AF668445

Recent Advances in Groundnut Production Technology

Recent Advances in Groundnut Production Technology

G. Krishna Reddy
P. V. R. M. Reddy
P. M. Reddy
K. Raja Reddy
and
T. Giridhara Krishna

BSP **BS Publications**
A unit of **BSP Books Pvt., Ltd.**
4-4-309/316, Giriraj Lane,
Sultan Bazar, Hyderabad - 500 095.

Recent Advances in Groundnut Production Technology
by G. Krishna Reddy, P. V. R. M. Reddy, P. M. Reddy, K. Raja Reddy and T. Giridhara Krishna

First Reprint 2018

Published by :

BSP BS Publications
A unit of **BSP Books Pvt., Ltd.**

4-4-309/316, Giriraj Lane, Sultan Bazar,
Hyderabad - 500 095.
Phone: 040 - 23445600, 23445688
e-mail: info@bspbooks.net
website: www.bspbooks.net

Printed and bound in India by Bhavish Print Solutions Pvt. Ltd.

ISBN: 978-93-5230-097-6 (Hardback)

CONTENTS

Preface.. (ix)

Contributors .... (xi)

1. Importance of Groundnut in Indian Economy
   D. M. Hedge 1
2. Groundnut in India
   Radhakrishnan T and Mishra GP 9
3. Reproductive Biology of Groundnut
   M. Subba Rao 19
4. Groundnut Improvement – Methods, Status and Issues
   SN Nigam.. 37
5. Breeding High Yielding Drought Resistant Varieties in Groundnut
   RP Vasanthi.. 84
6. Groundnut Varietal Characteristics and Recommended Varieties of India
   K. S. S. Naik.. 119
7. Biotechnological Tools for Groundnut Improvement
   Yellari Amaravathi 164
8. Quality Seed Production and Certification in Groundnut
   E. Venkata Ramana and B. Gopal Reddy.. 179
9. Role of Genomics in Groundnut Improvement
   D. M. Reddy, K. H. P. Reddy, M. Reddi Sekhar, N. V. Naidu and N. P. Eswar Reddy.. 194
10. DNA Fingerprinting in Groundnut Genotypes
    Yellari Amaravathi 207
11. Farm Mechanization in Groundnut Cultivation
    C. Ramana.. 219
12. Influence of Climate Change on Groundnut Productivity
    T. Prathima.. 236
13. Groundnut based Cropping Systems
    D. Sampath Kumar.. 247

14. Role of Nanoscale Nutrients for Improved Groundnut Production
T. N. V. K. V. Prasad, P. Sudhakar, K. V. Nagamadhuri, P. V. R. M. Reddy and T. Giridhara Krishna 255

15. Good Agricultural Practices (GAP) for Improving the Groundnut Productivity in Drylands
G. Krishna Reddy 267

16. Preparation of Biofertilisers and Methods of usage in Oilseeds in Particular to Groundnut
N. Trimurtulu and A. J. Suvarna Latha 304

17. Efficient use of Fertilizers in Groundnut Production
G. Prabhakara Reddy and V. Divya 318

18. Production Technology for Organic Groundnut
T. Giridhara Krishna 331

19. Micronutrient Management in Groundnut
P. V. R. M. Reddy, V. Naga Madhuri, G. Krishna Reddy, P. M. Reddy and P. Bala Hussain Reddy 343

20. Long-Term Effect of Fertilizers and Manures on Groundnut Productivity and Soil Properties
K. V. Naga Madhuri, P. V. R. M. Reddy, T. N. V. K. V. Prasad and T. Giridhara Krishna 363

21. Dry Land Technologies for Enhancing Groundnut Productivity under Rainfed Conditions
B. Sahadeva Reddy and A. Malliswara Reddy 384

22. Water Management Practices for Groundnut
Reddy S. R. and Deepthi Kiran Y. 415

23. Soil Profile Moisture Study – Hands on Experience
G. Krishna Reddy 474

24. Systems of Irrigation for the Enhanced Productivity in Groundnut
K. Madhusudhana Reddy, Y. Pavan Kumar Reddy and V. Siva Jyothi 479

25. Farm Pond Technology
T. Yellamanda Reddy 496

26. Drought Mitigation Strategies for Dryland Agriculture with Special Reference to Groundnut
P. Sudhakar, P. Latha and K. Sujatha 504

27. Integrated Weed Management in Groundnut
A. S. Rao ... 513

28. Soil and Water Conservation Measures in Rainfed Agriculture
A. Pratap Kumar Reddy and S. Hemalatha ... 523

29. Herbicide Formulations, Spray Fluid Calibration and Application Techniques
P. Maheswara Reddy, G. Krishna Reddy and P.V.R.M. Reddy ... 534

30. Integrated Pest Management in Groundnut
T. Murali Krishna, ARK Rao, K. Devaki and K. Raja Reddy ... 548

31. Seed Related Physiological and Quality Aspects of Groundnut
V. Umamahesh, V. Rajarajeswari, Y. Yohan and A. Reshma ... 564

32. Management of Viral Diseases of Groundnut
B. V. Bhaskara Reddy ... 593

33. Preparation of Microbial Biocontrol Agents for Insect Pest Management in Groundnut
K. Devaki, T. Muralikrishna and A. Ramakrishna Rao ... 608

34. Indigenous Technical Knowledge (ITK) in Groundnut Cultivation
E. Somasundaram ... 629

35. The Food and Processing Quality Characteristics of Groundnut
J. B. Misra ... 644

36. Kernel Qualitative Traits in Groundnut
Latha P ... 659

37. Problems and Prospects of Export Oriented Groundnut in India
V. Sumathi ... 670

38. Groundnut Seed Processing Technology
S. Kaleemullah ... 679

39. Value Addition to Groundnut
Swarnalatha and Veeraiah ... 706

40. Statistical Tools for Sustainable Agriculture
P. Lavanya Kumari ... 717

41. Correlation and Regression
B. Ravindra Reddy ... 753

42. Strategies for Mitigation of Phytonematode Problems in Groundnut
V. Krishna Rao ... 770

43. Fungal Disease Management in Groundnut
M. Reddykumar 783

44. Dissemination of Groundnut Research Technology through Farmers Call Centre: A Critical Analysis
S. Hemalatha and A. Pratap Kumar Reddy 797

45. Recent Advances in Agriculture Transfer of Technology
V. Anitha and R. Sudhakar Reddy 803

46. Technology Assessment and Refinement (TAR) in Groundnut – The Concept and Practice
K. Mohan 816

47. Planting Density –Impact on Groundnut Production
Mallu Srinivasa Reddy, P. V. M. Reddy, G. K. Reddy, P. M. Reddy and P. Bala Hussain Reddy 835

48. New Paradigms in Improvement of Drought Tolerance in Crop Plants
C. V. C. M Reddy, P. V. R. M Reddy and P. Bala Hussain Reddy 841

PREFACE

Winter school training programme was organized on "Recent Advances in Groundnut Production Technology" for 21 days at Regional Agricultural Research Station, the foot hills of Lord Sri Venkateswara, Tirupati, Andhra Pradesh from 25th September to 15th October, 2015.

Eminent Agricultural Scientists including Dr. J. B. Misra, Ex-Director DGR, Junagadh, Dr. Muralidharan, Ex-Director, NRCG-Junagadh, Dr. D.M. Hedge, Ex-Director, Indian Institute of Oilseeds Research, Hyderabad, SN Nigam, Former Principal Scientist (Groundnut), ICRISAT and Dr. S. R. Reddy, renowned Agronomist enriched the training program with their vast experience on groundnut production technology.

Major areas of Groundnut production technology viz., crop improvement, biotechnology, crop production, crop protection, farm machinery, biofertilisers, value addition, agricultural statistical tools and latest trends in technology transfer have been presented in 48 chapters.

Proceedings of the training programme have been published as "***Recent Advances in Groundnut Production Technology***" useful to the researchers of groundnut group.

The authors are deeply indebted to many individuals, organizations (ANGRAU and ICAR) for their support in making the manuscript.

- Authors

CONTRIBUTORS

D. M. Hedge
Ex-Director ICAR Indian Institute of Oilseeds Research, Rajendranagar, Hyderabad-510 030.

Radhakrishnan T, Mishra GP
ICAR-Directorate of Groundnut Research, Junagadh, 362 001, Gujarat, India.

M. Subba Rao
Principal Scientist (Millets) & Head, Agricultural Research Station, Perumallapalle, Tirupati - 517 505, A. P.

SN Nigam
Former Principal Scientist (Groundnut) and Expert-Agriculture for Development, 'Shiva-Parvati', Plot No. 125, Road No. 74, Jubilee Hills, Hyderabad 500 033, India.

R. P. Vasanthi
Principal Scientist (Pl.br), RARS, Tirupati.

K. S. S. Naik
Principal Scientist (Groundnut), ARS, Kadiri.

Dr. Y. Amaravathi
Institute of Frontier Technology, RARS, Tirupati.

Dr. E.Venktramana
Scientist (G.P.B.R), RARS, Tirupati.

D. M. Reddy
Dept. of. Genetics and Plant Breeding, S.V. Agril. College, Tirupati.

Dr. Yellari Amaravathi
Scientist (Biotechnology), Institute of Frontier Technology, RARS, Tirupati.

Dr. C. Ramana
Principal Scientist & University Head, RARS, ANGRAU, Tirupati.

Dr. T. Prathima
Senior Scientist, RARS, Tirupati.

Dr. D. Sampath Kumar
P S (Agro) & Co-ordinator, ANGRAU, DAATT Centre, Anantapuramu.

T. N. V. K. V. Prasad
P. Sudhakar, K.V. Nagamadhuri, P.V.R.M. Reddy and T.Giridhara Krishna.

Dr. G. K. Reddy
Principal Scientist (Agronomy), RARS, Tirupati.

N. Trimurtulu *Principal Scientist, Agricultural Research Station, ANGRAU, Amaravathi-522 020.*
Dr. G. Prabhakara Reddy *Department of Agronomy, S.V. Agricultural College, Tirupati -517 502, Andhra Pradesh.*
Dr. T. Giridhara Krishna *ADR, RARS, Tirupati.*
Dr. P. V. R. M. Reddy *Scientist (SS), RARS, Tirupati.*
K. V. Naga Madhuri *Senior Scientist(Soil Science), RARS, Tirupati.*
B. Sahadeva Reddy *Chief Agronomist, ARS, DCMS Buidings, Kamalanagar, Ananthapuramu.*
Dr. S. R. Reddy *Professor and Uni.Head(Rtd), ANGRAU S.V.Agril.Campus, Tirupati.*
Dr. G. K. Reddy *PS(Agronomy), RARS, Tirupati.*
K. Madhusudhana Reddy *Scientist(Agri.Engg), Agricultural Research Station, Anantapuramu, Andhra Pradesh.*
Dr. T. Yellamanda Reddy *Chief, Sustainable Agriculture, A.F. Ecology Centre, Anantapuramu, Former Dean of Agriculture, ANGRAU, Hyd.*
Dr. P. Sudhakar *Professor(Crop Physiology), S.V.Ag.College, Tirupati.*
Dr. A. S. Rao *Nodal Officer, Acharya N.G. Ranga Aricultural University, Rajendranagar, Hyderabad 500 030.*
Dr. A. Pratap Kumar Reddy *Principal Scientist (Agro), Farmers Call Centre, ANGRAU.*
P. Maheswara Reddy *Scientist (Agronomy) RARS, Tirupati.*
Dr. T. Murali Krishna *Principal Scientist (Ento), Regional Agricultural Research Station, Tirupati.*
Dr.V. Umamahesh *Assistant Professor (crop physiology), S.V.Ag. College, Tirupati.*
Dr. B. V. Bhaskara Reddy *RARS, ANGRAU, Tirupati.*

K. Devaki *Scientist (Entomology), IFT, RARS, Tirupati.*
Dr. E. Somasundaram *Professor and Head, Department of Sustainable Organic Agriculture, Tamil Nadu Agricultural University, Coimbatore- 641 003.*
J. B. Misra *Technical Advisor, Indian Oilseeds and Produce Export Promotion Council Mumbai.*
Dr. Latha *Scientist (Crop Physiology), RARS, Tirupati.*
Dr.V. Sumathi *Professor, Dept. of Agronomy, S. V. Ag. College, Tirupati.*
Dr. S. Kaleemullah *Professor & Head, Dept. of Agricultural Engineering, S.V. Agricultural College, Tirupati.*
Smt. Swarnalatha *RA (HSc), KVK, Utukur, YSR district.*
Dr. P. Lavanya Kumari *Scientist, Division of Statistics, RegionalAgricultural Research Station, Tirupati.*
Dr. B. Ravindra Reddy *Asst. Professor (Statistics &Maths) S.V. Ag. College, ANGRAU, Tirupati.*
V. Krishna Rao *Professor & Uni. Head (Plant Pathology) and Coordinator-Farmers Call Centre, PJ TS Agril. University, Rajendranagar, Hyderabad, Telangana.*
Dr. M. Reddykumar *Sr. Scientist (Pl. Pathology), IFT, RARS, Tirupati.*
S. Hemalatha *Principal Scientist(Agro), Farmers' Call Centre, ANGR Agril University, Seetaphalmandi, Telangana, India.*
R. Sudhakar Reddy *Co-ordinator, PJ TS Agril. University, Rajendranagar, Hyderabad.*

Chapter 1

Importance of Groundnut in Indian Economy

D. M. Hedge

Ex-Director ICAR Indian Institute of Oilseeds Research, Hyderabad.

1.1 Introduction

India is among the largest vegetable oil economies in the world next only to USA, China, Brazil and Argentina. In the agricultural economy of India, oil seeds one important next only to food grains in terms of heritage, production and value. Currently, India accounts for about 13-15% of world's oilseeds area, and 8-9% of world's oilseeds output and 10-11% of world's vegetable oils consumption. The diverse agro ecological conditions in the country are favorable for growing all the nine annual oilseeds, apart from a wide range of other minor oilseeds of horticultural and forest origin, including in particular coconut and oil palm. Oilseeds occupy about 13-14% of gross cropped and account for nearly 1.4% of gross national product and 8% of the value of all agricultural combustions. About 14 million formers and engaged in the production of oilseeds and action million in their processing.

Edible oil constitutes an important port of our daily diet, being source of energy, essential fatty acids like oleic and linoeic acids which are vital for our growth. Domestic consumption of vegetable also has increased substantially over the last two decades and has touched the level of (19.82 million tonnes in 2012-13) and is likely to increase further with enhancement in income and population. The growth in growth in production of domestic vegetable oils (4.22 million tonnes in 2012-13) has not been able to keep pace with the growth in consumption and the gap between production and consumption is being met through imports. The cost of import of 10.385 million tonnes of vegetable oils was Rs. 61106.43 crores during 2012-13 next only to crude oil and gold.

1.2 Groundnut as an Oilseed

Groundnut is the second most important annual oilseed crop after soybean. It is believed to be native of Brazil (South America) since many closely related species are found there. The crop spread from Brazil to Peru, Argentina and Ghana, from where it was introduced into Jamica, Cuba and other West Indies islands. The plant was introduced by the

Portuguese into Africa from where it was introduced into North America. It was then transported along the American west coast to Mexico and then across the Pacific to the Philippines, from where it spread to China, India, Malaysia and Indonesia.

The groundnut seed oil content varies from 44 to 50% depending upon the varieties and agronomic practices/ conditions. Its oil finds extensive use as a cooking medium both as refined refined oil and vanaspati ghee. It is also used in soap making, and in manufacture of cosmetics, lubricants olein, stearin and their salts. Kernels are also eaten raw, roasted or sweetened. They are rich in protein and vitamins A, B and some members of the B_2 group. Their calorific value is 349 per kg. The handpicked and selected (HPS) types of groundnut kernels are exported to other countries. The residual oilcake contains 7-8% N, 1.5% P_2O_5 and 1.2% K_2O and can be used as organic manure. It is an important supplement in cattle and poultry rations. It is also consumed as a confectionery product. The cake can be used for manufacturing artificial fiber. The haulms are fed (green, dried or silage) to livestock. Groundnut shell is used as fuel, for manufacturing coarse boards, coal substitutes, etc.,. Groundnut is also of value as a rotation crop. Being a legume with root nodules, it can fix atmospheric N and their by improves soil futility.

1.3 World Groundnut

The total world production of groundnut was 45.31 million tonnes from 25.46 million hectares with an average productivity of 1780 kg/ha during 2013-14 (Table 1.1) China, India, Nigeria, USA, Sudan, Myanmar, Indonesia and Senegal, are the major countries producing groundnut. Asia accounts for 55% of the global groundnut area and 67% of the global production while Africa constitutes 40% of global area and 26% of the production. The average yield of groundnut among the groundnut growing countries in the world varies between 300 kg and 5400 kg/ha

***Table 1.1** Area production and productivity of groundnut in major countries (2013-14)*

Country	Area (lakh.ha)	Production (lakh t)	Yield (kg/ha)
Cameroon	4.63	6.36	1373
China	46.82	169.19	3614
India	52.50	94.72	1804
Indonesia	5.14	11.50	2216
Myanmar	8.90	13.75	1545
Niger	7.20	2.80	389
Nigeria	23.60	30.00	1271
Senegal	7.70	7.10	922
Sudan	21.62	17.67	817
USA	4.21	18.93	4496
World	254.60	453.08	1780

During 1980-2006 periods, the global groundnut area grew by 1.0% yield by 2.4% and production by 3.4% annually during the same period, the growth rates for Asia were 0.6%, 3.2% and 3.8% respectively. For Africa these figures were 2.2%, 1.5%, and 3.7% respectively. In Asia, it was the yield which contributed to increased production but in Africa, the increased production came largely through area expansion.

Productivity and production of groundnut in Asia picked up substantially 1991 onwards with very little increase in area in Asia, China and Vietnam led the way. The advancement in groundnut production and productivity came along in a step by step manner with scientific achievements, government's encouragement and the hard work of the farming communities in these countries.

1.4 Groundnut in India

During 2013-14, India produced 9.67 million tonnes of groundnut from 5.53 million hectares with a productivity level of 1750 kg/ha. For the first time groundnut productivity in India nearly equaled the world average groundnut productivity (1780 kg/ha). Six states namely Andhra Pradesh, Gujarat, Karnataka, Maharashtra, Rajasthan and Tamil Nadu account for about 90% of the total groundnut area of the country (Table 1.2). Tamil Nadu has the highest productivity followed by Gujarat and Rajasthan. About 74% of the total produce is harvested from rainy season crop, while only 26% from the irrigated post-rainy season crop. The average productivity during the post-rainy season is as at two times more than rainy season.

Table 1.2 *State-wise area, production and yield of groundnut (2013-14)*

State	Area (lakh.ha)	Production (lakh t)	Yield (kg/ha)
Andhra Pradesh*	13.86	12.34	890
Gujarat	18.42	49.18	2670
Karnataka	7.25	6.58	908
Madhya Pradesh	2.00	1.98	990
Maharashtra	2.67	3.25	1217
Odessa	0.58	0.80	1379
Rajasthan	4.66	9.06	1944
Tamil Nadu	3.42	9.62	2813
Uttar Pradesh	0.96	0.86	896
West Bengal	0.79	2.02	2557
Others	0.66	1.04	1576
All India	55.27	96.73	1750

* Includes Telangana

The growth rates in area, production and productivity during pre-TMOP (1951-1986) and post-TMOP (1986-2012) period were 1.13, 1.59, 0.45% and 1.39, 0.20, 1.20% respectively. Groundnut acreage has declined in recent years largely because farmers switch over to more profitable or less risky crops such as Bt cotton and soybean. Groundnut contributes 0.4 to 0.6% of gross national product and 2 to3% of all agricultural commodities. In states like Gujarat and Andhra Pradesh groundnut contributes upto 4-6% of value of all agricultural commodities. India also earns sizeable foreign exchange through export of many groundnut products.

1.5 Record Yields in Groundnut

Groundnut is photo synthetically a highly efficient crop. Since of the examples of record pod yields are as follows:

- 10.5 t ha^{-1} over a small area under intensive cultivation in Shandong province, china (Yanhao et.al., 1996).
- 9.6 t ha^{-1} in large plots in Zimbabure (Hidebrand, 1996).
- 9.4 t ha^{-1} in a 0.2 ha plot in a summer groundnut crop in Maharashtra and 9.5 t ha^{-1} in a 3 cent plot in Andhra Pradesh, India (Nigam, 2000).

These figures represent the yield potential of groundnut for the respective agrological zones. Theoretical potential pod yield that can be researched in the crop in Shandong Province in China is 17.3 t ha^{-1} (Yanhao et.al., 1996).

1.6 What AILS Groundnut in India

Most of the groundnut in India is grown under four major agro ecologies:

1. Rainfed (kharif season, June – September)
2. Irrigated (rabi season, October – March)
3. Residual moisture (rabi season, October – March)
4. Irrigated (Summer / spring season, February – May)

In all the four major agro ecologies, various constraints gentle that limit groundnut productivity. Some of the constraints are common across all the four agro ecologies.

Groundnut is extensively grown in the semiarid tropics by resource – poor farmers where many Abiotic and biotic factors limit its productively and seed viability. The major Abiotic factors affecting groundnut production include drought, low availability of phosphorus especially under acidic soil condition, salinity and temperature stress (Both low and high). In calcareous soils, groundnut is also affected by iron-induced chlorosis. Rust, late leaf spot (LLS), and early leaf spot (ELS) are widely distributed foliar diseases. Groundnut rosette disease (GRD) is the most distructive disease of groundnut in Sub-Saharan Africa. Bacterial wilt of groundnut is prevalent in East Asian region, while peanut bud necrosis disease in South Asia. Unlike disease, insects are of localized importance. The only groundnut insect pests of significance are leaf miner in South Asia: Spodoptra in southeast Asia, termite in Africa, and cornear worm, lesser corn stock borer, and Sothern corn rootworm in USA. Nematodes are of importance in USA. Sucking pests such as aphids, Jassids, and thrips are not themselves considerable economically important but are carrier of virus diseases such as aphids for GRD (Dwivedi et.al., 2003).

Most of these stresses often occur in combinations and their severity and extent of distribution vary with cropping systems, growing seasons, and regions. Global yield loss caused by these stresses is enormous for example, yield loss done to rust and LLS together has been projected at US $ 1066 million of which nearly 50% can be realized through genetic enhancement.

1.7 Yield Gaps in Groundnut

The yield gap is difference between potential yield and actual yield. It can be decomposed into two parts, viz., yield gap-I and yield gap-II. Yield gap-I is the difference between experimental stations average attainable maximum yield and on farm experiments average maximum yield. This yield gap arises firm difference in environment that cannot be managed in the farmer's fields. Yield gap-II, which is of primarily concern is the difference between yield attained in on farm experiments and the average actual farm yield. This gap reflects the effects of biological, soil and water, physiological, genetic and socio economic constraints. This gap exists because farmers use sub optimal doses of inputs and cultural practices. It is manageable and narrowed by increasing efforts in research and extension services, as well as by appropriate government intervention, particularly in institutional issues. There gaps can also be classified as agronomic constraints, socio-economic constraints, institutional constraints management constraints etc.

Yield gaps calculated on the basis of the yields recorded under Front Line Demonstrations (FLD) of ICAR and average / highest yield recorded in the states is given in Table.1.3. It indicates that an average yield gap of 65% for kharif and 32% for rabi/summer groundnut over state average yield. However, yield gaps are higher in the states of Karnataka (72-101%) and Rajasthan (94%).

Table 1.3 *State-wise yield gap (kg/ha) groundnut*

State	Rabi 2011-12			Kharif-2012		
	SAY	FLD	Yield average (%)	SAY	FLD	Yield average (%)
Andhra Pradesh	1848	2947	70	533	1648	74
Gujarat	1914	2216	19	790	1446	42
Karnataka	917	2498	101	564	1694	72
Maharashtra	1446	2176	46	1037	2379	85
Odessa	-	-	-	1041	2424	88
Rajasthan	-	-	-	1549	3033	94
Tamil Nadu	3571	2027	-110	1928	1870	-4
West Bengal	1969	2962	63	-	-	-
Mean	1974	2471	32	1063	2078	65

SAY: State average yield, FLD : Frontline demonstration yield

Rajasthan and Tamil Nadu were identified as high potential-low gap states, while Gujarat was recorded under the high potential-high gap state. (Table 1.4). Potential districts for groundnut in important states are given in Table.1.5

***Table 1.4** Classification of states according to potential yield and yield gap in groundnut*

Gap	High potential			Low potential		
	State	Potential yield (kg/ha)	Yield Gap (%)	State	Potential yield (kg/ha)	Yield Gap (%)
High gap	Gujarat	2151	30	Karnataka	1677	28
Low gap	Rajasthan	2377	15	Maharashtra	1712	17
	Tamil Nadu	2209	10	Andhra Pradesh	1341	19

***Table1.5** Potential Districts for groundnut in important states*

State	Districts
Andhra Pradesh	Anantapuram, Chittoor, Kurnool, YSR Kadapa, Vizayagnagaram and Srikakulam.
Telangana	Mahabubnagar, Warangal and Nalgonda.
Gujarat	Junagadh, Jamnagar, Rajkot, Amreli, Bhavnagar, Porbandar, Kutch, Subarkantha, Banaskantha and Vadodara.
Karnataka	Chitradurga, Tumkur, Gadag, Bijapur, Belgaum, Bellary, Dharwad, Koppal, Raichur, Chickballapur, Bhagalkot, Kasargad and Haveri.
Maharashtra	Kolhapur, Satara, Pune, Sangli, Nasik and Dhale.
Rajasthan	Bikaner, Jaipur, Jodhpur, Sikar and Chitture.
Tamil Nadu	Thiruvannamalai, Villupuram, Vellore, Kancheepuram, Numakkal, Erode and Salem.
Odessa	Jajpur, Baragarh and Ganjam.

1.8 Prioritization of the Constraints

In spite of the moderate grains in productivity of groundnut made in India there remains on large gap in potential yield and yield realized at the farm level, particularly in rainfed agriculture. For to well defined and focused research agenda that will ensure sustainable yield grains in a consistent manner, the following is required. Identify production constraints including socio-economic factors operating in various agro ecologies. Rank them in order of their adverse impact on productivity. Identify best management (genetic or non-genetic) options for major constraints based in benefit cost ratio for return in research investment for each identical constraint. Identify research lags.

Indicate probability of success and research investment costs.

Research institutions should focus on the priority constraints of groundnut of their region. In many cases non-genetic option may be an easier and profitable solution of constraint alleviation. Socio-economic, policy and infrastructure related issues affecting groundnut productivity and production will have to be dealt with at government level removing impediments in reaping the benefits of research outputs by farmers.

1.9 Issues/Concerns in Groundnut

Some of the major issues/Concerns which affect the groundnut production, productivity and utilization are as listed below (Hedge, 2009)

- Continued dependence of groundnut on rainfed production system nearly 72% of groundnut area is under rainfed condition.
- Non availability of quality seeds of improved varieties due to weak seed chain. The seed replacement rate over in state like Gujarat is just about 3%.
- General tendency to promote only state varieties at the cost of nationally released varieties.
- Inadequate plant stand which is by for the most important constraint hindering higher productivity.
- New biotic threats like stem and bud necrosis has became major constraint in many regions.
- Inefficient processing as the crop is reserved for small scale sector lending to poor oil recovery and reduce earnings to farmers and higher cost to consumers.
- Inadequate marketing support and non remunerative prices often acting as a disincentive for greater investment on inputs in crop production.
- Unfavorable and unstable import policy lending to depress a domestic prices which adversely affects the farmers although consumers many benefit in the short term.

The country has already achieved an average productivity of 1.7 t/ha during 2013-14. If all the constraints ore addressed adequately there is every possibility of achieving a productivity level of 2-2.5 t/ha under rainfed conditions and 4-5 t/ha under irrigation lending to total national productivity of 2.5 to 3 t/ha under intensive cultivation, a productivity level of 6-7 t/ha is easily achievable.

References

1. *Dwivedi, S.L Crouch, J.H; Nigam, S.N; Ferguson, M.E. and Paterson, A.H 2003. Molecular breeding of groundnut for an enhanced productivity and food security in the semi-arid tropics. Opportunities and challenge. Advances in Agronomy, 86: 153-221.*
2. *Hedge, D.M 2009. Oilseeds production in India: Retrospect and prospect. National Academy of Agricultural Sciences Newsletter &(1):1-7*

3. *Hildebrand, F.L 1996. The status of technologies used to achieve high groundnut yields in Zimbabwe In: Achieving high groundnut yields, proceedings of an international workshop. 25-29 Ang 1995, Laixi city shanding, China (Renard, c; prowdu, C.L.L, Nigam, S.N and Johansen, C.Eds), Patamcheru -502324, Andhra Pradesh, India. International Crops Research Institute for the Semi Arid Tropics, pp 101-114.*

4. *Nigam. S.N 2000 Some strategic issues in breeding for high and stable yield in groundnut in India. Journal of Oilseeds Research, 17: 1-10.*

5. *Yanhao,S shvuxiang T and Caibin, W 1996. Theoretical foundation for high yield of groundnut in China. Jr: Achieving High Groundnut Yields: Proceedings of an International Workshop 25-29 Ang 1945, Laixi city, shanding, China [Renard, C; Gowda, C.L.L; Nigam S.N and Johansen, C.Eds). Patancheru-502324, Andhra Pradesh, India. International Crops Research Institute for the semi Arid Tropics, pp 129-139.*

Chapter 2

Groundnut in India

Radhakrishnan T. and Mishra GP

ICAR-Directorate of Groundnut Research, Junagadh, Gujarat, India.

2.1 Origin and Introduction in India

Groundnut (Arachis hypogaea, L.) also known as peanut is believed to have originated in the South America somewhere between southern Bolivia and north western Argentina. The state of Mato Grosso, Brazil is generally recognized as the primary centre of diversity while Africa is considered the secondary centre. Groundnut was introduced to India in the east coast of erstwhile state of Madras by the Spaniards in the first half of the 16th century while some believe that Portuguese introduced groundnut to the Malabar Coast of south-western India. The first large-scale (about 4000 ha) cultivation of groundnut has been reported to have occurred in the South Arcot district of Tamil Nadu in the year 1950-51.

2.2 Area, Production and Yield of Groundnut

Globally, though India ranks first in the cropped area, it ranks second after China in production. The average productivity in India is quite low («1200 kg/ha) compared to that of the USA (3800 kg/ha), China (3400 kg/ha), Argentina (2100 kg/ha), Indonesia (^1250 kg/ha) and the world average («1562 kg/ha).

***Table 2.1** Triennial averages of area, production and yield of major groundnut producing countries*

Country	Area (lakh/ ha)	Production (lakh/ tonne)	Yield (Kg ha-1)
China	46.7	166.3	3562
India	51.1	70.4	1367
Nigeria	23.7	30.1	1268
USA	5.0	22.0	4330
Myanmar	8.9	13.8	1560
Senegal	7.8	6.4	827

Table 2.1 ***Contd...***

Country	Area (lakh/ ha)	Production (lakh/ tonne)	Yield (Kg ha-1)
Indonesia	5.4	12.0	2234
Nigar	7.2	3.2	452
World	249.3	420.9	1688

*Triennial (2011-2013) average

2.3 The Indian Scenario

Since 1951, the largest area under this crop was 87.1 lakh ha (1989-90); the highest production was 96.6 lakh tonnes (1988-89); while the highest productivity was 1460 kg ha^{-1} (2007-08). Currently, six states *viz.* Gujarat, Andhra Pradesh, Karnataka, Tamil Nadu, Maharashtra, and Rajasthan account for nearly 90% of the total groundnut area of India. Whereas, Madhya Pradesh, Uttar Pradesh, Orissa, West Bengal, Chhattisgarh, Jharkhand, Punjab, Goa, NEH states and Kerala also contribute to national groundnut production to various extents. In India, groundnut is cultivated mainly in *kharif* season (rain-fed conditions) with low inputs and if available, with a few protective irrigations. In *kharif,* the pressures of diseases, insect-pests and weeds are quite high. The productivity largely depends on the quantity and distribution of rainfall over the season. An evenly distributed rainfall of 600-800 mm would generally assure a good productivity. However, dry spells which often occur during pod development phase would reduce the yields and hence *kharif* productivity shows wide variations over the years.

In *rabi,* the crop is grown under residual moisture of rice fallows or river beds under minimal or no irrigation and also in summer as an irrigated crop. Over the years, the area under rabi-summer groundnut system has expanded. The summer cultivation is practiced generally under high input conditions and the pressures of diseases, insect-pests and weeds are much low and hence the productivity is high. Spring groundnut, cultivated after the harvest of potato/foria, also gives high productivity.

2.4 Trends in Area, Production and Productivity

The area under groundnut in India during 1900-1910, was only 2.3 lakh ha which increased to 76.8 lakh ha by the end of century. During 1951-1960, the growth in area (17.2%) had the major contribution in increasing groundnut production in India. The area increased from 4.66 million ha during 1951-55 to an all-time high of 8.16 million ha during 1991-95. After hovering above 8 million ha from 1988-89 to 1993-94 period, it started declining and hit a low of 4.32 million ha in 2013-14 due to I various reasons, which include competition from other crops such as Bt cotton, soybean, sunflower, maize etc., availability of cheaper edible oil such as palm oil and frequent and persistent drought in major groundnut growing areas.

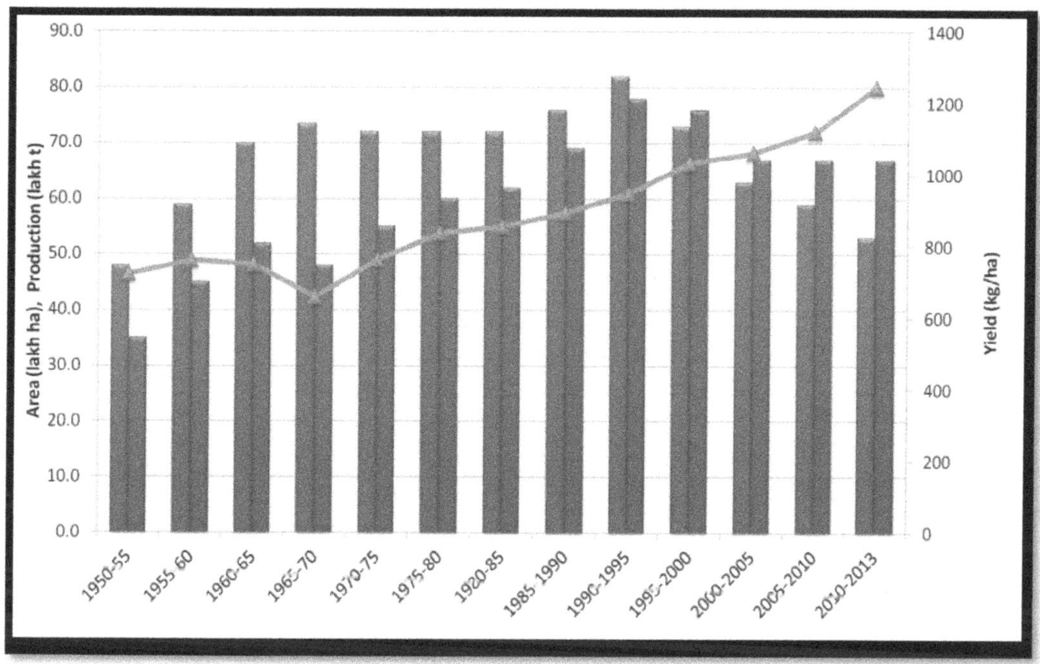

Fig. 2.1 Quinquennial averages of area, production and yield of groundnut in India

The productivity was below 800 kg/ha till 1975 but increased steadily to reach beyond 1000 kg/ha during 1996-2000. The increase in productivity was due mainly to release of high-yielding varieties and improved packages of practices, and also to a certain extent to cultivation in summer under high-input assured irrigation.

2.5 State-Wise Scenario of Area, Production and Productivity and Constraints

The triennial (2010, 2011, and 2012) averages of area and production of groundnut indicate that 91.5% of the cropped area is distributed across six states which together contribute 91.4% of total production. These states are Gujarat, Andhra Pradesh, Tamil Nadu, Karnataka, Maharashtra, and Rajasthan. Among other groundnut growing states are MadhyaPradesh, Odisha, Uttar Pradesh and West Bengal which together comprise nearly 7% of area and contribute about 7% of the total production.

Table 2.2 *Triennial (2010-11 to 2012-2013) averages of area, production and yield of groundnut in various states of India*

State	Area (lakh ha)	Production (lakh tonnes)	Yield (kg ha1)
Gujarat	16.5	22.8	1354
Andhra Pradesh	14.26	11.3	790
Tamil Nadu	3.7	9.1	2495
Karnataka	7	5.4	742
Maharashtra	2.8	3.6	1182
Rajasthan	3.8	7.1	1813
Madhya Pradesh	2	3.1	1543
Odisha	0.6	0.82	1199
West Bengal	0.6	1.26	2042
Uttar Pradesh	0.9	0.9	966
Chhatishgarh	0.2	0.4	1391
All India	52.8	66.4	1243

Source: Ministry of Agriculture, Govt. of India, India (Vanaspati producers association)

The average productivity of groundnut during the period 2001 to 2010, though ranged from 700-1460 kg/ha, the average productivity in three major groundnut growing states of the country, was in the range 510-2270 kg/ha in Gujarat, 300-1360 kg/ha in Andhra Pradesh, and 1150-1880 kg/ha in Tamil Nadu for *kharif* 'season and the corresponding values for the ra & i-summer season were 1473-2390 kg/ha in Gujarat, 1400-2130 kg/ha in Andhra Pradesh and 2100-3730 kg/ha in Tamil Nadu. The productivity of groundnut in NEH region and Goa, however, is more than that of national average.

2.6 Export of Groundnut

The major groundnut exporting countries of the world include China, India, USA and Argentina. Although during 1999 to 2002, the export of groundnut from India declined, but a considerable increase was registered subsequently. A record high export of 853 thousand tonnes was registered in 2011-12. Groundnut is exported both in-shell and kernels forms. For export purpose, two broad categories are defined- java and bold. The java types have smaller pods with constriction, kernels are oval with pink skin and the shell is thicker than the bold types. In bold types, the pods and kernels are bigger than the java type and the kernels are oblong with a red skin which becomes darker with ageing.

A very stringent maximum permissible limit (2 ppb) of aflatoxin contamination in groundnut has been set by the European Union. Aflatoxin contamination is now recognized as the chief non-tariff trade barrier for export of groundnuts from India as many a Indian consignments are rejected at the destination ports due to aflatoxin levels beyond the permissible limits.

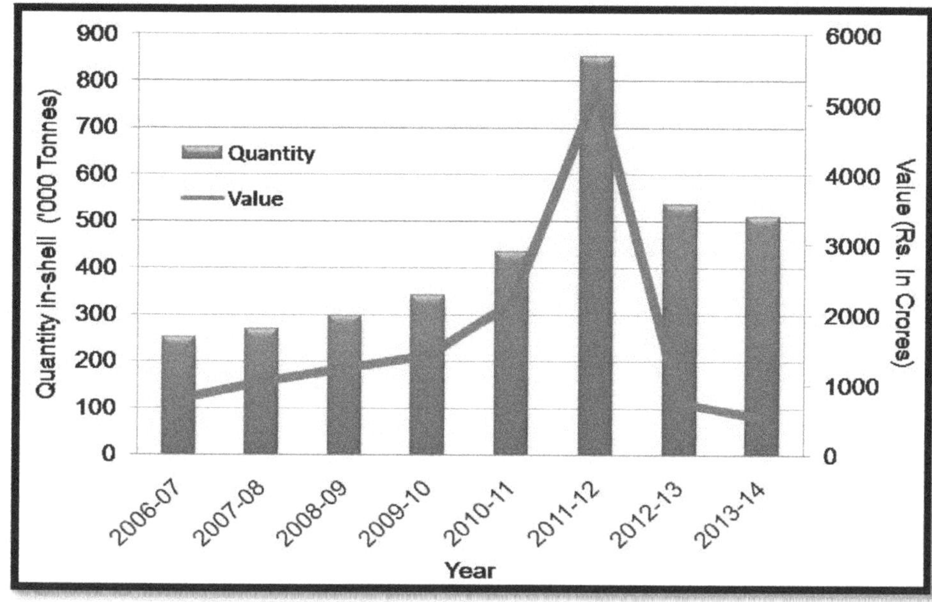

Fig. 2.2 Export of groundnut from India (source: DGCIS)

2.7 Current Utilization Pattern of Groundnut

The direct consumption of groundnut has been growing and its food value is being increasingly realized due to availability of other edible oils at economical prices. In 1960s, about 80% of the produce was crushed for oil expulsion and only about 6% was used for direct consumption, while the remaining 14% was used for seed. During the 1980s, about 81% of the total produced was used for oil expulsion and only 6% was used for direct consumption. However, in recent years, nearly 50% of the produce is used for direct consumption especially as value-added products like roasted, boiled and salted groundnut, *chikki* and peanut-butter. In the past two decades, the direct consumption (roasted and salted groundnuts, *chikki,* peanut butter and sauce) has gone up to 40% and currently about only 50% of the produce is crushed for oil expulsion.

2.8 Projected Demand of Groundnuts In 2050

The population of India is likely to touch a figure of 1,690 million by 2050. Providing food security to such a large population would be daunting task and also a challenge to agricultural scientists for developing production technologies to bridge the gap in the demand and supply on a sustainable basis. In times to come, groundnut would play a greater role as a supplementary food crop besides continuing to be an oilseed crop. The projected domestic demand of groundnut would be about 25 million tonnes. The area under groundnut may not shrink any further and would stay around 6-7 million ha.

Considering the current national average productivity of 1200 kg/ha, a growth rate of about 4-5% in productivity is called for.

Table 2.3 *Projected total oilseeds demand and groundnut as oilseeds and oils demand in India (in million tonnes)*

Year	Total oil seed required	Groundnut production	Production oil requirement
2020	50.4	14.8	3.5
2030	63.7	17.8	3.9
2040	79.6	21.6	4.3
2050	93.0	24.6	4.9

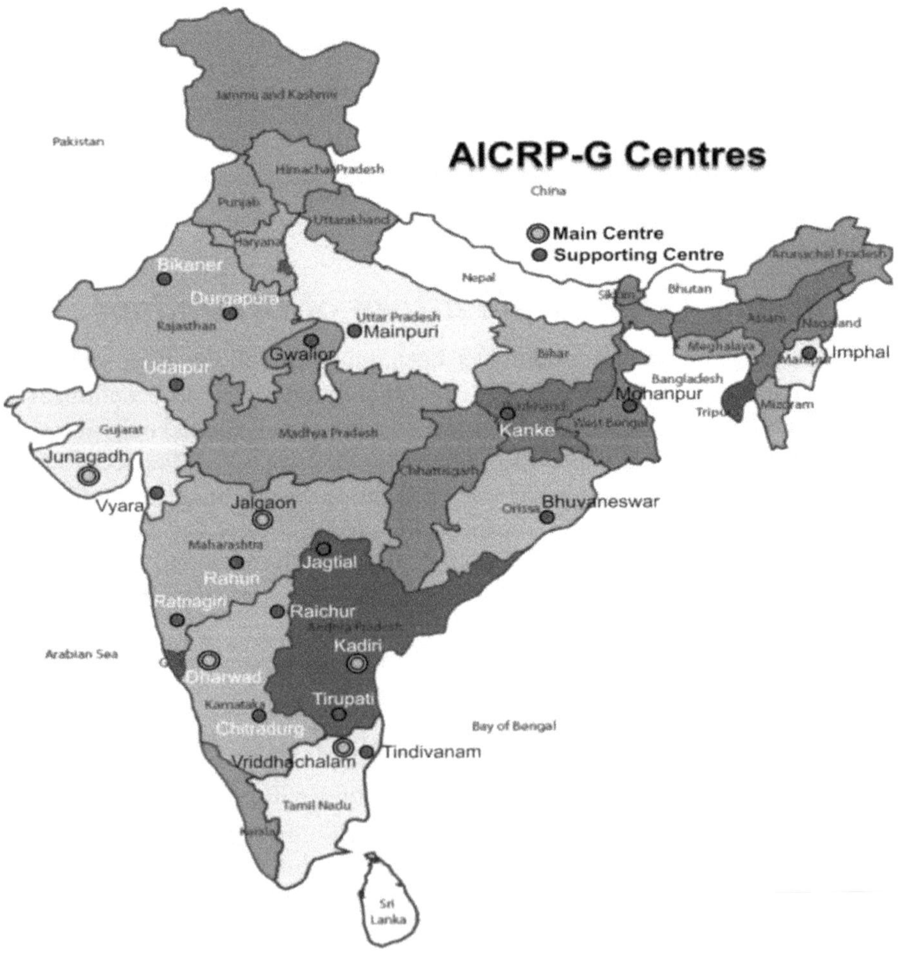

Fig. 2.3 *AICRP groundnut centres in India*

2.9 Icar-Directorate of Groundnut Research, Junagadh

The DGR was established in 1979 by the name National Research Centre for Groundnut (NRCG). The NRCG made its humble beginning in a small building with 18 ha farm land. Later, NRCG acquired another area of 105 ha land and the entire establishment was shifted to the new location. This was followed by shifting of headquarters of All India Coordinated Research Project on Groundnut (AICRP-G) from Akola to Junagadh in 1992 and bringing the AICRPG under the administrative control of NRCG. During the eleventh Five-Year Plan, the NRCG was upgraded from level of a 'NRC' to level of a 'Directorate'. The directorate is well equipped with human resource, equipment and infrastructure to conduct quality research. Since its inception, the Directorate of Groundnut Research has been addressing various issues concerning productivity of groundnut crop and has developed technologies to overcome the adverse effects of various biotic and abiotic stresses through multi-disciplinary approach. The efforts in future would be for further improving productivity, quality, and utilization of groundnut under changing scenarios of climate and global market.

The AICRP-G exists for multi-location evaluation of technologies developed by DGR and SAUs and a few ICAR institutes and KVKs. At present there are five main and seventeen supporting centres under AICRP-G. For the purpose of locating the centres, keeping in view the prevailing agro-climate and cropping density of groundnut, the country is classified into five zones for identifying the problems and prioritizing research under AICRP-G.

2.10 Breeder Seed Requirement and Production

Groundnut is a highly self-pollinated crop and as such the desirable seed replacement rate is 25%. To cover an area of 60 lakh hectares, the annual requirement of quality seed (truthfully labeled seed) comes out to be 9,600,000 q which can be produced from about 4688 q breeder seed as indicated below.

	Type of seed	Qty. (q)
A	Total seed requirement for 6 million ha	9,600,000
B	Certified seed (25% of A)	2,400,000
C	Foundation seed stage-II (one eighth of B)	300,000
D	Foundation seed stage-I (one eighth of C)	37,500
E	Breeder seed stage (one eighth of D)	4,688

The Directorate of Groundnut Research, Junagadh co-ordinates production of breeder seed to tune of 12,000 to 15,000 q every year and as such with a seed rate of 160 kg seed pods/ha and seed multiplication ratio of 1:8, the breeder seed produced in India is more than that is required for attaining desirable seed replacement rate.

***Table 2.4** Annual Production of breeder seed of groundnut in India*

Year	DAC Indent (q)	Production (q)
2009-10	22886	16406
2010-11	19679	15091
2011-12	25501	20075
2012-13	13075	12013
2013-14	12487	12996
2014-15	11310	10459

2.11 Groundnut Varieties

So far (till March 2013), 191 groundnut varieties have been released (139 state release and 52 central release). Of these 108 are Spanish bunch, 49 Virginia bunch, 29 Virginia runner, and 5 Valencia types. The varieties currently recommended for cultivation in various states of India are given below:

State	Kharif	Rabi-summer
Gujarat	GG 5, GG 7, TG 37A	GG 6, Dh 86, TPG 41
Andhra Pradesh	Kadiri-6, Kadiri-7, Narayani	TAG 24, Kadiri Harit Andhra, ICGV 00350
Rajasthan	Girnar 2, TG 37A, TBG 39	PM 1, PM 2
Karnataka	GPBD 4, ICGV 91114, Chinthamani 2	TAG 24, Dh 86, ICGV 00350
Tamil Nadu	VRI 2, VRI 6, TMV Gn 13, TNAU (Gn) Co- 6	VRI 2, VRI 6, TMV Gn 13
Maharashtra	JL 501, AK 159	JL 501, JL 220
Odisha	OG 52-1, Girnar-3	Dh 86, TG 51, TG 38B
West Bengal	Girnar-3, GPBD 5	Dh 86, TG 51, TG 38B

***Table 2.5** Some districts of India with pockets of high groundnut productivity*

State	Areas	Yield (kg/ha)
Gujarat	Halwad, Porbandar, Mangrol, Kutch	2500-3000
Andhra Pradesh	Cuddappa, Karnool, Mehaboobnagar, Nalgonda	2000-2500
Karnataka	Bagalkote, Dharwad, Belgaum, Udupi	3000-3500
Maharashtra	Kolhapur, Sangli, Satara, Shindudurg, Ratnagiri, Raigad	2500-3000
Odisha	Balasore, Bargarh, Ganjam, Jajpur, Malkangiri, Puri	2500-3000
West Bengal	Midnapur	2500-3000
Rajasthan	Bikaner, Jaipur	2000-2500
Tamil Nadu	Kanchipuram, Thiruvallur, Thnajvur Thiruvanamalai, Pattukottai	3000-3500

2.12 Reasons for Low Average Productivity of Groundnut in India

- Cultivated (85% area) by and large on poor soils under rainfed conditions
- Nearly 25% (^15-20 lakh ha) kharif area in and around Anantapur district is very dry (250-300 mm rainfall)
- Inefficient management of irrigation and rain-water
- Spread of improved varieties is too slow due to low seed multiplication ratio (1:8) and high seed rate (>100 kg seed/ha)
- Abiotic stresses: Chiefly drought and to some extent high-temperature and salinity
- Biotic stresses: Fungal (soil-borne and foliar) and viral diseases and insects pests mainly defoliators and sucking ones

2.13 Possible Technological Interventions for Increasing Productivity

- Popularization of new varieties and phasing out of inferior varieties
- Popularization of technology of rain water harvest and use of micro-irrigation systems
- Promotion of mechanization of cultivation
- Adoption of region specific IPM practices
- Promotion of application of region specific fertilizer mixtures (N, P, K and also of gypsum and micronutrients)
- Ensuring availability of quality biocontrol agents *(Trichoderma* and *Pseudomonas)* and bio-fertilizers (superior strains of *Rhizobiuma* and consortia of PGPR)

2.14 Strategic Researchable Issues

- Development of extra-early varieties for rice fallows and river beds for cultivation in Odisha, West Bengal and Goa
- Development of Spanish varieties with fresh-seed dormancy for cultivation in summer and spring
- Development of varieties resistant to soil borne (stem rot and collar rot) and viral
- (PBND and PSND) disease
- Refinement of fertigation technology for enhancing fertilizer and irrigation water use efficiency (through drip)
- Optimization of *in situ* moisture conservation technologies including application of hydrogel for combating drought

- Development of efficient delivery systems for micro-nutrients and bio-fertilizers
- Development of varieties and crop husbandry for potato fallows

2.15 Possible Policy Interventions

- Rejection of indents for old and inferior varieties
- Production of quality seed of new varieties to be taken up only in high productivity areas by promoting public private partnership
- Permitting seed multiplication through an increased number of steps by introducing an additional step at foundation seed stage:
- Nucleus seed-* Breeder seed — Foundation seed-I — Foundation seed-II — Certified seed
- Establishment of farmers participatory seed-villages for large-scale production of certified seed
- Incentive to farmers for using home grown seed of new varieties rather than providing seeds old varieties on subsidized prices

Chapter 3

Reproductive Biology of Groundnut

M. Subba Rao

Principal Scientist (Millets) & Head, Agricultural Research Station, Perumallapalle, Tirupati, A.P.

3.1 Introduction

Groundnut is the king of Vegetable oil seeds in India and occupies important position in the economy of national edible oils. It is also known as peanut, monkey nut, earthnut, goober pees, Jack nut and manila nut. Botanical name *Arachishypogaea* is derived from Greek words *Arachis* meaning legume and *hypogaea* meaning below ground, referring to the formation of pods in the soil.

Groundnut is grown in about 23.4 million hectares annually producing 37.2 million tons chiefly in India. China, Nigeria, USA, Brazil and Senegal. Groundnut grows well in places with moderate rainfall, abundant sunshine and relatively higher temperatures. The best quality of groundnut is secured in well drained, light sandy – loamsoils. It is highly susceptible to water logging.

3.2 Origin

There are two main schools of thoughts on the origin of groundnut, one supporting that it had in origin in Africa and the other to South America. However, recent botanical surveys and Archeological excavations have indicated South America as the most likely center of origin of groundnut (Seshadri, 1962). It appears that Portuguese introduced groundnut plant from Brazil in the Africa. Groundnut was introduced into India at the beginning of 16th century by Jesuit Fathers who followed Vasco De Gama shortly after was first launching in India.

3.3 Distribution in India

Groundnut accounts for 28 per cent of total area and 36 per cent of total oil seed production in the country. (Reddy, 2012). In India, groundnut is cultivated in 5.52 m.hawith production of 9.67 m.tons accounting to a productivity of 1750 Kg/ha.

Largest area under groundnut is in Gujarat followed by A.P. In Andhra Pradesh groundnut is grown in 11.76 lakh hectares with a production of 8.81 lakh tons accounting to productivity of 749 kg/ha.

3.4 Classification

Groundnut is a annual, herbaceous legume belonging to sub family *papilionaceae*of the family *Fabaceae* (formerly *Leguminosae*). Arachisconsistsof two large botanical groups Virginia and Spanish – Valencia on the basis of branching pattern as described by Richter (1899). *Arachishypogaea* consists of two subspecies each containing two botanical varieties. Variety subspecies *hypogaea* (Virginia type) Variety *hirsute,* Subspecies *fastigiata*

Variety *fastigiata* (Valencia types)

Variety v*ulgaris* (Spanish type)

Hypogaea plants have alternate branching, spreading or bunching habit, lack inflorescence on the main stem, possess appreciable fresh seed dormancy and a long maturation period. This group corresponds to Virginia type. The botanical Variety *hirsuta* is the Peruvian runner type.

Plants of subspecies *fastigiata* are upright, have sequential branching and inflorescence on main stem leaf axils, posses little fresh seed dormancy and are of shorter duration then *hypogeaea* variety it belongs to valencia type. Botanical Variety *vulgaris* corresponds to Spanish type (Gregory and Gregory, 1976).

Table 3.1*Comparative characters of Virginia spanish/valencia*

Character	Virginia *(hypogaea)*	Vaencia/Spanish *(fastigiata)*
Seed size	Medium to very large	Small to medium
Seeds per pod	Usually 2, occasionally 3	Usually 2-4, rarely 3
Seed dormancy	Moderate	Little to none
Branching	Moderate to profuse	Sparse to moderate
Flower bud on main axis	Alternate	Sequential
Flowering pattern	Alternate	Sequential
Primary lateral branches	Longer than main stem	Shorter than main stem
Growth habit	Spreading to semi spreading	Bunch
Growth period	Long (130-175 days)	Short (90-105 days)
Peak flowering	8 to 12 weeks	6 to 10 weeks
Leaf colour	Dark green	Pale green
Leaflet size	Small	Large
Leaflet apex	Pointed	Rounded
Pod distribution	Scattered	Close to the base
Ability to grow after seed setting	Retains growing point if pods are detached	Usually dies
Linoleic acid	Low	High

***Table 3.2** Comparative characters of Valencia spanish types*

Character	Valencia	Spanish
Secondary lateral branches	Absent	Present irregularly to primary laterals
Stem and petiole colour	Purple	Green
Hairiness	Less	More
Leaflet colour	Dark green	Light green
Inflorescence	Single	Compound
Seeds per pod	Usually 3 to 4	Usually 2
Constriction, reticulation and beak on the pod	Less prominent	More prominent
Shell thickness	Thick	Thin
Testacolour	Red to purple	Rose
Maturity	90-95 days	100-105 days

Knowledge on Botany and reproductive biology of the plant is very much essential for a breeder to accomplish the objective of development of improved cultivars (A.J. Norden, 1980). Flowering pattern and flower structure majorly decides the method of pollination which ultimately is very important to determine the breeding procedures to be followed for development of new crop varieties. Extent of selfing or out crossing also determines the isolation distance to be followed during large scale seed production. Knowledge on flowering and flower structure is essential to decide upon the method of emasculation to be followed and pollination to be done. A breeder should also be aware of anthesis time to attempt the pollinations. A brief description of botany of groundnut crop including its reproductive biology and reproductive efficiency in groundnut is described (Fig. 3.1).

3.5 Habit

Groundnut plant is a short statured plant with main axis being upright but major part of plant consists of primary branches. Secondary and tertiary branches are found in the Sami-spreading and spreading types (Virginia).

3.6 Root

Tap root with four series of spirally arranged lateral roots with abundant branching and usually heavily supplied with nodules. Root tip is without epidermis and without root hairs. The well-developed tap-root may penetrate to a depth of 130 cm but rarely goes beyond 90 cm (Intorzato and Tella, 1960). The root system is normally concentrated at a depth of 5 cm to 35 cm (Narasinga Rao, 1936) and root spread is confined to a radius of 12 to 14 cm. There are cultivar differences in the vigour of the root system (Mohammad and Khanna, 1932; Seshadri*et al.,* 1958). The spreading types are usually more vigorous than bunch types). It can be concluded that root elongation depends on the cultivar, the nature and the moisture content of the soil.

Root hairs are characteristically absent in groundnut both on the main and lateral roots. Maximum root development is seen within 15 cm of the soil. Hence there is necessity of giving good tilth to the first 15 cm of the soil while growing groundnut. Deep cultivation is not necessary. It is only during the second month that there is maximum development of the roots and this synchronizes with the period of the maximum shoot development. The short duration varieties do not have the root system as well developed as is the case in the long duration ones. The bunch varieties generally do not have well developed root system compared to the spreading varieties.

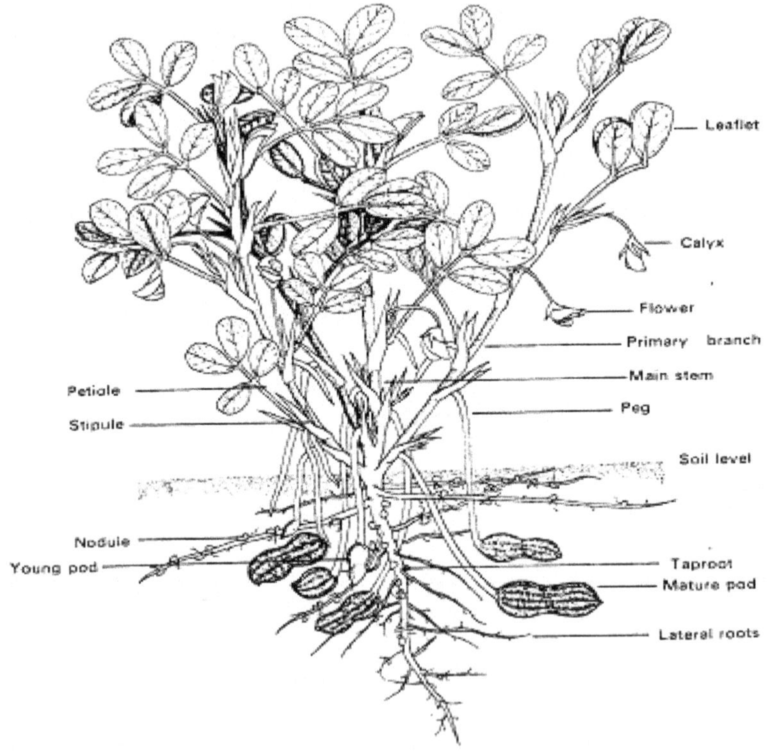

Fig. 3.1 Groundnut plant and its parts

3.7 Stem

The young stems are angular, usually pubescent and solid with large pith. As the plant ages, the stems become hollow and tend to be cylindrical and shed hairs. As *A.hypogaea*is a herbaceous annual, there is no indication of woody development. The main axis (n) develops from the terminal bud of the epicotyls (flanked by 2 opposite cotyledonary laterals (n+1). The height of the main axis is essentially cultivar-dependent, but is also influenced by environment to a great extent, and ranges from 12 to 65 cm. The thickness of the stem is highly variable although (Seshadri 1962) reported that generally the bunch types have thicker stems. Internodes are short and highly condensed at the base but are longer at the higher nodes.

3.8 Growth Habit

The cultivated groundnut plant has a distinct main stem and a varying number of lateral branches. The carriage of laterals is an important character which determines the growth habit of the plant. Two distinct forms of growth habits in groundnut – spreading (runner, trailing, procumbent, and prostrate) and erect (upright, erect bunch and bunch) – have long been recognized, and they provided the basis for several varietal classifications. The spreading form is usually characterized by an erect conspicuous main stem with procumbent or decumbent lateral branches. In the erect types the main axis becomes indistinguishable from the laterals. An intermediate semi-spreading form (spreading bunch, bunch runner, and runner bunch) is also recognized.

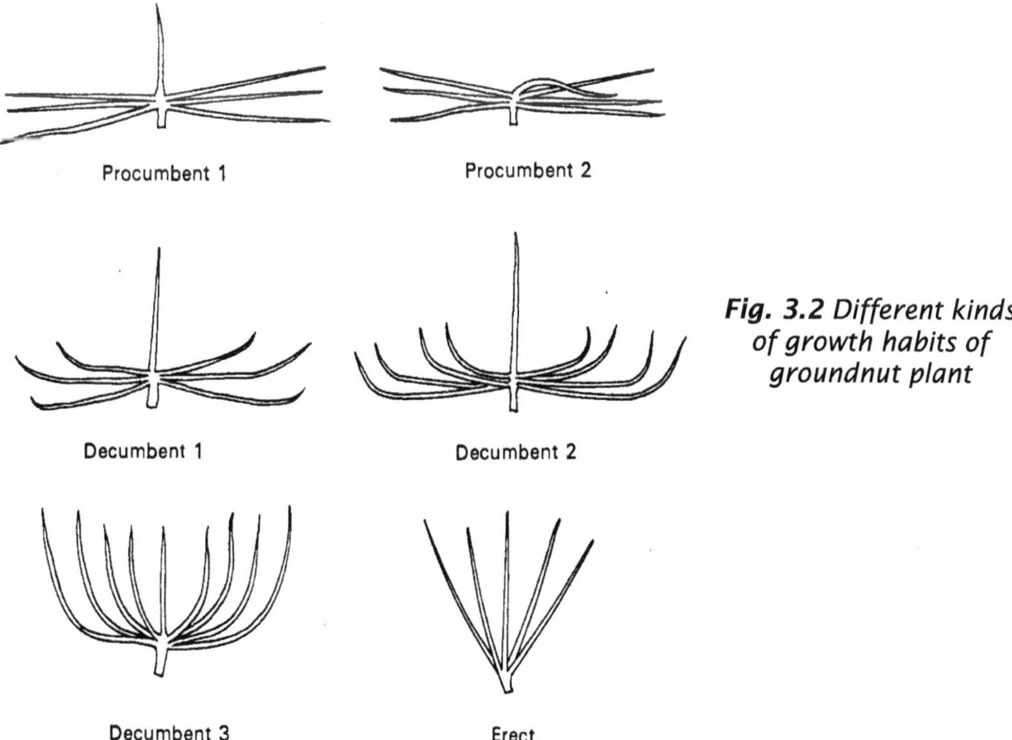

Fig. 3.2 Different kinds of growth habits of groundnut plant

Different kinds of growth habits have been observed in groundnut as described below (Fig 3.2).

1. **Procumbent1:**The main axis is erect and may vary in height. The lateral branches are prostrate.
2. **Procumbent 2:**This type is similar to Procumbent 1, but the main axis also tends to bend to the ground and continues to trail. This type is generally encountered in wild *Arachis.*The usual runner (=spreading, trailing and prostrate) habit group includes the above two forms.
3. **Decumbent 1:** The main axis is clearly distinct and the laterals, which normally trail on the ground, tend to become upright at distal nodes.
4. **Decumbent 2:** The main axis is clearly distinct and the laterals branches tend to become upright at the median nodes.
5. **Decumbent 3:** The main axis becomes indistinct because many lateral branches are almost upright and give the plant a bushy appearance.The above three forms can normally be included under spreading bunch (=semi-spreading, bunch runner, and runner bunch) from.
6. **Erect:**The lateral branches are at an acute angle to the main axis and give the plant an upright appearance. The main axis becomes almost indistinguishable.This form includes the erect bunch or bunch forms in the earlier classifications. As can be seen, the present system is based on only the carriage of plant and the position of primary branches in relation to the main axis (IBPGR/ICRISAT, 1990).

3.9 Branching and Flowering in Groundnut

The main axis of the plants is upright and from 15 cm to 35 cm long; but the greater part of the structure of the plant consists of or is derived from the primary branches.In all forms, the primary vegetative branches (the so-called n + 1 branches, the main axis being conventionally numbered (n) arise in the axils of the cotyledons and at a number of higher nodes on the main axis.

In the sequentially branched section, inflorescences are borne at the second and several subsequent nodes of the primary branches. The first node on such a branch may bear a secondary (n + 2) branch but it too often bears an inflorescence so that the first flowers are initiated very soon after germination. (In some varieties of this section, there may be a few n + 2 vegetative branches at high nodes of n + 1 branches). The first inflorescence is often borne below the surface, especially if the soil has been moved towards the plants by entirely below the surface and they are very short, with no vertical portion, so that the fruits lie very close to the tap root and the fruiting habit is consequently more or less compact. In the sequentially branched forms, inflorescences are also borne at some of the higher nodes of the main axis (Fig. 3.3).

In the *alternately branched* section, the first two nodes of n + 1 branches normally bear vegetative (n + 2) branches, the next two bear inflorescences, the nexttwo vegetative branches and so on. The same sequence is repeated on the n + 2 vegetative branches. The 2-2 alternation is not always exact, but groups of vegetative branches always alternate in this section with groups of inflorescences. In these forms, inflorescences are never borne on the main axis (Fig. 3.3).

Fig. 3.3 Branching and flowering in groundnut

3.10 Flowering and Anthesis

Flowering generally begins about 25 days after sowing. In the spreading types it may be a few days later. The flowers arise in the leaf axils either singly or in clusters of two to five. The dehiscence of the anthers takes place between 5 a.m. and 6 a.m. flowers open between 6 a.m. and 8 a.m.

For crossing purposes emasculation is done by opening the flower buds between 5 p.m. and 6 p.m. on the previous day, i.e., about twelve hours before the opening of the flowers. Artificial pollination of the emasculated flower is carried out the following day between 7 a.m. and 8 a.m., when the anthesis of flowers normally occurs. Bagging with muslin bags is not considered necessary since natural crossing is extremely rare.

However, bagging may be useful to protect the flowers from insect attack. Flowering initiated between 22 and 28 days in *kharif* (rainfed) and between 26 and 33 days in *rabi* (Irrigated) in different varieties. In general, cultivars belonging to Valencia group were early to flower followed by Spanish cultivars. Virginia cultivars were lateinto with regard to first flowering both under rainfed and irrigatedconditions (SubbaRao *et al.,* 1986).

General flowering occurred between 26 and 47 and 32 and 52 DAS under rainfed and irrigated conditions respectively. In general, Valencia Spanish group cultivars were early to general flowering followed by Virginia types. Both the Virginia bunch cultivars were late with respect to general flowering. (Smith, 1954;Subba Rao *et al.,* 1986). Flowering period was shorter in cultivars of Valencia group and relatively longer in Spanish cultivars. Virginia group cultivars had the longest flowering period. A marked difference in flowering period was observed between rainfed and irrigated crops.

A study of daily flowering pattern indicated that flowering in cultivars belonging to Spanish-valencia botanical group began after third week and attained maximum in about a week thereafter. The second flush in Spanish-valencia cultivars was of a lesser intensity as compared to first flush. In cultivars belonging to Virginia group, there were several flushes of varying intensities. However, the second flush was of relatively greater intensity as compared to other flushes. The second flush in Virginia cultivars was responsible for increased production compared to the first one in Spanish Valencia cultivars. (Divakar 1961 and Subba Rao *et al.,* 1986).

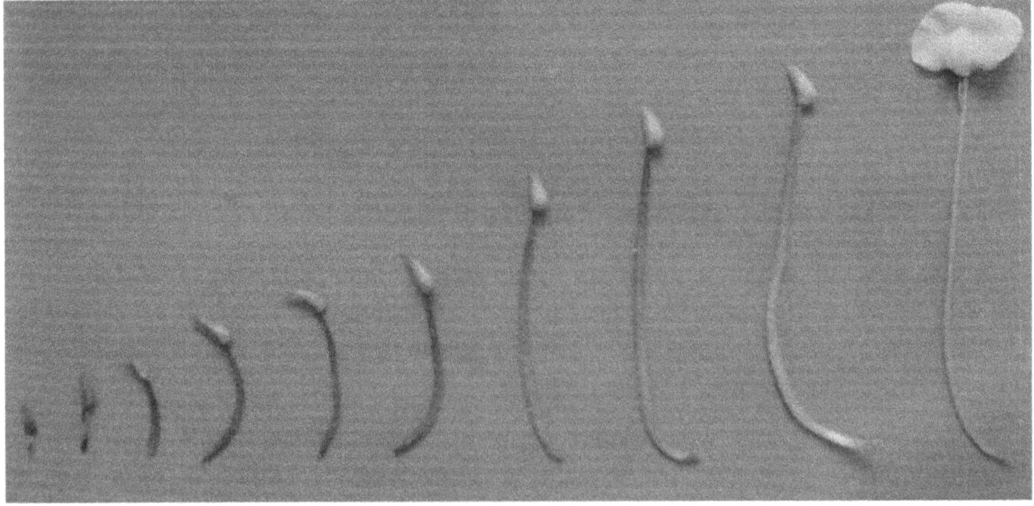

***Fig. 3.4** Elongation of hypanthium and development of buds in groundnut*

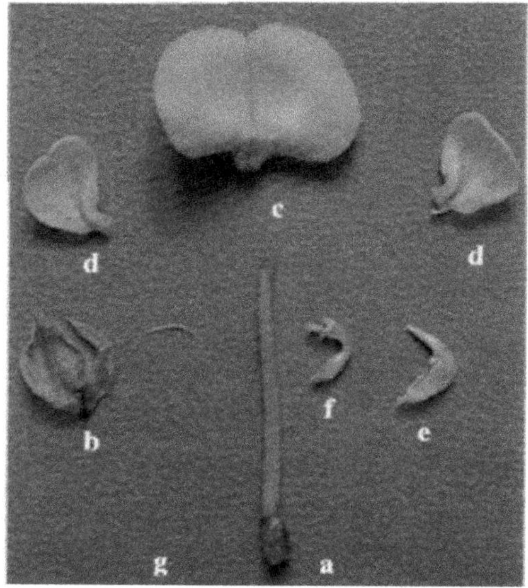

A dissected groundnut flower showing: (a) hypanthium, (b) single and fused sepals, (c) standard, (d) wings, (e) keel, (f) stamens, and (g) style and stigma.

Fig. 3.5 *Flowers structure and floral biology*

3.11 Flowers

The flowers are bisexual, zygomorphic, complete and sessile, flowers pea-like, enclosed between two bracts, one simple, subtending a very short peduncle, the other bifid, subtending the pedicel; sessile, but appear to be stalked after the growth of a tubular hypanthium just before anthesis (Fig. 3.4 and 3.5). The ovary is surrounded by the base of the hypanthium (perigynous), on the distal end of which are inserted two calyx lobes, one awl-like opposite the keel and the other broad and four-notched opposite the back of the standard. Petals are orange, yellow, cream or rarely white; inserted between the calyx lobes and the fused based of the anther filaments (staminal column). The standard is orange with red veins marking the more yellow, or yellow central face or brick-red by extension of the red veined area (Fig. 3.7). Wings yellow, or yellow at the base and orange apically, to brick-red; keel colorless to faintly yellow along its length. There are 10 Stamens, sterile filaments usually 2, anthers 8 (sometimes 9, rarely 10), 4 are globose, uniloculate, alternating with 4 oblong, 3 of which are biloculate and the one opposite the standard, uniloculate, dimorphic (Fig. 3.6). The tip of the ovary, bearing from 1-5 ovules, grows out from between the floral bracts, bearing with it the dried petals, calyx lobes and hypanthium; creating a unique floral structure-the peg.

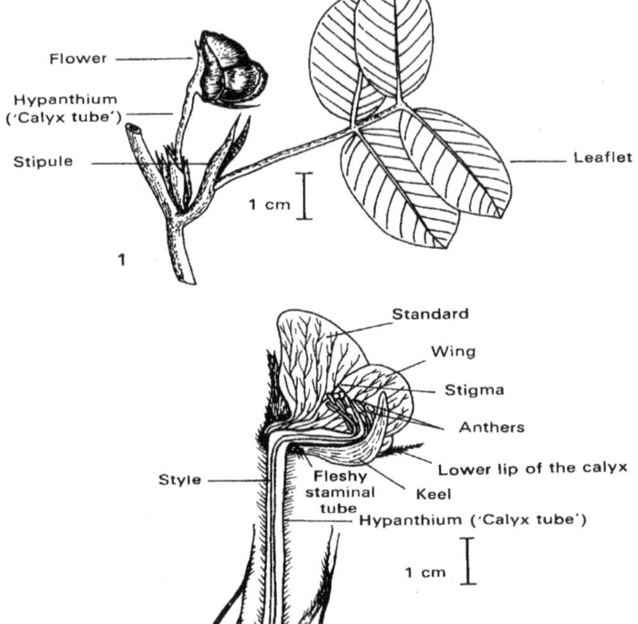

Fig. 3.6 *Groundnut flower and structure (Longitudinal view)*

Fig. 3.7 *Flower of groundnut crop*

3.12 Pollination and Fertilization

Self–pollinating, occasionally out-crossed by bees (Duke, 1981a). Pollination takes place early in the morning. As soon as the fertilization is complete, the flowers fade. After fertilization, an intercalary meristem becomes active at the base of the ovary above the point of attachment of the hypanthium, producing new tissue below itself and resulting in an elongated stalk, the peg. The pegs are positively geotropic, enter the soil and bend in a horizontal plane. Generally two, and occasionally one, three or four, fertilized ovules are borne at the tip of the peg which later swells to become the pod.

Groundnut is completely a self fertilised crop. The percentage of natural crossing in nature is practically negligible. The stigma protrudes above the anthers and is receptive at the time of dehiscence of anthers; cross-pollination does not take place because the stigma is still enclosed in the keel. Though the flowers are often said to be cleistogamous, they are never truly that. The flowers open and close within a short interval. After fertilization, the floral parts begin to wither and fall off.

3.13 Development of the Peg

The flowers which are sessile, after fertilization fade and fall off. The portion below the ovary as a result of the stimulus of fertilisation, begins to grow into a 'peg' like structure carrying the ovary at its apex. There is yet lot of differences of opinion about the morphology of the peg. It has been generally described as gynophores and by some as carpopodium. (Smith 1950) has reviewed the various interpretations and defines the peg as 'the young fruit during the stalk like phase of development'. The stalk like structure is not really the gynophores, but the basal portion of the ovary itself elongates by the growth of an intercalary meristematic region a little above the insertion of the calyx tube in the fertilized flower. The peg is visible about five to seven days after fertilization (Fig. 3.8).

Fig. 3.8 Photographs of peg and peg development

This peg grows towards the soil, i.e., it is positively geotropic. Though it has been proved to be positively geotropic, it is not negatively phototropic. The peg carrying the ovary pushes itself into the soil (Fig. 3.9). It is only after entering the soil that the ovary begins to develop and takes up a horizontal position. The flowers which are found in the basal branches very near the main stem alone develop pods. Flowers carried on the distal end of the branches and on the higher part of the plant either do not develop the pegs though fertilized or they develop pegs which do not reach the soil and hence the fertilized ovaries get aborted.

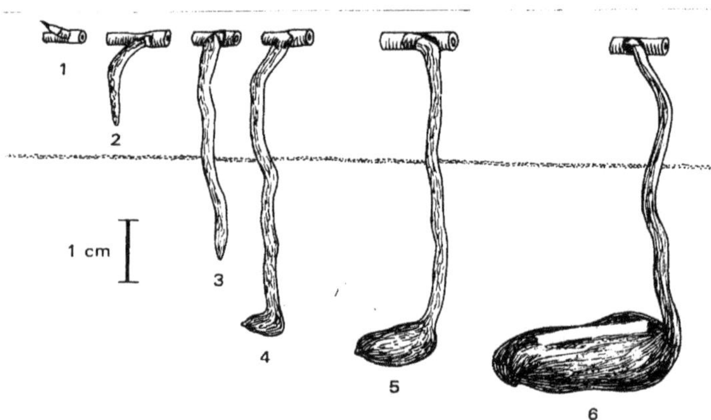

***Fig. 3.9** Stages of peg development*

3.14 The Fruit

Not all the pegs formed grow long enough to reach the soil and to develop into pods. Similarly not all the pegs that enter the soil may grow into mature pods. There seem to be significant cultivar differences in the proportion of pegs developing into pods (Seshadri, 1962; Hari Singh *et al.,*1981). The fruit of groundnut, normally referred to as a pod, is a lomentiform carpel, indehiscent and up to 10 cm in length. The mature pod normally contains up to 4 seeds, depending on the cultivar. Occasionally 5 or even 6 seeds per pod have been recorded. Single-seeded pods may be produced when all the ovules but the proximal abort. The pod size may range up to 6.0 cm X 2.7 cm. The fruit consists of 2 valves, structurally dehiscent but functionally indehiscent. When pressed the pods split along the longitudinal suture.

3.15 Development of the Pod

Before fertilization the ovule grows gradually up to flowering, and the growth continues after fertilization to some extent. The embryo is dormant during peg elongation and begins to develop 3-4 days after commencement of pod development (Murty*et al.,* 1980). The basal ovule develops first. When the peg has reached its maximum depth it becomes diageotropic and almost horizontal. The pod development is initiated when the peg ceases to elongate. The enlargement of the pod proceeds from base to apex. In general the pod expands rapidly underground by the development of a large parenchymatous tissue (endocarp) lying between the ovules and shell layers (Schenk, 1961). The endocarp recedes with the advancing growth of the ovules and disappears completely when the seeds associated with increased tannin content, and becomes very dark brown on maturation (Gregory *et al.,* 1973). Generally it takes about 60 days from the time of fertilization to full maturity (Patel. *et al.,* 1936). The dimensions of the mature pod and of the seeds are controlled by both internal and external factors such as genotype, soil, and method and time of cultivation (Shibuya, 1935). The normal podding zone is located 4-7 cm below the soil surface. At about 30°C the swelling of ovaries commences 2 days after and the pod attains its maximum size 3 weeks after the penetration of the peg into the soil. Higher or lower temperatures prolong the duration of pod development. The optimum soil temperature in the podding zone is 31-33°C (Ono, 1979).

3.15.1 Pod Beak

In groundnut the tip of the indehiscent fruit may end in an appendage called the beak. The prominence of the beak varies depending on the cultivar. This character has been used for the classification of groundnut cultivars (Gibbons *et al.,* 1972; Varisai Muhammad *et al.,* 1973a). Generally 4 grades of beak have been recognized in the past: not prominent, slightly prominent, prominent and very prominent. The presence or absence of a beak is an excellent diagnostic character. Five grades can be used to classify groundnut based on this character and they are: absent, slight, moderate, prominent and very prominent .There can be some variation between the pods of the same cultivar and the observation may be subjective; nevertheless, it is very useful. The shape of pod beak also varies (Fig. 3.11).

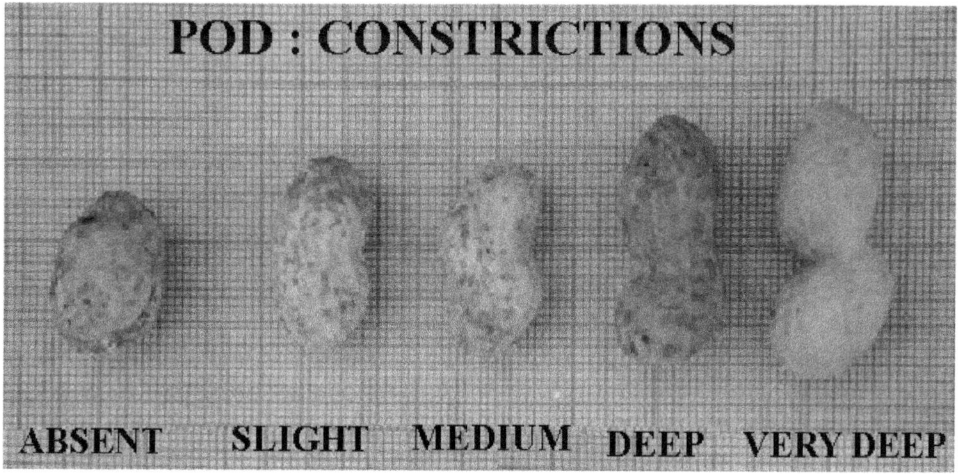

Fig. 3.10 Photographs of pods with different types of constricts

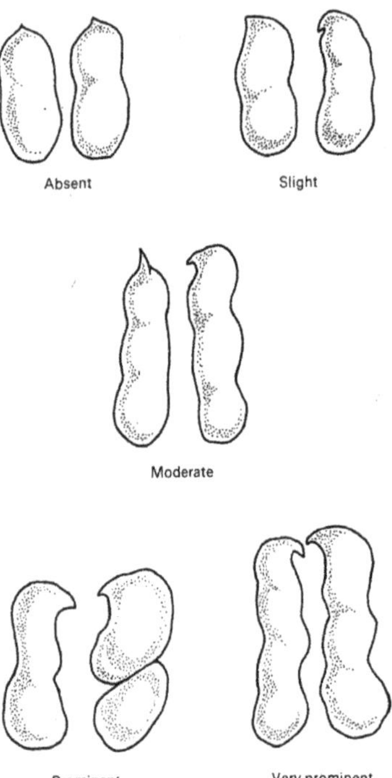

Fig. 3.11 Different types of beaks of groundnut Pods

3.15.2 Pod Constriction

The constriction of the pod is an interesting characteristic as it affects the developing seed. Non-constricted pods generally have seeds with flattened ends and are in contact with other seeds in the pod. The contact sometimes can result in damage to the embryo (Coffelt and Hammons, 1974). Such seeds also tend to split during shelling. On the other hand, pods with deep constrictions tend to carry soil on them. They also break during harvesting and shelling, reducing the market value. Most commercial cultivars have constriction between each seed; a few do not, especially the cultivars belonging to subsp. *fastigiata.* Under abnormal conditions (such as aerial fruiting), elongated pods with a filiformisthmus may be produced by a cultivar that is normally only slightly unconstructed (Smartt, 1976 a). In several wild species of groundnut a filiform isthmus is characteristic.

Pod constriction shows considerable variation both within and between cultivars (Fig 3.10). However, the range of variation is not great enough to prevent visual classification. Badami (1928) described 4 groups based on the depth of constriction. Varisai Muhammad *et al.,* (1973) recognized 3 grades of constriction: shallow, medium, and deep. All groundnut cultivars can easily be graded into 5 constriction groups: none, slight, moderate, deep and very deep.

3.15.3 Pod Reticulation

As stated earlier the mechanical tissue of the shell gives a reticulated or ridged appearance to the pod. Mostly the longitudinal ribbing is prominent, but vertical ridges appear to be prominent in some cultivars. Reticulation (Venation, ribbing, ridging) on the shell is a prominent visual characteristic. In general, most of the groundnut cultivars exhibit reticulation to different degrees. Typical var. *fastigiata* cultivars may be without any reticulation; however, some Peruvian forms show very prominent ribbing. As with constriction, reticulation also contributes to some extent the cleanliness with which pods come out of the soil at the time of harvest. (Varisai Muhammad *et al.,*(1973 a) recognized 4 grades of reticulation: non-distinct, slightly distinct, distinct, and very distinct. At the ICRISAT, 5 grades are distinguished: smooth, slight, moderate, prominent and very prominent.

3.15.4 Pod Size

(John *et al.* 1954) held that though the size of the pod is influenced to a certain extent by soil and other environmental conditions, it contributes to distinguish the cultivars. They classified the pod size arbitrarily into 5 grades. (Varisai Muhammad *et al.* 1973b) reported a range of 8-20 mm in diameter. They divided the groundnuts into 5 classes depending on the mean length and pod weight (g/100 pod): very small (less than 1.5 cm and 35-50 g), small (1.6-2.0 cm and 51-65 g), medium (2.1-2.25 cm and 65-105 g), big (2.6-3.0 cm and 160-155g), and very big (more than 3 cm and more than 155g). At the ICRISAT a range of 11-83 mm for pod length and 9-27 mm for pod diameter has been recorded.

3.16 Reproductive Efficiency

Studies made on percentage of flowers produced to other component factors such as peg, total pods and mature pods in groundnut revealed that the reproductive efficiency appeared to be a highly flexible process influenced by the climate, genotype, soil and other factors. Reproductive efficiency plays a major role in deciding the yield of groundnut varieties. Cultivars with higher reproductive efficiency may yield better due to higher number of mature pods per plant. According to (Beldose and Harries 1950), of the total flowers produced 58.7 percent produced gynophores and flowers producing fruits were only about 9.2 percent. The gynophores entering fruiting medium were about 64.3 per cent and of them 24.4 per cent produced fruits.

(Seshadri 1962) reported that the percentage of fertilization in groundnut was 4.9 to 67.5 per cent. In general this percentage was lower in spreading forms (4.8 to 58.9) than in bunch forms (21.9 to 67.5 per cent). Though total number of flowers produced in bunch types was low, the percentage of fertilization was high. It was also reported that at the time of harvest fertilized flowers were seen at different stages of development some as pegs, some as immature pods etc. The total percentage of the fraction failed to develop into mature pods was about 14.7 to 38.1 and was usually more in spreading forms than in bunch forms. Fruiting studies in bunch varieties revealed that the percentage of fertilization in groundnut was between 31.1 to 60.1 and pod setting was between 13.2 to 22.6 per cent of total flower production. Similar fruiting studies in spreading varieties indicated that the percentage of fertilization was between 14.0 to 49.6 and pod setting varied from 4.4 to 24.9 per cent of total flower production (Seshadri, 1962). It was concluded that though differences occurred in flowering pattern in virginia and bunch cultivars, the percentage of fruit set and immature pods were similar for all varieties (Goldin and Hartzook, 1966).

From the flowering and fructification studies it was confirmed by Wika layer (1975) that the varieties differed in number of flowers produced and the number of flowers that produced fruits varied from 11.2 to 19.5 percent. From the total flowers produced, 50 per cent developed into pegs and 22 percent produced mature pods (Rahman and Miah, 1978). The percentage of mature pods to flowers ranged from 12.34 to 17.02 and 11.50 to 19.37 during *kharif* and *rabi* respectively. The percentage of pegs to flowers ranged from 38.92 to 55.64 and 39.30 to 69.92 under rainfed and irrigated conditions respectively. A range of 24.65 to 43.23 and 16.86 to 46.25 under rainfed and irrigated conditions was observed for percentage mature pods to pegs. The percentage of total pods to flowers varied from 15.16 to 23.49 and 16.08 to 28.73 under rainfed and irrigated conditions respectively (Subba Rao and Raja Reddy, 1990).

References

1. *Badami V.K. 1928. Arachis hypogae (The groundnut). Ph.D. thesis (Vol.II Inheritance studies), Cambridge (England) University Library (unpublished) : 297-374.*
2. *Beldose R.W. and Harris C.H. 1950. The influence of Mineral deficiency on Vegetative growth flower and fruit production and Mineral Composition of Peanut Plant. Plant Physiol. 25: 63-77.*
3. *Coffelt T.A. and Hammons R.O. 1974. Inheritance of pod constriction in peanuts. Journal of Heredity 65: 94-96.*
4. *Divakar C.B. 1961. Flowering in groundnut varieties. Indian Oilseeds J. 5 (4): 264-273.*
5. *Duke J.A. (1981). Handbook of legumes of world economic importance. Plenum Press. New York.*
6. *Gibbons R.W., Buniting A.H. and Smartt J. 1972. The classification of varieties of groundnut (Arachishypogaea L.) Euphytica 21: 78-85.*
7. *Goldin E. and Hartzook A. 1966. Observations on the flowering and reproduction of groundnut (Arachishypogaea L.) in Inreal. Insrael. J. agric. Res. 16(1): 3-10.*

8. Gregory W.C. and Gegory M.P. 1976. Groundnut In : Simmons NW (Ed) Evolution of crop plants. Longman, New York.
9. Gregory W.C. Gregory M.P. and Krapovickas A. 1973. Structure, variation and evolution and classification in Arachis. In Advances in Legume Science (Summerfield RJ, Bunting AH, eds.). Royal Botanical Gardens, Kew, U.K. 469-481.
10. IBPGR/ICRISAT (1990). Classification of Arachis. In Report of a Workshop on Genetic Resources of Wild Arachis Species, Including Preliminary Descriptors for Archis IBPGR International Crop Network, Series 2. IBPGR, Rome, Italy.
11. Intorzato R. and Tella, R. 1960. Sistema radicular do amendoim. Bragantia 19: 119-23.
12. John C.M., Venkatanarayana, G. and Seshadri C.R. 1954. Varieties and forms of groundnut (Arachishypogaea Linn.) Their classification and economic characters. Indian Journal of Agricultural Sciences 24: 159-93.
13. Mohammad A.A. and Khanna K.I. 1932. Studies on germination and growth in groundnut. Agriculture and Live-Stock in India 3: 91-115.
14. Murty U.R., Rao N.G.P., Kirti P.B. and Bharati M. 1980. Fertilization in groundnut, Arachis hypogaea L. Oleagineux 36: 73-4.
15. Narasinga Rao V. 1936. A note on preliminary physiological observations on the groundnut. Proceedings of the Association of Economic Biologists, Coimbatore, Tamil Nadu : 1-6.
16. Norden, A.J. 1980. Proceedings of International Workshop on Groundnut, ICRISAT, Hyderabad : 58-60.
17. One, Yoshitaka 1979. Flowering and fruiting of peanut plants. JARQ 3: 226-9.
18. Patel J.S., John C.M. and Seshadri C.R. 1936. The inheritance of characters in the groundnut – Arachis hypogaea. Proceedings of the Indian Academy of Sciences 3: 214-233.
19. Rahman L. and Miah M.N.I. 1978. Selection for number of pods and oil content on the basis of flowering pattern and coefficients of fertility and productivity in groundnuts. Bangladesh Journal of Scientific and Industrial Research. 13 (1/4): 1-8.
20. Reddy, S.R. 2012. Agronomy of field crops, Kalyani Publishers, New Delhi, 348-382.
21. Richter, C.G. 1899. BeitragezurBiologie von Arachishypogaea, A. Schrieber (Ed.) Inaug. Diss. Kgl: 39.
22. Seshadri C.R. 1962. Groundnut. Indian Central Oilseed Committee, Hyderabad, India.
23. Seshadri C.R. and Srinivasulu N. 1958. Studies on the branching pattern in groundnut (Arachis hypogaea). Indian Journal of Agricultural Sciences 28 : 175-80.
24. Seshadri C.R. ed. 1962. Groundnut – A monograph. Indian Central Oilseeds Committee, Hyderabad: 50-67.

25. *Shibuya T. 1935. Morphological and physiological studies on the fructification of peanut (Arachis hypogaea L.) Memoirs of the Faculty of Science and Agriculture Taihoku Imperial University 17: 1-120.*

26. *Smartt J. 1976a. Comparative evolution of pulse crops. Euphytica 25: 139-42.*

27. *Smith B.W. 1950. Arachishypogaea L. Aerial flower and subterranean fruit. American Journal of Botany 34: 802-815.*

28. *Smith B.W. 1954. Arachishypogaea, reproductive efficiency. American Journal of Botany 41(8): 607-616.*

29. *Subba Rao M. and Raja Reddy C. 1990. Reproductive Efficiency in Groundnut (Arachis hypogaea L.)J. Res. APAU 18(1): 63-67.*

30. *Subba Rao M., Raja Reddy C. and Reddy C.D.R. 1986. Flowering pattern in Groundnut (Arachis hypogaea L.)The Andhra Agric. J. 33 (2) : 148-149.*

31. *Varisai Muhammad S., Ramanathan T., Ramachandran M. and Mohanbabu G. 1973. Variation in kernel weight and shelling out-turn in Arachishypogaea L. Madras Agricultural Journal 60:1394-8.*

32. *Varisai Muhammad S., Ramanathan T. and Ramachandran M. 1973a. Variation in morphological characters of Arachishypogaea L. Madras Agricultural Journal 60:1373-9.*

33. *Varisai Muhammad S., Ramanathan T. and Ramachandran M. 1973b. Classification of Arachishypogaea L. var. procumbens. Madras Agricultural Journal 60: 1403-8.*

34. *Wikalayer V.S. 1975. The biological features of flowering and fruiting of Arachis in Ginea. SelskokhozyaistVennaya. Biologia. 10(6): 933-936.*

Chapter 4

Groundnut Improvement – Methods, Status and Issues

SN Nigam

Former Principal Scientist (Groundnut) and Expert-Agriculture for Development, Hyderabad, India.

4.1 Introduction

Groundnut is a highly self-pollinated crop, though cross pollination can reach as high as 10% at locations and in seasons, where bee activity is high.

4.2 Methods of Cultivar Development

4.2.1 Standard Methods

Standard breeding methods, followed in groundnut for developing a cultivar, can be placed into two groups – (a) methods without hybridization, and (b) methods after hybridization. The former includes introduction, pure line/mass selection and mutation breeding and the latter bulk selection, pedigree selection, bulk pedigree selection, single seed descent method, back cross method and recurrent selection. Among these, pedigree and bulk selection methods are more commonly used by the breeders. Some breeders use a combination of breeding methods or make modification to conventional methods. (For detailed description of different breeding methods, readers are advised to consult any standard plant breeding book.)

4.2.1.1 Methods without Artificial Hybridization

These methods involve selection in existing (land races) or introduced (introductions) variability or variability created without sexual process (mutation breeding and genetic transformation).

Introduction: Plant introductions are useful in bringing in additional genetic diversity in a breeding program. Sometimes they can also be released directly as cultivars if found suitable. Indian cultivars TMV 2 and JL 24 have been introduced and released under different names in many countries in Southeast Asia and Africa.

4.2.1.2 Pure line/Mass Selection in Introduced Material

These methods involve selection and subsequent seed increase for release as cultivars of single plants (pure line) or group of plants (mass) selected from introduced material and landraces, etc. Many cultivars in Africa (Makulu Red, Apollo, Egret, Chalimbana, Mani Pintar and Malimba), India (JL 24) and the USA (New Mexico Valencia C) have been developed through pure line or mass selection in introduced material.

4.2.1.3 Mutation Breeding

X-rays, gamma-rays and various chemicals are used to break specific linkages or create variation for specific characters for further use in a breeding program. Gamma-rays are extensively used at BARC, Mumbai to create desired variation for further use in conjunction with other breeding methods. BARC released 15 groundnut varieties; some of these are TG 19, TAG 24, TG 37A,TG 38B, TGB 39, TPG 41, TLG 45, TG 51.

4.2.1.4 Methods after Artificial Hybridization

Hybridization provides opportunity to combine genes from different parents.Choice of parents is very critical for the success of a breeding program. Parents can be crossed in different manners to access desirable genes from them and recombine them (genes) in a single genotype - single cross, three-way cross, four-way cross, convergent cross, diallel mating, diallel selective mating.It usually takes 12-15 years after hybridization to develop a new cultivar. This period can be shortened by taking multiple crops in a year in a glasshouse under controlled conditions (and following single seed descent method) or raising off-season nursery at other locations where environmental conditions are favourable to raise a crop. In crosses between parents having equal or nearly equal number of alleles (k=0.5), selfing provides the greatest probability of recovering the desired plants. As k approaches 1, the probability of recovering desired plants by backcrossing to the better parent becomes high.

4.2.1.5 Bulk Selection

F_1 plants are space-planted to produce as much F_2 seed as possible, which is planted *en masse* the next generation. F_3 seed harvested in bulk (limited selection for easily identifiable traits (high heritability) such as plant type or seed size can be practiced on F_2 plants) and a sample planted the next generation with the process being repeated until F_6 generation. At this stage, a portion of the seed is space-planted, and individual plants selected for evaluation in progeny test. Undesirable plants are discarded from the F_6 derived progenies to make them uniform for replicated on-station and multilocation evaluation. Because of its low cost and little efforts required in generation advance, the method continues to be used in breeding programs. It can be used to select for desirable traits in segregating populations originating from inter-subspecific crosses between adapted and unadapted germplasm. But it is less effective in selection in intra-subspecific crosses. However, bulk selection may cause genetic shifts in segregating populations. It should be performed for few generations only.

4.2.1.6 Pedigree Method

It is a very commonly used method in spite of extensive record keeping of line descent or pedigree of each plant. Individual plants are selected based on desired traits in space-planted F_2 populations. Selected plants are then space-planted in progeny-rows in F_3 generation and again the best plants are selected and progeny-rowed in F_4. Selection process is repeated in F_5 and the selected plants are advanced to F_6 or F_7generation. When progenies become uniform (with each generation of selfing, the homozygosity increases), selection is made on plot basis. Selected genotypes (uniform experimental line) are then evaluated in replicated trials on-station and at multi location. Selection can also be carried out in a sequential manner at different locations to enhance adaptability and stability of the selected populations.The method was first used in the USA in (Florida in 1928). Most of the cultivars developed by pedigree method are F_5-derived lines.

4.2.1.7 Bulk Pedigree

This method, a combination of pedigree and bulk methods and often not described in breeding books, is practiced in groundnut breeding program at ICRISAT. It differs from bulk method that selected individual plants in F_2 population are bulked into different bulks based on morphological traits including pod and seed characteristics and their record of descent is kept.The method allows retention of enough subtle variation in the resultant uniform looking bulk populations for recipient breeders to exploit it while evaluating them under local conditions.

4.2.1.8 Single Seed Descent (SSD)

It is a modification of the bulk method that can be used to inbreed populations in a shorter length of time. F_2 seeds obtained from space-planted F_1 plants are dense planted and a single seed or pod or small quantity of F_3 seed is collected from each F_2 plant and bulked which is again grown densely and the process of selection is repeated until the desired number of inbreeding generations are achieved (generally F_6). At this stage, the final selected population is space-planted for selection of individual plants for evaluation in progeny test.Undesirable plants are discarded from the F_6 derived progenies to make them uniform for replicated on-station and multilocation evaluation.Following this method, a breeder can raise three generations in 12-14 months in a glasshouse and breaking seed dormancy before planting, if required.

4.2.1.9 Backcross

The method is mostly used for transferring desired traits from a donor source to existing cultivar. Occurrence of small number of simply inherited traits of economic importance in groundnut limits the use of this method. However, with increasing emphasis on disease resistance breeding, this method is gaining popularity among breeders. It was first used in Malawi (Southern Africa) to incorporate groundnut rosette virus disease resistance governed by two recessive genes into high-yielding cultivar Makulu Red. In the USA, it was used to transfer high oleic trait into acceptable agronomic backgrounds. For more details on two or three stages of crossing and selfing per cycle of selection, minimum number of plants required in each stage and cost effectiveness, consult Isleib, 1997 (*Crop Science 37: 139-144*).

4.2.1.10 Recurrent Selection

Recurrent selection has received little attention in groundnut breeding because of efforts needed to make a large number of crosses.The method allows incorporation of diversity in breeding programs while providing opportunities for recombination. It enhances the probability of incorporation of several desirable traits from different sources in a single genotype. Peanut breeding program at NC State University, NC, USA has extensively used this breeding method.

4.2.1.11 Diallel Selective Mating (DSM)

It is a form of recurrent selection, which allows use of a selected group of parents and provides opportunity for multiparental gene combination. Offspring of two-way crosses can be crossed to each other or can be grown out and examined for important characteristics and only selected plants are intercrossed. At any stage in the program, material can be intercrossed or carried through a pedigree, bulk or SSD selection program to obtain potentially new cultivars.

4.2.2 Issues Involved in Hybridization and Selection

4.2.2.1 Single Cross vs. Complex Crosses

Complex crosses (three-way, double cross, convergent crosses, etc.) are generally recommended for obtaining desirable combinations of traits from different sources. Some researchers found three-way crosses superior to single crosses, while others did not find any difference between pedigree selections from single and double crosses.

4.2.2.2 Inter-subspecific vs. Intra-subspecific Crosses

In addition to releasing greater diversity, inter-subspecific hybridization also results in deleterious albino, brachytic, dwarf and sterile plants in segregating generations. Inter-subspecific crosses generally show more heterosis than intra-subspecific crosses.

4.2.2.3 Reciprocal Cross Differences

Direction of the cross has a bearing on the types of segregants obtained. Reciprocal differences are observed due to cytoplasmic and maternal effects. These two need to be separated in reciprocal effects as cytoplasmic effects have a significant bearing and the maternal effects have very little bearing on breeding programs. Reciprocal differences often occur in inter-subspecific crosses (Spanish/Valencia types x Virginia types).

4.2.2.4 Maternal Effects

Maternal effects are observed particularly when studying seed characteristics (such as dormancy and fatty acid composition, etc.) as they are influenced by the mother plants on which seeds are borne. However, such maternal effects last for only one generation and have little consequence in breeding programs.

4.2.2.5 Early Generation Testing

Normally early generation testing is done to identify superior crosses and eliminate inferior crosses early in the breeding program, thus enhancing breeding efficiency. No clear and significant advantage in early generation testing is observed in groundnut as most breeding efforts deal with quantitatively inherited traits. Traits with higher heritability can only be selected effectively in early generations.

4.2.2.6 Index Selection

Index selection allows concurrent improvement of several traits in early generations which go into the making of selection index. A selection index may contain traits with higher heritability, which are related with other economically important desirable traits that have low heritability.

4.2.2.7 Natural Hybridization

Depending upon the level of bee activity, variety and weather conditions, the outcrossing rate may vary from < 1% to 10%. If natural hybridization occurs between the genotypes of two subspecies, it is easy to identify the natural hybrids in an otherwise uniform parental population. If it occurs between the genotypes within a subspecies, the identification becomes difficult. Natural hybridization causes genetic instability in the mother genotype as the resultant hybrids and their progenies genetically contaminate the mother genotype. Natural hybrids can also be exploited to develop new cultivars. However, the pollen parent in natural hybridization remains unknown.It was exploited as early as in 1923 to develop groundnut cultivars in Indonesia. At ICRISAT, cultivars ICGS 11, ICGS 44, ICGS 37, etc. originated from natural hybrid population of Robut 33-1.

4.2.2.8 Genotype x Environment Interaction (G × E)

Desirable quantitative traits usually have both genetic and environmental components. Separation of these components is necessary to achieve maximum efficiency in breeding. Significant G x E interaction has been observed for most of the quantitatively inherited traits confirming the need of extensive multiyear and multilocation testing prior to cultivar release.

4.2.2.9 Multiline Variety

A multiline variety consists of a bulk of phenotypically similar but genotypically dissimilar sister lines. Sister lines are maintained individually and bulked prior to each seed increase of Foundation seed. This method allows the breeder to improve the cultivar after its release by adding or dropping component lines. The main advantage of a multiline cultivar is in its wide genotypic variability and stability. The disadvantages are: less uniform, seed stocks difficult to maintain anda multiline cultivar is lower yielding than the best component line. It has been successfully used in Florida breeding program in the USA to develop Florunner and Florigiant multiline cultivars.

4.3 Farmer-Participatory Varietal Selection (FPVS) and Plant Breeding (FPPB)

These two approaches in a breeding program empower farmers to influence the outcome of the plant breeding research to suit their requirements. These also result in wider acceptability of new cultivars by the farmers.

4.3.1 FPVS

It allows farmers to select among advanced breeding lines or genetically stable populations through on-farm farmer-participatory varietal selection trials. Various models of FPVS trials are being pursued in on-farm trials (researcher-designed and researcher-managed, researcher-designed and farmer-managed and farmer-designed and farmer-managed).The trials canfollow various experimental designs based on their purpose and availability of seed material etc. Mother-Baby trials appear to be most popular design in FPVS. FPVS meets immediate need of better cultivars of the farmers as it identifies the material best suited to the farmers' need and socio-economic conditions and agroecological conditions prevailing in the locality. Also, it allows multiplication and wide distribution of seed of farmer-preferred varieties through on-farm FPVS trials. FPVS also helps to identify varieties which should be used as parents in future hybridization program. Recent example of FPVS in groundnut in India include - release of ICGV 91114 in 2006 in Andhra Pradesh, in 2008 in Odisha and in 2009 in Karnataka, of ICGV 87846 in 2010 as CO 6 and of ICGV 00351 in 2013 as CO 7 inTamil Nadu in a short-period of time in India.

4.3.2 FPPB

This approach allows farmers' participation in a crop improvement programat various levels, which is non-existent inconventional breeding. In complete participatory breeding model, farmers' involvement is there at all stages – selection of source germplasm, trait development (pre-breeding), cultivar development and varietal evaluation. In an efficient participatory breeding, farmers are involved at the selection of source germplasm and varietal evaluation stages. In traditional farmer breeding, all stages of crop improvement are conducted by the farmers themselves without any input from the scientists.FPPB consumes more resources if selection activities are carried out in farmers' fields. It should be used when the possibilities of FPVS have been exhausted. FPPB can be conducted at several locations with the same breeding populations to identify the best material suited to local conditions by exploiting G x E for the benefit of farmers. However, cultivation of several varieties enhances genetic diversity in farmers' fields but it may pose problem in seed production and certification program. FPPB is a 'demand driven' process whereas the conventional breeding is 'supply driven'. It is desirable to involve market players and processors also at appropriate stage in the evaluation of the varieties/breeding materials.

4.4 New Approaches and Methods

Groundnut has lagged behind in use of molecular genetic technology due to the shortage of genome infrastructure, tools and resources and low levels of molecular polymorphism observed in cultivated species. Molecular markers and dense genetic linkage map are necessary for the application of marker-assisted breeding in crop improvement. Infrequent and low polymorphisms have restricted the progress in the development and application of genetic maps in groundnut breeding except in cases where polymorphic chromosomal regions have been introgressed into *A. hypogaea* from diploid relatives. (Pandey et al. 2012) reviewed the advances in *Arachis* genomics. Their publication lists *Arachis* markers in public domain, main populations used in *Arachis* genomics research, details of some major genetic maps constructed in *Arachis* species and QTLs identified for some economically important traits in groundnut.

4.4.1 Marker-Assisted Selection (MAS)

MAS was proposed a plant breeding tool in 1923, based on principle – 'use of easy-to-score phenotypes to select difficult-to score or low heritability traits that are linked to them'. A good marker should permit separation of homozygotes from heterozygotes to allow more genetic gain per generation, have early expression in the plant to save time on waiting for the desired phenotype to develop, and not have interaction with other markers. Initially breeders depended on morphological markers but these had either dominance effects, late expression of phenotype, epistatic relationship or had deleterious effect on the plant and as such did not qualify as good markers. Efficiency of MAS is dependent on quality of mapping process, which requires substantially contrasting parents for target trait, precise phenotyping technique and large mapping population. Cost effective application of MAS requires one marker very close (or preferably within) to the gene of interest and two markers closely flanking both sides and that these markers are based on simple robust PCR-based marker-assays. Construction of a genetic linkage map is necessary to facilitate QTL analysis and gene tagging to apply marker-assisted selection in a crop.

4.4.1.1 Polymorphism

Polymorphism is a natural variation in a gene, DNA sequence, or chromosome that has no adverse effects on the individuals and occurs with fairly high frequency in general variation. It involves one of two or more variants of a particular DNA sequence. Most common polymorphism involves variation at a single base pair but can also be much longer in size and involve long stretches of DNA. Polymorphism is essential for marker-assisted molecular breeding. Abundant polymorphism occurs in wild *Arachis* species but only limited or moderate polymorphism occurs in cultivated groundnut.

Morphological traits are altered by one or a few major genes with expression influenced by modifiers and epistatic interactions and intense selection pressure under cultivation results in diversification. Variation for biochemical and molecular markers, which are not subject to direct selection, often decreases during domestication. Apparent lack of variation at the molecular level could be due to isolation of cultivated groundnut from other *Arachis* species soon after polyploidization of the single hybridization event between *A. duranensis* and *A. ipaensis* followed by successive selection during breeding efforts.

4.4.1.2 Quantitative Trait Loci (QTLs)

QTLs are stretches of DNA containing or linked to the genes that underline a quantitative trait. Identification and mapping of QTLs is required for marker-assisted breeding.

4.4.1.3 Molecular Markers

Molecular markers have great potential for increasing breeding efficiency because they have large number of polymorphisms,alternate alleles are rarely deleterious at the molecular or whole plant level, are often co-dominant, allow all genotypes to be distinguished in each generation, rarely segregate in epistatic ratio, scoring of molecular markers is not dependent upon gene expression, arenot affected by environment, an accurate genotype can be established using any plant tissue at any development stage, are useful in pyramidizing desirable genes and in tracking them through generations of backcrossing, and have reduced requirement of the time and space necessary to evaluate plant population. However, a large number of markers must be evaluated for association with various traits and the linkage between markers and desired traits must be known for effective breeding. Different types of markers are described below.

4.4.1.3.1 Isozymes

Any two proteins that catalyse the same biochemical reaction but differ in chemical composition are called isozymes. Isozymes were the earliest molecular marker systems used for plant analyses based on polymorphisms in enzyme mobility. They were not found useful in characterizing polymorphism in cultivated groundnut.

4.4.1.3.2 Restriction fragment length polymorphisms (RFLPs)

RFLPs were the first marker system that had a large number of polymorphisms. They were widely used both to create linkage maps and implement indirect selection strategies. They can be used to study both recessive genes and multiple alleles. Being co-dominant, they rapidly identify homozygous individuals. RFLPs are produced by digesting DNA with restriction endonucleases that recognize specific DNA sequence and then cleave the DNA strand in or near the sequence. Radioactivity is used to label the probes and bands are visualized in an auto radiograph. High cost and time involved in the assay discourage the use of these markers. Little variation is detected by RFLPs in cultivated groundnut but significant variation is observed among wild *Arachis* species.

4.4.1.3.3 Random amplified polymorphic DNA (RAPD)

RAPD is a polymerase chain reaction (PCR)-based marker system that detects only dominant markers. It requires only small amount of DNA to screen large number of markers, does not require radioactivity, is very sensitive to polymorphism and is easy and rapid to perform. It detected very little variation in cultivated groundnut but among wild *Arachis* species a large amount of genetic variation was detected.

4.4.1.3.4 Sequence characterized amplified regions (SCARs)

They are also a PCR-based marker system. SCAR is a genomic DNA fragment at a single genetically defined locus that is identified by PCR amplification using a pair of specific oligonucleotide primers. They are superior to RADP markers as they are less sensitive to reaction conditions, do not require radioactivity, and can be used as co-dominant genetic markers. Dominant SCARs may be used as a quick plus/minus assay for a particular product.

4.4.1.3.5 Amplified fragment length polymorphisms (AFLPs)

AFLPs are also a PCR-based dominant marker system that combines benefits of RFLPs and RAPDs. Like RAPDs no knowledge of nucleotide sequence is required. About 10-fold increase over RAPDs in number of loci screened observed with AFLPs. But, it showed less variation in cultivated groundnut.

4.4.1.3.6 Diversity arrays technology (DArT) markers

DArT is a DNA segment, present or absent in a defined genomic representation, depending upon the individual genotype. Many methods are available to produce a genomic representation. They are not very useful in genetics and breeding applications.

4.4.1.3.7 Simple sequence repeats (SSR) or microsatellite markers

SSR system is more variable than other marker systems. They are co-dominant and easily detected from relatively little amount of DNA after PCR amplification. Higher levels of polymorphism particularly with longer TC motif repeats are observed with these marker systems, which are transferable among related species. SSRs are currently the preferred markers of choice in groundnut as they work easily in the tetraploids. More than 3000 SSR markers are available; many of these SSR markers are developed from ESTs.

4.4.1.3.8 Single nucleotide polymorphism (SNP) markers

Most common type of genetic variation, each SNP represents a difference in a single nucleotide. These are highly amenable to high-throughput genotyping approaches.

4.4.1.3.9 Linkage maps

Dense genetic linkage maps are essential for wide spectrum of genetics and breeding applications such as linkage mapping or association analysis based trait mapping, marker-assisted breeding, map-based cloning and physical map alignment. *Reference genetic maps*, developed both in diploid AA and BB and in tetraploid genomes, created enabling comparison of different groundnut maps and even allowed alignment of maps with other legume species. Marker density, however, is not very satisfactory especially in the context of large genome size (2800 Mb/1C) and 20 linkage groups (LGs).Using marker segregation data for 10 RILs and one BC population from the international groundnut community, with the help of common markers across different populations, a *reference consensus genetic map* of cultivated groundnut has been developed. It is comprised of 913 marker loci including 911 simple sequence repeats (SSRs) and 2 cleaved amplified polymorphic sequence (CAPS) loci distributed on 20 LGs spanning a map distance of 3,607.97 cM with an average map density of 3.94 cM. This *reference consensus map* will serve as a reliable reference for aligning new genetic and physical maps, performing QTL analysis in a multi-populations design, evaluating the genetic background effect on QTL expression, and several other genetic and molecular breeding activities in groundnut. As more research groups are getting involved in genomics of groundnut, the linkage map is getting saturated fast.

4.4.1.3.9.1 Targeting induced local lesions in genomes (TILLING)

TILLING is a reverse genetic technique that requires knowledge of gene sequences since mutants are detected by screening for DNA sequence changes rather than phenotypic differences (forward genetics). It is used to find genes of interest in a mutant population of a species, to enhance genetic diversity available for exploitation and in functional genomic studies. A tilling population of over 3400 mutant lines from Tifrunner groundnut cultivar has been generated using chemical mutagenesis and screened for mutation in six genes.

4.4.1.3.9.2 *Expressed sequence tags (EST)*

ESTs are important transcriptome resources for major crops including crops with large genomes (such as groundnut) where genome sequence is not available, to enable gene discovery, microarray gene expression analysis and molecular marker development and genetic map construction. Many plant EST libraries have been sequenced as an alternative to whole genome sequencing where genome is complex and large. NCBI EST database contains 225, 264 ESTs from groundnut as of Nov 2011. ESTs will complement the whole genome sequence.

4.4.1.3.9.3 *BAC libraries*

Large insert genomic DNA libraries such as bacterial artificial chromosome (BAC) libraries provide a platform for physical mapping, map-based cloning of the genes for traits of interest, analysis of gene structure and function and genome sequencing. First BAC library with 182,784 clones was developed for tetraploid groundnut cultivar Florunner. Recently two BAC libraries for AA and BB genomes have been developed.

4.4.1.3.9.4 *Genome sequence database*

Peanut Genome Project (PGP) (http://www.peanutbioscience.com/peanutgenome project.html) has fully sequenced the groundnut genome in 2014 and is now engaged in the development genomic resources for use in groundnut improvement.

4.4.1.3.9.5 *Mapping populations*

Three primary types of populations are used for molecular mapping - F_2, backcross, and recombinant inbreds. These are created from F_1 lines that are derived from two parents that show differing phenotypes for a target trait. Members of a fixed mapping population contain differing amount of recombination and linkage disequilibrium between loci. Population is genotyped with molecular markers and linkage analysis is performed to estimate linkage map for population. F_2 and backcross populations are not eternal and will have to be created each time when DNA is needed. Recombinant inbreds are eternal as they are maintained by single seed descent method and are available any time to harvest DNA.

4.4.1.3.9.6 *Recombinant inbred lines (RILs)*

RILs are the offspring of two genetically distant parents (bi-parental population) that are inbred either through self-fertilization or by sibling mating. They have a little bit of each parent genome (due to recombination) but are homozygous at any given point in the genome so dominance effects do not need to be considered in subsequent genetic and phenotypic analyses.

4.4.1.3.9.7 *Multi-parents advanced generation intercross (MAGIC) population*

A MAGIC population consists of RIL soriginating from multi-parents crossing. It provides an useful germplasm resource with diverse allelic combinations for precise QTL mapping and for direct and indirect use in variety development. It allows parallel high resolution mapping of different complex traits in the same population.

4.4.1.3.9.8 Chromosome segment substitution lines (CSSLs)

CSSLs, which carry a particular chromosome segment from a donor line in the genetic background of another line, are useful populations for detection and mapping of QTLs for target traits. These can be used for detection of QTLs with small additive effects that are masked by QTLs with larger effects in primary populations such as F_2 and RIL populations. These are powerful QTL mapping populations to elucidate the molecular basis of interesting traits of wild species. These are developed by recurrent backcrossing followed by selfing (e.g. BC_3F_4).

4.4.1.3.9.9 Comparative mapping

Comparative genomics is an useful approach to define common attributes among species showing close phylogenetic relationship. Knowledge of genome structure and gene function gained from one species can be used to study the other related species. In legumes, the knowledge of genome structure and gene function gained from *Glycine*, *Medicago* and *Lotus* can be applied to other lesser studied legume species. This will allow breeders to mine desirable genes from germplasm collection rapidly and cost effectively.

4.4.1.3.9.10 Foreground and background selections

Foreground selection refers to using markers that are tightly linked to the gene of interest in order to select for the target allele or gene. Background selection refers to using markers that are not tightly linked to the gene of interest in order to select against other DNA from the donor parent (i.e., to select for recurrent parent alleles at other loci than the target).

4.5 Genetic Transformation

Genetic transformation creates new genetic diversity for exploitation in cultivar development and provides opportunity to combine conventional resistance with nonconventional resistance harnessed from unrelated organism. It can be adopted only when a reliable tissue culture system to regenerate plants exists. The regeneration is highly influenced by genotype, media, light, temperature and growth regulators. Now reliable regeneration protocols for groundnut are available. Gene promoters are the most critical element for obtaining gene expression in groundnut. Till date (2014) no groundnut transgenic cultivar has been released in the world. Public resistance to transgenic food crops, particularly in Europe, heavy cost of meeting regulatory requirements for the release of transgenic cultivars including food safety and issues related to patented technologies come in the way of field testing and release of transgenic cultivars. Product of all breeding approaches is a genetically modified genotype when compared to its predecessor genotype (GMO), the better term for product of genetic transformation is transgenics. The first transgenic food came in the market in 1996. In 2012, transgenic crops covered 1.5 billion ha globally. Most of soybean, cotton, maize and canola in the USA and other countries are transgenic.

4.5.1 Cisgenesis

The process of engineering a genotype with desired genes transferred artificially between organisms that could otherwise be conventionally bred is called cisgenesis. Unlike transgenics, genes are transferred between closely related organisms. Gene belongs to conventional gene pool, a cis plant contains no foreign gene. It has been used to transfer natural resistance genes to blight in potato and scab in apple.

4.5.2 Binary Vector

A pair of plasmids consisting of a binary plasmid and a helper plasmid is called binary vector. The two plasmids are used together to produce genetically modified plants. Both artificial vectors are created from T1 Plasmid found in *Agrobacterium tumefaciens*.

Agrobacterium-mediated transformation: Groundnut tissues susceptible to infection by *A. tumefaciens* include - leaf sections, cotyledonary nodes, longitudinal cotyledon halves, embryo axes, embryo leaflets, and hypocotyls. A target gene of interest is engineered into the T-DNA region of a disarmed plasmid and introduced into *A. tumefaciens*. Transgene within the T-DNA borders is further transferred into plant cells by co-cultivation of *A. tumefaciens* and wounded plant tissue in appropriate medium. PCR, which can amplify a few molecules of the gene from residual *Agrobacterium* because of its very sensitive nature and Southern Blot assay, where the digesting enzyme cuts at only one site or not at all within the introduced plasmid, should be used to confirm gene integration and stable transformation. It is an efficient and rapid technique for transformation due to circumvention of tissue culture step (4-5 months to obtain transgenic plants). However, technology is genotype specific and of limited use in groundnut.

4.5.3 Micro Projectile Bombardment-Mediated Transformation

This method directly transfers target genes into plant cells by delivering coated microprojectiles at high velocity. It is less dependent on host genotype. Microprojectile bombardment can be followed as long as groundnut can be generated from somatic tissues. It is low in efficiency, results in frequent infertility among tissue culture regenerants and takes longer (12-14 months) from induction of transformation event to plant maturity. It has been successfully used in transferring multiple genes for protein and RNAi-mediated gene silencing.

4.5.4 Steps Involved and Other Requirements in Field Testing and Release of Transgenics in India

Department of Biotechnology, Ministry of Science and Technology, Government of India has issued two publications – Recombinants DNA Safety Guidelines and Regulations in 1990 and Revised Guidelines for Research in Transgenic Plants and Guidelines for Toxicity Evaluation of Transgenic Seeds, Plants and Plant Parts in 1998 – which give guidelines and regulatory requirements to be followed by all those engaged in recombinant DNA research and product development. Department of Biotechnology provides an oversight to all DNA recombinant research in the country. There are different levels of transgenic research, which can be carried out only in laboratories following good practices and in specified greenhouse/glasshouse meeting prescribed standards of environmental safety. In addition to standard agronomic evaluation of transgenic material under controlled conditions on-station and in fields, biosafety tests are also required. Toxicity evaluation of transgenic seeds, vegetables and leaves involves oral toxicity tests using rats and goats and skin irritation tests involving rabbits and guinea pigs. Allergenicity tests are to be carried out in animal model.

4.8 Status of Crop Improvement

4.8.1 Status of Marker-Assisted Breeding in Groundnut

Molecular markers associated with resistance to rust, early leaf spot (ELS), late leaf spot (LLS), *Cylindrocladium* black rot, tomato spotted wilt virus (TSWV), nematode, aphids and drought, yield parameters, high oleic acid and other seed biochemical traitshave been identified. Many of these QTLs are not major (account for < 10% phenotypic variation). Major QTLs are identified for rust, LLS and nematode resistance. Genomic selection (GS) is now preferred over marker-assisted backcrossing (MABC) and marker-assisted recurrent selection (MARS) approaches for improving complex traits.

4.8.1.1 Root Knot Nematode

The first example of MAS in groundnut breeding is – transfer of a single dominant root knot-nematode resistance gene from *A. cardenasii* into a variety named Nema TAM in the USA.However, Nema TAM wasfound not suitable for cultivation in Southeastern USA due to its susceptibility to TSWV. Tifguard cultivar, which combines resistance to root-knot nematode and TSWV, was developed following conventional breeding approach.Tifguard High O/L, which combines resistance to root-knot nematode and TSWV and high O/L, was developed following three cycles of backcrossing using as pollen donors BC_nF_1 progenies selected with molecular markers for these traits. Use of co-dominant marker allowed backcrossing with heterozygous lines for high O/L ratio, which is a recessive trait, selfed BC_3F_2 plants yielded marker-homozygous individuals identified as Tifguard High O/L. The use of MAS backcross breeding compressed hybridization and selection phases to less than 3 years.

4.8.1.2 Foliar Diseases

With markers (RAPD, EST-SSR and SSR) and QTLsassociated with rust and LLS resistance already identified, marker-assisted breeding efforts are in progress for resistance to these diseases. Seven markers linked to LLS severity score, which explain 32-59% phenotypic variations, have been identified. Two EST-SSR markers closely linked to rust resistance have been identified. Soon LLS and rust resistant groundnut cultivars developed through MAS would become available for cultivation by farmers.

4.8.1.3 Aflatoxin Contamination

Six QTLs, each located on a different linkage group, for resistance to *A. flavus* invasion explaining 6.2-22.7% phenotypic variation are reported.

4.8.1.4 Drought

Identification of few major, many minor M-QTLs and QTL x QTL interactions indicates the complex and quantitative nature of drought tolerance in groundnut and suggests use of marker-assisted recurrent selection (MARS) or genomic selection (GS) instead of marker-assisted backcrossing approach in breeding to introgress a large number of QTLs associated with drought resistance.

4.8.1.5 Tomato Spotted wilt Virus

Two major QTLs for resistance to TSWV are identified.

4.8.1.6 Insect Pests

Three QTLs for component traits associated with bruchid resistance explaining 14-39% phenotypic variation are reported.

4.8.1.7 High oil/Oleic Acid Content

Three SSR alleles associated with high oil content in wild *Arachis* species are absent in cultivated groundnut. Using wild *Arachis* species the oil content in cultivated groundnut can be increased. One tightly linked marker each for high and low oil content is reported.

Tifguard High O/L cultivar was developed after three accelerated backcrossing combining nematode resistance from Tifguard and high O/L trait from Georgia-02C and Florida-07 following marker-assisted selection.

4.9 Status of Transgenic Material

First successful transformation and accompanying plant generation using microprojectile bombardment technique in groundnut was achieved in 1993 in the USA. In most cases, the level of resistance achieved through transgenic is more or less similar to that achieved through conventional breeding. However, transgenics do provide opportunity to combine conventional resistance with that of nonconventional resistance to improve the level of protection against pathogen or stress factors.

4.9.1 Virus Diseases

Virus diseases received more attention than fungal diseases in transgenic research. Coat protein-mediated resistance in transgenics, in general, offered only moderate protection against diseases. RNAi-mediated resistance is expected to offer higher levels of protection.

4.9.2 Peanut Bud Necrosis Disease (PBND)

Both, *A. tumefaciens*- and micro projectile-mediated genetic transformation approaches, using PBNV nucleocapsid gene encoding for viral coat proteinarepursued at ICRISAT. Transgenic events with $PBNV_{np}$ gene showed lower incidence and delayed onset of disease and also recovery from disease suggesting only a modest tolerance to PBNV. Currently, RNA i-mediated approach is being followed to counter the effect of non-structural silencing suppressor gene (NSs gene) in the PBNV genome.

4.9.3 Tomato Spotted wilt Virus Disease (TSWV)

Protection of transgenic plants against TSWV is under control of both RNA- and protein-mediated control. Nucleocapsid protein gene (NP) was introduced via microprojectile bombardment into New Mexico Valencia A cultivar and a runner cultivar *A. tumefaciens*-mediated transformation is also followed. AT 120 (with antisense nucleocapsid gene) and Marc 1 (with coat protein gene) cultivars have also been transformed.Expression of sense or antisense NP gene from TSWV delayed expression of symptoms and prevented systemic virus infection but did not provide complete resistance to the disease.This single gene resistance may be short-lived because of highly heterogeneous population of the virus.

4.9.4 Peanut Stripe Virus Disease (PStV)

Transgenic plants of Gajah and NC 7 cultivars containing one of the two forms of PStV coat protein gene (*cp* 2 or *cp* 4) exhibited high levels of resistance to PStV. Mechanism of resistance appears to be RNA-mediated.

4.9.5 Peanut Stem Necrosis Disease (PSND)

Transgenics produced following *A. tumefacien*-mediated transformation with TSV coat protein gene (TSV_{cp} gene) showed three symptoms – blockage of systemic movement of TSV within the plants, recovery from an initial infection and subsequent new growth devoid of TSV symptoms and susceptible reaction. Transgenic lines cv. JL 24 containing sense and antisense coat protein gene of TSV using *A. tumefaciens*-mediated transformation have been developed. However, these lines are yet to be tested for disease reaction at hotspot locations under field conditions.

4.9.6 Peanut Clump Disease

Transgenic lines having $IPCV_{cp}$ and $IPCV_{rep}$ genes of Indian peanut clump following *A. tumefaciens*-mediated transformation have been produced and tested under containment facilities at ICRISAT. Some events showed resistant phenotype where the virus titre declined with maturity.

4.9.7 Fungal Diseases

Several genes (glucanase, chitinase, SniOLP and Rs-AFP2 for late and early leaf spots, chitinase for rust, oxalate, glucanase and chitinase for sclerotinia blight and Stilbene synthase, glucanase, chitinase, mod 1, anionicperoxidase, synthetic peptide D4E1, LOX 1, Nonheme chloroperoxidase (cpo) and Pn LOX 3 for *A. flavus* infection and aflatoxin biosynthesis) have been used in genetic transformation. These genes suppressed the disease, delayed the onset of disease, enhanced resistance and decreased disease incidence. In the case of *A. flavus*, reduced aflatoxin production wasnoticed.

4.9.8 Insect Pests

Synthetic genes, *cry*1 EC against *S. litura,cry*1 X against *H. armigera* and *S. litura,* and *cry*1 Ac against lesser cornstalk borer showed good promise.

4.9.9 Drought and Salinity Tolerance

(a) A cis-acting transcription factor that binds to dehydration responsive element (DRE) from *Arabidopsis thaliana*, *AtDREB1 A*, under the control of a stress inducible promoter from the rd29A gene has been used to develop transgenic lines with drought tolerance. One transgenic line showed 40% increase in transpiration efficiency.

(b) Transgenic lines of groundnut transformed with *AtNHX1* showed drought and salt tolerance.

(c) Transgenic lines transformed with Isopentenyltransferase (IPT) gene driven by SARK promoter showed improved biomass retention and an average of 58% yield increase.

4.9.10 Nutritional Quality

4.9.11 Vitamin A

Zmpsy 1 gene from maize and beta-lycopene cyclase gene from tomatohave been used to enrich groundnut seeds with vitamin A. Second generation transgenic events showed many fold increase in vitamin A content.

4.9.12 Oleic/Linoleic Fatty Acid Ratio

Transgenic event with an FAD2 gene RNAi construct showed reduced content of linoleic acid and increased stability of groundnut oil.

4.9.13 Herbicide Tolerance

Transgenic groundnut expressing human Bcl-xL gene showed improved tolerance to paraquat.Transgenic groundnut over expressing pEGAD-EPSPS with altered kinetics of enzyme showed improved tolerance to glyphosate (round up).

4.10 Current Status of and Future Needs in Conventional Groundnut Breeding

From a large collection of more than 15,000 accessions of cultivated groundnut germplasm, only a limited number, mostly sources of resistance to diseases and insect pests, has been utilized in crop improvement. Main reasons for this limited utilization of resistance sources are their poor agronomic background causing linkage drag, absence of information on their reaction to other biotic and abiotic stresses and lack of information on their allelic relationship and combining ability. Keeping the requirement of consumers and processors with respect to taste and flavor and seed size and color, breeders in most applied breeding programs prefer to re-use improved breeding lines in hybridization.

4.10.1 Yield and Yield Related Traits

Newer high-yielding cultivars allocate a great proportion of dry matter to reproductive tissue early in the growth cycle with greater reproductive efficiency than the older cultivars. Newer releases in the USA also have more spreading growth habit and greater seed and pod weight than the earlier releases. In India also, the increase in yield in newer Spanish releases was largely due to gain in seed size, seed weight and number of pods.

Selection either for higher pod yield or for greater harvest index is essential for improvement of yield potential in future cultivars.Remobilization of reserves from vegetative dry matter to pods under conditions of source limitations (falling temperature, defoliation by pathogens or water stress) is likely to be significant in maintaining yields but may limit response to improved conditions specially with high partitioning. There still remains a huge gap between realized and potential yield of groundnut in most developing countries; Breeding programs in developing countries should aim at bridging this gap in yield through resistance breeding against biotic and abiotic stresses.

4.10.2 Resistance to Foliar Fungal Diseases

Infector-row technique is used to screen germplasm and breeding materials against foliar diseases in the field. Combining high levels of resistance to ELS and LLS into high yielding cultivars with acceptable market traits continues to be difficult. Resistance to leaf spots is generally associated with late maturity,longer incubation and latent periods, reduced sporulation, smaller lesion diameter, lower infection frequency and less defoliation in resistance sources.Level of resistance to foliar diseases is very high in wild *Arachis* species,interspecific derivatives confer higher levels of resistance to progenies.

4.10.2.1 ELS

ELS is more serious in southern Africa and the USA. Tolerant *A. hypogaea*genotypes have been identified in Malawi (4), western Africa (18) and India (15) with a disease score between 3.6 and 6.3 on a 1-9 scale where 1=no disease and 9=more than 81% foliage destroyed. Resistant sources reported from the USA (NC 3033, PI 270806, PI 259747 and PI 350680) are found susceptible in India and Malawi. Genotypes ICG 6284, ICG 6902, ICG 7878, ICG 10000, ICG 10948 and ICG 13917 show some resistance at more than one location. Rate-reducing resistance is quantitative in nature and controlled predominantly by additive gene effects. Resistant wild *Arachis*species among others include – *A. cardenasii*, *A. chacoense*, *A. stenosperma*, *A. appressipila*, *A. pusila*, *A. kempff-mercadoi*, *A. batizocoi*, *A. diogoi*, *A. hagenbeckii*, *A. glabrata*and *A. repens*.ELS tolerant cultivars reported, among others, in India are - ICGS 44, ICGS 76, M 335, BG 3, Somnath, CSMG 84-1, M 522, Prutha and GG 7 and in the USA are – Georgia-01R and Georgia-05E.

4.10.2.2 LLS

69*A. hypogaea* genotypes tolerant to LLS with disease score ranging between 3 and 5 on a 1-9 scale have been identified. 49 of these landraces belong to var. *peruviana* with low yield and poor shelling out turn. High levels of resistance/immunity in wild *Arachis* species – *A. cardenasii*, *A. chacoense*, *A. stenosperma*, *A. repens*, *A. appressipila, A. hagenbeckii, A. glabrata*, *A. batizocoi, A. duranensis*, *A. correntina*, *A. villosa*, *A. pusila*,*A. diogoi*, *A. kempff-mercadoi* and *A. paraguariensis* - are observed. Some tolerant cultivars released are: in India – RG 141, ICG(FDRS) 4, ICGV 86590, ICGV 86325, K 134, Girnar 1, GBPD 4, R 8808, ALR 1, ALR 2, ALR 3,BSR 1, VRI 5 and CSMG 84-1 among others, and in USA - Southern Runner, Florida MDR 98, C-99R, Georgia-01R and Georgia-05E.

4.10.2.3 Rust

169 A. hypogaea genotypes are reported resistant (a score of five or less on a 1-9 scale). Of these, 135 landraces belong to var.peruviana; many (ICG 7896, ICG 7897, ICG 7899, ICG 10014, ICG 10030, ICG 10052, ICG 10053, ICG 10067, ICG 10933, ICG 10939, ICG 10940, and ICG 10943) have a disease score of < 3 but are agronomically poor (low shelling outturn, thick pod shell, strong pod reticulation and unacceptable seed coat color). New sources of resistance – ICG 10056, ICG 10567, ICG 10925, ICG 10932, ICG 11108, ICG 12059, ICG 12112, and ICG 12113 and the interspecific derivatives involving A. batizocoi and A. duranensis – have high levels of resistance with good agronomic potential and resistance to other biotic stresses. Several other species belonging to section Arachis are either immune or highly resistant – A. cardenasii, A. chacoense, A. stenosperma, A. repens, A. appressipila, A. hagenbeckii,A. glabrata, A. batizocoi, A. duranensis, A. correntina, A. villosa, A. pusila, A. diogoi, A. kempff-mercadoi, A. paraguariensis, A. villosulicarpa, A. spegazzinii. Some resistant cultivars released in India are ICG(FDRS) 4, ICG(FDRS) 10, ICGV 86590, GBPD 4.

4.10.2.4 Multiple Resistance to Foliar Diseases

Resistance to rust and LLS is correlated (r=0.48-0.60).42 LLS resistant genotypes are also resistant to rust. ICG 1703, ICG 4995, ICG 10920 and interspecific derivative ICG 13917 [259-2 (red)] are useful in multiple resistance breeding, the last one being resistant to all three pathogens. Other useful sources of resistance to both LLS and rust with agronomic potential are: *A.hypogaea* genotypes – ICG 6330, ICG 7884, ICG 10023, ICG 10035 and ICG 11182, and interspecific derivatives – ICG 11312, ICG 11317 (also resistant to ELS), ICG 11321, ICG 11325, ICG 11337, ICG 13916, ICG 13917, ICG 13919, ICG 13920 and ICG 13922. Several wild *Arachis* species have very high levels of resistance or immunity to all the three foliar diseases.

4.10.2.5 Future Needs

Levels of resistance to leaf spots (both ELS and LLS) in cultivars need further improvement. As LLS and rust often occur together, a combined resistance to these pathogens is required. However, often with higher levels of resistance, crop duration is enhanced.Higher levels of resistance to foliar fungal diseases is required in short-duration cultivars without affecting their duration. Similarly, linkage drags (poor pod shape, unattractive seed coat color, thick shell, low partitioning, etc.) need to be eliminated completely in resistant cultivars. It is desirable to intermate second or third generation advanced breeding lines originating from different parents including interspecific derivatives to improve the level of resistance against foliar diseases without bringing in linkage drags.

4.10.2.6 Resistance to Soil-Borne Diseases

Breeding for resistance to soil-borne fungal diseases continues to be a difficult task as creating uniform disease pressure in disease screening nursery remains challenging. Among the soil-borne diseases, *A. flavus* infection and aflatoxin contamination receive the maximum attention.

4.10.2.6.1 A. Flavus/ A. Parasiticus seed infection and aflatoxin production

Three barriers exist to *A. flavus/A. parasiticus* seed infection and aflatoxin production - pod wall, seed coat and cotyledons.There are three resistance mechanisms – preharvest resistance, seed coat resistance (*in vitro* seed colonization (IVSC)) and cotyledon resistance (aflatoxin production). Their independent inheritance provides an opportunity for gene pyramiding.Sources resistant to preharvest infection (≤2%; 21 genotypes- ICG 1122, ICG 1173, ICG 1323, ICG 1326 (J 11)*, ICG 1859, ICG 1994, ICG 3263(U 4-47-7)*, ICG 3267, ICG 3336*, ICG 3700*, ICG 4589, ICG 4749 (PI 337394 F)*, ICG 4888, ICG 7633 (UF 71513)*, etc.; *consistent across locations), IVSC (≤15% seed colonized; PI 337394F*, PI 337409*, UF 71513, Ah 78223, J 11*, U $-47-7, Var 27, Faizpur, Monir 240-30 etc.; * consistent across locations and pathogen pressure) and aflatoxin production (< 0.7 μg per kg: ICG 10609, ICG 11682, ICG 10615, ICG 6760, ICG 9610, etc.) are available, but none of these is completely free from infection or aflatoxin production.Containment of preharvest infection is essential as once infected, the seed cannot be disinfected and the infection is carried forward. Seed coat resistance provides postharvest protection in storage. Recommended for use in breeding because of their multiple resistance are genotypes - ICG 1326, ICG 1859, ICG 3263, ICG 3336, ICG 3700, ICG 4749, ICG 7633, ICG 9407, ICG 9610, ICG 10094, etc. Drought predisposes groundnut to aflatoxin contamination,some drought tolerant lines also show low preharvest seed infection and aflatoxin production. Fatty acid composition and N_2 fixation and related traits are also reported to influence directly or indirectly aflatoxin contamination. *A. flavus* sick plot with imposed end-of-season drought during the dry season to avoid interference from rains is used for screening germplasm and advanced breeding lines at ICRISAT. In the USA, a large-scale field system at a desert location (Yuma, Arizona) with subsurface irrigation, which allows extended drought in pod zone without causing any plant mortality, was developed and used for screening against *A. flavus* infection and aflatoxin production. It is important to have more number of replications as plot to plot/plant to plant variation in preharvest infection could be large in spite of sufficient number of fungal propagules being present in the soil throughout the disease screening nursery. Screening for resistance to *in vitro* seed colonization and aflatoxin production is done in laboratory following protocols prescribed by various researchers. During field and laboratory screening, it is not unusual to find nil preharvest infection but presence of aflatoxin in the same genotype, the reverse is also observed. Conventional breeding alone does not ensure complete freedom from aflatoxin contamination, at best it is able to combine the level of resistance available in resistant parents with high yield and other agronomic characters. Level of resistance in breeding linesis similar to that of resistance sources. Gene pyramiding has also not changed the situation. Elite breeding lines giving good performance in India and Mali/Niger include – ICGV 88145, ICGV 89104, ICGV 91278, ICGV 91283, ICGV 91284, ICGV 87084, ICGV 87094 and ICGV 87110 and in China include ICGV 95440, ICGV 95422, ICGV 94435, ICGV 95435 and UF 71315.

Future Needs

Sampling procedures and screening methods including development of a sick plot need to be further refined to improve uniformity of infection, characterization and precision of estimation of infection and aflatoxin production in a genotype in a consistent manner. Allelic relationships among resistant sources need to be studied to ascertain if they possess different resistance genes.Marker-assisted selection is likely to improve efficiency of breeding. Similarly, genetic transformation also has potential to produce aflatoxin free genotypes.

4.10.2.7 White Mold

White mold or stem rot caused by *Sclerotium rolfsii*is wide spread in major groundnut growing areas in the world. Genetic variation for resistance to white mold exists in cultivated groundnut. Sources of moderate resistance are NC 2, NC Ac 18016, NC Ac 18416, ICG 15233, ICG 15234, ICG 15235, ICG 15236, ICGV 96590 and ICGV 87160 identified at ICRISAT and NRCGCS 47, NRCGCS 99, NRCGCS 131 and NRCGCS 319 identified at the Directorate of Groundnut Research, Junagadh in India. Cultivars with moderate resistance have been released in the USA- Southern Runner, Toalson, Pronto, Georgia Browne, Sunbelt Runner, Tamrun 96, Georgia-01R and Georgia-05E. Field screening is more consistent than greenhouse screening.Uniformity and level of inoculum in the sick plot can be enhanced by adding sterilized oat seed inoculum of *S. rolfsii*, but individual plants may still escape;'agar disk technique' is used to screen individual plants.

Future Needs

Soil-borne diseases are becoming increasingly important and they require more research attention.Higher levels of genetic resistance to all soil-borne diseasesare needed in released cultivars.

4.10.2.8 Resistance to Virus Diseases

Nine virus diseases are economically important in groundnut in different parts of the world - peanut mottle potyvirus (PMV), peanut stripe potyvirus (PStV), peanut clump furovirus (PCV), groundnut rosette disease (GRD) complex, tomato spotted wilt tospovirus (TSWV), peanut stunt cucumovirus (PSV), cowpea mild mottle carlavirus (CMMV), cucumber mosaic cucumovirus (CMV), peanut bud necrosis tospovirus (PBNV) and tobaco streak ilavirus (TSV). However, the breeding attention is focussed on PBNV in India, TSWV in the USA and GRD in Africa.

4.10.2.9 Peanut Bud Necrosis Disease (PBND)

PBND, caused by PBNV, is economically important in south and southeast Asia. It is transmitted by *Thrips palmi* – virus is acquired by the larvae but transmission is exclusively by adults. It is not seed transmitted. Several germplasm accessions with consistently low disease incidence in the field are ICG 848, ICG 851, ICG 852, ICG 862, ICG 869, ICG 885, ICG 2271, ICG 2306, ICG 2307, ICG 2323, ICG 2741, ICG 3042, ICG 3806, ICG 3873, ICG 5030, ICG 5024, ICG 5043, ICG 5044, ICG 6135, ICG 6317, ICG 6323, ICG 7676, ICG 7892 and others. Breeding lines, ICGV 86031 and ICGV 86388, are resistant to both vector and the virus. Other lines with resistance to PBND under field conditions include DRG 18, ICG 7812, ICG (FDRS) 10, ICGV 80325, JSSP 3, KNG 22 and PI 393516. Highly resistant wild species in section *Arachis* free from disease or virus in the field are *A. duranensis* (ICG 8199 (PI 468200), ICG 8956 (PI 468201), ICG 11552 (PI 475882), ICG 11553 (PI 475883) and ICG 11555 (PI 475885), *A. correntina* (ICG 8132 (PI 262808)), *A. monticola* (ICG 8189 (PI 468199)), *A. cardenasii* (ICG 11564, ICG 13164 and ICG 13165), *A. villosa* (ICG 8144 and ICG 13168). Resistant cultivars released are CO 3, ICGS 11, ICGS 44 (ICGV 87128), ICGS 37 (ICGV 87187), R 8808 (KRG 2), R 9251, K 134, DRG 12, RSHY 1, Kadiri 4 among others in India and Khon Kaen 6 in Thailand. Sources of resistance to vector are NC Ac 2242, NC Ac 2214, NC Ac 2243, NC Ac 2240, NC Ac 2232, NC Ac 2230 and others.

Future Needs

Combining virus resistance with vector resistance to improve the levels of resistance to virus diseases is essential. As coat protein gene in transgenics does not give complete protection against virus diseases, there is need to pursue RNAi approach to provide better protection against virus diseases.

4.10.3 Insect Pests Resistance

Breeding for resistance to insect pests is lagging behind due to difficulty in large scale screening of germplasm lines and breeding populations under sporadic and variable natural insect pests pressure.

4.10.3.1 Sucking Pests

These include mainly thrips (*Scirtothrips dorsalis, Thrips palmi, Frankliniella schultzei, Caliothrips indicus*), jassids (*Empoasca kerri*) and aphids (*Aphis craccivora*); thrips and aphids are also vectors of virus diseases. Genotypes resistant to thrips and jassids are reported and some are listed in sections on PBND and TSWV. High density, distribution and length of trichomes (NC Ac 2214, NC Ac 2230, NC Ac 2240) and thick leaf cuticle (NC Ac 2242, NC Ac 2243) are associated with resistance to thrips and jassids.Genotypes tolerant to jassid damage are: Breeding lines – ICGV 86388, ICGV 86252, ICGV 86393 and ICGV 85455, Cultivars –Georgia-01R and Georgia-05E. Antibiosis (reduced growth and fecundity) operates in the case of aphids resistance in NC Ac 343, ICG 5240 (EC 36892), ICG 12991and ICGV 86030 genotypes.Several wild *Arachis* species have very high levels of resistance to insect pests (thrips - *A. chacoense*; jassids - *A. cardenasii*, *A. duranensis*, *A. kempff-mercadoi*, *A. monticola*, *A. stenosperma*, *A. paraguariensis*, *A. pusila*, *A. triseminata*, *A. batizocoi*, *A. chacoense*).

4.10.3.2 Defoliators

Major defoliators in groundnut are leaf miner (*Aproaerema modicella*), tobacco caterpillar/armyworm (*Spodoptera litura*) and gram pod borer (*Helicoverpa armigera*). Screening for resistance to defoliators under natural field conditions is difficult because of variation in infestation in space and time. No-choice cage technique is used to screen for *S. litura*, nuclear insect culture is maintained on artificial diet, a known number of first- or third-instar larvae are released for varying period of time on 15-day old greenhouse grown plants which are kept inside a plastic jar cage with wire-mesh screen windows,observations on insect survival (# of surviving larvae and larval weight) and leaf area damage are recorded. For leaf miner, natural infestation is relied upon, which can be enhanced by planting soybean as an infester crop and creating prolonged drought, it is difficult to devise no-choice cage screening for leaf miner.

Breeding lines, ICGV 86031, ICGV 87154 and ICGV 87160 and germplasm accessions, ICG 2271 and ICG 1697, show resistance to both tobacco armyworm and leaf miner.Other genotypes which show promise against leaf miner are NC Ac 343, NC Ac 17090 and ICG (FDRS) 4. Several wild *Arachis* species also show resistance to these two insect pests - *A. chacoense*, *A. repens*, *A. appressipila*, *A. hagenbeckii*,*A. glabrata*, *A. batizocoi*, *A. correntina*, *A. villosa*, *A. diogoi*.

Future Needs

Resistance genes from wild *Arachis* species should be harnessed to improve resistance to insect pests in cultivated groundnut. Field screening techniques need to be improved to create desired and uniform insect pressure.

4.10.4 Nematode Resistance

4.10.4.1 Root Knot Nematode

Screening involves growing plants in glasshouse and inoculating them with *Meloidogyne arenaria* race 1 inoculum cultured on roots of tomato or egg plants and scoring the roots of 70 day old harvested groundnut plants for root galls and egg masses. In contrast to moderate level of resistance in cultivated groundnut, wild *Arachis* have a high level of resistance against this nematode. Accessions supporting less egg production of *M. arenaria*race 1 have been identified both in cultivated groundnut and wild species *A. cardenasii*, *A. batizocoi* and *A. diogoi*.

4.10.4.2 Kalahasti Malady

'Kalahasti disease' caused by stunt nematode (*Tylenchorhynchus brevilineatus*) was first noticed in 1975/76 in Kalahasti area in Andhra Pradesh, India. Thedisease is characterized by reduction in pod size and presence of small, brownish yellow lesions on pegs and young developing pods, later pod surface and roots become completely discoloured. From replicated screening of 1599 genotypes in a hot spot location in a farmer's field in Kalahasti during 1985/86 -1986/87, 14 resistant genotypes were identified. Most of these genotypes had undesirable pod/seed characteristics with the exception of TCG 1518, an advanced Virginia bunch breeding line, which was later released as Tirupati 3 for cultivation in disease-affected areas. In another screening of 39 genotypes during 1992-94, TCGS 307, TCGS 313 and TCGS 320 (released as Kalahasti) were also identified as resistant to the disease with the last two having pod yield exceeding 3 t/ha.

Future Needs

The resistance to nematodes needs to be combined with resistance to other major biotic stresses operating in a region to derive larger benefits from the resistance breeding efforts.

Drought tolerant cultivars that are also resistant/tolerant to a flatoxin contamination in different botanical backgrounds are needed. Marker-assisted recurrent selection (MARS) should be adopted to accumulate several QTLs with small effects in a single genotype when the dense linkage map in groundnut becomes available.

4.10.5 Physiological Traits

4.10.5.1 Drought

A majority of groundnut in developing countries is grown under rainfed conditions. Although considered a drought tolerant legume, it can suffer early-season, mid-season, end-of-season or intermittent droughts impacting adversely on yield and yield-related characters including quality of the produce due to reduced photosynthesis, N_2 fixation and calcium uptake by developing pods. Impact depends upon the timing of occurrence, duration and intensity of drought. A 20 to 26-day moisture stress soon after crop emergence is beneficial to the crop as it forces roots to go to deeper layers of soil and when the moisture stress is released it induces profuse flowering resulting in synchronized and uniform maturity and increased yield.The adverse effect of end-of-season drought can be overcome by developing short-duration varieties with their life cycle matching the period of soil moisture availability. Mid-season drought is the cause of concern as insufficient water at the time of flowering and fruiting reduces the yield significantly.

Direct selection for yield under drought is effective but is resource consuming and lacks repeatability across different environments. Passioura (1977) gave the following equation for yield under water-limited conditions: Pod yield = Transpiration (T) x Transpiration Efficiency (TE) x Harvest Index (HI).Drought tolerance can be enhanced by improvements in soil water extraction ability (T) or improvements in water-use efficiency (TE). Genetic variation for root system and transpiration efficiency (amount of dry matter produced for each unit of water transpired (g dry matter per kg of water transpired) is reported, but these traits are difficult to measure. Easily measurable surrogates for these traits are needed for use in a large scale breeding programs. Transpiration efficiency is negatively correlated with $\Delta^{13}C$ (carbon isotope discrimination) in leaves (r = –0.88 to – 0.92), which is rapid but expensive to measure. $\Delta\ ^{13}C$ is highly positively correlated with specific leaf area (SLA, ratio of leaf area to leaf dry weight), which is easy and inexpensive to measure. SLA has inverse relationship with relative leaf water content (RWC) and the low SLA types are drought tolerant as they are able to maintain higher RWC. However, SLA is influenced by the time of sampling and age of the leaf. SLA is inversely correlated with SPAD chlorophyll meter reading (SCMR) (r= 0.77), which, in turn, is positively correlated with TE. SCMR is measured by a hand held device, which is easy to operate and can rapidly record observations.Thus, for rapid screening, SCMR can be used in a large scale breeding programs aiming to improve drought tolerance in groundnut. SLA and SCMR can be recorded anytime after 60 days of crop growth, preferably under moisture deficit conditions. However, the utility of SLA and SCMR in screening for drought tolerance has been questioned in some studies. Sufficient variation for physiological traits such as SLA, WUE, T, TE and HI and in tolerance to mid-season and/or terminal droughts is reported.

Field screening at ICRISAT for tolerance to different droughts is done in dry season by withholding irrigation as follows: for early-season drought by with holding irrigation for 40 days after the first irrigation for germination and subsequently following normal irrigation schedule, for mid-season drought with holding irrigation after 40 days up to 80 days and before and after drought period following regular irrigation schedule, and for end-of-season drought with holding irrigation from 80 days onward till maturity and following normal irrigation schedule prior to imposition of drought. For germplasm screening, line-source sprinkler system of irrigation is used to create a gradient of soil moisture deficit and the genotypes are selected based on biomass and HI.Segregating populations are screened in the field under imposed drought conditions and selections are based on pod yield, pod number and pod filling. Surrogates SCMR or SLA can also be used along with pod yieldand other characters in selection of plants/populations. Both empirical (yield-based) and trait-based approaches are equally effective in selecting for drought tolerance. In the case of trait-based approach, TE is the major contributor to pod yield, which indicates more efficient utilization of available water. However, in the case of empirical approach, it is T which is a major contributor to pod yield, which indicates better mining of water from soil layers.However, better mining does not necessarily mean better utilization of water. In case of limited water availability, enough T may not occur thus impacting on pod yield. It is advisable to integrate surrogates of TE in the selection scheme for drought tolerance.

Some of the drought tolerant cultivars/genotypes are ICGS 11, ICGS 44, ICGS 76 and ICG (FDRS) 10 for mid-season drought and TPT 25 for both early- and mid-season drought in India and 55-437 (also tolerant to aflatoxin contamination), GC 8-35, 55-21, 55-33, TS 32-1, SRV 1-3, SR 1-96, ICGV 86024 (also tolerant to leaf spots), ICGV 86124 and ICGV-SM 87003 in West Africa.

When drought occurs, the temperature also rises; Drought and heat tolerance appear to be correlated. Heat tolerant genotypes include 796, 55-437, TMV 2, ICGS 11, ICG 1236, ICGV 86021, ICGV 87281, ICGV 92121, etc.*In vitro* pollen germination, pollen tube growth, membrane thermostability, growth rates, fruit set and partitioning are the traits used to record response of groundnut genotypes to high temperature.

4.10.5.2 Salinity

No targeted program is in progress to breed salinity tolerant groundnut. A pot culture technique has been standardized for screening for tolerance to salinity in groundnut. In limited studies, tolerant genotypes (cultivars and breeding and germplasm lines) based on plant survival and seed yield per plant have been identified and include NRCG 2588, NRCG 4659, NRCG 5513, NRCG 6131, NRCG 6450, NRCG 6820, NRCG 6919, NRCG 7206, TMV 2 NLM, TG 33, JNDS-2004-15, VRI 3, UF 70-103, TKG 19 A, S 206, Tirupati 4, M 522, Punjab 1, BG 3, Somnath and ICGV 86590.

4.10.6 Seed Dormancy

When rainfed groundnut is caught in rains at the time of harvest, it results in *in-situ* germination in Spanish and Valencia cultivars, thus causing significant losses in yield and quality of the produce. Incorporation of 2-3 weeks fresh seed dormancy in Spanish and Valencia cultivars will help to avoid these losses, which could reach up to 40%.

Different seed parts – seed coat, cotyledons and embryo – play a role in imparting dormancy. Fresh seed dormancy is more under control of testa than cotyledons. Complexity arises in studying inheritance of seed dormancy when both maternal (testa) and zygotic (cotyledons) tissues are involved in its control. Several Spanish breeding lines/cultivars with fresh seed dormancy are available now. Most of these originate from Virginia x Spanish/Valencia crosses. Instead of screening for seed dormancy in early generations, the advanced generation Spanish breeding lines are screened for fresh seed dormancy in laboratory and under field conditions.

4.10.7 Nitrogen Fixation

Nodulation and N_2 fixation are dependent on the interaction between the host cultivar and the nodulating *Bradyrhizobium* strain.Virginia types are superior to Spanish and Valencia types in N_2 fixation.Considerable variation exists for ability to nodulate and N_2 fixation, which are under genetic control of the host and are heritable. No directed breeding program currently is in progress to improve N_2 fixation in groundnut.Selection for high yield also indirectly improves N_2 fixation.

4.10.8 Adaptation Traits

Growth and development in groundnut is largely driven by temperature. Optimum temperature (T_o) for growth and development in groundnut ranges between 27 °C and 33 °C. Base temperature (T_b) in groundnut ranges between 9 °C and 13 °C below which the growth ceases.$T_b$varies for different phenological stages and among genotypes.Growth is slowed down at lower temperatures and it takes longer for crop to mature. Reverse is observed at relatively higher temperatures, but at temperatures exceeding 45 °C, the growth stops as protein gets denatured. At temperatures above the optimum, significant reduction in dry matter production and partitioning of dry matter to pods are observed but flower production is not affected. Photoperiod does not affect flowering in groundnut but it affects partitioning. However, these effects are genotypic specific. Irradiance, together with temperature, also plays a role in determining the crop duration.

4.10.9 Short-Duration

Early maturity is a relative term – in India early maturing varieties are less than 100-day duration whereas in China and USA a variety of 120-day duration will qualify as early maturing variety.Selection based on days to first flower alone is ineffective in identifying early maturing lines as other processes are also involved in reaching to maturity. Instead of calendar days, use of cumulative thermal time (CTT) measured in day degrees (°Cd) is recommended for selecting for early maturity at a given location. CTT is measured in day-degrees (˚Cd) above the base temperature and is calculated on successive days by subtracting the base temperature from the mean daily temperature and adding each value to the sub-total accumulated since the seed was sown ($CTT\ (°Cd) = \Sigma\ ((T_{max}+T_{min})/2 - T_b)$). In photoperiod-insensitive genotypes the CTT for maturity does not differ across environments barring the influence of environmental factors other than photoperiod. For photoperiod-sensitive genotypes, the CTT will vary with photoperiod over the photoperiod-sensitive range.

Based on the past records of temperatures (15-20 years), CTT for the desired crop duration at a given location should be worked out.When the desired CTT is achieved, the segregating breeding populations are uprooted and plants are selected visually in the field based on number of matured pods (internal pericarp color) and high pod yield for further laboratory evaluations for seed appearance, uniformity, and maturity, andplants, finally selected. They are advanced to the next generation. Advanced breeding lines are evaluated for 2–3 years in on-station replicated trials with staggered harvesting starting from a predetermined date based on the desired CTT. In addition to pod yield and pod and seed maturity, extent of further gains in pod yield, seed weight and shelling turn over in the following staggered harvesting are taken into account while selecting advanced breeding lines. Those with high pod yield and pod and seed maturity and minimum gains in pod yield, seed weight and shelling turnover over the preceding harvest are selected for further multi-location evaluation and selection.

Incorporating large seed size in short duration cultivars is unlikely to succeed as large seeds take more time to emerge on sowing, and to develop and mature. Similarly, combining higher levels of resistance to foliar diseases and short-duration will be difficult to achieve. On the other hand, a moderate level of resistance will have only limited influence on crop duration and would also stabilize productivity in a cropping system.

The following considerations in breeding strategy will help to achieve the objective of early maturity along with high yield: (i) Selection for low Tb and CTT for various phenological stages (ii) Selection for tolerance to high temperature (iii) Selection for photoperiod-insensitive genotypes (iv) Selection for high crop growth rate and partitioning (v) Selection for high water-use efficiency, and (vi) Evaluation in target environments/cropping systems.

Sources of Early Maturity

Chico, Gangapuri, JL 24, ICG 3540, ICG 3631, ICG 4558, ICG 4729, ICG 4890, ICG 9427, ICG 9930, ICG 11605, 91776, 91176, Dh 40, ALG(E) 57, TG 1E, TG 2E, TG 3E, etc.

Short-duration cultivars

Pronto and Spanco in the USA, Dh 40, TNAU 97, ALG(E) 57, GG 3, GG 5, GG 12, TG 26, R 9251, M 522, RS 138, VRI 3 and Co 4 in India and 55-437, TS 32-1, 73-30, KH 149A, KH 241D, Te 3 and Fleur 11 in Africa, among others, in different habit groups.

Future Need

Diversification of sources of earliness and studies on their genetic constitution and allelic relationships are needed to identify different genes for earliness which could be accumulated in a desired genotype.

4.10.10 Quality Traits

4.10.10.1 Oil Quality

Oxidative stability and shelf life of groundnut and its products can be enhanced by improving oleic to linoleic fatty acid ratio, which normally ranges between 0.8 and 2.5 in old commercial cultivars. These two fatty acids constitute about 80% (55% oleic acid (18:1) and 25% linoleic acid (18:2) of the oil content of groundnut. Of these two fatty acids, linoleic fatty acid is less saturated and less stable than oleic acid. In peanut breeding program at the University of Florida in 1987, two breeding lines originating from F 435, a high oleic acid spontaneous mutant, with 80% oleic and 2% linoleic fatty acid composition were identified. With simple inheritance (single recessive or two recessive genes and some possible modifiers depending upon the parents involved in the crosses), it is easy to transfer high-oleate trait to other genotypes through backcross breeding program.

Varieties developed with high O/L ratio in the USA through conventional breeding are SunOleic 95R, SunOleic 97R, Tamrun OL01, Georgia 04S, Andru II, Florida-07 and Hull, through chemical mutagenesis are Mycogen-Flavorunner and M 2-225 and through Gamma radiation are Georgia-02 C and Georgia Hi-high.

4.10.10.2 Biofortification

Large genetic variation exists for Zn (44-95 mg/kg) and Fe (33-68 mg/kg) contents in seeds among groundnut genotypes. Both Zn and Fe could be improved simultaneously as they are positively correlated. No trade-off between Zn and Fe contents and pod yield and oil content is observed. High yielding and stable genotypes across environments with high Zn and Fe contents are ICGV 06099 and ICGV 06040.

4.10.10.3 Fodder Quality

As groundnut fodder (haulms) is highly valued, dual purpose groundnut cultivars are preferred in developing countries. In spite of preference for dual-purpose cultivars, farmers are not willing to sacrifice pod yield for fodder yield. Studies have shown large variation in haulm yield, haulm nitrogen content, *in vitro* organic matter digestibility, metabolizing energy content and digestible haulm yield. High heritability for these traits and absence of any negative relationship between fodder quality traits and pod or haulm yield provide ample opportunities to improve fodder quality traits. In spite of the need to improve fodder quality and yield, the primary target trait in groundnut improvement remains pod yield. Due consideration to haulm yield and its quality parameters should be given in the selection scheme of cultivars as haulm is an economic component of produce of groundnut crop.

References

1. *Amin PW, Singh KN, Dwivedi SL and Rao VR. 1985. Sources of resistance to jassid (Empoasca kerri Purthi), thrips (Frankliniella shultzei (Trybom)) and termites (Odontotermes sp.) in groundnut (Arachis hypogaea L.). Peanut Science 12: 58-60.*
2. *Anderson WF, Patanothai A, Wynne JC and Gibbons RW. 1990. Assessment of a diallel cross for multiple foliar pest resistance in peanut. Oléagineux 45: 373-378.*
3. *Anonymous. 1988. USA Peanuts. National Peanut Council of America, Virginia, USA.*

4. Arunyanark A, Jogloy S, Akkasaeng C, Vorasoot N, Kesmala T, Rao RCN, Wright GC and Patanothai A. 2008. Chlorophyll stability is an indicator of drought tolerance in peanut. Journal of Agronomy and Crop Science 194: 113-125.

5. Arunyanark A, Pimratch S, Jogloy S, Wongkaew S, Vorasoot N, Akkasaeng C, Kesmala T, Patanothai A and Holbrook CC. 2012. Association between aflatoxin contamination and N_2 fixation in peanut under drought conditions. International Journal of Plant Production 6: 161-172.

6. Asif MA, Zafar Y, Iqbal J, Iqbal MM, Rashid U, Ali GM, Arif A and Nazir F. 2011. Enhanced expression of AtNHX1 in transgenic groundnut (Arachis hypogaea L.) improves salt and drought tolerance. Molecular Biotechnology 49: 250-256.

7. Bag S, Singh RS and Jain RK. 2007. Agrobacterium-mediated transformation of groundnut with coat protein gene of Tobacco streak virus. Indian Journal of Virology 18: 65-69.

8. Bandyopadhyay A, Nautiyal PC, Radhakrishnan T and Gor HK. 1999. Role of testa, cotyledon and embryonic axis in seed dormancy of groundnut (Arachis hypogaea L). Journal of Agronomy and Crop Science 182:37-41.

9. Banjara M, Zhu LF, Shen GX, Payton P and Zhang H. 2012. Expression of an Arabidopsis sodium/ proton antiporter gene (AtNHX1) in peanut to improve salt tolerance. Plant Biotechnology Reports 6: 59-67.

10. Baulcombe D. 1996. Mechanisms of pathogen-derived resistance to viruses in transgenic plants. Plant Cell 8: 1833-1844.

11. Bera SK, Kasundra SV, Kamdar JH, Ajay BC, Lal C, Thirumalasamy PP, Dash P and Maurya K. 2014. Variable response of interspecific breeding lines of groundnut to Sclerotium rolfsii infection under field and laboratory conditions. Electronic Journal of Plant Breeding 5: 22-29.

12. Bertioli, DJ, Seijo, G, Freitas, FO, Valls, JFM, Leal-Bertioli, SCM and Moretzohn, C. 2011. An overview of peanut and its wild relatives. Plant Genetic Resources: Characterization and Utilization 9: 134-149.

13. Bhatnagar-Mathur P, Devi MJ, Reddy DS, Lavanya M, Vadez V, Serraj R, Yamaguchi-Shinozaki K and Sharma KK.2007. Stress-inducible expression of At DREB 1A in transgenic peanut (Arachis hypogaea L.) increases transpiration efficiency under water-limiting conditions. Plant Cell Reports 26: 2071-2082.

14. Bhatnagar-Mathur P, Rao JS, Vadez V, Dumbala SR, Rathore A, Yamaguchi-Shinozaki K and Sharma KK. 2014. Transgenic peanut overexpressing the DREB1A transcription factor has higher yields under drought stress. Molecular Breeding 33: 327-340.

15. Bhatnagar-Panwar M, Bhatnagar-Mathur P, Bhaaskarla VV, Dumbala SR and Sharma KK. 2013. Rapid, accurate and routine HPLC method for large-scale screening of pro-vitamin A carotenoids in oilseeds. Journal of Biochemistry and Biotechnology. DOI 10.1007/s13562-013-0239-1

16. Bindu Madhava H, Sheshshayee MS, Shankar AG, Prasad TG and Udayakumar M. 2003. Use of SPAD chlorophyll meter to assess transpiration efficiency of peanut. Pages 3-9. In Breeding of drought-resistant peanuts. Cruickshank AW, Rachaputi NC, Wright GC and Nigam SN (eds.). ACIAR Proceedings No. 112. Australian Centre for International Agricultural Research, Canberra.

17. Bucher E, Sijen T, De Haan P, Goldbach R and Prins M. 2003. Negative-strand tospoviruses and tenuiviruses carry a gene for a suppressor of gene silencing at analogous genomic positions. Journal of Virology 77: 1329-1336.

18. Buiel, AAM. 1996. Quantitative resistance to peanut bud necrosis tospovirus in groundnut. Ph.D. dissertation, Wageningen Agricultural University, The Netherlands.

19. Burow MD, Simpson CE, Faries MW, Starr JL and Paterson AH. 2009. Molecular biogeographic study of recently described A- and B-genome Arachis species, also providing new insights into the origin of cultivated peanut. Genome 52: 107-119.

20. Burow MD, Simpson CE, Paterson AH and Starr JL. 1996. Identification of peanut (Arachis hypogaea L.) RAPD markers diagnostic of root-knot nematode (Meloidogyne arenaria (Neal) Chitwood) resistance. Molecular Breeding 2: 369-379.

21. Chander Rao S, Kumar PL, Reddy AS, Swamy Krishna T, Waliyar F, Nigam SN, Laxminarasu M and Sharma KK. 2006. Evaluation of transgenic peanut plants against peanut bud necrosis disease (PBND) under glasshouse and field conditions. Indian Journal of Virology 17: 135.

22. Chen CY, Nuti RC, Rowland DL, Faircloth WH, Lamb MC and Harvey E. 2013. Heritability and genetic relationships for drought-related traits in peanut. Crop Science 53: 1392-1402.

23. Chenault KD and Payton ME. 2003. Genetic transformation of runner-type peanut with the nucleocapsid gene of tomato spotted wilt virus. Peanut Science 30: 112-115.

24. Choi K, Burow MD, Church G, Paterson, AH, Simpson CE and Starr JL. 1999. Genetics and mechanisms of resistance to Meloidogyne arenaria in peanut germplasm. Journal of Nematology 31: 283-290.

25. Chu Y, Faustinelli P, Ramos ML, Hajduch M, Thelen JJ, Maleki SJ and Ozias-Akins P. 2008. Reduction of IgE binding and non-promotion of Aspergillus flavus fungal growth by simultaneously silencing Ara h2 and Ara h6 in peanut. Journal of Agriculture and Food Chemistry 56: 11225-11233.

26. Chu Y, Holbrook CC and Ozias-Akins P. 2009. Two alleles of ahFAD2B control the high oleic acid trait in cultivated peanut. Crop Science 49: 2029-2036.

27. Chu Y, Wu CL, Holbrook CC, Tillman BL, Person G and Ozias-Akins 2011. Marker-assisted selection to pyramid nematode resistance and high oleic trait in peanut. The Plant Genome 4: 110-117.

28. CIAT. Biology and agronomy of forage Arachis (eds. Kerridge PC and Hardy B). Centro Internacional de Agricultura Tropical (CIAT), Cali, Colombia.

29. Coffelt TA, Seaton ML and Van Scoyoc SW. 1989. Reproductive efficiency of 14 virginia-type peanut cultivars. Crop Science 29: 1217-1220.

30. Craufurd PQ, Prasad PVV, Kakani VG, Wheeler TR and Nigam SN. 2003. Heat tolerance in groundnut. Field Crops Research 80: 63-77.

31. Culbreath AK, Gorbet DW, Martinez-Ochoa N, Holbrook CC, Todd JW, Isleib TG and Tillman BL. 2005. High levels of field resistance to tomato spotted wilt virus in peanut breeding lines derived from hypogaea and hirsuta botanical varieties. Peanut Science 32: 20-24.

32. Culbreath AK, Todd JW and Brown SL. 2003. Epidemiology and management of tomato spotted wilt in peanut. Annual Review of Phytopathology 41: 53-75.

33. Davidson, JI Jr, Whitaker, TB and Dickens, JW. 1982. Grading, cleaning, storage, shelling, and marketing of peanuts in the United States. Pages 571- 623. In Peanut Science and Technology (eds. Pattee HE and Young, CT), American Peanut Research and Education Society, Inc. Yoakum, Texas 77995, USA.

34. Dawson WO.1996. Gene silencing and virus resistance: a common mechanism. Trends in Plant Science 1: 107-108.

35. Deshmukh, SN, Satpute, GN, Dabre, WM and Deshmukh, RG. 2001. The PKV method of groundnut seed production. International Arachis Newsletter 21: 23-24.

36. Devi MJ, Bhatnagar-Mathur P, Sharma KK, Serraj R, Anwar SY and Vadez V. 2011. Relationships between transpiration efficiency and its surrogate traits in rd29A: DREB 1Atransgenic lines of groundnut. Journal of Agronomy and Crop Science 197: 271-283.

37. Dodo HW, Konan KN, Chen FC, Egnin M and Viquez OM. 2008. Alleviating peanut allergy using genetic engineering: The silencing of the immunodominant allergen Ara h 2 leads to its significant reduction and a decrease in peanut allergencity. Plant Biotechnology Journal 6: 135-145.

38. Dwivedi SL, Amin PW, Rasheedunisa, Nigam SN, Nagabhushanam GVS, Rao VR and Gibbons RW. 1986. Genetic analysis of trichome characters associated with resistance to jassid (Empoascakerri Pruthi) in peanut. Peanut Science 13: 15-18.

39. Dwivedi SL, Crouch JH, Nigam SN, Ferguson ME and Paterson AH. 2003. Molecular breeding of groundnut for enhanced productivity and food security in the semi-arid tropics: opportunities and challenges. Advances in Agronomy 80: 153-221.

40. Dwivedi SL, Nigam SN, Reddy DVVR, Reddy AS and Ranga Rao GV. 1995. Progress in breeding groundnut varieties resistant to peanut bud necrosis virus and its vector. Pages 35-40. In AAM Buiel, JE Parlevliet and JM Lenne (eds.). Recent

studies on peanut bud necrosis disease: proceeding of a meeting. 20 March 1995, ICRISAT Asia Center, Patancheru 502 324, Andhra Pradesh, India. International Crops Research Institute for the Semi-Arid Tropics and PO Box 386, 6700 AJ Wageningen. The Netherlands: Dept. Plant Breed.

41. Entoori K, Sreevathsa R, Manoj KA, Kumar PA, Kumar ARV, Madhusudhan B and Udayakumar. 2008. A chimeric cry1X gene imparts resistance to Spodoptera litura and Helicoverpa armigera in the transgenic groundnut. EurAsian Journal of BioSciences 2: 53-65.

42. FAOSTAT 2013.

43. Farquhar GD and Richards RA. 1984. Isotopic composition of plant carbon correlates with water-use-efficiency of wheat genotypes. Australian Journal of Plant Physiology 11:539-552.

44. Gautami B, Foncěka D, Pandey MK, Moretzsohn MC, Sujay V, Qin H, Hong Y, Faye I, Chen X, BhanuPrakash A, Shah TM, Gowda MVC, Nigam SN, Liang X, Hoisington DA, Guo A, Bertioli DJ, Rami Jean-Francois and Varshney RK. 2012 a. An international reference consensus genetic map with 897 marker loci based on 11 mapping populations for tetraploid groundnut (Arachis hypogaea L.). PLoS ONE 7: 1-11.

45. Gautami B, Pandey MK, Vadez V, Nigam SN, Ratnakumar P, Krishnamurthy L, Radhakrishnan T, Gowda MVC, Narasu ML, Hoisington DA, Knapp SJ and Varshney, RK. 2012 b. Quantitative trait locus analysis and construction of consensus genetic map for drought tolerance traits based on three recombinant inbred line populations in cultivated groundnut (Arachis hypogaea L.). Molecular Breeding 30: 757-772.

46. Girdthai T, Jogloy S, Vorasoot N, Akkasaeng C, Wongkaew S, Holbrook and Patanothai A. 2010. Heritability of, and genotypic correlations between, aflatoxin traits and physiological traits for drought tolerance under end of season drought in peanut (Arachis hypogaea L.). Field Crops Research 118: 169-176.

47. Halward T, Stalker HT and Kochert G. 1993. Development of an RFLP linkage map in diploid peanut species. Theoretical and Applied Genetics 87: 379-384.

48. Hamidou F, Halilou O and Vadez V. 2013. Assessment of groundnut under combined heat and drought stress. Journal of Agronomy and Crop Science 199: 1-11.

49. Hanks RJ, Keller J, Rasmussen VP and Wilson GD. 1976. Line source sprinkler for continuous variable irrigation crop production studies. Soil Science Society of America Journal 40: 426-429.

50. Hapsoro D, Aswidinnoor H, Suseno R and Sudarsono J. 2005. Agrobacterium-mediated transformation of peanuts (Arachis hypogaea L.) with PStVcp gene. Journal of Tropical Agriculture 10: 85-93.

51. Hapsoro D, Aswidinnoor H, Jumanto, Suseno R and Sudarsono J. 2007. Resistance to peanut stripe virus (PStV) in transgenic peanuts (Arachis hypogaea

L.) carrying PStVcp gene was stable up to seven generations of selfing. Biota 12: 83-91.

52. Herselman L, Thwaites R, Kimmins FM, Courtois B, van der Merwe PJA and Seal SE. 2004. Identification and mapping of AFLP markers linked to peanut (Arachis hypogaea L.) resistance to the aphid vector of groundnut rosette disease. Theoretical and Applied Genetics 109: 1426-1433.
53. Higgins CM, Hall RM, Mitter N, Cruikshank A, and Dietzen RG. 2004. Peanut stripe potyvirus resistance in peanut (Arachis hypogaea L.) plants carrying viral coat protein gene sequences. Transgenic Research 13: 59-67.
54. Hildebrand, GL. 1996. The status of technologies used to achieve high groundnut yields in Zimbabwe. Pages 101-114 In Achieving high groundnut yields: proceedings of an international workshop, 25-29 Aug 1995, Laixi City, Shandong, China (eds. Renard, C, Gowda, CLL, Nigam, SN and Johansen, C), Patancheru 502 324, Andhra Pradesh, India. International Crops Research Institute for the Semi-Arid Tropics.
55. Holbrook CC, Anderson WF and Pittman RN. 1993. Selection of a core collection from the U.S. germplasm collection of peanut. Crop Science 33: 859-861.
56. Holbrook CC, Kvien CK, Ruker KS, Wilson DM, Hook JE and Matheron ME. 2000 a. Preharvest aflatoxin contamination in drought-tolerant and drought-intolerant peanut genotypes. Peanut Science 27: 45-48.
57. Holbrook CC, Matheron ME, Wilson DM, Anderson WF, Will ME and Norden AJ. 1994. Development of a large-scale field system for screening peanut for resistance to preharvest aflatoxin contamination. Peanut Science 21: 20-22.
58. Holbrook CC and Noe JP. 1992. Resistance to the peanut root-knot nematode (Meloidogyne arenaria) in Arachis hypogaea. Peanut Science 19: 35-37.
59. Holbrook, CC, Ozias-Akins, P, Chu, Y and Guo, B. 2011. Impact of molecular genetic research on peanut cultivar development. Agronomy 1: 3-17.
60. Holbrook, CC and Stalker, HT. 2003. Peanut breeding and genetic resources. Plant Breeding Reviews 22: 297-355.
61. Holbrook CC, Timper P and Culbreath AK. 2003. Resistance to tomato spotted wilt virus and root-knot nematode in peanut interspecific breeding lines. Crop Science 43: 1100-1113.
62. Holbrook CC, Timper P and Xue HQ. 2000 c. Evaluation of the core collection approach for identifying resistance to Meloidogyne arenaria in peanut. Crop Science 40: 1172-1175.
63. Holbrook CC, Wilson DM, Matheron ME, Hunter JE, Knauft DA and Gorbet DW. 2000 b. Aspergillus colonization and aflatoxin contamination in peanut genotypes with reduced linoleic acid composition. Plant Disease 84: 148-150.
64. Hong Y, Chen Y, Liang X, Liu H, Zhou G, Li S, Wen S, Holbrook CC and Guo B. 2010. A SSR-based composite genetic linkage map for the cultivated peanut (Arachis hypogaea L.) genome. BMC Plant Biology 10: 10-17.

65. Huang BY, Zhang XY, Miao LJ, Yan Z, Hai Y, Yi ML, Xu J and Chen ZK. 2008. RNAi transformation of Ah FAD2 gene and fatty acid analysis of transgenic seeds. Chinese Journal of Oil Crop Sciences 30: 290-293.

66. Huang L, Jiang H, Ren X, Chen Y, Xiao Y, Zhao X Tang M, Huang J, Upadhyaya HD and Liao B. 2012. Abundant microsatellite diversity and oil content in wild Arachis species. PLoS ONE 7 (11): e50002.

67. Huang L, Zhao X, Zhang W, Fan Z, Ren X, Liao B, Jiang H and Chen Y. 2011. Identification of SSR markers linked to oil content in peanut (Arachis hypogaea L.) through RIL population and natural population. ACTA AGRONOMICA SINICA 37: 1967-1974.

68. BPGR and ICRISAT. 1992. Descriptors for groundnut. Rome, Italy: International Board of Plant Genetic Resources; and Patancheru, Andhra Pradesh, India: International Crops Research Institute for the Semi-Arid Tropics.

69. Isleib TG, Pattee HE and Giesbrecht FG. 1995. Ancestral contribution to roasted peanut attribute. Peanut Science 22: 42-48.

70. Isleib, TG, Wynne, JC and Nigam, SN. 1994. Groundnut breeding. Pages 552-623. In The groundnut crop: A scientific basis for improvement. (ed. Smartt, J),Chapman and Hall, London.

71. Janila P and Nigam SN. 2013. Phenotyping for groundnut (Arachis hypogaea L.) improvement. Pages 129-167. In Phenotyping for plant breeding: Applications of phenotyping methods for crop improvement. (Panguluri SK and Kumar AA eds.). Springer Science+Business Media, New York.

72. Janila, P, Nigam, SN, Pandey, MK, Nagesh, P, and Varshney, RK. 2013. Groundnut improvement: use of genetic and genomic tools. Frontiers in Plant Science 4: 23. Doi: 10.3389/fpls. 2013. 00023.

73. Jayalakshmi V, Rajareddy C, Reddy PV and Nageswar Rao RC. 1999. Genetic analysis of carbon isotope discrimination and specific leaf area in groundnut (Arachis hypogaea L.). Journal of Oilseeds Research 16: 1-5.

74. Kakani VG, Prasad PVV, Craufurd PVV and Wheeler TR. 2002. Response of in vitro pollen germination, and pollen tube growth of groundnut (Arachis hypogaea L.) genotypes to temperature. Plant Cell & Environment 25: 1651-1661.

75. Khalfaoui JLB. 1991. Inheritance of seed dormancy in a cross between two Spanish peanut cultivars. Peanut Science 18: 65-67.

76. Khedikar YP, Gowda MVC, Sarvamangala C, Patgar KV, Upadhyaya HD and Varshney RK. 2010. A QTL study on late leaf spot and rust revealed one major QTL for molecular breeding for rust resistance in groundnut (Arachis hypogaea L.). Theoretical and Applied Genetics 121: 971-984.

77. Knauft DA, Moore KM and Gorbet DW. 1993. Further studies on the inheritance of fatty acid composition in peanut. Peanut Science 20: 74-76.

78. Knauft DA, Norden, AJ and Gorbet, DW.1987. Peanut. Pages 346-384. In Principles of cultivar development, Volume 2. Crop Species (ed. Fehr, WR). Macmillan Publishing Co., New York, USA.

79. Knauft, DA and Wynne, JC. 1995. Peanut breeding and genetics. Advances in Agronomy 55: 393-445.

80. Kochert G, Stalker HT, Gimenes M, Galgaro L, Lopez CR and Moore K. 1996. RFLP and cytogenetic evidence on the origin and evolution of allotetraploid domesticated peanut, Arachis hypogaea (Leguminosae). American Journal of Botany 83:1282-1291.

81. Li ZJ, Jarret RL and Demski JW. 1997. Engineered resistance to tomato spotted wilt virus in transgenic peanut expressing the viral nulceocapsid gene. Transgenic Research 6: 297-305.

82. Liang X, Zhou G, Hong Y, Chen X, Liu H and Li S. 2009. Overview of research progress on peanut (Arachis hypogaea L.) host resistance to aflatoxin contamination and genomics at the Guangdong Academy of Agricultural Sciences. Peanut Science 36: 29-34.

83. Lomonossoff GP. 1995. Pathogen-derived resistance to plant viruses. Annual Review of Phytopathology 33: 323-343.

84. Lopez Y, Smith OD, Senseman SA and Rooney WL. 2001. Genetic factors influencing high oleic acid content in Spanish market-type peanut cultivars. Crop Science 41: 51-56.

85. Lu M, Liang XQ, Dang P, Holbrook CC, Bausher MC, Lee RD and Guo BZ. 2005. Microarray-based screening of differentially expressed genes in peanut in response to Aspergillus parasiticus infection and drought stress. Plant Science 169: 695-703.

86. Lynch, RE. 1990. Resistance in peanut to major arthropod pests. Florida Entomologist 73: 422-445.

87. Magbanua ZV, Wilde HD, Roberts JK, Chowdhury K, Abad J, Moyer JW, Wetzstein HY and Parrott WA. 2000. Field resistance to tomato spotted wilt virus in transgenic peanut (Arachis hypogaea L.) expressing an antisense nucleocapsid gene sequence. Molecular Breeding 6: 227-236.

88. Manjunatha SB, Suma TC, Sreevathsa R, Devendra R, Udayakumar M and Prasad TG. 2008. Evaluation of advanced generation transgenic groundnut lines resistant to herbicide-glyphosate. Indian Journal of Weed Science 40: 162-165.

89. Mehan VK. 1989. Screening groundnuts for resistance to seed invasion by Aspergillus flavus and to aflatoxin production. Pp 323-334. In D McDonald and VK Mehan (eds.). Aflatoxin contamination of Groundnut. Proc. Int. Workshop, 6-9 Oct 1987, ICRISAT Center, India. Patancheru, AP 502 324, India. International Crops Research Institute for the Semi-Arid Tropics.

90. Mehan VK, Ba A, McDonald D, Renard JL, Rao RCN and Jayanthi S. 1991. Field screening of groundnuts for resistance to seed infection by Aspergillus flavus. Olėagineux 46: 109-118.

91. Mehan VK, Liao BS, Tan YJ, Robinson-Smith A, McDonald D and Hayward AC. 1994. Bacterial wilt of groundnut. Information Bulletin no. 35. Patancheru 502 324, Andhra Pradesh, India: International Crops Research Institute for the Semi-Arid Tropics. 28 pp

92. Mehan VK, Mayee, CD, Brenneman, TB and McDonald D.1995. Stem and pod rots of groundnut. Information Bulletin no. 44. Patancheru 502 324, Andhra Pradesh, India: International Crops Research Institute for the Semi-Arid Tropics and Griffin, GA 30223, USA: Peanut Collaborative Research Support Program. 28 pp

93. Mehan VK, McDonald D and Ramakrishna N. 1987. Resistance of peanut genotypes to seed infection by Aspergillus flavus in field in India. Peanut Science 14: 17-21.

94. Mehan VK, Reddy, DDR and McDonald D. 1993. Resistance in groundnut genotypes to Kalahasti malady caused by the stunt nematode, Tylenchorhynchus brevilineatus. International Journal of Pest Management 39: 201-203.

95. Melouk HA and Banks DJ. 1978. A method of screening peanut genotypes for resistance to Cercospora leafspot. Peanut Science 5: 112-114.

96. Mienie CMS and Pretorius AE.2013. Application of marker-assisted selection for ahFAD2A and ah FAD2B genes governing high oleic acid trait in South African groundnut cultivars (Arachis hypogaea L.). African Journal of Biotechnology 12: 4283-4289.

97. Milla SR, Isleib TG and Tallury SP. 2005. Identification of AFLP markers linked to reduced aflatoxin accumulation in A. cardenasii-derived germplasm lines of peanut. Proceedings of American Peanut Research and Education Society 37: 90.

98. Mondal S, Badigannavar AM and D'Souza SF. 2012. Development of genic molecular marker linked to a rust resistance gene in cultivated groundnut (Arachis hypogaea L.). Euphytica 188: 163-173.

99. Mondal S, Badigannavar AM and Murty SSS. 2008. RAPD markers linked to a rust resistance gene in cultivated groundnut (Arachis hypogaea L.). Euphytica 159: 233-239.

100. Mondal S, Hadapad AB, Hande PA and Badignnavar AM. 2014. Identification of quantitative trait loci for bruchid (Caryedon serratus L.) resistance components in cultivated groundnut (Arachis hypogaea L.). Molecular Breeding 33: 961-973.

101. Moore KM and Knauft DA. 1989. The inheritance of high oleic acid in peanut. Journal of Heredity 80: 252-253.

102. Moretzsohn MC, Leoi L, Proite K, Guimarães PM, Leal-Bertioli SCM, Gimenes MA, Martins WS, Valls JFM, Grattapaglia D and Bertioli DJ. 2005. A microsatellite-based, gene-rich linkage map for the AA genome of Arachis (Fabaceae). Theoretical and Applied Genetics 111: 1060-1071

103. Moretzsohn MC, Barbosa AVG, Alves-Freitas DMT, Teixeira C, Leal-Bertioli SCM, Guimarães PM, Pereira R, Lopez CR, Cavallari MM, Valls JFM, Bertioli DJ and Gimenes MA.2009. A linkage map for the B genome of Arachis (Fabaceae) and its synteny to the A-genome. BMC Plant Biology 9: 40.

104. Mozingo RW, Coffelt TA and Wynne JC. 1987. Genetic improvement in large-seeded Virginia-type peanut cultivars since 1944. Crop Science 27: 228-231.

105. Murthy, TGK and Reddy, PS.1993. Genetics of groundnut. Pages 144-269 in Cytogenetics and genetics of groundnuts. Intercept Ltd., UK.

106. Nageswar Rao RC and Nigam. 2001. Genetic options for drought management in groundnut. Pages 123-141. In Management of Agricultural Drought: Agronomic and Genetic Options. Saxena NP (ed.). Oxford and IBH, New Delhi.

107. Nageswar Rao RC, Talwar HS and Wright GC. 2001. Rapid assessment of specific leaf area and leaf N in peanut (Arachis hypogaea L.) using chlorophyll meter. Journal of Agronomy and Crop Science 189: 175-182.

108. Nageswar Rao RC, Williams JH, Wadia KDR, Hubick KT and Farquhar GD. 1993. Crop growth, water-use efficiency and carbon isotope discrimination in groundnut (Arachis hypogaea L.)genotypes under end-of-season drought conditions. Annals of Applied Biology 122: 357-367.

109. Nageswar Rao RC, Williams JH and Singh M. 1989. Genotypic sensitivity to drought and yield potential of peanut. Agronomy Journal 81: 887-893.

110. Nageswar Rao RC and Wright GC. 1994. Stability of the relationship between specific leaf area and carbon isotope discrimination across environments in peanut. Crop Science 34: 98-103.

111. Naidu, PH and Moses GJ. 2000. Identification and evaluation of sources of resistance to Kalahasti malady in groundnut. Indian Phytopathology 53: 395-398.

112. Narasimham, NV. 1999. Hand picked and selected (HPS) groundnut – Indian industry and trade. Agricultural Situation in India:179-190.

113. Nautiyal PC, Bandyopadhyay A and Ravindra V. 1999. Nature of fresh seed dormancy in peanut. ACIAR- Food Legumes Newsletter 30:5-6.

114. Nautiyal PC, Bandyopadhyay A and Zala PV. 2001. In situ sprouting and regulation of fresh seed dormancy in Spanish type groundnut (Arachis hypogaea L.). Field Crops Research 70: 233-241.

115. Nautiyal PC, Rachaputi NR and Joshi YC. 2002. Moisture-deficit-induced changes in leaf-water content, leaf carbon exchange rate and biomass production in groundnut cultivars differing in specific leaf area. Field Crops Research 74: 67-79.

116. Nautiyal PC, Ravindra V and Bandyopadhyay A. 1994. Peanut seed dormancy. ACIAR- Food Legumes Newsletter 22:2.

117. Nigam, SN. 2000. Some strategic issues in breeding for high and stable yield in groundnut in India. Journal of Oilseeds Research 17:1-10.

118. Nigam SN and Aruna R. 2008 a. Stability of Soil Plant Analytical Development (SPAD) Chlorophyll Meter Reading (SCMR) and specific leaf area (SLA) and their

association across varying soil moisture stress conditions in groundnut (Arachis hypogaea L.). Euphytica 160: 111-117.

119. Nigam SN and Aruna R. 2008 b. Improving breeding efficiency for early maturity in peanut. Plant Breeding Reviews 30: 295-322.

120. Nigam SN, Chandra S, Sridevi KR, Bhukta M, Reddy AGS, Rao RCN, Wright GC, Reddy PV, Deshmukh MP, Mathur RK, Basu MS, Vasundhara S, Varman PV and Nagda AK. 2005. Efficiency of physiology trait-based and empirical selection approaches for drought tolerance in groundnut. Annals of Applied Biology 146: 433-439.

121. Nigam SN, Dwivedi SL and Gibbons RW. 1991. Groundnut breeding: constraints, achievements and future possibilities. Plant Breeding Abstracts 61: 1127-1136.

122. Nigam SN, Nageswar Rao RC and Wright GC. 2003. Breeding for increased water-use efficiency in groundnut. Pages 305-318. In National seminar on stress management in oilseeds for attaining self-reliance in vegetable oils: Thematic papers. Rai M, Singh H and Hegde DM (eds.). Directorate of Oilseed Research, Rajendranagar, Hyderabad, Andhra Pradesh, India.

123. Nigam SN, Nageswara Rao RC, Wynne JC, Williams JH, Fitzner M and Nagabhushanam GVS. 1994. Effect and interaction of temperature and photoperiod on growth and partitioning in three groundnut (Arachis hypogaea L.) genotypes. Annals of Applied Biology 125: 541-552.

124. Nigam SN, Prasada Rao RDVJ, Bhatnagar-Mathur P and Sharma KK. 2012. Genetic management of virus diseases in peanut. Plant Breeding Reviews 36: 293-356.

125. Nigam SN, Upadhyaya HD, Chandra S, Nageswar Rao RC, Wright GC and Reddy AGS. 2001. Gene effects for specific leaf area and harvest index in three crosses of groundnut (Arachis hypogaea L.). Annals of Applied Biology 139: 301-306.

126. Nigam SN, Vasudeva Rao MJ and Gibbons RW. 1990. Artificial hybridization in groundnut. Information Bulletin no. 29. Patancheru, A.P. 502 324, India: International Crops Research Institute for the Semi-Arid Tropics.

127. Nigam SN, Waliyar F, Aruna R, Reddy SV, Lava Kumar P, Craufurd PQ, Diallo AT, Ntare BR and Upadhyaya HD. 2009. Breeding peanut for resistance to aflatoxin contamination at ICRISAT. Peanut Science 36: 42-49.

128. Norden AJ, Gorbet DW, Knauft DA and Young CT. 1987. Variability in oil quality among peanut genotypes in the Florida breeding program. Peanut Science 14: 7-11.

129. Norden, AJ, Smith, OD and Gorbet, DW. 1982. Breeding of the cultivated peanut. Pages 95- 122. In Peanut Science and Technology (eds. Pattee HE and Young, CT), American Peanut Research and Education Society, Inc. Yoakum, Texas 77995, USA.

130. Ntare BR, Williams JH and Dougbedji F. 2001. Evaluation of groundnut genotypes for heat tolerance under field conditions in a Sahelian environment using a simple physiological model for yield. Journal of Agricultural Sciences 136: 81-88.

131. Olorunju PE, Kuhn CW, Demski JW, Misari SM and Ansa OA. 1992. Inheritance of resistance in peanut to mixed infections of groundnut rosette virus (GRV) and groundnut rosette assistor virus and a single infection of GRV. Plant Disease 76: 95-100.

132. Olorunju PE, Ntare BR, Pande S and Reddy SV. 2001. Additional sources of resistance to groundnut rosette disease in groundnut germplasm and breeding lines. Annals of Applied Biology 159: 259-268.

133. Ozias-Akins P. 2007. Peanut. In Biotechnology in Agriculture and Forestry volume 61: 81-105, (Pua EC and Davey MR eds.). Springer-Verlag, Berlin, Heidelberg.

134. Ozias-Akins P, Schnall JA, Anderson WF, Singsit C, Clemente TE, Adang MJ and Weissinger AK. 1993. Regeneration of transgenic peanut plants from stably transformed embryogenic callus. Plant Science 93: 185-194.

135. Ozias-Akins P, Yang P, Culbreath AK, Gorbet DW and Weeks JR. 2002. Field resistance to Tomato spotted wilt virus in a transgenic peanut (Arachis hypogaea L.). Proceedings of American Research and Education Society 34:70.

136. Padgham DE, Kimmins FM and Ranga Rao GV. 1990. Resistance in groundnut (Arachis hypogaea L.) to Aphis craccivora(Koch.). Annals of Applied Biology 117: 285- 294.

137. Pandey MK, Monyo E, Ozias-Akins P, Liang X, Guimarães P, Nigam SN, Upadhyaya HD, Janila P, Zhang X, Guo B, Cook DR, Bertioli DJ. Michelmore R and Varshney RK. 2012. Advances in Arachis genomics for peanut improvement. Biotechnology Advances 30: 639-651.

138. Pang SZ, Slightom JL and Gonsalves D. 1993. Different mechanisms protect tobacco against tomato spotted wilt and impatients necrotic spot tospoviruses. Biotechnology 11: 819-824.

139. Patel M, Jung S, Moore K, Powell G, Ainsworth C and Abbott A. 2004. High-oleate peanut mutants result from a MITE insertion into the FAD2 gene. Theoretical and Applied Genetics 108: 1492-1502.

140. Pattee HE, Isleib TG and Giesbrecht FG. 1998. Variation in intensity of sweet and bitter sensory attributes across peanut genotypes. Peanut Science 25: 63-69.

141. Passioura JB. 1977. Grain yield, harvest index and water use of wheat. Journal of the Australian Institute of Agricultural Sciences 43: 117-120.

142. Pensuk K, Jogloy S, Wongkaew S and Patanothai A. 2004. Generation means analysis of resistance to peanut bud necrosis caused by peanut bud necrosis tospovirus in peanut. Plant Breeding 123: 90-92.

143. Pensuk K, Wongkaew S, Jogloy S, and Patanothai A. 2002. Combining ability for resistance in peanut (Arachis hypogaea) to peanut bud necrosis tospovirus (PBNV). Annals of Applied Biology 141: 143-146.

144. Poledate A, Laohasiriwong S, Jaisil P, Vorasoot N, Jogloy S, Kesmala T and Patanothai A. 2007. Gene effects for parameters of peanut bud necrosis virus (PBNV) resistance in peanut. Pakistan Journal of Biological Sciences 10: 1501-1506.

145. Qin H, Gu Q, Zhang JL, Sun L, Kuppu S, Zhang YZ, Burow M, Payton P, Blumwald E and Zhang H. 2011. Regulated expression of an isopentenyltransferase gene (IPT) in peanut significantly improves drought tolerance and increases yield under field conditions. Plant and Cell Physiology 52: 1904-1914.

146. Qin HD, Feng SP, Chen C, Guo YF, Knapp S, Culbreath A, He GH, Wang ML, Zhang XY, Holbrook CC, Ozias-Akins P and Guo BZ. 2012. An integrated genetic linkage map of cultivated peanut (Arachis hypogaea L.) constructed from two RIL populations. Theoretical and Applied Genetics 124: 653-664.

147. Rao MJV, Nigam SN and Huda AKS. 1991. Use of thermal time concept in selection for earliess in peanut. Peanut Science 19: 7-10.

148. Rao, M.J.V., S.N. Nigam, V.K. Mehan, and D. McDonald. 1989. Aspergillus flavus resistance breeding in groundnut: Progress made at ICRISAT Center. pp. 345-355. In D. McDonald and V.K Mehan (eds.) Aflatoxin contamination of groundnut. Proc. Int. Workshop, 6-9 Oct. 1987, ICRISAT Center, India. Patancheru, A.P. 502 324, India. International Crops Research Institute for the Semi-Arid Tropics.

149. Rathnakumar AL, Hariprasanna K and Lalwani, HB.2012. Genetic improvement in Spanish type groundnut. Journal of Oilseeds Research 27: 1-7.

150. Ravi K, Vadez V, Isobe S, Mir RR, Guo Y, Nigam SN, Gowda MVC, Radhakrishnan T, Bertioli DJ, Knapp SJ and Varshney, RK.2011. Identification of several small main-effect QTLs and a large number of epistatic QTLs for drought tolerance related traits in groundnut (Arachis hypogaea L.). Theoretical and Applied Genetics 122: 1119-1132.

151. Reddy PS and Murthy TGK. 1996. Current genetic research on groundnut in India. Genetica 97: 263-277.

152. Reddy PS, Zade VR and Desmukh SN.1985. A new Spanish bunch groundnut cultivar with fresh seed dormancy. Journal of Oilseed Research 2: 103-106.

153. Sanders, TH, Schubert AM and Pattee HE. 1982. Maturity methodology and post-harvest physiology. p. 624-654. In Peanut Science and Technology (eds. Pattee HE and Young CT), American Peanut Research and Education Society, Inc. Yoakum, Texas 77995, USA.

154. Seaton ML, Coffelt TA and Van Scoyoc SW. 1992. Comparison of vegetative and reproductive traits of 14 peanut cultivars. Oléagineux 47: 471-474.

155. Seijo JG, Lavia GI, Fernandez A, Krapovikas A, Ducasse D and Moscone EA. 2004. Physical mapping of the 5S and 18S-25S $_R$RNA gene by FISH as evidence that Arachis duranensis and A. ipaensis are the wild diploid progenitors of A. hypogaea (Leguminosae). American Journal of Botany 91: 1294-1303.

156. Shan ZH, Duan NX, Jiang HF, Tan YJ, Li D and Liao BS. 1998. Inheritance to bacterial wilt in Chinese dragon groundnuts. Pages 300-305. In Bacterial wilt

disease. (Prior P, Allen C and Elphinstone J eds.). Springler-Varlag, Berlin, Heidelberg.

157. *Sharma KK and Anjaiah VV. 2000. An efficient method for the production of transgenic plants of peanut (Arachis hypogaea L.) through Agrobacterium tumefaciense-mediated genetic transformation. Plant Science 159: 7-19.*

158. *Shoba D, Manivannan N, Vindhiyavarman P and Nigam SN. 2012. SSR markers associated for late leaf spot disease resistance by bulk segregant analysis in groundnut (Arachis hypogaea L.). Euphytica 188: 265-272.*

159. *Shoba D, Manivannan N, Vindhiyavarman P and Nigam SN. 2013. Identification of quantitative trait loci (QTL) for late leaf spot disease resistance in groundnut (Arachis hypogaea L.). Legumes Research 36: 467-472.*

160. *Shokes FM, Różalski K, Gorbet DW, Brenneman TB and Berger DA.1996.Techniques for inoculation of peanut with Sclerotium rolfsii in the greenhouse and field. Peanut Science 23: 124-128.*

161. *Singh AL, Hariprasanna K, Chaudhary V, Gor HK and Chikani BM. 2010. Identification of groundnut (Arachis hypogaea L.) cultivars tolerant of soil salinity. Journal of Plant Nutrition 33: 1761-1776.*

162. *Singh AL, Hariprassana K and Solanki RM. 2008. Screening and selection of groundnut genotypes for tolerance of soil salinity. Australian Journal of Crop Science 1: 69-77.*

163. *Singh AK and Nigam, SN. 1997. Groundnut. Pages 114-127 In Biodiversity in Trust (eds. Fuccillo D, Scars, L and Stapleton, P), Cambridge University Press, UK.*

164. *Singh AK, Mehan VK and Nigam SN. 1997. Sources of resistance to groundnut fungal and bacterial diseases: an update and appraisal. Information Bulletin no. 50. Patancheru, A.P. 502 324, India: International Crops Research Institute for the Semi-Arid Tropics. 48 pp.*

165. *Singsit C, Adang MJ, Lynch RE, Anderson WF, Wang A, Cardineau C and Ozias-Akins P. 1997. Expression of a Bacillus thuringiensis cry1A(c) gene in transgenetic peanut plants and its efficacy against lesser cornstalk borer. Transgenic Research 6: 169-176.*

166. *Songsri P, Jogloy S, Kesmala T, Vorasoot N, Akkasaeng C, Patonothai A and Holbrook CC. 2008 a. Heritability of drought resistance traits and correlation of drought resistance and agronomic traits in peanut. Crop Science 48: 2245-2253.*

167. *Songsri P, Jogloy S, Vorasoot N, Akkasaeng C, Patonothai A and Holbrook CC. 2008 b. Root distribution of drought-resistant peanut genotypes in response to drought. Journal of Agronomy and Crop Science 194: 92-103.*

168. *Srivastava N, Vadez V, Krishnamurthy L, Saxena KB, Nigam SN and Rupakula A. 2007. Standardization of a screening technique for salinity tolerance in groundnut (Arachis hypogaea L.) and pigeonpea (Cajanus cajan L.). Indian Journal of Crop Science 2: 209-214.*

169. Stalker, HT. 1997. Peanut (Arachis hypogaea L.). Field Crops Research 53: 205-217.

170. Stalker HT and Mozingo LG. 2001. Molecular markers of Arachis and marker-assisted selection. Peanut Science 28: 117-123.

171. Stalker, HT and Simpson, CE. 1995. Germplasm Resources in Arachis. Pages 14-53. In Advances in Peanut Science (eds. Pattee, HE and Stalker, HT), American Peanut Research and Education Society, Inc., Stillwater, OK 74078, USA.

172. Stalker, HT, Weissinger, AK, Milla-Lewis, S and Holbrook. 2009. Genomics: An emerging science in peanut. Peanut Science 36: 2-10.

173. Subrahmanyam P, Hildebrand GL, Naidu RA, Reddy LR and Singh AK. 1998. Sources of resistance to groundnut rosette disease in global groundnut germplasm. Annals of Applied Biology 132: 473-485.

174. Subrahmanyam P, McDonald D, Waliyar F, Reddy LJ, Nigam SN, Gibbons RW, Rao VR, Singh AK, Pande S, Reddy PM and Subba Rao PV. 1995. Screening methods and sources of resistance to rust and late leaf spot of groundnut. Information Bulletin no. 47. Patancheru, A.P. 502 324, India: International Crops Research Institute for the Semi-Arid Tropics. 24 pp.

175. Subrahmanyam P, Wongkaew S, Reddy DVVR, Demski JW, McDonald D, Sharma SB, Smith DH, Nigam SN and Sudini H. 2012. Field diagnosis of groundnut diseases. Information Bulletin no. 36 (revised). Patancheru, A.P. 502 324, India: International Crops Research Institute for the Semi-Arid Tropics. 88pp.

176. Sunkara S, Bhatnagar-Mathur P and Sharma KK. 2013. Transgenic interventions in peanut crop improvement: progress and prospects. Pages 179-216. In Genetics, Genomics and Breeding of peanuts. (Mallikarjuna N and Varshney RK eds.). CRC Press, Taylor and Francis Boca Ratob, Florida.

177. Taliansky ME, Robinson DJ and Murant AF. 1996. Complete nucleotide sequence and organization of the RNA genome of groundnut rosette umbravirus. Journal of General Virology 77: 2335-2345.

178. Tiwari S, Mishra DK, Singh A, Singh PK and Tuli R. 2008. Expression of a synthetic cry1EC gene for resistance against Spodoptera litura in transgenic peanut (Arachis hypogaea L.). Plant Cell Reports 27: 1017-1025.

179. Upadhyaya HD, Bramel PJ, Ortiz R and Singh S. 2002 a. Developing a minicore of peanut for utilization of genetic resources. Crop Science 42: 2150-2156.

180. Upadhyaya HD and Nigam SN. 1999. Inheritance of fresh seed dormancy in peanut. Crop Science 39: 98-101.

181. Upadhyaya, H.D., Nigam SN, Mehan VK and Lenne JM. 1997a. Aflatoxin contamination of groundnut- prospects of a genetic solution through conventional breeding, pp 81-85 In Aflatoxin contamination problems in groundnut in Asia: proceedings of the First working Group Meeting, 27-29 may 1996, Ministry of Agriculture and Rural Development, Hanoi, Vietnam (Eds. V.K. Mehan and C.LL. Gowda). Patancheru 502 324,

182. Upadhyaya HD, Nigam SN, Rao MJV, Reddy AGS, Yellaiah N and Reddy NS. 1997b. Registration of five Spanish peanut germplasm with fresh seed dormancy. Crop Science 37: 1027.

183. Upadhyaya, H.D., S.N. Nigam and R.P Thakur. 2002 b. Genetic enhancement for resistance to aflatoxin contamination in groundnut, pp 29-36 In Summary proceedings of the seventh ICRISAT regional groundnut meeting for Western and Central Africa, 6-8 December 2000, Cotonou, Benin. Patancheru 502 324, Andhra Pradesh, India: International Crops Research Institute for the Semi-Arid Tropics.

184. Upadhyaya HD, Ortiz R, Bramel PJ and Singh S. 2003. Development of a groundnut core collection using taxonomical, geographical and morphological descriptors. Genetic Resources and Crop Evolution 50: 139-148.

185. Upadhyaya, HD, Sharma, S and Dwivedi, SL. 2011. Arachis. Pages 1-19 In Wild Crop Relatives: Genomic and Breeding Resources, Legume Crops and Forages, (ed. C Kole). Springler-Verlag, Berlin Heidelberg.

186. Upadhyaya HD, Yadav D, Dronavalli N, Gowda CLLL and Singh S. 2010. Minicore germplasm collection for infusing genetic diversity in plant breeding program. Electronic Journal of Plant Breeding 1 (4): 1294-1309.

187. Utomo SD, Anderson WF, Wynne JC, Bouto MK, Hagler Jr. WM and Payne GA. 1990. Estimates of heritability and correlation among three mechanisms of resistance to Aspergillus parasiticus in peanut. Proceedings of the American Peanut Research and Education Society 22: 26. (Abstract).

188. Vadez V, Rao JS, Bhatnagar-Mathur P and Sharma KK. 2013. DREB 1A promotes root development in deep soil layers and increases water extraction under water stress in groundnut. Plant Biology 15: 45-52.

189. Varma TSN, Dwivedi SL, Pande S and Gowda MVC. 2005. SSR markers associated with resistance to rust (Puccinia arachidis Speg.) in groundnut (Arachis hypogaea L.). SABRAO Journal of Breeding and Genetics 37: 107-119.

190. Varshney, RK, Bertioli DJ, Moretzsohn MC, Vadez V, Krishnamurthy L, Aruna R, Nigam SN, Moss BJ, Seetha K, Ravi K, He G, Knapp SJ and Hoisington DA. 2009. The first SSR-based genetic linkage map for cultivated groundnut (Arachis hypogaea L.).Theoretical and Applied Genetics 118: 729-739.

191. Vasudeva Rao, MJ, Nigam, SN, and Huda, AKS.1992. The thermal time concept as a selection criterion for earliness in peanut. Peanut Science 19: 7-10

192. Vindhiyavarman, P and Gunasekaran, M. 2012.Participatory plant breeding theory and practices. Kalyani Publishers, India. 185 pp.

193. Wang M, Abbott D and Waterhouse PM. 2000. A single copy of a virus derived transgene encoding hairpin RNA gives immunity to barley yellow dwarf virus. Molecular Plant Pathology 1: 401-410.

194. Weber, JL. 1990. Informativeness of huan (dC-dA) (dG-dT) in polymorphism. Genomics 7: 524-530.

195. Wells R, Bi T, Anderson WF and Wynne JC. 1991. Peanut yield as a result of fifty years of breeding. Agronomy Journal 83: 957-961.

196. Wightman JA and Ranga Rao GV. 1994. Groundnut pests. Pages 395-479. In The groundnut crop: A scientific basis of for improvement. J Smartt (ed.). Chapman and Hall, London.

197. Wightman JA and Ranga Rao GV. 1993. A groundnut insect identification handbook for India. Information Bulletin no. 39. Patancheru, A.P. 502 324, India: International Crops Research Institute for the Semi-Arid Tropics. 64pp.

198. Williams JH and Boote KJ. 1995. Physiology and modeling-predicting the unpredictable legume. Pages 301-353. In Advances in peanut science. Pattee HE and Stalker HT (eds.). American Peanut Research and Education Society, Stillwater, USA.

199. Wright DL. 2008. Production of biofuel crops in Florida: Peanut. SS AGR 295, Agronomy Department, Florida Cooperative Extension Service, Institute of Food and Agricultural Sciences, University of Florida.

200. Wright GC, Nageswar Rao RC and Basu MS. 1996. A physiological approach to the understanding of genotype by environment interactions – A case study on improvement of drought adaptation in peanut. Pages 365-381. In Plant Adaptation and Crop Improvement. Cooper M and Hammer GL (eds.). CAB International, Wallingford, UK.

201. Wright GC, Nageswar Rao RC and Farquhar GD. 1994. Water-use efficiency and carbon isotope discrimination in peanut under water deficit conditions. Crop Science 34: 92-97.

202. Wynne, JC and Gregory, WC. 1981. Peanut breeding. Advances in Agronomy 34: 39-72.

203. Xing JY, Wang BZ, Jia KH, Wan SB, Meng JJ, Guo F and Li XG. 2011. Isolation of Arachis hypogaea Na+/H+ antiporter and its expression analysis under salt stress. African Journal of Biotechnology 10: 14302-14310.

204. Xue HQ. 2004. Evaluation of peanut (Arachis hypogaea L.) germplasm for resistance to aflatoxin production by Aspergillus flavusLink ex Fries. Ph D Thesis. North Carolina State University, Raleigh, NC, USA.

205. Xue HQ, Isleib TG, Payne GA, Novitzky WF and OBrian G. 2005. Aflatoxin production in peanut lines selected to represent a range of linoleic acid concentrations. Journal of Food Protection 68: 126-132.

206. Yang H, Singcit C, Wang A, Gonsalves D and Ozias-Akins P. 1998. Transgenic peanut plants containing a nucleocapsid protein gene of tomato spotted wilt virus show divergence levels of gene expression. Plant Cell Reports 17: 693-699.

207. Yanhao, S, Shouxiang, T, Caibin, W. 1996. Theoretical foundations for high yield of groundnut in China. Pages 129-139 in Achieving high groundnut yields: proceedings of an international workshop, 25-29 Aug 1995, Laixi City, Shandong, China (eds. Renard, C, Gowda, CLL, Nigam, SN and Johansen, C), Patancheru

502 324, Andhra Pradesh, India. International Crops Research Institute for the Semi-Arid Tropics.

208. *Yaw AJ, Richard A, Safo-Kantanka O, Adu-Dapaah HK, Ohemeng-Dapaah S and Agyeman A. 2008. Inheritance of fresh seed dormancy in groundnut. African Journal of Biotechnology 7: 421-424.*
209. *Yin DM, Tang HT, Tai GQ, Yang QY and Cui DQ. 2009. Expression analysis and establishment of regeneration system of oleate desaturase gene in peanut. Scientia agricultura Sinica 42: 1827-1832.*
210. *Zhang XQ, Shan L, Tang GY, Teng N and Bi YP. 2007. Transformation of RNAi suppressed expression vector containing of Δ^{12} fatty acid desaturase gene via Agrobacterium infection in peanut (Arachis hypogaea L.). Chinese Journal of Oil Crop Sciences 29: 409-415.*
211. *Zhou GY and Liang XQ. 2002. Analysis of major-minor genes related to resistance to infection by Aspergillus flavus in peanut. Journal of Peanut Science 31:11-14.*
212. *Zhou GY, Liang XQ, Li YC, Li XC and Li SL. 1999. Evaluation and application of introduced peanut cultivars for resistance to Aspergillus flavus invasion. Journal of Peanut Science 32: 14-17.*

Chapter 5

Breeding High Yielding Drought Resistant Varieties in Groundnut

RP. Vasanthi

Principal Scientist (Genetics and Plant Breeding), RARS, Tirupati.

5.1 Introduction

Groundnut is the major oilseed and food crop of India. It is being grown in over 5-52 million ha with a production of 61.1 million tones and productivity of 1105 kg ha. It is largely cultivated in kharif season in rainfed environment where major limiting factor for yield is moisture stress at different stages of crop growth variety from year to year with no fixed pattern. Though groundnut is an inherently drought tolerant crop with great resilence capacity, yield and its attributes are affected in rainfed situation due to moisture stress of varying intensities at different stages of crop growth. In kharif season, its productivity depends on quantity and distribution of rainfall during the season. Depending on the distribution, irrespective of the quantum of rainfall the yield levels fluctuate between 550 to 1350 kg/ha. In certain years, excess rainfall during early crop growth and later at pod development and pod maturity stage are equally detrimental to the crop.

Drought continues to be a challenge to agricultural scientists in general and plant breeders in particular, despite many decades of research. The development through breeding of cultivars with higher harvestable yield under drought would be a major breakthrough. Drought resistance is a complex trait, expression of which depends on action and interaction of different morphological (earliness, reduced leaf area, leaf rolling, wax content, efficient root system, leaf pubescence and stem hairiness and reduced branching), physiological (reduced transpiration, high water use efficiency, stomatal closure and osmotic adjustment) and biochemical (accumulation of proline, polyamines, trehalose, betaine, increased reductase activity and increased storage of carbohydrates) characters . The ability of some cultivars to give a higher economic yield under drought than others is a very elusive trait from genetic point of view. This is because severity, timing and duration of drought vary from year to year and cultivars successful in one dry year may fail in another (*Richards*, 1996). As there is no fixed pattern of dry spells in any given location, breeding programmes aim at comprehensively tackling the drought through consideration of various morphological and physiological mechanisms besides evaluation for yield in a given environment.

To formulate breeding programmes for breeding for drought resistance, it is essential to understand and know different morphological and physiological mechanisms that confer drought resistance i.e. trait identification. Establishing the importance of a particular trait is very difficult and time consuming. The nature of drought is such that its timing and intensity is unpredictable from year to year which means that the physiological responses to drought are complex and unpredictable. These make breeding for drought resistance slow and difficult, whether it can be in selection for yield or for the evaluation or identification of the important traits. Up to late 1970s, defining criteria for improving yield under drought was a haphazard affair. A shift in thinking occurred in the late 1970s which changed the basis for identifying criteria for improving yield under drought. There is a switch away from thinking how plants protect themselves against drought to thinking how to maximize the economic product when water is limited (*Passioura*, 1977). He proposed that when water is limited, the yield of biomass is a function of the amount of water transpired by the crop (T) and how efficiently the crop uses this water i.e., the water use efficiency (WUE),

Where $\text{biomass} = T \times WUE$

$$\text{Economic yield} = T \times WUE \times HI$$

Using this model, it is possible to assess whether a particular trait is likely to increase any one of the components thereby increasing yield under drought. This provided a framework to more critically identify important traits for yield improvement under drought condition.

In the normal situation, yield potential is explained by the physiological model proposed by (Duncan et al. 1978)

As per the model. $Y_R = C^*D_R^*\ P$ where, Y_R – Yield potential or the genotypes C – Mean crop growth rate, D_R – Duration of reproductive growth and P = Mean fraction of C partitioned towards the reproductive organs. C is defined as dry matter produced per unit land area per unit time ($gm^{-2}\ day^{-1}$). It is the integration of intercepted radiation and radiation use efficiency. D_R and P are the integration of reproductive growth and ability of partitioning of photosynthates to pods. The crop duration is generally fixed for a given location or cropping system. It is determined by soil moisture availability and prevailing temperatures during the cropping season. D_R can be increased to some extent by reducing the duration of the vegetative phase by selecting for early emergence and profuse early flowering. Thus, in fixed crop duration, C and P are the major determinants of the final yield (Nigam, 2004).

5.2 Drought Resistance Mechanisms in Groundnut

It is important to identify the specific physiological and morphological attributes contributing to their adaptability to the moisture stress pattern in a given target environment. This would assist the breeders to selectively combine these attributes into the high yielding cultivars for rainfed cropping systems. (Rosielle and Hamblin, 1981). In genetic sense, the mechanisms of drought resistance can be grouped into three categories viz. drought escape, drought avoidance and drought tolerance (Levitt, 1972). However crop plants use more than one mechanism at a time to resist drought. Drought escape is defined as the ability of plant to complete its life cycle before serious soil and plant water deficits develop. This mechanism involves rapid phenological development, developmental plasticity and remobilization of pre- anthesis assimilates to grain (Turner 1979). Drought avoidance is the ability of plants to maintain relatively high tissue water potential despite a shortage of soil moisture, where as drought tolerance refers to the ability of the plant to withstand water deficit with low tissue water potential. Mechanisms for improving water uptake, storing in plant cell and reducing water loss confer drought avoidance. The responses of plant to tissue water deficit determine their level of drought tolerance (Subba Rao et.al, 1995).

Drought avoidance is performed by maintenance of turgor through increased rooting depth, efficient root system and increased hydraulic conductance and by reduction of water loss through reduced epidermal (stomatal and lenticular) conductance, reduced absorption of radiation by leaf rolling or folding (Tode and Moya, 1978) and reduced evapo-transpiration (Turner, 1986). These help to survive under drought condition by doing a balancing act between maintenance of turgor and reduction of water loss (Shashidhar et al, 2000). The mechanisms of drought tolerance are maintenance of turgor through osmotic adjustment, increase in elasticity in cell and decrease in cell size and desiccation tolerance by protoplasmic resistance (Sullivan and Ross, 1979).

Most of these mechanisms/traits adaptive to drought have disadvantages. A genotype of short duration usually yields less compared to that of normal duration. The mechanisms that confer drought resistance by reducing water loss usually result in reduced assimilation of carbon dioxide. Osmotic adjustment increases drought resistance by maintaining plant turgor, but the increased solute concentration responsible for osmotic adjustment may have detrimental effect in addition to energy requirement for osmotic adjustment. Hence, breeders should aim at moulding genotypes with a moderation in each of the traits that confer drought tolerance and yield traits.

5.2.1 Drought Escape

Drought escape is an important strategy of matching phenological development with the period of soil moisture availability to minimize the impact of drought stress on crop production in environments where the growing season is short and terminal drought stress predominates (Turner, 1986). Plant breeding programmes should aim at development of high yielding genotypes with phonological patterns to match seasonal soil moisture availability (Fischer etal, 1982).

Land races mostly belonging to Virginia bunch and Virginia runners face terminal moisture stress and because of this season and difficulty in harvesting they were almost phased out from cultivation. Short duration Spanish bunch variety, TMV2 subsequently occupied major area. This was subsequently high yielding Spanish bunch varieties, JL24, later on followed by Narayani, Kadiri 6. Virginia bunch/ Virginia runners have better tolerance to early and mid season moisture stress in comparison with presently popular Spanish bunch varieties. Hence, breeders aim at developing high yielding varieties with short-duration early maturing nature with tolerance attributes of Virginia lines without sacrificing yield. Breeding for shorter duration is major objective, not only to match phenology to season length, but also for other reasons such as to fit crops/genotypes in to existing rainfed system.

However various penalties are associated with reducing crop duration to better fit likely soil moisture patterns. Primarily, earliness ultimately reduces the potential yield of the crop by reducing dry matter at flowering and number sites of post flowering and pod filling. The degree of earliness required is a generally compromise between development of sufficient biomass without reducing soil water to a level that will limit reproductive growth (Fischer, 1979). Simulation studies in soybean have shown that early genotypes yield more than those maturing later if late drought reduces the yield of later maturing cultivars by at least 40% (Muchow and Sinclair, 1986)

Mid season drought which occurs during the pod filling stage is important in groundnut as it limits the productivity. Mid season drought has been shown to significantly reduce pod yield of groundnut (Nageswara Rao. et al, 1989), while terminal drought can cause not only a reduction in yield but also an increase in the incidence of the aflatoxin contamination (Holbrook and Stalker, 2003, Arunyanark et al, 2009, Girdthai et al, 2010). Drought stress can also occur during the early flowering stage, but early season drought stresses is not detrimental to peanut yield and it sometimes actually increases yield of peanut (Nageswar Rao. et al 1985, Nautiyal, 1999, Puang but et al. 2009). Limited moisture between emergence and flowering allows the plants to 'acclimate' and become better prepared for a drought season (Balota, 2012)

5.2.2 Dehydration Avoidance

Crop species have evolved several mechanisms to maintain plant water status within reasonable limits for normal metabolic functioning under limited water supply or when the evaporative demand of the atmosphere become excessive. These can be broadly divided into: maximize water uptake through improving the capacity of the root system to acquire water and optimization of the use of the absorbed water for the production of dry matter.

5.2.2.1 Root Attributes

Root size, morphology, depth, length, density, hydraulic conductance and function are basic to meet the transpirational demands of the short (Passioura, 1982). For maximizing extraction of moisture from the soil, the requirements are 1.Deep penetration of roots 2. Adequate root density through the soil profile and 3.Adequate longitudinal conductance in main roots (Fischer et al, 1982). The water uptake pattern across the depth of rooting zone is not uniform. In general, nearly 40% of the total water uptake occurs over the first one fourth of the root zone, 30% over the second, 20% over the third, and 10% over the last fourth of the total rooting depth (Doorenbos and Pruitt, 1975; Nageswara Rao and Wright, 1994). Crop plants, often maintain higher rot length densities than are required in the surface layer: 1.To facilitate rapid uptake of recent rain before it evaporates 2. To provide reserve capacity in case of disease or pest damage 3. To extract relatively immobile nutrients like P; and to compete with weeds or other neighboring plants for both water and nutrients (Passioura, 1983).

5.2.2.1.1 Rooting depth

In the rainfed environments, the depth of rooting and the ability to sustain an un-interrupted supply of water are important factors (Gregory, 1988). In groundnut, substantial genotypic variation in rooting depth, root volume and water extraction pattern at different depths has been reported (Ketring; 1984, Mathews etal.1988 a, Wright et al. 1991, Chapman et al. 1993 a)

5.2.2.1.2 Root length density

Genotypic variation for root length density was reported in groundnut (Wright et al. 1994 and Nageswara Rao & Wright 1994). Screening and selection for rooting depth on a large scale is expensive and laborious (Blum, 1988). Thus, this type of evaluation is restricted to a few promising germplasm lines or cultivars and in choosing potential parents in a breeding program (Fischer et al. 1982)

5.2.2.2 Shoot Attributes

A number of shoot attributes play important roles in regulating water use of crop plants when grown under moisture-limiting environments. These include development, structure and surface properties of the canopy, ability to adjust the leaf area according to moisture availability and functional attributes such as osmotic adjustment (Subba Rao et al, 1995).

5.2.2.2.1 Canopy structure

Canopy structure is determined by leaf size, leaf shape, leaf surface characteristics and reflectance properties, leaf angle and the geometrical arrangement of leaves in the canopy. These traits determine the light extinction co-effficient (K) and radiation use efficiency (RUE) of the crop and canopy structure plays an important role in controlling water loss from the canopy. Under field conditions, the boundary layer over the crop canopy cause gas exchange to be less dependent on stomatal conductance, and thus can influence transpiration efficiency (TE) (Jarvis and Mc Naughton, 1986). Boundary layer resistance is a function of the thickness of the unstirred air boundary layer adjacent to the leaf which depends on leaf size (Parkhurst and Loucks, 1972) and canopy architecture. The aerodynamic resistance of the canopy determines the relative importance of stomatal conductance to TE. If the canopy resistance to heat and water vapour diffusion is large, an increase in stomatal conductance (gs) would tend to cool and humidify the air in the boundary layer, thus lowering the leaf-air vapour pressure deficit (vpd) and increasing TE (Farquhar et al. 1989)

With closed canopy, solar radiation is attenuated downward with cumulative leaf area index (LAI) in accordance with Beer's law, with a 'K' characteristic of canopy (Lawn, 1989). Crop cultivars with more erect and narrow leaves, and lower K values and hence higher critical leaf area index (CLAI- Leaf area index that intercepts 95% of the incoming solar radiation), generally have higher crop growth rates (Duncan, 1971). The advantages of narrow, vertically oriented leaves with a resultant higher CLAI and RUE has been well demonstrated in groundnut using mutants with variable leaf size and shape (Nageswara Rao, 1992). The narrow leaf mutant, TMV2-NLM has CLAI values around 5 to 6 compared with its parental line; TMV2 with CLAI values around 2 to 3, although in both cases, LAI reaches 5 at the onset of flowering. The RUE of the mutant is consistently higher than that of the parent (Nageswara Rao, 1992). Also, narrow leaves are considered to be an adaptive trait to stress conditions (Blum, 1980). In addition to this, because of aerodynamic implications, narrow leaves are usually stress less (Gates, 1968). Many legumes produce leaf area beyond CLAI which results in inefficient use of water and radiation for dry matter production and yield. In groundnut, the LAI often reaches more than 6, although the CLAI is only about 2to 3 (Williams etal, 1986; Nageswara Rao, 1992). Leaf losses of upto 50% can be tolerated by many legumes including groundnut without any major effect on yield. Thus, genotypes that are conservative in leaf area development beyond CLAI and also have narrow leaves that permit better utilization of radiation and water should be advantageous in water-limited environments. Significant differences were found in the values of LAI from 36 to 72 DAS. Leaf area index (LAI) describes the size of assimilatory apparatus of a crop or the crop canopy expansion. LAI increased from 36 to 72 DAS. (Duncan *et al.* 1978) reported that maximum light interception occurred when the LAI was three. In the present study, LAI values greater than three were observed in all the cultivars at 72 DAS. The cultivar, JL-24 showed maximum LAI at both 36 and 72 DAS. It showed fairly high LAI at 54 DAS also. The cultivar, K-134 showed maximum LAI at 54 DAS and also at 36 and 72 DAS (Suneetha, 2007).

5.2.2.2.2 Leaf movements

Once leaf area development is complete, an important mechanism by which legumes can adapt to drought stress is through changes in leaf angle (Begg, 1980). Leaf lets orient perpendicular to incident light in the absence of the water deficit but parallel to it during water deficits (Squire, 1990). This can effectively reduce the radiation load on water stressed leaves when less water is available to dissipate the energy as latent heat, thus minimizing heat damage (Ludlow and Bjorkman, 1984, Forseth and Teranura, 1986). Advantages in ability to change leaf angle or orientation (paraheliotropism) are reversibility, rapidity of recovery on the relief of water deficit, and a minimum reduction in yield during water deficit (Shackel and Hall, 1979). The leaf orientation will allow maximum radiation interception when evaporative demand is low and WUE is high (Turner, 1982, Muchow, 1985 b). Genotypic variation was observed in groundnut in paraheliotropism response to radiation and water deficit (Law, 1982) a, Ludlaw and Bojork man, 1984, Mathews et al, 1988 b). However, contribution of paraheliotropism to genotypic performance would be limited under severe moisture deficit environment.

5.2.2.2.3 Leaf surface characteristics

A smooth leaf is likely to lose more water than a crinkled leaf which tends create small pockets of still air (Rosenberg, 1978). The increase in pubescence and waxiness observed under stress increases leaf reflectance and reduces water loss (Ehleringer, 1980). Leaf pubescence may have an adaptive value in water limited environments as the hairs allow the leaf to fix more carbon, to avoid potentially lethal high leaf temperatures and lose less water daily, which allows the plant to extend its growth for a longer period into the drought (Ehleringer, 1980).

5.2.2.2.4 Stomatal and cuticular characteristics

By adjusting their apertures, stomata achieve the best compromise between the requirement for Co_2 and the need to conserve water (Berninger and Hari, 1993). Stomata play a major role in regulating water loss so as to match the evapotranspirational demand to the water supplying capacity of the roots, but this comes in to operation when the fraction of transpirable water falls to 0.3 or below (Sinclair and Ludlow, 1986). This is essential in maintaining internal plant water status above a critical threshold level, thus contributing to dehydration avoidance strategies (Turner et al. 1984). Genetic variability for stomatal characteristics, such as stomatal density, aperture size, sensitivity to changes in internal and external water was observed (Ciha and Brun, 1975, Markhart, 1985). The heritability for stomatal characteristics is high, indicating the feasibility for genetic manipulation (Buttery et al, 1993).

Morphological features such as a thick cuticle or wax deposits on the leaf surface can reduce evaporational water losses from the leaf surface and thus minimize residual transpiration rate (Jefferson et al. 1989). Genotypes with lower residual transpiration rates usually have a functional advantage during moisture- limiting environments as this leads to efficient water use (Walker and Miller, 1986, Paje et al. 1988).

5.3 Osmotic Adjustment

Osmotic adjustment (OA) can be defined as the active accumulation of solutes within the plant tissue (either in roots or shoots) in response to a lowering of soil water potential

(WP_s) (Morgan, 1984). This could lead to lowering of osmotic potential (OP), which provides the driving force for extracting water from low WP_s. Osmotic adjustment can play a major role in determining the drought resistance of given genotype by: (1) maintaining turgor over fluctuating soil water potentials, (2) maintaining stomatal conductance and thus photo synthesis (3) maintaining growth, (4) increasing de hydration tolerance and (5) increasing the extraction of soil water (Turner and Jones, 1980; Wright et al., 1983; Flower and Ludlow, 1986; Ludlow, 1987).

***Table 5.1** Values and ranges of Osmotic Adjustment (OA) in grain Legumes (Subba Rao et al, 1995)*

Legume species	QA (MPa)	Ref
Soybeans	-0.3 to -1	Muchow, 1985c; Cortes and Sinclair, 1986, 1987; Oosterhuis and Wullschleger, 1988
Pigeon pea	-0.12 to -1.28	Muchow, 1985c; Flower, 1985; Flower and Ludlow, 1987; Lopez et al., 1987
Groundnut	-0.3 to -1.58	Bennett et al., 1981, 1984; Erickson and ketring, 1985; Black et al., 1985; Stirling et al., 1989; Ketring, 1986
Green gram (V. radiata)	-0.2 to 0.44	Zhao et al., 1985, Muchow, 1985c
Black gram (V. unguiculata)	-0.05 to 0.5	Muchow, 1985c; Sinclair and Ludlow, 1986 Ashraf and Karim, 1991
Cowpea (V. unguculata)	0 to 0.40	Shackel and Hall, 1983; Muchow, 1985c; Sinclair and Ludlow, 1986; Lopez et al., 1987
Lablab beans (Lablab purpureus)	-0.2	Muchow, 1985c
Lupin	-0.14 to -0.48	Turner et al., 1987
QA in roots		
Pea	-0.3 to -0.8	Greacen and Oh, 1972

5.3.1 Dehydration Tolerance

The ability of cells to continue metabolizing at low WP_L is termed as dehydration tolerance. Yield reduction due to water deficit becomes important before severe desiccation occurs. Enhancement of dehydration tolerance results in continued leaf growth and decreased senescence during mild or moderate drought. It would have a positive effect on agricultural production. Sinclair and Ludlow (1986) considered critical relative water content as a most meaningful index for identifying genotypes with contrasting differences in dehydration tolerance. High dehydration tolerance is usually associated with slow rates of growth and development and most plants with good dehydration tolerance lack dehydration avoidance strategies (Ludlow, 1980). I Moderate dehydration tolerance is desirable in semi-arid environments (Steponkus et al 1982). Dehydration tolerance is related to cellular and sub-cellular processes. The stability of cell membranes under stress is central to cell performance. Measurements of membrane stability can be easily done with the electrolyte leakage test (Sullivan and Eastin, 1974) after subjecting plant tissue to stress. Chlorophyll fluorescence is also being used for this purpose.

5.4 Integrated Traits Associated with Crop Performance

Traits such as seedling establishment early growth vigor and leaf area maintenance are determined by a number of functionally integrated mechanisms that may contribute to both dehydration avoidance and tolerance aspects

5.4.1 Seedling Establishment

In arid and semi arid environments soil moistures in the seed bed can be sub optimum which causes reduced germination and emergence and results in low yields (Saxena, 1987). Rapid root development and growth facilitate successful establishment of seedlings and soil moisture is sub optimal after sowing (Asay and Johnson 1983)

5.4.2 Early Growth Vigor

Early development of the crop canopy is necessary to optimally utilize the production environment (i.e. like water and conducive temperatures). Rapid early growth and canopy development will reduce surface soil evaporation thus increasing the moisture availability for transpiration

5.4.3 Leaf Area Maintenance

Leaf area expansion is more sensitive to drought than photosynthesis and transpiration (Hoogenboom et al. 1987). When intermittent drought patterns are common by ability to retain leaf area during the period of moisture deficit place an important role in determining productivity of the genotype. Large genotypic differences and leaf area reduction have been observed under drought conditions (Nageswara Rao and Wright 1994).The ability to retain leaf area is an integrity function of many lower level traits such as deep root system, other root traits related to moisture extraction, stomatal control, osmotic adjustment, dehydration tolerance and leaf movements. Thus, genetic improvement of crop performance under drought conditions may be based on genotypes that retain leaf area during drought and undergo the usual senescence at maturity (Boyer 1983). This leaf retention is easy to visually assess among large number of plants in segregating population in breeding programme.

Significant differences were observed in LAD values from 54 to 90 DAS. The cultivars JL-24, Tirupati-3, Tirupati-4, K-134 and TCGS-717 maintained consistently higher LAD values throughout the crop period from 54 DAS to 90 DAS. In some cultivars *i.e.* TCGS-645 (79.28), TIR-23 (66.53), Abhaya (64.78), JL-501 (63.16),

R-2001-2 (55.01), TCGS-653 (55.01), Prasuna (54.25), TIR-9 (52.62) and TCGS-913 (52.09) the rate of increase in LAD from 72 DAS to 90 DAS was very high that ranged from 52.09 to 79.28. The increase in LAD in these cultivars coincides with pod filling phase. The maintenance of high LAD values during pod filling phase helps in sustained production of photosynthates and is a desirable attribute. Studies of Enyi (1977) on growth aspect of two varieties of groundnut, a sequential branching type, Natal common and an alternate type, 'Domodo Edible' indicated greater grain to leaf ratio (grain weight at harvest/LAD between pod initiation and final sampling) in spreading types than bunch types. In the present study, the cultivars that showed high LAD values during pod filling phase were mostly derivatives of infra-specific crosses. (Sunnetha, 2007)

5.4.4 Transpiration Efficiency

TE is defined as aerial dry matter produced for unit of water transpired by the crop accurate estimates of TE for different genotypes of a crop for entire growing season (season –long TE) are difficult and laborious to obtain (Tanner and Sinclair 1983). Carbon isotope (^{13}c) discrimination in c_3 plants was found to be negatively correlated with TE (Farquhar and Richards 1984). Genotypic variation in Δ has been reported in groundnut (Hubick et al. 1986; Wright et al. 1984). An increase of 1 % in Δ corresponds to a decline of 17 % in TE in groundnut. The variation in TE was 65 % in groundnut (Hubick et al. 1986). It is a difficult and expensive to measure. $\Delta\ ^{13}c$ is highly positively correlated with specific leaf area (SLA), ratio of leaf area to leaf dry weight which is easy and inexpensive to measure. SLA has inverse relationship with relative water content (RWC) and the low SLA types are drought tolerant as they are able to maintain higher RWC. However, SLA is influenced by the time of sampling and age of the leaf. SLA is inversely correlated with SPAD chlorophyll meter reading (SCMR) (r=0.77) which is positively correlated with TE. SCMR is measured by a handheld device which is easy to operate and large number of observations can be recorded rapidly.

5.4.5 Developmental Plasticity

Developmental plasticity can be defined as the ability to adjust the duration of different growth phases to suit soil moisture availability during the growing season. Developmental plasticity includes an ability to adjust canopy development pattern according to soil moisture availability this included leaf area adjustment, leaf movements, stomatal control and ability to produce new leaf on relief of moisture deficit it also includes plasticity of root system. Flexibility of reproductive development patterns is also required, with ability to adjust maturity according to soil moisture availability, and ability to produce new flushes of flowers and pods when the moisture situation improves (e.g., as in pigeon pea [Chauhan et al.1987]). In groundnut, peg initiation and elongation ceases when soil moil moisture is depleted to 80% of the extractable limit (Chapman, 1989). However, those pegs initiated prior to or during the drought period had the capacity for renewed elongation after rewatering. This plasticity in pod development apparently plays an important role in the adaptation of groundnut to intermittent drought situations (Harris et al. 1988). Genotypic variation in rooting pattern- rooting depth and root length distribution pattern and other components such as canopy structure, rate of phenological development (Carter et al. 1982 and Fischer, Turner 1978) have been reported. Plant breeders are quite successful in incorporating various degrees of phenological plasticity in groundnut. This has led to the development of cultivars that can best fit particular production environments.

5.4.6 Mobilization of Pre Flowering Stored Reserves

Remobilization of starch results stored in stems contributes significantly to yield of legumes (Constable and Hearn 1978). These carbohydrate reserves act as a buffer against availability of current photosynthates particularly during grain filling (schnyder 1993). Genotypic variation has been reported the ability, store and mobilize carbohydrates for seed filling during terminal moisture stress (Wright et al. 1991). Harvest index indicates the extent of dry matter partitioning into economic products *i.e.* pods, seeds and oil in groundnut. The cultivar, R-2001-2 exhibited highest harvest index (63.66 %) followed by TCGS-888 (62.54%0, TG-47 (58.59%) and TCGP-6 (53.45%) respectively. Duncan *et al.* (1978) stated that most of the yield variation among the five groundnut cultivars grown in Florida was due to differences in the partitioning of assimilates between vegetative and reproductive parts, the length of grain-filling period and rate of fruit establishment. Estimates of partitioning to fruit ranged from 41% in the first cultivar released in US to 98% in the latest cultivar though CGR did not vary much among different cultivars. In the present study, harvest index values ranged from 31.70% in Tirupati-3 to 66.66% in R-2001-2 during rainy season. In a set of 11 genotypes which consisted of released varieties and advanced breeding lines, (Vasanthi *et al.* 2003) observed a range of 41% (TCGS-647) to 53% (JL-24) in harvest index in post-rainy season.

5.4.7 Symbiotic Nitrogen Fixation

The delicate balance between the host plant and the symbiont demands an optimum soil moisture environment for maximum efficiency of N_2 fixation (Swaraj 1987). Most tropical legumes require soil moisture levels of atleast 70-80 % of the field capacity for optimum function of legume-rhizobium symbiosis (Swaraj 1987). Identification of genotypes capable of forming functional nodules and deeper soil layers are desirable as well as rhizobial strains capable of thriving at these depths. Nitrogenase activity in groundnut following a drought induce decrease in nodule fresh weight of 40% (Rao and venkateswarlu 1987). Nitrogen fixing ability declines in many legumes as the crop approaches maturity (Lawn and Brun 1974). Nearly 60-70 % of leaf nitrogen is diverted to the pod filling process and this leads to early cessation of photosynthetic ability termed as self destruction process (Sinclair and deWit 1975). Terminal moisture deficit accelerates this destruction (Cure et al 1985). Therefore, the ability to extend nitrogen fixation activity into the reproductive growth phase under moisture limited environments plays a major role in improving productivity in these environments.

5.5 Inheritance Studies on Drought Related Traits in Rars, Tirupati

Genetic parameters, variability, heritability and genetic advance as percentage of mean were studied among genotypes as well as in segregating generations of different cross combinations so that it would be useful in making choice of suitable selection method to be followed to arrive at useful selections. Genotypic co-efficient of variation (GCV), heritability(h^2) and genetic advance were estimated in a set of 29 genotypes in 2007. Leaf area duration (LAD) at 54, 72 and 90 DAS showed high to moderate GCV, PCV, high heritability and high genetic advance as per cent of mean (GAM). Crop growth rate (CGR) at 54 and 72 DAS also exhibited high GCV, PCV, heritability and GAM. LAI at 36, 54 and 72 DAS registered high GCV, PCV, heritability and GAM. SLA at 18, 36 and 54 DAS showed low genetic variability with moderate to high heritability and moderate genetic gain except at 36 DAS. SLA at 72 DAS showed moderate GCV coupled with high heritability and high GAM indicating that SLA at 72 DAS was least influenced by environment compared to other stages in the crop growth and is governed by additive gene action. The characters, seedling vigour index, harvest index and pod yield per plant recorded moderate to high GCV, heritability and high GAM indicating the role of additive gene action in the inheritance of these characters. Low GCV, moderate heritability and low GAM were observed for SCMR and germination percentage indicating that these characters were under the control of non-additive gene action (Suneetha, 2007).

Broad sense heritability estimates were high for plant height, plant dry weight, total plant dry weight, harvest index, SCMR and SLA in a study involving F1 of crosses during rabi 2002-03 (Personal communication). Moderate values of heritability GCV, PCV and genetic advance were obtained for SCMR, dry haulms weight per plant, mature pods plant, primaries & secondaries/plant total weight of the plant, pod weight per plant, sound mature kernel percentage & immature pods/plant in a study involving 34 F1s and 11 parents (Vasanthi et al.2003). The traits that confer WUE, high SPAD and low SLA showed moderate to high h^2 i.e. 40.7 and 66.0 and moderate to lower GAM i.e., 27.4 and 14.7% respectively. Genetic variability was moderate for SCMR (20.7%) while it was low for SLA (6.5% GCV). In the same study, expression of WUE traits in F1s was studied. Among F1s, the SCMR varied between 37 (Vri 2 x Jug 43) to 46 (TCGP6 x JAL30). SCMR in F1s was either equal to high SCMR parent or in between the two parents involved in the cross. The SLA values ranged from 142 (K-134 x TCGS-647) to 190 (F1 of JL-24 x Jug 43). F1's SLA values were lower than high SLA parent indicating the dominant nature of the trait, SLA (low) i.e., thickness over thinness of leaf.

Table 5.2 *Estimates of genetic parameters in different studies for traits related to drought tolerance in groundnut*

Trait/ character	Generation	GCV	h^2	GAM	Reference
SCMR	F2	-3.70 to 8.50	-56.08 to 66.92	3.65 to 14.21	Seethaladevi (2004)
	F3	0.03 to 6.54	0.00 to 32.70	0.00 to 7.72	Sirisha (2005)
	F4	2.06 to 61.5	4.83 to 43.07	0.93 to 9.29	Sirisha (2005)
		3.83	34.90	4.66	N.Suneetha (2007)
		2.90 to 5.11	13 to 36	2.18 to 6.29	Shreya et al (2014)
SLA	F3	11.46 to 59.61	35.54 to 76.24	11.48 to 31.96	Sirisha (2005)
	F4	4.69 to 10.19	11.05 to 37.42	10.37 to 32.14	Sirisha (2005)
		8.58	43.26	11.63	N.Suneetha (2007)
		4.59 to 24.02		2.42 to 40.52	Shreya et al (2014)
Harvest index (HI)	F2	-7.67 to 25.55	-5.93 to 55.35	-3.84 to 36.48	Seethaladevi (2004)
	F3	12.29 to 27.14	9.02 to 33.58	7.60 to 32.40	Sirisha (2005)
	F4	8.93 to 27.32	6.51 to 32.49	2.69 to 32.08	Sirisha (2005)
		16.05	88.20	31.06	N.Suneetha (2007)
		10.07 to 14.09	31 to 52	11.63 to 20.91	Shreya et al (2014)
Pod yield	F2	-13.31 to 44.89	-8.58 to 37.09	-8.03 to 79.99	Seethaladevi (2004)
	F3	20.31 to 47.91	11.99 to 67.39	14.49 to 81.03	Sirisha (2005)
	F4	14.52 to 43.43	10.12 to 42.37	9.59 to 58.24	Sirisha (2005)
		15.54	40.86	20.46	N.Suneetha (2007)
Dry haulm wt	F2	13.35 to 46.34	23.73 to 86.41	13.11 to 88.74	Seethaladevi (2004)
Root length	F3	12.54 to 13.43	19.85 to 29.40	11.51 to 15.01	Sirisha (2005)
	F4	8.60 to 14.98	29.08 to 40.17	26.49 to 27.30	Sirisha (2005)
Root mass	F3	13.98 to 41.82	16.71 to 57.69	11.77 to 65.44	Sirisha (2005)
	F4	11.79 to 119.07	8.42 to 91.45	7.05 to 234.58	Sirisha (2005)

Inheritance of leaf chlorophyll content (in terms of SPAD chlorophyll meter reading) was studied in five crosses of groundnut involving common male parent, TCGS-647 and female parents, TCGS-320, Tirupati-4, K-134, URI-2 and TCGS-29 advanced breeding lines released varieties which are from ANGRAU/TNAU analysis of F2 frequency distribution of five groundnut crosses indicated that the leaf chlorophyll content may be a quasi-quantitative trait with modifiers acting either to increase or decrease the expression of the trait in different genetic backgrounds. SCMR of F1s was intermediate between the parents involved with a slight shift towards parent of high SCMR (Vasanthi et al 2005). In a similar study involving JAL 30, an advanced breeding line with high SCMR (45) and the same set of female parents, same inferences were drawn (Babitha et al 2006). Genotypic co-efficient of variability was low with genetic advance as % of mean below 20% for SCMR.

5.6 Breeding Approaches

Defining the traits that improve adaptation in drought condition and evaluating the lines generated keeping these as selection criteria together or alone takes substantial breeding effort. A more pragmatic approach was proposed by (Acevedo and Ceccarelli 1989) who suggested crossing genotypes diverging in four or five of the most important stress resistance traits. Individual F_2 plants are harvested and from then onwards divergent selection are carried out of the traits under study. Contrasts can be made for single traits in combination with other traits. The traits are considered valuable, alone or in combination, if they increase the selection efficiency when compared to yield alone as the selection criteria by regressing the values of F_4 on the F_3 families. Estimates of realized heritability are generated. Final yield testing is done in F_5 or F_6 using contrasting environments. In this process, sources of stress resistance as well as stress resistant traits can be identified.

The sources of drought resistance may be available in (i) cultivated varieties, (ii) land races, (iii) related wild species or may be, (iv) introduced by genetic engineering. Amongst these, the first priority should be the cultivated species, since it is least problematic of all sources. Undesirable linkages may pose some problem due to undesirable association, though the magnitude of such problems may be limited in case of land races. Wild relatives as sources of resistance have two serious limitations; (i) various attributes of wild species may be of little advantage to a crop variety and (ii) crossability and linkage-drag.

Three breeding approaches for drought have been evolved. The first is to breed for high yield under optimum (water- stress-free) condition. As the maximum genetic potential of yield is expected to be realized under optimum condition and a high positive correlation exists between performance in optimum and stress conditions, a genotype superior under optimum level will also yield relatively well under drought condition. This is the basic philosophy of this approach.

However, the concept of expression of maximum genetic potential in optimum condition is debated as genotype- environment interaction which may restrict the high yielding genotype to perform well under drought. Thus, the second approach i.e. to breed under actual drought condition has been suggested by (Blum 1972). The second approach suffers from the problem that the intensity of drought is highly variable from year to year and as a consequence, environmental selection pressure on breeding materials changes drastically from generation to generation. This problem confounded with low heritability of yield (Blum and Naveh, 1976) makes it complicated and results in slow progress.

An alternative approach to the above two would be to improve drought resistance in high yielding genotypes through incorporation of morphological and physiological mechanisms of drought resistance. However, transferring drought resistance into high yielding genotypes is complicated due to due to its complex nature and involvement of many traits, physiological, biochemical and morphological and lack of understanding of physiological and genetic basis of adaptation in drought condition. In contrast, improving the yield potential of an already resistant material may be a more promising approach, provided there is genetic variation within such a material (Ritchie, 1980). Simultaneous selection in non-stress environment for yield and for stability in stress condition may be useful in achieving the desired goal of evolving drought resistant genotype with high yield.

5.6.1 Breeding Methods

The breeding methodology to be followed for drought resistance is the same as that for improving other traits. In general, pedigree and bulk method could be used for self pollinated crops like groundnut. For instance, in an approach for combining yield potential and drought resistance traits e.g., cuticular wax and appropriate cross to combine these traits could be affected. F_1 is advanced to F_4 under optimal moisture. In F_4, a large number of individual plants are selected and harvested separately. In F_5, individual plant progenies are grown under moisture stress; and progenies are evaluated for cuticular wax only: superior progenies are harvested separately. The selected progenies are grown under optimal moisture in F_6 and the selection is based on yield, yield attributes and quality traits. It may often be necessary to repeat the operations of F_5 and F_6 one or more times to identify desirable progenies.

5.6.1.1 Different Approaches Followed in Breeding Drought Tolerant High Yielding Genotypes at RARS, Tirupati and their Outcome

Regional Agricultural Research Station, Tirupati, situated at an altitude of 182.90 m above mean sea level, 13° N latitude and 79° E longitude. The soil is sandy clay loam with medium fertility.

5.6.1.2 Drought Avoidance Trait, Earliness Exploited in Initial Stages of Breeding for Drought Resistance

In the initial stages of breeding at RARS, Tirupati, efforts were made to develop early maturing varieties and exploited the drought avoidance trait, earliness. After the release in1940 TMV2 , a mass selection from Gudiyatham bunch was widely grown in Andhra Pradesh besides land races in Virginia bunch (Local Peddagutti) and Virginia runner types. In the beginning of 1980s, spanish bunch variety, JL 24 released from Jalagaon centre of Maharashtra was quickly adopted and became popular in Andhra Pradesh in kharif rainfed situation. The long duration Virginia bunch and Virginia runner land races that were cultivated were slowly phased out from cultivation due to different reasons major reason being difficulty in harvesting due to insufficient soil moisture due to failure of rains from north-east monsoon frequently at the stage of crop maturity .

From germplasm, EC 106983/3, an early maturing germplasm line with high yielding ability and drought tolerance in comparison with TMV 2 was selected and was evaluated in yield trials in different stages as TCG 1074 and released as Tirupati 1 in 1989 by state varietal release committee. The special features of Tirupati 1 (TCG 1704) are given hereunder

Fig. 5.1 Groundnut varieties

The variety, JL24 was highly susceptible to moisture stress at the pod filling stage. Hence, with the objective of evolving an early maturing variety that could circumvent the moisture stress at the pod formation and the pod development stages, massive hybridization programme was undertaken involving sources of earliness, Ah 316/S (red testa line), EC 21137-1 and germplasm lines, EC21083A etc., as male parents and varieties, TMV 2, J 11, Local peddagutti and JL 24 as female parents. The material was advanced by mass pedigree method of breeding with emphasis in selection for earliness and high pod and kernel yields. About 200 lines were brought into yield trials during 1987. After evaluation in yield trials in different stages, three lines, TCGS 29, TCGS 30 and TCGS 37 were identified as high yielding lines with earliness. Both TCGS 29 and TCGS 37 were red testa lines only difference in between these two being in pod reticulation i.e. TCGS 29 pods were reticulated while TCGS 37 pods were smooth like JL 24 while TCGS 30 had rose testa kernels and reticulated pods. As there was preference for rose testa, TCGS 30 was released as Tirupati 4 in 1995 and later on TCGS 29 was released as Narayani in 2002 by state varietal release committee. Tirupati 4 matures in 105 days and possesses midseason moisture stress tolerance besides high yield potential and attractive medium bold kernels with rose testa.

Narayani (TCGS 29) is early by a week compared to JL24. Yield evaluation was carried out at research station from 1988 to 1992 as well as on farmers holdings during 1995 and 1996 rainy season and 1998/99 post-rainy season. During rainy season, the yield advantage was 13% over JL 24 with pod yield of 1675 kg/ha and during post-rainy season, the yield advantage was 23% over JL24 with average pod yield of 2637 kg/ha. In minikit trials, the average pod yield was 1236 kg/ha with yield advantage of 43% over locally grown spanish bunch cultivars. In post-rainy season, It recorded average pod yield of 3883 kg/ha with 16% hjgher yield over the other locally grown red-kernelled variety "Pollachi Red".

***Table 5.3** Characteristic features of Tirupati 1, Tirupati 4 and Narayani*

Variety	Pedigree	Year of release	Duration	Pod yield/kg/ha		Shelling %	100-seed weight (g)	Sound mature kernel %	Oil content (%)	Special features
Tirupati-1	E.C-106983/3	1989	100	18-20	30-40	79	33-36	90-95	49	Early maturing and drought tolerant Spanish bunch groundnut variety, leaf lets narrow with ashy coating.
Tirupati-4	JL-24XAh316/S	1995	105	20-25	35-45	76	43-47	85-90	48-50	High yielding, mid-season moisture stress tolerant, early maturing variety, attractive light rose testa
Narayani	JL 24/Ah316/S	2002	100	20-25	40-45	76	40-45	90-95	49	Early maturing (100days) drought tolerant performance highly stable in different types of soils and situations, popular among farmers of Rayalaseema and Telangana districts of Andhra Pradesh , replaced to an extent of 20%, the existing Spanish bunch varieties TMV 2 and JL24

The leaflets of TCGS 29 are long, elliptical and green. The stem is angular with light greenish purple pigmentation. Its growth habit is determinate and erect. It possesses four primary branches (very rarely five) and secondary branches are more. It has medium sized pods (100-pod mass of 90-99g and 100- seed mass of 42-45g) with moderate reticulation and moderate constriction. Seeds have light red testa with oil content of 47-49% and shelling out-turn of 74-76% . The other important desirable attribute of TCGS 29 is synchronous maturity of all pods in a plant. Early vigour and fast production of biomass are other good traits that are necessary for a genotype to produce hjgher yield in rainfed situation

5.6.1.3 Selection for Drought Resistance and Higher Yield through above - Ground Correlated Traits to Water-use Efficiency (SCMR and SLA)

Drought resistant breeding in groundnut got a fillip with handling of two ACIAR funded projects “Selection for water use efficiency in food legumes” from 1-7-1993 to 30-6-1996 and “More efficient breeding of drought resistant peanuts in India and Australia from 1-7-1997 to 30-6-2001. It led to more scientific understanding of drought related resistant traits and variability among the genotypes, their heritability, mode of inheritance were understood through many studies. It helped in breeding of drought resistant varieties in a more scientific way with yield evaluation in rainfed environment, under artificially created stress situation (Rain-out shelters) and post-rainy season evaluation under stress situation. In the second project of ACIAR- ICAR- ANGRAU collaborative program. Three crosses in common and one location specific cross were made at each collaborative centre (ICRISAT, Junagadh, Jalagaon, Tirupati centers) in total eight crosses were made i.e. ICGS 76 × CSMG 84-1, ICGC 44 × CSMG 84-1, ICGV 86031 × TAG 24, ICGS 44 × ICGS 76, JL 220 × TAG 24, GG2 × ICGV 86031, K 134 × TAG 24 and TAG 24 × ICGV 86031 and progenies selected in F6 generation were evaluated across all the four locations and promising selections in terms of water-use efficiency traits were identified. The parents were selections after vigorous exercise from the earlier project “Selection for water use efficiency in food legumes”. Two selection approaches i.e. trait based and empirical were followed. Top ten genotypes in terms of kernel yield in rainy season, 2000 and post- rainy season 2000-01 were viz rainy season: ICR-17, JAL 30, TIR 44, TIR 42, JAL 03, ICR 12, TIR 17, JAL 14, JUG 18 & ICR 26 and Post rainy season: JAL 24, JAL 06, TIR 16, ICR 45, JAL 02, JUG 06, JUG 03, JAL 28, JAL 18 AND JAL 29.

In the trait- based approach (selection index involving three traits, HI, T & TE), there was a major gain for HI (6-50%) compared to T (< 0-12%) and TE (< 0-25%) was observed. Among the top 10 genotypes yield wise as well as in T & TE there is no significant difference among the selections made through trait and empirical approaches. One of the selections, TPT 25, a short statured line with high water-use efficiency and high yielding ability and drought tolerance was released as Abhaya by state varietal release in 2006. Many of the selections were found to be deficient in acceptable agronomic attributes and could not be promoted further in yield trials. However, these genotypes formed the base material for another major project "Physiological and biochemical approaches for breeding drought tolerant groundnut genotypes" funded by AP-Netherlands biotechnology programme. In this project the methodology followed and outcome of the project are given hereunder:

- 287 germplasm lines obtained from RARS, Tirupati (obtained from earlier ACIAR funded programmes), ICRISAT, ARS, Vridhachalam, ARS, Jalagaon, DGR, Junagadh and BARC, Trombay were screened for physiological traits, SCMR, SLA, RWC, SLN, Chlorophyll stability index etc. Among the top fifty genotypes for each trait, five lines with a good combination of majority of the physiological traits were selected as donor parents for water-use efficiency traits and drought tolerance.
- Water-use efficiency of the lines selected was verified by pot culture studies. Among the selected lines, significant positive relationship was observed between SCMR and WUE (r^2=0.44).

Table 5.4 *Total dry wt. (g), cumulative water transpired (kg), water-use efficiency (g/kg) and SPAD chlorophyll meter reading of parental lines*

Genotype	Total dry wt/pl	CWT	WUE	SCMR
TCGS-320	3.29	1.050	3.41	45.7
JL-24	4.77	1.268	3.79	43.9
CHICO	3.71	1.186	3.12	43.6
ICGV86031	3.39	0.822	4.16	45.6
TCGS-29	7.40	2.017	3.67	46.8
VRI-2	6.43	1.795	3.70	46.2
K-134	4.60	1.134	4.00	46.4
TPT-4	4.96	1.299	3.87	40.5
TCGS-647	2.14	0.488	4.37	50.8
JAL-30	1.58	0.346	4.57	56.9
TCGP-6	4.62	1.100	4.19	45.5
JUG-43	6.41	1.526	4.20	52.8
TIR-45	3.80	0.912	4.21	48.2

- Hybridization programme was undertaken involving six high yielding varieties and one advanced breeding line as female parents and identified five advanced breeding lines with high WUE as male parents. One parent, TCGP 6 is used both as male and female parent. Details of parents along with their characteristic features are given here under

Table 5.5 *Male parents used and their important attributes*

1	TIR-45	High SPAD, high heat tolerance, high KHI, high S.P%
2	JAL-30	High SPAD, high S.P%
3	TCGP-6	High SPAD, moderate heat tolerance
4	TCGS-647	High SPAD, moderate KHI, insect pest tolerant
5	JUG-43	High SPAD, high S.P%

***Table 5.6** Female parents used and their important attributes*

1	TCGS-320	Released as kalahasti from ANGRAU, it is a high yielding variety with resistance to kalahasti malady.
2	TIRUPATI-4	High yielding variety released from ANGRAU. It has tolerance to mid season moisture stress.
3	VEMANA	High yielding variety released from ANGRAU. Has tolerance to moisture stress and foliar diseases.
4	VRI-2	High yielding variety released from TNAU.
5	TCGP-6	Drought resistant genotype with high yield potential but with low shelling %
6	TCGS-29	Released as Narayani. High yielding variety released from ANGRAU. With earliness and mid season moisture stress tolerance.
7	JL-24	High yielding variety released from Jalagaon of Maharashtra

- A total of 5846 pollinations and 1435 F_1 plants were realized i.e. 24.6 %. In total, 34 cross combinations were made with on an average 42 F_1s in each cross. Crosses were affected in line x tester fashion.
- The segregating material was handled by modified pedigree method. In total, 62 crosses segregating material was handled.
- F_2 population size in each cross ranged from 500-1500. In F_2 generation, heritability was high for SCMR, total dry matter/plant, kernel weight /plant and plant height. Best cross combinations were identified taking into consideration mean values of pod and kernel yield and WUE traits in F_1 and F_2 generations.
- In F_2 generation, single plants were selected considering yield, yield attributes and colour of the foliage. Number of selections in each cross ranged from 9 (JL 24 × Jug 43 and JL ×JAL 30) to 99 (Narayani× TIR 45). In total, 1313 single plants were made from 34 crosses.
- In F_3 generation, 60 uniform bulks and 543 single plant selections were made.
- 689 single plant progenies and 67 uniform bulk selections derived from 34 single crosses were grown in F_4 generation during kharif, 2004. All the progenies were visually judged at the time of harvest and progenies that were phenotypically similar in plant and pod characters were selected. In the progenies that were not phenotypically similar, individual plants were selected. The leftover 707 single plant selections made in F_2 generation during kharif 2003 were studied in F_3 generation during kharif 2004. From these two generations, 219 uniform bulks and 57 single plant selections were made.
- From this, 23 uniform bulks and 149 single plant selections were grown during rabi 2004-05 in F_4 and F_5 generations. From this, material, 58 single plants and 16 uniform progenies and 32 single plants and 54 uniform progenies were selected in F_4 and F_5 generations respectively.
- In F_5 and F_6 generations, 140 single plants and 105 uniform bulk selections were made in kharif, 2005. 57 selections were evaluated in initial varietal trial in three sets.

- During kharif 2006, 93 single plant progenies and 48 uniform bulks were grown out of which 73 single plant selections and 52 uniform bulks were made. 43 uniform bulk selections were evaluated in initial varietal trial trial in two sets.18 advanced breeding lines promoted from initial varietal trials were evaluated in advanced varietal trial. 133 uniform bulk selections were promoted to initial varietal trial.
- During kharif 2007, in total, 133 uniform bulk selections were evaluated in four sets in initial varietal trial. 41 entries were promoted to advanced varietal trial and evaluated in kharif, 2008. Among these, 22 entries were promoted to advanced varietal trial (II year). Groundnut cultures, TCGS 1043 and 1045 were found to be high yielders under rainfed conditions with high mean pod yield (3 years mean, 2007, 2008 and 2009 kharif season) of 1579 kg/ha and 1665 kg/ha with 84% and 94% higher pod yield over the best check Narayani (858 kg/ha) and 34% and 42% over Abhaya (1176 kg/ha).

5.6.1.4 Identification of TCGS 1043, a High Yielding Drought Tolerant Culture

The genotypes developed in the project were evaluated in three seasons which represented different rainfed environments viz. favorable, not-so favorable and unfavorable environment –kharif, 2007, kharif, 2008 and kharif, 2009 respectively as discussed here under.

During kharif, 2007 highly favorable environment was observed for groundnut crop. In RARS, Tirupati, crop was sown in first fortnight of July with the onset of south-west monsoon in time. There were no dry spells in the month of August which is usually a dry month. During the month of September, rainfall distribution was uniform with neither excess rainfall nor significant dry spell which contributed for better pod filling. Average temperature was around 25^0c throughout the crop period. Incidence of pests and diseases was also at minimum. Relative humidity and sun-shine hours were also optimum range. The total amount of rainfall received during the crop period (July to October) was 691 mm.

During kharif, 2008, there was deficit rainfall of 17.5% with intermittent dry spells of 15- 30 days. Crop experienced moisture stress during pod filling phase. Total rainfall received during crop period (July to October) was 372 mm.

Kharif, 2009 was highly unfavorable to groundnut crop. Total rainfall received from July to October was 274 mm. Though crop was sown in time with receipt of meager rainfall during July, crop experienced moisture stress in different stages- at seedling stage, flowering and pegging, pod filling stages. Duration got extended and in the month of November, 183.6 mm rainfall was received at the time of harvest which affected the yield and quality through in-situ germination.

In kharif, 2009, though results were non-significant and co-efficient of variation for pod and kernel yields was high in advanced varietal trial, TCGS 1043 was the only line which recorded high pod and kernel yield with 12% higher pod yield and 31% higher kernel yield with shelling out-turn of 71%, 100-kernel weight of 39g,. Haulms yield was moderate (2908 kg/ha) over the best check variety, Abhaya (Table). The mean pod and kernel yields over three seasons, Kharif, 2007, 2008 and 2009 were 1579kg/ha and 1199kg/ha with yield advantage of 84 and 30% respectively over Narayani (Table). This was further promoted to testing in minikit, multilocation trial and AICRP system. Its superiority in terms of yield and drought tolerance was proved beyond doubt and was released as Dharani on 11th July, 2012 by state varietal release committee for general cultivation.

The characteristic features of Dharani (TCGS 1043) are given in Table. In Dharani (TCGS 1043), Aflatoxin contamination was found to be low in comparison with popular varieties. Holbrook et al (2000) reported lower levels of Aflatoxin contamination in drought tolerant genotypes, PI 145681, 259639, 269106, 295722, 298836, 315628 and Tifton8. A similar relationship between drought tolerance and reduced Aflatoxin contamination was observed in the drought tolerant cultivar, 'Streeton' in Australia (Cruickshank et al, 2000) , Physiological studies (Nageswara Rao et al (2000) have shown that the lower aflatoxin incidence is associated with better root water uptake resulting in better maintenance of crop water status during severe end-of-season drought.

Fig. 5.2 Groundnut variety dharani in different forms

Table 5.7 *Performance of groundnut genotypes in advanced varietal trial (AVT-2nd year) during Kharif 2009, the worst drought year*

Entry	Days to harvest	Final plant stand ('000)	Pod yield (Kg/ha)	%of check (Abhaya)	Kernel yield (kg/ha)	%of check (Abhaya)	100-pod weight (g)	100-kernel weight (g)	Shelling per cent	SMK (%)	Haulms weight (kg /ha)
TCGS-983	109	200	910	92	510	85	80	33	56	73	4280
TCGS -985	109	217	682	69	375	62	82	35	55	71	5045
TCGS -986	109	207	930	94	558	93	93	32	60	66	1736
TCGS -987	109	221	1022	104	552	92	80	37	54	78	2165
TCGS -997	109	235	841	85	462	77	90	37	55	78	1644
TCGS -999	109	259	1008	102	666	111	78	36	66	78	3104
TCGS -1001	109	230	1042	106	615	102	82	31	59	84	2528
TCGS -1039	109	236	626	63	363	60	78	31	58	77	1602
TCGS -1042	109	190	479	49	254	42	73	28	53	75	2994
TCGS -1043	109	220	1108	112	786	131	81	39	71	76	2908
TCGS -1045	109	220	1116	113	558	93	87	28	50	50	1540
TCGS -1051	109	230	941	95	574	95	81	32	61	81	4649
TCGS -1058	109	185	892	90	598	99	69	33	67	58	1186
TCGS -1070	109	216	979	99	607	101	75	35	62	69	1638
TCGS -1071	109	223	910	92	510	85	79	30	61	60	1430
TCGS -1072	109	206	783	79	556	92	82	38	71	61	1522
TCGS -1073	109	203	830	84	548	91	93	39	66	87	3969
TCGS -1074	109	199	852	86	545	91	79	37	64	76	2030
TCGS -1075	109	211	829	84	497	83	78	31	60	90	2664
TCGS -1076	109	186	802	81	513	85	74	29	64	55	3720
TCGS -1082	109	170	831	84	540	90	78	31	65	71	1732
TCGS -1087	109	232	802	81	433	72	82	32	54	81	1996
Narayani	103	223	882	89	564	94	81	33	64	67	3026
Abhaya	110	202	987	100	602	100	78	30	61	77	4199
K-134	109	198	764	77	489	81	71	30	64	73	7539
GM			931		568						
CV %			26.61		27.85						
CD at 5%			NS		NS						

Table 5.8 *Mean performance of top ten groundnut genotypes in yield trials over three years*

Genotype	Pod yied kg/ha			Mean	% of the check	Kernel yield kg/ha			Mean	% of the check
	2007	2008	2009			2007	2008	2009		
TCGS-1043	2562	1066	1108	1579	184	1973	799	826	1199	130
TCGS-1045	2422	1454	1118	1665	194	1623	771	586	993	108
TCGS-1047	2464	1217	3068	2250	262	1750	682	3111	1848	200
TCGS-1058	2047	1326	894	1422	166	1535	848	894	1092	118
TCGS-1070	1980	1231	979	1397	163	1406	788	981	1058	115
TCGS-1071	2362	956	1063	1460	170	1654	622	1063	1113	121
TCGS-1072	2367	1239	784	1463	171	1657	855	784	1098	119
TCGS-1073	2207	1300	832	1446	169	1589	871	832	1097	119
TCGS-1075	2067	929	830	1275	149	1426	631	830	1128	122
TCGS-1076	2033	1226	803	1354	158	1423	834	803	1028	111
Narayani	1800	889	884	858		1278	604	883	922	

5.6.1.5 Selection for Drought Resistance and Higher Yield Considering Root Traits and above - Groundnut Correlated Traits to Water-use Efficiency (SCMR and SLA)

- For phenotyping of root traits, specially designed raised beds (15x2x2m-LxBxH) with soil characteristics similar to natural soil were constructed during kharif, 2009. Phenotyping for root traits, WUE traits for selected 70 genotypes was carried out. Top ten genotypes for SCMR (SPAD chlorophyll meter reading) an indirect measure of WUE are ICGV 06040 (53.8), TCGS 986 (52.7), TCGS 933 (50.9), TCGS 1058 (50.3), TCGS 985(49.3),ICGV 07045 (48.9), TCGS 645(48.9), AVK-2007-8 (48.8), TCGS 1015 (48.1) and ICGV 07356 (47.9).Top ten genotypes for root length are : TCGS-985 -61cm, ICGV-86590 -59 cm, ICGV-87846 -57 cm, ICGV-06045 -55 cm, ICGV-07072, -50 cm, ICGV-07227 -47.5 cm, ICGV-08020 -47.5 cm,TCGV-1071 -47.5 cm, ICGV-05062 -47 cm and ICGV-06122 -47 cm. SCMR ranged from 37.0 -53.8. Root length (cm) ranged from 8.5-61.0. Shoot length (cm) ranged from 6.5 -23.5. Root dry weight (g) ranged from 0.16-1.54.Shoot dry weight (g) ranged from - 2.69-17.42. Root volume (ml) ranged from – 1.0 -3.0. Root- shoot ratio ranged from – 1:4-1:26.
- Traits related to WUE i.e. photosynthetic rate stomatal conductance and the rate of transpiration were estimated in 70 genotypes through photosynthesis system during Kharif, 2009. Highest photosynthetic rate (μ mole Co_2 / m^2 /sec) was observed in the genotypes Jyothi (38.75) and TCGS 941 (36.35) followed by IVK-06-12 (34.17) and IVK-06-10 (34.07) among released varieties, it varied between 20.25 (Tirupathi-4) to 30.65 (Greeshma and Rohini). Stomatal conductance ranged from 0.19 (IVK -6-08)

to Jyothi (7.76). The rate of transpiration was lowest in IVK-6-08 and highest in (4.5) TCAS 1039 (14.47).

- To develop mapping populations for root traits, ten crosses were affected during rabi 2010-11 taking Tirupati 4 and Kadiri 6 as female parents and ICGV 07072, 07132, TCGS 1043, TCGS 1039 and Jyothi as male parents.
- F_1s were raised during kharif, 2011. F_1 s and parents were evaluated for root triats. F_1 of K6 x TCGS 1043 recorded highest root length (93 cm) and high root dry weight (2.94 g) followed by TPT4x TCGS1073.
- F_2 generation material of ten crosses was studied during rabi 2011-12. Part of F_2 material of the cross, Kadiri 6 x TCGS 1043 was grown in specialized root structures to phenotype for root traits. Based on SCMR, SLA, root length, root mass and yield, starting from 363 single plants in F_2 generation 25 stabilized selections were finally brought to yield trials in F_6 generation.
- In 2014 kharif season, they were evaluated in randomized block design with three replications under rainfed situation. Genetic parameters, genotypic co-efficient variation (GCV), heritability and genetic advance as % of the mean were computed from data collected on 14 attributes. Analysis of variance revealed significant differences for all the traits studied. GCV was high (>20) for the traits viz number of primaries and haulms weight per plant. It was moderate (10-20) for the traits, plant height, mature pods, immature pods, 100-pod weight and 100-kernel weight. While genetic advance as % the mean was high (>20) for plant height, secondaries and haulms weight per plant indicating the scope for phenotypic selection for bringing about improvement in these traits. For SCMR, both GCV and genetic advance as % the mean were though the heritability was high indicating non additive nature of inheritance while it was moderate for SLA indicating moderate scope for improvement through phenotypic selection. .
- In initial varietal stage, one of the genotypes, TCGS 1616 recorded significantly higher pod yield over Dharani (TCGS 1043) with acceptable pod and seed traits with high shelling percent of 76%.Now, the material is in advanced varietal trial stage.
- The material in other crosses is in initial varietal trial stage.

To develop varieties suitable for rainfed environment where growing period is short and dry spells occur intermittently, it s necessary to take into consideration improvement of different traits contributing for improved performance together in a wholistic way or improve the genotypes step by step which takes long time when we look into the traits involved which is definitely a challenging task. In Dharani (TCGS 1043), most of the attributes required for better performance in rainfed situation have come together. It is necessary to still reduce the duration by one week. This is being aimed at and among the progenies being evaluated in initial and advanced trials, there is hope of getting genotypes as good as Dharani in yield and drought tolerance along with slightly reduced duration of 100 days.

Variety	Pedigree	Year of release	Duration	Pod yield (q/ha)		Shelling %	100 seed weight (g)	Sound mature kernel percentage	Oil content (%)	Special features
				Kharif	Rabi					
Abhaya	K-134X TAG-24	2006	105-110	20-25	40-45	76-79	38-40	85-90	52	Drought tolerant, LLS tolerant with fresh seed dormancy, short compact plant type, suitable for high rainfall areas
Greeshma (TCGSAPNL-888)	TIR46X JUG37	2009	95-100	20-25	40-45	76	40-45	85-90	50	Early (95-100days), water use efficient, short compact plant type, LLS tolerant
Rohini (TCGS-APNL-913)	Tirupati 4 x TIR 45	2010	90-95	20-25	40-45	75	38-42	85-90	49	Ultra-early (90-95days), water use efficient, short compact plant type, LLS tolerant
Dharani (TCGS-1043)	VRI-2 x TCGP-6	2012	100-105	16-26	37-43	75-77	40-43	90	49	Drought tolerant – withstands up to 35 days dry spell, uniform maturity, high SMK%, attractive pods, moderate stature and tolerant to low light conditions

5.7 Genomic Approaches in Breeding for Drought Tolerance

Genomics provides novel opportunities to unravel the genetic basis of drought tolerance and manipulate its genetic determinants (QTLs) through marker assisted selection to obtain improved drought tolerant cultivars. The marker assisted breeding reduces or eliminates altogether the reliance on environmental conditions during selection phase, a major hindrance to conventional breeding when dealing with traits influenced by drought (flowering time, pod filling etc., in groundnut). During the past decade, an increasing number of studies have strived to map QTLs affecting drought- related traits and yield in major crops exposed to water deficit (for an updated list, see www.plantstress.com). In a limited number of cases, marker-assisted selection (MAS) has been deployed to manipulate the target traits and to ameliorate tolerance to drought (Mackill et al. 1999; Crouch and serraj 2002; Ribaut et al. 2002; Saxena and O' Toole 2002). More recently, bioinformatics (Bray 2002) and the deluge of information generated by the post-genomics platforms (Tuberosa et al.2002 a; Hazen and Kay 2003) have added new dimensions for deciphering the role and function of genes governing the response to drought. Despite all these impressive and fascinating technological breakthroughs, the overall impact of MAS and genomics on the release of drought –resilient cultivars has so far been disappointingly negligible. Herein, we critically analyze the reasons for this limited applicative impact and review the potentially of a number of emerging post- genomics platforms for more effectively selecting drought- tolerant cultivars.

5.7.1 Marker-Assisted Selection for Improving Drought Tolerance

(Ribaut et al. 2002) pointed out that QTL manipulation to improve germplasm for polygenic traits has a number of limitations, the most distinct being the inability to predict the phenotype of any given genotype based on its allelic composition, a constraint particularly important when epistatic interactions influence the target trait (Podlick et al. 2004). Another limitation of MAS pertains to the high cost associated with QTL discovery and validation, as well as the release of superior lines. At CIMMYT, a new strategy to overcome these limitations has been implemented for improving the cost-effectiveness of MAS and its capacity to deliver improved germplasm (Ribaut and Betran 1999). This new method, named single large-scale MAS (SLA-MAS), combines conventional breeding with the use of MAS conducted at early generations for selecting plants with fixed, favorable alleles at target QTLs. The first step of SLS-MAS entails the selection of a number of elite lines with a high level of allelic complementarity (e.g. belonging to different heterotic groups) and optimal expression of the target traits. This choice can be based on phenotypic data for the target traits obtained with conventional approaches (e.g. performance per se and in diallel crossing) complemented with molecular data on the genetic distances (e.g. finger printing profiles) of the lines. In the second phase, agronomically valuable QTL alleles are identified through the evaluation of mapping populations derived from the cross of the elite lines and testers with poor performance for the target traits. Subsequently, the lines are intercrossed to develop large segregating populations from which plants homozygous for desirable alleles at target QTLs are selected in the F_2-F_3 generations. Due to the large number of plants analyzed, the cost-effectiveness of this approach relies on the utilization of PCR-based markers (Ribaut et al. 1997a). During the third and final step, pedigree selection based on the characteristics of the target environment and local needs is carried out to develop new elite lines. The distinctive feature of SLS-MAS is that MAS is deployed only once and at an early generation to fix the desired alleles at target QTLs. Although this approach reduces the chances of undesired recombination between the QTL and the flanking markers, it requires the selection of a minimum number of plants to guarantee a sufficient allelic variability at non-selected loci.

5.7.2 Association Mapping

QTL discovery through association mapping (Flint-Garcia et al. 2003) relies on the molecular and phenotypic characterization of unrelated accessions. The analysis evaluates the difference in allele frequency in case-control samples, or preferably when dealing with complex traits, the change in the mean of the investigated traits caused by allele substitution. In plants, the interest lies in the possibility of performing QTL analysis and cloning without the costly and time-consuming production of large experimental populations (Morgante and Salamini 2003). Factors influencing the applicability of association mapping are the level of LD, availability and cost of molecular makers, and knowledge of population structure. Strategies of association mapping are strongly influenced by the level of LD present in the target population. Populations characterized by a higher level of LD are more suitable for gene/QTL discovery, particularly when the panel of accessions has been profiled at a limited number of loci. The validation of a candidate gene, particularly if its sequence is available, requires the utilization of panels with much lower LD, hence a much higher level of genetic resolution. A major point to consider is related to the fact that the chromosome region underlying a particular QTL may be a non-coding sequence, in some cases separated by several kb from the effector gene modulating the QTL effect, as recently shown in maize for vgtl, a QTL for flowering time (Salvi et al.2003). As recognized by (Rafalski and Morgante 2004), the identification of regulatory regions often quite distant from the effector genes indicates that the selection of a candidate sequence to be tested for association mapping with a phenotype is not a trivial undertaking if the genomic scan aims to be comprehensive. Indeed, it is likely that a relevant portion of the genetic variability observed for drought-related traits lies in regulation of gene expression, rather than in functional differences of the encoded products.

5.7.3 "Omics" and Beyond

Once a major QTL is mapped, the next logical step is to identify suitable candidate sequences accounting for its effects, validate their role and proceed with a more direct manipulation using the gene itself as marker for MAS within the same species (Tuberosa and Coraggio 2004). The identification of candidate genes and the elucidation of their sequences can be facilitated combining QTL analysis with different sources of information and technological platforms (Pfleiger et al. 2001; Wayne and McIntyre 2002). The recent progress in the mass-scale profiling of the transcriptome, proteome and metabolome (i.e. the so-called "omics") offers the possibility of investigating the concerted response of thousands of genes to drought and other stresses. High-throughput profiling of mRNAs has been applied to investigate the changes in gene expression in response to dehydration in Arabidopsis (Oono et al. 2003; seki et al.2003; Kawaguchi et al. 2004), barley (Ozturk et al.2002) and maize (Zinselmeier et al. 2002; Yu and setter 2003). Additionally, it should be noted that the functional basis of a number of cloned QTLs may in some cases be possible through a direct profiling approach applied to suitable genetic materials (Wayne and McIntyre 2002; Hazen and Kay 2003). An interesting application of transcriptome analysis is the identification of the so-called eQTLs, namely QTLs that influence the level of expression (hence the "e") of a particular gene. In this case, the analysis of the level of gene expression carried out on each progeny of a mapping population will identify eQTLs influencing the observed variability among progenies in mRNA level of the profiled genes (Tuberosa et al. 2003). Clearly, the cost for profiling a large number of RNA samples required to identify eQTLs is still prohibitive to conceive the routine application of this approach. Instead, transcriptome profiling is better suited for studies involving a limited number of samples extracted from congenic strains differing at key genomic regions (e.g. NILs, BDLs, etc) and/or bulked samples obtained from the tails of mapping populations. Besides providing clues for the identification of candidate genes, transcriptome profiling of congenic QTL strains offers the opportunity of enriching the region with additional functional markers, provided that the different levels of expression of the investigated genes are caused by a sequence polymorphism (e.g. SNPs or insertion/deletions) that can be utilized for producing a useful marker.

5.7.4 Future Challenges and Opportunities

Genomics and bioinformatics allow us to investigate sequence colinearity in the main crops (Paterson et al. 1996; Bowers et al. 2003) and to compare their gene order and content with those of model species like *Arabidopsis* whose genome have been sequenced. This is particularly true for the identification of genes/QTLs controlling the perception of dehydration and the activation of the signal-transduction cascade that leads to the adaptive response to drought. Additionally, significant variability has been detected for WUE (Nienhuis et al. 1994) and drought- adaptive characteristics of *Arabidopsis* (Meyre et al. 2001).

In terms of experimental materials utilized for gene/QTL discovery, growing attention will be devoted to the exploitation of progenies derived from multiparental crosses and large panels of germplasm accessions with different levels of LD. The mapping populations so far utilized for QTL discovery have three major shortcomings: their size is usually small, only two alleles are explored at each single QTL locus in the biparental crosses and most QTL loci goes undetected due to the presence of functionally similar alleles. Exploring more than two genotypes will increase the chances of identifying functional variability due to QTLs and selecting most beneficial allele in terms of agronomic performance. On a similar line, useful information can also be obtained through the construction of comparative consensus maps integrating the information of anchor markers and the results of different mapping populations within the same species (Khavkin and Coe 1997; Tuberosa et al. 2002; Ribaut et al. 2004) and based on synteny information, from different species (Paterson et al. 1996; Gale and Devos 1998). Preliminary, albeit limited, evidence suggests the possibility that clusters of QTLs may govern developmental features (Khavkin and Coe 1997), some of which influence adaptation to drought conditions. Finally,the importance of an accurate phenotyping based upon good field practices and appropriate experimental designs coupled with the application of the most advanced software cannot be over emphasized for optimizing the discovery of QTLs and the characterization of their effects in different environments.

In groundnut, based on extensive phenotyping data and updated genotyping data, a large number of QTLs have been identified using CIM and EIA approaches. Phenotypic variation explained by all these QTLs was found to be small; this suggested that drought tolerance in groundnut is governed by several M-QTLs and E-QTLs each with a small phenotypic variation. Therefore, genome wide marker approaches, such as MARS and GS should be more effective approaches as compared to marker-assisted backcrossing (MABC) in case of groundnut to develop the varieties with enhanced drought tolerance through molecular breeding (Ravi et al., 2011).

With the aim of understanding the genetic basis and identification of quantitative trait loci (QTL) for drought tolerance, two new recombinant inbred line (RIL) mapping populations, namely ICGS 76 × CSMG 84-1 (RIL-2) and ICGS 44 × ICGS 76 (RIL-3), were used. After screening of 3,215 simple sequence repeat (SSR) markers on the parental genotypes of these populations, two new genetic maps were developed with 119 (RIL-2) and 82 (RIL-3) SSR loci. Together with these maps and the reference map with 191 SSR loci based on TAG 24 × ICGV 86031 (RIL-1), a consensus map was constructed with 293 SSR loci distributed over 20 linkage groups, spanning 2,840.8 cM. As all these three populations segregate for drought-tolerance-related traits, a comprehensive QTL analysis identified 153 main effect QTL (M-QTL) and 25 epistatic QTL (E-QTL) for drought-tolerance-related traits. Localization of these QTL on the consensus map provided 16 genomic regions that contained 125 QTL. A few key genomic regions were selected on the basis of the QTL identified in each region, and their expected role in drought adaptation is also discussed. Given that no major QTL for drought adaptation were identified, novel breeding approaches such as marker-assisted recurrent selection (MARS) and genomic selection (GS) approaches are likely to be the preferred approaches for introgression of a larger number of QTL in order to breed drought-tolerant groundnut genotypes (Gautami etal, 2012). The consensus map developed in this study consists of a total of 293 SSR marker loci with a total distance of 2,840.8 cM on 20 LGs (Fig. 2). The LGs in the consensus map ranged from 6.3 (LG_AhXX) to 293.37 cM (LG_AhIV) with a mean of 142.04 cM. The number of markers per LG ranged from two (LG_AhXX) to 31 (LG_AhVII) (ESM 4). Out of 293 mapped loci, 65.19% (191 loci) of marker intervals were smaller than 10 cM, 26.90% (79 loci) were between 10 and 30 cM, and 7.85% (23 loci) were greater than 30 cM (ESM 13).

In Regional Agricultural Research Station Tirupati, research on breeding for drought resistance has been going on from 1979 and in this process, many lines were developed and most promising ones are released as varieties. Lines with defects /advantages in traits related to drought tolerance and varying levels of different traits are available as genetic stocks and using these lines, it is possible to identify genes/QTLs associated with different traits conferring drought tolerance. It is being attempted now in RKVY funded project "Identification of SSR markers linked to drought related traits in advanced breeding lines of groundnut".

References

1. *Acevedo, E.and Ceccarelli S1989 Role of physiologist-breeder in a breeding program for drought resistance. In: Drought Resistance in Cereals. (pp.117-139), Baker, F.W.G, Ed, CAB International U.K.*
2. *Berninger F. And Hari P (1993) optimal regulation of gas exchange: evidence from field data. Ann. Bot. 71:135-140.*
3. *Begg J.E (1980) morphological adaptations of leaves to water stress. In: adaptations of plants to water and high temperature stress. (pp. 33-42).*
4. *Blum A (1980): Breeding programmes for improving crop resistance to water stresses. In: Crop reactions to water and temperature stresses in humid, temperature climates. (pp .263-275).*
5. *Chapman S.C, Ludlow M.M, Blamcy F.P.C and Fischer K.S (1993a) Effect of drought during early reproductive development on growth of cultivars of groundnut.I. Utilization of radiation and water during drought. Field crops Res.32:193-210.*
6. *Ciha A. J and Brun W.A (1975) Stomatal size and frequency in soyabeans. Crop Sci. 15:309-313.*
7. *Duncan W.G (1971): leaf angle, leaf area and canopy photosynthesis. Crop Sci. 11:482-485.*
8. *Doorenbos J, and Pruitt W O (1975) Guidelines for predicting crop water requirements. FAO Irrigation and Drainage paper no. 24, FAO, Rome, Italy.*

9. Duncan, WG, McCloud DE, McGraw R and Boote K.J 1978 Physiological aspects of peanut yield improvement. Crop science 18:1015-1020.
10. Ehleringer J. (1980) leaf morphology and reflectance in relation to water and temperature stress. In: adaptation of plants to water and high temperature stress. (pp.295-308).
11. Flower D.J. and Ludlow M.M (1986) contribution of osmotic adjustment to the dehydration tolerance of water stressed pigeonpea leaves. Plant cell Environ. 9:33-40.
12. Forseth I.N and Teranura A.H (1986) leaf energy budget and caluculated transpiration: the influence of leaflet orientation. Ecology.67:564-571.
13. Fischer, R.A 1979 Growth and water limitation to dryland wheat yield in Australia : A physiological framework J.Aust.Inst.Agric.science 45:83-94.
14. Fischer K S, Johnson E C and Edmeades G O(1982) Breeding and selection for drought resistant in tropical maize, In :Drought resistance in crops with Emphasis on Rice (PP.377-399), IRRI. Philippines.
15. Farquhar GD Ehleringer JR and Hubick KT 1989 Annu.Rev. Plant Physiol. $0:503-537.
16. Girdthai T, Jogloy S, Vorasoot N, Akkasaeng C, Wongkaew S, Holbrook and Patanothai A. 2010. Heritability of, and genotypic correlations between, aflatoxin traits and physiological traits for drought tolerance under end of season drought in peanut (Arachis hypogaea L.). Field Crops Research 118: 169-176.
17. Gregory PJ: (1988) Root growth of chickpea, fababean, lentil and pea effects of water and salt stress. In: World crops: cool season Food Legumes. (pp. 857-867), Summerfield, R.J, Ed, Kluwer Academic Publishers, London, U.K.
18. Gates D.M (1968): transpiration and leaf temperature. Annu. Rev. Plant physiol. 19:211-238.
19. Gautami B.M. K. Pandey, V. Vadez, S. N. Nigam, P. Ratnakumar, L. Krishna murthy T. Radha krishnan, V. C. Gowda, M. L. Narasu, D. A. Hoisington, S. J. Knapp, R. K. Varshney 2012 Quanitative trait locus analysis and construction of consensus genetic map for drought tolerance traits based on three recombinant inbred line populations in cultivated groundnut (Arachis hypogaea L.)Molecular breeding 2012, 30 (2): 757-772.
20. Holbrook, CC and Stalker, HT. 2003. Peanut breeding and genetic resources. Plant Breeding Reviews 22: 297-355.
21. Holbrook C C, GuoB.Z, Wilson D.M, and Timper.P 2009 The U.S . Breeding program to develop peanut with drought tolerance and reduced aflatoxin contamination Peanut science 36:50-53.
22. Jarvis P.G and Mc Naughtan K.G(1986) : Stomatal control of transpiration: scalling up from leaf to region. Adv.Ecol. Res. 15:1-49
23. Ketring D.L (1984) root diversity among peanut genotypes crop sci. 24:229-232.

24. *Ludlow M.M and Bjorkman O (1984):paraheliotropic leaf movement in siratro as a protective mechanism against drought induced damage to primary photosynthetic reactions: damage by excessive light and heat, planta, 161:505-518.*
25. *Ludlow M.M (1987) contribution of osmotic adjustment to the maintenance of photosynthesis during water stress. In: Progress in photosynthesis Research. Vol.4.(pp.161-168).*
26. *Lawn R.J (1982 a) responses of 4 grain legumes to water stress in south eastern queensland. I.physiological response mechanisms. Aust.J.Agri.Res. 33:481-496.*
27. *Lawn R.J(1989): Agronomic and physiological constraints to the productivity of tropical grain legumes and prospects for improvement .Exp. Agric.25:509-528.*
28. *Muchow R.C (1985b) canopy development in grain legumes grown under different soil water regimes in a semi-arid tropical environment. Field crops Res.11:99-109.*
29. *Muchow , R.C and Sinclair T.R 1986 water and nitrogen limitations in soybean grain production. II Field and model analysis. Field crops Res.15:143-156.*
30. *Mathewa R.B , Harris D, Nageswararao R.C, Williams J.H and Wadia K.D.R (1988a) the physiological basis for yield differences between 4 genotypes of groundnut in response to drought. I. Dry matter production and water use.exp.Agric.24:191-202.*
31. *Mathewa R.B , Harris D, Nageswararao R.C, Williams J.H (1988b) the physiological basis for yield differences between 4 genotypes of groundnut in response to drought. II. Solar radiation interception and leaf movement. Exp. Agric.24:203-213.*
32. *Markhart A.H.(1985) comparative water relations of phaseolus vulgaris L. And Phaseolus acutifolius gray. Plant Physiol.77:113-117.*
33. *Morgan J.M (1984) Osmoregulational and water stress in higher plants. Annu. Rev.Plant Physiol. 35:299-319.*
34. *Nageswara Rao, R.C, Williams J.H and Singh M 1989 Genotypic sensitivity to drought and yield potential of peanut Agronomy J.81:887-893.*
35. *Nageswara Rao, R.C 1992 Some crop physiological approaches for groundnut improvement J.oilseeds Res. 9: 286-296.*
36. *Nageswara Rao,R.C and Wright GC 1994 Stability of the relationship between specific laf area and carbon isotope discrimination across environments in peanut crop sci 34(1):98-103.*
37. *Nautiyal PC, Bandyopadhyay A and Ravindra V. 1999. Nature of fresh seed dormancy in peanut. ACIAR- Food Legumes Newsletter 30:5-6.*
38. *Passioura JB. 1977. Grain yield, harvest index and water use of wheat. Journal of the Australian Institute of Agricultural Sciences 43: 117-120.*
39. *Passioura JB (1982) the role of root system characteristics in the drought resistant of crop plants. In: Drought resistant in crops with Emphasis on Rice (PP.71-82), IRRI, Philippines.*
40. *Parkhurst D.F and Loucks O.L.(1972): optimal leaf size in relation to environment. J.Ecol, 60:505-537.*

41. *Paje M.C.M, Ludlow M.M and Lawn R.J (1988) variation among soyabean accessions in epidermal conductance of leaves. Aust. J. Agric. Res. 39:363-373.*
42. *Rosielle,A.A and Hamblin J (1981)Theoretical aspects of selection for yield in stress and non- stress environments. Crop science 21:943-946.*
43. *Rosenberg N.J (1978) North American droughts.AAAS selected synopsism, American society of Agronomy.*
44. *Ravi K, VadeZ V, Isobe S, Mir, RR, Guo Y, Nigam SN, Gowda MV, Radhakrishnan T, Bertioli DJ, Knapp SJ and Varshney RK 2011 Identification of several small main-effect QTLs and a large number of epistatic QTLs for drought tolerance related traits in groundnut (Arachis hypogaea L.). Thoer.applied genetics122:6 1119-1132.*
45. *Subbarao GV, Johansen C, Slinkard A.E, Nageswara Rao R.C, Saxena N.P and Chauhan Y.S 1995 Stratagies for improving drought resistant in Grain Legumes: In Critical Reviews in plant sciences 14(6):469-523.*
46. *Sponchiado B.N, White J.W, Castillo J.A and Jones P.G (1989) root growth of commom bean cultivars in relation to drought tolerance in environments with contrasting soil types Exp . Agri. 25:249-257.*
47. *Squire G.R (1990) the physiology of tropical crop production Shackel K.A and Hall A.E (1979) reversible leaf movements in relation to drought adaptations of cowpeas. Aust.J.plant physiol. 6:265-276.*
48. *Sinclair T.R and Ludlow M.M (1986) influence of soil water supply on the plant water balance of 4 tropical grain legumes. Aust. J. Plant physiol. 13:329-341.*
49. *Seethala Devi, G. 2004. Genetic studies on certain morphological and physiological attributes in 10 F_2 populations of groundnut (Arachis hypogaea L.) M.Sc. (Ag.) Thesis, Acharya N. G. Ranga Agricultural University, Hyderabad. India.*
50. *Shreya, R.P. Vasanthi, M, Shanthi Priya and P. Sudhakar 2014 Genetic parameters for drought tolerance related traits in groundnut (Arachis hypogaea L.) Electronic Journal of Plant Breeding, 5(4): 834-838.*
51. *Sirisha P 2005 Genetic studies in F3 and F4 generations of five crosses in groundnut. Thesis submitted to Acharya NG Ranga Agricultural University p232.*
52. *Suneetha N 2007 Morphological and RAPD based gene diversity studies among released cultivars and pre-release cultures of Groundnut. Thesis submitted to Acharya NG Ranga Agricultural University p251.*
53. *Tuberosa R and Salvi S (2004) Markers, genomics and post-genomics approaches-will they assist in selecting for drought tolerance? In "New directions for a diverse planet". (pp1-19).*
54. *Turner N.C (1979) drought resistance and adaption to water deficit in crop plants. In: stress physiology in crop plants. (Pp.343-372).*
55. *Turner N.C and Jones M.M (1980) Turgor maintenance by osmotic adjustment: A review and evaluation. In: Adaptations of plants to water and high temperature stress. (pp.87-103).*

56. *Turner N.C (1982) the role of shoot characteristics in drought resistance of crop plants. In: drought resistance in crops with emphasis on rice. (pp.115-134).*

57. *Turner N.C, Schulze E.D. and Gollan T. (1984) the responses of stomata and leaf gas exchange to vapour pressure deficits and soil water content. I.species comparisons at high soil water contents. 63: 338-342.*

58. *Tanzerella O.A depace C and Filippetti A (1984) Stomatal frequency and size in vicia faba L.Crop Sci. 24:1070-1076.*

59. *Upadhyaya, H. D., Shivali Sharma., Sube Singh and Murari Singh. 2010. Inheritance of drought resistance related traits in two crosses of groundnut (Arachis hypogaea L.). Euphytica. 177:55-66.*

60. *Vishnuvardhan, K.M., Vasanthi, R.P., Hariprasad Reddy, K and Bhaskar Reddy, B.V.2012.Genetic variability studies for yield attributes and resistance to foliar disease in groundnut (Arachis hypogaeaL.). International Journal of Applied Biology and Pharmaceutical Technology.3(1):390-394*

61. *Vasanthi, R .P., Babitha, M., Sudhakar, P., Reddy, P. V and John, K. 2004. Heritability studies for water use efficiency traits in groundnut (Arachis hypogaea L.). Proceeding of National Seminars: Physiological Interventions for Improved Crop Productivity and Quality: Opportunities and Constraints. 12-14 December, 2003, S.V. Agricultural College, ANGRAU, Tirupati.*

62. *Vasanthi, R.P, Seethala Devi G., Babitha M and Sudhakar P. 2005. Inheritance of leaf chlorophyll content (in terms of SPAD chlorophyll meter reading) in groundnut (Arachis hypogaea L.). Indian Journal of Genetics and plant Breeding.65: 196-198.*

63. *Wright G.C, Smith R.C.G and Mc William J.R (1983) differences between two grain Sorghum genotypes in adaptation to drought stress.III.Physiological responses. Aust. J.Agric. Res. 34:637-651.*

64. *Wright Gc, Hubick KT and Farquhar 1991 Physiologcal analysis of peanut cultivars esponse to timing and duration of drought stress Aust. J. Agric. Res. 42: 453-470.*

65. *White J.W and Castillo J.A studies at CIAT on mechanism of drought tolerance in beans. In: Research on drought tolerance in common bean. (pp 146-151).*

66. *Walker D.W and Miller J.C (1986), Jr; rate of water loss from detached leaves of drought resistant and susceptible genotypes of cowpea. Horti Sci.21:131-132.*

Chapter 6

Groundnut Varietal Characteristics and Recommended Varieties of India

K. S. S. Naik

Principal Scientist (Groundnut), ARS, Kadiri.

6.1 Introduction

Botanical description of Groundnut (Arachis Hypogaea L.)

Family: Leguminosae.

sub-family: Papilionaceae

S. Name: Arachis hypogaea L.

Two subspecies each containing two botanical varieties.

Subspecies hypogaea

variety hypogaea (The Virginia Group) Kadiri 1, 3, Tirupati 3, Kadiri 7 & 8 -variety hirsuta Subspecies fastigiata

variety fastigiata (the Valencia Group) MH2

variety vulgaris (the Spanish Group) TMV2, JL24, Kadiri 4, 5, 6, 9, KH, Anantha, Tirupati 1, 2 &4

6.2 Quantitative Groundnut Descriptors

Table 6.1

S. No.	Characteristics	States	Note	Example varieties	Stage of observation	Type of assessment
1	2	3	4		6	7
1.	Plant: Growth habit	Erect	1	ALR 1	60-70	VG
(*)		Semi- spreading	2	Chitra		
(+)		Spreading	3	Punjab 1		
2.	Leaflet: Size	Small	3	CSMG 84-1	50-60	MS

Table 6.1 ***Contd...***

S. No.	Characteristics	States	Note	Example varieties	Stage of observation	Type of assessment
(*)	(fully developed	(<4.0 cm)				
	basal leaflet)	Medium	5	ALR 2		
		(<4.0 -6.0 cm)				
		Large	7	GG 3		
		(>6.0 cm)				
3.	Leaflet Colour	Light green	1	GG 3	50-60	VG
		Green	2	M 145		
		Dark green	3	ALR 2		
4.	Stem: Pubescence	Absent	1	M 13	60-70	vs
(+)		Sparse	3	IGGV 86325		
		Medium	5	Girnar-1		
5.	Flower: Presence	Absent	1	ICGV 86325	60-70	VG
(*)	on main axis	Present	9	ALR 2		
(+)						
6.	Flower:	Sequential	1	AK 12-24	60-70	VG
(*)	Arrangement on	Alternate	2	ALR 2,		
(+)	side branches	Irregular	3	BAU 13		
7.	Inflorescence	Simple	1	GG 3	80-90	VG
(+)		Compound	2	ICGV 86325		
8.	Tune of maturity	Very early	1	Chico	90	VG
(*)	(For curing)	(< 90 days)				
		Early	3	MH 4		
		(90-100 days)				
		Medium	5	JL 24		
		(101-110 days)				
		Late	7	Somnath		
		(111-120 days)				
		Very late	9	Punjab 1		
		(>120 days)				
9. (*) (+)	Pod: Constriction	Absent Shallow Medium Deep	1 3 5	ICG(FDRS)4 ICGV 86325 JL 24 BPZ 691	100	VG
10. (*) (+)	Pod: Reticulation	Absent Medium Prominent	1 3 5	JL 24 GG20 BAU 13	100	VG
11.	Pod: Number ot	>60% 2 seeded	1	GG 2	100	MS
(*)	kernels (on 100 pod	>60% 3 seeded	3	ICGV 86590		
	basis)	>60% 4 seeded	5			

Table 6.1 *Contd...*

S. No.	Characteristics	States	Note	Example varieties	Stage of observation	Type of assessment
12. (*)	Pod: Presence of	Absent	1	SG 84	100	VS
(+)	beak	Present	9	ICGV 86590		
13.	Pod: Shelling	Low	3	BAU 13	100	MO
	percentage	(<66)				
		Medium	5	GG 1 1		
		(66-75)				
		High	7	Tirupati 1		
		(>7S)				
14.	Testa: Colour	Uniform	1	GG20	100	VG
<*>		Variegated	9	TMV I0		
15	Kernel: Colour of	White (1 A 1)	1	SPA 411 White	100	VG
(*)	testa (varieties, with	Off white (1 A 2)	2	PEARL 4		
	monochrome testa	Tan (12 E 4)	3	TMV 1		
	only)	Rose (Grayish red 8 B	4	Girnar 1		
		Purple (14 F 4)	5	AH 7835		
		Dark purple	6	SAM COL S3		
		(14 F ~)				
		Salmon (6 A 4)	7	M 13		
		Red (IO B 7)	8	Gangapuri		
		Dark red (11 C 8)	9	M 145		
16. (+) (*)	Kernel: Sliape	Spheroid Cylindrical Fusiform	1 2 3	GG 2 GG 20 RS 138	100	VG
17.	Kernel: Weight of	l-ow	3	AK 12-24	100	MG
(*)	lOO kernels (about	(-36 g)				
	9°-'o moisture)	Medium	5	ALR 1		
		(36-50 g)				
		High (51-65 g)	7	TKG 19A		
		Very hrgh	9	TPG41		
		C>65 g)				
18.	Seed: Fresh seed	Absent	1	GG 2	90	MG
<*>	dormancy	Present	9	Punjab 1		

Table 6.1 ***Contd...***

S. No.	Characteristics	States	Note	Example varieties	Stage of observation	Type of assessment
19.	Kernel: Oil	Low	3	China	100	MG
	percentage	(<45%)				
		Medium	5	TG 26		
		(45-48%)				
		High	7	TMV 2. GG 13.		
		(49-52%)		TAG 24		
		Very liigli	9			
		(>52%)		TMV IO		

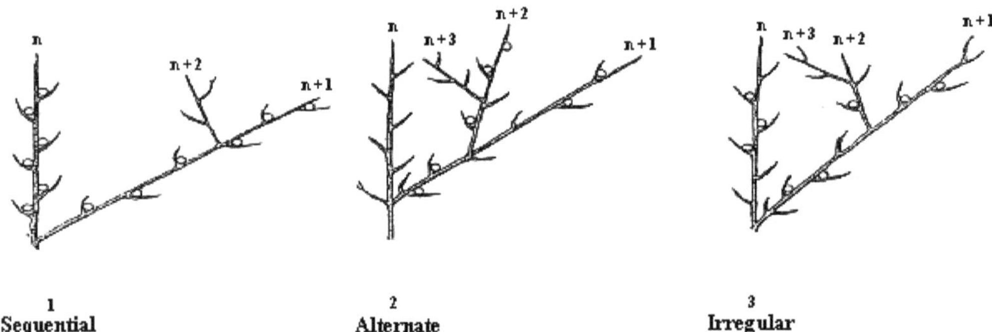

Fig. 6.1 *Flower arrangement on side branches*

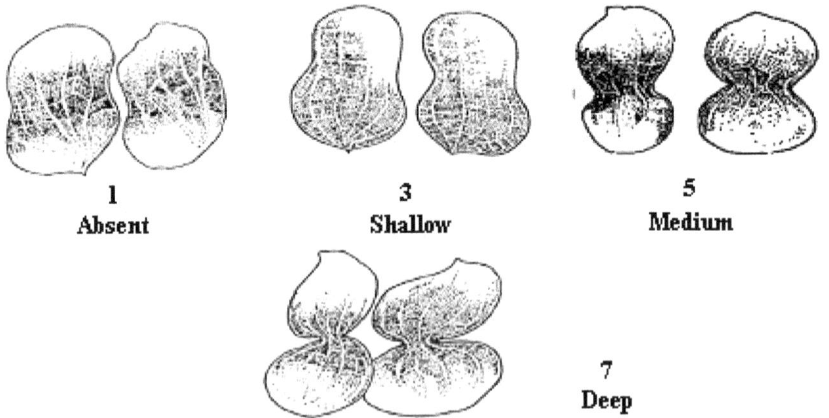

Fig. 6.2 *Constrictions of pod (to be observed within one month of harvest)*

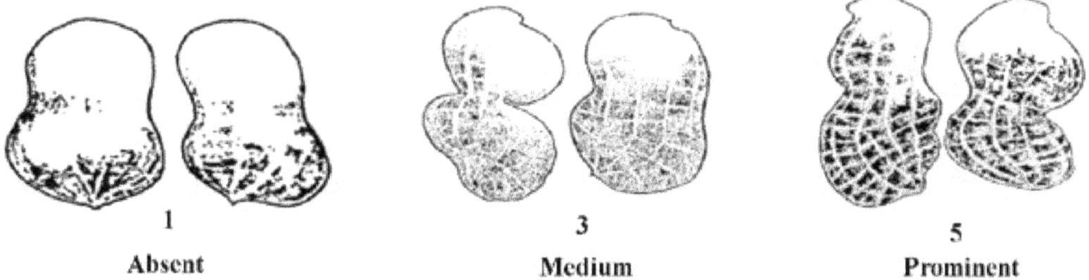

Fig. 6.3 *Pod Reticulation (to be observed within one month of harvest)*

Fig. 6.4 *Pod Presence of beak (to be observed within one month of harvest)*

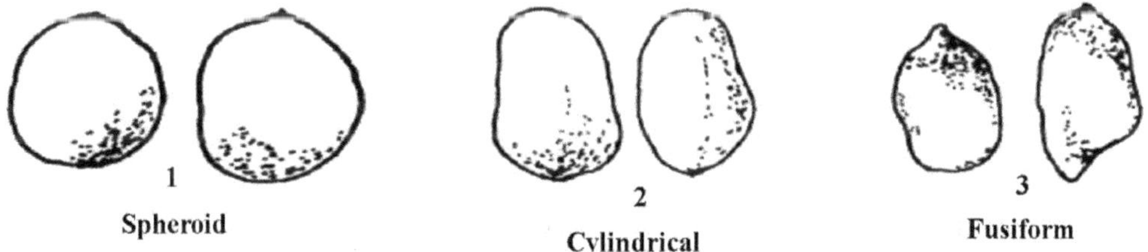

Fig. 6.5 *Kernel shape (to be observed within one month of harvest)*

Table 6.2 *Qualitative characters of groundnut*

S. No.	Character	S. No.	Character
1	Plant: growth habit (*) (+)	12	Pod texture (reticulation) (*)(+)
2	Prostrate varieties only: Main stem: growth habit (+)	13	Pod: Number of kernels (on 100 pod basis)
3	Prostrate varieties only: Side branches: growth habit (+)	14	Pod: Prominence of beak (*) (+)
4	Leaflet size (fully developed basal leaflet)	15	Pod: Shape of beak (*)
5	Leaflet colour	16	Kernel : colour (*)

Table 6.2 *Contd...*

S. No.	Character	S. No.	Character
6	Stem : Pubescence	16a	Varieties with monochrome testa: Kernel: Colour of testa (*)
7	Flowering pattern on side branches (+) (*)	17	Kernel: Shape (+)
8	Inflorescence (+)	18	Kernel : weight of 100 Kernel (7% moisture) (*)
9	Time of maturity (for curing) (*)	19	Seed dormancy (*)
10	Pod constriction (*)(+)	20	Kernel : percentage of shell

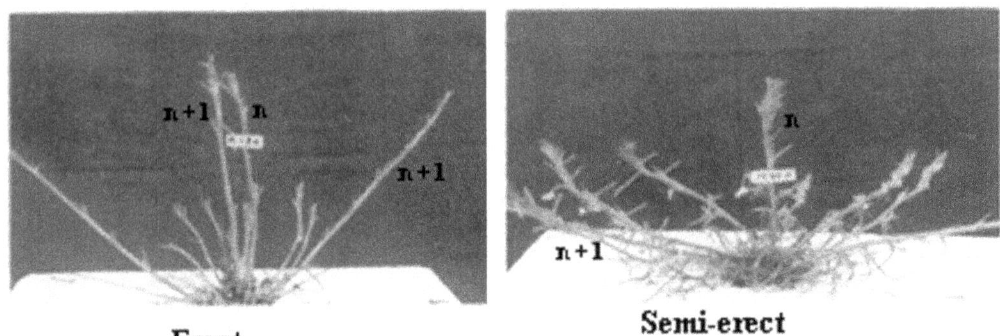

Fig. 6.6 Growth habit

Fig. 6.7 Leaf let size

Fig. 6.8 *Leaf let colour*

Fig. 6.9 *Flowering pattern*

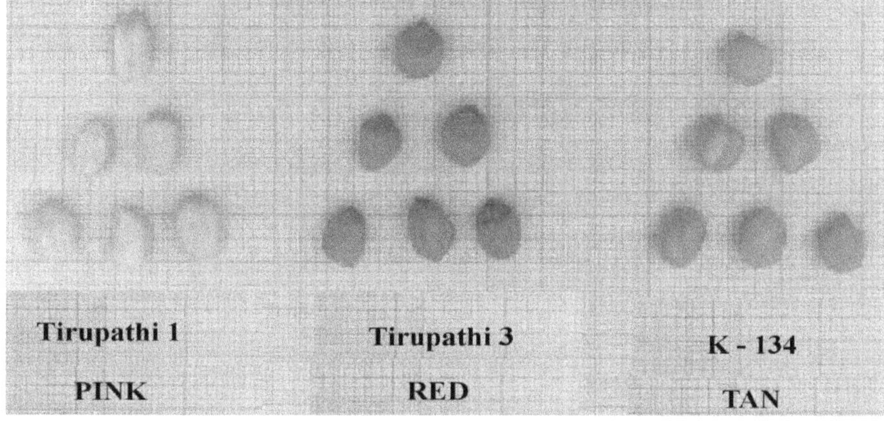

Fig. 6.10 *Kernel colour*

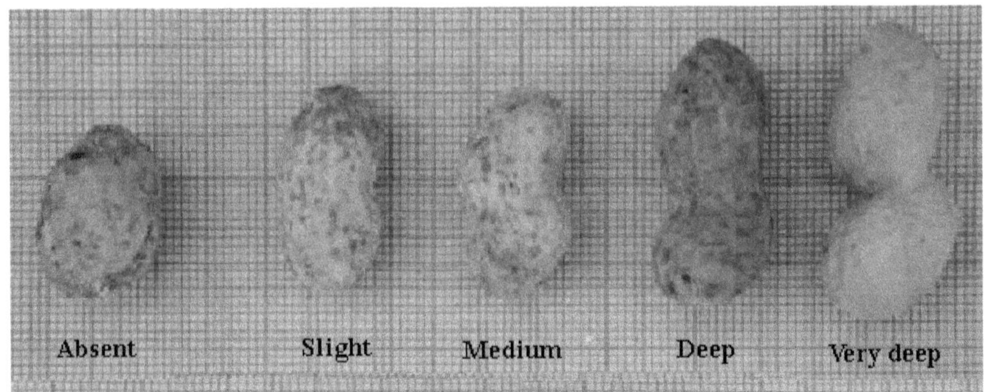

Fig. 6.11 Pod Constrictions

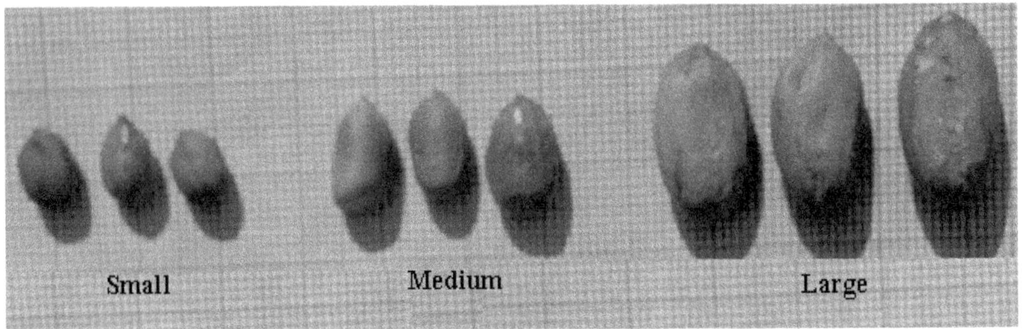

Fig. 6.12 Kernel size

***Table 6.3** Groundnut test illustrating different degrees of staining with NaOH AND KOH*

Control after 1 h	NaOH @ 2% after 1 h	KOH @ 5% after 3 h	Colour of kernel
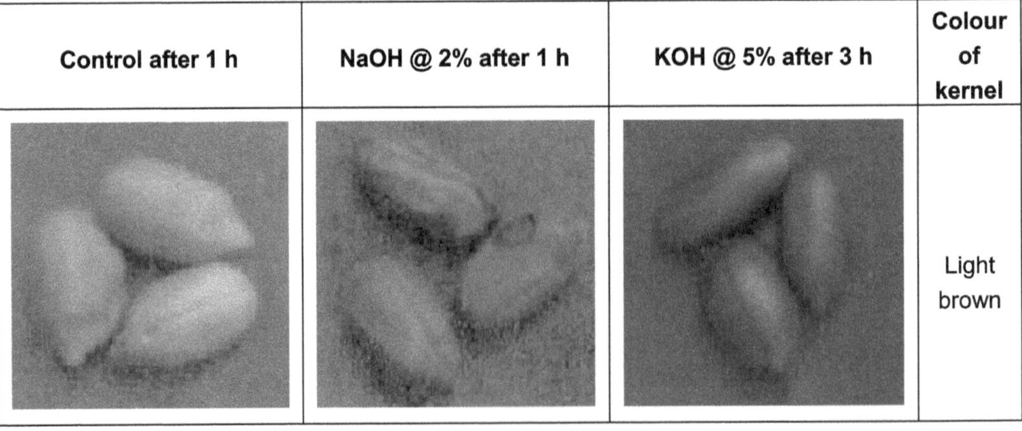			Light brown

Table 6.3 ***Contd...***

Control after 1 h	NaOH @ 2% after 1 h	KOH @ 5% after 3 h	Colour of kernel
	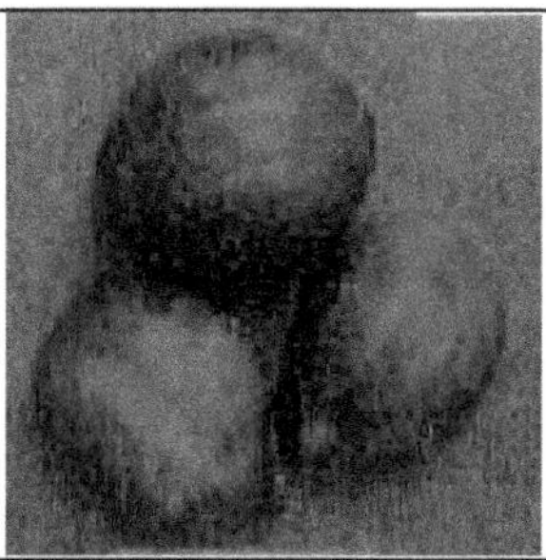	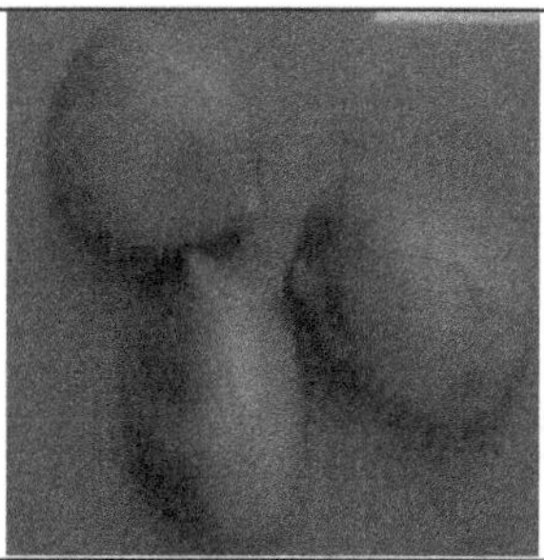	Brown
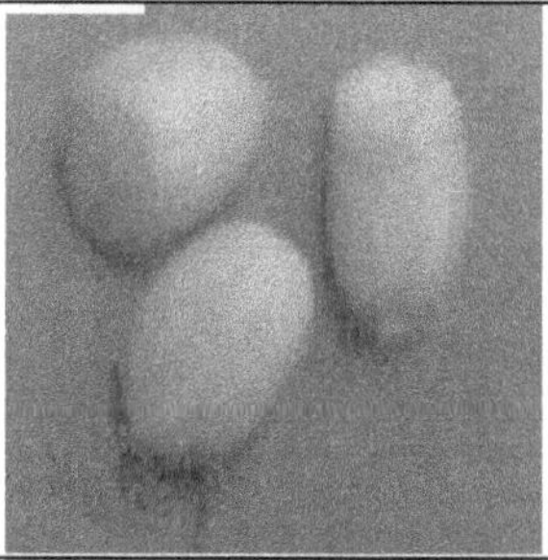	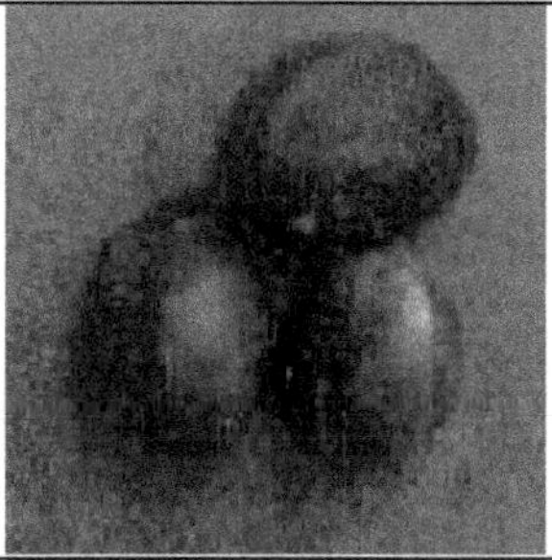	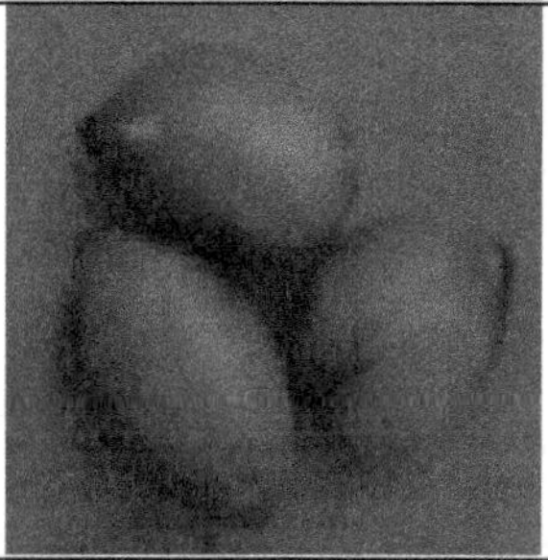	Dark brown
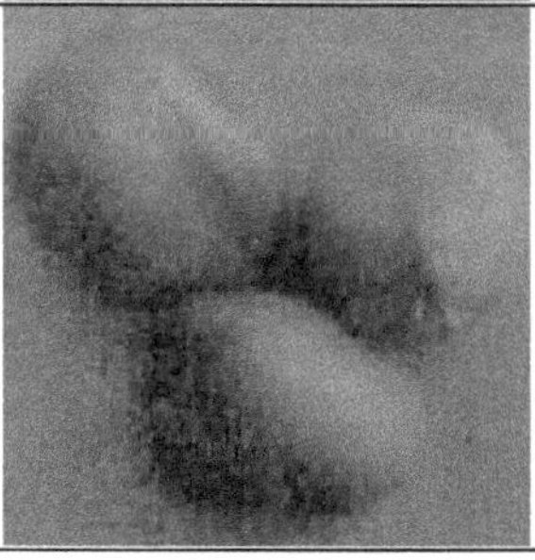	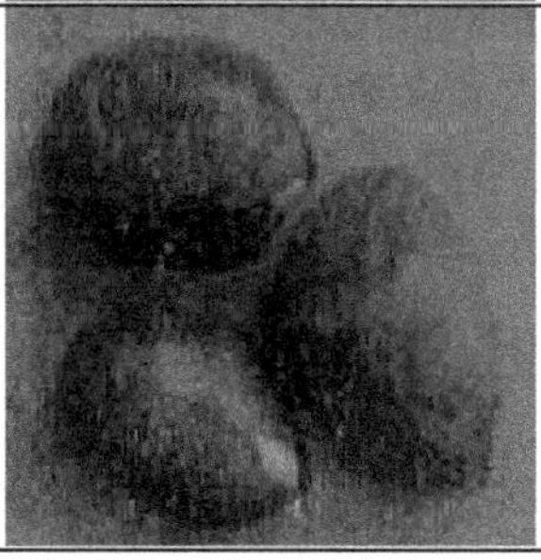	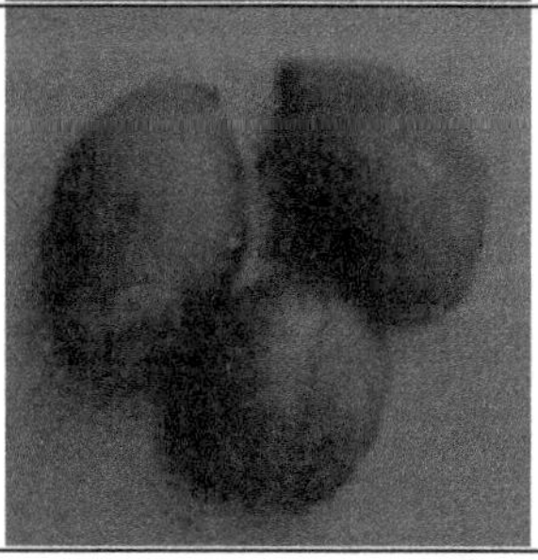	

Under these studies, the varieties showed polymorphism for plant growth, branching pattern, leaf let size and colour, flowering pattern, pod characters and testa colour. The biochemical test viz. NaOH (2%), KOH (5%), GA_3 (25 ppm) and 2-4 D(5ppm) did not show much variation for colour reaction. The varieties can be classified for showing dark brown and brown colour reaction with NaOH and KOH, while reaction to GA_3 and 2,4 -D showed high medium and low response to seedling growth. Majority of the varieties showed the common mode of medium response.

The total protein analysis by electrophoretic banding (SDS- PAGE) of 88 cultivars showed large variation in banding pattern *viz.* number of bands ranging from 9-23, intensity and presence or absence of bands. The compiled data has been published in the manual 'Morphological, chemical and electrophoretic descriptors of groundnut varieties.'

Table 6.4 *Zone-wise released groundnut-varieties and their important characteristics up to 1999*

S. No.	Name of the Variety	Pedigree	Habit Group	Year of release	Originating Centre	Area for which released	Days to maturity	Pod yield kg/ha	Shelling (%)	Oil (%)	HKW (g)	Disease reaction	Pests reaction	Salient Features
ZONE I														
1.	Gangapuri	--	Valencia	--	Khargone	Madhya Pradesh and entire country	--	2000	59.7	49.5	36.8	--	Susceptible to aphid and jassids	3-4 seeded pod with purple testa
2.	Ak 12-24	Selection from local	Spenish Bunch	1940	Akola	Vidharbha region Maharashtra Orissa, M.P. & Rajasthan	--	1250	75.0	48.5	29.4	--	--	Wider adaptability, suited for medium to heavy soils
3.	RS 1	Selection from local collection	Virginia Runner	1953	Dugrapura	Rajasthan	--	1300	71.0	45.0	40.0	--	--	Kernel medium bold
4.	T 28	Sel. from Bombayl collection	Virginia Bunch	1960	Mainpuri	Uttar Pradesh	--	1900	74.0	47.5	32.6	--	--	Pod 3 seeded with prominent beak & reticulation, kenel medium spherical wihith light red testa

Table 6.4 *Contd...*

S. No.	Name of the Variety	Pedigree	Habit Group	Year of release	Originating Centre	Area for which released	Days to maturity	Pod yield kg/ha	Shelling (%)	Oil (%)	HKW (g)	Disease reaction	Pests reaction	Salient Features
5.	RSB 87	Sel. from Brazillianl collection	Virginia Bunch	1961	Durgapura	Kota division & other heavy soil area of Rajasthan	--	1750	66.0	50.0	49.7	--	--	
6.	C 501	D 3 x Ah 477	Virginia Bunch	1961	Ludhiana	Punjab state	--	1750	68.0	48.0	40.4	--	--	1-3 seeded for sandy loam soil under irrigation
7.	T 64	Selection from EC 1664	Virginia Bunch	1966	Mainpuri	Uttar Pradesh state	--	2100	70.0	48.0	37.2	--	--	Bold pods HPS catagory with prominent reticulation and beak light red testa
8.	M 145	A 1-1 x D 3	Virginia Bunch	1968	Ludhiana	Punjab state	--	2200	76.0	50.5	40.6	--	--	Light green leaves 1-3 seeded pods purple coloured testa smaller lateral spread

Table 6.4 *Contd...*

S. No.	Name of the Variety	Pedigree	Habit Group	Year of release	Originating Centre	Area for which released	Days to maturity	Pod yield kg/ha	Shelling (%)	Oil (%)	HKW (g)	Disease reaction	Pests reaction	Salient Features
9.	Punjab 1	Selection	Virginia Runner	1953	Ludhiana	Punjab but grown in Gujarat & Rajasthan	--	1900	69.0	49.0	46.5	--	--	Wider adaptability
10.	M 13	Selection from M 13	Virginia Runner	1972	Ludhiana	Entire country	--	2750	68.0	49.0	58.1	--	--	Leaves drak green with waxy coating. Large size pod HPS category tion and medium constriction tolerant to leafspot
11.	TG 1 (Vikaram)	X-Ray mutant from spanish improved	Virginia Bunch	1973	BARC Trombay	Entire country	--	2695	68.0	46.5	72.0	--	--	Bold pod HPS Grade profuse branching but late maturity
12.	Chandra	Selection from Ah 114	Virginia Runner	1977	Mainpuri	Uttar Pradesh	--	2500	70.0	47.7	40.2	--	--	Bold pod kernel light rose big oval shaped

Table 6.4 *Contd...*

S. No.	Name of the Variety	Pedigree	Habit Group	Year of release	Originating Centre	Area for which released	Days to maturity	Pod yield kg/ha	Shelling (%)	Oil (%)	HKW (g)	Disease reaction	Pests reaction	Salient Features
13.	MH 1	Selection from Faizpur 1-5	Spanish Bunch	1975	Hissar	Haryana state	--	2000	70.0	50.0	40.8	--	--	Light green leaves and light rose kernels small pods
14.	MH 2	Selection from Gujarat dwarf mutant (GDM)	Valencia	1978	Hissar	Haryana	--	3000	72.0	49.0	27.2			Vary dwarf plant pod 1-4 seeded red testa responsive to closer spacing 15 x 15 cm
15.	BG 1	X-Ray mutant of 41-C	Virginia Bunch	1979	Kanke	Bihar state	--	2000	70.0	48.0	50.0	--	--	Bold pod SMK 93-95%
16.	BG 2	X-Ray mutant of 41-C	Virginia Bunch	1979	Kanke	Bihar state	--	2200	69.0	49.3	51.0	--	--	Bold pod SMK 93-95%
17.	M 37	A 1 x C 6-4-7-2	Virginia Runner	1980	Ludhiana	Punjab	--	1650	69.0	50.5	47.4	--	--	Pod two seeded with small beak kernels with light brown testa
18.	M 197	C 501 x U 4-7-2	Virginia Runner	1982	Ludhiana	Punjab state	--	1800	67.4	51.2	48.7	--	--	Leaves dark green pod with smooth vanation earlier than C 501. Bold kernel.

Table 6.4 *Contd...*

S. No.	Name of the Variety	Pedigree	Habit Group	Year of release	Originating Centre	Area for which released	Days to maturity	Pod yield kg/ha	Shelling (%)	Oil (%)	HKW (g)	Disease reaction	Pests reaction	Salient Features
19.	Chitra	Spanish 5B-1 x EC 1688	Virginia Runner	1984	Mainpuri	Uttar Pradesh state	--	2000	71.0	49.0	32.0	--	--	Leaves dark green testa varigated on rose back ground.
20.	Kaushal (G 201)	Selection from T28	Virginia Runner	1984	Mainpuri	All India	108-112	1700	71.0	49.0	40.4	--	--	Compact plant dark green leaves even at early maturity1-3 seeded pod.
21.	M 335	M 13 x F 7	Virginia Runner	1986	Ludhiana	Sandy areas of Punjab under irrigated condition during Kharif	120-125	2300	67.0	49.0	71.0	Tolerant to early and late leaf spots	Tolerant to leaf webber	Plant has less lateral spread leaflets large and dark green colour, pod bold with 1-2 seeded and rarely 3 seeded. Reticulation prominent, constriction medium, kernels bold with light brown testa.

Table 6.4 ***Contd...***

ZONE I														
1.	SG 84	Selection from ICGS 1	Spanish Bunch	1986	Ludhiana	Loamy soil of Punjab under irrigation for spring season	115-125	2062	64.0	50.0	49.0	--	--	Plant grows erect with medium sized dark green leaves seeded with medium constriction and slight beak brown
2.	BG 3	Early Runner X Asiriya Mwit – unde	Virginia Bunch	1986	BAU. Kanke	Bihar	115-125	2500	72.0	49.0	52.0	Fairly resistant to early and late leaf spots	--	Virginia Bunch variety with early maturity
3.	MA 16	Selection from EC 16664	Virginia Bunch	1986	Mainpuri	Uttar Pradesh	120 -125	(1500 R)	--	--	--	Moderately resistant to leaf spots	--	Bold seeded suitable for HPS, Moderately resistant to leafspot
4.	RG 141	Robut 33-1 X NCAC 2821	Spanish Bunch	1989	RAU Durgapura	Black soils of Rajasthan	110-115	2100 (K)	72.0	50.8	42.0	Tolerant to late leaf spot and BND	--	Variety with dark green foliage
5.	RS 138	Sel. from Braz- ilian culture	Spanish Bunch	1989	RAU Durgapura	Rajasthan Durgapura	105-115	2100	72.0	50.0	56.0	--	--	Variety resembling RSB 87.

Table 6.4 ***Contd...***

ZONE I													
6.	Somnath	TG 18 A (Induced mutant of TG 18A X M 13	Virginia Runner	1990	BARC Bombay	Gujarat and Rajastan	120-125	1900 (K)	72.3	50.1	60.0	Tolerant to early and late leaf spot	--	High yielding , early maturing bold seeded, suitable fo export purpose
7.	ICGS 1	Selection from Robut 33-1	Spanish Bunch	1990	ICRISAT	U.P. Bihar Haryana Punjab and Rajasthan	110-115	2300 (R)	2300	70.0	51.4	Tolerant leaf spots	Tolerant to leaf webber	Suitable for Spring season
8.	CSMG 84-1	Selection from MA 10	Virginia Runner	1992	CSAU A&T Mainpuri	Utter Pradesh Rajasthan and Haryana	125-135	2704 (K)	64.0	50.4	48.0	Resistant to rust	Tolerant to thrips, termites	Rose varigated kernel colour , bold two seeded pod
9.	ICGS 5	(Robut 33-1 x NCAC 316)F 2	Virginia Bunch	1992	ICRISAT	Uttar Pradesh	110-120	2385 (K)	68.0	50.9	38.0	Resistant to rust	--	Pod one to two seeded and tolerant to drought.
10.	BAU 13	BAU 6 x M 13	Virginia Bunch	1993	BAU Kanke	Bihar, U.P. Rajasthan Gujarat Maharashtra and Tamil Nadu.	125-135	2191 (K)	63.0	48.0	71.0	Tolerant to leaf spots colar rot and BND	--	Higher yield with bold seeded HPS for export trade

Table 6.4 *Contd...*

ZONE II													
11.	J-11	Ah 4218 x Ah 4354	Spanish Bunch	1964	Junagadh	Gujarat state suitable for other state also	--	1300	75.0	49.0	36.7	--	--	collar rot and *Aspergillus flavus* invasion. Suitable for summer season also
12.	M 13	Selection from M 13	Virginia Runner	1972	Ludhiana	Entire country	--	2750	68.0	49.0	58.1	--	--	Leaves drak green with waxy coating. Large size pod HPS category tion and medium constriction tolerant to leafspot
13.	GAUG 1	Ak 10 x Ak 12-24	Spanish Bunch	1973	Junagadh	Gujarat state	--	1500	76.0	50.0	34.0	--	--	Small thin shelled pods having prominent beak
14.	TG 1 (Vikaram)	X-Ray mutant from spanish improved	Virginia Bunch	1973	BARC Trombay	Entire country	--	2695	68.0	46.5	72.0	--	--	Bold pod HPS Grade profuse branching but late maturity
15.	GAU 10	G 224-3 x GO 343	Virginia Runner	1973	Junagadh	Gujarat	--	1800	74.0	50.0	45.4	--	--	Plant well branched medium sized pod 2-1 seeded with rose testa
16.	GG 2	J 11 x EC 16659	Spanish Bunch	1983	Junagadh	Saurastra region of Gujarat for summer season	--	3100	70.5	49.0	36.0	--	--	Early flowering dark green leaves, reticulated 2 seeded pod

Table 6.4 *Contd...*

ZONE II													
17.	Kaushal (G 201)	Selection from T28	Virginia Runner	1984	Mainpuri	All India	108-112	1700	71.0	49.0	40.4	--	--	Compact plant dark green leaves even at early maturity 1-3 seeded pod.
18.	GG 11	M 13 x GAUG 10	Virginia Runner	1984	Junagadh	Entire Gujarat	--	2050	72.6	48.6	52.0	--	--	Leaflet and pod bigger than GAUG 10 4.9 % higher yield than GAUG 10 and 14.3 % higher yield than M 13.
19.	ICGS 44	Selection from Robut 33-1	Spanish Bunch	1988	ICRISAT	Gujarat Northern Maharastra & M.P.	120-125	2500 R/S	70.0	49.0	60.0	Tolerant to early leaf spot and BND	--	Suitable for Rabi /Summer
20.	Somnath	TG 18 A (Induced mutant of TG 18A X M 13	Virginia Runner	1990	BARC Bombay	Gujarat and Rajastan	120-125	1900 (K)	72.3	50.1	60.0	Tolerant to early and late leaf spot	--	High yielding , early maturing bold seeded, suitable for export purpose
21.	ICGS 37	Selection from Robut 33-1	Spanish Bunch	1990	ICRISAT	Gujarat	110-125	3000	70.0	48.0	53.0	Moderately resistant to late leaf spots and rust	Tolerant to BND	Sequential flowering.
22.	TG -26		Spanish Bunch	1995	BARC Trombay	Rabi/ Summer G'nut producing area of Gujarat, North Maharastra andMadhya Pradesh	100-110	1596 (K) 2425 (R)	65.0	49.0	35.0		--	High yielding, earliness and fresh seed dormancy

Table 6.4 ***Contd...***

ZONE III													
23.	Gangapuri	--	Valencia	--	Khargone	Madhya Pradesh and entire country	--	2000	59.7	49.5	36.8	--	Susceptible to aphid and jassids	3-4 seeded pod with purple testa
24.	Kopergaon 1	Selection from local	Semi Spreeding	1933	Kopergaon	Sangli and Kolhapur District of Maharashtra	--	1250	72.0	47.0	44.8	--	--	Pod medium in size
25.	Kopergaon 3	Selection from local	Valencia	1933	Kopergaon	Western Maharashtra	90	1900	69.0	47.0	35.0	--	--	Early maturity, three seeded red testa preferred as a table type
26.	Ak 12-24	Selection from local	Spenish Bunch	1940	Akola	Vidharbha region Maharashtra Orissa, M.P. & Rajasthan	--	1250	75.0	48.5	29.4	--	--	Wider adaptability, suited for medium to heavy soils
27.	Ak 12-24	Selection from local	Spenish Bunch	1940	Akola	Vidharbha region Maharashtra Orissa, M.P. & Rajasthan	--	1250	75.0	48.5	29.4	--	--	Wider adaptability, suited for medium to heavy soils
28.	Spanish Improved)	Sel. from Spenish Peanut	Spenish Bunch	1905	Dharwad	Tract of bombay & Karnataka	--	2400	67.0	44.0	29.6	--	--	Less Prominent beak & constriction, kernels with round end
29.	Karad 4-11	Selection from local	Virginia Runner	1957	Karad	Satara, Sangli, Kolhapur and Pune District	--	1000	72.0	48.0	42.4	--	--	Medium to long pod 1-3 seeded

Table 6.4 *Contd...*

ZONE III														
30.	SB XI	Ah 4213 x Ah4354	Spanish Bunch	1965	Originally at Junagadh but finally tested at Jalgaon	Maharastra state	--	1300	75.0	49.0	35.2	--	--	Suitable for cultivation in summer season also.
31.	Jyoti	Selection from exotic 7	Spanish Bunch	1971	Khargaon	Madhya Pradesh state		1600	77.8	53.3	31.5			Less incidence of root rot and collar rot
32.	M 13	Selection from M 13	Virginia Runner	1972	Ludhiana	Entire country	--	2750	68.0	49.0	58.1	--	--	Leaves drak green with waxy coating. Large size pod HPS category tion and medium constriction tolerant to leafspot
33.	GG 3	GAUG 1 x JL 24	Spanish Bunch	1994	GAU Junagadh	Madhya Pradesh and Northern Maharashtra	100-105	1283	68.7	51.0	32.0	--	--	Early maturity, foliage remians green at harvest stage.
34.	TG 1 (Vikaram)	X-Ray mutant from spanish improved	Virginia Bunch	1973	BARC Trombay	Entire country	--	2695	68.0	46.5	72.0	--	--	Bold pod HPS Grade profuse branching but late maturity
35.	TG 17	Dark green mutant x TG 1	Spanish Bunch	1982	BARC Trombay	Maharastra state for summer season	--	2000	69.0	49.0	--	--	--	Bold poded pinkish kernel high harvest index. Fresh seed dormancy upto 30 days

Table 6.4 ***Contd...***

ZONE III														
36.	Kaushal (G 201)	Selection from T28	Virginia Runner	1984	Mainpuri	All India	108-112	1700	71.0	49.0	40.4	--	--	Compact plant dark green leaves even at early maturity 1-3 seeded pod.
37.	UF 70-103	Intro-duction from USA	Virginia Runner	1984	Akola	Vidharbha and western region of Maharashtra state	--	2000	71.3	52.0	31.2	--	--	Suitable for summer.
38.	ICGS 11	Selection from Robut 33-1	Spanish Bunch	1986	ICRISAT	Zone III comprising parts of M.P. and Maharashtra and Zone V comprising Andhra Pradesh Karnataka and part of Maharashtra for Rabi/ Summer cultivation	110-115	2000	70.0	48.7	42.0	Tolerant to PBND	--	Dark green leaflet , compct plant , pod 2 seeded constriction and reticulation less prominent and beak absent
39.	Girnar 1	X 14-4-B-19-B X NCAC 17090	Spanish Bunch	1988	NRCG Junagadh	Western Maharastra T.N. A.P.	95-105	(2000R)	70.9	50.0	33.0	Resistant to leaf spots and rust	Resistant to jassids	Early maturing multiple resistant variety suitable to rainfed.
40.	ICGS 44	Selection from Robut 33-1	Spanish Bunch	1988	ICRISAT	Gujarat Northern Maharashtra & M.P.	·20-125	2500 R/S	70.0	49.0	60.0	Tolerant to early leaf spot and BND	--	Suitable for Rabi /Summer

Table 6.4 *Contd...*

ZONE III													
41.	ICGS 76	TMV 10 x Chico	Virginia Bunch	1989	ICRISAT A.P.	Southern Maharastra and Karnataka	115-125	1300 (K)	72.0	42.5	44.0	Resistant to early leaf spots	--	A variety with two seeded pods
42.	TAG 24	Selection from TGS 2	Spanish Bunch	1991	BARC Trombay	Maharastra	100-105	1683	70.0	51.0	38.0	Tolerant to leaf spots	Tolerant to BND, Jassids	Early with 2-3 seeded, tolerant to BND, leaf spot, jassid
43.	BAU 13	BAU 6 x M 13	Virginia Bunch	1993	BAU Kanke	Bihar, U.P. Rajasthan Gujarat Maharashtra and Tamil Nadu.	125-135	2191 (K)	63.0	48.0	71.0	Tolerant to leaf spots colar rot and BND	--	Higher yield with bold seeded HPS for export trade
44.	B 95	M 13 x Sulmith	Virginia Bunch	1993	MPKV Rahuri	Southern Maharashtra.	115-125	3345	68.0	48.0	90.0	Resistant to late leaf spots and BND	--	Suitable for Summer cultivation, extra bold seeded, HPS grade
45.	ICGV 86325	ICGS 20 x G 201	Virginia Bunch	1994	ICRISAT	Southern Maharashtra Andhra Pradesh Tamil Nadu Karnataka & Kerala.	115-125	2700	70.0	45.0	35.0	Tolerant torust, late leaf spot and BND	--	High yielding under rainfed

Table 6.4 *Contd...*

ZONE IV													
46.	Ak 12-24	Selection from Ical	Spenish Bunch	1940	Akola	Vidharbha region Maha-rashtra Orissa, M.P. & Rajasthan	--	1250	75.0	48.5	29.4	--	--	Wider adaptability, suited for medium to heavy soils
47.	M 13	Selection from M 13	Virginia Runner	1972	Ludhiana	Entire country	--	2750	68.0	49.0	58.1	--	--	Leaves drak green with waxy coating. Large size pod HPS category tion and medium constriction to lerant to leafspot
48.	TG 1 (Vikaram)	X-Ray mutant from spanish improved	Virginia Bunch	1973	BARC Trombay	Entire country	--	2695	68.0	46.5	72.0	--	--	Bold pod HPS Grade profuse branching but late maturity
49.	Kisan	Spanish Improved X B 31	Spanish Bunch	1980	Chiplima	Orissa state	--	1600	70.8	49.9	35.2	--	--	Small pod with prominent reticulation
50.	Jawan	J 11 x Asiriya mwitunde	Spanish Bunch	1983	Chiplima	Orissa state	--	1230 (Rainfed) 2000 (Irrigated)	66.3	47.6	36.5	--	--	Elongated pod with slight larger beak, medium rose kernel.

Table 6.4 ***Contd...***

ZONE IV														
51.	Kaushal (G 201)	Selection from T28	Virginia Runner	1984	Mainpuri	All India	108-112	1700	71.0	49.0	40.4	--	--	Compact plant dark green leaves even at early maturity 1-3 seeded pod.
52.	RSHY 1	GDM x TMV 2	Spanish Bunch	1991	IARI Hyderabad	Orissa and Coastal Andhra Pradesh	110-120	2480 (R)	73.0	50.0	27.0	Moderately resistant to late leaf spots and rust	Tolerant to BND	Suitable for Rabi redidual moisture
ZONE V														
53.	Spenish Improved	Sel. from Spenish Peanut	Spenish Bunch	1905	Dharwad	Tract of bombay & Karnataka	--	2400	67.0	44.0	29.6	--	--	Less Prominent beak & constriction, kernels with round end
54.	TMV- 1	Mass Sel. from salom	Virginia Runner	1940	Tindivanum	Tamil Nadu State	--	1450	73.5	50.0	32.3	--	--	Pod 1-3 seeded
55.	TMV- 3	Selection from Bassi	Virginia Runner	1943	Tindivanum	Tamil Nadu State	-	1450	76.7	49.6	36.6	--	--	Pod short 1-3 seeded with reticulation and constriction

Table 6.4 ***Contd...***

ZONE V														
56.	TMV - 4	Selection from North Carolina	Virginia Runner	1947	Tindivanum	Tamil Nadu State	--	1450	75.5	50.0	36.6	--	--	High percentase of 3-4 seeded pod
57.	TMV 5	Cross between native Tangani yaka & Gudhia-tham bunch	Spanish Bunch	1961	Tindiva nam	Tamil Nadu State	---	1250	74.6	50.5	33.3	--	--	Mature in 110 days dormant
58.	TMV 6	Selection from Virginia Bunch	Virginia Bunch	1961	Tindiva nam	Tamil Nadu State	--	950	73.0	48.0	40.3	--	--	Suited for table purpose1-3 seeded with shallow to medium constriction reticulation
59.	TMV 7	Selection from Tenesse white	Spanish Bunch	1967	Tindivanam	Tamil Nadu state	--	1400	75.2	49.6	35.8	--	--	Bold poded dark green foliage
60.	Pollachi 1	Selection from Malayadian Bunch	Spanish Bunch	1968	Pollachi	Pollachi tract of Tamil Nadu	--	1450	76.6	48.2	31.1	--	--	Fit for HPS type.

Table 6.4 ***Contd...***

ZONE V													
61.	TMV 8	Selection from Manapuri local	Virginia Bunch	1968	Tindivanam	Southern and central region of Tamil Nadu state	--	1600	74.4	50.1	34.0	--	--	Mostly 3 seeded pod, dormant upto 60 days.
62.	S 230	Selection from Tandur local	Spanish Bunch	1969	Raichur	Northern Karnataka	--	1280	70.0	50.0	41.2	--	--	--
63.	S 206	Selection from Manvi local	Spanish Bunch	1969	Raichur	Northern Karnataka		1900	73.0	49.5	33.8			Reticulated pod with slight beak and constriction
64.	TMV 9	Ah 3490 x Ah 477	Spanish Bunch	1970	Tindivanam	Tamil Nadu state	--	1150 (Rainfed) 2000 (Irrigated)	75.6	51.4	33.3	--	--	Dormancy upto 15 days.
65.	TMV 10	Natural mutant from Argentina	Virginia Bunch	1970	Tindivanam	Tamil Nadu but suitable for entire country	--	1700	77.7	54.4	54.0	--	--	Testa varigated content
66.	Kadiri 71-1	Selection from Ah 288	Virginia Runner	1971	Kadiri	Rayalseema region of Andhra Pradesh	--	1390	74.0	47.0	38.0	--	--	Efficient in fixing atmos-pheric nitrogen

Table 6.4 *Contd...*

ZONE V														
67.	TMV- 2	Mass Sel. from "Gudhitham bunch"	Spenish Bunch	1940	Tindivanum	Tamil Nadu currently grown in A.P. Karnataka	--	1025	76.7	49.0	32.7	--	--	Widely adapted, well suited for Summer season
68.	M 13	Selection from M 13	Virginia Runner	1972	Ludhiana	Entire country	--	2750	68.0	49.0	58.1	--	--	Leaves drak green with waxy coating. Large size pod HPS category tion and medium constriction tolerant to leafspot
69.	Pollachi 2	Pollachi Red x Ah 2105	Spanish Bunch	1973	Pollachi	Coimbatore Salem & Trichi districts of Tamil Nadu	--	1500 (Rainfed) 2700 (Irrigated)	78.6	48.6	--	--	--	High shelling percentage
70.	TG 1 (Vikaram)	X-Ray mutant from spanish improved	Virginia Bunch	1973	BARC Trombay	Entire country	--	2695	68.0	46.5	72.0	--	--	Bold pod HPS Grade profuse branching but late maturity
71.	Dh 3-30	Spanish Improved x US 4	Spanish Bunch	1975	Dharwad	Northern Karnataka	--	2800	77.6	46.5	40.0			Pod with prominant beak and constriction category

Table 6.4 *Contd...*

ZONE V														
72.	TMV 11	Selection from	Valencia	1977	Tindivanam	Tamil Nadu	--	1200	73.0	50.2	33.6			Big pod 1-3 seeded thick shelled purple testa.
73.	Kadiri 2	Nigerian culture MK 374	Virginia Bunch	1978	Kadiri	Srikakulam Vishakha-patnam Chittoor district of A.P.	115-120	1800	78.0	47.0	50.9	--	--	Efficient in fixing atmospheri c nitrogen Bold seeded brownish testa.
74.	Kadiri 3	Selection from Robut 33	Virginia Bunch	1978	Kadiri	Andhra Pradesh but adopted to entire country	105-110	2100	75.0	49.0	41.0	--	--	Clustered short pod, suitable for summer season also.
75.	TMV 12	Pure line sel. from ugandan culture	Spanish Bunch	1978	Tindivanam	Tamil Nadu	--	1250 Rainfed) 2100 (Irrigated)	72.0	51.2	35.3	--	--	Medium to bold pods shallow constri-ction.
76.	Co 1	Ah 6279 x TMV 3	Spanish Bunch	1979	Coimbatore	Tamil Nadu	--	1300 (Rainfed) 2100 (Irrigated)	47.0	50.4	34.0	--	--	High harvest Index suited Rabi/ Summer cultivation also.

Table 6.4 ***Contd...***

ZONE V														
77.	KRG 1	Selection from Argentina	Spanish Bunch	1981	Raichur Bellary &	Raichur Bellary & Gulbarga districts of Karnataka	--	1200 (Rainfed) 2230 (Irrigated)	73.0	49.0	42.0	--	--	Medium sized pod 2 seeded suitable for summer season also.
78.	Co 2	EMS mutant from pol 1	Spanish Bunch	1983	Coimba tore	Tamil Nadu state	--	1650 (Rainfed) 2700 (Irrigated)	77.0	51.4	40.0	--	--	Medium pod 1-2 seeded plumpy rose colour testa.
79.	Dh 8	Selection from RS 144	Spanish Bunch	1984	Dharwad	Transitional tract of northern Karnataka	--	3200	71.7	46.0	33.0	--	--	Dark green leaves Compact plant, tolerant late leaf spot, pod small, smooth kernel rose rounded at one end and sharply pointed at the other.
80.	UF 70-103	Introduction from USA	Virginia Runner	1984	Akola	Vidharbha and western region of Maharastra state	--	2000	71.3	52.0	31.2	--	--	Suitable for summer.

Table 6.4 ***Contd...***

ZONE V														
81.	TG 3	Mutant of Spanish Improved	Spanish Bunch	1985	BARC Trombay	Kerala	--	1200 (R)	--	48.5	--	Tolerant to tikka and pod borer	--	Suitable for kharif and summer season .
82.	ICGS 11	Selection from Robut 33-1	Spanish Bunch	1986	ICRISAT	Zone III compris ing parts of M.P. and Maharastra and Zone V comprising Andhra Pradesh Karnataka and part of Maharastra for Rabi/ Summer cultivation	110-115	2000	70.0	48.7	42.0	Tolerant to PBND	--	Dark green leaflet , compct plant , pod 2 seeded constriction and reticulation less prominent and beak absent
83.	VRI 1	TMV 7 x FSB 7	Spanish Bunch	1986	RRS. Vriddha chalam	Tamil Nadu	95-105	(1590K) (1877R)	74.0	48.0	36.0	--	--	Suitable for rainfed and irrigated condition with high shelling, fresh seed dormancy for a week.

Table 6.4 *Contd…*

ZONE V													
84.	ALR 1	Pol 2 x PPG 4	Virginia Bunch	1987	ARS Aliyarnagar	Pollachi tract of Tamil Nadu	115-120	(1840K) (2000R)	70.0	49.0	33.0	Resistant to leaf spots and rust	--	Suitable for rust endemic areas (Pollachi tract).
85.	Girnar 1	X 14-4-B-19-B X NCAC 17090	Spanish Bunch	1988	NRCG Junagadh	Western Maharastra T.N. A.P.	95-105	(2000R)	70.9	50.0	33.0	Resistant to leaf spots and rust	Resistant to jassids	Early maturing multiple resistant variety suitable to rainfed.
86.	ICGS 76	TMV 10 x Chico	Virginia Bunch	1989	ICRISAT A.P.	Southern Maharastra and Karnataka	115-125	1300 (K)	72.0	42.5	44.0	Resistant to early leaf spots	--	A variety with two seeded pods
87.	VRI 2	JL 24 x Co 2	Spanish Bunch	1989	RRS Vriddha -chalam	Tamil Nadu Vriddha-chalam	100-105	1500 (K) 2000 (R)	74.0	48.0	50.0	--	--	Bold poded with medium constriction suitable to both rainfed and irrigated.
88.	ICG (FDRS) 10	Ah 65 x NCAc 17090	Spanish Bunch	1990	ICRISAT	Peninsular India	105-115	2200 (K)	67.0	47.8	36.0	Tolerant to late leaf spot and BND	--	Suitable for Kharif season where rust and LLS diseases are problem.

Table 6.4 *Contd...*

ZONE V														
89.	VRI 3	J 11 x Robut 33-1	Spanish Bunch	1990	RRS Vriddha -chalam	Tamil Nadu	90-95	1668	75.0	48.7	33.0	--	---	Pod and seed are small, pod have slight to moderate constriction with little or no beak. Highly suitable for situation of delused South West monsoon.
90.	RSHY 1	GDM x TMV 2	Spanish Bunch	1991	IARI Hyderabad	Orissa and Coastal Andhra Pradesh	110-120	2480 (R)	73.0	50.0	27.0	Moderately resistant to late leaf spots and rust	Tolerant to BND	Suitable for Rabi redidual moisture
91.	Tirupati 2	GAUG 1/NCAC FLA-14	Spanish Bunch	1991	APAU Tirupati	Andhra Pradesh	105-110	2100 (K) 3500 (R)	77.0	49.0	34.0	Tolerant to nematode incidence Kalahasti-melody	--	Suitable for both season, possesses high pg strength.

Table 6.4 ***Contd...***

ZONE V														
92.	ICGV 86590	x 14-4-B-19-B x PJ 259747	Spanish Bunch	1991	ICRISAT	Peninsular India	105-110	1785 (K)	62.0	41.6	33.0	Resistant to rust tolerant to late leaf spots	--	Multiple disease and insect pest resistant.
93.	Vemana (K-134)	Kadiri-3 x JL-24	Spanish bunch	1993	ANGRAU, Kadiri	AP, Karnataka, Tamilnadu, Orissa	105-110	1800-2300(K) 3500-4000®	76.0	49	34.0	Resistant to BND, tolorent to late leaf spot	Tolorent to Heliothis and sucking pests	Early maturing with tolorence to leaf spot and drought. Remains green till harvest.
94.	BAU 13	BAU 6 x M 13	Virginia Bunch	1993	BAU Kanke	Bihar, U.P. Rajasthan Gujarat Maharashtra and Tamil Nadu.	125-135	2191 (K)	63.0	48.0	71.0	Tolerant to leaf spots colar rot and BND	--	Higher yield with bold seeded HPS for export trade
95.	TKG 19A	TG 17 x TG 1	Virginia Bunch	1993	K.K.V. Dapoli	Konkan region for R/S and Kharif	120-125	2260 (K)	63.0	46.0	60.5	--	--	Bold and attractive kernels seed qualifies for HPS grade.

Table 6.4 *Contd...*

ZONE V													
96.	ICGV 86325	ICGS 20 x G 201	Virginia Bunch	1994	ICRISAT	Southern Maharashtra Andhra Pradesh Tamil Nadu Karnataka & Kerala.	115-125	2700	70.0	45.0	35.0	Tolerant to rust, late leaf spot and BND	--	High yielding under rainfed.
97.	Kadiri-4 (K-150)	Dh-3-30 x NCAC 2230	Spanish bunch	1997	ANGRAU, Kadiri	Andhra Pradesh for rabi & rice fallows	100-105	3500-4000 ®	74.0	48.0	40-45	Tolorent to BND, LLS	-	Non dormant, short plant, light green foliage, round tip, recommended for rice fallows.

Table 6.5 Important varieties released and suitable for different states of India

Variety	Year of release	Releasing centre	Yield Potential	Oil content (%)	Recommended for (state/region)	Specific features
VRI (Gn) 5	2001	Vridhachalam TNAU	2133 (R) 2384 (I)	50	Tamil Nadu	Resistant to rust and Late Leaf spot (LLS); suitable for both *kharif* and *rabi*-summer seasons
Co (Gn) 4 (TNAU 269)		Coimbatore TNAU	1500 (R) 1950 (I)	53	Tamil Nadu	Resistant to rust and LLS; suitable for all seasons
GG 7 (J-38)	2001	Junagadh GAU	2149	49	Gujarat & Southern Rajasthan	Tolerant to LLS; suitable for *kharif* season
AK 159	2002	PKV, Akola,MH	1606	51	Maharashtra and Madhya Pradesh	Early (105-110 days) maturity; recommended for *kharif* season

Table 6.5 *Contd...*

Variety	Year of release	Releasing centre	Yield Potential	Oil content (%)	Recommended for (state/region)	Specific features
Kalahasti (TCGS 320)	2002	Tirupati ANGRAU	3764	52	Andhra Pradesh	Tolerant to Peanut Bud Necrosis Disease PBND) and Jassids; suitable for *rabi* season in kalahasti malady endemic areas and also for *kharif* in north coastal and north Telangana regions
Narayani (TCGS29)	2002	Tirupati ANGRAU	3764	48	Andhra Pradesh	Tolerant to mid-season moisture stress conditions; recommended for both *kharif* and *rabi*-summer seasons
Sneha*	2002	Vellayani, KAU	-	-	Kerala	Recommended for *kharif* season
Snigdha*	2002	Vellayani, KAU	-	-	Kerala	Recommended for *kharif* season
GG 6	2003	Junagadh, JAU	2782	50	Gujarat	Suitable for cultivation in summer
GG 7 (J-38)	2001	Junagadh,GAU,	2149	49	Gujarat & Southern Rajasthan	Tolerant to LLS; suitable for *kharif* season
AK 159	2002	PKV, Akola	1606	51	Maharashtra and Madhya Pradesh	Early (105-110 days) maturity; recommended for *kharif* season
Kalahasti	2002	Tirupati ANGRAU	3764	52	Andhra Pradesh	Tolerant to Peanut Bud Necrosis Disease(PBND) and Jassids; suitable for *rabi* season in kalahasti malady endemic areas and also for *kharif* in north coastal and north Telangana regions
(TCGS 320)						
Narayani (TCGS29)	2002	Tirupati ANGRAU	3764	48	Andhra Pradesh	Tolerant to mid-season moisture stress conditions; recommended for both *kharif* and *rabi*-summer seasons

Table 6.5 *Contd...*

Variety	Year of release	Releasing centre	Yield Potential	Oil content (%)	Recommended for (state/region)	Specific features
Sneha*	2002	Vellayani, KAU	-	-	Kerala	Recommended for *kharif*
Snigdha*	2002	Vellayani, KAU	-	-	Kerala	Recommended for *kharif*
GG 14 (JSP 28)	2003	Junagadh, JAU	2159	52	Punjab, Haryana, Uttar Pradesh	Northern Rajasthan,Tolerant to thrips, *Spodoptera and* leaf miner; recommended for *kharif*
TPG 41	2004	BARC, Mumbai	2008	49	All India	Moderately resistant to rust; bold-seeded (HSM >60g); high O/L ratio; recommended for *kharif*
TG 37A	2004	BARC, Mumbai	1900	48	All India	Tolerant to collar rot, rust and late leaf spot; suitable for both *kharif* and *rabi*-summer seasons; possess fresh seed dormancy up to 15 days
TPG 41	2004	BARC, Mumbai	2008	49	All India	Moderately resistant to rust; bold-seeded (HSM > possess fresh seed dormancy up to 15 days
Vikas (GPBD4)	2004	Dharwad, UAS	1900-2200	49	All India	Resistant to LLS and rust; recommended for *kharif*
TLG 45	2004	Latur, MAU	1506	51	Maharashtra	Large-seeded (HSM = 59 g); medium maturity (114 days); recommended for *kharif*
SG 99	2004	Ludhiana, PAU	2501	52	Punjab	Tolerant to bud necrosis disease; possess long fresh seed dormancy (30days); suitable for summer
Phule Unap (JL286)	2004	Jalgaon, MPKV	2231	49	Maharashtra	Tolerant to LLS, rust and stem rot; also tolerant to thrips, leaf miner and Spodoptera
Prutha (Dh 86)	2005	Dharwad, UAS	4022	48	Recommended for All India	Tolerant of LLS and sucking pests; suitable for rabi-summer

Table 6.5 *Contd...*

Variety	Year of release	Releasing centre	Yield Potential	Oil content (%)	Recommended for (state/region)	Specific features
Kadiri 5	2005	Kadiri , ANGRAU	1800-2200	48	Andhra Pradesh	Tolerant to leaf spots and drought; recommended for kharif
Kadiri 6 (K 1240)	2005	Kadiri , ANGRAU	1800-2400	49	Andhra Pradesh	Tolerant to leaf spots; recommended for Kharif
Pratap Mugphali2 (ICUG 92195)	2005	Udaipur MPUA&T,	1800-2800	49	Rajasthan	Tolerant to Early Leaf Spot (ELS), LLS and PBND; tolerant to Spotdoptera,, leaf miner and
Co (GN) 5	2005	Vridhachalam, TNAU	1585	54	Tamil Nadu	Tolerant to rust, PBND; tolerant to leaf miner and Spodoptera; recommended for kharif
Pratap Mugphali1 (ICUG 92035)	2005	Udaipur, MPUA&T	2500.-3000	49	Rajasthan	Tolerant to rust, PBND; tolerant to leaf miner and Spodoptera; recommended for kharif
Ratneshwar (LGN 1)	2005	Latur, MAU	1487	51	Maharashtra	Moderately resistant to LLS, stem rot, rust and PBND; tolerant to sucking pests; recommended for kharif season
Utkarsh (CSMG 9510)	2005	Mainpuri, CSAUAT	21.92	49	Uttar Pradesh, Punjab, Northern Rajasthan	Resistant to rust, possess fresh seed dormancy up to 40-45 days; recommended for kharif
GG 21(JSSP 15)	2005	Junagadh, JAU	1843	53	Uttar Pradesh, Punjab, northern Rajasthan	Recommended for kharif
Durga (RG 382)	2005	RAU, Durgapura	2203	55	Rajasthan	Suitable for sandy and loamy soils; recommended for kharif
TMV (Gn) 13	2006	Coimbatore, TNAU	2580	50	Tamil Nadu	Tolerant to early and mid season moisture deficit stress conditions; recommended for kharif
GG 8 (J 53)	2006	JAU, Junagadh	1716	46	Northern Maharashtra and Madhya Pradesh	Moderately tolerant to PBND, collor rot and stem rot diseases

Table 6.6 *Varieties released from Agricultural Research Station, Kadiri*

S. No.	Name of the Variety	Year of release	Pedigree	Recommend area	Days to maturity	Plant habit group	Resistant to stem rot	Resistant to rust	Fresh seed dormancy (Days)	Pod characters	Shelling %	Leaf let color
1.	Kadiri-1	1971	A selection from Ah-228	For Kharif in rainfed tract of Rayalaseema districts of A.P.	135-140	Virginia Runner	Susceptible	Moderately tolerant	40 - 45	Medium size, pod reticulation medium, beak is less prominent, pod with prominent constriction	65-68	Dark green
2.	Kadiri-2	1972	A selection from Nigerian culture MK-374	For Kharif in rainfed tract of Rayalaseema districts of A.P.	115-120	Virginia Bunch	Susceptible	Moderately tolerant	30 - 40	Big, prominent constriction, reticulation and beak	70-72	Dark green
3.	Kadiri-3	1978	A selection from Rahout-33	Cultivation in Kharif and Rabi throughout A.P. in groundnut growing tracts and in certain parts of the country	105-110 (Kharif) 125-130 (Rabi)	Virginia Bunch	Susceptible	Susceptible	35 - 40	Medium, slight constriction and reticulation, beak is less prominent	70-72	Dark green
4.	Vemana	1993	A cross derivative of Kadiri-3 X JL-24	Groundnut growing tracts throughout A.P and in certain parts of the country	105-110	Spanish Bunch	Susceptible	Susceptible	10 - 15	Medium, moderate constriction, slight reticulation and beak,	72-75	Green

Table 6.6 *Contd...*

S. No.	Name of the Variety	Year of release	Pedigree	Recommend area	Days to maturity	Plant habit group	Resistant to stem rot	Resistant to rust	Fresh seed dormancy (Days)	Pod characters	Shelling %	Leaf let color
5.	Kadiri-4	1995	A cross derivative of Dh3-30 X NCACC-2230	Recommended for Rabi/ Summer cultivation under less irrigation water facilities throughout A.P.	95-100	Spanish Bunch	Susceptible	Susceptible	-- Nil --	Medium bold, less prominent constriction, prominent reticulation and beak	70-72	Green
6.	Kadiri-5	2002	A cross derivative of JL-24 X VG-55(E)-7	Recommended for Kharif/Rabi cultivation in groundnut growing parts	105-110	Spanish Bunch	Susceptible	Susceptible	-- Nil --	Medium bold, constriction prominent, prominent reticulation and beak	72-74	Light green
7.	Kadiri-6	2002	A cross derivative JL24 XAh316 / S	Recommended for both Kharif & Rabi groundnut tracts of A.P.	95-100	Spanish Bunch	Susceptible	Susceptible	-- Nil --	Bolder than Kadiri-5, prominent constriction, reticulation and beak	70-72	Light green

Table 6.7 Varieties released from Agricultural Research Station, Kadiri and RARS, Tirupati after 2002

S. No.	Variety	Year	Days	Yield (kg/ha)		Specific attributes	Distinguishing feature
				Kharif	Rabi		
1	Kadiri-5 (S.B)	2002	110	1800-2200	3500 - 4000	Tol.to LLS & Drought.	Broad oval leaves, medium attractive pods.
2	Kadiri -6 (S.B)	2002	100	1800-2400	3800 - 4200	Early with bolder & long shiny attractive kernels	Broad long leaves, bold, beaked reticulated, constricted pods
3	Kadiri -7 bold (VG)	2009	125	2500–3000	4000 - 4500	Confectionary HPS Type, Tol. to thrips, field tol. to PSND & PBND HKW 65-75g. protein 29%, TSS 14%.	Pointed dark green small leaves. Bold round attractive pods. More primaries & secondary branches.
4	Kadiri -8bold (VG)	2009	120	2500–3000	4000	Confectionary HPS Type, Tol. to thrips, field tol. to PSND & PBND HKW 65-75g. Protein 29%, TSS 14%.	Pointed dark green small leaves. Long bold round attractive pods. More primaries & secondary branches.
5	Kadiri -9 (S.B)	2009	110	2000–2500	3500 - 4000	Res. to Drought, Leaf spot, Sucking pests, dry rootrot, thrips, PSND & PBND, Kalahasti malady, 30 days fresh seed dormancy.	Dark green oval hairy leaves with sturdy plant, medium reticulated stumpy pods. Kernel medium round Tan in colour
6	Anantha (S.B)	2010	110	2000–2500	3500 - 4000	Tol. to Drought, Leaf spot and Sucking pests	Medium dark green leaves with hairy and wavy margins. Medium smooth attractive pods like Vemana. Remains green till harvest
7	Kadiri Harithandra (S.B)	2010	110	2000–2500	3500 - 4000	Early, res. to LLS(2 score), Rust, Thrips and Jassids	Dark Green Pointed Leathery Foliage and Rose Testa Colour With Stumpy Humped Pod. Remains green till harvest
8	Narayani (S.B)	2002	105	2000-2500	4000 - 4500	Early, uniform maturity, Tol.to Drought.	Light green broad leaves like JL-24. Medium smooth clustered attractive pods. Red oval shiny kernel
9	Kalahasti (S.B)	2002	110	---	4500 - 5000	Tol to. Kalahasthy malady. Oil 52%	Medium green pointed leaves. Red bolder shiny kernels.
10	Abhaya	2006	105	2250-2500	3500 - 4000	Tol.to LLS & Drought.	Light green foliage. Medium short plant with mostly three seeded pods
11	Prasuna	2006	115	---	3375 - 3500	Tol to. Kalahasthy malady.	-

***Table 6.8** Charateristics of high yielding groundnut varieties*

S. No.	Variety	Charateristics
1	 VEMANA	Vemana (1993)(SB)(Kadiri 3xJL-24) Duration (days):105-110 Pod yield (q/ha): 18-26q/ha(k), 35-40q/ha (R) Shelling out-turn(%): 75 Oil content (%):49 100 kernel weight(g): 45-47 Special characters: Tolerant to LLS and PBND and also to *Heliothis*. Early with Tolerant to Drought and leaf spot, collar rot, remains green till harvest.
2.	 Kadiri 4	Kadiri-4 (DH3-300x NcAc2230) (1997) (SB) Duration (days):95-105 Pod yield (q/ha): 35-40q/ha (R) Shelling out-turn(%): 74 Oil content (%): 48 100 kernel weight (g): 45-47 Special characters: Tolerant to leaf spot and PBND. Suitable for late sowings recommended to Rabi/Summer, Rice fallows.
3.	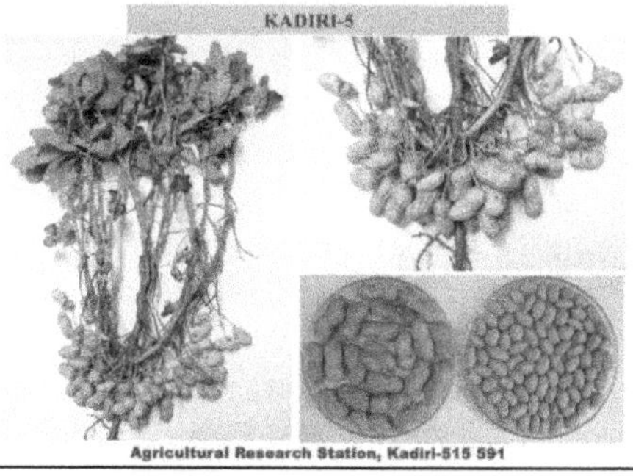 KADIRI-5 Agricultural Research Station, Kadiri-515 591	Kadiri-5 (JL 24xVG55-7)(2002): Duration (days): 105-110 Pod yield (q/ha): 18-22q/ha (k), 34-40q/ha (R) Shelling out-turn(%): 75 Oil content (%): 48 100 kernel weight (g): 45-47 Special characters: Tolerant to leaf spot and drought
4.	 Kadiri 6	Kadiri-6 (JL 24xAh316/s) (2002) Duration (days): 100-105 Pod yield (q/ha): 18-24q/ha (k), 38-42q/ha (R) Shelling out-turn(%): 75 Oil content (%): 48 100 kernel weight(g): 45-47 Special characters: Large seed, Early and high yield than JL24

Table 6.8 *Contd...*

S. No.	Variety	Charateristics
5.		Kadiri Harithandhra (2008) 91/57-2 X PI476177(ICGX930181P3) Released at National level for zone III A Duration 110-120 days Pod yield ranged from 3606 to 5506 kg/ha during rabi Oil content: 48%. It is resistant to rusts and leaf spot and Tolerant to sucking pests such as thrips, jassids and it is less susceptible to Spodoptera.
6.		Kadiri-7 Bold (2009) [{ICGV86522XICG(FDRS)10}XICGV91172]: Released for cultivation under high input management both in *Kharif* and *Rabi* conditions, Yield potential: 50 to 75 Q/ha, Duration 125 days in *Kharif* and 135 days in *Rabi* Hundred kernel weight 75-80 gms with bold round pods and kernels , Protein content : 27% Total Soluble Sugars: 14% and Oil content: 49% Tolerant to LLS, sucking insects such as Thrips and Jassids
7.		Kadiri-8 Bold (2009) [{ICGV86522XICG(FDRS)10}XICGV91172]: Released for cultivation under high input management both in *Kharif* and *Rabi* conditions, Yield potential of 50 to 65 Q/ha, Duration of 125 days in *Kharif* and 135 days in *Rabi,* Hundred kernel weight of 65-75 gms with bold long pods and kernels, Protein content: 27% Total Soluble Sugars: 14.5 % Oil content: 49 % This variety is tolerant to LLS and sucking insects such as Thrips and Jassids

Table 6.8 ***Contd...***

8.		Kadiri-9 (2009) (Kadiri 4x Vemana): Highly drought resistant variety (can withstand dry spell up to 50 days) Duration :110-115 days Stable yielding (kharif 800-1000, rabi 1400-1600 kg/acre) Under severe drought conditions its yield is not less than1200kg/ha Fresh seed dormancy: 30 days Multiple tolerant for diseases (ELS, LLS, Rust, PSND,PBND, Dry root rot), Multiple tolerant for pests (Jassids, Thrips, Defoliators, Mite and Nematodes) High shelling: 78% High oil content: 53% High W U E with more photosynthetic activity, most suitable for end season drought.
9.		Anantha (2010) (VemanaX Girnar): Variety can withstand dry spell up to 30 days. matures in 100-105. It gives a pod yield of 22-26 Q/ha under normal conditions and 14-18 Q/ha under drought conditions. It is tolerant to foliar diseases and sucking pests. It has a shelling Percent of 75 and an oil content of 48%. Its drought tolerance is attributed to its high Proline, high Relative Water Content and quick regenerative capacity
10.		Tirupati-4 (1985)(JL24 x Ah316/S) Duration (days): 105 kharif, 110 rabi Pod yield (q/ha): 20-25 kharif, 40-45 rabi Shelling out-turn(%): 76 Oil content (%): 48 100 kernel weight(g): 42-45 Special characters: Tolerant to mid season drought, rose testa variety
11.		Narayani (JL24 x Ah316/S)(2002) Duration (days) :100 kharif, 105 rabi Pod yield (q/ha):20-25 kharif, 40-45 rabi Shelling out-turn(%): 75-76 Oil content (%): 49 100 kernel weight(g) : 42-45 Special characters: Uniform maturity, suitable for low to medium rainfall areas, tolerant to drought, high frequency of mature kernels (95%), red testa variety

Table 6.8 *Contd...*

S. No.	Variety	Charateristics
12.		Kalahasti (TCG1709 x TCG 1518)(2002) Duration (days) :105-110 rabi Pod yield (q/ha): 45-50 rabi Shelling out-turn(%): 76 Oil content (%): 53 100 kernel weight (g) : 46-47 Special characters: Suitable for rabi, resistant to kalahasti malady, red testa variety with high oil content
13	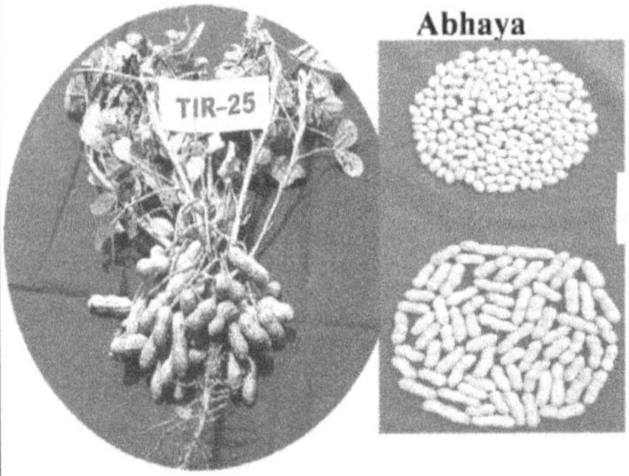 Abhaya TIR-25	Abhaya (K 134 x TAG 24)(2006) Duration (days): 105 kharif, 115 rabi Pod yield (q/ha): 20-25 kharif, 40-45 rabi Shelling out-turn(%): 77-78 Oil content (%): 50-52 100 kernel weight (g): 40-42 Special characters: High water use efficiency, tolerant to drought, late leaf spot, high frequency of three seeded pods (75%) and tolerant to cloudy weather, rose testa variety
14.	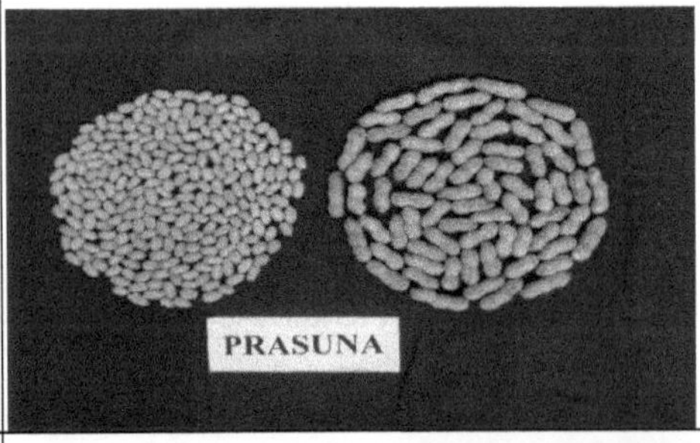 PRASUNA	Prasuna (TCG 1717x TCG 1518)(2006) Duration (days):105-110 rabi Pod yield (q/ha): 40-45 rabi Shelling out-turn(%): 75-76 Oil content (%): 48 100 kernel weight(g): 45-47 Special characters: Suitable for rabi, resistant to Kalahasti malady, rose testa variety
15.		Greesma (2009) Duration (days): 105-110 rabi Pod yield (q/ha): 40-45 rabi Shelling out-turn(%): 75-76 Oil content (%): 48 100 kernel weight(g): 45-47Special characters: Suitable for rabi, resistant to Kalahasti malady, rose testa variety

Table 6.8 ***Contd...***

S. No.	Variety	Charateristics
16.		Rohini (2010): Duration (days) :90-95 kharif, 100 rabi Pod yield (q/ha) :20-25 kharif, 40-45 rabi Shelling out-turn(%): 75 Oil content (%): 50 100 kernel weight: 36-40 Special characters: High water use efficiency, tolerant to drought, late leaf spot, rose testa
17.		Bhima(2010) Duration (days): 110-115 Pod yield (q/ha): 25-30 Shelling out-turn(%): 72 Oil content (%): 45 100 kernel weight(g): 69 Protein content (%): 14 Sucrose content (%): 7 Special characters: Early, bold seeded confectionery variety
18.		TAG-24 (TGS2xTGE-1) Duration (days):95-90 Pod yield (q/ha): 40-45q/ha(R) Shelling out-turn(%): 70 Oil content (%): 48 100 kernel weight (g): 45-47

Chapter 7

Biotechnological Tools for Groundnut Improvement

Yellari Amaravathi

Scientist (Biotechnology), Institute of Frontier Technology, RARS, Tirupati.

7.1 DNA Isolation and Purification from Groundnut Seedlings

Access to high quality genomic DNA for downstream analyses requires isolation of DNA from plant tissue and most commonly used method for DNA isolation is by Saghai-Maroof *et al.* (1984). Fresh young leaves of groundnut are preferred tissue for DNA extraction as they contain low concentration of metabolites and plysaccharides. The disruption of the tissue is the first step in the process of DNA isolation. Plant cells contain a rigid cell wall which needs shear forces to break it down and to release DNA in the extraction buffer. Usually tissue disruption by mortar and pestle is done in the presence of liquid nitrogen which minimizes oxidation of polyphenols, inhibits nuclease activity and aids the grinding process. After disrupting the plant cell wall, cell membrane also needs to be degraged in order to allow the release of the cellular constituents including DNA in the extraction buffer.

7.1.1 DNA Extraction Buffer

The 100 ml of extraction buffer contains 2g of CTAB, 10 ml of 1 M Tris(pH-8.0), 14 ml of 5M NaCl, 2 ml of 0.5 M EDTA(pH-8.0) and 1ml 2-Mercapto ethanol dissolved in doubled distilled water and finally make up the volume to 100 ml.

7.1.2 Cetyltrimethyl Ammonium Bromide (CTAB) as Detergent

Detergents allow the release of membrane-bound DNA. These assist with the disruption of tissues by removing the lipid molecule from the cell membrane and allowing DNA and other cellular material to be released into the solution. The most commonly used detergents are cetyltrimethyl ammonium bromide (CTAB), sodium dodecyl sulphate (SDS), Triton X-100 and comercial biological laundary detergent. CTAB detergent not only releases DNA from cellular membranes and proteins but also separates polysaccharides from the extracted DNA molecule. At low salt concentration, the positively charged CTAB solubilizes the cell membrane and forma a complex with negatively charged DNA, later the DNA can be precipitated from this complex mixture by increasing the salt concentration. Higher concentrations of CTAB are recommended for polysaccharide-rich plant tissues, but at the cost of little yield of the DNA.

7.1.3 Tris HCl as Buffering Agent

This is needed to control the pH of the extraction solution. Usually 100 mM Tris HCl (pH 7.0-8.0) is used as a buffering agent in most of the DNA extraction protocols.

7.1.4 Ethylene diamine tetraacetic Acid (EDTA) as Chelating Agent

Chilating agent is required to inactivate the nuclease activity. Nuclease activity can cause shearing of DNA molecules. Keeping tissue frozen during or prior to homogenization reduces the risk of nuclease-activated degradation of DNA molecules. Activities of endogenous nuclease are dependent on the pH of the extraction buffer. The pH of the extraction buffer needs to be monitored carefully, especially after the addition of the plant material. Magnesium (Mg^{2++}) ions acts as a cofactor for most of the nuclease activity, therefore addition of bivalent ion chelator EDTA in extraction buffer prevents DNA degradation.

7.1.5 2-Mercaptoethanol (2-ME) as Reducing Agent

2-mercaptoethanol (2-ME) to prevent peroxidase or polyphenoloxidase. Polyphenols form complexes with DNA very quickly and we need to include polyphenol absorbents, reductants and oxidants. Maintaining the freezing temperatures during tissue homogenization minimizes the polyphenol oxidation. 2-ME prevents DNA from quinones, disulphides, peroxidases and polyphenol oxidases and therefore retards polyphenol oxidation. Activated charcoal has also been reported as an effective endogenous polyphenol absorbent.

7.1.6 Phenol, Chlorofrom and Isoamylalcohol as Seperating Agents

DNA purification is usually conducted a two-step process, the first step involves treating the cell extracts with an organic agent like phenol or chloroform or mixtures of phenol, chlorofrom and isoamylalcohol. The second step involves centrifugation of the lysis mixture with the purpose of separating the DNA-containg aqueous phase from the organic phase, containing denatured protein and lipid contaminants. Phenol denatures proteins and eliminates cellular debris; chloroform also denatures proteins, and isoamylalcohol facilitates the separation from aqueous phase to organic phase and reduces foaming during centrifugation.

7.1.7 Ethanol as Precipitating and Purification Agent

Following the organic extraction step, DNA is precipitated by ethanol or isopropanol in the presence of sodium chloride, sodium acetate, or ammonium acetate. When the yield is high the precipitated DNA can be spooled out from the solution using a glass rod. Centrifugation is the preferred method when small amounts of DNA are precipitated. The DNA is usually rinsed with 70% ethanol to remove traces of salt and then dried.

7.1.8 NaCl as Macromolecular Separating Agent

Polysaccharides in plants are diversified compounds which vary with species and with different developmental stages and cause the majority of problems in regard to purity of extracted DNA. Carbohydrates in plants are clear and sticky in nature and can easily be identified by their viscous consistency in the solution. Younger tissues are preferred because they contain fewer polysaccharides than older or mature tissues. Exclusion of polysaccharides by using high concentrations of NaCl is an easy, quick and inexpensive method for plant DNA isolation.

7.1.9 RNAse as Purifying Agent to free DNA from RNA

During the DNA isolation a considerable amont of RNA is often extracted. RNA contamination can cause overestimation of the amount of DNA extracted when quantified. It usually appears as low molecular weight DNA in agarose gel electrophoresis. As RNA may cause suppression of PCR amplification and lead to mis-priming of DNA templates during thermal cycling, its removal from the samples is essential. For this the DNA samples are generally dissolved in low concentration TE buffer and then treated with RNAse at 37°C from 30 min to 1 h.

7.2 Equipments Needed

- Pestle and mortar
- Vortexer
- Liquid nitrogen container
- High speed centrifuge
- Auto-pippets (0.5-10 µl, 20-200 µl, 200-1000 µl)
- Waterbath
- Deep freezer (-20°C)
- Refrigerator

- Flourometer
- Gel documentations systems or UV transilluminator
- Gel electrophoresis appratus
- Uninterupted Power supply

Buffers, Reagents and Accessories needed

- 2mL centrifuge tubes
- Liquid nitrogen
- 0.5 and 1.5mL microcentrifuge tubes
- Tips (2-20 µL, 20-200 µL, 200-1000 µL)
- CTAB Buffer
- Isopropanol
- Saturated phenol
- Chloroform : isoamylalcohol (24:1) mixture
- 10:1 TE
- R-Nase A (10mg / ml)
- 70% ethanol
- 50x TAE buffer, pH 8.0

7.2.1 Procedure of CTAB Method (Saghai-Maroof et al. 1984)

- Weigh 1 g of clean young leaf tissue. Freeze the tissue rapidly in liquid nitrogen and grind to a powder with a mortar and pestle. It is important not to let the tissue thaw once frozen, until it is added to the DNA extraction buffer.
- Transfer the frozen powder to 2 ml centrifuge tube with 1ml CTAB buffer maintained at 60° C in a water bath. Mix vigorously or vortex.
- Incubate at 60° C for 30min with intermittent mixing.
- Fill the tube with chloroform: isoamyl alcohol (24:1). Mix gently by inverting for 5 min.
- Spin at 13,000 rpm for 10 min in a centrifuge at 25° C.
- Transfer aqueous phase to a fresh centrifuge tube. Add 0.7 vol of isopropanol (stored at -20°C) and let the DNA to settle down for 20 min.
- Spool out the DNA. Drain out the excess chemicals with a pipette.
- Add 0.5 ml of 70% ethanol. Mix gently and incubate for 30 min. Decant and repeat the 70% ethanol treatment. Decant off and dry the pellet.

- Dissolve DNA in minimum volume of 10:1 TE or distilled deionized water.
- Add RNAse (1 μl) and incubate at 37° C for one hour.
- Add equal volume of phenol: chloroform (1: 1), mix properly for at least 2 min and spin for 5 min. Take out the DNA supernatant and after this perform two chloroform: isoamyl alcohol extractions as before. Spin after each extraction.
- Precipitate DNA by adding 1/10 volume of 3M NaOAc and 2.5 times of the total volume chilled ethanol. Mix and spool out the DNA. Remove extra salts by two washings with 70% ethanol. Dry under vaccum.
- Add minimum volume of TE (10:1). Dissolve at room temperature. Store frozen at – 20° C.

Caution: the samples should be handled gently to avoid shearing of DNA

7.3 Quantity Assessment of DNA

DNA quantification is an important step in many procedures where it is necessary to know the amount of DNA that is present when carrying out restriction digests or performing different techniques such as PCR for RAPD or AFLPs or SSRs etc. There are several methods for quantifying DNA , the most widespread being: i) the comparison of an aliquot of the extracted sample with standard DNAs of known concentration using gel electrophoresis; ii) spectrophotometric determination, and iii) flourimetric determination

7.3.1 Electrophoresis of a DNA Sample of Unknown Concentration with a Known Standard

- Set the electrophoresis apparatus with combs in such a way that 2 mm gap is maintained between the bottom of the gel and the comb tip.
- Prepare a 0.8% (w/V) agarose gel suspension by heating in 1 x TAE buffer. Allow the gel to cool and then pour it onto the gel tray and allow it to solidify for at least 30 min.
- Remove the combs. Place the solidified gel into the electrophoresis tank and pour 1x TAE until the gel is completely covered.
- Mix 4 μl distilled deionized water, 1μl loading dye and 1μl DNA and load onto the gel.
- Load 2 ml of DNA standard of known concentration (25 ng, 50 ng, 100 ng) into the side wells.
- Connect the apparatus to power supply and run the gel at 60-70 V for 1 h or until the dye moves 3-4 cm from the wells.

- Stain the gel with ethidium bromide (stock of 10mg/mL) solution for 30-45 min. Alterntively, ethidium bromide to a concentration of 0.5 µg/ mL can be added in step 2 above.
- View the gel under UV light and photograph under UV light.
- Compare the intensity of the DNA bands of the samples with the intensity of the standard DNA bands. As the amount of DNA present in each standard DNA bands is known, the amount of DNA of each sample can be calculated by comparing the fluorescent yield of the sample with that of the standard.

7.3.2 DNA Quantification by Nanodrop Spectrophotometer

- Clean up the upper and lower optical surface of the microspectrophotomete sample retension system by 1 to 2µi of clean deionized water. Close the lever arm and tap it few times. Lift the lever arm and wipe off the surface.
- Open the nanodrop software and select the nucleic acid module.
- Initialize the spectrophotometer by placing 1µl clean water on to the lower optic surface and initialize in the nanodrop software (10 sec).
- Perform blank measurementby loading 1µl of buffer (TE) and select BLANK.
- Measure the nucleic acid sample by loading 1µl and select measure.
- Once the measurement is complete, clean both optical surfaces and repeat fpr other samples.

7.4 Polymerase Chain Reaction

The advent of Polymerase Chain Reaction (PCR) by Kary, B. Mullis revolutionized molecular biology. The PCR is a rapid procedure for in vitro enzymatic amplification of a specific segment of DNA. PCR has infinite applications which include direct cloning from genomic DNA or cDNA, in vitro mutagenesis and engineering of DNA, genetic fingerprinting of forensic samples, genotype identification, assays for the presence of infectious agents, prenatal diagnosis of genetic diseases, analysis of allelic sequence variations, analysis of RNA transcript structure, genomic footprinting, and direct nucleotide sequencing of genomic DNA and cDNA.

7.4.1 Composition of the PCR Reaction Mixture

A typical PCR reaction mixture contains template DNA, primers, $MgCl_2$, dNTPs and *Taq* DNA polymerase each of which is described below.

7.4.1.1 Template DNA

Usually the template DNA amount is in the range of 0.1-1µg for genomic DNA, for a total reaction mixture of 50µl. Higher template DNA amounts usually increase the yield of nonspecific PCR products, but if the fidelity of synthesis is crucial, maximal allowable template DNA quantities in conjunction with limiting number of PCR cycles should be used to increase the percentage of "correct" PCR products. Nearly all routine methods are suitable for template DNA purification. Although even trace amounts of agents used in DNA purification procedures strongly inhibit *Taq* DNA polymerase, ethanol precipitation of DNA and repetitive treatments of DNA pellets with 70% ethanol is usually effective in removing traces of contaminants from the DNA sample.

7.4.1.2 Primers

- PCR primers are usually 15-30 nucleotides in length. Longer primers provide sufficient specificity.
- The GC content should be 40-60%. The C and G nucleotides should be distributed uniformly within the full length of the primer. More than three G or C nucleotides at the 3'-end of the primer should be avoided, as nonspecific priming may occur.
- The primer should not be self-complementary or complementary to any other primer in the reaction mixture, in order to avoid primer-dimer and hairpin formation.
- The melting temperature of flanking primers should not differ by more than 5°C, so the GC content and length must be chosen accordingly.
- All possible sites of complementarity between primers and the template DNA should be noted.
- If primers are degenerate, at least 3 conservative nucleotides must be located at the primer's 3'-end
- Annealing temperature should be approx. 5°C lower than the melting temperature. If the primer is longer than 25 nucleotides, the melting temperature should be calculated using specialized computer programs where the interactions of adjacent bases, the influence of salt concentration, etc. are evaluated.

7.4.1.3 Magnesium Chloride ($MgCl_2$)

Since Mg^{2+} ions form complexes with dNTPs, primers and DNA templates, the optimal concentration of $MgCl_2$ has to be selected for each experiment. To few Mg^{2+} ions result in a low yield of PCR product, and too many increase the yield of non-specific products and promote misincorporation. Lower Mg^{2+} concentrations are desirable when fidelity of DNA synthesis is critical. The recommended range of $MgCl_2$ concentration is 1-4 mM, under the standard reaction conditions specified. If the DNA samples contain EDTA or other chelators, the $MgCl_2$ concentration in the reaction mixture should be raised proportionally.

7.4.1.4 Deoxy Ribonucleotide Tri Phosphates (dNTPs)

The concentration of each dNTP in the reaction mixture is usually 200µM. It is very important to have equal concentrations of each dNTP (dATP, dCTP, dGTP, dTTP), as inaccuracy in the concentration of even single dNTP dramatically increases the misincorporation level. When maximum fidelity of the PCR process is crucial, the final dNTP concentration should be 10-50 µM, since the fidelity of DNA synthesis is maximal in this concentration range. In addition, the concentration of $MgCl_2$ should be selected empirically, starting from 1mM and increasing in 0.1mM steps, until a sufficient yield of PCR product is obtained.

7.4.1.5 *Taq* DNA Polymerase

Usually 1-1.5 units of *Taq* DNA polymerase are used in 50 µl of reaction mix. Higher *Taq* DNA polymerase concentrations may cause synthesis of nonspecific products. However, if inhibitors are present in the reaction mix (e.g., if the template DNA used is not highly purified), higher amounts of *Taq* DNA polymerase (2-3 units) are helpful in obtaining a better yield of amplification products.

7.4.2 Temperature Cycling

There are nucleic acid segments viz., double-stranded DNA to be amplified and two single-stranded oligonucleotide primers flanking it. In addition, there is a a DNA polymerase, deoxyribonucleoside triphosphates (dNTPs), a buffer, and salts to assist in the process. The primers are added in vast excess compared to the DNA to be amplified. They hybridize to opposite strands of the DNA and are oriented with their 3' ends facing each other so that synthesis by DNA polymerase (which catalyzes growth of new strands 5→3) extends across the segment of DNA between them. One round of synthesis results in new strands of indeterminate length which, like the parental strands, can hybridize to the primers upon denaturation and annealing. These products accumulate only arithmetically with each subsequent cycle of denaturation, annealing to primers, and synthesis.

However, the second cycle of denaturation, annealing, and synthesis produces two single-stranded products that together compose a discrete double-stranded product which is exactly the length between the primer ends. Each strand of this discrete product is complementary to one of the two primers and can therefore participate as a template in subsequent cycles. The amount of this product doubles with every subsequent cycle of synthesis, denaturation, and annealing; accumulating exponentially so that 30 cycles should result in a 2^{28}-fold (270 million–fold) amplification of the discrete product.

7.4.2.1 Initial Denaturation

The complete denaturation of the DNA template at the start of the PCR reaction is of key importance. Incomplete denaturation of DNA results in the inefficient utilization of template in the first amplification cycle and in a poor yield of PCR product. The initial denaturation should be performed over an interval of 1-5 min at 95°C if the GC content is 50% or less. This interval should be extended up to 10 min for GC-rich templates. If the initial denaturation is no longer than 3 min at 95°C, *Taq* DNA polymerase can be added into the initial reaction mixture. If longer initial denaturation or a higher temperature is necessary, *Taq* DNA polymerase should be added only after the initial denaturation, as the stability of the enzyme dramatically decreases at temperatures over 95°C.

7.4.2.2 Denaturation

Usually 0.5-2 min denaturation at 94-95°C is sufficient, since the PCR product synthesized in the first amplification cycle is significantly shorter than the template DNA and is completely denatured under these conditions. If the amplified DNA has a very high GC content, denaturation time may be increased up to 3-4 min. Alternatively, additives facilitating DNA denaturation - glycerol (up to 10-15 vol.%), DMSO (up to 10%) or formamide (up to 5%) - should be used. In the presence of such additives, the annealing temperature should be adjusted experimentally, since the melting temperature of the primer-template DNA duplex decreases significantly. If additives are used, the amount of *Taq* DNA polymerase in the reaction mix should be increased, because DMSO and formamide, at the suggested concentrations, inhibit the enzyme approximately 50%. Alternatively, a common way to decrease the melting temperature of the PCR product is to substitute dGTP with 7-deaza-dGTP in the reaction mix.

7.4.2.3 Primer Annealing

Usually the optimal annealing temperature is 5°C lower than the melting temperature of primer-template DNA duplex. Incubation for 0.5-2 min is usually sufficient. However, if nonspecific PCR products are obtained in addition to the expected product, the annealing temperature should be optimized by increasing it stepwise by 1-2°C.

7.4.2.4 Primer Extension

Usually the extension step is performed at 70-75°C. The rate of DNA synthesis by *Taq* DNA polymerase is highest at this temperature (2-4 kb/min), and a 1min extending time is sufficient for the synthesis of PCR fragments up to 2 kb. When larger DNA fragments are amplified, the extension time is usually increased by 1min for each 1000 bp.

7.4.2.5 Number of PCR Cycles

The number of PCR cycles depends on the amount of template DNA in the reaction mix and on the expected yield of the PCR product. For less than 10 copies of template DNA, 40 cycles should be performed. If the initial quantity of template DNA is higher, 25-35 cycles are usually sufficient.

7.4.2.6 Final Extension

After the last cycle, the samples are usually incubated at 72°C for 5-15min to fill-in the protruding ends of newly synthesized PCR products. Also, during this step, the terminal transferase activity of *Taq* DNA polymerase adds extra A nucleotides to the 3'-ends of PCR products. Therefore, if PCR fragments are to be cloned into T/A vectors, this step can be extended to up to 30 min.

7.4.3 PCR-based molecular marker techniques

This section describes some of the basic molecular marker techniques like RAPD, ISSR, SSR and gene specific amplification that make use of PCR. After PCR one needs to separate the amplified DNA fragments using gel electrophoresis which is described separtely in next chapter.

7.4.3.1 Random Amplified Polymorphic DNA (RAPD)

Three main techniques fall within the category of PCR-based markers using arbitrary primers: RAPD, DAF and AP-PCR. MAAP is the acronym proposed, but not commonly used to encompass all of these closely related techniques. The random amplified polymorphic DNA (RAPD) technique is a PCR-based method that uses a short primer (usually 10 bases) to amplify anonymous stretches of DNA. With this technique, there is no specific target DNA, so each particular primer will adhere to the template DNA randomly. As a result, the nature of the obtained products will be unknown. The DNA fragments generated are then separated and detected by gel electrophoresis. Polymorphism of amplified fragments are caused by: (1) base substitutions or deletions in the priming sites, (2) Insertions that render priming sites too distant to support amplification, or (3) insertions or deletions that change the size of the amplified fragment.

7.4.3.1.1 Requirements

10X *Taq* poymerase buffer, 25 mM magnesium chloride, 2 mM dNTP's (mixture of dATP, dCTP, dGTP and dTTP), *Taq* poymerase (5 Units/µL), 10 µM RAPD primers, template or genomic DNA (10 ng/µL), sterile double distilled water, TAE buffer, agarose, gel loading dye, DNA size standard, ethidium bromide, tips, 0.5mL and 1.5 mL micro-centrifuge tubes, 0.2 mL PCR tubes, staining trays

7.4.3.1.2 Protocol

- Take out $MgCl_2$, dNTPs, primers and buffer from the deep freezer and let these thaw at room temperature. *Taq* DNA polymerase needs to taken out only when it is to be added and return immediately to into the deep freezer after use.
- For setting up PCR reaction, prepare master mix in 1.5 µL centrifuge tube of the following components except template DNA:

S. No.	Component	Stock concentration	Required concentration	Amount required for one reaction
1	Taq polymerase buffer	10x	1x	2.5 µL
2	$MgCl_2$	25 mM	3 mM	3.0 µL
3	Primer	5 µM	0.5 µM	2.5 µL
4	dNTPs	25 mM	0.2 mM	0.2 µL
5	Taq DNA polymerase	5 U/ µL	1 U	0.2 µL
6	Template DNA	10 ng/ µL	50 ng	5.0 µL
7	Sterile double distilled water (to make 10 µL volume)	-	-	11.6 µL
	Total			25.0 µL

Note: This RAPD cocktail is standardized for PCR with groundnut leaf DNA samples.

- Dispense 20 µL of above master mix in each of the 0.2 mL PCR tube.
- Add 5 µL of template DNA and mix thoroughly and gently by pipetting 3-4 times.
- Turn on the thermocyler and place samples in it.
- The following general thermocycling steps are followed:
- Step1: Initial denaturation at 94°C for 5.0 min
- Step2: Denaturation at 94° C for 1.0 min
- Step3: Primer annealing at 45-55° C for 1.0 min
- Step4: Primer extension at 72° C for 2.0 min
- Step5: Go to 2, 39 times
- Step6: Final extension at 72° C for 10 min
- Step 7: 4° C ∞
- After the run is completed turn off the machine and remove the samples.
- Add 2.5 µL gel loading dye in each sample and carry out electrophoresis or store samples at 4°C until electrophoresis.

7.4.3.2 Simple Sequence Repeats (SSRs)

PCR-based assays for SSRs have become the most popular and powerful of the current methods for genotype identification as they reveal highly polymorphic Mendelian markers. A SSR loci consists of reiterated short sequences (usually di-, tri-, or tetranucleotides) that are tandemly arrayed at a particular choromosomal location with variation in repeat copy number. A microsatellite array can be considered a reminiscent of a minisatellite array, except that each of its repeat units is much shorter. Therefore alleles at SSR loci can be distinguished in acrylamide (rather than agarose) gels.

7.4.3.2.1 Requirements

10X *Taq* poymerase buffer, 25 mM magnesium chloride, 2 mM dNTP's (mixture of dATP, dCTP, dGTP and dTTP), *Taq* poymerase (5 Units/ µL), 10 µM RAPD primers, template or genomic DNA (10ng/µL), sterile double distilled water, TAE buffer, agarose, gel loading dye, DNA size standard, ethidium bromide, tips, 0.5 mL and 1.5 mL micro-centrifuge tubes, 0.2 mL PCR tubes, staining trays

7.4.3.2.2 Protocol

- Take out $MgCl_2$, dNTPs, primers and buffer from the deep freezer and let these thaw at room temperature. *Taq* DNA polymerase needs to taken out only when it is to be added and return immediately to into the deep freezer after use.
- For setting up PCR reaction, prepare master mix in 1.5 µL centrifuge tube of the following components except template DNA.

S. No.	Component	Stock concentration	Required concentration	Amount required for one reaction
1	Taq polymerase buffer	10 x	1 x	1 µL
2	$MgCl_2$	25 mM	1 mM	0.4 µL
3	Primer(Forward Reverse)	20 pM	20 pM	1F+1 R µL
4	dNTPs	25 mM	0.2 mM	0.2 µL
5	Taq DNA polymerase	5000 U/ µL	0.15 U	0.03 µL
6	Template DNA	20 ng/ µL	40 ng	2.0 µL
7	Sterile double distilled water (to make 10 µL volume)	-	-	4.37 µL
	Total			10 µL

Note: This SSR cocktail is standardized for PCR with groundnut leaf DNA samples.

- Dispense 23 µL of above master mix in each of the 0.2 mL PCR tube.
- Add 2.0 µL of template DNA and mix thoroughly and gently by pipetting 3-4 times.
- Turn on the thermocyler and place samples in it.
- The following general thermocycling steps are followed:

  Step1: Initial denaturation at 95°C for 5.0 min

  Step2: Denaturation at 94° C for 1.0 min

  Step3: Primer annealing at suitable annealing temp. for 1.0 min

  Step4: Primer extension at 72° C for 2.0 min

  Step5: Go to 2, 39 times

  Step6: Final extension at 72° C for 8 min

  Step7: 4° C ∞
- After the run is completed turn off the machine and remove the samples.
- Add 2.5 µL gel loading dye in each sample and carry out electrophoresis or store samples at 4°C until electrophoresis.

7.5 Agarose Gel Electrophoresis

Agarose gel electrophoresis is the most effective way of separating DNA fragments of varying sizes ranging from 100 bp to 25 kb. Agarose is isolated from the seaweed genera *Gelidium* and *Gracilaria*, and consists of repeated agarobiose (L- and D-galactose) subunits. During gelation, agarose polymers associate non-covalently and form a network of bundles whose pore sizes determine a gel's molecular sieving properties. The use of agarose gel electrophoresis revolutionized the separation of DNA. Prior to the adoption of agarose gels, DNA was primarily separated using sucrose density gradient centrifugation, which only provided an approximation of size. To separate DNA using agarose gel electrophoresis, the DNA is loaded into pre-cast wells in the gel and a current applied. The phosphate backbone of the DNA (and RNA) molecule is negatively charged, therefore when placed in an electric field, DNA fragments will migrate to the positively charged anode. Because DNA has a uniform mass/charge ratio, DNA molecules are separated by size within an agarose gel in a pattern such that the distance traveled is inversely proportional to the log of its molecular weight. The leading model for DNA movement through an agarose gel is "biased reptation", whereby the leading edge moves forward and pulls the rest of the molecule along. The rate of migration of a DNA molecule through a gel is determined by the following: 1) size of DNA molecule; 2) agarose concentration; 3) DNA conformation; 4) voltage applied; 5) presence of ethidium bromide, 6) type of agarose and 7) electrophoresis buffer. After separation, the DNA molecules can be visualized under uv light after staining with an appropriate dye.

7.5.1 Requirements

Horizontal electrophoresis unit,TBE buffer, sterile double distilled water, agarose, gel loading dye, DNA size standard, ethidium bromide, tips, 0.5mL and 1.5 mL micro-centrifuge tubes, 0.2 mL PCR tubes, staining trays

7.5.2 Protcol

- To prepare 100 mL of a 1.5% agarose solution, measure 1.5g agarose into a glass beaker or flask and add 100 ml 1X TAE.
- Microwave or stir on a hot plate until agarose is dissolved and solution is clear.
- Allow solution to cool to about 55°C before pouring. (ethidium bromide can be added at this point to concentration of 0.5 μg/mL)
- Place the comb in gel tray.
- Pour 50° C gel solution into tray to a depth of about 5mm. Allow the gel to solidify for about 20 min at room temperature.
- To run, gently remove the comb, place the tray in electrophoresis chamber, and cover (just until wells are submerged) with electrophoresis buffer (the same buffer used to prepare the agarose).

- Load 25 µL of DNA per well. Load also the DNA size standards along side ISSR reactions.
- Connect the electrodes to the power pack and carry out electrophoresis at 50-150 Volts until the bromophenol blue dye has reached three fourth of the gel length.
- Stain the gel with ethidium bromide solution.
- Examine the gel under UV light in the gel documentation system.

Note: *Ethidium bromide is a mutagen and a probable carcinogen. Wear gloves when working with ethidium bromide solutions. Also use care not to contaminate the work area with the solution. UV light is damaging and must be used with caution. UV light causes burns and can damage the eyes.*

7.6 Polyacryl Amide Gel Electrophoresis

Polyacrylamide gels are chemically crosslinked gels formed by the reaction of acrylamide with bifunctional crosslinking agent such as N,N′-methylenebisacrylamide (Bis). The composition of the gel is given by % T, the total (w/v) concentration of acrylamide plus crosslinker, and % C, the (w/w) percentage of crosslinker included in % T. Polyacrylamide gels are polydisperse in structure, because Bis polymerizes with itself more rapidly than with acrylamide. For this reason, polyacrylamide gels consist of highly crosslinked, Bis-rich nodules linked together by sparsely crosslinked, relatively acrylamide-rich fibers.

7.6.1 Requirements

Vertical electrophoresis unit, TBE buffer, sterile double distilled water, 30% Acrylamide solution, 0.8% Bisacrylamide solution, 10XTBE buffer, 0.1% Ammonium persulphate, TEMED, Sterile double distilled water, gel loading dye, DNA size standard, ethidium bromide, tips, 0.5mL and 1.5 mL micro-centrifuge tubes, 0.2 mL PCR tubes, staining trays

7.6.2 Protocol

- Clean both plates outer (large) glass plate (Bio-RAD unit) and inner (small) glass plate thoroughly with warm water, detergent and finally rinse them with deionized water.
- Assemble the glass plates with spacers in gel caster.
- Prepare the gel solution with the desired polyacrylamide 9% percentagent to make 30 ml by adding 9 ml 30% Acrylamide/ Bisacrylamide solution (29 :1), 4.5 ml TBE buffer (10X), add freshly prepared 180 µl of Ammonium persulphate(10%), 40 µl of TEMED and 16.5 ml diodnized water and mix the contents gently by swirling to avoid bubbles

- Work quickly after addition of TEMED to complete the gel before the acrylamide polymerizes.
- Immediately insert the appropriate comb into the gel, being careful not to allow air bubbles to become trapped under the teeth. The tops of the teeth should be slightly higher than the top of the glass. Clamp the comb in place with bulldog paper clips. If necessary, use the remaining acrylamide gel solution to fill the gel mold completely. Make sure that no acrylamide solution is leaking from the gel mold.
- Allow the acrylamide to polymerize for 30-60 minutes at room temperature.
- Mix the DNA samples with the appropriate amount of gelloading buffer. Load the mixture into the wells using a micropipette equipped with a drawn-out plastic tip.
- Run the gel until the marker dyes have migrated the desired distance. Turn off the electric power, disconnect the leads, and discard the electrophoresis buffer from the reservoirs.
- Detach the glass plates. Lay the glass plates on the bench. Use a spacer or plastic wedge to lift a corner of the upper glass plate. Check that the gel remains attached to the lower (white) plate. Pull the upper plate smoothly away. Remove the spacers.
- Stain gels with EtBr or silver staining and xamine the gel under UV light or white ligh trespectively in the gel documentation system.

References

1. *Caetano-Anollés G, Bassam BJ and Gresshoff PM. 1991b. DNA amplification using very short arbitrary oligonucleotide primers. Bio/Technology 9:553-557.*
2. *Litt M and Luty JA. 1989. A hypervariable microsatellite revealed by in vitro amplification of a dinucleotide repeat within the cardiac muscle actin gene. Am. J. Hum. Genet. 44:397-401.*
3. *Saghai-Maroof MA Soliman KM, Jorgensen RA and Allard RW. 1984. Ribosomal DNA spacer-length polymorphism in barley: Mendelian inheritance, chromosomal location, and population dynamics, Proc Natl Acad Sci USA 81: 8014-8018.*
4. *Tautz D.1989. Hypervariablity of simple sequences as a general source of polymorphic DNA markers. Nucleic Acids Res. 17: 6463-6471.*
5. *Weber JL and May PE. 1989. Abundant class of human DNA polymorhisms which can be typed using the polymerare chain reaction. Am. J. Hum. Genet. 44: 388-396.*
6. *Welsh J and McClelland M. 1990. Fingerprinting genomes using PCR with arbitrary primers. Nucleic Acids Res. 18:7213-7218.*

Chapter 8

Quality Seed Production and Certification in Groundnut

E. Venkata Ramana and B. Gopal Reddy

Scientist (Plant Breeding), RARS, Tirupati and Professor and University Head (Agronpmy, RARS, Nandyal.

8.1 Introduction

Groundnut is the most important oilseed crop in the world. The production of groundnut is concentrated in Asia and Africa. It is rich in oil (36-54%) and protein (21-36%) and has a high energy value of 2363 KJ.100 g^{-1}. Seeds are used for oil extraction, as food, and as an ingredient in confectionery products. After extraction, the residual cake is processed largely for use as animal feed for milch cattle. The haulms are valued as fodder.

In groundnut, seed is the costliest and basic input and its quality is of vital importance which has a bearing on the economic yield obtainable on field scale. The quality of seed used by farmers determines the status of agriculture they practice and the fate of the farmer. Most of the groundnut varieties released during the past 20 years have not reached the farmers due to problems in the seed production chain. Saturation of vast area with quality groundnut seed (certified seed) under the present seed production system is impossible even at 5 per cent replacement rate. During 1995-96, the requirement of breeder seed was 7513 q but only 6097 q of seed could be produced. At a growth rate of 2.2 per cent, the requirement 12,666 q of breeder seed in 2020 AD is visualized. The public seed production network will not be able to meet this demand. It is also unlikely that private seed companies will enter into groundnut seed business because of profit margin. A possible solution is to involve the progressive farmers in quality seed production system. There is an urgent need to unburden the need of producing breeder seed of varieties which are old but are still in demand.

For maximum gain in breeder seed production in groundnut, use of both improved integrated agronomic crop management practices and then maximum purity of the improved varieties are required. They contribute to increased productivity individually, and they also act synergistically. As the new high yielding varieties do not reach the farmers immediately, the demand persists even for breeder seed of old varieties also. Hence, two different approaches have to be adopted in breeder seed production - one for the old varieties and the another for the new varieties. The full chain of seed categories can be dispensed with for the old varieties. Quality seed of these varieties can directly be produced in farmer's fields under the cooperation and supervision of the local and concerned groundnut breeders and all the agencies associated with groundnut seed production and distribution. The seed village concept can greatly reduce the burden of producing quality groundnut seed on the public sector system.

The non-availability of improved groundnut variety seed pods is a major constraint in most of the groundnut growing countries including our India. With little interest of private sector in groundnut seed enterprise due to reasons such as low seed multiplication ratio, bulky nature of the produce, quick loss of seed viability, high cost of transportation, low profit margin, self pollinated nature of groundnut crop, the task of making the seed pods of improved groundnut varieties available to farmers in required quantities and at the right price lies with the public sector seed agencies. Unfortunately, the public sector seed agencies have not been able to meet the demand of good quality seed of improved varieties of groundnut in many countries. There remains a large gap between the groundnut seed demand and seed supply resulting in low area coverage by the improved varieties in this crop. The problem is further accentuated by the low seed multiplication ratio in groundnut. The vertical seed replacement rate for groundnut in India during 1999-2000 was 6.15% (Tiwari, 2002). Unless enterprising and progressive farmers and NGOs come forward to take up formal/informal seed production, this situation is likely to remain unchanged. The seed crop requires a treatment different from that of a commercial crop hence knowledge on groundnut seed production practices and attendant crop husbandry is essential.

Cultivated groundnut (*Arachis hypogaea* L.) belongs to genus *Arachis* in subtribe *Stylosanthinae* of tribe *Aeschynomeneae* of family *Leguminosae*. It is a self-pollinated, tropical annual legume. At locations where bee activity is high, some cross-pollination can occur (Nigam et al., 1983). Cultivated groundnut has two subspecies, *hypogaea* and *fastigiata*, which in turn have two (var. *hypogaea* and var. *hirsuta*) and four (var. *fastigiata*, var. *vulgaris*, var. *peruviana*, and var. *aequatoriana*) botanical varieties. Each of these botanical varieties has different plant, pod, and seed characteristics (Krapovickas and Gregory, 1994). However, most of the commercially cultivated varieties belong to var. *hypogaea* (common name/market type: virginia or runner), var. *fastigiata* (valencia), and var. *vulgaris* (spanish) botanical variety groups. The main characteristics of these three botanical varieties are described below.

***Variety hypogaea*:** No floral axes or branches on main stem; alternating pairs of vegetative and reproductive axes on branches (alternate branching); inflorescence simple; vegetative branches moderate to profuse; primary branches longer than main stem; growth habit spreading, intermediate, or erect; usually two seeds per pod; pod beak not very prominent; seed size medium (runner market type) to large (virginia market type); testa color generally tan (red, white, purple, or variegated also exist); cured seed dormancy moderate; maturity medium to late.

***Variety fastigiata*:** Floral axes on main stem; irregular pattern of vegetative and reproductive branches with reproductive branches predominating on branches (sequential branching); inflorescence usually simple; vegetative branches sparse; primary branches shorter than main stem; growth habit upright; two to four seeds per pod; pod beak absent, slight, or prominent; seed size small to medium; testa color tan, red, white, yellow, purple, or variegated; cured seed dormancy little.

***Variety vulgaris*:** Floral axes on main stem; irregular pattern of vegetative and reproductive branches with reproductive branches predominating on branches (sequential branching); inflorescence compound; vegetative branches moderate; primary branches shorter than main stem; growth habit upright; mostly two seeds per pod (three seeds are rare); beak absent or present; seed size small to medium; testa color tan, red, white, or purple; cured seed dormancy little. New groundnut cultivars, resulting mostly from hybridization within and between subspecies, sometimes, have overlapping characteristics described above.

Groundnut seed is very delicate and is highly sensitive to different kinds of stresses before, during, and after harvest including storage. Factors affecting yield and seed quality in groundnut include cultural practices, diseases, insect pests, weeds, drought, temperature, and curing and drying after harvest. Prolonged cloudiness and high humidity during day times while drying is not desirable also becomes important in storage for seed viability. Seed quality is established in the field and cannot be improved after harvest. Setting up of management plans and cultural practices that produce high quality seed is very important in the groundnut seed enterprise. Currently, the seed multiplication ratio in groundnut ranges between 1:5 and 1:10. Under assured growing conditions and high input management, this multiplication ratio can be further increased. It is important to get a high pod yield in a seed crop to ensure lower seed prices for the purchasing farmers and more profit to seed growers.

8.2 Classes of Seed

In a formal seed production system, the following five classes of seed are recognized.

8.2.1 Nucleus Seed

The Nucleus seed is produced from the basic seed stock, Nucleus seed, or Breeder seed available with the originating breeder or institution under the direct supervision of the originating or a sponsored plant breeder following the progeny row method. True to type (representing diagnostic characteristics of a released variety selected for Nucleus seed production) plants are selected individually from the space planted basic seed stock (or Nucleus or Breeder seed plot). The number of selected plants will depend upon the quantity of Nucleus seed to be produced taking the multiplication ratio into account. These selected plants are studied for plant characteristics during the growing period in the field and for pod and seed characteristics after harvest only those plants which fully conform to the diagnostic characteristics of the variety under multiplication are retained individually. In the following season, these plants are space-planted in progeny rows and each progeny is again studied carefully during pre- and post-harvest, for diagnostic characteristics of the variety under multiplication. Any progeny deviating from these diagnostic characteristics is rejected. The selected progenies are then bulked to form Nucleus seed stock.

8.2.2 Breeder Seed

Nucleus seed is used to produce the Breeder seed, which is done under the direct supervision of the originating or a sponsored plant breeder. It is used to increase the Foundation seed and is not available for general cultivation. Because of its low SMR in groundnut, two stages of Breeder seed production are permissible in India. The Nucleus seed is multiplied to obtain Breeder seed Stage I, which in turn is multiplied to obtain Breeder seed Stage II.

However, it is essential to use only duly certified Breeder seed Stage I to produce Breeder seed Stage II. The Breeder seed crop is sown at normal recommended plant density.

8.2.3 Foundation Seed

This is the offspring of Breeder seed or occasionally the progeny of Foundation seed. The breeder and originating institution help to maintain genetic purity and identity of the Foundation seed conforming to the standards prescribed for this class of seed.

8.2.4 Registered Seed

This is the offspring of Foundation seed and is produced under agency regulations to maintain varietal purity and identity.

8.2.5 Certified Seed

This is the offspring of Foundation, Registered or occasionally Certified seed and is available to farmers for general cultivation. Certified seed produced from Certified seed is not eligible for further seed increase under certification. Seed of only notified varieties is eligible for certification. The agricultural universities, public and private sector seed producing agencies, authorized farmers' organizations, and registered individual farmers generally undertake Foundation, Registered, and Certified seed production programs. Each class of seeds has its own prescribed seed standards, which are to be met by the seed growers for certification. In an informal seed sector, enterprising farmers and small seed traders produce/procure seed of varieties released in the public domain and sell it to the farming community as 'Truthful seed'. The selling of 'Truthful seed' is allowed legally in India, but its purity and quality is not assured. In leguminous crops, where seed multiplication ratio is low and seed is bulky (such as groundnut, chickpea, pigeonpea, and others), the informal seed sector plays a significant role in making the seed available to local farmers.

8.3 Seed Certification Standards

A regulatory authority sets the seed certification standards for various classes of seed of different crops in each country. These standards could vary depending on the local situations. Nucleus seed represents the highest degree of purity and stringent standards, which relax as the seed category moves down to Certified seed. Breeder seed does not have any prescribed certification standards. However, as per the Indian Minimum Seed Certification Standards, "Breeder seed should be genetically so pure as to guarantee that in the subsequent generation i.e. certified Foundation seed conform to the prescribed standards". The following are some of the seed certification standards set by various regulatory authorities for groundnut in their jurisdiction. In developing countries, where non-availability of seed continues to remain a major constraint to promoting improved groundnut cultivars among the farmers, it may be advisable to relax some of these standards to stimulate the seed production chain in the formal seed sector.

8.4 Monitoring and Inspection

The Nucleus seed and Breeder seed do not come under the purview of a seed certification scheme. As such, there is no prescribed monitoring/inspection procedure for them. However, the breeder responsible for Nucleus seed production should ensure full conformity to diagnostic characteristics of the variety under Nucleus seed production and the highest purity standards of the seed. The breeder responsible for Breeder seed production should carry out a thorough inspection of the crop before and after flowering and at harvest to eliminate any unhealthy, abnormal and off-type plant. This will ensure genetic purity of the Breeder seed, which in the next generation would conform to the standards of Foundation class of seed. Production should ensure full conformity to diagnostic characteristics of the variety under Nucleus seed production and the highest purity standards of the seed. The breeder responsible for Breeder seed production should carry out a thorough inspection of the crop before and after flowering and at harvest to eliminate any unhealthy, abnormal and off-type plant. This will ensure genetic purity of the Breeder seed, which in the next generation would conform to the standards of Foundation class of seed. Monitoring/inspection is mandatory for certification of other classes of seed (Foundation, Registered, and Certified). A duly authorized seed certification agency of a State organizes the field and post-harvest inspections by a team of technically qualified personnel. A seed analysis report and results of a growout test, wherever prescribed, are taken into account before issuance of a certificate by the seed certificating agency.

8.5 Cultural Management of A Seed Production Crop of Groundnut

A seed crop requires more care and attention than a food crop. A healthy crop meeting the prescribed standards of seed certification is essential for a successful seed production program.

Season for seed production: In many countries, there is only one season suitable for groundnut cultivation in a year. However, in major groundnut growing areas in India and southeast Asia at least two groundnut crops can be grown in a year. In such a situation, it would be better to grow a groundnut seed crop in a season where productivity and seed quality would be high. In India, crop season × seed germination of the produce interactions has been reported in the past. Generally, the produce of the post rainy (Oct/Nov-April/May) season, when high temperatures prevail at the time of maturity, is reported to have lower germination. However, in a study at 30 locations in Tamil Nadu, (Natarajan 1996) did not find any germination difference between the kharif (rainy season, June/July-Sept/Oct) and rabi (post rainy) season produce of groundnut variety VRI 2. If proper field drying and curing procedures are followed (avoiding direct exposure of the pods to the sun when field drying), the germination of the produce will not be affected.

Selection and care of parental seed: The class of parental seed will depend on the class of seed one has opted to produce. It is essential that the appropriate class of certified parental seed is used to produce the desired class of certified seed. The parental seed should be obtained from the authorized sources and the attached certification seed labels should be preserved till the next seed crop is certified. The seed procured for sowing should be stored in a cool and dry place. The seed bags should be handled carefully. Rough handling and dropping of bags on the floor can damage the seed radicle (protruding part of the seed which gives rise to the primary root) resulting in seed decay, poor germination, and abnormal seedling growth. If the seed is not treated with fungicide and insecticide, it should be treated with them just 1 or 2 days prior to sowing. The seed treatment with recommended fungicide and insecticide should be carried out gently so as to avoid injury to seed radicle. Seed treatment ensures good germination and establishment of healthy seedlings, which leads to higher productivity in the crop. The seed rate depends on the recommended plant spacing for the region and 100-seed weight of the variety under multiplication. It could vary from 80 kg ha-1 to 150 kg ha-1 under Indian conditions.

Selection of field: A healthy field, well drained free from soil borne diseases like kalahasti malady and insect pests root grub and weeds, is essential for a seed production program. The field should be leveled and well drained and should preferably have sandy loam soil. It should also have irrigation facilities. Unless supplementary irrigation is available, seed production under rainfed conditions should be avoided. Do not select a field in which groundnut was grown in the past two seasons. This will avoid contamination of the seed crop from volunteer plants of previous crops. If this is impossible to follow, the variety under seed multiplication should be the same as was grown in the previous seasons with equivalent or higher class of certified seed. However, it is not advisable to grow groundnut after groundnut in the same field as it could lead to build up of diseases and insect pests in the soil. The field should also have easy accessibility for monitoring/inspections.

***Isolation distance*:** In general, natural cross pollination in groundnut is almost absent. However, at locations where bee activity is intense, depending on the variety and season, natural cross pollination can occur. Therefore, it is essential to have adequate isolation distance between varieties in seed production fields to help prevent contamination with pollen from other varieties and mechanical mixtures. In Florida and Georgia, the isolation distance for Certified seed is a minimum 15 m between two groundnut plots. For Foundation seed in Georgia (0.25 – 6.16% natural cross pollination), the minimum isolation distance required is 30 m. In India where natural cross pollination is almost negligible, an isolation distance of 3 m between varieties is required for all classes of Certified seeds. Isolation distance should be determined for each location, season, and variety depending upon the extent of natural cross pollination. It would be advisable to carry out natural cross pollination studies periodically, particularly with newly released varieties, to make necessary adjustments in isolation distance, if required.

***Shelling*:** Shelling of groundnut pods for seed should preferably be done manually. Manual shelling can avoid damage and splitting of seeds, which can happen in mechanical shelling. At the time of shelling, any seed, which is infected, damaged, or does not conform to shape, size, and color of the variety under seed production, should be removed. The seeds should be treated with appropriate fungicides and insecticides before sowing.

***Land preparation and sowing of the crop*:** The land should be well prepared to give a fine tilth required by the groundnut crop. Although, groundnut is sown on flat beds in many countries, it is preferable to grow the crop on raised beds as they permit ease of field operations, better water management, and avoid compaction of seedbeds. The recommended plant spacing and sowing depth should be followed to raise the crop. However, only one seed per hill should be sown so as to permit easy rouging, if needed. The crop should be sown at the optimum time and followed with irrigation to ensure uniform and better crop establishment. A field with uneven crop emergence may fail to qualify for seed certification.

***Nutrients application*:** Groundnut does well in soils which are rich in organic matter. The fertilizer doses would depend on the results of soil tests and the productivity levels targeted for the crop. They could vary from field to field. In addition to major nutrients N, P, and K, calcium (Ca) is a critical nutrient in producing high quality groundnut seed.

There must be enough Ca in the top 8-12 cm of soil after the peg enters it. Characteristics of Ca deficiency include a dark plumule, slow germination and emergence, low seedling survival, poor growth, and a weakened plant. In tests, seed germination averaged 92% when the Ca level was greater than 420 ppm but decreased to 52% when the Ca level decreased to 200 ppm. It is important to maintain near to neutral soil pH levels (as high or low levels may create either deficiency or toxicity of micro-nutrients) and Ca to K ratios of less than 3 to 1. Excessive K in the podding zone interferes with Ca uptake and results in pod rot and pops (unfilled pods). Regardless of soil tests, it is advisable to apply 150-200 kg ha-1 Ca to small-seeded varieties and 300-400 kg ha-1 to large-seeded varieties at the time of peak flowering. The gypsum/lime (depending upon soil pH) should be applied as side placement followed by light inter-cultivation to mix it with soil and remove weeds. It is also important to monitor Mg, B, and Zn levels closely in seed production fields. They can interfere with the availability of other nutrients and their toxicity or deficiency can affect plant growth and production.

***Water management*:** Appropriate water management is essential to achieve early germination, uniform plant establishment. If soil moisture is insufficient, the field should be irrigated before or soon after sowing to ensure uniform and rapid crop emergence. A light soil moisture stress during the vegetative phase (up to 3-4 weeks after sowing) does not affect yield but it can reduce vine growth enough to minimize diseases later. However, moisture stress during flowering can delay or inhibit flower formation. After flowering, peg penetration requires adequate moisture. Once active pegging and pod formation have begun (about 50-60 days after sowing), the pegging zone (the top 8-12 cm of soil) should be kept moist. This can be achieved by applying frequent but small amount of irrigation. The presence of moisture in the podding zone improves Ca uptake, which is essential for proper pod and seed development. A lack of water in this zone during pod addition and development results in more pops, more one-seeded pods, a less mature crop, and a lower Ca content in the seed, which in turn affects germination and seed quality. Water deprivation can reduce germination by 40%, particularly in Virginia types.Too much water also causes problems by promoting excessive vine growth, diseases, peg deterioration, low shelling out-turn and non-uniform maturity late in the season.

***Weed management*:** A seed production field should be weed free as weeds not only affect productivity and other field operations but also interfere with roguing and field inspections by the monitoring team.

***Plant protection*:** Diseases and insect pests affect crop productivity and the quality of produce (poor pod filling, low shelling out-turn, small seed size, shriveled seed, seed discoloration, seed damage, low germination etc). They also interfere with recording of proper field observations. Recommended plant protection measures against diseases and insect pests should be regularly followed during the cropping season.

***Roguing*:** Minimum two (preferably three) roguings should be carried out before harvest to remove off-type groundnut plants in the seed production field. At the seedling stage, weak, distorted, variegated, diseased, and out of the row alignment seedlings should be removed and destroyed. At the flowering stage, variants, not conforming to flower morphology, branching pattern, growth habit, and other diagnostic characteristics of the variety under seed multiplication should be removed from the field. Similarly, at the podding stage, based on peg morphology and other vegetative characters, the remaining off-types including late flowering plants should be removed. The last roguing is done on the harvested plants to remove plants with diseased pods and off-types based on pod and seed characteristics.

***Harvesting*:** The timing of harvesting is very critical as it can significantly affect the economic yield and the quality of seeds. Both pre-mature harvesting and over maturity can be harmful. When the crop is mature, the leaves start yellowing. The mature pods become reticulated and within it, seed is separated from the shell of the pod and the inside of the shell becomes dark in color. The crop should be harvested when (i) 70-75% (sometime this percentage can be slightly lower in the case of alternate branching types) of the pods are mature, (ii) leaves start yellowing and old leaves start shedding, (iii) pods becoming hard and tough with dark tannin coloration inside the shell, (iv) Kernel surface becoming smooth and (v) testa developing seed colour characteristics of the genotype. A few representative plants in the field should be uprooted and their pods should be studied to determine the optimum time for harvesting. After harvest, the plants should be left in rows for final roguing. Any off-type plant (based on pod and seed characteristics) and plants with diseased pods should be removed.

***Drying and curing*:** Initial curing of plants and pods is usually done by allowing them to dry in windrows under ambient temperature conditions. At the time of harvest, groundnut pods generally contain 35-60% moisture. If the highly moist (>30% moisture) and fresh pods are exposed to hot ans strong sunlight (temperature >40^0c) even for one day, the seed viability is affected. Until the moisture is reduced to <10%, the pods are prone to mould attacks. Plants are allowed to remain in the windrows until whole pod moisture drops down to 18-20% for mechanical threshing and to 15% for hand threshing. In places where high temperatures prevail at the time of harvest (post rainy season crop in India), direct exposure of pods to the sun should be avoided. This can be achieved by arranging the harvested plants in circular heaps with pods facing inside if a thresher is used to strip pods, it should be thoroughly cleaned before use for each variety. All the possibilities of mechanical admixing at the time of threshing should be eliminated. After threshing, the pods should be dried in a bin by forcing low humidity air to evaporate the excess moisture. Or else, they could be dried in the shade (but it would take longer). Exposure to high temperatures (above 45^0C) may affect the seed quality. The moisture content in unshelled groundnut should be brought down to 8% (in case of seed 6%) for storage. Only sound, mature, clean, and well-filled pods should be selected for seed.

Preliminary consideration in groundnut processing is to dry the seed pods to safe moisture level which would not impair seed viability during storage. The dried seed pods are cleaned in the processing plant by sieving, the soil and under size pods are small enough easily separated and removed. If any chaffy material and ill filled pods are present they are easily blown away and thus separated by pre-cleaning operations. The cleaned seed lot is then processed for size grading. Specific gravity separator is used to separate pods according to the weight of pod and size.

Packaging and storage: Except for the lot size, there are no specifications on packaging of groundnut seed pods in India. The cleaned and well-dried seed pod produce should be packaged in polythene-lined gunny bags with appropriate certification tags. Each breeder seed bag should be properly stitched along with the information on package of practices furnished in pamphlet kept inside the bag and then sealed. The details printed on different seed tags used in India are given in Fig . The color of the tag for Breeder seed is golden yellow, for Foundation seed white, and for Certified seed, azure blue. The seed producer should maintain a proper record of use of seed tags. Groundnut stores better in pods than in seeds. It is known to loose seed viability quickly if stored improperly. The storage temperature should be low. In general, the lower the temperature, the longer the expected storage life. Temperatures below 13°C inactivate most insects and arrest growth and influence of other seed quality deteriorating factors. Higher temperature >45^{0}C prevailing for longer periods leads to decrease in seed moisture consequently it will result in death of embryo.

Class of Seed BREEDER SEED

007894

A TAG Nº 309809

APSSCA HYDERABAD

Kind ....

Variety ....

Lot No ....

"Use of the seed after expiry of the validity period by any person is entirely at his risk and the holder of the certificate shall not be responsible for any damage to the buyer of seed. No one should purchase the seed if seal of the certification tag has been tampered with."

Name and Full Address of Certified seed Producer

CERTIFIED SEED

Class of Seed FOUNDATION ....

Certificate No. ....

Date of Issue of Certificate

Date of Test ....

Certificate Valid up to ....

(Provided seed is stored under cool and dry environment)

Seed Certification Officer (Name and Place)

Validity of Certificate further extended up to ....

Seed Certification Officer (Name and Place)

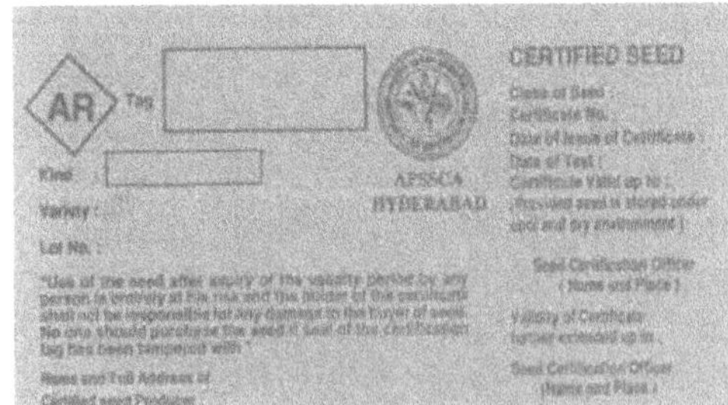

AR

CERTIFIED SEED

APSSCA HYDERABAD

(*Source:* Nigam SN, Giri DY and Reddy AGS. 2004).

The relative humidity (RH) should be between 65% and 70%. Mould growth is encouraged at higher RH levels. However, at RH below 65%, the groundnut pods loose weight and the seeds become brittle and split during handling. The RH, rather than seed moisture content per se, is the key criterion for many physiological events and microfloral associations during seed storage. If proper storage conditions (low temperature and low humidity) are not available, well dried, unshelled groundnut should be packaged in polythene-lined bags, and fused calcium chloride @ 250 g 20 kg pods^{-1} in a wide-mouthed bottle covered with thin muslin cloth should be placed in the center. The bags should be stored in a ventilated, clean, and dry store in such a way that the mouth of the bottle is upward in direction (Basu and Reddy,1989). The bags should be placed over wooden planks to avoid dampness from below and should not be stacked too high. In case of pest outbreak in storage, fumigation with Celphos (tablets) should be carried out strictly following the recommended procedure. It is important to follow proper sanitation in seed stores. Groundnut seed should not be carried over more than a year if it is to be sold as seed.

8.6 SEED Systems in India

***Formal seed system*:** The system of official release of crop varieties in India was formalized in 1964 with the establishment of the Central Variety Release Committee (CVRC) at the national level and the State Variety Release Committee (SVRC) at the state level in various states. The Government of India enacted the Seeds Act in 1966, which was implemented in 1968, after framing the Seed Rules (Yadav and Singhal, 2003). In 1969, the functions of CVRC were taken over by the then newly constituted Central Seed Committee (CSC), who advises the Central and State governments on matters arising from implementation of the Seeds Act. The Central Seed Sub-Committee on crop standards, notification, and release of varieties, appointed by the CSC, scrutinizes identification and release proposals forwarded by the Variety Identification Committee constituted by ICAR during the annual crop workshops. It recommends varieties to CSC for their release at the national or state level and notification. The State level varietal system of testing and release is almost identical to that of the Central system. The State Seed Sub- Committee can release a variety at the State level, but only the CSC can issue a notification of such releases. Only notified varieties can be certified under the formal seed production program. As per the Seed Rule of 1988, varieties of foreign origin may get provisional notification after one year of testing at 15-20 locations in a season. For regular notification however, two additional years of testing is required. As per the Seeds Act, certification (pre-marketing seed control) of seed is not essential as the Act states, "Any person selling, keeping for sale, offering to sale, bartering or otherwise supplying any seed of any notified kind or variety, may, if he desires to have such seed certified by Certification Agency, apply to the Certification Agency for the grant of a certificate for the purpose". However, for seed quality law enforcement (marketing control), labeling in the prescribed manner is essential. The person whose name appears on the mark or label is responsible for accuracy of the information required to appear on the mark or label. The Indian seed program largely adheres to the limited generation system for seed multiplication: i. Breeder seed, ii. Certified seed (Foundation seed and certified seed). Seed certification agency notified under the Indian Seeds Act or any certification agency established in any foreign country, provided the certification agency has been recognized by the Government of India through notification in the Official Gazette, can certify the Foundation and Certified seeds.

***Breeder seed production*:** The availability of Breeder seed in required quantities is essential to promote and sustain the seed production chain to popularize newly released and notified varieties among the farmers. Seed corporations, cooperative societies, and related organizations with the State Department of Agriculture can place their indent for Breeder seed requirement with the Seed Division, Department of Agriculture, Ministry of Agriculture and Cooperation, Government of India. Private seed companies and individuals can submit their requirements to the Seed Association of India who consolidate the requirements and place the indent with the Seed Division. These indents are then screened and complied variety-wise, state-wise, agency-wise and sent to the Deputy Director General (Crops), Indian Council of Agricultural Research (ICAR). The Deputy Director General (Crops), ICAR sends these consolidated indents to Project Coordinators of the respective crops for making arrangements for the production of Breeder seed. The assignment for production of Breeder seed is discussed in annual crop workshops of ICAR and then allotted to different research institutions/centers for production. Each participating center keeps the Project Coordinator and the Seed Division informed of the progress in Breeder seed production by sending information in prescribed Breeder seed production (BSP) proformas as described below.

***Table 8.1** Proforma for breeder seed production*

S. No.	Content	Issued by
1	Allocation of Breeder seed production (only notified varieties)	Project Coordinator of the crop
2	Schedule of production and expected availability of Breeder seed	Submitted by concerned producing breeder
3	Monitoring team report	Submitted by concerned producing breeder
4	Variety-wise Breeder seed production – Final report	Submitted by concerned producing breeder
5	Supply of Breeder seed to seed producing agencies identified by the Seed Division	Submitted by concerned producing breeder

(Adapted from Shanmugam, 2003)

Although the Breeder seed does not come under the purview of seed certification, a multidisciplinary team consisting of a plant breeder, a plant pathologist, an entomologist, and a scientist from the National Groundnut Research Centre, Junagadh, form the monitoring/inspection team for Breeder seed production. One representative each from the National Seed Corporation and the State Seed Corporation is also invited to join the team. At least two inspections from flowering to harvesting period are prescribed. The monitoring team generally inspects the Breeder seed crop 2-3 weeks before harvest to certify the conformity of the crop to diagnostic characteristics of the variety under multiplication and other prescribed standards of crop husbandry and isolation distance. Due to low seed multiplication ratio, sometimes two stages of Breeder seed multiplication (Stage 1 and Stage 2) are permissible for groundnut.

***Foundation and certified seed production*:** Foundation seed is generally produced at their own farms by the seed producing agencies. Like Breeder seed, two stages of Foundation seed multiplication are also permissible in groundnut. The Certified seed is generally produced in farmers' fields. The Foundation seed producing agencies identify farmers and enter into an agreement with them to produce Certified seed. These fields are regularly inspected by the seed producing agencies, who are responsible for getting the produce certified by the State Seed Certification Agency. An individual farmer can also produce Foundation and Certified seed at his/her farm by obtaining the appropriate class of parent seed (Breeder or Foundation seed) from authorized sources, maintaining the prescribed field and crop standards and getting the field inspected and produce certified by the State Seed Certification Agency. Farmers should retain the original tags of parent seed until all the inspections by the Seed Certification Agency are completed.

***Informal seed systems*:** The yearly requirement of Certified seed of groundnut in India is 1.5 million metric t. The public sector seed producing agencies meet only a small fraction (0.065 million metric t) of this requirement (Tiwari, 2002). Because of non-availability of seed of newly released varieties in required quantities, the old varieties such as TMV 2, and others continue to occupy in some parts of groundnut farming in India. Most of the farmers either save their own seed for the next season or buy it from local farmers and traders which often sell nondescript or mixed seed; thus resulting in low productivity of the crop. Unless farmers, NGOs, and the private sector come forward to procure breeder seed pods in a big way for further multiplication, the demand for Certified seed of groundnut will remain unfulfilled. As a consequence, crop productivity will remain low due to nonrealization of full returns on the investment made in groundnut improvement research. Groundnut scientists and policy makers have suggested several schemes to promote the informal seed sector to overcome the shortage of good quality seed and hasten the diffusion of improved varieties of groundnut among the farming community.

8.7 Groundnut Varietal Descriptions: Diagnostic Characteristics

A. Morphological traits

1. Growth habit
2. Branching pattern
3. Plant height and breadth (mm)
4. Stem pigmentation
   - (i) Present
   - (ii) Absent

5. Stem surface
   - (i) Glabrous
   - (ii) Sub-glabrous
   - (iii) Moderately hairy
   - (iv) Very hairy
   - (v) Woolly
6. No. of primary and secondary branches
7. Leaf character
   - (i) Size (small, medium or large)
   - (ii) Shape
   - (iii) Color
   - (iv) Hairiness of young leaflets
8. Flower color
   - (i) Standard petal
   - (ii) Standard petal markings
9. Peg pigmentation
   - (i) Present
   - (ii) Absent
10. Pod characters
    - (i) Pod beak
    - (ii) Pod constriction
    - (iii) Pod reticulation
    - (iv) Pod ridge
    - (v) Seeds/pod
    - (vi) Pod length (mm)
    - (vii) Pod breadth (mm)
11. Seed characters
    - (i) Seed length (mm)
    - (ii) Seed breadth (mm)
    - (iii) 100-seed mass (g)/ Seed size
    - (iv) Seed color
    - (v) Seed shape

B. Quality traits

1. Oil (%)
2. Protein (%)
3. Oleic / Linoleic (O/L) fatty acid ratio

C. Reaction to biotic stresses

1. Foliar diseases
2. Soil borne diseases
3. Virus diseases
4. Insect pests

D. Reaction to abiotic stresses

1. Drought
2. Others

E. Post-harvest seed dormancy

1. Present
2. Absent (specify duration)

F. Growing season and area of adaptation

References

1. *Basu MS and Reddy PS. 1989. Technology for increasing groundnut production. National Research Centre for Groundnut (NRCG), Indian Council of Agricultural Research (ICAR), P.O. Timbawadi, Junagadh 362 015, Gujarat. Publications & Information Division, ICAR, New Delhi. pp 18.*
2. *Krapovickas A and Gregory WC. 1994. Taxonomia dol genero Arachis (Leguminosae). Bonplandia VIII: 1-187.*
3. *Natarajan S. 1996. Influence of season and provenance on quality of groundnut seed (Arachis hypogaea L.). M.Sc. (Ag.) Thesis, Tamil Nadu Agricultural University, Coimbatore 641 003, India. pp. 160.*
4. *Nigam SN, Rao V Ramanatha and Gibbons RW. 1983. Utilization of natural hybrids in the improvement of groundnuts (Arachis hypogaea). Expl Agric., 19:355-359.*
5. *Shanmugam, C. 2003. Regulatory system for Breeder seed production in India. Pages 65-72 in Hybrid seed production in field crops (Principles and Practices) (Singhal NC ed.). Kalyani Publihsers, New Delhi.*
6. *Nigam SN, Giri DY and Reddy AGS. 2004. Groundnut seed production manual. Patancheru 502 324, Andhra Pradesh, India: International Crops Research Institute for the Semi-Arid Tropics. 32 pp.*
7. *Tiwari S Prakash. 2002. Moving towards integrated seed supply systems in India. Pages 61-71, in Proceedings of the XI National Seed Seminar on Quality Seed to Enhance Agricultural Profitability (Shekhargouda M and Naik L Krishna eds.), 18-20 January 2002, University of Agricultural Sciences (UAS), Dharwad. UAS, Dharwad.*
8. *Yadav SP and Singhal NC. 2003. Indian regulatory system for plant variety testing, release and notification. Pages 43-50 in Hybrid seed production in field crops (Principles and Practices) (Singhal NC ed.). Kalyani Publihsers, New Delhi.*

Chapter 9

Role of Genomics in Groundnut Improvement

D. M.Reddy, K. H. P. Reddy, M. Reddi Sekhar, N. V. Naidu and N. P. Eswar Reddy

Dept. of Genetics and Plant Breeding, S.V. Agril. College, Tirupati-517 502.

9.1 Introduction

Groundnut, also called as peanut (*Arachis hypogaea* L. 2n = 4x = 40), is the third most important oilseed crop in the World. This crop is cultivated extensively throughout the semi-arid tropics of Asia, Africa and Latin America in an area of 25.5 million ha with its annual production of 45.3 million tons and productivity of 1780 kg/ha. In India it is grown in an area of 5.3 million ha with a production of 9.5 million tons and productivity of 1804 kg/ha(Anonymous, 2013-14). Groundnut is generally used for oil, confectionary purpose and as a fodder for cattle. It contains 40 to 50% oil, 20-40% protein and 10-20% carbohydrates. In addition to these, groundnuts are also a good source of calcium, phosphorus, iron, zinc, boron, vitamin E, small amounts of vitamin B complex, antioxidants, biologically active polyphenols, flavonoids, and isoflavones. Realising its importance in the indian diet, plant breeders through conventional breeding approaches made significant contributions in breeding several improved varieties with high yield, different maturity durations, resistance to various diseases and pests, tolerance to drought, enhanced oil content, improved quality traits etc., However, keeping in view the future demands of groundnut production and predicted problems like decreasing arable land, increasing trend of pest and diseases, intolerance to abiotic stresses (as this crop is grown mainly in Semi Arid Tropics with 70% of its total production), quality problems and climatic change etc., there is an immediate need to increase the productivity levels using the advances in the molecular biology and genetics. In addition, as this crop also suffers with low genetic variability and the reservoir of new alleles present in wild species remains under-utilized and expected progress could not be achieved, either due to lack of reliable, precise and cost effective high through put phenotyping or due to fertility barriers that hampers harnessing the genetic variation present in secondary and tertiary gene pools. Hence, there is an urgent need to exploit various options to genetically improve the groundnut crop suitable for various niches and objectives, which requires thorough understanding of the groundnut genome. Among several viable options, genomics could be considered as the most important one (Varshaney et al., 2014). The recent understanding on the genome structure of cultivated and wild *Arachis* has brought new insights for gene occurrence and evolution in the genome context. In parallel, functional genomics has elucidated expression profiling and regulation of *Arachis* genes in a global genome network context. This knowledge will facilitate and broaden the use of the wild *Arachis* extensive genetic diversity for peanut improvement, by either introgression using interspecific hybridization or transgenic approaches (Miranda Brasileiro *et.al.,* 2014). Despite, several advances in genomics have great potential to play a pivotal role in crop improvement, some of the important issues have been discussed in the present discussion.

9.2 Groundnut Genome

Groundnut genome is an allotetraploid with an AB type genome ($2n = 4x = 40$). The origin of cultivated groundnut was probably through a few or even a single hybridization event between two diploid wild species, *A. duranensis* (A-genome) and *A. ipaensis* (B-genome), followed by a spontaneous chromosome duplication (Halward et al. 1993 and Seijo et.al., 2007) and has very low DNA polymorphism, a characteristic that has hampered genetic studies. Consequently, homeologous A and B genic sequences have very high sequence identity. Genetically, cultivated peanut behaves as a diploid. Due to its tetraploid character, peanut has been reproductively isolated from other wild species, which are mostly diploid, thus showing a limited genetic variability. Peanut is an under represented crop in terms of genome sequencing and structural genomics studies comparing to model plants or economically important crops, such as *Arabidopsis*, rice and soybean. Fortunately, a recent international joint initiative (The International Peanut Genome Initiative – IPGI (*http://www. peanutbioscience.com*) was succeeded in releasing the draft genome sequence from *A. hypogaea* and its diploid parental (*A. duranensis* and *A. ipaensis*) genomes. This effort will increase the genome structural knowledge of *Arachis* genus as a whole (Miranda Brasileiro *et.al.,*2014). The genome of peanut is estimated at about 2.8 Gb and with a high repetitive DNA content. Where as *A. duranensis* and *A. ipaensis* are estimated to have a similar size, which is almost half (1.3-1.4 Gb) of the peanut genome. Scott Jackson, director of the University of Georgia Institute of Plant Breeding, Genetics and Genomics at the College of Agricultural and Environmental Sciences, serves as chair of the International Peanut Genome Initiative.

9.3 Genetics to Genomics

Ever since, understanding the basic principles of Genetics-" the science which deals with the study of heredity and variation" in the early 20th century, plant breeding methodologies like introduction, selection, intra and inter specific hybridization and mutation breeding etc., made significant advances through release of several elite cultivars for groundnut crop improvement. However, the limited variability in groundnut remained a bottleneck for full exploitation of the genetic potential. Fortunately, the advances in molecular biology coupled with computing technology made paradigm shifts in biology research and helped to originate new subject called genomics – "the branch of molecular biology which deals with the structure, function, evolution, and mapping of genomes". It provided almost a holistic understanding of genes and proteins in different organisms instead of single traits or genes and simultaneously led to tremendous advances in understanding the crops in totality. Besides understanding the crops in totality, this subject has also provided new tools and methodologies to the breeders for accelerated plant breeding. Therefore, within a "genome to gene" approach, structural and functional genomic data integration will allow not only gene discovery, but the understanding of gene structure, function, modulation and their importance for the control of agronomical traits. While, classical genetics revolutionized plant breeding at the beginning of the 20th century, genomics is leading to a new revolution in plant breeding at the beginning of the 21thcentury (Perez-de-Castro, 2012).

9.4 Role of Genomic Approaches for Groundnut Improvement

9.4.1 Strengthening Germplasm

For any crop improvement programmes, gene pools play a pivotal role. Hence enrichment of germplasm through genetic enhancement or pre-breeding activities is highly essential. As cultivated groundnut is originated very recently in evolutionary time, and has a narrow genetic variation as compared to other crop species, tapping available genetic diversity is essential for crop improvement (Pandey et al., 2012). Broadening the genetic diversity in *Arachis* species, though conventional breeding approaches made reasonable progress, is hampered due to limited genetic variability. Application of genomic approaches through gene introgressions from wild species or from alien sources paved the way for bringing new genes in to the existing gene pools.

In addition to germplasm collections that represent naturally-occurring variation, induced mutant collections for tetraploid peanut also are being developed. One of such technique is called TILLING (Targeting Induced Local Lesions In Genomes), which is a reverse genetic approach used to identify rare mutations for understanding gene functions. In peanut over 3400 mutant (M_2) lines generated by ethylmethane sulfonate (EMS) treatment have been used to screen for mutations in six allergen genes (the homeolog pairs *Ara h 1.01/1.02*, *Ara h 2.01/2.02* (encoding major allergen proteins) and *FAD2A/B* (encoding delta-12-desaturase, the enzyme primarily controlling oleic to linoleic acid ratio in seeds) using the technique of TILLING (Knoll et al. 2011). Similarly, (Yufang Guo et. al., 2015) used TILLING by sequencing approach in peanut root cDNA library and identified three candidate LOX genes. The gene AhLOX7 was selected for TILLING due to its high expression in seeds and roots. By screening 768 M_2 lines from the TILLING population, four missense mutations for AhLOX7, three missense mutations for AhPLD, one missense and two silent mutations for Ara h 1.01, three silent and five missense mutations for Ara h 1.02, one missense mutation for AhFAD2B and one silent mutation for Ara h 2.02 were identified. With these findings the authors opiened that these mutations identified could be used to further study gene function and have potential usefulness in breeding programmes.

Advances in genomics with clear understanding of the genomes, generation of synthetic genomes were also made possible. Cultivated groundnut is thought to be of monophyletic origin, harboring relatively little genetic diversity. To introduce variability from diploid wild species into tetraploid cultivated *Arachis hypogaea*, a synthetic amphididiploid was used as donor parent to generate a backcross population of 78 progenies (Burow et al., 2001).

9.4.2 Discovery of New Genes or Candidate Genes that Could Confer Important Traits

Advances in functional genomics made possible for identifying new genes or candidate genes for traits of varied interests. However, the use of functional genomics for discovering genes that confer new interesting traits in cultivated groundnut is not an easy task due to several factors such as polyploidy, large genome size and abundance of repetitive sequences (Stalker et.al. 2009). Therefore, the use of diploid wild relatives as source of genetic variability has a strategic role to play as they can offer new unexploited genes useful for the improvement of peanut cultivars (Miranda Brasileiro, 2014). When compared with model species or major crops, peanut and its wild relatives have little transcriptome information and so for, the identification and studies on genes of agronomic interest have been quite limited to groundnut from the 274,910 ESTs currently available on public databases, of which 205, 442 are from *A. hypogaea*; 35,291 from *A. duranensis*; 32,787 from *A. ipaensis*; 6, 264 from *A. stenosperma;* 750 from *A. magna and* 400 from *A. appressipila.*(http://www.ncbi.nlm.nih.gov/; verified on 12th july, 2015). Most of the initial functional studies in *Arachis* have concentrated on the generation of cDNA libraries and their subsequent partial (ESTs) or complete (full length cDNA) sequencing of which expressed sequences were used for select directly candidate genes or to produce oligonucleotide microarrays. Some of these studies aimed at identifying plant genes preferentially expressed in specific organs or plant tissues (Miranda Brasileiro, 2014). (Wu et al., 2010), used subtraction at mRNA level to identify uniquely expressed genes in peanut immature pod and leaf, whilst (Li et al., 2010), sequenced full-length cDNA libraries to compare the expression of acetyl-CoA carboxylase, an enzyme involved in fatty acid biosynthesis, in developing seeds and other tissues.

Another advantage of having a large number of ESTs is to construct the microarrays, which could be used to study differential expression of genes globally at various growth stages in a large set of genotypes. For example, (Payton et al., 2009) used micro array expression profiling in a variety of peanut tissues to understand the regulatory role of microRNAs (miRNAs), which are known to control gene expression at the post-transcriptional level in peanut, and identified 14 novel miRNA families as well as 75 conserved miRNAs. Similarly genomic techniques were used to identify genes involved in resistance to late leaf spot *P. personata* (Luo et al., 2005a); differentially regulated genes in response to aflatoxin-producing fungus *A. parasiticus* (Luo et al., 2005); differentially expressed genes, including those involved in ethylene and jasmonic acid signal transduction pathways, from both peanut leaves and roots challenged with the bacterial pathogen *Ralstonia solanacearum* (Huang et al. 2012); drought tolerant genes (Ranganayakulu et al.,2012 and Dang et al., 2013) and genes for root knot nematode tolerance (Tirumalaraju et al., 2011). As these transcriptome studies by generation of EST's have become routine in the recent past, it further paved the way to play a key role in *in silico* gene discovery and annotation and is highly demanding for the present day groundnut breeding programmes. For example, (Mondel et al., 2012), mined 5,184 EST's from NCBI databse and found EST's closely linked to rust resistance [R] gene in groundnut and further developed several genic molecular markers linked to rust resistance in groundnut.

9.4.3 Identification and Development of New Class of DNA Markers

In groundnut crop improvement programmes, conventional breeding programmes which involves crossing whole genomes followed by selection of superior transgressive segregants helped to made significant success. But the whole success of these efforts were completely dependent on the availability of the morphological markers that are limited in nature and highly dependent on environment. However, advent of DNA marker technology offered several types of new markers to the plant breeders to limit these problems faced during conventional breeding progrmmes besides helping in preparation of saturated linkage maps. These markers are now widely being used in ground nut to track loci and genome regions in several crop-breeding programmes. Though hybridization-based markers such as restriction fragment length polymorphism (RFLPs) were used in the beginning, with the advances in PCR-based technologies, several markers like random amplification of polymorphic DNA (RAPD), amplified fragment length polymorphism (AFLP) and microsatellite or simple sequence repeats (SSRs) have become popular and became the choice of the breederes as these are simple and fast to operate with. However, with the advances in genomics and next-generation sequencing technologies sequence-based markers like genic molecular markers (Varshaney et al., 2007), single nucleotide polymorphism (SNP), DArT (Diversity Arrays Technology) and genotype-by-sequencing techniques etc., are now becoming more popular in ground nut crop improvement programmes. The details of these markers have been discussed elaborately in seperate chapter in the same series.

9.4.4 Development of Transgenics

Advances in the genomics has helped to identify new genes across the boundaries of the living organisms, which could be highly useful not only for cultivar development/ improvement, but also to import new diversity in to the crop since ground nut suffers with low polymorphism due to highly conserved genome. Several workers attempted to incorporate a wide variety of genes from various sources through recombinant DNA technology and achieved success in effectively addressing the current problems in ground nut crop improvement programmes. Thus the transgenic development in groundnut proved to be very useful for the development of groundnut varieties with superior performance. But, till now there is no transgenic groundnut released in India for commercial cultivation. Despite, several debates over pro and against GM crops, transgenic groundnuts have been made with resistance to various biotic and abiotic stresses besides improved grain quality. Nevertheless, such examples of gene transfers are endless to mention, some are presented in the Table 1.

Table 9.1 *List of some tansgenes used for development of transgenics in groundnut*

S. No.	Transgene	Trait	Reference
1.	AtDREB1A	Drought and Salinity stress	Tanmoy Sarkar et.al., 2013.
2.	Mustard defensin	Late Leaf spot resistance	Swathi Anuradha et al., 2008.
3.	Synthetic cry1EC gene	Resiatance aganest *Spodoptera litura*	Tiwari et al., 2008.
4.	sodium/proton antiporter gene (AtNHX1)	Salt tolerance	Banjara et.al., 2012.
5.	Bcl-Xl gene	Herbicide paraquat tolerance	Chu et al., 2008.
6.	viral nucleocapsid gene	Peanut bud necrosis virus	Rao et al., 2013.
7.	Coat protein-mediated gene	Peanut stem necrosis disease	Mehta et al., 2013.
8.	PR genes	Late leaf spot disease	Vasavirama and Kirthi, 2012.
9.	Tobacco ß 1–3 glucanase	Resistance to *Cercospora arachidicola* and *Aspergillus flavus*	Sundaresha et al., 2010.
10.	Bacterial *mtlD* gene coding for mannitol 1-phosphate dehydrogenase.	Drought tolerance through accumulation of mannitol	Tengale et al., 2014.
11.	SbASR-1 Gene	Salinity and Drought Endurance	Tiwari et al., 2015.

9.4.5 Aids in Construction of Genome Maps

Genome maps are pre-requisites for any crop improvement programmes and are useful for scanning the genomes for the trait of interest. In recent years, many efforts have been made to construct linkage maps, which can form the genetic basis for quantitative trait locus (QTL) analyses and marker assisted selection of important and complex traits.

9.4.5.1 Linkage Maps or Genetic Maps

In groundnut using RFLP markers, first genetic map was developed by (Halward et al., 1993) from F_2 population developed from a cross of *A. stenosperma* × *A. Cardenasii.* Later, with the advances in PCR techniques and genomics tools, SSR markers (Varshney et al., 2007 and Ravi et al., 2011) became the choice of inkage map construction. However, the maps constructed using RFLP and SSRs generally present a low density and are difficult to provide precise information on the QTLs controlling the traits of interest. Hence, of late, with the advances in the NGS (Next Generation Sequencing) technologies, SNP markers are also being used extensively for construction of genetic maps. (Zhou et al., 2014).

9.4.5.2 Physical Maps

Recently international joint initiative (The International Peanut Genome Initiative – IPGI) succeeded in declaring the draft sequence of the entire genome of *A. hypogaea* and its diploid parents *A. duranensis* and *A. ipaensis*. This effort will increase the knowledge of *Arachis* genus as a whole and in the coming years is expected to bring low-cost genotyping to the groundnut community which will facilitate use of modern genetics and breeding approaches to breed more productive and resilient groundnut varieties.

9.4.6 Mapping QTLs and Genes

9.4.6.1 QTL Mapping

In groundnut, molecular marker technologies greatly helped to prepare linkage maps followed by mapping genes and QTLs for several traits of economic importance. However, the advances in genomics and sequencing technologies will aid in discovering new types of markers which will play important role in mapping QTLs and genes. In groundnut QTL analysis was used for identification of QTLs and genes for several important traits such as yield, yield contributing traits, disease and pest resistance, drought tolerance related traits, oil and nutritional quality traits etc., which are reviewed in detail by Varshaney et.al., 2007 & 2014, (Holbrook et al., 2011) and Pandey et al., 2012. Confirmation of markers tightly associated to the important traits and genes will enable the integration of molecular breeding in groundnut improvement.

9.4.6.2 Association Mapping

Association mapping is a method of mapping quantitative trait loci (QTLs), which involves the correlation of phenotypes to genotypes in unrelated individuals. The linkage maps, which discussed above though utilized for mapping QTLs, banks upon family-based mapping efforts and generally helps in identification of few quantitative trait loci

(QTLs) for simply inherited traits with major phenotypic effects. However, as several QTLs for complex traits with low phenotypic effect are most commonly observed in crop species like groundnut, family-based trait mapping approach has several limitations such as inability to address multiple agronomic traits using single population, time-consuming population development process, use of low density genetic maps, low QTL resolution and over estimation of phenotypic effect of QTLs (Gupta et.al., 2005). Moreover, the association mapping techniques are relatively more rapid and cost-effective than the traditional linkage mapping. Since majority of the agronomically important traits are quantitative in nature, association studies with genomewide marker coverage which allow high resolution mapping of such traits by exploiting historical recombination may enhance the efficiency of candidate gene identification and facilitate genomics-assisted breeding (GAB) for complex traits (Varshney et.al., 2012). Moreover, the association mapping techniques are relatively more rapid and cost-effective than the traditional linkage mapping. Pandey et al. 2014, generated high resolution genotyping and multiple season phenotyping data for 50 important agronomic, disease and quality traits on the 'reference set' of peanut and reported comprehensive analyses of allelic diversity, population structure, linkage disequilibrium (LD) decay and marker-trait association (MTA) in peanut. Detailed analysis from their study identified a total of 524 highly significant MTAs (pvalue .2.1610–6) with wide phenotypic variance (PV) range (5.81–90.09%) for 36 traits. They also opined that these MTAs after validation might be deployed in improving biotic resistance, oil/seed/nutritional quality, drought tolerance related traits, and yield/yield components.

9.4.7 Increased Selection Efficiency

9.4.7.1 Marker Assisted Selection

The use of Marker-assisted selection (MAS) i.e. utilization of molecular markers which can be traced to breed for traits of interest that are difficult to observe phenotypically.

The MAS is very important molecular breeding approach for crop improvement. The first groundnut variety developed through integrated marker-assisted selection (MAS) in the USA is NemaTAM, a root-knot nematode-resistant variety (Simpson et al., 2003). Later several other traits viz., disease resistance, pest resistance and quality traits have been incorporated through this approach in groundnut, which have been reviewed elaborately by several authors. (Pasupuleti et.al., 2013 and Pandey et al., 2014). This approach is also highly effective for gene pyramiding in groundnut.

9.4.7.2 Genomic Selection

Genomic selection (GS) was proposed by (Meuwissen et al., in 2001). Genomic selection is a form of MAS (Marker Assisted Selection), where the whole genome covered by genetic markers are used for estimating the breeding value of selection candidates for a quantitative trait. By applying genomic selection, the selection cycle for selecting elite lines can be shortened and it results in greater gains/unit time and cost (Heffner et al., 2010).

The use of marker assisted selection (MAS) in groundnut crop improvement as discussed in the earlier section showed promising results and will no doubt continue to help the breeders in the future to breed the traits that are difficult to observe them phenotypically. However, the MAS methods, hat are being used currently are better suited for manipulating a few major effect genes than many small-effect genes. But, the fact is that most of the polygenic traits like yield etc., which are pivotal for success of improving the crop varieties are controlled by these small effect genes. Two primary limitations to MAS are (i) the biparental mapping populations used in most QTL studies do not readily translate to breeding applications and (ii) statistical methods used to identify target loci and implement MAS have been inadequate for improving polygenic traits controlled by many loci of small effect. The application of genomic selection (GS) proposed by (Meuwissen et al., 2001) to breeding populations using high marker densities is emerging as a solution to address both of these deficiencies. In groundnut also, to exploit the recent advances in groundnut genomics to improve complex traits such as drought tolerance and seed yield, efforts have also been initiated at ICRISAT to apply GS using about 300 advanced breeding lines (Pasupuleti et.al., 2013).

9.4.8 Bio-informatics Resources and In-silico Prediction Analysis

Genomic technologies are generating an extraordinary amount of biological information, unprecedented in the history of biology, which led to emerge a new scientific discipline called bioinformatics, at the intersection between biology and computation. Bioinformatics addresses the specific areas in data acquisition, storage, analysis and integration that research in genomics generates. These resources once developed are used for genome analysis, genome comparisions, identification and prediction of genes, gene models, proteins and markers etc., and such knowledge would be greatly utilized while handling the complex traits in crop improvement programmes. Peanut Base (*http://www.peanutbase.org*) is one of such resource supporting with genetic and genomic data aimed for rapid crop improvement in ground nut.

9.5 Summary

In groundnut, the conventional plant breeding methods made significant contributions in breeding several improved varieties with high yield, different maturity durations, resistance to various diseases and pests, tolerance to drought, enhanced oil content, improved quality traits etc., However, keeping in mind the future demands of groundnut production and predicted problems like decreasing arable land, increasing trend of pest and diseases, intolerance to abiotic stresses, quality problems and climatic change etc., there is an immediate need to exploit non-conventional options to genetically improve the groundnut crop suitable for various niches and objectives. Among several viable options, genomics could be considered as the most important one. Recent advances in the area of crop genomics have offered molecular tools to assist breeding. The recent release of the draft genome of cultivated and wild *Arachis* has brought new insights for crop improvement. The genomics knowledge would play a pivotal role in understanding the genomes, genome maps; to identifying new genes, markers and to design improved selection procedures. Furthermore, with the rapid advances in the next generation sequencing technologies, molecular genetics and bioinformatics, the field of genomics provide novel breeding approaches (also referred to as "genomics-assisted breeding" approaches) for "holistic" groundnut crop improvement. However, novel computing methodologies and bioinformatics resources are highly essential to exploit full potential of genomics in groundnut crop improvement programmes.

References

1. *Anonymous 2014. Status paper on oil seeds, Ministry of Agriculture, Govt. of. India.*
2. *Banjara M, Zhu L, Shen G, Payton P, Zhang H. Expression of an Arabidopsis sodium/proton antiporter gene (AtNHX1) in peanut to improve salt tolerance. Plant Biotechnol Rep. 2012; 6: 59-67.*
3. *Burow D M, Charles E. Simpson, James L. Starr and Andrew H. Paterson. Transmission Genetics of Chromatin From a Synthetic Amphidiploid to Cultivated Peanut (Arachis hypogaea L.): Broadening the Gene Pool of a Monophyletic Polyploid Species. Genetics. 2001; 159: 823–837.*
4. *Chu Y, Deng XY, Faustinelli P, Ozias-Akins P. Bcl-xL transformed peanut (Arachis hypogaea L.) exhibits paraquat tolerance. Plant Cell Rep. 2008; 27: 85-92.*
5. *Dang PM, Chen CY, Holbrook CC. Evaluation of five peanut (Arachis hypogaea) genotypes to identify drought responsive mechanisms utilising candidate-gene approach. Funct Plant Biol. 2013; 40: 1323- 33.*
6. *Gupta PK, Rustgi S, Kulwal PL. Linkage disequilibrium and association studies in higher plants: present status and future prospects. Plant Mol Biol.2005; 57: 461–485.*
7. *Halward T M, Stalker H T, Lame E A and Kochen G. Development of RFLP linkage map in diploid peanut species. Theoretical and Applied Genetics. 1993; 87: 379- 384.*
8. *Heffner E.L., Lorenz A.J., Jannink J.-L., Sorrells M.E. Plant Breeding with Genomic Selection: Gain per Unit Time and Cost. Crop Sci. 2010; 50:1681-1690.*
9. *Holbrook CC, Peggy Ozias-Akins, Ye Chu and Baozhu Guo. Impact of Molecular Genetic Research on Peanut Cultivar Development. Agronomy 2011; 1, 3-17.*
10. *Huang J, Yan L, Lei Y, Jiang H, Ren X, Liao B. Expressed sequence tags in cultivated peanut (Arachis hypogaea): discovery of genes in seed development and response to Ralstonia solanacearum challenge. J Plant Res. 2012; 125: 755-769.*

11. Knoll, J.E., Ramos, M L., Zeng, Y., HolBrok, C.C., Chow, M., Chen, S., et.al.TILLING for allergen reduction and improvement of quality traits in peanut (Arachis hypogaea L.). BMC Plant Biology, 2011; 11:81.
12. Li M-J, Xia H, Zhao C-Z, Li A-Q, Li C-S, Bi Y-P, et al. Isolation and characterization of putative Acetyl-CoA Carboxylases in Arachis hypogaea L. Plant Mol Biol Rep. 2010; 28: 58-68.
13. Liang XQ, Holbrook CC, Lynch RE, Guo BZ. beta-,3-Glucanase Activity in Peanut Seed (Arachis hypogaea) is Induced by Inoculation with Aspergillus flavus and Copurifies with a Conglutin-Like Protein. Phytopathology. 2005; 95: 506-511.
14. Luo M, Dang P, Bausher MG, Holbrook CC, Lee RD, Lynch RE, Guo BZ. Identification of Transcripts Involved in Resistance Responses to Leaf Spot Disease Caused by Cercosporidium personatum in Peanut (Arachis hypogaea). Phytopathology. 2005; 95: 381-387.
15. Mehta R, Radhakrishnan T, Kumar A, Yadav R, Dobaria JR, Thirumalaisamy PP, et al. Coat protein-mediated transgenic resistance of peanut (Arachis hypogaea L.) to peanut stem necrosis disease through Agrobacterium-mediated genetic transformation. Indian J Virol. 2013; 24: 205-213.
16. Meuwissen THE, Hayes BJ, Goddard ME. Prediction of total genetic value using genome wide dense marker maps. Genetics. 2001; 157:1819-29.
17. Miranda Brasileiro AC, Guerra Araujo AC, Leal-Bertioli SC, Guimarães PM. Genomics and Genetic Transformation in Arachis. Int J Plant Biol Res. 2014; 2(3): 1017.
18. Mondal S, Badigannavar AM, D'Souza SF. Development of genic molecular markers linked to a rust resistance gene in cultivated groundnut (Arachis hypogaea L.). Euphytica. 2012; 188: 163-73.
19. Pandey M. K., Monyo E., Ozias-Akins P., Liang X., Guimarães P., Nigam S. N., et al. Advances in Arachis genomics for peanut improvement. Biotechnol. Adv. 2012; 30: 639–651.
20. Pandey MK, Upadhyaya HD, Rathore A, Vadez V, Sheshshayee MS, et al. Genomewide Association Studies for 50 Agronomic Traits in Peanut Using the 'Reference Set' Comprising 300 Genotypes from 48 Countries of the Semi-Arid Tropics of the World. PLoS ONE. 2014, 9(8): 105-228.
21. Pasupuleti Janila, S. N. Nigam, Manish K. Pandey, P. Nagesh and Rajeev K. Varshney. Groundnut improvement: Use of genetic and genomic tools. Frontiers in plant science, 2013; 4:1-16.
22. Payton P, Kottapalli K, Rowland D, Faircloth W, Guo B, Burow M, et al. Gene expression profiling in peanut using high density oligonucleotide microarrays. BMC Genomics. 2009; 10:265.
23. Perez-de-Castro A.M, Vilanova S, Cañizares J, Pascual L, Blanca J.M, Díez, Prohens M.J and Picó B. Application of Genomic Tools in Plant Breeding. Current Genomics. 2012; 13: 179-195.

24. Ranganayakulu GS, Chandraobulreddy P, Thippeswamy M, Veeranagamallaiah G, Sudhakar C. Identification of drought stress-responsive genes from drought-tolerant groundnut cultivar (Arachis hypogaea L. cv K-134) through analysis of subtracted expressed sequence tags. Acta Physiol Plant. 2012; 34: 361-77.
25. Rao SC, Bhatnagar-Mathur P, Kumar PL, Reddy AS, Sharma KK. Pathogen-derived resistance using a viral nucleocapsid gene confers only partial non-durable protection in peanut against peanut bud necrosis virus. Arch Virol. 2013; 158: 133-143.
26. Ravi, K., Vadez, V., Isobe, S, Mir, R.R., Guo, Y, Nigam, S.N, Gowda, M.V.C, Radhakrishnan, T, Bertioli, D.J, Knapp, S.J., et al. Identification of several small main-effect QTLs and a large number of epistatic QTLs for drought tolerance related traits in groundnut (Arachis hypogaea L.). Theor. Appl. Gen. 2011, 122: 1119-1132.
27. Seijo G, Lavia GI, Fernández A, Krapovickas A, Ducasse DA, Bertioli DJ, Moscone EA. Genomic relationships between the cultivated peanut (Arachis hypogaea, Leguminosae) and its close relatives revealed by double GISH. Am J Bot. 2007; 94: 1963-1971.
28. Simpson,C.E., Starr,J.L., Church,G. T., Burrow,M.D and Paterson,A. H. Registration of NemaTAM peanut. CropSci. 2003; 43, 1561.
29. Stalker HT, Weissinger AK, Milla-Lewis S, Holbrook CC. Genomics: An evolving science in peanut. Peanut Sci. 2009; 36: 2-10.
30. Sundaresha S, Manoj Kumar A, Rohini S, Math SA, Keshamma E, Chandrashekar SC, et al. Enhanced protection against two major fungal pathogens of groundnut, Cercospora arachidicola and Aspergillus flavus in transgenic groundnut over-expressing a tobacco ß 1–3 glucanase. Eur J Plant Pathol. 2010; 126: 497-508.
31. Swathi Anuradha T, Divya K, Jami SK and Kirti PB. Transgenic tobacco and peanut plants expressing a mustard defensin show resistance to fungal pathogens. Plant Cell Rep. 2008; 27(11):1777-86.
32. Tanmoy Sarkar, Radhakrishnan T., A. Kumar, G.P. Mishra and J.R. Dobaraia. AtDREB1A transgenic groundnut shows tolerance to drought and salinity stresses. Proceedings of National Conference of Plant Physiology (13–16 December, 2013), Junagadh- Gujarat, India. pp. 383-384.
33. Tengale Dipak Bhauso, Thankappan Radhakrishnan, Abhay Kumar, Gyan PrakashMishra, Jentilal Ramjibhai Dobaria, Kirankumar Patel, and Manchikatla Venkat Rajam.. Overexpression of Bacterial mtlD Gene in Peanut Improves Drought Tolerance through Accumulation of Mannitol. 2014. The scientific world journal. 2014:1-10.
34. Tirumalaraju SV, Jain M, Gallo M. Differential gene expression in roots of nematode-resistant and -susceptible peanut (Arachis hypogaea) cultivars in response to early stages of peanut root-knot nematode (Meloidogyne arenaria) parasitization. J Plant Physiol. 2011; 168: 481-92.
35. Tiwari V, Chaturvedi AK, Mishra A, Jha B (2015) Introgression of the SbASR-1 Gene Cloned from a Halophyte Salicornia brachiata Enhances Salinity and Drought

Endurance in Transgenic Groundnut (Arachis hypogaea) and Acts as a Transcription Factor. PLoS ONE. 2015; 10:7.

36. *Varshney RK, Thudi Mahendar, Ramesh K. Aggarwal and Andreas Borner. Genic molecular markers in plants: Development and applications. R.K. Varshney and R. Tuberosa (eds.), Genomics-Assisted Crop Improvement: Vol. 1: Genomics Approaches and Platforms. 2007; 13–29.*
37. *Varshney RK, Ribaut J-M, Buckler ED, Tuberosa R, Rafalski JA, et al. Can genomics boost productivity of orphan crops?. Nat Biotech. 2012; 30: 1172–1176.*
38. *Varshney RK, Terauchi R, McCouch SR Harvesting the Promising Fruits of Genomics: Applying Genome Sequencing Technologies to Crop Breeding. PLoS Biol. 2014; 12(6):1-8.*
39. *Vasavirama K and Kirti PB. Increased resistance to late leaf spot disease in transgenic peanut using a combination of PR genes. Funct Integr Genomics. 2012; 12: 625-634.*
40. *Wu N, Matand K, Williams S. A novel mRNA level subtraction method for quick identification of target-orientated uniquely expressed genes between peanut immature pod and leaf. Biol Proced Online. 2010; 12: 44-55.*
41. *Yufang Guo, Brian Abernathy, Yajuan Zeng and Peggy Ozias-Akins. TILLING by sequencing to identify induced mutations in stress resistance genes of peanut (Arachis hypogaea).* BMC Genomics *2015, 16:157.1-13.*
42. *Zhou Xiaojing , Youlin Xia, Xiaoping Ren, Yulin Chen, Li Huang, Shunmou Huang, Boshou Liao, Yong Lei, Liyin Yan and Huifang Jiang. Construction of a SNP-based genetic linkage map in cultivated peanut based on large scale marker development using next-generation double-digest restriction-site-associated DNA sequencing (ddRADseq). BMC Genomics. 2014; 15:351.*

Chapter 10

DNA Fingerprinting in Groundnut Genotypes

Yellari Amaravathi

Scientist (Biotechnology), Institute of Frontier Technology, RARS, Tirupati.

10.1 Introduction

India is a rich source of biodiversity andcentre of origin for important crops like rice, pigeon pea, chickpea, cowpea, mung, brinjal, cucumber, Indian radish, noble canes, G. arboreum, Mango, orange, coconut, banana, etc. A large number of land races, farmer's varieties and scientifically developed cultivars have been replaced by new and better varieties. Inorder to contain the loss of genetic diversity neede to be properly chararized, conserved and utilized in the breeding programmes to sustain agricultural growth. In this direction, the "Plant Varieties Protection and Farmers'Rights Act (PVP&FR)" was passed by Parliament in 2001 is a singnificant leap farward. It facilitates the growth of seed industry and makes available high quality seed and planting material to farmers. The World Trade Organisation (WTO) requires its member countries to accord protection to plant varieties under Trade-Related Intellectual Property Rights (TRIPs) agreement. The protection, intended to buoy up innovation, can be provided by patents, by an effective *sui generis* system, or by any combination thereof. Convention on Biological Diversity (CBD) provides sovereign rights to nations over their genetic resources. Therefore, a perfect system to identify parental lines, landraces and wild relatives along with released varieties is the most fundamental requirement to enforce the propriety over plant varieties and germplasm. Variability in morphological and biochemical characters has been useful in this regard because the scoring is easy, economical and practical. Morphological characteristics cannot always be interpreted so that they provide the correct assessment of genetic differences. However, morphological and biochemical markers suffer from drawbacks such as low variability, environmental influence, epistasis, and their quantitative nature and complex inheritance pattern. Molecular markers, on the other hand, exhibit a high degree of non-tissue-specific polymorphism and simple inheritance patterns with minimal influence of environment and epistasis. The utility of marker system largely depends on genetic polymorphism i.e., simultaneous occurance of a trait in the same population of two or more discontinuous variants or genotypes. Several techniques have been developed for the past few years for distinguishing the individuals at DNA sequence level.

10.1.1 DNA Fingerprinting

The term DNA fingerprinting was coined by (Alec Jeffery in 1985) to describe bar-code like DNA fragment patterns generated by electrophoretic seperation of genomic DNA fragments. DNA fingerprinting is the application of molecular marker techniques for the identification of cultivars which can unambiguously distinguish one individual from another with much higher probability. Recently, the term DNA fingerprinting or profiling is used to describe the combined use of several single locus detection systems and are being used for investigating various aspects of plant genome. These include characterization of genetic variability, genome fingerprinting, genome mapping, gene localization, analysis of genome evolution, population genetics, taxonomy, pedigree analysis and to effect selection in plant breeding etc., DNA fingerprinting can also be used to in identification and linkage of disease loci and cloning and tranformation of resistance gene analogues to counter the disease. Conversely, plant DNA fingerprinting deals with more number of species with different mating systems and genetic structure, and in-depth genetic studies for many crops (except for a few such as rice, maize and tomato) are yet to begin.

10.2 Methods of DNA Fingerprinting

DNA fingerprinting can be broadly grouped into two categories:

(i) Hybridization based methods like RFLP, VNTR

(ii) PCR based methods like RAPD, AFLP, SSR, SNP etc

10.2.1 Hybridization based Methods

DNA profiles are visualized by hybridizing the restriction enzyme-digested DNA, to a labeled probe, which is a DNA fragment of known origin or sequence

10.2.1.1 Restriction Fragment Length Polymorphism (RFLP)

This technique is mainly based on the special class of enzyme i.e. Restriction Endonucleases. They have their origin in the DNA, occur due to evolutionary processes, point mutations within the restriction enzyme recognition site sequences, insertions or deletions within the fragments, and unequal crossing over (Schlotterer & Tautz, 1992). In RFLP analysis, restriction enzyme-digested genomic DNA is resolved by gel electrophoresis and then blotted (Southern, 1975) on to a nitrocellulose membrane. Specific banding patterns are then visualized by hybridization with labeled probe. These probes are mostly species-specific single locus probes of about 0.5– 3.0 kb in size, obtained from a cDNA library or a genomic library.Different sizes or lengths of restriction fragments are typically produced when different individuals are tested. Such a polymorphism can be used to distinguish plant species, genotypes and, in some cases, individual plants (Karp et al., 1998).

10.2.1.2 Variable Number of Tandem Repeats (VNTR)

The loci contain tandem repeats that vary in the number of repeat units between genotypes and are referred to as variable number of tandem repeats (VNTRs) (i.e. a single locus that contains variable number of tandem repeats between individuals) or hypervariable regions (HVRs). Minisatellites are a conceptually very different class of marker. They consist of chromosomal regions containing tandem repeat units of a 10–50 base motif, flanked by conserved DNA restriction sites. Variation in the number of repeat units, due to unequal crossing over or gene conversion, is considered to be the main cause of length polymorphisms. Due to the high mutation rate of minisatellites, the level of polymorphism is substantial, generally resulting in unique multilocus profiles for different individuals within a population.

10.2.2 PCR based Methods

PCR is a versatile technique which involve in vitro amplification of particular DNA sequences or loci, with the help of specifically or arbitrarily chosen oligonucleotide sequences (primers) and a thermostable DNA polymerase enzyme. PCR is extremely sensitive and operates at a very high speed. Its application for diverse purposes has opened up a multitude of new possibilities in the field of molecular biology.

10.2.2.1 Random Amplified Polymorphic DNA (RAPD)

RAPD is a PCR-based technology based on enzymatic amplification of target or random DNA segments with arbitrary primers. These oligonucleotides serve as both forward and reverse primer, and are usually able to amplify fragments from 1–10 genomic sites simultaneously. Amplified products (usually within the 0.5–5 kb size range) are separated on agarose gels in the presence of ethidium bromide and view under ultraviolet light (Jones et al. 1997) and presence and absence of band will be observed. This procedure detects nucleotide sequence polymorphisms in DNA by using a single primer of arbitrary nucleotide sequence. In this reaction, a single species of primer anneals to the genomic DNA at two different sites on complementary strands of DNA template. If these priming sites are within an amplifiable range of each other, a discrete DNA product is formed through thermo cyclic amplification. On an average, each primer directs amplification of several discrete loci in the genome, making the assay useful for efficient screening of nucleotide sequence polymorphism between individuals (Williams et al.1995).

10.2.2.2 AFLP (Amplified Fragment Length Polymorphism)

Amplified fragment length polymerphism (AFLP), which is essentially intermediate between RFLPs and PCR. AFLP is based on a selectively amplifying a subset of restriction fragments from a complex mixture of DNA fragments obtained after digestion of genomic DNA with restriction endonucleases (Vos et al 1995). Polymorphisms are detected from differences in the length of the amplified fragments by polyacrylamide gel electrophoresis (PAGE) (Matthes et al. 1998) or by capillary electrophoresis. The technique involves four steps: (1) restriction of DNA and ligation of oligonucletide adapters (2) preselective amplification (3) selective amplification (4) gel analysis of amplified fragments. AFLP is a DNA fingerprinting technique, which detects DNA restriction fragments by means of PCR amplification.

10.2.2.3 Simple Sequence Repeat (SSR)

Simple Sequence Repeats (SSRs) are sections of DNA, consisting of tandemly repeating mono-, di-, tri-, tetra- or penta-nucleotide units that are arranged throughout the genomes of most eukaryotic species (Powell et al. 1996). Microsatellite markers, developed from genomic libraries, can belong to either the transcribed region or the non transcribed region of the genome, and rarely is there information available regarding their functions. Microsatellite sequences are especially suited to distinguish closely related genotypes; because of their high degree of variability, they are, therefore, favoured in population studies (Smith & Devey 1994) and for the identification of closely related cultivars (Vosman et al. 1992). Microsatellites and their flanking sequences can be identified by constructing a small-insert genomic library, screening the library with a synthetically labelled oligonucleotide repeat and sequencing the positive clones. Alternatively, microsatellite may be identified by screening sequence databases for microsatellite sequence motifs from which adjacent primers may then be designed. Polymerase slippage during DNA replication, or slipped strand mispairing, is considered to be the main cause of variation in the number of repeat units of a microsatellite, resulting in length polymorphisms that can be detected by gel electrophoresis. Other causes have also been reported (Matsuoka et al. 2002).

10.2.2.4 Single Nucleotide Polymorphism (SNP)

A noval class of DNA markers namely single nucleotide polymorphism in genome (SNPs) has recently become highly preffered in genomic studies. The fact that in many organisms most polymorphisms result from changes in a single nucleotide position (point mutations), has led to the development of techniques to study single nucleotide polymorphisms (SNPs). Analytical procedures require sequence information for the design of allele specific PCR primers or oligonucleotide probes. SNPs and flanking sequences can be found by library construction and sequencing or through the screening of readily available sequence databases. Once the location of SNPs is identified and appropriate primers designed, one of the advantages they offer is the possibility of high throughput automation. To achieve high sample throughput, multiplex PCR and hybridization to oligonucleotide microarrays or analysis on automated sequencers are often used to interrogate the presence of SNPs. SNP analysis may be useful for cultivar discrimination in crops where it is difficult to find polymorphisms, such as in the cultivated tomato. SNPs may also be used to saturate linkage maps in order to locate relevant traits in the genome. For instance, in Arabidopsis thaliana a highdensity linkage map for easy to score DNA markers was lacking until SNPs became available (Cho et al. 1999). To date, SNP markers are not yet routinely applied in genebanks, in particular because of the high costs involved.

Anumber of markers are now available and each marker system has its own merits and demerits. Depending on the objective the reseacher can choose the best marker systemwhich is economical and most suitable to achieve the objectives. Table 10.1 summerizes the comparative account of all marker systems.

10.3 DNA Fingerprinting of Groundnut (arachis hypogaea l.) Varieties of Tirupati using SSR Markers

As more and more number of cultivars are being developed from few elite parents, it becomes more difficult to distinguish varieties solely based on morphological characters. This necessitated the development of an authentic method for varietal identification viz., DNA fingerprinting to protect plant varieties by appropriate characterization and documentation. Compared with other crops, groundnut is lagged behind in molecular characterization and the reason being meagre polymorphism detected by protein or DNA based markers like RAPD and AFLP (Paik-Ro et al., 1992; Stalker et al., 1994). The analysis of simple sequence repeat (SSR) loci is the current method of choice for fingerprinting groundnut as the allelic diversities are extremely high in SSR markers (He et al., 2003). SSRs are ideal markers for fingerprinting in groundnut as they are hyper variable, multi-allelic, distributed throughout the genome in high abundance (Ahmed et al., 2013). These markers have been shown to be highly informative because of their variability, ease of use, accessibility, reproducibility and co-dominant with greater informativeness, (Jannati et al., 2009).

We developed unique fingerprints for twelve varieties released over a period of 25 years from Regional Agricultural Research Station, Tirupati (Table 10.2) of Acharya N. G. Ranga Agricultural University using SSR markers. The unique fingerprints will safeguard plant varieties in the form of breeders' rights in the long run. Fifteen SSR markers were used for fingerprinting, out of which four were monomorphic and the remaining 11 markers generated 44 polymorphic loci (Table 10.3). Representative fingerprint pattern of twelve groundnut varieties with SSR markers IPAHM 23 and PM377 were shown in Fig. 1. The polymorphic primers generated alleles ranging from 2 to 7 with an average of four per locus. When one primer would not distinguish individual variety from others, additional markers needed to be tested to get unique profiles. All the twelve genotypes were successfully discriminated among themselves with 11 SSR markers. The genotypic data from all the loci provided unique SSR allelic fingerprints which help in varietal identification of groundnut (Fig. 2). The groundnut fingerprint data can unambiguously distinguish one individual from another with much higher probability than DUS characters alone. Recently, varietal fingerprinting is mandate for variety registration which helps in avoiding misappropriation and for protection of plant breeder's and farmers' rights.

Polymorphism information content (PIC) provides an estimate of the discriminatory power of the marker taking in to account both the number and relative frequency of the alleles (Smith et al., 1998) and the values ranged from 0 (monomorphic) to 1 (Highly polymorphic). In this study, PIC values ranged from 0 to 0.85. PIC values in polymorphic markers ranged from 0.49 (PM384) to 0.85 (TC1A02) with 0.67 as average (Table 3). This set of markers had the enormous potential to identify most of the groundnut genotypes. The core set of SSR markers identified in the present study are highly useful in discrimination of varieties bred from Tirupati can be employed for genotyping varieties bred in other stations and thereby the efficiency of the markers can be further verified.

10.4 Conclusion

The diagrammatic representation of DNA fingerprint pattern generated can serve as a guide for visual comparison of any additional genotypes as and when they become available. The finger printing of the groundnut genotypes provide information about phylogenetic relationships and assists groundnut breeders in varietal registration and protection of intellectual property rights. Further, the SSR analysis will help the breeders as additional tool in bringing together useful traits into single genetic background.

***Table 10.1** Comparative summery of commonly used DNA markers*

S. No.	Feature	RFLP	RAPD	AFLP	SSRs	SNPs
1	Quantity of DNA required (μg)	10	0.02	0.5-1.0	0.05	0.05
2	DNA quality	High	High	Moderate	Moderate	High
3	PCR based	No	Yes	Yes	Yes	Yes
4	No. Of polymorphic loci analyzed	1-3	1.5-50	20-200	1-3	1
5	Ease of use	Not easy	Easy	Easy	Easy	Easy
6	Amenable for automation	Low	Moderate	Moderate	High	High
7	Reproducibility	High	Low	High	High	High
8	Developmen cost	Low	Low	Moderate	High	High
9	Cost per assay	High	Low	Moderate	Low	Low

***Table 10.2** Groundnut varieties from regional agricultural research station, tirupati used for DNA fingerprinting*

S. No.	Variety	Pedigree	Year of release	Duration (Days)	Season recommended	Salient features	Reference
1.	Tirupati 1	Selection from E.C-106983/3	1989	100	Rainy and Post rainy seasons	Spanish bunch variety, Early maturing, drought tolerant	Vasanthi *et al.* (1998).
2.	Tirupati 2	GAUG-1 X Nc.Ac.FLA.14	1989	100-105	Post rainy season	Spanish bunch variety, Moderately tolerant Kalahasthi malady, suitable for black soils	Vasanthi *et al.* (1997).
3.	Tirupati 3	Selection from TMV-10	1991	125-130	Rainy and Post rainy seasons	Virginia bunch variety, red testa colour and resistant to Kalahasthi malady	Vasanthi *et al.* (1998).
4.	Tirupati 4	JL-24 X Ah316/S	1995	105	Rainy and Post rainy seasons	Spanish bunch Early maturing variety light rose testa, high yielding, tolerant to midseason moisture stress,	Reddy *et al.* (2000).
5.	Narayani	JL-24/Ah316/S	2002	100	Rainy and Post rainy seasons	Spanish bunch variety with red testa colour, early maturing and drought tolerant	Vasanthi *et al.* (2003).
6.	Kalahasti	TCG1709/ TCG1518	2002	105-110	Post rainy season	Spanish bunch variety with red testa colour, early maturing and resistant to Kalahasthi malady	Vasanthi *et al.* (2003).
7.	Prasuna	TCG1717/ TCG1518	2006	105-110	Early Rainy and Post rainy seasons	Spanish bunch variety, medium bold seeds and rose testa, tolerant Kalahasthi malady	Vasanthi *et al.* (2011).

Table 10.2 *Contd...*

S. No.	Variety	Pedigree	Year of release	Duration (Days)	Season recommended	Salient features	Reference
8.	Abhaya	K-134 X TAG-24	2006	105-110	Early Rainy and Post rainy seasons	Spanish bunch Drought tolerant variety with fresh seed dormancy and late leaf spot tolerant(LLS)	Vasanthi *et al.* (2006).
9.	Greeshma	TIR46 X JUG37	2009	95-100	Early Rainy and Post rainy seasons	Spanish bunch early maturing variety, LLS tolerant and water use efficient	Vasanthi *et al.* (2012).
10.	Rohini	Tirupati4 X TIR45	2010	90-95	Early and post rainy seasons	Spanish bunch variety, ultra early and LLS tolerant	Vasanthi *et al.* (2012).
11.	Bheema	TAG24 X TG19	2010	110-115	Post rainy season	Bold seeded Spanish bunch variety with low oil and high sucrose suitable for table purpose	Vasanthi *et al.* (2012).
12.	Dharani	VRI-2 X TCGP-6	2012	100-105	Earl, rainy and post rainy seasons	Spanish bunch variety, drought resistant, stem rot tolerant	Vasanthi *et al.* (2012).

Table 10.3 *Alleles and PIC of polymorphic SSR markers in 12 groundnut varieties*

S. No.	SSR marker	No. of Alleles	PIC
1	PM3	3	0.67
2	PM36	3	0.57
3	PM375	3	0.57
4	PM377	6	0.80
5	PM384	2	0.49
6	Seq5D5	4	0.67
7	Seq16G8	4	0.60
8	IPAHM23	4	0.71
9	GM1489	4	0.73
10	TC1A02	7	0.82
11	TC5AO6	4	0.72

A)

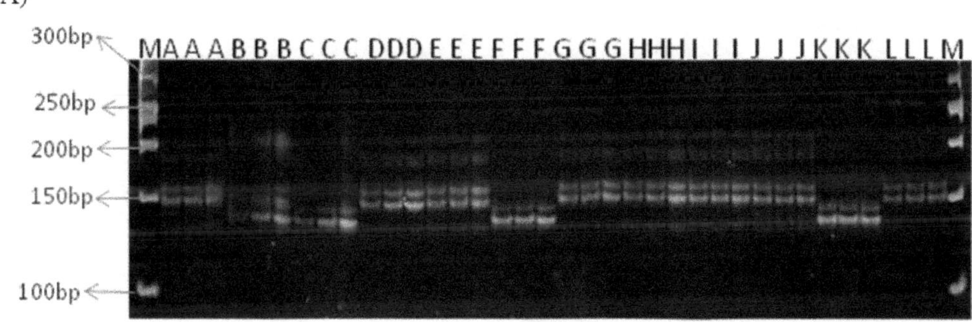

B)

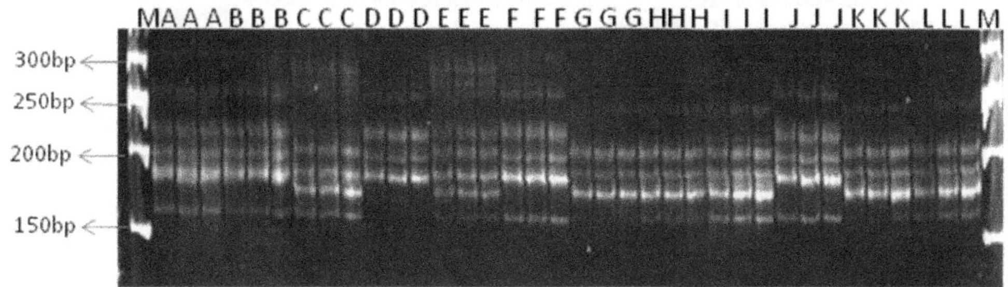

Fig. 10.1 *DNA profiles of groundnut cultivars with SSR markers A) IPAHM 23 and B) PM377 where M: 50bp ladder; A: Abhaya; B: Kalahasthi; C: Dharani; D: Greeshma; E: Prasuna; F: Bheema; G: Narayani H: Tirupati-4; I: Rohini; J: Tirupati-3; K: Tirupati-1 and L: Tirupati-2*

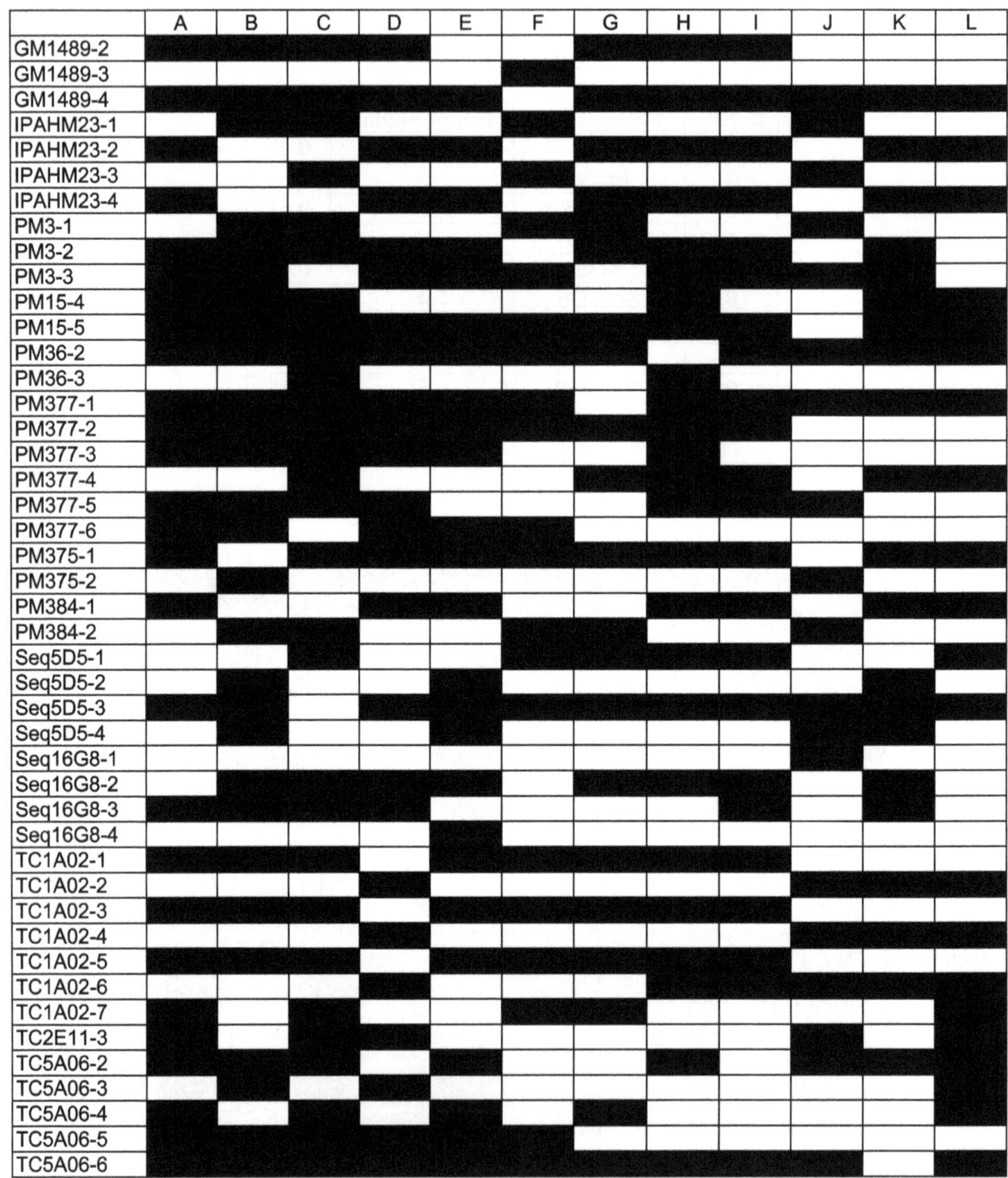

A: Abhaya; B: Kalahasthi; C: Dharani; D: Greeshma; E: Prasuna; F: Bheema; G: Narayani H: Tirupati-4; I: Rohini; J: Tirupati-3; K: Tirupati-1 and L: Tirupati-2

Fig. 10.2 *Bar-code-like DNA fingerprint pattern generated with 11polymorphic SSR markers after electrophoretic separation of DNA fragments.* ▬ *Presence of allele* ▭ *Absence of allele*

References

1. Ahmed, M.M., Guo, H., Huang, C., Zhang, X. And Lin, Z. 2013. Selection of core SSR markers for fingerprinting upland cotton cultivars and hybrids. Australian J. of crop Science. 7(12): 1912-1920.
2. Cho R.J., Mindrinos M., Richards D.R., Sapolsky R.J., Anderson M., Drenkard E., Dewdney J., Reuber T.L., Stammers M., Federspiel N., Theologis A., Yang W.H., Hubbell E., Au M., Chung E.Y., Lashkari D., Lemieux B., Dean C., Lipshutz R.J., Ausubel F.M., Davis R.W. and Oefner P.J. (1999) Genome-wide mapping with biallelic markers in Arabidopsis thaliana. Nature Genet 23:203–207
3. He, G., Meng, R., Newman, M., Gao, G., Pittman, RN. And Prakask, C.S. 2003. Microsatellites as DNA markers in cultivated peanut (Arachis hypogaea L.). BMC plnt biology.
4. Jannati, M., Fotouhi, R., Pourjan, A.and Salehi, A.Z. 2009. Genetic diversity analysis of Iranian citrus varieties using microsatellite (SSR) based markers. J. Hortic For 1: 120–125.
5. Jeffreys, A., Wilson, V. and Thein, S. (1985). "Individual-specific 'fingerprints' of human DNA". Nature **316** (6023): 76–79.
6. Jones C.J., Edwards K.J., Castaglione S., Winfield M.O., Sala F., Wiel C. van de, Bredemeijer G., Vosman B., Matthes M., Daly A., Brettschneider R., Bettini P., Buiatti M., Maestri E., Malcevschi A., Marmiroli N., Aert R., Volckaert G., Rudea J., Linacero R., Vazquez A. and Karp A. (1997). Reproducibility testing of RAPD, AFLP and SSR markers in plants by a network of European laboratories. Mol Breed 3: 381–390
7. Karp A., Isaac P.G. and Ingram G.S. (1998). Molecular Tools for Screening Biodiversity: Plants and Animals. Chapman & Hall, Thompson Sci., London
8. Matsuoka Y., Mitchell S.E., Kresovich S., Goodman M. and Doebley J. (2002). Microsatellites in Zea— variability, patterns of mutations, and use for evolutionary studies. Theor Appl Genet 104:436– 450
9. Paik-Ro, O.G., Smith, R.L. and Knauft, D.A. 1992. Restriction fragment length polymorphism evaluation of six peanut species within the Arachis section. Theor. Appl. Genet. 1992, 84:201-208.
10. Powell W., Machray G.C. and Provan J. (1996). Polymorphism revealed by simple sequence repeats. Tren Plant Sci 1:215–222
11. Schlotterer, C., and Tautz, D. (1992). Slippage synthesis of simple sequence DNA. Nucl. Acids Res. 20(2):211-215
12. Smith D.N. and Devey M.E. (1994) Occurrence and inheritance of microsatellites in Pinus radia. Genome 37: 977–983
13. Smith, J.S.C. and Register, J.C. 1998 Genetic purity and testing technologies for seed quality: a company perspective. Seed Sci. Res. 8: 285–293.
14. Southern E.M. (1975). Detection of specific sequences among DNA fragments separated by gel electrophoresis. Journal of Molecular Biology 98: 503–517.

15. Stalker, H.T., Phillips, T.G., Murphy, J.P. and Jones, T.M. 1994. Diversity of isozyme patterns in Arachis species Theor. Appl. Genet. 87:746-755.

16. Vasanthi, R.P., Venkataramana, E. Bhaskara Reddy, B.V. Thanuja Rani, I. Sivaprasad, Y. and Raja Reddy K. 2011. Identification of SSR Markers Associated with Resistance to Nematode Tylenchorhynchus brevilineatus Infection in Groundnut (Arachis hypogaea L.). 2011. AAJPSB.10-16.

17. Vasanthi, R.P. Reddy, J. R. Rajagopal, N. Reddy, P.V. Prasanthi, L.. John, K. Venkateswarlu O. and Chenchu Reddy. B. 2003. Groundnut variety Narayani suitable for cultivation in Andhra Pradesh, India. International Arachis News letter. 23:16-17.

18. Vasanthi, R.P. 1998. Testa colour inheritance in groundnut (Arachis hypogaea L.). Indian journal of genetics. 58(4): 433-437.

19. Vasanthi, R.P. Reddy, J. R. Rajagopal, N. Harinath Naidu, P. Prasanthi, L. John, K. Venkateswarlu O. and Chenchu Reddy, B. 2003. A Kalahasti Malady Resistant Groundnut Variety Suitable for Postrainy Season Cultivation in Andhra Pradesh, India.International Arachis News letter. 23:17-18.

20. Vasanthi, R.P., John, K., Mohan Vishnuvardhan, K., Purushotham, M., Sirisha, K., Thejaswi, Y. And Giridhar Krishna T. 2012. In: Groundnut varieties developed/ Released by RARS, Tirupati (AP). Pp.17-26.

21. Vasanthi, R.P., Reddy, P.V., Jayalakshmi, V., Sudhakar, P., Asalatha, M., Sudhakar Reddy, P. Harinatha Naidu, P., Muralikrishna, T., Venkateswarlu, O., John, K., Basu, M.S., Nigam, S. N. RC. Nageswara Rao and GC wright. 2006. A High-yielding Drought tolerant Groundnut Variety, Abhaya. IAN 26: 15-16

22. Vos P., Hogers R., Bleeker M., Reijans M., Lee van de T., Hornes M., Frijters A., Pot J., Peleman J., Kuiper M. and Zabeau M. (1995). AFLP: a new technique for DNA fingerprinting. Nucl Acids Res 23:4407–4414

23. Vosman, B., Arens, P., Rus-Kortekaas, W., and Smulders, M.J.M. 1992. Identification of highly polymorphic DNA regions in tomato. Theor. Appl. Genet. 85: 239–244

24. Williams J.G.K., Kubelik A.R., Livak K.J.., Rafalski J.A. and Tingey S.V. (1990) DNA polymorphisms amplified by arbitrary primers are useful as genetic markers. Nucl Acids Res 18:6531–6535.

Chapter 11

Farm Mechanization in Groundnut Cultivation

C. Ramana

Principal Scientist & University Head, RARS, ANGRAU, Tirupati.

11.1 Introduction

Advent of tractorization and new improved agricultural production technologies forces the farmer to resort for market available machinery without sound knowledge over the subject. The neglected popularization of operation specific or area specific machinery caused dent in usage of correct machinery in Indian farms. Some of farmers still under the impression that tractor with trailer/cultivator is whole about farm mechanization. Moreover changed agricultural scenario acute shortage of work force and incompatible machinery is neither meeting the demands of high production technologies nor able to complete the agricultural operation in time. Moreover the improved life standards of rural India, is neither allowing farmer to follow traditional agriculture nor empowered with new machinery management. This mismatch between demand, available machinery and farmer understanding is sometimes forcing a farmer to renounce the agriculture.

The timeliness of operations has assumed greater significance in obtaining optimal yields of groundnut, which is possible only by mechanizing. More over improved Farm mechanization technologies not only reduces the cost of cultivation but also to bring significant improvement in groundnut productivity. Farm Power Availability and Productivity of Food Grains in India (1951-2011) Farm power availability in 1951 was 0,25 kw/ha and increased to 1.68 in 2011 productivity improved from 0-55 to 1-92 t/ha during same period, hence agricultural productivity has a positive correlation with farm power availability (C. R. Mehtha, 2013). More importantly farm mechanization can improve operational comfort and rural house income with minimum risk i.e., timely completion of operations and improved productivity in agriculture and infuses better working environment there by Social Status.

India being a very important agriculture dependent country and farmers are entrepreneurial in adopting new varieties and latest technologies but still majority of the farmer are under impression that cultivator is lone equipment available for land preparation and can be used for all operations. The main reason for this is extension gap. Technical officers who are involved in the process have little or no knowledge on the machinery available. Hence it is dire essential to have orientation on operation specific machinery to solve the problem of completing timely operation. Farm machinery broadly classified in to primary, secondary, intercultural and harvesting equipment. The energy requirement for a semi-mechanized production system of groundnut in Kiashahr region, north of Iran. Data were gathered from 62 groundnut farms using a face to face questionnaire method. Results showed that the energy input of the diesel fuel had the highest share (43.51%) in the total energy inputs followed by the chemical fertilizers (29.11%). Total energy input was found to be 20164.36 MJ/ha and total output energy was determined as 79252.02 MJ/ha. Energy output-input ratio, specific energy, energy productivity, and net energy gain were computed as 3.93, 4.74 MJ/kg, 0.212 kg/MJ, and 59087.66 MJ/ha, respectively. Renewable and non-renewable energy were 19.27% and 80.73% in the total input energy, respectively. The direct and indirect energy shares for semi-mechanized groundnut production also were found to be 53.80% and 46.20% MJ/ha of the total input energy, respectively. (Saeed Firouzi* and Hashem Aminpanah, 2012).

11.2 Primary Tillage for Groundnut

Groundnut crop requires better tilh and congenial environment to extend roots proliferation, but majority of farmers till (land preparation) the land with cultivator irrespective of crop grown and some of the farmers do with rotovatos which are basically secondary/intercultural equipment. The operation by these equipment cannot break open hard-pan (consolidated soil mass formed due to weathering and machinery movement) crop grown under such environment will restrict its growth to the extent of loose soil. Hence primary tillage is dire essential in groundnut production system. To break open the consolidated soil up to the depth of 30-45cm and to provide soft, friable soil base for progress of plant growth use of primary tillage implements is essential. Blevins et al. (1971) indicated till treatments had higher volumetric moisture content to a depth of 60 cm during most of the growing season. The largest differences occurred in the upper 0 to 8 cm depth. However, (Blevins et al. 1971) also suggested soil moisture curves indicated two different water withdrawal patterns under the two contrasting methods of tillage, indicated crop type may have influenced results. (Chang and Lindwall 1990) indicated from a literature review, soil property changes due to tillage are related to several things. Those things include soil type, type of tillage equipment, tillage depth, soil conditions such as moisture content at the time of tillage and climatic conditions. There are several forces, natural and man-induced, that compact a soil. This force can be great, such as from a tractor, combine or tillage implement, or it can come from something as small as a raindrop.

11.2.1 Raindrop Impact

This is certainly a natural cause of compaction, and we see it as a soil crust (usually less than 1/2 inch thick at the soil surface) that may prevent seedling emergence. Rotary hoeing can often alleviate this problem. **Tillage operations** – Continuous moldboard plowing or disking or cultivator tilling at the same depth will cause serious tillage pans (compacted layers) just below the depth of tillage in some soils. This tillage pan is generally relatively thin (1-2 inches thick), may not have a significant effect on crop production, and can be alleviated by varying depth of tillage over time or by special tillage operations.

Primary tillage is focused on soil fracturing and, while loosening the soil, it will destroy soil structure and bury a great deal of crop residue--particularly those fragile ones such as soybean residue--, leaving the soil exposed to erosion. Primary tillage operations should be the last choice when improving soil productivity. Knowing your tillage implements and their settings can go a long way in establishing a good seed bed, sufficient seed-to-soil contact, and hardier and more profitable crops. These techniques require fewer trips through the field, lower the levels of waste of fertilizers, and leave the soil better prepared to withstand erosion (Mahdi Al-Kaisi Mark Hanna *and* Michael Tidman, 2004).

11.2.2 Mould Board Plough

Is useful in preparing field up to 45 cm, turned and pulverized. The MB plough enters in to the soil due to suctions in configuration and is 300 % efficient when compared to Bullock drawn country plough or cultivator. The implement can efficiently be used in stone free, non sticky soils. To avoid undulations in the field, two way reversible Mould board ploughs are available which can be operated with hydraulic shift lever.

Fig. 11.1 Mould board Plough

Fig. 11.2 Disc Plough

11.2.3 Disc Plough

In stony stumpy soils, where MB plough cannot be administered Disc plough can be successfully administered. This plough has got rolling plough bottom, hence it is useful for any type of soils and plough enters in to the soil due to self weight of the implement. If required, dead weights can be added to the implement for more depth up to 45cm even in dry soils. The rolling discs orientation like disc angle and tilt angle be changed for getting better quality soil tilt.

The research results on root development studies infers that groundnut can extend its roots up to 2.1 m (Sudhakar, et.al., 2010) under favorable root environment system. Hence to provide favorable root growth environment the field must be prepared to the possible depth. This not only envisages better root growth but also aids in in-situ moisture conservation. Similarly studies on the effect of primary tillage (MB plough and rotovator) on groundnut growth parameters revealed that there is an increase of 14.5 percent yield growth over farmers' practice i.e., cultivator and rotovator used for preparatory cultivation (Ramana. C 2015). Moreover, depth of cut and disturbance of soil will provide better exposure to nutrients and moisture therein productivity enhancement is made easy

11.2.4 Sub Soiler

Sub soiling i.e., cutting soil strata up to the depth of 40-75 cm on the field, makes vertical cut and administered once in three years to manipulate lower layers of soil strata and aids in adding new soil with old soil on top layers. But the sub soiling (vertical tillage) is also creates vertical trough for entry of rain water received in that field and lateral movement of excess water as runoff is restricted. The entry of rain water in deep layers through vertical cuts will moist soil below surface and be intact (as reservoir) without getting evaporated. Moisture below surface layers help crop to sustain even in long periods of dry spell and insures the crop. This phenomenon of entry of rain water in to the deep layers minimizes surface runoff, thereby minimizing transportation of fertile top-soil, remains in the same field (protecting soil against erosion). Deep layer tilling through sub-soiling also envisages the plant to develop deep roots and accessed to preserved moisture and more volume of soil nutrients. Experiment results also shown that there is less disease (collar root rot) instance, even in prolonged dry spell with sudden down pour (precipitation) recorded in sub-soiled field, but whereas complete damage due to collar root rot in un-treated fields (shallow ploughing was done) of groundnut, this is mainly due to moisture presence in the field soil where Sub-soiling is administered (Ramana. C, 2014). There have been a lot of discussion and research lately, regarding the benefits of deep rippers. If deep ripping is used for treating real soil compaction, the use of a ripper in dry soil conditions is recommended. However, because best results occur in dry soils, this operation may be best suited as a fall practice rather than a spring tillage investment. With very narrow blades or shanks that penetrate the soil from 12 to 14 inches, deep rippers loosen the soil in fractured planes running diagonally from the tip of the tool to the soil surface, break the compacted soil layer, and allow for the development of dense root systems. Heavily rooted plants gain greater access to subsoil moisture and nutrients throughout the soil profile (erosion (Mahdi Al-Kaisi Mark Hanna and Michael Tidman, 2004).

Allen et al. (1980) found conservation tillage increased fallow season soil water storage and resulted in larger crop yields in dryland wheat and sorghum crops. Javadi and Shahidzadeh (2005) developed a combined equipment with chisel plow behind of moldboard plow and indicated that the combined plow enabled to break plow pan during plowing in one pass and would avoid un-necessary sub-soiling in deep soil layers separately.

Fig. 11.3 Sub Soiler in operation

Hence usage of primary tillage during summer not only improves the productivity but also reduces pest complex, field preparation with primary tillage implements can improve production of up to 12% in the rain fed and 15-18% in irrigated dry crop situation.

11.2.5 Secondary Tillage

Act of primary tillage, ploughing to a depth of 45 cm will cut , invert and to some extent pulverizes soil, which handles more mass of soil and makes the field undulating (with soil clods). It is not congenial for sowing process and germination cannot be initiated in the undulating field conditions. Hence to make the field ready for seed bed (soil) need to be further processed/pulverized and requires a special type implements like disc harrow, spike toothed harrow, blade harrow etc.

Rotovator: this is an active implement and gets the power from PTO shaft of the tractor and rotates horizontally mounted central shaft on which soil working tynes are mounted and cuts the soils clods. The action of Rotovator on the already ploughed land makes friable and soft. This action of pulverizing soil automatically fills pores with air (pockets) in to the soil system and be useful for plant & root growth. Since the Rotovator consists of door in the rear to adjust clod size of the soil in the operation and it also helpful in making field level and clod free.

Fig. 11.4 Rotovator

11.2.6 Power Hoe

Fig. 11.5 Power hoe

This is an active tool and gets the power from the tractor PTO to rotate the vertically (independently) mounted tynes in the housing. These tynes are made of high carbon steel and sharpened to generate cutting action on the soil. The rotation of the tynes makes clods in to friable soil. Since the tynes are located vertically the bottom portion of the field will not become hard and puffing action is created. The soil will become more of air pockets and be very useful for root crops like groundnut, potato, onion etc. The implement is also useful for churning and mixing up of the soil. The rotary hoe also will have door in the rear to make the field level one and be immediately used as seed bed for the crops. The power hoe can complete 0.75-1 acre/hour.

11.2.7 Disc Harrow

Disc harrow consists of rotating discs as soil working components, it is a passive tool. The implement will have number of gangs on which number of discs mounted with spool and scraper so as to mix the soil thoroughly. The numbers of gangs are 2 or 4 and a disc on the gang varies from 6-18 depending on the size of the equipment. The gangs are provided with two different discs usually in front gang notched type discs and plain discs in rear gang. These discs are mounted on square centered shaft and with some disc, angle (20-23^0) to the line of travel. The arrangement inculcates the rotation when the implement dragged with tractor. The rotation of discs will crush the clods and mixes with two opposite disc gang arrangements. The capacity of the implement depends on number of gangs and discs per gang. Normally it can cover 1 ac/hr.

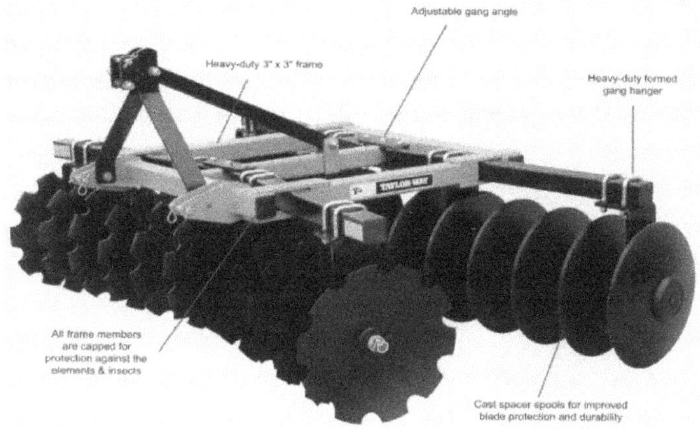

***Fig. 11.6** Disc harrow*

11.2.8 Cultivator

Spring tyne cultivators also be used as secondary tillage operations for crushing clods and breaking top soil crust. Since all farmers are familier and used to operate it also can be used as secondary tillage purpose. To make it more effective, use of 'S' type tynes are helpful both to increase depth of operation and reducing soil clod aggregate. To add some more emphasis and reducing number of passages and cost of operation, combining two implements like disc harrow or 'S' type cultivator roller/Cambridge clod crusher is effective. Moreover soil uniformity was significantly improved by using combined machine. It can be concluded that there were some benefits of using new machine in the secondary tillage operation. In case of similar effect, it can be also recommended that combined machine would be preferable reducing the number of pass, time and costs.

***Fig. 11.7** Spring tyne cultivator*

(Kamal Mohamed Adam), 1990, Moisture content at time of tillage has a significant effect on aggregate size distribution. The maximum force at which the aggregates crush was significantly affected by moisture content at the time of tillage and by implement type. The energy to crush is significantly affected by the moisture content at the time of tillage. However, Implement type showed no significant effect on crushing energy.

11.2.9 Spike Tooth Harrow

This harrow is used for clod crushing and combing up operation after primary tillage operation. This is a passive implement and does good mixing and crushing with mounted tynes. Some of the spike tooth harrow are mounted with spring tynes which can enter to better depths than ordinary tynes. This tynes are mounted on frame, this implement is better used for cleaning of crop & weed residues from the fields.

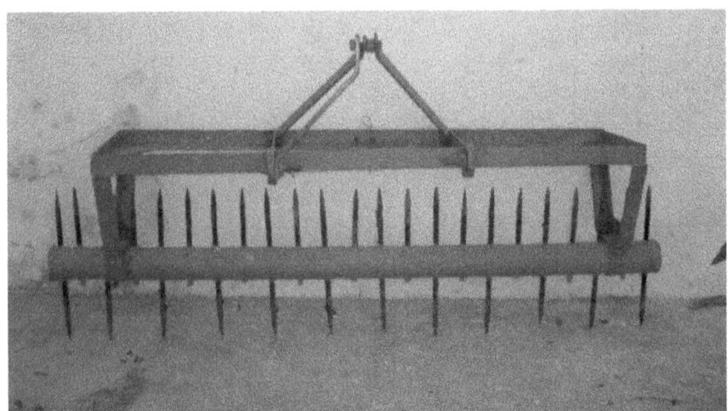

Fig. 11.8 Spike tooth harrow

11.2.10 Blade Harrow

This is a passive tool and being attached with 3 point linkage system of tractor as integral part. This is used after primary tillage and at the end of secondary tillage operation Blade harrow is fixed with a single blade or double blades to the vertical drops (standards) of frame. Frame consists of the provision to mount on a tractor as a mounted implement. The blade of the implement will cut undulating soil portions beneath the ploughed (or) tilled soil. The scraping of these undulating soil lumps will greatly help in uniform irrigation in entire field and avoid dry patches of crop in the field. The cutting of hidden bumps under soil is very much essential and saves irrigation water up to 10% and provides uniform crop growth & maturity which improves water productivity.

Fig. 11.9 Blade harrow

11.3 Seed Treatment Equipment

11.3.1 Seed Treatment Drum

Seed treatment drum is used for thorough mixing of pesticides with groundnut seed before sowing as a plant protection measure. It consists of a drum mounted on a frame at 40^0 with horizontal and operated manually by using a crank handle. It saves 33% labour, operating time and cost of operation over conventional method of mixing manually with land. Its overall dimensions are 900 X 700 X 400 mm and weigh about 25 kg. The drum capacity and mixing capacity are 10 kg and 100 kg/h respectively with mixing efficiency up to 90%.

Fig. 11.10 Seed treating drum

11.3.2 Sowing Equipment

Sowing is a very important operation which has to be performed timely especially in irrigated dry and rain-fed cultivation. The sowing also very important for making complete mechanization like intercultural operation, spraying and harvesting. Since the groundnut kernels are very soft and care need to be taken (not to crush damaged) during sowing. The seed rate also needs to be accurate for maintaining groundnut population in turn productivity.

11.3.3 Sathi Bullock Drawn Groundnut Planter

To mechanize the sowing operation by using bullock drawn groundnut planter was designed and developed at Agricultural Research Station, ANGRAU, Ananthapur. It is provided with a trough type seed metering mechanism for seed placement in the row. This covers four rows at a time with row to row distance of 30 cm and maintains seed to seed distance of 10 cm in a row at 4-5 cm depth of sowing. The recommended seed rate i.e. 100 Kg/ha can be maintained. It can also be used for other crops like Bengal gram, castor and red gram by changing the row to row spacing and disc in seed metering mechanism. The field capacity is in between 1.5 to 2 ha/day. The capacity of hopper is 8 kg for groundnut.

Fig. 11.11 Sathi groundnut planter

11.4 Seed Cum Fertilizer Drill

11.4.1 (Gujarat Model)

This seed cum fertilizer is provided with seed and fertilizer boxes along with seed metering mechanism (trough feed) and mounted on 9 tynes cultivator (Rigid and spring tyned optional based on soil type). The depth control system was provided to maintain uniform depth through two guage wheels. The row spacing of the sowing can be adjusted as per the season/requirement. The covering device is placed behind the implement to close the furrows immediately after sowing (with rear plank). Similarly the same seed drill can be used for any type of seed sowing for which seed metering scoop wheels need to be changed. Fertilizer drilling qualities also can be monitored by changing sliding door at the bottom of fertilizer box and beginning of the fertilizer spout.

Fig. 11.12 *Seed cum fertilizer drill (Gujarat model)*

11.4.2 Tractor Drawn Anantha Groundnut Seed Cum Ferti Drill

Tractor drawn groundnut seed cum ferti drill–row tractor operated groundnut seed drill with row to row spacing of 30 cm fir timely sowing with mechanical advantage and intercropping fertilizer facility is provided. This is provided with a hopper and seed metering mechanism as the main components. The hopper is divided into boxes each can accommodate 5 kg of seed (total 40 kg). the inclined plate seed metering mechanism gives correct seed to seed distance of 10 cm in a row and maintains the recommended seed rate of 100 kg/ha with optimum plant population of 33 per square meter area. Placement of seed is at proper depth of 4-5 cm. the seed damage is negligible and the field capacity is 6 to 7 ha/day and facilitates coverage of large area before the soil moisture is dried up. A 5 cm width covering blade is also attached behind the furrow openers to cover the furrows opened after seed placement. The intercropping of redgram or castor can also be possible using Anantha planter along with groundnut sowing.

The spring type cultivator frame of this planter facilitates to work even in stony soils. The cost of Anantha planter is approx. Rs.55, 000.

11.4.3 Intercultivation Equipment

The study conducted in two major groundnut-growing states, viz. Gujarat and Andhra Pradesh, has revealed that farmers employ more human labour in weeding and harvesting operations in groundnut than in other operations. The practise of manual decortication and stripping is followed by a larger number of farmers in Andhra Pradesh than in Gujarat, indicating less mechanization in the former. For weeding, though, the human labour-use in weedicide + bullock intercultivation + hand weeding technique is almost half of that of the bullock inter-cultivation + hand weeding, only 13 per cent of the farmers practise this labour-saving technique and hence this method should be disseminated in both the regions to reduce human-labour demand. In Andhra Pradesh, for all the operations in groundnut cultivation except harvesting, the cost as well as labour-use has been reduced substantially due to use of partial/complete mechanization methods. Hence, the necessary infrastructure (labour-saving machineries) should be created at the village or block level to reduce the human-labour demand (G. Govindaraj* and A.P. Mishra, 2011).

11.4.4 Bullock Drawn Inter-Cultivation Implement

It is a 4 row bullock drawn inter culture implement used for removal of shallow depth weeds in between rows of groundnut crop. It consists of 4 straight blades, frame, handle and beam to attach with a pair of bullocks. The blades are fixed to the frame to which handle is attached. The blades are the working components which are made from medium carbon steel or mild steel for more strength to resist soil friction and to have long life. The width of each blade is 15 cm. for operation, the weeder is passed in between the rows of crop so the blades cut and uproot the weeds. Its field capacity is 2.0 ha/day.

Fig. 11.13 Bullock drawn intercultivation equipment

11.4.5 Tractor Drawn Inter-Cultivation Implement

This is a 8 row tractor operated inter-culture implement used for weeding in groundnut crop developed at Agricultural Research Station, ANGRAU, Ananthapure. Its frame is provided with 8 tynes each tyne attached with T or V-shape sweeps to work in between 30 cm row spacing of the crop without any plant damage. Two small width pneumatic tyres of 8.3 X 28 size need to be fitted to the rear axle of the tractor to run in between rows of the crop instead of normal size tyres to prevent trampling of plants under the tyres. The size of the seeps range from "4 to 6". The cost of inter-culture implement and pneumatic tyres with sweeps is approx. Rs.45, 000. Its field capacity is 4 to 5 ha/day.

Fig. 11.14 Boom sprayer

Fig. 11.15 Tractor drawn inter-cultivation implement

11.4.6 Boom Sprayer

It is a sprayer with larger width of operation and covers more area in a lesser compared to power sprayers. It consists of a pump, one plastic or fiber glass made tank of 400 litres capacity, control valves, relief valve and a spray boom fitted with nozzles. The pump is operated by the PTO shaft of the tractor at the high pressure of around 20-55 kg/cm^2. The complete sprayer is mounted on 3-point linkage of the tractor. The boom may be of flexible hose pipe on which nozzles are mounted to meet crop row spacing. The bottom is fixed with a rigid beam by clamps. Inlet liquid supply to boom is provided at two points for even distribution of liquid. Hole are provided on the frame to lower or raise the beam to adjust the height of spray. This boom has 13-15 triple action nozzles and can cover 7-8 m width. Two small width pneumatic tyres of 8.3 "X28" size need to be fitted to the rear axle of the tractor to run in between rows of the crop instead of normal size tyres to prevent trampling of plants. Its field capacity is 8 ha/day at the operating speed of 3.5 Km/hr. application rate is 400 L/ha and initial cost about 1, 10, 000.

11.4.7 Harvesting Equipment

Harvesting in groundnut is very important and requires favourable condition of soil moisture and disease free coditios during harvest. A 45-65 hp tyer tractor was used for drawing this implement. The parameters evaluated were, soil moisture content, effective field capacity, theoretical field capacity, field efficiency, digging efficiency, percentages of damaged pods, exposed pods loss, unexposed pods loss, undug pods loss and total of pods loss. The experimental plots were laid side by side in a randomized complete block design (RCBD). Results of the main field test indicate that using the tractor drawn groundnut digger only performed better than using the tractor drawn groundnut digger/shaker combination with highest average digging efficiency of 43.74% at a soil moisture content of 5.60% (db), field efficiency of 85.93% and effective field capacity of 0.65 ha/hr both at soil moisture content of 4.63% (db). While the tractor drawn groundnut digger/shaker combination recorded the highest average digging efficiency of 39.73% at a soil moisture content of 3.02% (db), field efficiency of 60.16% and effective field capacity of 0.41 ha/hr both at soil moisture content of 3.02% (db). Results obtained from this study also showed that digging efficiency and percentage of total pod loss are inversely related to one another signifying that at lower digging efficiency there would be high percentage of total pod loss and vice versa (Chinsuwan, W.1998,). Output 150 – 200 kg/h, few unpicked and broken fruits. The machines are simply structured, easily produced with low capital, which permits massproduction. In addition, the Institute have also carried out research and design of one-step groundnut harvester (Ademiluyi, Y.S, 2004).

The use of machines is one of the inputs which, apart from giving relief from drudgery, influences the efficacy of the other inputs either directly or indirectly. Timely and correct placement of seeds, optimal harvesting procedures and timing, adequate shelling and related items are examples where machines could influence the production efficiency. Cost of production would also be influenced by the degree of usage of machines (Kaul, 1978).

11.4.7.1 ANGRAU Blade Guntaka

ANGRAU blade guntaka designed and developed at Agricultural Research Station, ANGRAU, Anantapur is used for digging of groundnut crop after maturity at the soil moisture range of 8-15%. It is provided with main components frame, 3-point linkage and a straight blade. All the components are made of MS material. The blade is the working component for digging of crop and its length and width are 135 and 8 cm respectively. It has the working width of 135 cm and covers 4-covers 4-5 ha in a day at the recommended speed of operation of 2-3 km/hr. the digging efficiency is 90-95%.

***Fig. 11.16** ANGRAU blade guntaka*

11.4.7.2 Groundnut Digger, Shaker Cum Windrower

It is also used for harvesting of groundnut crop at soil moisture levels of 8-15% and operated with above 45 H.P tractor. It has the working width of 120 cm and covers 4 rows of groundnut crop at row to row spacing of 30 cm. its overall dimensions are 1700 X 1000 X 1050 mm provided with soil loosening tool of sweep type, a pick conveying mechanism and gatherer windrower. The soil engaging tool is made of high strength mild steel. At the rear, a gatherer windrower the conveying crop. While conveying, soil get removed from crop due to shaking action. The field capacity is 0.8-1.0 ha/h at the recommended speed of operation 2-3 km/h. harvesting and soil separation efficiencies are 96 and 95% respectively. Saving in labor cost and time are 50 and 95% respectively compared to manual harvesting. Its cost is about Rs.1, 80, 000.

***Fig. 11.17** Groundnut digger, shaker cum windrower*

11.5 Threshing Equipment

11.5.1 Groundnut Fresh Pod Stripper

It is suitable for stripping of groundnut pods from harvested crop and consists of a wire spike type cylinder powered with 2 H.P electric motor. Stripping is done by holding the portion of a bunch manually over spiked cylinder. Three persons can work at a time. It is also provided with a blower and sieve for separation of pods from plant stalk, leaves etc. it saves 40% labour, 50% operating time and 30% cost of operation and it also results in 4% reduction in losses compared to conventional method of stripping. It was developed at ANGRAU, Hyderabad. Output capacity is 120 kg/ha with 100% stripping efficiency and 98% cleaning efficiency.

***Fig. 11.18** Fresh pod stripper*

11.5.2 Groundnut Fresh Pod Stripper

It is a throw-in type thresher used for separating pods from plants immediately after harvesting of groundnut crop. The farmer no needs to wait for drying of crop after harvesting for threshing. It is operated with PTO shaft (Speed 540 RPM) of 35-45 H.P tractors. It consists of frame, feed hopper, drum type threshing cylinder, concave, oscillating sieves and a blower. Total construction sits on the main frame. The threshing cylinder has the diameter and length of 50 c, and 90 cm respectively and working speed of 320 RPM. The cylinder surface is provided with flat pegs arranged in 6 rows such that each row has 7-8 pegs (length of peg 10 cm). A concave is provided under the threshing cylinder for rough separation of pods and stripped plants. An outlet is provided at the rear portion of cylinder for stripped plants. In order to separate all the unwanted material after threshing from the pod, two sieves have been provided below the concave. The top sieve has holes of 50 X 17 mm size and the bottom sieve has holes of 25 X 9 mm size. A centrifugal blower with spiral casing has been provided in between the two cleaning sieves for blowing of light weight plant material coming along with threshed pods from the concave. It has the feed rate of 750 Kg/h and output capacity of 300 kg/h with 96% threshing efficiency. It is very useful where the influence of north east monsoon will be more and continuous rains occur at harvesting time. Initial cost is around Rs. 1,80,000. It can easily transported from one place to another place as it is provided with pneumatic tyres.

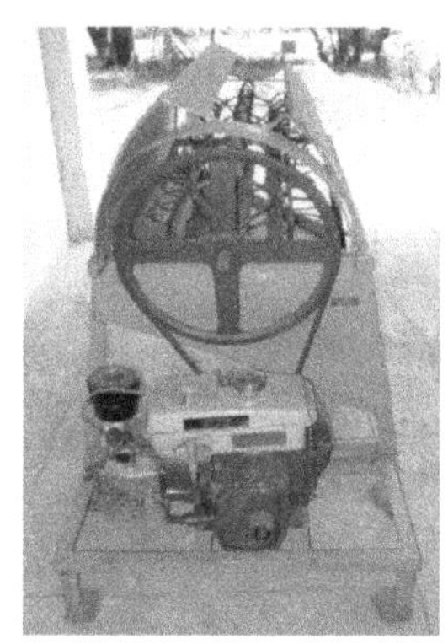

Fig. 11.19 Fresh pod thresher

11.5.3 Dry Pod Thresher

This is also a throw-in type thresher for groundnut crop having moisture content of 15-17% crop harvested needs to be dried before threshing. It is operated either with 10 H.P diesel engine or electric motor. It consists of frame, feed hopper, hammer type threshing cylinder, concave, oscillating sieves and a blower. A concave is provided under the threshing cylinder for rough separation of pods and stripped plants. In order to separate all the unwanted material after threshing from the pod, two sieves have been provided below the concave. A centrifugal blower with spiral casing has been provided in between the two cleaning sieves for blowing of light weight plant material coming along with threshed pods from the concave. It has the feed rate of 500 kg/h and output capacity of 200 kg/h with 95% threshing efficiency. Its initial cost is around Rs. 1, 10, 000 including 10 H.P diesel engine (M.A.Helmy,).

11.6 Decorticating Equipment

The shelling efficiency ranged from 95.44 to 95.27 and 98.92-98.86% at feed rate ranges of 80-120 and 140-180 kg/h at box speed of 1.4m/s and moisture content about 17.12 %d.b. The cleaning efficiency ranged from 99.01 to 98.79 and 99.17 to 98.93% at feed rate ranged from 80 to 120 and 140 to 180 kg/h at box speed of 1.4 m/s, air velocity of 8.37m/s and moisture content about 17.12 %d.b. for the Sheller

11.6.1 Hand Operated Decorticator

Hand operated groundnut decorticator can be used to shell groundnut pods and to separate kernels. It consists of and oscillating sector with sieve bottom and a handle. Number of hard rubber or cat iron lined assemblies are fitted in the oscillating sector unit. The groundnut pods are shelled between the oscillating sector and the fixed perforated concave screen by rubbing action. The decorticated shells and kernels fall down through the perforated concave sieve. The kernel and shells are collected at the bottom of the unit and separated manually. Clearance between the concave and oscillating sector is adjustable to suit the different varieties and concave sieves are also replaceable depending upon the pod size. Its overall dimensions are 600 X 350 X 700 mm. the capacity and efficiency of the unit are 50 kg/ha and 98% respectively.

Fig. 11.20 Hand operated decorticator

11.6.2 Power Operated Groundnut Decorticator

It is used to shell groundnut pods and to separate kernels. This is operated with 2 H.P single phase electric motor. It consists of feed hopper, rasp bar cylinder with hard rubber linings, perforated concave screen, tow oscillating sieves and a blower. The groundnut pods are shelled between the rubber linings of cylinder and fixed perforated concave screen by rubbing action. The decorticated shells ad kernels fall down on oscillating sieves through the perforated concave screen. A centrifugal blower with spiral casing provided in between the perforated concave screen and oscillating sieves separates the light weight shells from kernels. Oscillating sieves pods from stalk, leaves and other foreign material. Its capacity is 250-300 kg/h and its cost is around Rs. 40, 000 along with electric motor. he bigger capacity machines also available for commercial industrial use.

Fig. 11.21 Power operated decorticator

References

1. Allen, R.R. J.T. Musick, P.W. Unger and A.F. Wiesse. 1980. Soil, Water and Energy Conserving Tillage - Southern Plains. Proceedings of the ASAE Publication 7-81 Conference of Crop Production with Conservation in the 80's. 99-101.
2. Blevins, R.L., D. Cook, S.H. Phillips and R.E. Phillips. 1971. Influence of No Tillage on Soil Moisture. Agronomy Journal (63):593-596.
3. Chang, C. and C.W. Lindwall. 1990. Comparison of the Effect of Long Term Tillage and Crop Rotation on Physical Properties of a Soil. Canadian Agriculture Engineering (32):53-55.
4. C.R. Mehtha., 2013: CSAM Regional Forum 2013 (un-csam.org) Page 11 of 31
5. Javadi, A. and M. Shahidzadeh, 2005. Investigating possibility to break plow pan during plowing. Agric. Eng. Res. J., 24: 45–59
6. Arzhang javadi1 and Ali Hajiahmad 2006,Effect of a New Combined Implement for Reducing Secondary Tillage Operation; international journal of agriculture & biology 1560–8530/2006/08–6–724–727
7. Kamal Mohamed Adam,1990, Secondary tillage tool effect on soil aggregation, Iowa State University Digital Repository @ Iowa State University
8. Ademilyi, Y.S., James, D., Ozumba, I. C. and Olowonibi, M. M. 2004. Performance Evaluation of a Tractor Drawn Groundnut Digger/shaker for Agricultural Productivity Ozumba National Centre for Agricultural Mechanization (NCAM)Agricultural Engineering Department, University of Ibadan. Pg. 4
9. Chinsuwan, W., Wongpichet, S., Sudajan, S., Chakkapak, C., Krishsanaserani, S., Mongkolthanatas, J., and Thongsawatwong, P. 1998. Final Report Groundnut Mechanization Project: A Report Submitted to the International Development Research Centre (IDRC, Canada). IDRC File No. 3-P-87-0041
10. G. Govindaraj* and A.P. Mishra , 2011 Agricultural Economics Research Review Vol. 24 (Conference Number) 2011 pp 423-428 * Author for correspondence, Email: mggraj@yahoo.co.in Labour Demand and Labour-saving Options: A Case of Groundnut Crop in India
11. Mahdi Al-Kaisi, Mark Hanna and Michael Tidman, 2004, Extension Agricultural Engineer This article originally appeared on page 24 of the IC-492 (4) -- April 19, 2004 issue.
12. Sudhakar, P. Latha 2011, Annual report of RARS, results of Department Technical programmes
13. System Saeed Firouzi* and Hashem Aminpanah,2012, Production in a Semi-mechanized Cultivation Department of Agronomy, Agricultural College, Rasht Branch, Islamic Azad University, Rasht, Iran
14. M.A. Helmy, A.Mitrroi, S.E. Abdallah,Mohamed, A. Basiouny “Modification and evaluation of a Reciprocating machine for shelling peanut,” Misr J. Ag. Eng., 24(2): 283-298.

Chapter 12

Influence of Climate Change on Groundnut Productivity

T. Prathima

Senior Scientist, RARS, Tirupati.

12.1 Introduction

Groundnut (*Arachis hypogaea* L.) is an important food and oilseed crop suitable for cultivation in tropical areas of the world. In India the crop is grown by small and marginal farmers in diverse agro-climatic environments in India. Groundnut is believed to be a native of South America. Groundnut is the third largest oilseed produced in the world and second largest in India. India occupies the second position in terms of production and first in terms of area. China is the single the largest producer as well as consumer of groundnut in the world. Groundnut seeds are rich source of edible oil (43-55%) as well as protein (25-28%). About two thirds of world production is crushed for oil and the remaining one third is consumed as nuts. The cake or meal is used as animal feed and the shell sometimes used as fodder.

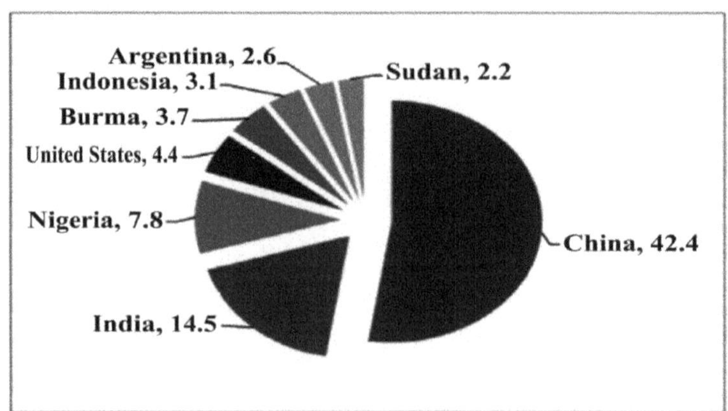

Fig. 12.1 Share of major groundnut producing countries

12.1.1 World Scenario

China is the largest producer of groundnut in the world accounting for more than 40% followed by India with about 15% share. Though, India is the largest cultivator of groundnut crop in terms of acreage, low yields kept her in the second place in terms of output. Further, production of groundnut in India has persistently been declining from mid 2000s.

12.1.2 Domestic Scenario

India has remained in the second among the top groundnut producing countries despite having the largest area under groundnut cultivation in the world owing to the low yield. In addition, area under cultivation has been declining steadily. Groundnut used to be largest produced oilseed in the country but its share has declined particularly in 2000s due to sharp rise in soybean production. Nevertheless, groundnut output remained largely stable in the recent years moving in a narrow range. Although the estimates from the GoI and the USDA differ widely their long term trend appears to be similar. The estimates of the GoI have been in the range of 7 to 8 million tons during the past four years while that of USDA stood in the range of 5 to 6 million tons in the same period. Nevertheless, the trend of relative change from year to year observed to in the same direction though the magnitude of change or fluctuation was small in case of USDA estimates.

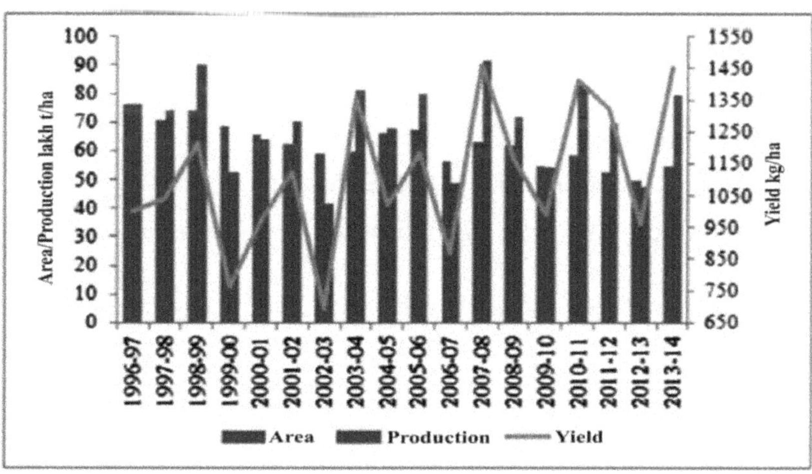

Fig. 12.2 Trends in area, production and yield

12.2 Overview of Climate Change and its Impact on Crop Growth

Increasing concentration of greenhouse gases in the atmosphere is warming the globe, which is causing climate change in terms of higher air temperatures, changing precipitation patterns and increased frequency of extreme weather events such as floods, heavy storms and droughts (IPCC, 2007). Depending upon the location on the globe, the plant growth and yield of crops will be negatively or positively affected by the climate change (Howden *et al.*, 2007). Increasing temperatures affect growth and development of crops, thus influencing potential yields.

Changes in the temperature, solar radiation and precipitation will have an effect on crop productivity and livestock agriculture. Climate change will also have an economic impact on agriculture, including changes in farm profitability, prices, supply, demand, trade and regional comparative advantages. The magnitude and geographical distribution of such climate induced changes may affect our ability to expand the food production area as required to feed the burgeoning population of more than 10,000 million people projected for the middle of the next century.

Agriculture is sensitive to short-term changes in weather and to seasonal, annual and longer-term variations in climate. For the long-term changes, agriculture is able to tolerate moderate variations in the climatic mean. Changes beyond these bands of tolerance may require shifts in cultivars and crops, new technologies and infrastructure or ultimately conversion to different land uses. Crop yield is the culmination of a diversified range of factors. The variations in the meteorological parameters are more of transitory in nature and have paramount influence on the agricultural systems, although other parameters, like soil characteristic, seed genetics, pest and disease and agronomic practices also do impact crop yields. Among these factors, pest and diseases cause a significant loss to world food production under different climatic conditions.

12.3 Influence of Climate on Crop Growth Management

Climate change impacts ron crop yield depends on the latitude of the area and precipitation.

1. The crop yield can be increased with increase in precipitation during the crop growth.
2. Crop yield is more sensitive to the precipitation than temperature.
3. If water availability is reduced in the future, soil of high water holding capacity will be better to reduce the frequency of drought and improve the crop yield (Popova and Kercheva, 2005).
4. With climate change, the growing period will reduce, and the planting date also needs to change for higher crop production.
5. Climate change can decrease the crop rotation period, so farmers need to consider crop varieties, crop densities and fertilization levels when planting crops (Cuculeanu *et al.*, 2002).

The positive effects of climate change on agriculture are concerned with the CO_2 concentration augment, crop growth period increases in higher latitudes and montane ecosystems; the negative effects include the increasing incidence of pests and diseases, and soil degradation owing to temperature change (Lal, 2005). This has urged scientists to develop more crop varieties suiting the changing climate and degrading soil in order to obtain sufficient yield for the increasing population. Quantification of climate uncertainty is an important indicator for crop yield variation in future climate scenarios.

***Table 12.1** Optimal range of temperatures and rainfall required at various phenological stages of Groundnut*

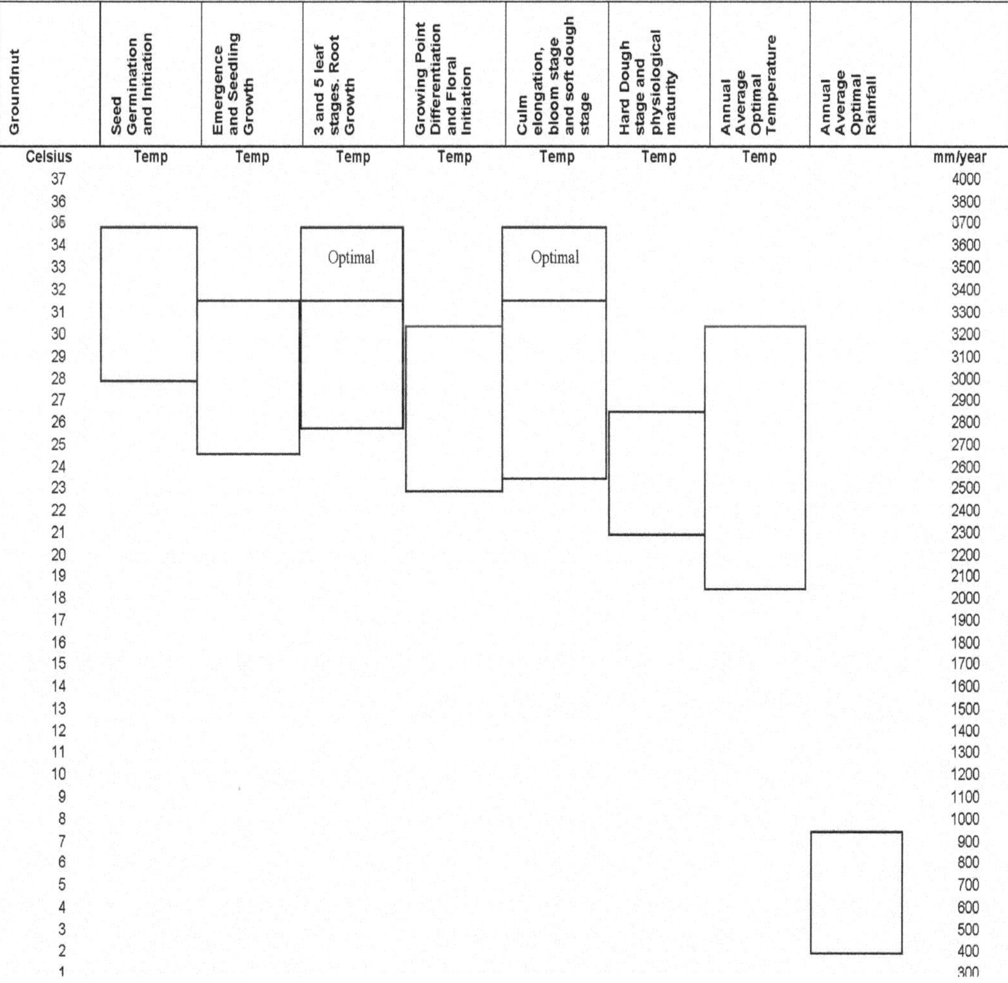

Table 12.2 *Influence of climate change on groundnut at various growth phases*

	Vegetative Stages												Reproductive Stages											
	Germination and Seedling emergence				Stem elongation and Root Development				Stem Growth and Leaf Development				Infloresce nce emer-gence			Flowering				Pod / Seed Ripening and Harvest				
Climatic Variables	Rainfall		Temperature		Rainfall		Temperature		Rainfall	Temperature			Rainfall		Temperature	Rainfall		Temperature		Rainfall		Temperature		
Assumed change from normal condition	Increased	Decreased	Warmer than Normal	Cooler than Normal	Increased	Decreased	Warmer than Normal	Cooler than Normal	Increased	Decreased	Warmer than Normal	Cooler than Normal	Increased	Decreased	Warmer than Normal	Cooler than Normal	Increased	Decreased	Warmer than Normal	Cooler than Normal	Increased	Decreased	Warmer than Normal	Cooler than Normal
Impact on Ground nut productivity	+	-	0	0	0	-	-	-	0	-	-	0	-	0	-	-	0	-	-	-	0	-	-	-

Legend
+ Favourable conditions expected to increase productivity of crops (Increase pest/disease attacks on crops)
0 Little or no influence or impact expected
- Non- favourable conditions expected to decrease productivity of crops (Decrease pest/disease attacks on crops)

12.4 Case Studies of Climate Change and its Impact on Groundnut Productivity

12.4.1 Climate Change in Terms of Temperature and its Effect on Groundnut Productivity

A study was conducted by (Paira Singh *et al.*, 2014). where Climate change was projected to alter the growing conditions of groundnut crop differently in different regions of India. The CROPGRO-Groundnut model was used to quantify the impact of climate change on the productivity of groundnut at three sites (Anantapur, Mahboobnagar and Junagadh) in India. Increase in temperature significantly ($p<0.05$) decreased pod yield of groundnut at all the sites by 2050. But the net effect of changes in temperature, rainfall and CO_2 was 4% decreaser in yield at Anantapur and 11% increase at both Mahboobnagar and Junagadh. A number of agronomic practices evaluated under climate change at Anantapur showed that the maximum increase in yield was simulated with supplemental irrigation, followed by delay in sowing and growing a longer maturity variety. at Mahboobnagar,

Table 12.3 *Influence of climate change in terms of pest and disease attack on growth of groundnut*

	Germination and Seedling emergence				Stem elongation and Root Development				Stem Growth and Leaf Development				Inflorescence emergence				Flowering				Pod / Seed Ripening and Harvest			
Climatic Variables	Rainfall		Temperature		Rainfall		Temperature		Rainfall		Temperature		Rainfall		Temperature		Rainfall		Temperature		Rainfall		Temperature	
Assumed change from normal condition	Increased	Decreased	Warmer than Normal	Cooler than Normal	Increased	Decreased	Warmer than Normal	Cooler than Normal	Increased	Decreased	Warmer than Normal	Cooler than Normal	Increased	Decreased	Warmer than Normal	Cooler than Normal	Increased	Decreased	Warmer than Normal	Cooler than Normal	Increased	Decreased	Warmer than Normal	Cooler than Normal
Yellow witch weed ***Alectra vogelii***	+	-	+	-	+	-	+	-	+	-	+	-	+	0	+	0	0	0	0	0	0	0	0	0
Peanut pod nematode Ditylenchus africanus	+	0	+	0	+	0	+	0	0	0	+	0	0	0	+	0	0	0	+	0	0	0	0	0
Fungal diseases																								
Groundnut rust ***Puccinia Arachidas***	0	0	0	0	+	-	+	-	+	-	+	-	+	-	+	+	0	0	0	0	0	0	0	0
Early leaf spot ***Cercospora arachidicola***	0	0	0	0	0	0	0	0	+	-	+	-	+	-	+	-	+	0	+	0	0	0	0	0
Late leaf spot ***Cercospora personatum***	0	0	0	0	0	0	0	0	+	-	+	-	+	-	+	-	+	0	+	0	0	0	0	0
Viral diseases																								
Groundnut Rosette Virus (GRV)	0	0	0	0	0	0	0	0	-	+	+	-	-	+	+	-	-	+	+	-	0	0	0	0

the maximum yield gain was with delayed sowing, followed by growing a longer maturity variety, supplemental irrigation and application of crop residues. At Junagadh, the yield increase was the maximum with supplemental irrigation, followed by application of crop residues. It is concluded that the relative contribution and prioritization of agronomic practices to increase groundnut yield under climate change varied with the region and the CROPGRO-Groundnut model was useful in quantifying such benefits.

12.4.2 Climate Change in Terms of Rainfall and its Effect on Groundnut Productivity

In another study conducted for understanding the relationship between climatic variables and groundnut production in different farming systems in Uganda by (Aizhen *et al.*, 2013), showed a negative rate of technological change revealing productivity losses over the time period studied. In this study alternative production function models are estimated using pooled cross-sectional time series data at the district level. The models incorporate land area, indicators for farming systems, technological change, and either rainfall or the El Niño–Southern Oscillation (ENSO) effect as variables to account for climatic conditions. The results suggested that the Climatic conditions, measured by rainfall, have a non-significant effect; but, when the ENSO phenomenon is used instead a significant negative effect is detected particularly for the warm phase.

12.4.3 Climate Change in terms of Pests/Diseases/ Weeds and its Effect on Groundnut Productivity

(Gadgil *et al.*, 1999) conducted a study i.e. a heuristic model to see the indirect impact of climate variability via the triggering of pests/diseases/weeds for rainfed groundnut over the Anantapur region. A simple hydrological model was used to determine the soil moisture for the rainfall pattern, in any given year. The criteria for determining when specific farming operations, such as ploughing and sowing, are performed are defined in terms of the soil moisture, on the basis of the farming practices -in the region. The events that trigger the growth and incidence/infestation of the major pests/diseases, *viz.* wet/dry spells are also defined in terms of the soil moisture and rainfall. The PNUTGRO model, which can incorporate the direct impact of climate on growth and development as well as the indirect impact via triggering of pests and diseases can be used for understanding the response of the groundnut yield to climate variability and in decision support systems for the region was used. The probabilities of the occurrence of the different pests/diseases incorporated in the model are then calculated by running the model for eighty years with daily rainfall data of Anantapur observatory. The model has been validated by comparison of the results with the observations of the incidence/infestation of specific pests/diseases at the Anantapur agricultural station.

Two kinds of decisions can be concluded from the analysis conducted. The first conclusion out of that is with management options after a specific variety is sown i.e. there are options such as the choice of the strategy (e.g. to spray or not to spray pesticides), while the second, with the choice of the sowing date and/or variety. On the basis of probabilities of the attack of certain pests/diseases estimated by the heuristic model, the cost/benefit of each of the two strategies can be estimated and hence a decision about spraying pesticides could be taken objectively[20]. For example we find that that the probability of a dry spell which promotes leaf miner over this region is above 93%. In about 30% of such years, a drenching shower occurs, implying that no spraying is necessary. Hence once leaf miner populations have started growing, the probability that a pesticide will be useful is 63%. This information could be used in deciding the optimum strategy. The optimum strategy may involve spraying as soon as the leaf miner populations cross the threshold or a 'wait and see' strategy depending on the climatic variability of the region 21.

The second conclusion of the study was for tailoring crop varieties/operations to be optimal for the climate variability of the region (or the expected variability for a given year if reliable predictions are available). Such tailoring is essential in diseases such as pod-rot for which there are no remedial measures available. For such tailoring, an easy. option available to the farmer is the choice of the sowing date and choice of best suitable cultivars. Although choosing an alternative sowing date may not involve any additional costs, it could involve losses, if the farmer chooses to skip the first opportunity and the second one does not arise.

12.5 Conclusion

Anthropogenic greenhouse gas emissions and climate change have a number of implications for agricultural productivity, but the aggregate impact of these is not yet known and indeed many such impacts and their interactions have not yet been reliably quantified, especially at the global scale. An increase in mean temperature can be confidently expected, but the impacts on productivity may depend more on the magnitude and timing of extreme temperatures. Mean sea-level rise can also be confidently expected, which could eventually result in the loss of agricultural land through permanent inundation, but the impacts of temporary flooding through storm surges may be large although less predictable. Freshwater availability is critical, but predictability of precipitation is highly uncertain and there is an added problem of lack of clarity on the relevant metric for drought - some studies including IPCC consider metrics based on local precipitation and temperature such as the Palmer Drought Severity Index, but this does not include all relevant factors. Agricultural impacts in some regions may arise from climate changes in other regions, owing to the dependency on rivers fed by precipitation, snowmelt and glaciers some distance away. Drought may also be offset to some extent by an increased efficiency of 2020.

Climate change impacts on crop yield are often integrated with its effects on water productivity and soil water balance. Global warming will influence temperature and rainfall, which will directly have effects on the soil moisture status and groundwater level. Crop yield is constrained to crop varieties and planting areas, soil degradation, growing climate and water availability during the crop growth period. With temperature increasing and precipitation fluctuating, water availability and crop production will decrease in the future. If the irrigated areas are expanded, the total crop yield will increase; however, food and environmental quality may degrade. Soil evaporation and plant transpiration will be changed with climate change; thus, water use efficiency may decrease in the future. Improving water productivity and keeping stable relations with global food suppliers will be vital for food security.

A critical variable is the number of days a crop is exposed to temperatures above specific thresholds at critical growth stages, i.e. flowering, pollination or grain-filling, thus reducing the quantity and quality of economic yield (Prasad *et al.*, 2003). Free air carbon enrichment (FACE) experiments indicated crop productivity to increase in the range of 15-25% for C3 crops, like wheat, rice and groundnut, with the increase in CO_2 concentration in the atmosphere (Tubiello *et al.*, 2007). The positive effects of enhanced CO_2 will continue alongside the negative effects of supra-optimal temperatures until the temperature thresholds are reached agronomic techniques are available to help farmers adapt.

These early-stage adaptations include varying sowing dates and cultivars, fertilization, irrigation scheduling, soil management such as reduced tillage and residue management, changing to better-adapted alternative crops and increasing crop diversity (Aggarwal, 2008 and Howden *et al.*, 2007).

Because agriculture will not experience the same magnitude of vulnerability to climate change in all regions, site-specific cropping systems and management practices will be needed to match yield potential with inputs, soil fertility and the range of climate variability in each area. Research technologies and management tools that can accelerate the adaptation of cropping system simulation modelling and remote sensing (Boote *et al.*, 2011).

It is well known that the sowing date has a significant impact on the yield. For example, Singh *et al.,* 6 have shown that under biotic stress-free conditions, the simulated yield increased from 0.4 to 1.2 tons/hectare when the sowing date was changed from 12 July to 12 August in 1988. In addition, it may also be possible to choose varieties which differ in the length of the growing season. For a given variety, the choice of sowing date determines the life history stages at which the different critical events such as wet spells/dry spells occur and hence the extent of moisture and biotic stress suffered. It also determines whether wet/dry spells will occur during harvest and lead to operational difficulties.

By detailed analysis of the variation in yield with sowing dates from the climate variability of the region, it would be possible to come up with prescriptions of the optimum sowing date for the region, as well as suggestions of traits (like duration, etc.) of a variety which is best suited for the climate variability of the region.

Supplemental irrigation and change in date of sowing are the major agronomic management options discussed in scenario of climate change which will be attractive to farmers and helps to address the financial risk associated with climate variability in groundnut.

References

1. *Aggarwal, P.K.(2008). Global climate change and Indian agriculture: impacts, adaptation and mitigation. Indian J. Agric. Sci.,* ***78****: 911-919.*
2. *Aizhen, Li., Boris, E., Bravo-Ureta., David, K., Okello., Carl, M., Deom and Naveen-Puppala, 2013. Groundnut Production and Climatic Variability: Evidence from Uganda. Zwick Center for Food and Resource Policy. Working Papers Series No. 17.*
3. *Boote, K.J., Ibrahim, A.M.H., Laftte, R., McCulley, R., Messina, C., Murray, S.C., Specht, J.E., Taylor, S., Westgate, M.E., Glasener, K., Bijl, C.G.and Giese, J.H.(2011). Position statement on crop adaptation to climate change. Crop Sci., v****51****: 2337-43.*
4. *Cuculeanu, V., Tuinea, P and Baltoanu, D. 2002. Climate change impacts in Romania: vulnerability and adaptation options. Geol. J.* ***57****: 203–9.*
5. *Gadgil, Sulochana, Abrol, Y. P. and Seshagiri, P. R. 1999. Curr. Sci.* ***76****:548-556.*
6. *Howden, S.M., Soussana, J.F., Tubiello, F.N., Chhetri, N., Dunlop, M., and Meinke, H.(2007). Adapting agriculture to climate change. Proc. National Acad. Sci.,* ***104****: 19691-19696.*
7. *IPCC,(2007).Climate change 2007: the physical science basis. Contribution of working group I to the fourth assessment report of the intergovernmental panel on climate change. Cambridge University Press, Cambridge, UK, pp. 996.*
8. *Lal R. Climate change, soil carbon dynamics, and global food security. 2005. In: Lal R, Stewart B, Uphoff N, et al., editors. Climate change and global food security. Boca Raton (FL): CRC Press: p. 43-113.*
9. *Popova Z and Kercheva M. 2005. CERES model application for increasing preparedness to climate variability in agricultural planning-risk analyses. Phys Chem Earth* ***30****:117–24.*
10. *Prasad, P.V.V., Boote, K.J., Allen, L.H. Jr., and Thomas, J.M.G. (2003). Supra-optimal temperatures are detrimental to peanut (Arachishypogaea L.) reproductive processes and yield at ambient and elevated carbon dioxide. Global Change Biology,* ***9****: 1775-87.*

11. *Singh, P., Boote, K.J., Kumar, U., Srinivas, K., Nigam, S.N., and Jones, J.W.(2012). Evaluation of genetic traits for Improving productivity and adaptation of groundnut to climate change in India. J. Agron. Crop Sci ,* ***198****: 399-413.*

12. *Singh, P., Singh, N.P., Boote1, K. J., Nedumaran, S., Srinivas, K. J. and Bantilan, C. 2014. Management options to increase groundnut productivity under climate change at selected sites in India. June 2013] Management option to increase groundnut productivity under climate change. J. Agrometeoro.* ***16*** *(1): 52-59.*

13. *Tubiello, F.N., Soussana, J., and Howden, S. M. (2007). Crop and pasture response to climate change. Proc. National Acad. of Sci.* ***105****: 19686-90.*

Chapter 13

Groundnut based Cropping Systems

D. Sampath Kumar

Principal Scientist (Agronomy) & Co-ordinator, ANGRAU, DAATT Centre, Anantapuramu.

13.1 Introduction

Groundnut (*Arachis hypogea* L.), an unpredictable legume, may continue to be an important commercial crop in rainfed areas. The crop contributes about 40 percent to the total oilseeds production in the country. At this level of contribution, the projected demand for groundnut by 2020 AD will be about 14 mt. The present production level is around 8.2 mt. Therefore, a gap of about 5.8 m t has to be bridged. This calls for a growth rate of about 2.2 percent per annum, in production. This growth has to come mainly from the increase in productivity. Productivity of groundnut has been showing a positive growth since eighties. At present, average yield in the kharif is around 1000 kg/ha and that of in the rabi/summer is around 2000 kg/ha. A fair projection for enhanced productivity with sustainability by 2020 AD will be about 1200 kg/ha for kharif groundnut and 2500 kg/ha for Rabi/ summer groundnut. But both production and productivity especially of kharif groundnut have shown highly fluctuating trends. The erratic trend in productivity, relatively high cost of cultivation, the associated financial risk and non availability of quality seed in sufficient quantities, have led to a decline in area and production in some traditional groundnut belts. Good management of cropping system where in the groundnut has been involved may increase the productivity levels of the system as whole and also the groundnut.

A cropping system refers to growing a combination of crops in space and time.

13.2 An Ideal Cropping System

- Should use natural resources efficiently
- Provide stable and high returns
- Do not damage the environment

13.3 Commonly Practiced Cropping Systems

- Intercropping systems
- Mixed cropping systems
- Crop rotation practices

Rain dependent groundnut based production systems have been and may continue to be the mainstay of the future of groundnut agriculture in India. Not more than 20 per cent of the total groundnut area is expected to be under irrigation by 2020 AD, from the existing 17 percent. Adverse weather conditions like delay in the onset of rains and/or failure of rains for few days to weeks some time or other during the crop period is very common in the rain fed groundnut growing areas. Adverse weather conditions result in economic losses to the farmers due to the partial or total failure of groundnut crop. To over come this situation there is need to adopt or follow groundnut based cropping systems like intercropping or mixed cropping in rain fed groundnut growing areas. Intercropping refers to growing more than one crop in the same land area in rows of definite proportion and pattern. Intercropping seems to be a viable proposition, since it is known to use the growth resources better with greater stability of production across different seasons, minimizing the risk of crop production, acting as an insurance against total crop failure, besides enhancing the cropping intensity. Farmers usually intercrop pigeonpea with groundnut in different row proportions, to meet the domestic needs in the rainfed red soil areas of Andhra Pradesh. Some farmers grow castor, pearlmillet, sorghum as intercrops in groundnut, without any definite row proportion and recently sunflower is also gaining importance in cropping systems, due to its short duration and photo insensitivity.

13.4 An Ideal Intercropping Should Aim To

- Produce higher yields per unit area through better use of natural resources;
- Offer greater stability in production under adverse weather conditions and with disease and insect infestation;
- Meet the domestic needs of the farmer;
- Provide an equitable distribution of farm resources.

The following intercropping practices were found to be remunerative than sole crop of groundnut in AP.

Groundnut + Red gram (pigeon pea)	7:1 ratio
Groundnut + Castor	7:1 ratio
Groundnut + Sorghum	6:2 ratio
Groundnut + Pearl millet	6:2 ratio

(7:1 ratio indicates 7 rows of groundnut and one row of the other crop)

Groundnut + Sorghum inter cropping system is recommended to farmers to meet the fodder needs of cattle and milch animals.

The following intercropping practices were found to be remunerative in other groundnut growing States in India.

State	Crop combination
Maharastra	Groundnut + Red gram (6:1/4:1) Groundnut + Soybean (6:2) Groundnut + Sunflower (6:2/3:1)
Gujarat	Groundnut + Castor (9:2/3:1) Groundnut + Sunflower (3:1/2:1) Groundnut + Red gram (4:1)
Tamil Nadu	Groundnut + Bengal gram (4:1) Groundnut + Pearlmillet (4:1) Groundnut + Greengram (2:1)
Karnataka	Groundnut + Chilli (2:2) Groundnut + Sunflower (3:1) Groundnut + Red gram (4:1)

13.5 Groundnut Intercropped with Long Season Annual Crops

No growth studies have been reported for combinations of groundnuts with any of the long season annuals. However, it is evident from the general growth patterns of the crops that considerable temporal complementarily of growth occurs. The groundnuts can give reasonably efficient use of resources during the early period when the long season annuals are slow to establish; after groundnut harvest, the long season annuals are able to make use of later resources, especially of the residual soil moisture.

13.6 Groundnut/Pigeonpea Intercropping

This combination is particularly prevalent on red soils of the southern States of India. A common practice here is that if rains commence at the normal time a groundnut/sorghum or groundnut/millet intercrop is grown, but if rains are delayed groundnut/pigeonpea is grown. Pigeonpea rows are usually wide-spaced up to 5 m apart with up to 10 groundnut rows in between. This traditional practice helps to obtain high yields of the groundnut cash crop but the overall advantage of intercropping may not be high because pigeonpea is too sparsely distributed to make efficient use of late season resources and produce a worthwhile yield contribution. Most studies have examined this predominantly groundnut situation.

(John et al. 1943) reported from a 3-year study that groundnut/pigeonpea in 8:1 proportion was 43% more profitable than sole groundnut. Similar results were reported from studies at Tindivanam over a 7-year period during 1942-49 (Seshadri et al. 1956). (Veeraswamy et al. 1974) and (Appadurai et al. 1974) showed that the arrangement of 6 groundnut: 1 pigeonpea was more economical than 8:1; groundnut gave 99% of its sole crop yield and pigeonpea 37% of its sole crop yield, totaling an advantage of 36%.

Sampath Kumar et al. (2010) reported that, intercropping of groundnut + castor or groundnut + pigeonpea system in 7:1 row proportion would be the best system for the rainfed alfisols of Andhra Pradesh in view of their higher productive capacity and efficiency of cropping system as indicated by increased cropping intensity. The highest Land Equivalent Ratio (LER) was recorded with groundnut + castor 7:1 system (1.66) and groundnut + pigeonpea 7:1 system (1.65) indicating an yield advantage of 66 and 65 per cent, respectively with groundnut + castor and groundnut + pigeonpea, during the two years of study. Higher yield advantage in these intercropping systems was due to higher yield of component crops as a result of minimum competition between the base and intercrops, due to differential growth habits.

At the other extreme, an alternate row arrangement at ICRISAT gave an LER of 1.53 comprising 95% pigeonpea and 58% groundnut (Rao and Willey 1980). This may not be ideal economically because of the reduced groundnut contribution, but it illustrates that higher yield advantages can be obtained with higher proportions of pigeonpea.

13.7 Groundnut/Castor Intercropping

(Reddy et al. 1965) reported that growing castor mixed with groundnut was better than raising a pure crop of castor, and monetary returns were 61.9% higher than pure castor. They also reported that the yield of castor was more when it was grown mixed with groundnut compared to castor grown mixed with greengram, cowpea, *Setaria,* millet or sorghum. (Tarhalkar and Rao 1975) reported that intercropping of castor/groundnut gave monetary returns up to Rs 4394 per hectare compared with Rs 3317 per hectare obtained from a pure castor crop. Similarly, Sampath Kumar et al. (2010) also reported highest pooled mean groundnut pod equivalent yield with groundnut + castor system at 7:1 row arrangement (2636 kg/ha) which was however, comparable with groundnut + pigeonpea system (2541 kg/ha).

13.8 Groundnut/Pearl Millet Intercropping

(M.S. Reddy et al. 1980) reported that, groundnut produced about the same yield per plant in intercropping with pearlmillet as in sole cropping, while the much more dominant millet approximately doubled its yield per plant in intercropping. The actual LERs were 0.71 for groundnut and 0.55 for millet, giving a total LER of 1.26, indicating an overall yield advantage of 26% for groundnut + perlmillet intercropping system at 3:1 row arrangement. Similarly (Sampath Kumar et al. 2001) reported that, LER was highest

(1.60) with groundnut/pearlmillet intercropping at 3:1 row proportion which was comparable with 2:1 arrangement (1.53), indicating an yield advantage of 60 and 53 per cent over sole crops respectively.

Mixed cropping refers to simultaneously growing more than one crop in the same land area as a mixture. Unlike in intercropping system, in mixed cropping the crops are grown without any definite proportion or pattern. Mixed cropping is practiced in traditional subsisting farming to meet the domestic needs of the farmer's family. Thus, the number of crops grown mixed varies depending on the family needs. Even though crops in the mixed cropping meets the farmer's family needs, the yield of crops will be very low due to the competition between the crops for water, light, nutrients etc. A better cropping system will be adopting intercropping system involving the major crops. However to meet the family needs, the minor crops could still be grown on field bunds, on the field borders, and in the house back yards.

13.9 Groundnut based Sequence Cropping Systems

In addition to the intercropping and mixed cropping systems, crop rotations also plays vital role in the productivity levels of groundnut production system. Growing the same crop year after year in the same field is called mono cropping. In contrast to mono cropping, in crop rotation practice crops are grown in a planned sequence from season to season within a year or from year to year.

This planned rotation sequence may be

- Within a year – double cropping in two seasons
- For a two- year period – two year crop rotation
- Three-year period – three year crop rotation.

13.10 Some of the General Purposes of Crop Rotations

- Improve or maintain soil fertility,
- Reduce erosion,
- Reduce the build-up of pests,
- Spread the workload on family labor, use of bullocks and farm equipment, hired labor etc.
- Reduce risk of weather damage,
- Reduce reliance on agricultural chemicals,
- Increase net profits.

When selecting a rotation, the long term viability of that rotation to reduce weed, insect and disease pressure, as well as its economic viability must be considered. A well developed plan which can be altered when necessary should always be followed. Otherwise, the desired crop sequence may be interrupted and the maximum benefits of the rotational effect will not be obtained. The following are important considerations when designing a crop rotation, regardless of location.

13.11 How will the Previous Crop Affect Subsequent Crop Production?

- Disease, insect, weed control, and residue management
- Soil moisture and fertility utilization
- Seed bed preparation
- Planting and harvesting schedule
- Gross economic returns

Generally, as a rainfed crop, groundnut is grown year after year in a mono cropping system. This practice may lead to the crop failure due to adverse weather conditions during crop growth period. The groundnut in general should be rotated with cereals like maize, wheat, sorghum (jowar), pearl millet (bajra) or minor millets. It was also found that groundnut does well after wheat, maize, sorghum (jowar), pearl millet (bajra) and tobacco. The crop sequences recommended for different groundnut growing states in India are as follows:

State	Rain fed-Two-Year Rotation	Residual Moisture Double Cropping
Andhra Pradesh	Groundnut-Sorghum Groundnut-Millet Groundnut-Tobacco	Groundnut-Bengalgram Groundnut-Saflower Groundnut-Sesame
Gujarat	Groundnut-Sesame	Groundnut-Sorghum (Fodder) Groundnut-Mustard
Karnataka	Groundnut-Sorghum Groundnut-Ragi	Groundnut-Saflower
Maharastra	Groundnut-Sorghum	Groundnut-(Saflower) Groundnut-Maize(Fodder)
Madhya Pradesh	Groundnut-sorghum	Groundnut(Saflower)
Orissa	Groundnut-Sorghum/ Pearlmillet-bajra	Groundnut-Bengal gram Groundnut-Sesame Groundnut-Sorghum(Fodder) Groundnut-Mustard

Contd...

State	Rain fed-Two-Year Rotation	Residual Moisture Double Cropping
Punjab	Groundnut-Sorghum/ Pearlmillet-(bajra)	Groundnut-Mustard
Rajasthan	Groundnut-Pearlmillet (bajra)	Groundnut-Barley
Tamil Nadu	Groundnut-Sesame Groundnut-Cotton	Groundnut-Sesame
Uttar Pradesh	Groundnut-Sorghum/ Pearlmillet-Bajra	Groundnut-Mustard

(Sampath Kumar et al. 2010) reported that, there is adequate scope for groundnut based sequence cropping in rainfed alfisols in scarce rainfall zone of Andhra Pradesh. Pooled mean total groundnut pod equivalents of the sequence cropping system as whole was highest with groundnut – clusterbean (vegetable) system at sandy loam soils of Kadiri. This was followed by groundnut – cowpea, groundnut – horsegram and groundnut – greengram, which were comparable among them, but significantly superior to groundnut – sunflower, groundnut – pearlmillet and groundnut – sorghum. The lowest pooled mean economic yield was recorded with groundnut – sorghum sequence cropping system. Production efficiency and land use efficiency were highest with groundnut – clusterbean system, indicating that productivity per unit area per unit time as well as the efficiency with which the land was used for intensive crop production would follow pattern of economic yield of the sequence cropping system. (Reddi e*t al.,* 1981) and (Soundara Rajan *et al.,* 1990) also confirmed the possibilities of double cropping in rainfed alfisols of Chittoor district (Andhra Pradesh) provided there is there is extended rainfall due to north-east monsoon.

13.12 Conclusions

There is good evidence that, intercropping groundnut with cereals and other long duration crops always shows yield advantages over their sole crops and gives more productivity per unit area. There is sufficient research results on the yield and other complementarily benefits in groundnut intercropping with pigeonpea. More rapid early growth of the cereals and late maturity of groundnut compared with early cereals may also be an important factor in giving complementarily between groundnut and cereals and allowing them to better utilization of resources viz., moisture and light. Similarly intercropping groundnut with long duration crops, also have same sort of better use of resources. Growth of long duration crops like pigeonpea, castor will be slower in early stages and groundnut completes its duration by the time of these long duration crops comes to reproductive stage and there by better temporal and spatial advantages. Further, there is much scope for groundnut based cropping sequence in rainfed areas where there is benefit of bimodal distribution of rainfall. Groundnut followed by clusterbean for green pods or cowpea or greengram are resulted in higher production and land use efficiency. However, much research needed in this direction to establish profitable groundnut based cropping systems.

References

1. *Appadurai, R., and Selvaraj, K. V. 1974. A note on groundnut — redgram mixture in lower Bhawani Project area. Madras Agricultural Journal 61: 803-804.*
2. *John, C. M., Seshadril, C. R., and Bhavani Shankar Rao, M. 1943. Mixed cropping of groundnut. Madras Agricultural Journal 31: 191-200.*
3. *Rao, M. R., and Willey, R. W. 1980. Preliminary studies of intercropping combinations based on pigeonpea or sorghum. Experimental Agriculture 16: 29-39.*
4. *Reddy, G. P., Rao, S. C, and Reddy, R. 1965. Mixed cropping in castor. Indian Oilseeds Journal 9(4): 310-316.*
5. *Reddy, M. S., and Willey, R. W. 1980. Growth and resource use studies in an intercrop pearl millet/groundnut. Field Crops Research: 133-141*
6. *Reddi, G. H.S., Soundara Rajan, M.S. and Venkataramana Reddy, V. 1981. Sequence cropping in drylands. Crop management to meet new chanllenges: 1-10.*
7. *Sampath Kumar, D., Reddy, C.R. and Reddy, S.R. 2001. Production potentials of dryland intercropping systems under different row proportions of groundnut, sunflower and pearlmillet. Journal of Oilseeds Research. 18(2): 275-278.*
8. *Sampath Kumar, D., Reddy, C.R. and Reddy, S.R. 2001. Production potentials of dryland intercropping systems under different row proportions of groundnut, sunflower and pearlmillet. Journal of Oilseeds Research. 18(2): 275-278.*
9. *Sampath Kumar, D., Reddy, D.S. and Reddy, T.Y. 2010. Productivity of groundnut (Arachis hypogaea L.) based intercropping systems under rainfed conditions. Current Biotica. 3(4): 490-499.*
10. *Sampath Kumar, D., Reddy, D.S. and Reddy, T.Y. 2010. Production potential and economic feasibility of groundnut based sequence cropping systems in rainfed alfisols. Current Biotica. 4(1): 73-81.*
11. *Seshadri, C. R., Aiyadurail, J. G., and Srinivasulu, N. 1956. Groundnut-mixed cropping experiment. Madras Agricultural Journal 43: 496-504.*
12. *Soundara Rajan, M. S., Sankaranarayana Rao, G. M. and Usharani, N. R. 1990. Profitability of efficient cropping systems in drylands. Indian Farming. 40(6): 21-22.*
13. *Tarhlakar, P.P., and Rao, N.G.P. 1975. Changing concepts and practices of cropping systems. Indian Farming, June 3-7.*
14. *Veeraswamy, R, Rathnaswamy, R and Palaniswamy, G. A. 1974. Studies in mixed cropping of red - gram and groundnut under irrigation. Madras Agricultural Journal 61: 801-802.*

Chapter 14

Role of Nanoscale Nutrients for Improved Groundnut Production

T. N. V. K. V. Prasad, P. Sudhakar, K. V. Nagamadhuri, P. V. R. M. Reddy and T. Giridhara Krishna

Nanotechnology laboratory, IFT RARS, ANGRAU, Tirupati -517 502, A.P., India.

ABSTRACT

Nanotechnology is been a recent addition to agricultural crop production technologies. Nanoscale nutrients (<100 nm size in at least one dimension) showed both detrimental and positive effects on plant growth and productivity across a variety of agricultural crops tested so far. Though the research on application of nanoscale materials as nutrients in agriculture is at its infancy, but no crop is runaway from the use of nanoscale materials either in its productivity or protection. However, there is an urgent need to initiate the research in a big way on large scale production of non-toxic nanoscale materials for diverse applications in agriculture sector. This chapter deals with the performance of the nanoscale materials on agricultural crops with the special reference to groundnut crop.

14.1 Introduction

FAO predicted that (2009) the global grain production is required to increase by 70% to meet the food demands of 9.6 billions of population by 2050 (UN, 2013). To increase food production, it is necessary to use the different technologies in agriculture. Nanotechnology has been described as a set of technologies relating to materials, systems and processes which operate at a scale of 100 nanometers or less (Ditta, 2012; Mousavi and (Rezai, 2011). Nanoscale materials show novel physical, chemical and biological properties, which are completely distinct from their bulk counter parts and individual molecules (Li *et al.*, 2001). Nanoparticles have large surface to volume ratio and possess strong affinity to biological targets such as proteins (Kumar *et al.*, 2010). With the limited natural resources such as land, water and soil fertility, demand for food has increased tremendously as it is required to feed the population of 9 billion by 2050. The cost of chemical fertilizers, pesticides and other production inputs has been drastically increased due to limited reserves of natural gas and petroleum (Ditta, 2012). There is a need to overcome these constraints with the help of precision farming practices and effective application of nanotechnology to agriculture (Chowdappa *et al.*, 2013).

Nanotechnology has provided new solutions to problems in plants and food science (post-harvest products) and offers new approaches to the rational selection of raw materials, or the processing of such materials to enhance the quality of plant products. The heart of nanotechnology lies in the ability to compress the tools and devices to the nanometer range and to accumulate atoms and molecules in to bulkier structures while the size remains very small (Chowdappa *et al.*, 2013). In developing countries, nanotechnology has got important application for enhancing agricultural productivity, along with other emerging technologies such as biotechnology including genetic s, plant breeding, disease control, fertilizer technology, precision agriculture, and other allied fields (Chowdappa *et al.*, 2013).

Nanotechnology can be used for combating the plant diseases either by controlled delivery of functional molecules or as diagnostic tool for disease detection. Nanosensors and other field sensing devices can be used in detection and measurement of crop nutrient status, insects, pathogens, weeds, moisture level, soil fertility, temperature, etc. which helps in real time monitoring of the crop growth and provide essential data for precision farming practices leading to minimize agricultural inputs and maximizing resource output and yield (Scott and Chen, 2003). Nanosensors can also provide information about optimal times for planting and harvesting crops and provides useful information for timely application of agrochemicals like fertilizers, pesticides and herbicides (Chowdappa *et al.*, 2013).

14.2 Plant Production

Plant-based agricultural production is the basis of broad agriculture systems providing food, feed, fiber, fire (thermal energy), and fuels through advancements in materials sciences, and biomass conversion technologies. While the demand for crop yield will rapidly increase in the future, the agriculture and natural resources such as land, water and soil fertility are finite. Other production inputs including synthetic fertilizers and pesticides are predicted to be much more expensive due to the constraints of known petroleum reserve. Precision farming is hence an important area of study to minimize production inputs and maximize agricultural production

Outputs for meeting the increasing needs of the world sustainability. Given that nanotechnology may allow for the precise control of manufacturing at the nanometer scale, a number of novel possibilities in elevating the precision farming practices are possible (Chen *et al.*, 2011).

Many nanoscale carriers, including encapsulation and entrapment, polymers and dendrimers, surface ionic and weak bond attachments and other mechanisms may be used to store, protect, deliver and release by control of intended payloads in crop production processes. One of the advantages of nanoscale delivery vehicles in agronomic applications is its improved stability of the payloads against degradation in the environment, thereby increasing its effectiveness while reducing the amount applied. The nanoscale delivery vehicles may be designed to "anchor" to plant roots or the surrounding soil structures and organic matter if molecular or conformational affinity between the delivery nanoscale structure and targeted structures and matters in soil could be utilized (Johnson *et al.*, 2010).

14.3 Nanoparticles as Macronutrient Nanofertilizers

In the process of achieving the sustainability in agriculture, applying innovative technologies, nanotechnology in particular, in agriculture including fertilizer development is considered to be one of the promising approaches to significantly increase crop production. It is evident that the efficiency of conventional fertilizers is as low as 30-50% and leaving a few management options to enhance the productivity necessitates the application of nanotechnology to fertilizer research and development. A few reports are available in the literature on nanofertilizer research but directly-related research is a draw back. (Ghormade *et al.*, 2011; Nair *et al.*, 2010). Huge quantities of macronutrient fertilizers (mainly N P and K fertilizers) are used for increasing the production of food, fiber, and other essential commodities. Total global macronutrient fertilizer ($N + P_2O_5 + K_2O$) consumption was 175.7 million ton (Mt) in 2011 and is projected to increase to 263Mt in 2050 (Alexandratos and Bruinsma, 2012). It was estimated that N fertilizers have contributed a roughly 40 percent increase in per-capita food production in the past 50 years, indicating the critical role of these macronutrient fertilizers in global food production (Smile, 2002). Moreover, due to the low efficiency (30–50%) and heavy application of these macronutrient fertilizers, significant amounts of these nutrients (N and P) are transported into surface and groundwater bodies, disrupting aquatic ecosystems and threatening health of human and aquatic life. Therefore, an urgent and practically essential research direction is to develop highly efficient and environmentally-friendly macronutrient (N and P) nanofertilizers to replace the conventional N and P fertilizers and to ensure the sustainable food production while protecting the environment. A brief discussion on some promising macronutrient nanofertilizer is given below (Ruiqiang Liu *et al.*, 2015).

14.3.1 Phosphorus (P) NPs

Hydroxyapatite ($Ca_5(PO_4)_3OH$) NPs of ~16 nm in size were synthesized by (Liu and Lal 2014) and evaluated the fertilizing effect on soybean (Glycine max) through a greenhouse experiment. The data showed that application of NPs increased the growth rate and seed yield by 33% and 20%, respectively, compared to those with a regular P fertilizer ($Ca(H_2PO_4)_2$. It is evident from the data that soybean roots can absorb hydroxyapatite NPs as an effective P nutrient source and maintain healthy growth and high yield (Ruiqiang Liu *et al.*, 2015). In addition to P, the NPs may have also supplied the plants with nutrient Ca. Using apatite NPs as a new class of P fertilizer can potentially enhance agronomical yield, increase efficiency of P application, and reduce risks of eutrophication. For example, the apatites NPs have much weaker interactions with soil components than charged ions of PO_4^{3-}, HPO_4^{2-}, H_2PO_4-, or Ca^{2+}. Therefore, a significant portion of NPs remain in the soil solution for roots to absorb while most of the charged phosphate ions from conventional P fertilizers are often absorbed by the soil particles and not available for plant uptake (Ruiqiang Liu *et al.*, 2015).

14.3.2 Calcium (Ca) NPs

The most critical element in the production of groundnut with sound mature kernels is calcium (Meena *et al.*, 2007). Calcium is an immobile element in phloem. Calcium moves upward (xylem) in the peanut plant but does not move downward (phloem). Thus, calcium does not move through the peg to the pod and developing kernel. (Amanda and Alexander, 1989). (Skelton and Shear, 1971) stated that calcium is poorly translocated *via* the phloem of the gynophores and must be absorbed by the developing pod. However, groundnut crop is mostly cultivated in rainfed conditions, where moisture cannot be assured at the time of gypsum application, continued to suffer due to calcium deficiency.

Through applying $CaCO_3$ NPs (20–80 nm, 160 mg L^{-1} as Ca) as a Ca nutrient to peanut (*Arachis hypogaea*) seedlings grown in sand with Hoagland solution for 80 days, (Liu *et al.*, 2005) observed that Ca-NPs significantly improved the seedling growth in comparison with that under control (no Ca application). Plant roots could absorb Ca-NPs as Ca source and transport Ca from roots to shoots (Ca content in seedling stem and roots were 3.04 and 1.58%, respectively, as compared with 0.58 and 0.43% for no-Ca control), satisfying the plant demand for Ca. (Liu *et al.*, 2005) also reported that applying Ca-NPs and humic acids (1000 mg L^{-1}) together achieved the maximum seedling growth (dry biomass reached 5.78 g per plant, a 30% increase over that for the control and a 14% over that for the Ca-NPs alone). Therefore, Ca-NPs have a large potential as Ca fertilizers for field crops (Prasad *et al.*, 2014).

Results of the hydroponics experiment using solution culture technique revealed that foliar application of n-CaO at different concentrations (10,50,100,500,1000 ppm) on groundnut plants confirmed the entry of calcium into leaves and stems through phloem compared to bulk source of calcium sprayed (CaO and $CaNO_3$). After spraying of n-CaO, calcium content in roots, shoots and leaves significantly increased. Based on visual scoring of calcium deficiency correction and calcium content in plant parts, we may establish the fact that nanoscale calcium oxide particles (size 69.9 nm), could move through phloem tissue in groundnut. This is the first report on phloem transport of nanoscale calcium oxide particles in plants (Prasad *et al.*, 2014) and this result points to the use of nanoscale calcium oxide particles as calcium source to the plants through foliar application, agricultural crops in particular, as bulk calcium application through foliar nutrition is restricted due to its non mobility in phloem.

14.4 Micronutrient Nanofertilizers

Plant micronutrients include iron (Fe), manganese (Mn), zinc (Zn) and copper (Cu). Only trace levels of micronutrients are required for healthy growth of crops and other plants as is shown in the composition of Hoagland solution (Hoagland and Arnon, 1950). Micronutrients are often added to the N, P, and K fertilizers (collectively called composite fertilizers) at low rates ($<5 mgL^{-1}$) as soluble salts for crop uptake. Micronutrients in these composite fertilizers usually provide enough nutrients and cause little environmental risks. However, plant-availability of the applied micronutrients may become low and micronutrient deficiency may occur in some soils of alkaline pH, coarse texture, or those containing low soil organic matter (SOM) (Fageria, 2009). Presumably, micronutrient nanofertilizers may enhance bioavailability of these nutrients to plants even under these worst-case scenarios. Since development and application of nanofertilizers are still at initial stages, there are few if any specific researches or systemic studies on the effects and advantages of applying micronutrient nanofertilizers under field conditions (Ruiqiang Liu *et al.*, 2015).

14.4.1 Nanoscale Zinc and Zinc Oxide Particles

Zinc plays a fundamental role in protecting and maintaining structural stability of cell membranes (Welch *et al.*, 1982; Cakmak, 2000). Zn is used for protein synthesis, membrane function, cell elongation and tolerance to environmental stresses (Cakmak, 2000). Plants emerging from seeds with low Zn have poor seedling vigor and field establishment on Zn-deficient soils (Yilmaz *et al.*, 1998). (Rengel and Graham, 1995) reported from pot culture experiments on wheat plants that increasing seed zinc content from 0.25 μg per seed to 0.70 μg per seed significantly improved root and shoot growth under Zn deficiency. Hence, it may be concluded that high Zn content in seed could act as a starter fertilizer. (Ajouri *et al.*, 2004) reported that seed priming with Zn was very effective in improving seed germination and seedling development in barley. (Prasad *et al.*, 2012) reported on effects of nanoscale Zinc Oxide particles on the germination, growth and yield of peanuts. The results indicated that high Zn concentration in seeds has very important physiological roles during seed germination and early seedling growth. (Slaton *et al.*, 2001) reported that treating rice seeds with Zn greatly increased grain yield and concluded that this type of Zn application method is a very economical alternative to more expensive broadcast Zn fertilizer applications. Studies revealed that, treating groundnut seeds with nanoscale ZnO particles with a concentration of 1000 ppm has shown significant increment in germination, shoot length, root length and vigor index over other concentrations of the same material and varying concentrations of another material (chelated zinc sulfate) tested. The exact reason for these effects is not known but it is likely to be due to the higher concentrations of zinc in the seed when treated with nanoscale ZnO particles (Prasad *et al.*, 2012).

Foliar fertilization is more effective than soil application. Foliar Zn application significantly increased grain Zn concentrations of wheat, indicating high mobility of Zn within plants. Spraying with 0.5% $ZnSO_4$ gave significantly higher peanut pod yield compared to no spraying. However, soil application of 10 kg ha^{-1} $ZnSO_4$ at sowing gave yield on par with no

$ZnSO_4$ application. This indicates that groundnut responds to foliar spray but not to soil application (Channabasavanna and Setty, 1993). The effectiveness of various synthetic and natural chelates has been widely investigated (Alvarez and Gonzalez, 2006; Gonzalez *et al.*, 2007; Prasad and Sinha, 1981). Apart from their effectiveness, application of chelates is generally expensive and may result in potential leaching risk because the more mobile the chelate, or the less biodegradable the carrier, the greater the risk of leaching (Gonzalez *et al.*, 2007). Zinc sulfate, which is highly soluble, can easily be taken up by plants but is known to fall off quickly. The retention time in the plant system is low. So the bioavailability of nutrients for long period was not sure with the use of $ZnSO_4$. If the plants are soft or sensitive and the conditions are harsh like high temperatures, $ZnSO_4$ has a large salt index, which may burn the plant. Moreover, the zinc content in the mixture is usually very low (9-12%). But studies revealed that ZnO in the nanoscale form is absorbed by plants to a larger extent unlike bulk $ZnSO_4$. These particles proved effective in enhancing plant growth, development and yield. A lower dose of foliar application is proved to be significantly productive. The post harvest leaf and kernel samples analysis revealed a significant increment in zinc content in leaves (42%, 29%) and kernels (42%, 36.6%) when supplied with nanoscale ZnO (Prasad *et al.*, 2012) compared to chelated $ZnSO_4$ (in *Rabi* seasons 2009 and 2010 respectively). Similarly, nanoscale nutrients at high concentrations are detrimental just as the bulk nutrients. Similar results were observed by (Racuciu and Creanga 2007) when they analyzed the influence of magnetic nanoparticles coated with tetramethyl ammonium hydroxide on the growth of *Zea mays* plant in early ontogenetic stages.

The mechanism of foliar uptake pathway for aqueous solutes and water–suspended nanoparticles was well discussed by (Eichert *et al.*, 2008) in the context of *Allium porrum* and *Vicia faba* (L). The results suggest that the stomatal pathway differ fundamentally from the cuticular foliar uptake pathway. Low penetration rates in thick leaves, rapid drying of spray solution, limited translocation within the plant, and leaf damage are the problems of concern and most foliar applied micronutrients are not efficiently transported towards the roots. Concentrated liquid suspensions of ZnO are used for foliar application but their performance is strongly determined by the size range specification of the ZnO particles present in the formulation (Moran, 2004). Leaf water repellency of adaxial or abaxial surface is a main limiting factor, which can affect the Zn uptake through spray application processes (Watanabe and Yamaguchi, 1991; Holder, 2007). The permeability of the cuticle to water and to lipophilic organic molecules increases with mobility (distribution co-efficients) and solubility (partition co-efficients) of these compounds within the transport-limiting barrier of the cuticles. Ions being highly water soluble might have some hindrance in penetrating the lipophilic cuticle. This may be acting as a limiting factor in the case of chelated $ZnSO_4$. But our custom-made nanoscale ZnO, which is having less hydrophilicity and being more dispersible in lipophilic substances compared to the ions, can penetrate through the leaf surface (Da Silva *et al.*, 2006) compared to $ZnSO_4$. Also the mobility of the nanoparticles is known to be very high which ensures the phloem transport and ensures the nutrient to reach all parts of the plant. The presence of nanoparticles both in the extracellular space and within some cells in the living plant *Cucurbita pepo* was reported (Gonzalez-Melendi *et al.*, 2008). The bioavailability of the nanoparticle because of its size and lower water solubility (which inhibit rapid falling off compared to ionic supplements) can also be higher compared to chelated $ZnSO_4$. The inherent small size and the associated large surface area of nanoscale ZnO fertilizer may increase the uptake as reported earlier. This enhanced uptake of Zn was seen in the EDAX analysis of the seeds also. All these factors may be responsible to give higher yields for nanoscale ZnO compared to chelated $ZnSO_4$. The promotory effects of nanoscale ZnO at cellular level has to be understood by further in depth investigations (Prasad *et al.*, 2012).

14.4.2 Nanoscale Iron Particles

The impact of using Fe-NPs was similar to that of an effective Fe source for the plants - Fe-EDTA at concentrations by 45 mg L^{-1} as Fe. (Delfani *et al.*, 2014) reported that a foliar application of 500 mg L^{-1} Fe-NPs to black-eyed peas significantly increased the number of pods per plant (by 47%), weight of 1000-seeds (by 7%), Fe content in leaves (by 34%) and chlorophyll content (by 10%) over those of the controls. (Ghafariyan *et al.*, 2013) reported that low concentrations of super paramagnetic Fe-NPs significantly increased the chlorophyll contents in sub-apical leaves of soybeans in a greenhouse test under hydroponic conditions, suggesting that soybean could use this type of Fe-NPs as source of Fe and reduce chlorotic symptoms of Fe deficiency. Application of Fe-NPs also improved crop performance more than that by application of a regular Fe salt. The abovementioned parameters were increased by 28%, 4%, 45%, and 12%, respectively, under the Fe-NPs treatment compared with these under treatment with a Fe salt. In addition, Fe-NPs significantly improved the beneficial effect of another nanofertilizer (Mg- NPs) on black-eyed peas (Ruiqiang Liu *et al.*, 2015). For an optimal growth, most plants generally require 1–5mg L^{-1} Fe in soil solution (Hoagland and Arnon, 1950).

14.4.3 Nanoscale Manganese Particles

(Pradhan et al., 2013) observed that Mn-NPs enhanced growth of mung bean (*Vigna radiata*) and augmented its photosynthesis. The mung bean seedlings were incubated for 15 days in an inert media (perlite) with Hoagland solution in growth chambers. Application of Mn-NPs at 0.05 mgL^{-1} produced the maximum growth enhancement over that of the controls (without Mn) in root length (by 52%), shoot length (by 38%), number of rootlets (by 71%), fresh biomass (by 38%), and dry biomass (by 100%). Interestingly, application of $MnSO_4$exhibited an inhibitory effect on the plant growth at concentration of 1 mg L^{-1} while the response was still positive at this level with the application of Mn-NPs. Plants often require 0.5 mg L^{-1} of Mn in the soil solution for healthy growth (Ruiqiang Liu *et al.*, 2015).

14.4.4 Nanoscale Copper Particles

(Nekrasova *et al.*, 2011) reported on a 3-day incubation study using a type of waterweed (Elodea. Densa planch) and the results indicated that low concentrations (<0.25 mg L^{-1} as Cu) of Cu-NPs stimulated plant photosynthesis rate by 35% compared with that of the control (without Cu). (Shah and Belozerova, 2009) reported that soil amended with metallic Cu-NPs (130 and 600 mg kg^{-1}) significantly increased 15-day lettuce seedling growth by 40% and 91%, respectively. But higher concentrations (200 to 1000 mg L^{-1}) of metallic Cu-NPs exhibited toxic effects on seedling growth of mung bean, wheat (*Triticum aestivum*), and yellow squash (*Cucurbita pepo subspecies ovifera cv yellow crookneck*) (Lee *et al.*, 2008; Musante and White, 2012). (Stampoulis *et al.*, 2009) reported that application of 1000 mg L^{-1} of metallic Cu-NPs reduced biomass of zucchini (*Cucurbita pepo cv costata romanesco*) by 90% compared to that of the control (without Cu) after the seedlings were incubated in Hoagland solution for 14 days. However, aqueous Cu concentration optimal for plant growth is only 0.02 mg L^{-1}, and phytoxicity can occur at higher Cu levels (Ruiqiang Liu *et al.*, 2015).

14.5 Mechanism of Uptake of Nanoscale Nutrients by Plants

The complete mechanism of uptake and absorbance of NPs is not known clearly yet, however, it is possible for NPs to directly enter plant cells through the sieve like cell wall structures if the particle sizes are smaller than the sizes of cell wall pores (5 to 20 nm). But penetration of bigger NPs (even 100nm) through cell membrane and interactions with cytoplasm needs a thorough check of quantum effects of NPs because the quantum confinement of the nanoscale materials can only give clues to better explanation of the NPs mobility and their interaction with the plant cells. However, no research has excluded it as one of the primary mechanisms in this regard that the nutrient elements could be absorbed by plant root system through nanoparticle dissolution in water/soil solution. In other words, NPs simply dissolve in solution and release the nutrient(s) as soluble ions. Plants absorb the soluble nutrient ions indiscriminately as those from the dissolved conventional fertilizers. But, the dissolution rate and extent of NPs in water/soil solution should be higher than those of the related bulk solids because of the much smaller particle sizes and higher specific surface areas of the former (Ruiqiang Liu *et al.*, 2015).

14.6 Conclusion

Nanotechnology application in agriculture and food systems is still at the nascent stage and a lot more applications can be expected in the years to come. Nanotechnology applications have the potential to change agricultural production by allowing better management and conservation of inputs to plant production. Nanotechnology holds the promise of controlled delivery of agrochemicals to improve disease resistance, plant growth enhancement and nutrient utilization. Even though the toxicity of nanomaterials has not yet clearly understood, it plays a significant role in crop protection because of its unique physical and chemical properties. Nevertheless, application of nanoscale materials for the productivity enhancement in groundnut is promising but more research need to be done on all plant nutrients and plant protection agents to know the full potential of nanotechnological interventions in groundnut productivity.

References

1. *Ditta A (2012). How helpful is nanotechnology in agriculture. Adv. Nat. Sci.: Nanosci. Nanotechnol 3:10.*
2. *Mousavi S.R and Rezaei M (2011). Nanotechnology in agriculture and food production. J. Appl.*
3. *Environ. Biol. Sci 1(10): 414–419.*
4. *Li L.S, Hu J, Yang W, and Alivisatos A.P (2001). Band gap variation of size- and shape-controlled colloidal CdSe quantum rods. Nano Letters 1: 349-351.*
5. *Kumar R, Sharon M, and Choudhary A.K (2010). Nanotechnology in agricultural diseases and food safety. Journal of Phytology 2: 83.92.*
6. *Chowdappa P, and Shivakumar gowda (2013). Nanotechnology in crop protection: Status and scope, Review article. Pest Management in Horticultural Ecosystems 19: (2)131-151*
7. *Chen H, and Yada R (2011). Nanotechnologies in agriculture: New tools for sustainable development. Trends in Food Science & Technology 22: 585-594.*
8. *Johnston C.T (2010). Probing the nanoscale architecture of clay minerals. Clay Minerals 45: 245-279.*
9. *Ruiqiang Liu, Rattan Lal (2015). Review: Potentials of engineered nanoparticles as fertilizers for increasing agronomic productions. Science of the Total Environment 514: 131–139.*

10. Ghormade V, Deshpande M.V, and Paknikar K.M (2011). Perspectives for nanobiotechnology enabled protection and nutrition of plants. Biotechnol. Adv 29: 792 803.

11. Nair R, Varghese S.H, Nair B.G, Maekawa T, Yoshida Y, and Kumar D.S (2010). Nanoparticulate material delivery to plants. Plant Sci 179: 154 163.

12. Alexandratos N, and Bruinsma J (2012). World agriculture towards 2030/2050: the revision. ESA Working Paper No. 12-03. FAO, Rome.

13. Amanda Webb J and Alexander Hansen P (1989). Histological Changes of the Peanut (Arachis hypogaea) Gynophore and Fruit Surface during Development, and their Potential Significance for Nutrient Uptake.Annals of Botany.64 (3):351-357.

14. Skelton B.J and Shear G.M (1971). Calcium translocation in the peanut (Arachis hypogaea L.). Agronomy Journal, 63: 409-412.

15. Deepa M , Sudhakar.P, Nagamadhuri K.V, Balakrishna Reddy K, Giridhara Krishna K, Prasad T N V K V. (2014). First evidence on phloem transport of nanoscale calcium oxide in groundnut using solution culture technique, Applied Nanoscience, DOI 10.1007/s13204-014-0348-8.

16. Welch R.M, Webb M.J, and Loneragan J.F (1982). Zinc in membrane function and its role in phosphorus toxicity. In: Proceedings of the Ninth Plant Nutrition Colloquium, ed. A. Scaife, Wallingford, UK: CAB International 710–715.

17. Cakmak I (2000). Role of zinc in protecting plant cells from reactive oxygen species. New Phytologist 146: 185–205.

18. Yilmaz A, Ekiz H, Gultekin I, Torun B, Barut H, Karanlik S, and Cakmak I (1998). Effect of seed zinc content on grain yield and zinc concentration of wheat growth in zinc-deficient calcareous soils. Journal of Plant Nutrition 21: 2257–2264.

19. Rengel Z, and Graham R.D (1995). Importance of seed zinc content for wheat growth on zinc-deficient soil. I. Vegetative growth. Plant and Soil 173: 259–266.

20. Ajouri A, Asgedom H, and Becker M (2004). Seed priming enhances germination and seedling growth of barley under conditions of P and Zn deficiency. Journal of Plant Nutrition and Soil Science 167: 630–636.

21. Prasad T.N.V.K.V, Sudhakar P, Sreenivasulu Y, Latha P, Munaswamy V, Raja Reddy K, Sreeprasad T.S, Sajanlal P.R, and Pradeep T (2012). Effect of nanoscale zinc oxide particles on the germination, growth and yield of peanut, Journal of Plant Nutrition 35:905–927

22. Slaton N.A, Wilson C.E, Ntamatungiro S, Norman R.J, and Boothe D.L (2001). Evaluation of zinc seed treatments for rice. Agronomy Journal 93: 152–157.

23. Channabasavanna A.S, and Setty R.A (1993). Effect of nitrogen, ferrous sulfate and zinc sulfate on groundnut yield in deep black soils. Indian Journal of Agronomy 38: 329–330.

24. Alvarez J. M, and Gonzalez D (2006). Zinc transformations in neutral soil and zinc efficiency in maize fertilization. Journal of Agricultural and Food Chemistry 54: 9488–9495.

25. Gonzalez D, Obrador A, and Alvarez J.M (2007). Behavior of zinc from six organic fertilizers applied to a navy bean crop grown in a calcareous soil. Journal of Agricultural and Food Chemistry 55: 7084–7092.

26. Prasad B, and Sinha M.K (1981). The relative efficiency of zinc carriers on growth and zinc nutrition of corn. Plant and Soil 62: 45–52.

27. Racuciu M, and Creanga D (2007). TMA-OH coated magnetic nanoparticles internalized in vegetal tissues. Romanian Journal of Physics 52: 395–402.

28. Eichert T, Kurtz A, Steiner U, and Goldbach H.E (2008). Size exclusion limits and lateral heterogeneity of the stomatal foliar uptake pathway for aqueous solutes and water suspended nanoparticles. Physiologia Plantarum 134: 151–160.

29. Moran K (2004). Micronutrient Product Types and Their Development. International Fertilizer Society – Proceeding 545. York, England: International Fertilizer Society.

30. Watanabe T, and Yamaguchi I (1991). Evaluation of wettability of plant leaf surfaces. Journal of Pesticide Science 16: 491–498.

31. Holder C.D (2007). Leaf water repellency of species in Guatemala and Colorado (USA) and its significance to forest hydrology studies. Journal of Hydrology 336: 147–154.

32. Da Silva L.C, Oliva M.A, Azevedo A.A, and De Araujo M.J (2006). Response of resting plant species to pollution from an iron pelletization factory. Water, Air and Soil Pollution 175: 241–256.

33. Gonzalez-Melendi P, Fernandez Pacheco R, Coronado M.J, Corredor E, Testillano P.S, Risueno M.C, Marquina C, Ibarra M.R, Rubiales D, and Perez-De-Luque A (2008). Nanoparticles as smart treatment-delivery systems in plants: Assessment of different techniques of microscopy for their visualization in plant tissues. Annals of Botany 101: 187–195.

34. Ghafariyan M.H, Malakouti M.J, Dadpour M.R, Stroeve P, and Mahmoudi M (2013). Effects of magnetite nanoparticles on soybean chlorophyll. Environ. Sci. Technol 47: 10645–10652.

35. Delfani M, Firouzabadi M.B, Farrokhi N, and Makarian H (2014). Some physiological responses of black-eyed pea to iron and magnesium nanofertilizers. Commun. Soil Sci. Plant Anal 45: 530–540.

36. Hoagland D.R, and Arnon D.I (1950). The water-culture method for growing plants without soil. Calif. Agric. Exp. Stat. Circ 347: 1–32.

37. Pradhan S, Patra P, Das S, Chandra S, Mitra S, and Dey K.K (2013). Photochemical modulation of biosafe manganese nanoparticles on Vigna radiata: a detailed molecular, biochemical, and biophysical study. Environ. Sci. Technol 47:13122–13131.

38. *Nekrasova G.F, Ushakova O.S, Ermakov A.E, Uimin M.A, and Byzov I.V (2011). Effects of Copper (II) ions and copper oxide nanoparticles on Elodea densa Planch. Russ. J. Ecol 42: 458–463.*
39. *Shah V, and Belozerova I (2009). Influence of metal nanoparticles on the soil microbial community and germination of lettuce seeds. Water Air Soil Pollut 197:143–148.*
40. *Lee W, An Y, Yoon H, and Kweon H (2008). Toxicity and bioavailability of copper nanoparticles to the terrestrial plants mung bean (Phaseolus radiatus) and wheat (Triticum aestivum): plant agar test for water-insoluble nanoparticles. Environ. Toxicol. Chem 27:1915–1921.*
41. *Musante C, White J.C (2012). Toxicity of silver and copper to Cucurbita pepo: differential effects of nano and bulk-size particles. Environ. Toxicol 27: 510–517.*
42. *Stampoulis D, Sinha S.K, and White J.C (2009). Assay-dependent phytotoxicity of nanoparticles to plants. Environ. Sci. Technol 43: 9473–9479.*

Chapter 15

Good Agricultural Practices (GAP) for Improving the Groundnut Productivity in Drylands

G. Krishna Reddy

Principal Scientist (Agronomy), RARS, Tirupati.

15.1 Introduction

Groundnut is an important self pollinated leguminous oilseed crop grown in India. Globally India ranks first in area and second in production. India accounts for 31 % of the total groundnut area in the world (24.6 m.ha) and 22 % of the total production (35.7 mt.) Cultivated in nearly 100 countries in areas ranging from latitude 40°N to 40°S. In Andhra Pradesh it is mainly cultivated in Rayalaseema districts viz., Anantapur, Kadapa, Kurnool and Chittoor Districts followed by Telangana and coastal districts. In Andhra Pradesh it is cultivated in an area of around 10.59 lakh ha in kharif with a production of about 5.64 lakh tones and a productivity of 533 kg/ha (2012-13). In *Kharif,* it is mainly grown as rainfed crop.

15.2 Soils

Groundnut is grown on all types of soils such as well drained loose and friable sandy, sandy loams and heavy black soils. It is seen to thrive best on sandy loams. Heavy and stiff clay soils are not desirable as they tend to become hard during dry weather thereby interfering with peg penetration into the soil and also makes the harvest extremely difficult (Seshadri 1962). Most suitable soils for groundnut production are well drained light sandy loams with an ample supply of calcium and moderate organic matter (Gibbons, 1980). Groundnut is one of the most acid tolerant crops with a critical pH range of 5 to 5.5. Ideal pH is 6.5 to 7.0. It is considered moderately sensitive to soil salinity (Adams 1981). A salinity threshold (salinity at initial yield decline) of 3.2 dSm^{-1} had been reported (Maas and Hoffman 1977). Groundnut is highly susceptible to water logging.

15.2.1 Tillage

Land preparation for groundnut depends on the soil type and onset of monsoon for rainfed crop and on the previous crop grown for irrigated crop. Set line cultivation In black soils, instead of ploughing, blade-harrow is worked several times after the receipt of South - West monsoon rains. After the land is turned deep with a mould board plough equipped with a litter burial device, the final seedbed is prepared to accomplish a completely flat level bed with wheel tracks established by tracking off the land immediately after turning or by using precision power - driven rotary tiller. Only the rows are ploughed and inter - row spacing is cleared of weeds by harrowing. The available organic residues, farm - yard manure and fertilizers are ploughed into the row before sowing groundnut. The set - line cultivation has been found to increase infiltration rate in the row and thus more moisture is conserved. The general fertility of soil in the row increases. The inter - row spacing may be reduced up to 60 cm for virginia - runner types and 45cm for Spanish - bunch type groundnut. Groundnut crop requires a thorough preparation of the soil before sowing. Higher pod yields and lower loss of pods were reported by using mould board plough for land preparation as compared with the disc - harrow. Also higher groundnut yields were obtained when land was ploughed with a mould-board plough in the fall (December) than when ploughed in the spring (March), though the yields were not influenced by the method of land preparation. A depth of soil preparation beyond 15 - 20 cm is generally not considered for groundnut cultivation, since deep ploughing tends to form pods in deeper layers, rendering the harvesting of groundnut crop more difficult.

In command areas and under well irrigation, land preparation for irrigated groundnut after Kharif rice is a problem if the soils are relatively heavy textured. The land frequently ploughed at high soil moisture content for sowing the groundnut in optimum time. This results in soil compaction leading to poor crop stand and low yield. If the soil is not well prepared, soil bulk density beyond 1.6 g cm^{-3} limits the peg penetration. Under such situations, it is desirable to give one light ploughing soil moisture content is brought down considerably, the field can be harrowed 3 to 4 times at 2 to 3 days interval for taking up groundnut sowing.

15.3 Varieties

15.3.1 Varietal Improvement

According to (Seshadri 1962), there are four varieties under cultivation in india by 1960, all of which were introductions: Coramandal, Bold, Peanuts and Red Natal. They were marketed under different names in different parts of the country. The first two were spreading types and the later two are bunch types.

Table 15.1 *Recommended groundnut varieties for Andhra Pradesh (Annual Report, 2014)*

Season	Suitability	Varieties	Duration in days
Kharif/ Rabi	For Scarce Rainfall areas	Kadiri – 9* Kadiri Harithandra* Kadiri – 4 Kadiri – 5 Narayani Abhaya and Dharani ICGV-91114	105-110
	For excess rainfall conditions	Tirupati , Kadiri – 7 Bold , Kadiri – 8 Bold Vemana, Kadiri -5 Abhaya, Kadiri – 9	125-130 105-110
	For assured rainfall areas with supplementary irrigations	Kadiri – 7 Bold Kadiri – 8 Bold Vemana, Kadiri -5, K - 6, K — 9, Kadiri Harithandra, Tirupati - 4,Greeshma, Narayani, Abhaya, ICGV-91114	125-130 105-110
	Varieties having tolerance to leaf spot	Kadiri – 9, Kadiri Harithandra, Greeshma, Vemana, Abhaya, Kadiri - 5	105-110
	Nematodes infected areas	Tirupati – 3, Kalahasti and Prasuna	125-130 100-105
	For delayed monsoons	Kadiri – 4, Kadiri -5	90-100
	Varieties having drought tolerance	Kadiri – 9, Kadiri - 5, Vemana, Abhaya, ICGV-91114 *Dharani	105-110
	For rice fallow conditions	Kadiri – 4, Kadiri - 5, Kadiri – 6, Greeshma Kadiri Harithandra, TAG - 24, Rohini, Bheema	95-100
	Coastal sands	*Dharani, Vemana, Kadiri – 4, Kadiri – 6, Tirupati - 4, Greeshma, TAG - 24, Narayani and Rohini	105-110

(* Released in 2012)

Table15.2 *Recommended groundnut varieties for different states (DOR 2011)*

S. No	State	Kharif	Rabi – Summer
1	Tamilnadu	VRI(Gn) 5, Co(Gn) 4, TG 37 A CO (Gn) 5, GG16, TMV (Gn)13, Vikas (GPBD 4) AK 265, Ajeya (R 2001-3), ICGV – 00348,VRI (Gn) 7, Mallika (ICGH- 00440)	VRI (Gn) 5,6, Co (GN) 4, TG 37-A, VRI (Gn) 6, Mallika (ICHG 00440), Vijetha (R 2001-2).
2	Karnataka	GG 16 (JSP 39), AK 265, Ajeya (R 2001-3), ICGV 00348, VRI (Gn) 6, Mallika (ICHG 00440), TGLPS 3 (TDG 39,) Vijetha (R 2001-2)	VRI (Gn) 6, Mallika (ICHG 00440), TGLPS 3 (TDG 39,) Vijetha (R 2001-2), K 1319
3	Chatishgarh/ Madgya Pradesh	AK 159, GG 8 (J 53), JGN 23, Mallika (ICHG 00440).	Mallika (ICHG 00440).

Table15.2 Contd...

S. No	State	Kharif	Rabi – Summer
4	Maharashtra	AK 159, Ratneshwar (LGN 1,) Phule Unap (JL 286), GG8 (J 53) GG 16 (JSP 39), AK 265, AK 303, Ajeya (R 2001-3), TG 37A, Vikas (GPBD 4), TLG 45, ICGV 00348, VRI (Gn) 6, Mallika (ICHG 00440), Vijetha (R 2001 -2)	VRI (Gn) 6, Mallika (ICHG 00440), Vijetha (R 2001 -2) TG 37A, K 1319
5	Gujarat	GG7, GG 6, Prutha (Dh 86), Mallika (ICHG 00440), JL 501.	GG 6, Prutha (Dh 86), Mallika (ICHG 00440)
6	Rajasthan	TPG 41, GG14, Pratap Mugphali 1, Pratap Mugphali 2, Utakarsh (CSMG 9510), GG 21 (JSSP 15), Durga (RG 382), Girnar 2 (PBS 24030), Mallika (ICHG 00440), JL 501, HNG 69.	Prutha (Dh 86), Mallika (ICHG 00440)
7	Uttar Pradesh	GG14, TPG41, Utakarsh (CSMG 9510), GG21 (JSSp 15), Girnar 2 (PBS 24030), VL-Moongphali-1 (for Uttarakhand), Mallika (ICHG 00440), HNG 69.	Mallika (ICHG 00440)
8	Punjab	GG14, TPG 41, Utakarsh (CSMG 9510) GG21(JSSP 15), M548, Girnar 2 (PBS 24030), Mallika (ICHG 00440), HNG 69,	SG 99, Mallika (ICHG 00440)
9	Jharkhand	Mallika (ICHG 00440), Vijetha (R 2001 -2)	TG 38B (TG 38), TG 51, Vasundhara (Dh 101), Mallika (ICHG 00440), Vijetha (R 2001 -2)
10	Orissa	Devi (ICGV 91114), Mallika (ICHG 00440), Vijetha (R 2001 -2), Girnar 3	Prutha (Dh 86), TG 51, TG 38B (TG 38), Vasundhara (Dh 101), Mallika (ICHG 00440), Vijetha (R 2001 -2)
11	West Bengal	Mallika (ICHG 00440), Vijetha (R 2001 -2), Girnar 3.	Prutha (Dh 86), TG 51, TG 38B (TG 38), Mallika (ICHG 00440), Vijetha (R 2001 -2)
12	NEH region	Mallika (ICHG 00440), Girnar 3, Vijetha (R 2001 -2),	Prutha (Dh 86), TG 51, TG 38B (TG 38), Vasundhara (Dh 101), Mallika (ICHG 00440), Vijetha (R 2001 -2)

Some of the Recently (2011 Onwards) Released Improved Groundnut Varieties are Listed Below

Pratap Raj Mungphalli: For *kharif* and summer seasons in Rajasthan.

ALG 06-320: Suitable for *rabi*/summer in Vidharbha and southern MP.

RG 510: Ideal for *kharif*/summer in Punjab and Rajasthan.

RG 425: Suitable for *kharif* in Rajasthan.

RHRG 6021: Suitable for *kharif* in western Maharashtra.

Divya: Ideal for *kharif*/summer in UP.

HNG 123: Suitable for *kharif* and summer in Rajasthan and Punjab.

GJG 17: Spreading groundnut suitable for *kharif* in Gujarat.

GJG 22: Semi-spreading groundnut ideal for *kharif* in Saurashtra, and south Gujarat.

ICGV 00350: For all the seasons in AP and TN.

Improved Groundnut Varieties Developed at the Bombay Atomic Research Center (Barc) After1990 Are Listed Below

TAG 24: For *kharif* and *rabi* in AP, TN, WB, Maharashtra, Karnataka, Rajasthan.

TAG 26: MP. For summer in Gujarat and Maharashtra.

TG 37A: For *kharif* in Punjab, Haryana, Rajasthan, Gujarat, UP and WB.

TG 38: For Assam, Odisha and WB in *rabi*/summer.

TLG 45: For *kharif* in Maharashtra.

TPG 41: Ideal for summer throughout the country.

TBG 39: For *rabi*/summer in Rajasthan and Karnataka.

TG 51: Spanish bunch for irrigated conditions of Bihar, Odisha and WB.

TG 47: For irrigated conditions in Andhra Pradesh.

15.4 Seasons

The crop growing season should be ideal for growth, development of crop and for optimum yield. India has a unique distinction of growing groundnut in four major seasons: Kharif (80% area), rabi (11% area), summer (5% area) and spring (4% area), Thus, Kharif groundnut (June-July to October-November, depending upon the south-west monsoon) is the important crop for the country.

15.4.1 Kharif

Around 80 percent of the area under groundnut is grown during *kharif* in rainfed conditions. Depending on the onset of monsoon, the season extends from April to August.

In Pollachi area of Tamil Nadu, which gets the benefit of premonsoon showers, Spanish bunch type is sown in April. In Ramanathapuram, and Tirunelveli areas, sowings are taken up in August-September due to late receipt of rains. In other areas of south India, sowings are done from June to August. Other parts of the county complete the sowings in June-July. Earlier the sowings in June, higher will be the yield.

15.4.2 Rabi

Groundnut cropping during rabi is limited to areas where winter is not severe and temperatures do not grow below 15^0 C. Rabi groundnut crop is usually after *kharif* rice where irrigation water is inadequate for *rabi* rice. Sowing period extend from October to December. In general, second half of November is the ideal sowing time.

15.4.3 Summer

Groundnut yields are higher during summer due to ideal climate and irrigation althrough the crop period. This crop is largely confined to south India states. Sowings commence during second fort night of December and continue up to February first fortnight. Mid-Dec. to Mid-Jan. is ideal for high yield.

15.4.4 Early Kharif

Time of sowing influenced groundnut pod yield significantly. Among different dates of sowing, the highest pod yield of 3397 kg/ha was recorded when sowing was done on May, 16. Lowest pod yield of 1736 kg/ha was obtained when sown on June, 16 (2011). Significant differences were also observed among the dates of sowing. Sowing of groundnut crop on May 16th recorded the highest pod yield of 3689 kg/ha and was significantly superior over rest of dates of sowing (2012). Pooled data of 2011 and 2012 showed that among the different dates of sowing May 16th sown crop recorded significantly superior mean pod yield of 3544 kg/ha while, it was the lowest with June 16th sown crop (1557 kg/ha)as reported by (G.K. Reddy et.al., 2014).

***Table15.3** Performance groundnut varieties under irrigated condition during early kharif, 2011*

Treatment	Pl ht. (cm)	100 pod wt (g)	100 kernel wt (g)	Shelling %	Filled pods/pl	ill filled pods/pl	Pod yield (Kg/ha)
Variety							
Greeshma	39.0	53.5	34.5	68.4	12.5	2.1	2514
Dharani	51.6	63.1	37.5	71.3	12.9	2.5	2620
CD at 5 %	4.38	2.29	1.2	1.9	NS	NS	NS

Table 15.3 Contd...

Treatment	Pl ht. (cm)	100 pod wt (g)	100 kernel wt (g)	Shelling %	Filled pods/pl	ill filled pods/pl	Pod yield (Kg/ha)
Date of sowing							
April 1	39.9	57.7	35.9	68.3	13.1	2.2	2706
April,16	46.6	60.5	37.4	69.6	12.8	1.9	2574
May,1	47.2	65.3	39.6	69.1	13.1	2.7	2760
May,16	38.1	53.0	33.8	73.6	13.9	2.3	3397
June,1	46.9	56.5	34.5	69.6	11.8	2.7	2229
June,16	53.0	57.0	34.8	68.9	11.5	2.1	1736
CD at 5 %	7.59	3.96	2.1	3.3	1.59	NS	285
V x D	NS	NS	NS	NS	2.29	1.35	Significant

***Table15.4** Performance groundnut varieties under irrigated condition during early kharif, 2012*

Treatment	Pl ht. (cm)	Filled pods/pl	ill filled pods/pl	100 pod wt (g)	100 kernel wt (g)	Shelling %	Pod yield (Kg/ha)
Variety							
Greeshma	32.3	8.9	1.3	89.3	40.1	70.7	2321
Dharani	40.0	10.4	1.2	91.6	40.7	73.5	2640
CD at 5 %	2.98	NS	NS	NS	NS	1.8	79
Date of sowing							
April 1	32.5	6.6	1.8	86.8	39.1	73.1	2001
April,16	35.2	10.6	1.3	92.0	39.6	73.6	2220
May,1	41.9	13.1	0.7	95.6	45.2	74.0	3394
May,16	34.5	14.1	0.2	104.3	44.9	74.6	3689
June,1	31.4	7.5	1.4	91.8	42.3	72.6	2199
June,16	41.7	5.9	2.2	72.2	31.4	64.3	1378
CD at 5 %	5.2	1.95	0.6	7.40	2.3	3.17	136
V x D	NS	NS	0.85	NS	NS	NS	193

Years	2011	2012	Pooled
2011	---	---	2567
2012	---	---	2480
CD at 5 %	---	---	86.14
Variety			
Greeshma	2514	2321	2418
TCGS 1043	2620	2640	2630
CD at 5 %	NS	79	86.74
Date of sowing			
April 1	2153	2774	2464
April,16	2375	2398	2387

Table15.4 Contd...

Years	2011	2012	Pooled
May,1	2947	3207	3077
May,16	3452	3636	3544
June,1	2205	2025	2115
June,16	1374	1741	1557
CD at 5 %	285	136	122.6
CV			7.44

15.5 Seeds and Seeding

Quality seed is essential for establishing adequate plant population for efficient use of growth resources leading to economic pod yield.

15.5.1 Seed

Germination less than 85 percent is not considered satisfactory. It is a practice among formers to set apart good quality pods for seed purpose. Selected pods are thoroughly cleaned, dried, packed preferably in gunny bags and stored in a well ventilated place.

Groundnut pods for seed purpose are usually hand shelled about a week in advance of sowing. The viability of seed will be last if shelled long before seeding. Shellers are also used for shelling the pods to minimize the cost of groundnut production. After shelling, shriveled and damaged seeds are rejected by hand picking. Seed retained on 5 mm sieve (100 kernel weight above 27g) germinates early and produce vigorous plants for high pod yield.

Seed treatment against seed and soil born diseases is essential for stand establishment by preventing damage to seeds and seedlings emerging from soil. Seed treatment with Thiram (3g kg^{-1}), Carbandazim (2g kg^{-1}) or Dithane M 45 (3g kg^{-1}) is effective for about 20 days form sowing. Most of the chemicals used for seed treatment against fungal and bacterial diseases also affect *Rhizobium,* thus rendering the inoculation ineffective for a short time. When both seed treatment and inoculation are essential, seeds may be treated with fungicides and *Rhizobium* culture sprayed into seed rows and covered with soil. Granulated *Rhizobium* strain can also be sown with seed.

15.5.2 Optimum Time of Seeding

A large number of trails have been conducted all through the groundnut growing areas for determining optimum time of sowing for *kharif, rabi* and *summer* groundnut. In spite of fixing an ideal date for seeding *kharif* groundnut in different places, there is no scope for adhering to the recommended dates because of years variation in the onset of monsoon. Rabi and summer groundnut sowings largely depends on the harvest of proceeding crops in the intensive cropping systems.

In general, June seeding appears to the ideal for *kharif* groundnut, November for *rabi* and mid-December to mid-January for summer crop.

***Table15.5** Optimum time of sowing groundnut in different states (Reddy 1982).*

State	Kharif	rabi	Summer
Gujarat	15 June - 15 July	---	Jan to Feb
Andhra Pradesh			
North costal	First fort night of June	Second fortnight of November	15 Dec to 15 Jan
Telangana	First fort night of June	Second fortnight of November	15 Dec to 15 Jan
Rayalaseema	15 June - 7 July	Second fortnight of November	15 Dec to 15 Jan
Tamil Nadu			
Pollachi area	April - May	October	15 Dec to 15 Jan
Northern region	15 June - 15 July	October	15 Dec to 15 Jan
Southern region	July -August	October	15 Dec to 15 Jan
Karnataka	---	---	---
Bunch type	First fort night of June	Second fortnight of November	Northern region second fort night of Jan
Spreading	15 June - 15 July	Second fortnight of November	Others 15 Dec to 15 Jan
Maharastra	25 June to 7 July	---	Jan - Feb
Madhya Pradesh	25 June to 7 July	---	---
Punjab	June 15 - July 10	---	---
Haryana	25 June to 7 July	---	---
Bihar	Last week of June	---	---
Utttar Pradesh	Last week of June	---	---
Orissa	15 June - 15 July	---	15 Dec - 15 Jan
West Bengal	15 June - 15 July	---	15 Feb - 15 Mar

15.5.3 Spacing and Seed Rate

Inadequate crop stand is one of the major factors limiting the productivity of groundnut Optimum plant population depends on plant type (bunch or spreading), soil fertility, soil moisture and management practices. There is no agreement on optimum spacing and plant population for groundnut in India. The crop is seeded with spacing between 15 × 10 cm and 45 × 20 cm corresponding to plant densities of 6.6 and 1.1 Lakh plants ha^{-1}.

Often, significant variations in pod yield of groundnut are not observed due to differences in plant population, indicating plasticity. Higher plant population reduce the number of branches per plant while lower plant population increase the number of branches per plant. In groundnut, number of pods per plant is the major yield component determining the final pod yield. Number of pods per plant increase with increasing number of branches per plant. As such there is no significant variation in pod yield due to differences in plant population. In general, irrespective of initial plant population above 3.33 lakh plant ha^{-1}, the final plant population at harvest tends to be around 3 lakh plant ha^{-1}. As such, a spacing of 30 x 10 cm (3.3 lakh plant ha^{-1}) appears to be the ideal for *kharif* bunch types under most situations. A spacing of 22.5 x 10 cm is optimum during *rabi* and *summer* seasons. Under conditions of Africa, USA and Israel, the row spacing even for bunch types ranges from 50 to 75 cm with intra-row spacing between 10 and 20 cm. For spreading types, the spacing vary between 75 x 20 cm and 100 x 20 cm.

Of the total operating cost of production, cost of seed alone is around 40 percent. Seed rate recommended for groundnut (about 120 kg ha^{-1}) in India is more than twice to that in Africa, USA, Israel etc. There is need for reexamine the seed rates recommended for different types of groundnut to bring down the production cost considerably.

Seed rate depends upon the test weight (100 kernel weight), inter -row and intra-row spacing. Most common spacing recommended for bunch types is 30 cm x 10 cm. At this spacing, seed rate workout to about 120 kg ha^{-1} to obtain a plant population 3.3 lakh plants ha^{-1}. In the case of spreading types, the most common spacing is 30 x 15 cm, requiring a seed rate of 100 kg ha^{-1} obtain a plant population of 2.2 lakh plant ha^{-1}. Bold seeded varieties require an additional seed rate of 20 kg ha^{-1}. In Andhra Pradesh, recommended seed rates for bunch types are 120 and 150 kg ha^{-1} during khrif and rabi respectively. Corresponding seed rates for semi spreading types are 150 and 180 kg ha^{-1}. These seed rates appear to be relatively higher that the actual requirement under most situations. Recommended spacings (ICAR) in major groundnut growing states in India are given in table.

***Table15.6** Recommended spacing (cm) for groundnut in India (Reddy 1982).*

State	Spanish bunch (cm)	Virginia bunch and runner (cm)
Gujarat	45 x 10	60 x 10
Andhra Pradesh		
Rain fed	30 x 10	22.5 x 10
Irrigated	30 x 15	22.5 x 10
Karnataka	30 x 10	45 x 15
Maharstra	30 x 10	45 x 15
Utttar Pradesh	---	30/45 x 15
Punjab	---	30 x 22.5
Rajasthan	30 x 10	---
Tamil Nadu	30 x 8	30 x 20
Tindivanam	15 x 15	22.5 x 10
Alianagar	30 x 10	22.5 x 10
Madhya Pradesh	37 x 10	---

A closer spacing of 22.5 cm × 10 cm is recommended to Spanish bunch under irrigation, especially in Andhra Pradesh. Gap filling if necessary, should be completed taking advantage of the soil moisture at the time of sowing the seed within a week. If gap filling is done taking advantage of subsequent late rain, plants from the first sowing smother the later gap filled plants. As such, there may not be any yield advantage due to gap filling in groundnut.

Determination of Seed Rate for a Given Plant Density

Seed rate depends on row spacing, plant population, test weight of seed and percentage germination. For experiments, it is desirable to increase the seed rate by around 20 percent to ensure adequate stand establishment.

Given:

100 seed weight = 32 g

Plant population = 333000 plants ha^{-1}

100 percent germination

Find:

Seed rate ha^{-1} and 15 m^2 plot with eight 5 m rows?

Solution:

Seed rate per ha:

333000 plants ha^{-1} = 33.3 seed m^2

32 g per 100 seed = 0.3 g per seed

$$33.3\ seeds\ m^{-2} \times 0.32\ g\ per\ seed = 10.65\ g\ seed\ per\ m^2\ or$$

$$\frac{10.65\,\text{g} \times 10000\ \text{m}^2}{1000\ \text{g kg}^{-1}} = 106.5\ \text{kg ha}^{-1}$$

Add 20 percent to meet unforeseen losses due to birds, insects or diseases during germination.

106.5 kg ha^{-1} + 20% or 106.5 kg ha^{-1} x 21.3 = 127.8 kg ha^{-1}

1 m^2 requires:

$$\frac{127.8\ \text{g} \times 1000\,\text{g kg}^{-1}}{10000\ \text{m}^2} = 12.78\ \text{g m}^{-2}$$

A plot of 15 m^2 will require 12.78 g m^2 x 15 m^2 = 191.7 g

Seed per row:

A plot of eight rows requires 191.7 g

Hence, one row requires:

$$\frac{191.7 \text{ g}}{8 \text{ rows}} = 23.96 \text{ g per row}$$

If the germination percentage is 80 percent, then the seed rate will be:

If 80 seeds out of 100 seeds germinate, required seeds for 100 plants will be.

$$\frac{100 \times 100}{80} = 125 \text{seeds should be sown.}$$

This means for each gram of seed, 1.25 will be the multiplying factor to compensate for germination loss.

Then the seed requirement would be:

$$125 \text{ kg ha}^{-1} \times 1.25 = 160 \text{ kg ha}^{-1}$$

$$192 \text{ g per plot} \times 1.25 = 240 \text{ g per plot}$$

$$24 \text{ g per row} \times 1.25 = 30 \text{ g per row.}$$

15.5.4 Method and Depth of Seeding

Mechanical or bullock drawn seed drills are used for sowing the groundnut seed. The other method is dropping the seed in plough furrow behind country plough. Hand dibbling is also adapted to a limited extent.

Groundnut is mostly sown under rainfed conditions with a bullock drawn seed drill in Andhra Pradesh. Karnataka and Maharashtra. The number of cultures for seed drill varies from 3 to 6 with 15 to 60 cm between cultures. One ha of land can be covered with a drill of 6 cutlers drawn by bullocks in 4 hours. Tractor drawn mechanical seed drills are also used if the area to be covered is large. Sowing with indigenous bullock drawn or mechanical seed drill is the best method of seeding for uniform stand establishment. Inter cultivation and harvesting can also be easily carried out with implements resulting in considerable saving in production expenditure.

Common method of sowing in Tamil Nadu and adjoining areas of Karnataka and Andhra Pradesh is ***dropping the seed in furrows behind the country plough***. It is often argued that in this method, the seeds are placed at uneven depths and often on the ridges of the furrows leading to uneven germination and poor stand establishment. It appears that this method is more ideal when the soil moisture content is relatively low at the time of sowing since the seed is dropped deeper in soil than that with drilling. There is not problem for using inter cultivation implements since the row spacing can be maintained as in the case of drilling.

Groundnut seed can emerge even from around 20 cm depth at soil bulk densities more than 1.6 g cm^{-3}. However, sowing deeper than 8.0 cm is not desirable. In light soils, the seeds are usually sown to a depth of 5-7 cm and in heavier soils to a depth of 4-5 cm. Depth of sowing should be less if the soil moisture content is higher than when it is relatively dry. It is desirable to work a wooden plank after seeding to establish adequate soil-seed contact to ensure quicker and higher percentage of germination. A blade harrow along the line of sowing is more effective than working a wooden plank since the former operation can slightly compact the soil besides uprootting the emerging weeds is any.

15.6 Nutrient Management

Oilseeds are energy-rich crops and hence the requirement of major nutrients as well as secondary and micronutrients is very high. The nutrient removal varies considerably, depending upon crop productivity and soil fertility (Hegde, 2000). Groundnut removes fairly large quantities of nutrients from the soil. It depletes the soil nutrients rapidly unless the soil is adequately manured. Adequate manuring does not only improve the yield but also maintains the soil health and sustain the productivity (Lourduraj, 1999). (Ghosh *et al.* 2002) further stressed that proper fertilizer management of groundnut crop with right kind of nutrients at right time adapting right method of application has significant effect on yield and quality.

For every 1.0 t of pods and 2.0t of haulms about 63 N, 11 P_2O_5, 46 K_2O, 27 CaO and 14 MgO kg ha^{-1} are removed by the groundnut crop (Collins and Morris 1941). Of these about 50 percent nitrogen, 80 to 90 percent phosphorous, potassium, calcium and magnesium are found in haulms. On sandy loam soils, for the production of 100 kg pods 4.38 kg N, 0.92 kg P_2O_5, 3.12 kg K_2O. 1.25 kg Mg and 4.0 g Zn are required (Raghavaiah 1982). A balanced fertiliser progrmme with particular emphasis on P,K,Ca and Mg is essential for optimum pod yield.

15.6.1 Deficiency Symptoms

Nitrogen deficiency is characterized by a general chlorosis of leaves, which become light yellow or nearly white in some cases. The stems become distinctly red besides poor nodulation. Lack of phosphorus leads to dark blue-green leaves with reduced size. Characteristic symptom is the deep red or purple colouration of the stem in later stage, older leaves yellow and drop down. Potassium deficiency leads to stunted growth, with drying up of leaf margins. The stem gets a reddish colour at the tip of branches.

Calcium deficiency is characterised by development of localized pitted areas on lower surface of leaves. Later on, large necrotic spots are founds on both the leaf surfaces, which give the leaf a bronze colour. The youngest foliage presents a distorted appearance. The basal stem cracks and dieback of the shoots occurs during the late stages. Calcium deficiency leads to poor filling of pods (pops). Sulphur deficiency produces pale green colour in leaves. Magnesium deficiency leads to interveinal chlorosis, which starts, from leaf margins and advances towards midrib.

Iron deficiency produces stunted growth and characteristic chlorosis symptoms. Small and distorted terminal chlorotic leaflets with a few yellowish white spots indicate iron deficiency. Boron deficient leaves resemble calcium deficient leaves except that in the former necrotic areas are localized near leaf margins instead of being distributed over the entire leaf surface.

15.6.2 Nutrient Concentration

The primary function of plant analysis is to diagnose problems or to monitor the nutrient status during crop growing season for timely correction of problems, if any. Concentrations of N, P, K, Cu, Mn and Zn in groundnut leaves, generally, decrease with increasing age. The concentration of Ca markedly increases with leaf age and Mg concentration tends to increase. Adequate concentration of different nutrients is given in Table 15.7.

Table15.7 *Adequate concentration of nutrients in upper stem and leaves of groundnut at early pegging (Small and Ohirogge 1973).*

Nutrient	Adequate concentration (g kg^{-1})	Nutrient	Adequate concentration (mg kg^{-1})
N	35.0 - 45.0	Fe	100 - 250
P	2.0 - 3.5	Mn	100 - 350
K	17.0 -30.0	Zn	20 - 50
Ca	12.5 - 17.5	B	20 - 50
Mg	3.0 - 8.0	Cu	10 - 50
S	2.0 - 3.0	Mo	1 - 5

15.6.3 Nitrogen

The necessity for fertiliser nitrogen to groundnut is reduced because of atmospheric nitrogen fixation in root nodules. The contribution of biologically fixed nitrogen has not been adequately quantified. However, there are indications that around 200 kg N ha^{-1} can be fixed under ideal conditions, reducing the need for fertiliser nitrogen application. Groundnut, as member of the cowpea cross-inoculation group, is nodulated by a large number of diverse Rhizobial strains. However, not all Rhizobial strains are efficient in fixing nitrogen through symbiosis, rainfed groundnut is often subjected to soil moisture stress, which also adversely affects the efficiency of nitrogen fixation. This is one of the reasons why response of groundnut to applied nitrogen is not consistent. Efficient Rhizobial strains used in groundnut are NC 92, Tal 1000 and THA 205.

Nitrogen in general is the major structural constituent of the plant cell. It plays an important role in plant metabolism by virtue of being an essential constituent of metabolically active components like amino acids, protein, nucleic acid, flavins, purines and pyrimidines, nucleotides, enzymes and alkaloids. The biological role of N is evidenced through the chlorophyll in harvesting solar energy, phosphorylated compounds in energy transformation, nucleic acid in the transfer of genetic information and regulation of the cellular metabolism and biological catalysts (Puntankar and Bathkal 1988).

Growth and development of crops depend largely on the development of root system. Nitrogen is the key plant nutrient that stimulates root and shoot growth (Jana *et al.,* 1990). Nitrogen is closely linked to control the vegetative growth of plant and hence determines the fate of reproductive cycle (Wojnowska *et al.,* 1995). (Srinivas *et al.* 2005) stated that groundnut being a leguminous crop depended on two major sources of nitrogen for their growth viz., atmospheric nitrogen and mineral nitrogen benefit from the appropriate complementary operation of both biological nitrogen fixation by the species and by nitrate reduction. In general, the contribution of nitrogen in the legume would be 25 Kg ha^{-1}.

Groundnut showed a significant increase in plant height with increasing levels of N from 0 to 40 kg ha^{-1} in soils with low N status (Jakhro, 1984 and Barik *et al.,* 1994). (Sukanya *et al.* 1995) observed that increasing level of nitrogen increased the nodule number, nodule mass, total dry mass, total nitrogen content, pod yield and harvest index in groundnut.

(Barik *et al.* 1998) reported that dry matter production, LAI and plant height were increased significantly with the enhanced rate of nitrogen supply at various stages of growth up to harvest and the highest value was observed with 40 kg ha^{-1}. However, the results of the research experiment conducted by (Edna Antony *et al.* 2000) revealed that leaf area duration, leaf area index and leaf net assimilation rate increased with an increase in nitrogen dose in all genotypes studied and concluded that 25 kg N ha^{-1} was necessary for optimal yield. Yield of groundnut tended to decrease with higher dose of N beyond 25 kg ha^{-1}.

(Kandil *et al.* 2007) reported that the increasing nitrogen levels increased number of leaves, stems, total pods and pod dry weight per plant, number of pods per plant, weight of pods per plant, number of seeds per plant, weight of seeds per plant, 100-pod weight, 100-seed weight, pod yield, straw yield, seed protein content and NPK contents. However, numbers of pods per plant and seed oil content were decreased by increasing nitrogen levels.

(Cox *et al.* 1992) opined that application of nitrogen in the form of urea in two equal splits at the time of sowing and at 30 DAS significantly increased the dry matter with each increment levels. Studies on the response of three levels of nitrogen *viz.,* 0, 15 and 30 kg ha^{1} to groundnut during summer season was made and the results revealed that application of N at higher dose did not significantly improve the dry matter accumulation and yield attributes at harvest (Chawale *et al.,* 1993).

(Reddy *et al.* 1992) observed considerable increase in pod as well as haulm yields with the application of 40 kg N ha^{-1} as compared to 20 kg N in alfisols having low availability of N. (Yakadri *et al.* 1992) observed that 100 kernel weight was significantly increased with the application of 30 kg N ha^{-1} over unfertilized control in red sandy loam soils in Southern Telengana zone of Andhra Pradesh.

Application of 40 kg N ha^{-1} significantly increased the number of pods per plant, kernel and oil yield by 16.6, 18.8 and 24.7 per cent, respectively (Patra *et al.,* 1995). (Gogoi *et al.* 2000) compared the response of different levels of N *viz.,* 0, 20, 40, 60 and 80 kg ha^{1} to groundnut and found that increased level of nitrogen application upto 80 kg ha^{-1} increased the number of branches, pegs, pods per plant and shelling percentage. However, significant increase in yield and yield attributes were noted only up to application 40 kg ha^{-1}. Similarly, (Deka *et al.* 2001) indicated that increasing the level of nitrogen up to 40 kg ha^{1} increased the nutrient uptake and resulted in significantly higher kernel and haulm yields of groundnut.

(Chavan and Patil 1995) studied the response of four groundnut varieties to three levels of nitrogen and observed that the variety UF 70103 showed a consistent and significant response to increased levels of nitrogen up to 40 kg ha^{-1}, while variety JL 24 showed a significant increase in yield in response to nitrogen levels up to 25 kg ha^{-1}. Whereas SB XI was not consistent in its response pattern.

Increasing nitrogen levels up to 60 kg N ha^{-1} significantly increased the pod yield of groundnut and it did not respond to N beyond 60 kg N ha^{-1} (Singh and Singh, 2001).

^{15}N labelled fertilizer application in groundnut

There are several concepts that are key to conserving nutrients in a soil-plant system. It is important to make nutrient applications in line with crop uptake and utilization patterns. Therefore, the time (stage of crop growth), method, and rates of application are all very important in achieving optimum crop uptake and utilization of the applied nutrients. ^{15}N as a tracer can be used to understand the plant demand on N during different phenological stages, as well as its distribution within the plant (Wallace *et al.,* 2007).

(Bloom *et al.* 1988) observed that the apparent recovery of N by winter wheat (measured as the difference between uptake from fertilized and unfertilized crops) varied from 40 to 88 per cent. The recovery of applied ^{15}N by the crop differed substantially depending on the rate and method of N application (Recous *et al.,*1988).

Rhizobium Inoculation Procedure

Rhizobium or Bradyrhizobium inoculation is a cheap and effective way of providing nitrogen to legumes. Cultures of bradyrhizobium like NC 92, IGR 6, IGR 40, TAL 1000 and TNAU 14 are recommended for groundnut. Several factors influence the success of Rhizobium inoculation and efficiency of nitrogen fixation. Intercropping with cereals leads to reduction in nitrogen fixation by groundnut; deep sowing leading to elongated hypocotyls, poor rooting, poor nodulation and poor nitrogen fixation.

Bradyhizobium population of 10^2 to 10^4 cells g^{-1} soil are usually detected in many soils. A minimum of 10^6 cells per seed are needed for successful field inoculation

(Nambiar 1990). Use of higher rates (10^6 to10^8 cells per seed) for initial inoculation, may help early establishment of inoculants strains. Common inoculation procedures are:

Direct Seed Inoculation

Seed inoculation with inoculum pasts: Seeds are inoculated by direct coating with Rhizobium (10^6 cells per seed). Seed inoculation is planned 48 hours prior to sowing. Before coating with Rhizobium, seeds may be moistened or paste of inoculum may be used. Carrier in the paste-like inoculums helps the Rhizobial cells to stick to the seeds. Jiggery or sugar solution (10-15%) may be used as inoculums carrier.

Mix the inoculums and seed thoroughly. It is better to add half the required inoculums to the seed, mix and then add the remaining half and mix again. This will ensure black specks on each seed, indicating the presence of inoculums.

The amount of inoculums to be used is usually indicated on the container of the commercial product. Nine ml of slurry containing 4.4 g of peat inoculums per kg see has been suggested for seed of the size of soybean.

***Pelleting*:** In this method, appropriate Rhizobial culture is added to the seed at the factory or main distribution point. Seed is enclosed in a clump of lime or phosphate rock along with inoculums. This implies protection of the bacteria and seed, plus amelioration of adverse conditions in the soil adjacent to the seed.

Material used in this procedure is lime stone or rock phosphate as a base with stickers such as gum.-arabic (15% water solution) and methyl-ethyl -cellulose (1% solution). The procedure involves inoculation of seed using gum-arbic (40%) or methyl-ethyl-cellulose (50%); while the seed is still moist tumble it in an appropriate container with the chosen coating material like rock-phosphate or fine limestone powder.

Field Inoculation

Groundnut seed is often treated with fungicides, which may be toxic to Rhizobum, prior to sowing to control seedling diseases. Application of inoculum in the seed furrow in peat-based slurry form appears effective. Mixing of peat in water (0.7 g per liter) and pouring the mixture below the seed (4-5 ml per seed) into the furrow, to give a population of more than 10^6 Rhizobial cells per seed appears effective (Nambiar 1990)

Results of 200 trials in different agroecological zones in the country revealed that application above 22 kg N ha^{-1} is not beneficial. (Mann 1965). Results of 436 trials on cultivator's fields in groundnut growing states revealed that rainfed groundnut responds to 30 kg N ha^{-1} if phosphorous and potassium are also applied along with nitrogen (Kanwar 1978). Rainfed groundnut on black soils responded to 20 kg N ha^{-1} at different locations. On shallow red soils of Rayalaseema, rainfed groundnut responded to 10 kg N ha^{-1} during drought years. However, under conditions of adequate rainfall, it responded to 30 kg N ha^{-1}. Instance of groundnut responding to nitrogen up to 60 kg N ha^{-1} on light soils and up 30-40 kg N ha^{-1} on medium to deep soils are not uncommon.

From the results of experiments, it is clear that response of groundnut to nitrogen largely depends on soil type and distribution of rainfall during crop season. In general, 20 kg N ha^{-1} for rainfed groundnut and 30 kg N ha^{-1} for irrigate groundnut can be recommended for economic yields under most crop growing situation. For a rainfed crop, entire dose of nitrogen should be applied by placement at seeding. A fertiseed drill could be effective for placement of fertiliser along with groundnut seed. For an irrigated crop, nitrogen may be applied in two equal splits at seeding and 30 days after seeding, especially for the crop on light soils with high infiltration rates to minimize leaching losses of applied fertiliser.

15.6.4 Phosphorus

Phosphorus is required in small amount than N for plant growth but is equally important for crop growth (Chen *et al.,* 1994).

Phosphorus is one of the major limiting plant nutrients (Nandwa 1998; Rao *et al.* 2004) in the tropical and sub-tropical soils. Agasimani and Babalad (1991) reported that response to P could be obtained when the available P status in the soil was less than 35 kg P_2O_5 ha^{-1}. (Kulkarni *et al.* 1986) reported that phosphorus application up to 50 kg ha^{-1} increased the number pods per plant and dry matter production accumulation in the plant.

(Patel *et al.* 1990) revealed that phosphorus is the most important nutrient which affects the yield and quality of leguminous crops including groundnut. P fertilization particularly at flowering and pod formation stages were beneficial (Singh *et al.,* 1991). Groundnut responds to P application by increasing shelling percentage, oil yield and nodulation. (Pushpendra Singh *et al.* 1994) reported that phosphorus application brought about significant increase in biological yield in calcareous soils at Udaipur in Rajasthan. However, they observed significant differences at 40 and 60 kg over 20 kg P_2O_5 ha^{-1}.

(Mudalagiriyappa *et al.* 1995) observed positive effect on pod yield due to P application @ 50 kg P_2O5 ha^{-1} in the form of single super phosphate in different proportions in vertisols. Mehta and Ram Mohan Rao (1996) reported that application of 50 kg P_2O_5 ha^{-1} registered significantly higher number of pods per plant and 100 kernel weight. The increasing trend in pods and haulm yield was noticed upto 75 kg P_2O5 ha^{-1}.

Sharma and Yadav (1997) reported that phosphorus plays a beneficial role in legume growth promoting extensive root development and thereby ensuring a good yield. (Barik *et al.* 1994) observed that the plant height increased linearly by the application of P and the highest value was observed at 80 kg P_2O_5 ha^{-1}. Intodia *et al.* (1998) reported that application of 60 kg P_2O_5 significantly increased number of pods per plant, shelling percentage, pod yield, haulm yield, harvest index and oil yield of groundnut. Phosphorus fertilization was investigated by several workers and recommended varied doses of P_2O_5 kg ha^{-1} for increasing the yield and its attributes, *i.e.* 25 kg (Bhatol *et al.,* 1994); 33 kg ha^{1} (Kumar and Ray Chudhuri, 1997); 50 kg ha^{1} (Patel *et al.,* 1995); 60 kg ha^{-1} (Yakadri *et al.,* 1992) and about 114 kg ha^{-1} (El-Far and Ramadan, 2000).

Rath *et al.* (2000) reported that application of 75 kg P_2O_5 ha⁻ produced the highest pod yield (21.51q ha^{-1}), whereas the highest oil yield was obtained with 50 kg P_2O_5 ha^{-1} due to higher shelling percentage and oil content.

(Bhatol *et al.* 1994) concluded that application of 25 kg P_2O_5 ha^{-1} markedly increased the pod yield over control, but a higher rate of 50 kg P_2O_5 ha^{-1}, the yield was significantly decreased. Similarly (Choudhary *et al.* 1991) reported that deletion of P application had significantly reduced the pod and haulm yields of groundnut.

The total amount of phosphorus uptake by groundnut plant is relatively small compared to nitrogen and potassium. On sandy loan soils of Tirupati (AP), 20 kg P_2O_5 of available phosphorus in the soil was adequate for pod yield of 1.0 t under rainfed conditions (Raghavaiah 1982). Though, the phosphorus requirement is small, large quantity of fertilizer has to be applied as the efficiency of uptake from fertiliser is low.

From the results of 200 trials under difference situations in the country, it is evident that rainfed groundnut responded to around 30 kg P_2O_5 ha^{-1}. The responses, in general, are highest on red soils than on black soils (Mann 1965). Simple fertiliser trails in the country revealed necessity for 40 kg P_2O_5 ha^{-1} for rainfed groundnut (Mahapatra *et al* 1973). There are also results indicating that 60 kg P_2O_5 ha^{-1} is needed for irrigated groundnut. From 20 to 60 kg P_2O_5 ha^{-1} for realizing economic yields under different situation. In general, rainfed groundnut should be given 40 kg P_2O_5 ha^{-1}, whereas irrigated crop need 50 kg P_2O_5 ha^{-1}, especially on red soils. Theses recommendation holds good provided recommended dose of nitrogen is applied. If only phosphorus is applied to groundnut, optimum doses are 30 to 40 kg P_2O_5 ha^{-1} for rainfed and irrigated crops respectively.

Common source of phosphorus is single super phosphate (SSP), which also supplies calcium to groundnut. Response to applied phosphorus is usually evaluated based on phosphorus content of SSP, ignoring the effect of calcium and sulphur content of SSP. On equal P_2O_5 basis, response to SSP is higher than that due to diammunium phosphate (DAP) due to calcium, sulphur and trace elements in SSP. For irrigated groundnut, DAP is the most common source of phosphorus as it almost meets the needs of both nitrogen and phosphorus.

Entire recommended dose of phosphorus should be applied at sowing along with nitrogen by placement, preferably using ferticumseed drill. Several studies indicate beneficial effect of foliar application of phosphorus on pod yield of groundnut. However, its implementation at field level is limited due to practical problems.

15.6.5 Potassium

Groundnut crop responds well for potassium (K) application and addition of K increased its concentration at all stages in groundnut crop. The concentration of K was high in initial stages and declined in the later stage, indicating that groundnut absorbs K rapidly in early stages (Madkour *et al.,* 1992). Potassium is also required in large amount by oil seed crop (Singh, 2004).

Additional dose of 12 kg ha^{-1} apart from the recommended basal application of 54 kg ha^{-1} gave 10 per cent higher yield (CSM, 1990). Application of potassium at 100 kg ha^{-1} significantly increased the plant height, nodule weight, pod number, pod and haulm yields of groundnut (Singh and Vidya Chaudhari, 1996). (Kankapure *et al.* 1994) observed beneficial effect of K on growth characters including dry matter of groundnut.

(Jana *et al.* 1990) observed that K up to 49.8 kg ha^{-1} had increased the number of pods per plant, 100 seed weight, pod yield and oil yields. The response was quadratic and also influenced the K content of seeds. The pod and haulm yields of groundnut increased significantly with application of 40 kg K_2O ha^{-1} over lower dose and further increase beyond this level did not increase the yield. Oil content in kernel increased with graded levels of K and the effect was marked to the higher at 60 kg K_2O ha^{-1}. However, increase in protein content and protein yield was only upto application of 40 kg K_2O ha^{-1} (Deshmukh *et al.,* 1992).

Application of K also resulted in higher uptake of NPK and also increased the pod and kernel yield. Quality parameters were also enhanced by the application of K which reflected in oil and protein content (Lakshmamma *et al.,* 1996).

(Hameed Ansari *et al.* 1993) found that application of potassium up to 45 kg ha^{-1} significantly improved the pod yield (3392 kg ha^{-1}) and its contributing characters compared to lower dose of 15 and 30 kg Kha^{-1}.

Groundnut plant takes up potassium throughout the growth period. Though, its uptake is high, groundnut in general will not show any significant response to added potassium as most of the Indian soils are rich in potassium. Response to its application can be expected only when the available potassium content is less than 40 kg K_2O ha^{-1}. There is a mutual antagonistic effect on the uptake of K, Ca and Mg. The ratio of K:Ca:Mg is more important than total amount of any of them. High concentration of potassium in fruiting zone is harmful, especially at low levels of Ca.

Results of the experiments in groundnut growing areas of the country indicated no response to applied potassium at several situation. Even if there is response, it is limited to 20-40 kg K_2O ha^{-1} both under irrigated and rainfed conditions. From the available information, there is no necessity for potassium application to rainfed groundnut yielding around 1.0 t ha^{-1}. For an irrigated crop, a does of 40-60 kg K_2O ha^{-1} is necessary provided nitrogen is applied at recommended rate.

On equal K_2O basis, all potassium fertilisers are equally effective. Entire dose of potassium should be applied along with phosphorus at sowing, preferably using a fertiseed drill. Marginal increase in yield due to foliar application of potassium has been reported. However, it may not be a viable practice, especially under rainfed conditions.

15.6.6 Balanced Fertiliser Use

Application of all the three major nutrients in a balanced proportion increases the pod yield of groundnut both under rainfed and irrigated conditions. Combined application of 10 N, 20 P_2O5, and 20 K_2O kg ha^{-1} resulted in 150 percent increase in pod yield of groundnut over control and significantly higher pod yield due to N, P and K applied singly or when only two of them were applied together on light and medium texture soils (Singh and Pathak 1969). Experiments in farmers fields indicted that application of NPK in 1:2:2 or 1:2:3 ratio will be economical (Maini and Ramamurthy 1969). Results of 436 trails conducted all over the country revealed that combined application of 30 N and 40 P2O5 double the yield of groundnut compared with 30 kg N ha^{-1} alone and addition of 20 kg K_2O ha^{-1} to 30 N and 40 P_2O_5 kg ha^{-1} improved the response considerably (Kanwar 1978). Similarly application of 20 N, 40 P2O5 and 40 K_2O kg ha^{-1} was optimum for rainfed groundnut on sandy loam soils of Tirupati (AP).

Review of work done in India on response of groundnut to fertilisers for about two decades conclusively proved the need for balanced fertiliser use for realizing economic groundnut yields. Table givens recommended fertilsers schedules (ICAR) for different states.

Table15.8 Fertiliser schedules recommended for different states in India (Reddy 1982)

State	Rainfed (kg ha^{-1})			Irrigated (kg ha^{-1})		
	N	P_2O_5	K_2O	N	P_2O_5	K_2O
Gujarat	13	25	0	15	50	0
Andhra Pradesh	20	40	20	30	40	50
Karnataka	15	30	25	25	75	25
Tamil Nadu	10	20	40	10	10	75
Madhya Pradesh	20	40	20	-	-	-
Punjab	15	40	25	10	15	25
Orissa	10	15	15	-	-	-
Rajasthan	20	60	0	30	60	0
Uttar Pradesh	15	30	45	-	-	-

Recommended fertiliser schedule for rainfed groundnut in Andhra Pradesh, at present is 20 N, 40 P_2O_5 and 50 P_2O_5 kg ha^{-1}. For an irrigated crop additional 10 N kg ha^{-1} is recommended as top dressing at 30 DAS.

Recommended fertiliser schedules are based on several experiments conducted under different agroecological conditions. It is important to note that earlier to the introduction of fertilisers as the source of nutrients supply to the crops, farmers depended on farm yard manure (FYM) as a source of nutrient supply. Framers used to apply around 15 t ha^{-1} of FYM for groundnut crop once in 2 to 3 years. The FYM under rainfed conditions, besides supplying nutrients aids in improving water receptivity of soils, which is more important than supply of nutrients. Application of FYM alone used to give pod yields around 1.0 t ha^{-1}, which is not much different from that due to fertilisers as a source of nutrient supply to groundnut. It is not uncommon to observe farmers depending on FYM alone for groundnut cultivation, especially when the area is relatively small. As such, it is desirable to apply at lest around 5.0 t ha^{-1} of FYM only to groundnut productivity besides improving soil health. The traditional practice is to apply FYM only to groundnut crop in cropping systems such as groundnut - sorghum/pearlmillet/cotton/maize (2 or 3 years) and grow the subsequent crops on the residual effect of farm yard manure applied to groundnut.

NPK content and uptake by groundnut

(Rao and Narayan 1990) noticed that many groundnut cultivars had higher uptake and accumulation of N and P due to application of higher doses of N and P.

The difference in potassium uptake could only be due to differences in the pod and haulm yields because the crop received uniform potassium doses (Prameela Rani *et al.,* 1991). Application of K, in general increased N, P and K content in all the plant parts at harvest stage. On an average 137.3, 16.6 and 63.34 kg N, P and K ha^{-1}, respectively were removed by groundnut crop (Deshmukh *et al.,* 1992).

High nitrogen levels increased nitrogen and phosphorus contents and concentration of all plant components in early growth stage only (Loubser and Human, 1993). (Manoharan *et al.* 1994) reported that the uptake of N, P and K increased with increasing levels of nitrogen. (Zharare 1996) demonstrated that potassium application increased the nitrogen uptake but did not affect the uptake of phosphorus and potassium.

(Sarmah *et al.* 1995) found that nitrogen uptake was maximum with 80 kg ha1. An experiment conducted in sandy loam soils of Bhuvaneswar indicated that nitrogen up to 60 kg ha-1 increased the uptake of N, P and K (Mishra *et al.,* 1995). Increase in available N might be due to the direct addition of N through inorganic fertilizers to the available pool as reported by (Bellakki and Badanur 1997). Combined application of N and K in equal split doses as basal and on 45 DAS coinciding with peg formation increased the uptake and availability of nutrients as well as groundnut production (Balasubramanian, 1997). Enhanced P content and uptake in groundnut leaf, stem and kernel could be attributed to increased availability of P in soil due to application of organic manures (Lupwayi *et al.,* 1999).

(Deka *et al.* 2001) studied the effect of lime and N on nutrient uptake in groundnut and the results revealed that with each successive increase in the dose of nitrogen, nitrogen uptake increased significantly up to 40 kg ha^{-1}, which was however at par with 60 kg ha^{-1}.

Badole *et al.* (2003) studied the effect of organic and inorganic fertilizer on the uptake of groundnut at Parbani during summer season in the clay loam soil and found that integrated nutrient supply system (50 per cent of recommended nutrient content through organic and remaining by inorganic fertilizers) significantly improved the uptake of groundnut compared to control.

***Table 15.9** Nutrient uptake/removal in groundnuts (kg ha-1)*

Plant part	Yield	N	P	K	Ca	Mg	S
Pods	3 t ha-1	120	11	18	13	9	7
Haulms	5 t ha-1	72	11	48	64	16	8
Total		192	22	66	77	25	15
Source: Gascho, 2001							

(Dutta *et al.* 2003) reported that potassium content both in kernel and haulm was significantly increased by increased potassium application and the highest was observed at 50 kg K_2O ha^{-1}. Application of graded levels of potassium produced notable difference in uptake of N, P and K and the increase was significant at higher doses of potassium application (50 kg K_2O ha^{-1}).

(Laxminarayana 2004) recommended that integrated application of organic and inorganic fertilizers showed higher uptake of NPK compared to sole application of organic manures due to increased nutrient availability and improvement in the physical condition of the soil. (Chitdeshwari *et al.* 2007) reported that increasing levels of the respective nutrients (34:64:108 kg NPK ha^{-1}) increased the nutrient content in the crop and its uptake.

15.6.7 Calcium and Sulphur

These two nutrient are absorbed by pegs and developing pods and the common source of supply is gypsum. Adequate calcium is essential in root and pod zones for yields and quality of kernels. Calcium deficiency leads to unfilled pods (pops). Sulphur is directly involved in the biosynthesis of oil.

About 1.0 meq 100 g^{-1} of soil in the root zone depth and 3.0 meq 100 g^{-1} of soil in the pod zone are considered threshold values of calcium sufficiency (Virmani 1973). About 100 ppm of heat soluble sulphur is the critical limit of available sulphur for groundnut (Karwar et al 1983).

Supply of calcium and sulphur to groundnut crop through CaO or gypsum has been observed to increase the yield by more than twice depending on its availability in the soil. Application of gypsum near pegging zone at flowering (30DAS) appears to be ideal. On sandy loam soils, application of gypsum at 250 kg ha^{-1} for irrigated crop and 500 kg ha^{-1} for rainfed crop resulted in highest yields (Raghavaiah 1982) by meeting the calcium and sulphur needs of groundnut. Response of groundnut to gypsum ranges from 250 to 100 kg ha^{-1} in different situations.

Gypsum is recommended at top dressing at flowering (30DAS) if soil test shows less than 400 Ca ha^{-1} and /or the ratio of calcium to potassium is less than 3:1 for Virginia groundnut (Smith 1995). For Spanish types, gypsum is necessary for soils testing less than 225 kg ha^{-1} or the ratio of calcium to potassium is less than 3:1. If heavy rains occur within tow weeks after application a second lighter application of gypsum may be necessary around three weeks after first application.

15.6.8 K:Ca:Mg Ratio

Groundnut is rather sensitive to imbalanced nutrient supply (Harris and Bledsoe 1951). The ratio K:Ca:Mg is more important than the total amount of any of them (Reid and Cox 1973). Increase in concentration of Mg in nutrient solution decreased the uptake of K by groundnut. Similarly, uptake of Ca deceased with increasing concentration of Mg. There is amutual antagonistic effect on the uptake of K, Ca and Mg. The ideal of K:Ca:Mg for groundnut is 4:4:2: (Comber 1959). Under field conditions, a ratio of 4:2:0 will be optimum both for rainfed and irrigated crops (Subba Rao et al 1980).

15.6.9 Zinc, Boron and Iron

Zinc deficiency is common on sandy and sandy loam soils. Zinc deficiency affects flower production and reduces the number of pegs penetrating into the soil. Results of experiment all over the country indicated increase in pod yield ranging from 12 to 13 percent due to application of zinc sulphate at 10 to 50 kg ha^{-1}. application of zinc sulphate to soil 25 kg ha^{-1} once in two years corrects the deficiency. If zinc deficiency is noticed in the standing crop, foliar application of 0.2 percent zinc sulphate along with 0.2 percent lime can correct the deficiency.

Boron deficiency lads to hollow heart. Boron deficiency hs been reported in Punjab and Tamil Nadu. The threshold level for boron is 0.25 ppm. Deficiency can be corrected by soil application of 5-10 kg ha^{-1} of boron depending on the level of deficiency. Excess born inhibits iron uptake resulting in typical burning of leaf margin. Its deficiency in standing crop can be corrected by 0.1 percent borax spray.

Iron chlorosis is largely due to its reduced availability in the soil. Immobilization of iron in the soil may be due to high levels of lime, high pH (more than 7.6) or high level of bicarbonates in soil or irrigation water. Spraying 1.0 percent ferrous sulphate mixed with1. Percent ammonium citrate around 50 DAS corrects the iron deficiency.

15.6.10 Integrated Nutrient Management on Groundnut

Continuous application of inorganic fertilizer decreased the C: N ratio of soil and also improved the activities of microorganisms responsible for nitrogen mineralization (Sarkar and Rathore, 1992). (Bocchi and Tano 1994) reported that positive interaction between the combination of organic manures and urea as nitrogen source.

Integrated nutrient use of both organic manures and inorganic fertilizers simultaneously is probably the most effective method to maintain healthy sustainable soil system while increasing crop productivity (Janssen, 1993). (Sudha 1999) also reported the beneficial effect of integration of organic and inorganic fertilizers.

(Toor and Bahl 1997) showed that the combined addition of poultry manure and fertilizer phosphorus had a synergistic effect and increased phosphorus uptake compared with fertilizer P alone.

(Jagdev Singh and Singh 2000) reported that application of FYM combined with NPK also improved the soil environment, which encouraged proliferous root system resulting in better absorption of water and nutrients from lower layers and thus resulting in higher yield and nutrient uptake. (John *et al.* 2004) had advocated for integral use of organic manure and inorganic fertilizers for the supply of adequate quantities of plant nutrients required to sustain maximum crop productivity and profitability, while minimizing environmental impact from nutrient use.

(Desai *et al.* 1999) found a positive interaction between organic manures and urea as nitrogen sources. Groundnut crop responded favourably to organic and inorganic fertilizer application. Nutrient management of the soil on sustainable basis includes fertilizer types, amounts, time and methods of their application. Proper amount and time of fertilizer application is considered a key to the bumper crop (Jan and Khan, 2000).

(Geethalakshmi *et al.* 1993) reported that integration of all the sources of nutrients (NPK + FYM + *Rhizobium* + gypsum) registered the highest pod yield of groundnut compared to NPK fertilizer alone. Similarly (Deshmukh *et al.* 1995) reported that conjunctive use of inorganic fertilizer along with FYM significantly increased the number of pods and shelling percentage of groundnut. Higher shelling percentage, harvest index, oil yield and protein content were registered with application of 34:17:54 kg NPK ha^{-1} +12.5 t FYM ha^{-1} (basal) +17 kg P_2O_5 ha^{-1} (30 DAS) + 400 kg gypsum ha^{-1} (40 DAS). The recommended level of NPK (17:34:54 kg NPK ha^{-1}) alone recorded lower shelling percentage, harvest index, oil yield and protein content. (Kumaran, 2000).

(Kadwe *et al.* 2003) reported that application of 75 per cent RDF along with press mud @ 5 t ha^{-1} delayed days to 50 per cent flowering, peg formation and days to maturity and significantly increased plant height, number of branches, number and weight of nodules per plant, yield and harvest index.

(Laxminarayana and Patiram 2005) reported that the integrated use of inorganic and organic manures by applying 40 kg N, 26 kg P and 33 kg K ha^{-1}, in combination with farmyard manure @ 15 t ha^{-1} gave the highest pod and haulm yields, followed by NPK + pig manure @ 5 t ha^{-1} and NPK + poultry manure @ 5 t ha^{-1}.

(Subrahmaniyan *et al.* 2000) reported that the application of combined nutrients (Diammonium phosphate 2.5 kg + ammonium sulphate 1.0 kg + borax 0.5 kg + NAA (naphthalene acetic acid) @ 40 ppm) on 25th and 35th day after sowing (DAS), followed by foliar spray of NAA @ 40 ppm alone on 45th and 55th DAS significantly gave the highest values of growth and yield attributes and pod yield of groundnut. (Ali and Mowafy 2003) observed that foliar spray of Zn (2 per cent) improved the groundnut yield attributes and yield as well as quality.

15.6.11 Economics

Subrahmaniyan *et al.* (1998) reported that the foliar application of combined nutrients @ 40 ppm on 25th and 35th DAS followed by the foliar spray of NAA @ 40 ppm alone on 40th and 55th DAS + 100 per cent NPK gave the highest pod yield of 2040 kg ha^{-1} with a net return of Rs. 17320 ha^{-1} and BCR (2.32) as compared to 100 percent NPK alone. (Subrahmaniyan *et al.* 2000 a) found that application of FYM @ 15t ha^{1} enhanced the pod yield of groundnut (2890 kg ha^{-1}) and net return (Rs. 28607 ha^{-1}) compared to lower level of FYM and control under rainfed conditions.

15.7 Water Management

Irrigated groundnut accounts for over 20 percent of the total area under the crop in the country. Yields around 4.2 t ha^{-1} have been reported under irrigation in several situations.

15.7.1 Critical Stages for Irrigation

Adequate available sol moisture is essential throughout the growth duration of the crop for normal growth and development. However, certain growth and development stages are more sensitive to water deficits than other stages leading to low pod yield of groundnut.

Studies on critical stages for soil moisture stress indicated that peak flowering (45-60 DAS) is the most sensitive stage (Klepper 1973). Very early growth phase (up to 20 DAS) is less sensitive. Flowering, peg penetration and early pod development stages are the three moisture sensitive stages for pod yield (Sankara Reddi 1976). Several other studies indicated that the period from peak flowering to early pod development (50-70 DAS) us very sensitive to soil moisture stress.

Since flowering is almost a continuous process in groundnut from about 25 DAS to physiological maturity, there is overlap of flowering, pegging and pod development. As such, it is difficult to pinpoint exact stage most sensitive to soil moisture stress. Therefore, it is reasonable to state that the period from peak flowering to early pod formation is the most critical phase for water deficits in groundnut.

15.7.2 Scheduling irrigation

On sandy loam soils, scheduling irrigations at 25 percent DASM throughout the growth period results in higher pod yield. Irrigating the crop at 25 percent DASM from pegging to early pod development and at 50 percent DASM at other stages appears to be ideal for high water use efficiency without significant reduction in pod yield. An IW/CPE ratio of 1.0 at moisture sensitive stages and 0.6 during other stages leads to high water use efficiency (Prihar and Sandhu 1987).

Keeping the total quantity of irrigation water applied constant, high frequency irrigation leads to high pod yield of groundnut on sandy loam soils (Bhaskara Reddy et al 1979). At several situations, 10-12 irrigations at 10 days interval resulted in highest pod yield. On sandy loam soils, seven irrigations: pre sowing irrigation, 25, 35, 45, 55, 70 and 90 DAS appears to be ideal at times of water scarcity.

Based on several experimental results throughout the country, it is evident that a good crop of groundnut can be grown with 8 irrigations. There is no necessity for irrigation up to 25 DAS. After pre sowing irrigation, the next irrigation should be 25 DAS followed by four irrigations at 10 days interval. Later two irrigations can be at 15 days interval. For bunch types, there is no need for any irrigation after 90 days of sowing. If the soil is very hard for harvesting the crop, a high irrigation may be necessary a day or two before pulling out the plants.

Depending on the soil types, evapotranspiration during the crop period and crop duration, water requirement of groundnut, on an average, ranges between 450 and 650 mm. The water utilisation efficiency (pod yield per unit of water (evapotranspired) is 0.6 to 0.8 kg m^{-3}. The crop is usually irrigated by check basin method. However, border strip method is more suitable than other methods. ***Sprinkler irrigation*** is ideal for groundnut crop on sandy soils.

15.8 Weed Management

Weed problem is severe in early stage of groundnut because of its slow growth. The competition is from both grasses and broad leaf weeds. Important weed flora in the groundnut crop is indicated below:

Cyperus rotundus	*Digitaria sanguinalis*
Chloris barbata	*Commelina benghalnsis*
Cynodon doctylon	*Celosia argentena*
Amaranthus viridis	*Clemoe viscose*
Protulaca oleracea	*Trichodesma indicum*
Boerhaavia diffusa	*Eclipta alba*

Weed competition is critical up to 35DAS. Yield losses to the extent of 70 percent are not uncommon, especially under rainfed conditions. When once pegging beings (40 DAS), there should not be only disturbance to pegs through manual or mechanical weeding.

15.8.1 Cultural Management

Most common methods of weed control in groundnut crop are hand weeding and harrowing. Line sowing with a row sparing around 30 cm facilitates working with different types of bullock drawn implements. Intercultivation usually starts around 10 days after emergence and continues up to 35 DAS at 7-10 days interval till pegging begins. Weeds within the row are removed byhand weeding. Hand weeding is done twice, first around 20 DAS and the second about 35 DAS. There should not be any intercultivation or hand weeding from 40 DAS.

15.8.2 Use of Herbicides

Cultural methods alone may not provide complete weed free environment for groundnut crop and repeated cultural practices may be costly or labour may not be available for timely weeding. Under such situation, herbicides use is the only option. Any of the following herbicides are effective against many weed species.

PPI herbicides:

Fluchloralin (1.0 - 2.0)
Nitralin (0.5 - 1.0)
Pendimethalin (1.0 -2.0)
Pronamid (1.5 - 2.5)
Trifluralin (0.5 -1.0)
Metolachlor (1.5 -2.0)
Imazethapyr (50 -70 g)

These herbicides can be PPI or PPS (preplant surface). They will be effective only if rain or irrigation follow their application. Herbicides can be applied about 5 DAS (before crop emergence).

PRE herbicides:

Nitrofen (3.0 - 4.0)
Simazin (1.0 - 1.5)
Alachlor (1.5 - 2.5)
Oxadiazon (1.5 - 2.0)
Prometryn (0.5 - 1.0)
Oxyfluorfen (0.25 - 0.50)

A mixture of Oxadiazon and Dinoseb eah at 1.7 kg ha-1 gives excellent control of weeds besides reducing stem rot in groundnut. Fluazfop (150-250g) is a promising POST herbicides for controlling grasses, particularly *Cynodon doctylon*, 35-40 DAS.

- Crop must be weed free upto 45 days after sowing.
- Intercultivation at 20 and 40 DAS followed by one hand weeding

- The crop should not be disturbed by weeding or inter cultivation after 45 DAS.
- Pre-emergence application of Butachlor/Metalachlor/Pendimethalin @ 2.5 to 3.1./ ha. or Oxyflourfen 1.5 to 2.0 l/ha followed by one intercultivation and one hand weeding will effectively control the weeds.
- Wherever, pre-emergence herbicides could not apply, weeds can be controlled by post-emergence herbicides by spraying imazethaphyr @ 750 ml/ha or Quizalofop ethyl @ 1.0 l/ha at 20 DAS.
- In heavy soils higher dose of chemical should be used.

15.9 Harvesting

Flowering extends over two months in bunch types and over three months in spreading types. It is, therefore, usual to find pods at different stages of maturity at harvest. Several methods to predict optimum harvesting date are adopted in western countries. Such methods are not common in India.

When good percentage of pods is well developed and is fairly intact with line, the plant turns yellow and leaf shedding starts. Maturity of pods is determined by inside of the shell turning dark, kernel reaching its normal size and the test having normal colour of the variety. Pulling, out few plants in the field at random and examining for maturity of pods determine proper harvest stage. When around 75 percent of pods mature, harvesting has to be commenced.

Besides the above symptoms, the following procedures can be used for determining maturity in groundnut (Sanders et al 1982).

Shell out method: Groundnut pods from several plants are picked and cracked or cut open to determine maturity. The percentage of pods with tan to brown colour inside the hull and pink to dark pink seed coats is worked out. Harvesting is recommended when mature pos range from 60 to 80 percent, depending on the variety, presence of dormancy and environmental factors.

Seed weight method: Mean seed weight per plant is determined to estimate maturity at successive intervals. It reaches a constant value when the crop is mature. If two to three plants harvested on 10 November give an average seed weight of 30 g and the other harvested on 11th or 12th November also gives the same weight, then the crop is mature.

Seed/hull ratio (Maturity index): All pods for selected plants for evaluation are removed, washed, towel-dried and opened. Those pods at or beyond maturity stage are characterised by cracks in the white internal pericarp. Such pods are separated into seed and hull. Less mature pods without such cracks are placed with the hulls. A fresh mass ratio (FMI) is determined by seed mass divided by the hull mass. The grouped kernels and hulls are forced air-dried or dried at room temperature to determine the dry mass ratio (DMI). When FMI or DMI of the two samples is constant, harvesting is recommended.

After a variety attains 50 percent maturity a close monitoring is required to determine the date of harvest.

Harvesting consists of two operations; pulling or lifting the plants (vines) from the soil with pods intact and separation of pods from vines. Manual labor is employed for pulling out the vines, if soil moisture is adequate for easy pulling. If the soil is dry bullock or tractor drawn blades are used for lifting the vines with pods.

Several methods are followed for stripping the pods from vines. In bunch type, if the vines are still green, the plants are knocked against

- Should be done at right stage of maturity
- At the time of 70-80% leaves and stems turn yellow
- When the inner side of the shell turn black
- When sufficient moisture is available in the root zone

15.10 Storage

- Seed should not contain more than 9% moisture for storage
- Prefer poly ethylene/gunny bags for storage
- Spray Malathion 5 ml/liter of water once in 2-3 weeks on storage bags against storage pests.

15.11 Tips for Increasing Groundnut Seed Production

- Deep summer ploughing once in three seasons
- Adoption of quality seed of HYVs
- Seed treatment
- Adoption of recommended seed rate
- Adoption of Ferti-cum-seed drill to ensure right placement of seed and fertilizer
- Ensure optimum population
- Adopt recommended fertilizer dose
- Apply Gypsum and SSP to provide Calcium and Sulphur
- Avoid inter cultivation/weeding after 45 DAS, but earthing up should be done.
- Adopt IPM package
- Practice crop rotation and intercropping
- Use mechanization for sowing, inter cultivation, harvesting and stripping to reduce cost of cultivation.

References

1. Agasimani, C.A., and H.B. Babalad. 1991. Recent advances in agronomy of groundnut (Arachis hypagaea L.). J. Oilseeds Res. 8:133 - 158.
2. Annual Report 2012, Basic record of Agronomy, Regional Agricultural Research Station, Tirupati.
3. Badole, S.B., S.N. More, P.B. Adsul, A.K. Shaikh, and A.L. Dhamak. 2003. Residual effect of Integrated Nutrient ManagementSystem of cotton on yield and nutrient uptake of summer groundnut. J. Soil and Crop. 13(1): 77-80.
4. Balasubramanian, P. 1997. Integrated nutrient management in irrigated groundnut (Arachis hypogaea L.). Indian J. Agron. 42(4):683-687.
5. Barik, A., P.K. Jana, G. Sounda, and A.K. Mukherjee. 1994. Influence of Nitrogen, phosphorus and potassium fertilization on growth, yield and oil content of kharif groundnut. Indian Agric. 38(2): 105 - 111.
6. Barik, A.K., A.K. Mukherjee, and B.K. Mandal. 1998. Growth and yield of sorghum and groundnut grown as sole and intercrop under different nitrogen regimes. Indian J. Agron. 43(1): 27-32.
7. Bellakki, M.A and V.P. Badanur. 1997. Long-term effect of Integrated nutrient management on properties of Vertisol under dry land agriculture. J. Indian Soc., Soil Sci. 45: 438- 442
8. Bhaskara Reddy, G.Reddy SR and Reddi GHS 1979. Frequency and depth of irrigation for groundnut. Agric Water Manage 3:45-51.
9. Bhatol, D.P., N.A. Patel, and R.P. Pavaya. 1994. Effect of nitrogen, phosphorus and zinc application on yield and uptake of nutrients by groundnut. Indian J. Agric. Res. 28 (3): 209-213.
10. Bloom, T.M., R. Sylvester Bradley, L.V. vaidyanathan, and A.W.A. Murray. 1988. Apparent recovery of fertilizer nitrogen by winter wheat. In nitrogen efficiency in agricultural soils (Eds D.S. Jenkinson and K.A. Smith). pp. 27-37.
11. Bocchi, S., and F. Tano, 1994. Effect of cattle manure and components of Pig slurry on maize growth and production. European J.Agron. 3: 235-41
12. Chavan, L.S., and B.P. Patil. 1995. Response of groundnut varieties to fertilizer nitrogen in medium black soils of the Konkan region of Maharashtra, India. International Arachis Newsletter 15: 76-77.
13. Chawale, V.V., G.M. Bharad, S.K. Kohale ,and M.B. Nagdeve. 1993. Effect of nitrogen and farmyard manure levels on growth and yield of summer groundnut. Indian J. Agron. 38(3): 500-502.
14. Chen, M. L, X. L. Jaing, B. Y. Wang, B. Y. Zhov, Z. Y Zheri and G. S. Xu. 1994. athematical models and best combination of high yield cultivation techniques for rapseed variety Zhenyouyom-2. Acta Agriculturae Zhejiangensis 6: 22-26.

15. Chitdeshwari, T. P. K. Selvaraj, and P. M. Shanmugam. 2007. Influence of levels and split application of fertilizers on the yield and nutrient uptake by groundnut. Agricultural Science Digest 27(2): 91-94.
16. Choudhary, T.S.D., C.S. Vaidya, and A.C. Shekar. 1991. Effects of graded doses of phosphorus and sulphur on the growth, yield and oil content of groundnut. Ann. Pl Soil. Res, 8(2): 106-11.
17. Collins H and Morris HD 1941. Soil fertility studies with peanut. Bull North Carolina Agric Exptl St, USA.
18. Comber RS 1959. Effect on the groundnut of variation in supply of K, Ca and Mg. Nature 184:961:1003.
19. Cox F.R., F. Adams, and B.B. Tucker. 1992. Liming, fertilization and mineral nutrition. In: Pattee H. and Young C. (eds), Peanut Science and Technology. American Peanut Research and Education Society. pp. 138-159.
20. CSM (Oils seeds). 1990. Crop scientists meet on oilseeds. In Proc. of Crop Scientists Meet on Oilseeds. 25-26 April, Tamil Nadu Agricultural University. Coimbatore
21. Deka, N.C., R. Dutta and P.K. Gogoi. 2001. Effect of lime and nitrogen on nutrient uptake and residual soil fertility in groundnut. Legume Res. 24(2): 118-120.
22. Desai B.B., P.M. Kotecha, and D.K. Salunkhe. 1999. Science and Technology of Groundnut: Biology, Production, Processing and Utilization. Naya Prokash, Calcutta, pp.677.
23. Deshmukh, V.N., R.T. Warokar, and B.T. Kanakpure. 1992. Yield, quality and nutrient uptake by groundnut as influenced by potash fertilization and time of application. J. Potassium Res. 8(4): 367-370.
24. Deshmukh, V.N., S.M. Bhoyar, and A.D.Portikar. 1995. Response of groundnut to levels of FYM, N and P fertilization. J. Soils and Crops. 5(1): 53-56.
25. Dilip Kumar Majumdar, 2009 Irrigation water management principles and practice
26. DOR 2011. Recent varieties/hybrids of oilseed crops recommended for different state (2001-2010). Directorate of Oilseeds Research (ICAR), Rejendranagar, Hyderabad – 500 030.
27. Dutta, R., P.K. Gogoi and N.C. Deka. 2003. Effect of levels of lime and potassium on nutrient uptake and residual soil fertility ingroundnut. Crop Res. 26(3): 420-423.
28. Edna Antony, M.B., Doddamani, U.V. Mummigatti, and M.B. Chetti. 2000. Canopy characteristics and its relation with yield in groundnut. Haryana J. Agron. 16(1&2): 7-10.
29. El-Far, I.A., and B.R. Ramadan. 2000. Response of yield, yield components and seed quality of groundnut (Arachis hypogaea L.) to plant density and PK fertilization in sandy calcareous soil. Proc. 9th Conf. Agron., Minufiya Univ., Egypt. 1-2 Sept. 2000: 453466.
30. Geethalakshmi, V., A. Christopher Lourduraj, A. John Joel, and K. Rajamanickam. 1993. Nutrient Management in Groundnut. Madras Agric. J. 80(7): 412-414.

31. Gogoi, P.K., R.M. Choudhury, R. Dutta, and N.C. Deka. 2000. Effect of levels of lime and nitrogen on production of groundnut (Arachis hypogaea L.). Crop Res. 20: 274-278.

32. Hameed Ansari, A., S.M. Qayym, and M. Usman Usmani Khail. 1993. Impact of row spacing and NPK fertilizer levels on the growth, seed yield and oil content in pea nut. Oil crops Newsletter 10: 50-53.

33. Harris HC and Bledsoe RW 1951. Physiology and mineral nutrition. In: peanut – unpredictable legume. Nat Fert Assoc, Washington, DC.

34. Hegde, D. M. 2000. Nutrient management in oilseed crops. Fert. Res. 45(4): 31-38 & 41.

35. Intodia, S.K., S.C. Mahnot, and M.P. Sahu. 1998. Effect of organic manures and phosphorus on growth and yield of groundnut (Arachis hypogaea L.). Crop res. 5(1): 22-26.

36. Jagdev Singh, and K.P. Singh. 2000. Effect of Aztobater FYM and fertility levels on yield nitrogen recovery and use efficiency in spring sunflower. Haryana University journal of Agronomy. 16: 57 - 60

37. Jakhro, A.A. 1984. Growth, nodulation and yield of groundnut as affected by nitrogen rates. Planter 60(6): 149-153.

38. Jan, M.T., and S. Khan. 2000. Response of wheat yield components of N-fertilizer their levels and application time. Pak. J.Bid. Sci. 3: 1227-1230.

39. Jana, P.K., S. Ghatak, A. Barik, and B.C. Biswas. 1990. Response of summer groundnut to nitrogen and potassium. Indian. J.Agron. 35: 137-143.

40. Janssen, B.H. 1993. Integrated nutrient management: the use of organic and mineral fertilizer. In: Van Reuler, H., Prins, W.H.(Eds.), The Role of Plant Nutrients for Sustainable Food Crop Production in Sub-Saharan Africa. VKP, Ledschendam, TheNetherlands, pp. 89-105.

41. John, L.W., D.B. Jamer, L.T. Samuel, and L.W. Warner. 2004. Soil Fertility and Fertilizers: An Introduction to Nutrient Management, Pearson Education, India. pp: 106- 53.

42. Kadwe, S.B., C.N. Hatmode, R.D. Deotale, C.N. Chore, and Archana Thorat. 2003. Effect of different levels of fertilizers and pressmud on morphophysiological parameters and yield of groundnut. J. Soils and Crops. 13(2): 305-308.

43. Kandil, A. A., A. K. A. El-Haleem, M. A. Khalafallah, S. F. El-Habbasha, N. S. Abu-Hagaza, and T. G. Behairy. 2007. Effect of nitrogen levels and some bio-fertilizers on dry matter, yield and yield attributes of groundnut. Bulletin of the National ResearchCentre (Cairo). 32(3): 341-359

44. Kankapure, B.T., R.T. Warokar, and V. N. Deshmukh. 1994. Effect of potassium levels and its time of application on groundnut in Vertisol. J. of Maharashtra Agric. Univ. 19 (1): 122-123.

45. Kanwar, J.S., H.L. Nijhawan, and S.K. Raheja. 1983. Groundnut nutrition and fertilizer responses in India. ICAR, New Delhi.

46. *Kanwar JS 1978. Trends in consumption and productions of fetilisers. Proc FAI – IFC Semi 1977.*

47. *Klepper B 1973. Water relations of peanut. In: Peanuts culture and use. Am peanut Res Ed Assoc. Inc, Stillwater, Oklahome.*

48. *Krishna Reddy G, Muneendra Babu A, Maheswara Reddy P, and Giridhara Krishna T 2014 Growth and pod yield of Groundnut as influenced by cultivars and times of sowing under irrigated conditions during Early Kharif. The Andhra Agricultural Journal 61(4):744-747.*

49. *Kulkarni, J.H., P.K. Joshi, and V.K. Sojitra. 1986. Influence of phosphorus and potassium application on nodulation, nitrogen accumulation and pod yield of groundnut. Legume Res. 7: 6-9.*

50. *Kumar, K., and M. Ray Chudhuri. 1997. Differential response of groundnut varieties to phosphorous nutrition in a typic Kanhaplohumult. Annals of Agric. Res. 18(4): 415 - 419.*

51. *Kumaran, S. 2000. Role of organic manure, fertilizer levels, split application of phosphorus and gypsum application on shelling percentage, harvest index, pod and oil yield of irrigated groundnut. Res. on Crops. 1: 344-347.*

52. *Lakshmamma, P., A. Shivraj, and L.M. Rao. 1996. Effect of cobalt, phosphorus and potassium on yield and nutrient uptake in groundnut. Ann. Agric. Res. 17(3): 335-336.*

53. *Laxminarayana, K. 2004. Effect of organic and inorganic manures on yield and nutrient uptake of groundnut (Arachis hypogaea L.) in ultisols of Mizoram. Journal of Oil Seeds Research 21(2): 280-283.*

54. *Laxminarayana, K., and Patiram. 2005. Influence of inorganic, biological and organic manures on yield and nutrient uptake of groundnut (Arachis hypogaea) and soil properties. Indian Journal of Agricultural Sciences 75 (4): 218-221.*

55. *Loubser, H.L., and J.J. Human. 1993. The effect of nitrogen and phosphorus fertilization on the nitrogen and phosphorus absorption by sunflowers. J. Agronomy and Crop Science. 171: 206-215.*

56. *Lourduraj, A. C. 1999. Nutrient management in groundnut (Arachis hypogaea L.) cultivation - A review. Agric. Rev. 20 (1): 14 -20.*

57. *Lupwayi, N.Z., I. Haue, A.R. Saka, and D.E. Siaw. 1999. Lucaena hedgerow intercropping and cattle manure application in Ethiopian highlands: II. Maize yields and nutrient uptake. Biol. Fertil. Soils. 28: 196 - 203.*

58. *Madkour, M.A., I. Salwa, El-Mohandas, and A.M. El-Wakil. 1992. Effect of row spacing, phosphorus, potassium and boron application on some groundnut cultivars. Egypt J.Agron. 17(1-2): 127 - 140.*

59. *Mahapatra IC, Prasad R, Krishnan KS, Goswamy NN and Bapat SR 1973. Response of rice, jowar, maize, bajra, groundnut and castor to fertilizer under rainfed conditions on farmer's fields. Fert News 18:18:21.*

60. Mann HS 1965. Response of paddy, wheat, sugarcane, potato, ground-nut and cotton to fertiliser and other agronomic factors on alluvial black, red and lateritic soil of India, Indian J Agro 10: 1-9.

61. Manoharan, S., S. Senthivel, and K. Balakrishnan. 1994. Effect of different levels of nitrogen on sunflower. Madras Agric. J. 81(6): 344 - 345.

62. Mehta, A.K., and D.S. Ram Mohan Rao. 1996. Effect of Rhizobium inoculation nitrogen and phosphorus application on yield attributes of groundnut. Legume Res. 19(3): 151-154.

63. Mishra, A., P. Dash, and P.K. Paikaray. 1995. Yield and nutrient uptake by winter Sunflower as influenced by nitrogen and phosphorus. Indian J. Agron. 40(1): 137-138.

64. Mondal, S.S., and S.B. Goswami. 1991. Effect of split application of potassium on rainfed groundnut. Journal of Potassium Research 7(4): 304-308.

65. Mudalagiriyappa, C.A., Agasimani, and M.N. Sreenivasa. 1995. Response of groundnut to sources of phosphate and phosphate solublizers in vertisols. J. Oilseeds Res. 12(1): 141 - 145.

66. Nambiar PTC 1990. Nitrogen nutrition of groundnut in Alfisols. Information Bulletin No.30. ICRISAT, AP, India.

67. Nandwa, S.M. 1998. Overview of soil fertility in Tropical Africa. A lecture presented at the International Training Programme on development of fertilizer recommendations for optimum crop production. Holiday Inn Hotel, Nairobi, Kenya. April 20-25 1998.

68. Patel, M.P., V.B. Shelke, and D.K. Shelke. 1990. Response of groundnut to weed management and phosphate in pre-monsoon season. J. Maharashtra Agric. Univ, 15: 313-316.

69. Patra, A.K. 1995. Effect of planting variables and water management practices on seed production of groundnut varieties. Ph.D. (Ag.). Thesis submitted to Bidhan Chandra Krishi Viswavidyalaya. Mohanpur, Nadia, West Bengal.

70. Prameela Rani, B., K. Subrahmanyam, and S. Rami Reddy. 1991. Effect of Straight and complex fertilizers on growth and yield of groundnut (Arachis hypogea L.). Intern. J. of Trop. Agric. 9(1): 25-32.

71. Prihar SS and Sandhu BS 1987. Irrigation on field crops – Principles and practices, Indian Council of Agricultural Research, New Delhi.

72. Puntankar, S.S. and B.G. Bathkal. 1988. Influence of N, P and K fertilizers on composition, growth and yield of groundnut. Indian J. Agron. 12: 344-350

73. Pushpendra Singh., B.S. Varma, and M.P. Sahu. 1994. Effect of level and source of phosphorus and bio-regulators on groundnut (Arachis hypogaea L.). Indian J. Agron. 39(1): 66 - 70.

74. Raghavaiah R 1982. Studies on the effect of Ca and S on growth, yield and nutrient uptake of groundnut. PG Thesis submitted to APAU, Hyderabad, India.

75. *Rao, J.S., and A. Narayan. 1990. Physiological aspects of iron deficiency in groundnut and blackgram. Madras Agric. J. 77: 151157.*
76. *Rath, B.S., R.K. Paikary, and K.C. Barik. 2000. Response of Rabi groundnut to sources levels of phosphorus in red and lateritic soils of Inland districts of Orissa. Legume Res. 23(3): 167 -169.*
77. *Recous, S., J.M. Machet, and B. Mary. 1988. The fate of labeled 15N urea and ammonium nitrate applied to a winter wheat crop. I. Plant uptake and N efficiency. Plant and Soil, Dordrecht, 11(2): 215-224.*
78. *Reddy PS 1982. Production technology for increasing groundnut yield in India. Ann Kharif Oilseeds Workshop, Bangalore 7-11 May, 1982.*
79. *Reddy S., K., V.M.M. Srirama Murthy, and M.V. Shantaram. 1992. Nitrogen, phosphrous and sulphur nutrition of groundnut (Arachis hypogaea L.) grown in alfisols. J. Oilseeds Res. 9(2): 189 - 195.*
80. *Reid PH and Cox FR1973 Soil properties, mineral nutrition and fertilization practices .In: Peanuts culture and use. Am peanut Res and Edu Assoc, Inc, Stillwater, Oklahome.*
81. *Sanders TM, Schubert AM and Pattee HE 1982. Maturity methodology and postharvest physiology. In: Pattee HE and young CT (eds) Peanut Science and Technology. Yoakum, Texas, USA: American Peanut Research and Education Society.*
82. *Sankara Reddi GH 1976. Management of irrigation water, Tech Bull No.33, APAU, Hyderabad, India.*
83. *Sarkar, S., and T.R. Rathore. 1992. Soil properties changes associated with wheat straw management. Indian Agriculturist 43:177-183.*
84. *Sarmah, P.C., S.K. Katyal, and A.L. Bhola. 1995. Nutrient uptake and quality of spring sunflower as influenced by fertility levels and plant population. Haryana Agricultural University Journal Research 25: 21-28.*
85. *Seshadri CR 1962 Groundnut. Indian Central Oilseed Committee, Hyderabad,India*
86. *Sharma, B.M., and J.S.P. Yadav. 1997. Availability of phosphorus to grain as influenced by phosphatic fertilization and irrigation regimes. Indian J. Agric. Sci. 46: 205-210.*
87. *Singh A.L, M. S. Basu, and N. B. Singh. 1991. Mineral disorders of groundnut. National Research Center for Groundnut, Junagadh, India. p 85.*
88. *Singh C and Pathak SS 1969. Groundnut responds to nitrogen, phosphorus and potassium. Fert Nees 14:26-27.*
89. *Singh, A.C., and S.P. Singh. 2001. Growth and yield of spring sunflower and groundnut as influenced by different cropping systems and rates of nitrogen. Ann. Pl. Soil Res. 3 (2): 222-225.*
90. *Singh, A.L. 2004. Mineral nutrient requirement, their disorders and remedies in groundnut. Groundnut Research In India, National Research Centre for Groundnut, Junagadh, India. pp. 137-159.*

91. Singh, A.L., and Vidya Chaudhari. 1996. Interaction of sulphur with phosphorus and potassium in groundnut nutrition in calcareous soil. Indian J. Plant Physiol. New series (1): pp 21-27.

92. Small HG and Ohlrogge AJ 1973. Plant analysis is an aid in fertilizing Soybean and penuts. IN: Walsh LM and Baton JD (eds) Soil testing and plant analysis. Soil Sci Soc Am, Madison, Wisconsin.

93. Smith CW 1995. Crop production. John Wiley and sons Inc, New York.

94. Srinivas, M., Shaik Mohammed, and A. Sairam. 2005. Yield components and yield of castor (Ricinus communis) as influenced by different planting geometries and row proportions of intercropped groundnut or peralmillet. Crop Res. 30 (3): 349-354.

95. Subba Rao R, Rajan MSS and Reddi GHS 1980. Effect of K, Ca and Mg ratio on field attributes and yield of irrigated groundnut. Indian Potash J 5:2-6.

96. Subrahmaniyan, K., N. Arulmozhi, and P. Kalaiselvan. 1998. Effect of irrigation layout, irrigation and fertilizer levels on the yield of rainfed groundnut (Arachis hypogaea L.). Crop Res. 18: 19-21.

97. Subrahmaniyan, K., P. Kalaiselvan, and N. Arulmozhi. 2000. Studies on the effect of nutrient spray and graded level of NPK fertilizers on the growth and yield of groundnut. Intern. J. Trop. Agric. 18 (3): 287-290.

98. Sudha, K.N. 1999. Response of rainfed groundnut to sand mulching and organics in verticincepti soils. M.Sc. (Agri.) thesis, University of Agricultural Sciences, Dharwad, India.

99. Sukanya, D.H., M.V.C. Gowda, and S.G. Hedge. 1995. Response of erect bunch groundnut genotypes to nitrogen fertilizer. International Arachis Newsletter 15: 77-79.

100. Toor, G.S., and G.S. Bahl. 1997. Effect of solitary and integrated use of poultry manure and fertilizer phosphorus on the dynamics of P availability in different soils. Bioresour. Technol. 62: 25-28.

101. Virmani SM 1973. Some edaphic factors affecting yield of groundnut in acid brown soils. Indian J Agric Sci. 43:119-122.

102. Wallace, A., Z.I. Zidan, R.T.Mueller, and C.P. North. 2007. Translocation of nitrogen in citrus trees. Proceedings of the American Society for Horticultural Science, 64, pp.87- 104

103. Wojnowska, T. H. Panak, and S. Siekiewiez. 1995. Reaction of winter oil-seed rape to increasing level of nitrogen fertilizer application under condition of Ketrzyn chernozem. Rosling Oleiste, 16: 173-180.

104. Yakadri, M. M.M. Husain, and V. Satyanarayana. 1992. Response of rainfed groundnut (Arachis hypogaea L.) to potassium with varying levels of nitrogen and phosphorus. Indian J. Agron. 37(1): 202 - 203.

105. Zharare G.E. 1996. Research priorities for groundnut (Arachis hypogaea L.) nutrition- A scientific basis for manipulating soil fertility to optimize groundnut yields. Agronomy Institute Annual Review and Planning W orkshop, 6-7 August 1996. Department of Research and Specialist Services, Harare, Zimbabwe.

Chapter 16

Preparation of Biofertilisers and Methods of usage in Oilseeds in Particular to Groundnut

N. Trimurtulu[*] and A. J. Suvarna Latha

Agricultural Research Station, ANGRAU, Amaravathi.

16.1 Introduction

In India, groundnut crop is grown predominantly in rainfed situations in marginal lands with suboptimal external nutrient supply. Even with the adoption of advanced technologies and high yielding varieties, the productivity of groundnut is around 1270 kg/ha. Being a legume crop most of its Nitrogen requirement should be met from biological nitrogen fixation. With an average pod yield of 2t/ha, there is a requirement of about 250 kg N/ha. In rainfed and irrigated conditions, it is recommended to apply only 12.5 kg and 25 kg N/ha, respectively, and thus rest of the requirement of N is being fixed by biological nitrogen fixation. Groundnut is the one which is often nodulated by all strains of rhizobia present in the tropic and subtropic soils (Brahmaprakash and Pramod kumar, 2012). Most of the times, ineffective native rhizobia strongly compete for nodule occupancy because of this there is highly erratic biological nitrogen fixation in groundnut crop. As a result, there is acute imbalance in the nitrogen availability in groundnut, thus affecting the productivity. Most of the groundnut growing soils are deficient in available phosphorus and an external application of P_2O_5 @ 40 kg/ha is recommended for groundnut cultivation. It is well documented that in acid soil, P is fixed as Fe and Al phosphates and in alkaline soil it is fixed as tri-calcium phosphate (Illmer and Schinner, 1995). In such soils, the groundnut productivity recorded to be low. To improve the phosphorus nutrition, inoculation with phosphate solubilizing microorganisms is much encouraged. The marginal soils are also often deficient in micronutrients in different parts of the country, specially Fe, Mo, Zn. etc. which were corrected through application of special group of biofertiliser organisms called plant growth promoting rhizobacteria and AM fungi etc.

16.2 Biofertilisers in Agriculture

Intensive commercial farming involves excessive use of agricultural land area through multiple cropping and uses of chemical fertilizers and pesticides as well. It is feared that practice of use of chemical fertilizers and pesticides continually would result in gradual decrease in quality of soils especially in terms of fertility. The quantity of application of chemical fertilizers is increasing over the years in Indian Agricultural System. In the last five decades, the rate of nitrogen, phosphorus and potassium (NPK) fertilizer application has increased tremendously. As per International Fertilizer Association records India is standing second in position by consuming more than 21.65 million tons of NPK fertilizers in the world. The challenge therefore is to continue agricultural productivity in a way that minimizes harmful environmental effects of fertilizers.

The role of bio-fertilisers in sustainable agriculture assumes special significance particularly in the present context of negative nutrient balance and prohibitive cost of chemical fertilizers. Bio-fertilisers cannot totally replace conventional chemical fertilizers but their usage along with organic manure can sustain crop production. Biofertilisers are apparently environment friendly, low cost, non-bulky agricultural inputs which could play a significant role in plant nutrition. For most crops and in most soil conditions up to 20–25% of nitrogen requirement can be met through biofertilisers. This situation coupled with cost advantage provides vast scope for introduction of alternative sources of nitrogen particularly biofertilisers (Katyal et al. 1994). Thus, there is a need to enhance the efficiency of the externally applied nutrients present in the soil. The latter becomes possible by the application of efficient strains of the rhizosphere microorganisms that can influence the nutrient availability to plants in many ways including transformation of the unavailable forms of minerals to the available forms.

In order to increase the unit area productivity of agricultural land, role of different crop nutrients in contributing increased crop yield play a vital role. Among the crop nutrients, nitrogen as well as phosphorus plays an important role in increasing the crop productivity. Further, the nitrogenous chemical fertilizers are manufactured industrially using non-renewable petroleum whose cost affects the cost of chemical fertilizers. On the other hand when increased doses of chemical fertilizers are used to increase crop production it causes environmental pollution and is also toxic to soil as it kills the beneficial microorganisms. The plants remove many nutrients from soil in modern intensive cultivation and needs replenishment. Under such conditions microbes offer good alternative technology to replenish crop nutrients.

In agricultural eco-system, microorganisms are having vital role in fixing/solubilizing/mobilizing/recycling nutrients. These microrgansims occur in soil naturally, but their populations are often scanty. The beneficial microbial populations occurring in soils always may not be supporting for higher crop yield. In order to increase the crop yield, the desired microbes from rhizosphere are isolated and cultured in adequate count and mixed with suitable carriers. These are known as biofertilisers or microbial inoculants. It includes *Rhizobium, Azotobacter, Acetobacter, Azospirillum*, Phosphate solubilizing microorganism (PSM), Arbuscular mycrorrhal fungi (AM fungi) , Plant growth promoting rhizobacteria (PGPR), micronutrient mobilizing bacteria like *Thiobacillus* sp. etc.

These biofertilisers are microbial inoculants which contain living cells of efficient nitrogen fixing microorganism, that fix atmospheric nitrogen either symbiotically with host plant or free living, phosphate solubilizing/mobilizing microorganisms as well as potassium mobilizers release or mobilize P_2O_5 and K_2O from soil. These biofertilisers are available in market as carrier based inoculants or as liquid inoculants. Biofertilisers are inexpensive and eco-friendly. Many State Agricultural Universities, Govt. Agriculture/ Forest Departments and good number of commercial units in public and private sectors are producing and distributing biofertilisers.

16.3 Nitrogenous Biofertilisers in Groundnut

Biological nitrogen fixation (BNF) is considered to be an important process which determines nitrogen balance in soil ecosystem. Legumes can potentially fix about 80% of their own nitrogen and in addition can contribute to the yield of subsequent crops. However, the potentiality is always flexible due to one or the other abiotic or biotic constraints. Abiotic constraints include moisture stress, salinity, acidity, high pH, P-deficiency due to fixation in calcareous soils, molybdenum and cobalt deficiency in acid soil etc.

16.3.1 Rhizobium in Groundnut

Rhizobium is a symbiotic bacterium that infects groundnut roots and form nodules. Within such nodules this bacterium reduces the nitrogen to ammonia which is readily utilized by plant to produce proteins, vitamins etc. The success story of rhizobium inoculants depends on the groundnut rhizoial strain having ability to overcome antagonism and competition from other soil microorganisms and other native rhizobial strains. Excess or insufficient soil moisture reduces N fixation. Application of cobalt has been reported to increase the nodulation in high pH soils. Increased application of nitrogen fertilizers decreases the rate of nitrogen fixation. Application of seed dressing chemicals are detrimental to the coated rhizobium survival hence indirect method of furrow application may be adopted in that case.

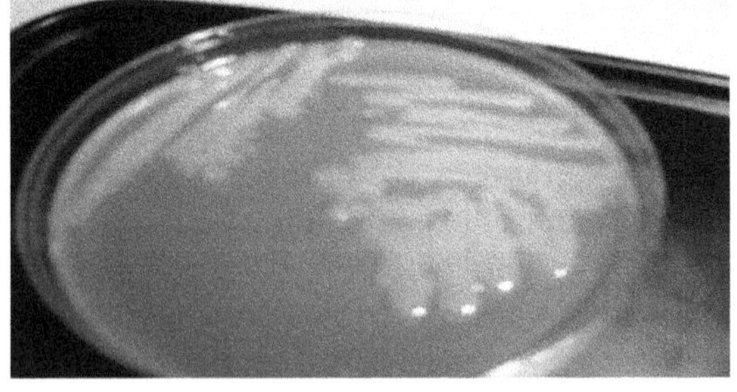

Fig. 16.1 Groundnut Rhizobial strain GN714

At ARS, Amaravathi we found that carbofuran 3G, Thiram, and Carbandizam doesn't harm the Rhizobial strain of GN 706, GN708 and GN714 (Fig. 16.1). Soil with low Fe levels are found to be low in nodule number and nitrogen fixation. Siderophore producing strains of rhizobium like NC92 was found to be good in such low Fe soils. Presence of high bicarbonates also adversely affected the Nitrogen fixation in groundnut.

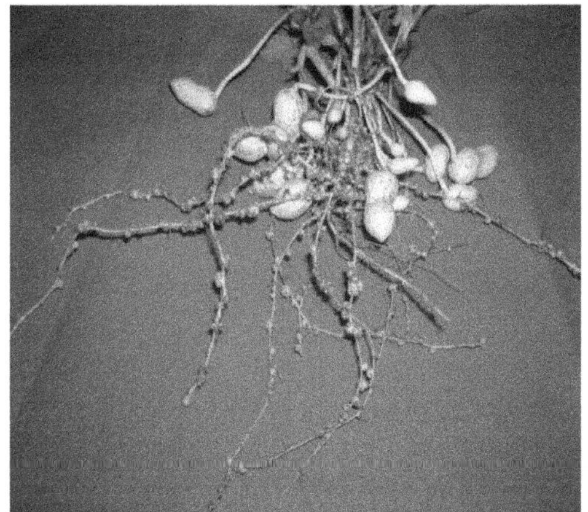

Fig 16.2 Uninoculated control

Fig. 16.3 Inoculated with Rhizobium GN714

Identification of strains with high saprophytic competence is essential for groundnut. Development of competitive strains in highly promiscuous crop like groundnut is challenging because the native flora are generally more competitive than the inoculated strain. Lack of response to inoculation and low yield in groundnut are probably due to competition from ineffective strains in the soil. Inoculation trials conducted with efficient strain GN 714 at ARS, Amarvathi showed significantly higher nodule number (Fig. 16.2 and 16.3). Experiments conducted by ICRISAT at Patancheru showed that inoculation with an effective Rhizobium strain NC91 increased the yield of groundnut cultivar "Robut 33-1" in such fileds where ineffective native strains were present. In Tamil Nadu, a strain of Rhizobium TNAU 14 and Tt 9 were used in commercial production.

16.4 Phosphatic Biofertilisers in Groundnut

Since, P solubilising microorganisms are heterotrophic in nature they need large amount of organic matter before they can excrete organic acids. Even if phosphate is solubilised, phosphate ions are incorporated into the microbial biomass, so roots can not absorb enough of them. The addition of a large amount of organic matter makes phosphate solubilising microorganisms proliferate and solubilise bound phosphates (Fig. 16.4) (Illmer and Schinner, 1995). Solubilised phosphates are incorporated into the microbial biomass during other microbial multiplication using organic matter. Once the organic matter becomes exhausted the microbial biomass cell disintegrates and releases phosphate into soil. The death of the microbial biomass can be accelerated by various soil treatments, including tillage, drying, liming and sterilization.

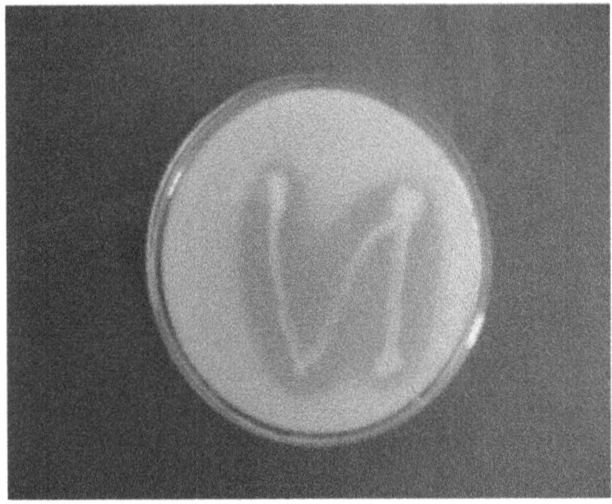

***Fig. 16.4** PSB biofertiliser AMT 1005 of ARS, Amaravathi*

The absorption of phosphate by groundnut plants can be accelerated by inoculation with Arbuscular Mycorrhizal Fungi (AMF). Arbuscular mycorrhizal fungus is a novel soil microorganism which associates with the roots of 95% of the crop plants and enhances nutrient uptake especially phosphorus, potassium, zinc, iron, boron and molybdenum etc. In addition to stimulating the nutrient uptake by plants, it is hoped that AMF will do the mining of water under moisture deficit situations and prevent the infection by root pathogens. Our survey revealed that groundnut is the one which is heavily colonized by Arbuscular Mycorrhizal fungi in the AP soils. Further, found that Alfisols are much more predominant with AM fungal spores than Vertisol soils. Hence, inoculation is a mandate practice in Vertisol soils of groundnut grown areas in AP.

16.5 Mechanism of Improved Plant Growth Due to Am Fungi

- Nutrient uptake especially P and micronutrients.
- Production of growth promoting substances
- Inducing drought tolerance and disease resistance
- Can reduce nearly 25 to 50% phosphatic fertilizers consumption in groundnut.

Researchers, demonstrated that different strains may sometimes vary greatly in their ability to infect the roots of groundnut plants. Therefore AMF products composed of multiple strains of Glomus intraradicis, Glomus etunicatum, Gigaspora etc., all with confirmed infection abilities. Spores and mycelia produced by the cultivation of host (Fig. 16.5 & 16.6) are packed with mineral carriers or peat moss. Studies on AM fungi in groundnut revealed that inoculated plants showed significant increase in root colonization over uninoculated control. AM fungi significantly increased the root volume nutrient uptake in groundnut crop. By application AMF the final pod yields of groundnut was reported to be increased by 11-14% (Wani and Lee, 1992).

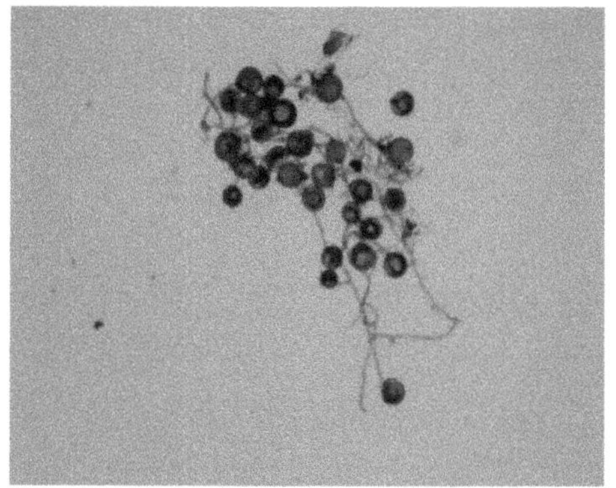

Fig. 16.5 AM fungal inoculum

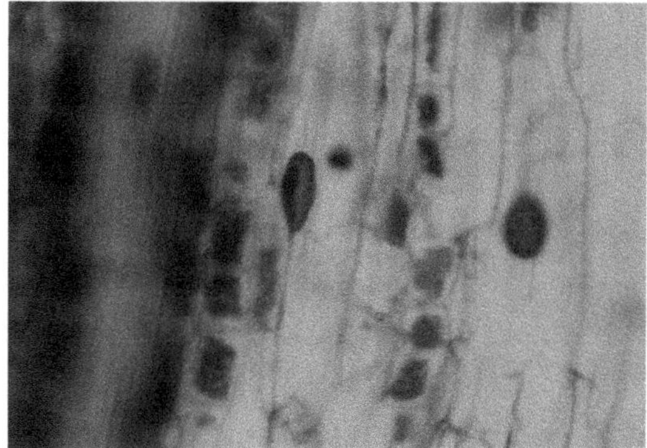

Fig. 16.6 AM fungi infected groundnut root

16.6 Plant Growth Promoting Rhizobacterial Biofertilisers for Groundnut

Rhizosphere bacteria which are known to stimulate the plant growth are usually referred to as plant growth promoting rihzobacteria or PGPR. They are mainly fluorescent pseudomonads and Bacillus sps. The mechanisms of PGPR mediated plant growth enhancement and yields are not yet fully understood. In groundnut these organisms facilitate plant growth, phytohormones like IAA, and gibberellins, by producing siderophores for chelating iron and synthesis of low molecular weigh compounds or enzymes that can modulate plant growth and development. PGPR are also reported to produce antibiotics that suppress deleterious rhizobacteria/plant pathogenic fungi or through some other unidentified mechanisms and thus provide healthy environment for better root growth (Adesemoye and Kloepper, 2009). Much of their beneficial effects may be a PGP/bio-control effect mediated by variety of ways including P-solubilisation as one but not the exclusive cause. In groundnut much of the scientific literature indicates that antagonism against soil borne phytopathogenic fungi like *Aspergillus niger*, *Sclerotium rolfsi* and *Aspergillus flavus* is

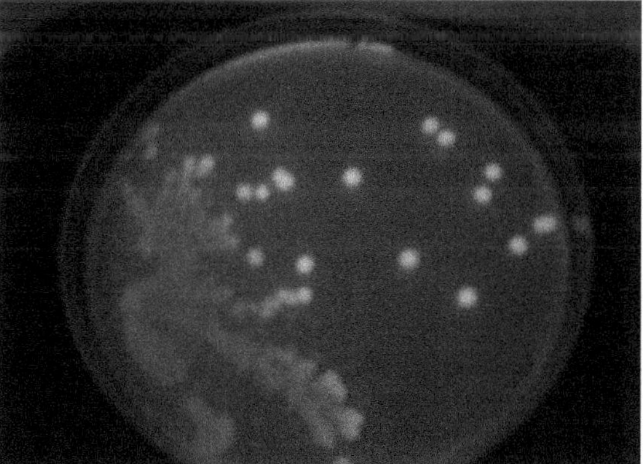

Fig. 16.7 PGPR organism of PSF1, ARS, Amaravathi

because of producing siderophores, B- 1,3-glucanases, Chitinases, antibiotics, fluorescent pigment, volatile substances and cyanide etc (Vessey, 2003).

Further these PGPR inoculants are reported for allevation of biotic and abiotic stresses by modulating the functional environment of plant tissues. Seed bacterization with PGPR organisms (Fig. 16.7) significantly enhanced the pod yield (23-28%), haulm yield and nodule dry weight over control at ARS, Amaravathi. Seed bacterization with PGPR strains suppressed the soil borne fungal diseases like collar rot of groundnut caused by *Aspergillus niger* and stem rot caused by *Sclerotium rolfsi* (Dey and Pal, 2014).

16.7 Why Biofertilisers to be Applied in Groundnut Cultivation?

- To enhance the efficiency of the externally applied nutrients.
- For transformation, mobilization and availability of the fixed or unavailable forms of nutrients.
- To sustain crop production, maintain soil health and soil biodiversity in the long run
- To integrate biofertilisers, biocontrol agents and organic materials in farming system to provide stability and sustainability in production.
- To promote eco-friendly agriculture and organic farming.

Excessive and indiscriminate use of chemical fertilizers may lead to:

- Decrease in organic carbon
- Decrease in population and diversity of beneficial soil microflora
- Increase in soil acidity, alkalinity and hardening
- Excessive use of N fertilizers may pollute ground water and leads to deterioration of soil health
- Imbalance in soil nutrient status

Though the effect of microorganisms is slow, it is sustainable in long term basis and would provide benefits like increasing crop yield up to 15% by reducing external application of fertilizers by 25%. Further they play a key role in stimulating plant growth, activating soil biological system, restoring natural fertility, maintaining soil biodiversity, providing protection against drought and some soil borne fungal pathogens etc.

16.8 Types of Biofertiliser Formulations Available for Groundnut

Formulation comprises aids to preserve organisms and to delivering them to their target fields and to improve their activities. A technical concentrate of an organism that has been achieved by a particular process is called as a formulation, or a product, which may be stored and put on sale. A wide range of formulations with additives is available in market.

16.8.1 Dry inoculum Products

These formulations comprises dusts, granules and briquettes based on particle size. They mainly contain 30% of an organism by weight.

16.8.2 Granules, Pellets, and capsules

Granules are discrete masses 5-10 mm^3. Pellets are more than 10 mm^3. In this, carriers include clay minerals, starch polymers and ground plant residues. Soft carriers, eg. Bentonite disburse quickly to release the organism. The concentration of organisms in granules is 20-30%.

16.8.3 Wettable Powders

The formulations contains lignite, vermiculite blended with 3% gum to make them stable during storage on the shelf and readily stick with seeds.

16.8.4 Liquid Formulations

These are special formulations those contain not only the desired microorganisnm, but also special cell protectants or substances that encourage formation of resting bodies or cysts for longer shelf life and tolerance to adverse conditions (Fig. 16.8).

Fig. 16.8 Liquid biofertilisers released from ARS, Amaravathi

Temperature is an important factor for the shelf life of microbial product and it can affect activity before or after application. In general colonization proceeds at field temperatures in the cropping season, but slows at lower or higher temperatures. Further, biofertiliser organism needs the surface to be wet in order to establish and perform its activity. These needs can be overcome by liquid formulations as the product contain humectants. To counter harmful effects of sunlight, sunscreens are added to liquids formulations. The pH of the product plays a vital role in liquid inoculum preparations. A buffer therefore is maintained by adding some addititives, which render better shelf life in liquids. The population of bacteria is also more in liquid. These formulations are very good for drip irrigation systems. The shelf life of these liquid biofertilisers are 12 to 18 months (Trimurtulu and Rao 2014).

16.9 Methods of Application of Biofertiliser in Groundnut

16.9.1 Seed Treatment

The seeds can be coated with either carrier based or liquid based inoculants in Groundnut. Generally, biofertilisers of groundnut can be mixed with the seeds in a plastic container as the size of seed is larger. The use of 10% sugar or 4% gum Arabic in the suspending solution facilitate the survival of rhizobia on seed. Carrier-based inoculants are mixed with minimum amount of water to form a slurry so as to coat uniformly on the surface of the seed. The seeds are dried in shade and sown immediately (Deaker, et al., 2004).

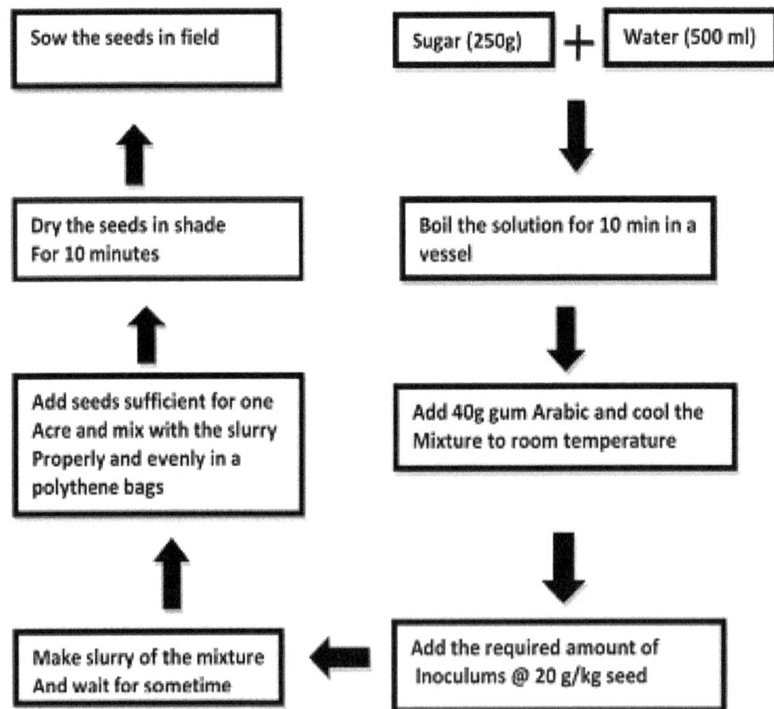

Fig. 16.9 Inoculation of biofertilisers in groundnut

16.9.2 Soil Application

For even distribution and spread over a large area, biofertilisers can be applied by mixing with FYM/ or any other organic manure, but sufficient moisture must be maintained to prevent drying up of inoculum other wise the inoculum load will fall. In groundnut, FYM enriched with biofertilisers can be applied in set furrows and then seeds are sown in the furrows and irrigation will be given.

16.9.3 Through Drip System or Irrigation Water

Liquid biofertilisers diluted in sufficiently can also be applied in irrigation channels or through drip system. This will also increases the viability of the cells and prevent desiccation/death of live bacteria.

16.10 New Formulations in Biofertilisers for Groundnut Crop

Formulations like mixing N-fixers, P-solubilisers and PGPRs in the same Biofertiliser packet (mixed microbial consortium) has been evaluated at ARS, Amaravathi and found very good results. Co-inoculation of *Rhizobium.PSB and* AMF has been found to be significantly better than their single inoculation. Mixed Bio Fertilizer formulations involving diazotrophs and phosphate solubilisers/PGPRs are the new concepts and should be encouraged. Largely the local isolates are more effective than the best introduced isolates showing the importance of the ecological adaptation of the strains in a given habitat and the difficulty of introducing an organism in an alien environment. Inoculation, of course, is not required if sufficient effective biofertiliser organism is already present in the soil. In the absence of simple indicators of the need for inoculation, many farmers inoculate as a form of insurance.

Single biofertiliser organism does not have all the characters to enhance the crop yields. Moreover there is a chance of failure of inoculation of single organism. Thus application of a mixed biofertilisers having mutually compatible strains of rhizobia, PSB, PGPR and AM fungi would be best available choice to provide healthy growth of ground crop under field conditions. The consortium developed at ARS, Amaravathi with a combination of Rhizobium (GN714), PSB(AMT1005)+PGPR 706 with 100% RDF gave significantly higher pod yields than other treatments and it was 18.8% higher than 100% RDF treatment in Vertisols of AP.

Table 16.1 *Recommended dose of biofertilisers for groundnut*

S. No.	Name of the Biofertil izer	Recommended dose of carrier inoculum
1	Rhizobium	20-25 g per kg seed
2	PSB	5 kg/ha
3	PGPR	5 kg/ha
4	AM fungi(VAM)	12.5 kg/ha

16.10.1 Method of Application

Take suitable biofertiliser organism specified groundnut crop of one hectare dose (Table 16.1) and check for their compatibility and inoculate all the biofertilisers to 500 kgs of organic manure one day before application and apply the biofertilisers mixture at the time of sowing. Care should be taken that the biofertiliser mixture should fall below the seed or nearer to seed in the main field. At the time of biofertilisers application field should be with a minimum of 30-40% moisture.

The advantages of application of consortium of Biofertilisers

- Enhance the biological nitrogen fixation by more nodule occupancy and efficient N fixation.
- Enhances the phosphate solubilization and uptake from soil.
- It will enhance the mobilization and uptake of N, P, K Fe and Zn etc.
- It produces several growth promoting substances like IAA, siderophores, vitamins etc.
- Easy to apply in large scale.
- Low cost
- Can be used for both rainfed and irrigated situations.

Precaution

- Farmers should not apply biofertilisers along with chemical fertilizers and pesticides
  until unless they are compatible.
- In case of dry sowing practice farmer may apply at the time of first inter-culturing time.
- Use right combination of biofertilisers.
- Biofertilisers are live products and they require care in storage.
- Use mixed biofertilisers consortium for best results in groundnut.
- Biofertilisers are not replacement of chemical fertilizers but can supplement their requirement.

16.11 Usage of Biofertilisers in Groundnut Crop

Legumes can fix about 80% of their own requirement of nitrogen and in addition can contribute to the yields of subsequent crop. However, the potentiality is seldom realized in groundnut due to one or other constraints including biotic and abiotic stresses. But identifying competitive and efficient nitrogen fixing strains rhizobia in groundnut can solve the problem of ineffective nodulation by native rhizobia. The most widely adopted and released strains of different biofertiliser organism across Indian subcontinent has been given in Table 16.2 (Rao, 2008).

***Table 16.2** Influence of different biofertilisers on the yields of groundnut in India (Rao, 2008)*

S. No.	Name of the Biofertiliser	Isolates/strains used	Yield improvement	Increase in nutrient uptake
1	Rhizobium/ Bradyrhizobium	NC 92, GN706, GN708, GN714 TNAU14, NRCG9	10-19%	N
2	PSB	AMT 1001, AMT 1005, BM8	7-12%	P
3	PGPR	PSF1, PGPR706 PGPR4, PGPR8	12-16%	N,P, Fe, Zn etc.
4	AM fungi	Glomus intraradicis, G.mossae , Gigaspora etc.	6-10%	P, and other micronutrients
5	Mixed biofertiliser consortium	Rhizobium+ PSB+PGPR+AMF	13-20%	N, P and other micronutrients

In a three year study on vertisols by AICRP-BNF, Parbani, Rhizobium inoculation of groundnut grown with recommended dose of NPK (25:50:30) increased the pod yield by 3.9 q/ha while FYM alone @ 5t/ha increased it by 1.5 q/ha, combined application of FYM and Rhizobium increased it by 7.3 q/ha (Rao, 2008). The inoculation technology has been successfully demonstrated through various extension programmes on front line demonstrations in farmers fields carried out for the last decade to popularize the use of biofertilisers along with organic manures in the groundnut crop.

16.12 Quality of Biofertilisers

Because most extension advisors lack any knowledge of microbial products, confusion and trouble frequency occur. It is now well established and documented that introduction of standards and quality testing services has improved the situation regarding inoculants in at least in developed countries.

Evaluation of inoculant quality by enumeration of viable cells is an accurate index of inoculating potential. It was observed that substantial proportions of the inoculants examined appear unsatisfactory for farmer use because of low populations of required organism or high numbers of microbial contaminants. The dirt cheap cost of biofertilisers, the perceived ease of their production often in a cottage industry style employing skilled labor with non-microbiologists supervision, has contributed not only a poor image for the product but also cases the failure of technology Improper handling at the time of application by even technical personal has further worsened the problem. The enthusiasm of the biofertiliser researcher is not shared by farmers or planners due to inconsistent response or exaggerated claims. Transfer of biofertilisers technology to the benefit of small farmers through the agricultural extension services, IPNS practices should be economically viable, sociologically acceptable and environmentally safe.

16.12.1 Quality Control of Biofertilisers

As per the Bureau of Indian Standards the following are the relevant specifications for biofertilisers.

1. The inoculants shall be a carrier based
2. It should contain a minimum of 10^8 viable CFU per gram of carrier on dry mass basis within 15 days of manufacture and 10^7 within 15 days before expiry date when the inoculants stored at 25-30^{0}C
3. The inoculants should have a maximum shelf life of 6 months from the date of manufacture.
4. The pH of the inoculants shall be between 6.5 to 7.5
5. The inoculants shall show effective nodulation on all the crops listed on the packet.
6. Carrier should be a neutralized powder.
7. The manufacturer should check the quality of broth and maintain record of tests.
8. The inoculants should be packed in 50-75 u low-density polyethylene packets.

16.13 Frontline Demonstrations in Groundnut

Frontline demonstrations were conducted during summer season of 2003-2007 on farmers field to evaluate the effect of Rhizobium and PSB and their dual inoculation on Groundnut var Tag-24. Rhizobium and PSB inoculation were used @250g /10 kg seed. Highest pod yield of 4338 kg/ha was recorded in inoculated groundnut during year summer 2003. Rhizobium with PSB inoculation gave 22.0% increase in pod yield over uninoculated (Table 16.3).

Table 16.3 *FLDs conducted in tamilnadu on groundnut crop with biofertilisers*

Year	Pod Yield(kg ha^{-1})		%
	Uninoculated	Inoculated	increase
2003	3423	4338	27
2004	2536	3073	21
2005	1674	1988	18
2006	1640	1877	15
2007	1567	1917	23
Mean	2168	2639	22

16.14 Conclusion

Application of biofertilisers in groundnut is good to improve the soil health and for better transformation and mobilization nutirnts in the soil. With the growing awareness of organic agriculture demand for biofertilisers is high because of their major stand. Thus, to enhance the productivity of groundnut, application of highly competitive strains of groundnut rhizobia, phosphate solubilizing microorganisms, plant growth promoting rhizobacteria (PGPR), AM fungi and other beneficial bacteria like endophytes would help in realizing good yield in sustainable manner. Application of these biofertilisers is a long term sustainable perspective and should not be thought of a short term gain. Quality of the biofertiliser is the major constraint in India which ultimately affect the benefits of these novel inputs in groundnut production.

References

1. *Adesemoye, A. O. and Kloepper, J. W. 2009. Plant-microbes interactions in enhanced fertilizer- use efficiency. Appl. Microbiol. Biotechnol. 85:1-12.*
2. *Brahmaprakash, G. P. and Pramod Kumar, S. 2012. Biofertilisers for Sustainability. J of Ind. Insti. of Sci. 92(1): 37-62.*
3. *Deaker, R., Roughley, R. J., and Kennedy, I. R. 2004. Legume seed inoculation Technology – a review. Soil Biol. Biochem. 36 (8): 1275-1288.*
4. *Dey, R and Pal, K. K. 2014. Biofertilisers for enhancing groundnut productivity, ICAR- Directorate of Groundnut Research, Ivnagar Road PB. No.5, Junagadh, Gujarat, India, 32P.*
5. *Illmer, P and Schinner, F. 1995. Solubilization of in organic calcium phosphates–solubilization mechanisms. Soil Biol. Biochem. 27(3): 265-270.*
6. *Katyal, J. C., Venkateswarlu, B., and Das, S. K. 1994 Biofertilisers for nutrient supplementation in dryland agriculture. Fert. News. 39: 27-32.*
7. *Rao, D. L. N. 2008. Biofertilisers Research Progress (2004-2007), All India Network Project on Biofertilisers, IISS Bhopal. pp105.*
8. *Trimurtulu, N., and Rao, D. L. N. 2004. Liquid Microbial Inoculants and their Efficacy on Field Crops. ANGRAU, Agricultural Research Station, Amaravathi pp 54.*
9. *Vessey, K.J. 2003. Plant growth promoting rhizobacteria as biofertilisers. Pl. Soil., 25: 557-586.*
10. *Wani S. P. and Lee, K.K. 1992. Role of Biofertilisers in upland crop production. In. Fertilizers of Organic Manures Recycle Wastes and Biofertilisers. (Ed. HLS Tondon), Fertilizer Development and Consultation Organization, New Delhi, pp 96-107.*

Chapter 17

Efficient Use of Fertilizers in Groundnut Production

G. Prabhakara Reddy and V. Divya

Dept of Agronomy, S. V. Agricultural College, Tirupati, Andhra Pradesh.

17.1 Introduction

Oilseed crops have been backbone of agricultural economy of India from time immemorial. Groundnut, besides being a premier oilseed crop, is also a valuable high quality protein in India and presently grown in about 260 districts in various states. Among the 9 annual oilseeds cultivated in India, in about 25 million ha, groundnut holds 32% share of the total area under oilseeds and contributes 42% of the total oilseed production. Gujarat, Andhra Pradesh, Tamil Nadu, Karnataka and Maharashtra are the five major groundnut growing states account for about 85% of the total groundnut area and contribute nearly 85% of the total groundnut production. West Bengal, Assam and NE Hill Regions are the potential Non-traditional areas where groundnut is grown in both rainy (uplands/terraces) and post rainy seasons. The majority of the groundnut area (5.5-6.0 m ha) is under rainfed where productivity fluctuates between 500 - 1500 kg ha^{-1} depending upon the monsoon. Rabi-summer groundnut on the other hand is grown in about 1.5-2.0 m ha under assured moisture condition with a productivity of 2000 - 3000 kg ha^{-1}.

17.2 Causes for Poor Yields of Groundnut

Yields of groundnut are low due to cultivation under rainfed conditions, poor soil fertility, inadequate plant population under field conditions, nutrient deficiencies, inadequate and imbalanced nutrient application, pests and diseases etc.,

17.3 Importance of Nutrient Management in Groundnut

The range of groundnut grown soils is so wide that some are having exceptionally low or excess amount of certain nutrients. Groundnut requires 13 mineral elements for its growth and development. The essential mineral nutrients, based on their requirement by plant, are classified as macro-nutrients (N, P, K, Ca, S and Mg) and micro-nutrients (Fe, Mn, Zn, Cu, B, Mo and Cl). In addition, nickel (Ni), cobalt (Co) and sometimes Al are also beneficial for groundnut crop. In groundnut, the macronutrients are largely involved in structure, however, the micronutrients act as catalytic and regulatory role in enzyme and are required relatively in small quantities. Unlike other plant, the groundnut nutrition is unique as the pod develops under soil and most of the seed nutrition is directly through pod rather than those transported from root, shoot and back to the seed.

The mineral nutrient disorders in groundnut limit crop production. As it requires all the essential nutrients in balanced proportions, deviation from this results in mineral disorders, which may be due to deficiency or the toxicity of a particular nutrient or multi-nutrients. India is the world's largest producer of groundnut where nutritional disorders cause yield reductions from 30-70% depending upon the soil types, soil nutrient status and groundnut varieties.

Groundnut grows in an environment facing various climatic conditions and soil types and following factors are responsible for mineral deficiencies and toxicities:

- Continuous withdrawal and inadequate supply of nutrients in the soil;
- Leaching, runoff and nutrient fixation in the soil;
- Edaphic factors (soil, water, temperature and environment) preventing absorption of nutrients by plants;
- Changes in soil physico-chemical conditions such as pH and EC;
- Imbalanced use of fertilizers;
- Induced deficiency (Al induced P and Ca deficiencies in acid soils and Ca-induced iron-deficiency in calcareous soil);
- Interaction between minerals during uptake and antagonism. Low or excess of any element influences the uptake of other nutrients also that may be synergistic (beneficial) or antagonistic (detrimental) effects. At low levels of K, the Ca and Mg uptake are more in groundnut plant tissues except seeds which decreases with increasing the levels of K. Similarly, the concentrations of K and Mg in plant tissues are high at low levels of Ca and decrease with increase of Ca levels;
- Use of intensive nutrient (responsive) requiring and nutrient in-efficient genotypes;
- Biotic (disease and pests) factors;
- The water availability, light intensities and duration and extreme temperature alter the concentrations of elements;
- The temperature influences the movements, translocation and utilization of elements by plants;

- Increasing light exposure reduces the N, P, K concentrations while shading increases the P, K, Al, Ca, Fe and Mn concentrations and decreases the Cu, Mg and Zn concentrations in leaves;
- The humidity influences the rate of transpiration and thus indirectly affects the nutrient content in plants;

The prominent mineral nutrient deficiencies in the major groundnut growing areas and their causes are furnished in Table 17.1.

***Table 17.1** Nutrient deficiencies in different groundnut growing areas*

States and regions	Soil types	Nutrient deficiencies	Causes for nutrient deficiencies
Gujarat (Saurashtra)	Coastal alluvial ,medium black calcareous, sandy to clayey	P, S, Fe, Zn and B	Calcareous soil, low rainfall and excessive irrigation
Gujarat (Kutch)	Coastal alluvial, sandy	P, K, S, Zn and B	Sandy soil, low fertility, low rainfall
Andhra Pradesh	Red and mixed red	N, P, Ca, K and B	Low soil fertility, and drought
Tamil Nadu	Red and mixed red	Ca, N, P, K and B	Low soil fertility due to leaching, Al-toxicity
Karnataka	Laterite , red and mixed red soils	Ca, Mg, N, P, K and B	Excess and low rainfall, low fertility
Maharashtra	Black calcareous, mixed red	P, K, S, Fe, Zn and B	Calcareous soil, low fertility,
Orissa	Laterite, red and mixed red	Ca, Mg, N, P, K and B	Excess and low rainfall, low soil fertility,
Rajasthan	Desert, red yellow and mixed red	P, K, S, Fe, Zn and B	Low fertility, sandy and calcareous soil, low rainfall and drought
Punjab and Haryana	Alluvial, alkaline	P, K, S, Fe and Zn	Salinity, alkalinity, excess irrigation
Madhya Pradesh and Chhattisgarh	Red and mixed red, medium black	N, Ca, P and K	low soil fertility and low rainfall
Uttar Pradesh	Alluvial, red and mixed red	Ca, P, K and B	Salinity, alkalinity,
Bihar & Jharkhand	Alluvial, red and mixed red	Ca, P, K and B	High rainfall and leaching

Table 17.1 *Contd...*

States and regions	Soil types	Nutrient deficiencies	Causes for nutrient deficiencies
Assam & NE	Red, mixed red and laterite soils on upland , Tilla, and foot hill and alluvial in medium land	Ca, Mg, P, K, B and Mo	High rainfall, leaching, crusting erosion and compaction, Al - toxicity
West Bengal	Alluvial, red, mixed red and laterite	Ca, Mg, P, K, B and Mo	High rainfall, leaching, compaction and Al - toxicity

Groundnut is an exhaustive crop and depending upon the yield, it removes large amount of macro and micronutrients A groundnut crop, with an economic yield of 2.0 to 2.5 t ha^{-1} removes 160-180 kg N, 20-25 kg P, 80-100 kg K, 60-80 kg Ca, 15-20 kg S, 30-45 kg Mg, 3-4 kg Fe, 300-400 g Mn, 150-200g Zn, 140-180 g B, 30-40 g Cu and 8-10 g Mo which is quiet high and to harvest 5 t ha^{-1} pod in coming 20 years, it will harvest double the amount. Among these Ca, K, P, S, Fe and B are involved in the kernel filling and oil synthesis and are always limiting hence need more emphasis. The average worldwide yield increases due to application of calcium in the pegging zone is 24%.

Inadequate and imbalanced use of nutrients is one of the major factors responsible for low yields in groundnut. Providing the adequate quantity of nutrients can double the production and thus meeting the challenges of the future 20 years. Optimization of the mineral nutrition through proper nutrient management with balance doses of all the nutrients is the key way to optimize the production of groundnut, as it is a high nutrient requiring crop and being an unpredictable legume, if proper doses of fertilizers are not there, many a times there is either no response of any fertilizer or negative response of a single fertilizer. Management of N, P, K, S, Ca and Mg for higher yields of groundnut is discussed below.

17.4 Nitrogen Management

Nitrogen requirement of groundnut is much higher than cereals because of its high protein content, but most of the soils where groundnut is grown are deficient in nitrogen. The fact is that groundnut is capable of meeting its nitrogen requirements both from symbiotic nitrogen fixation (60-80%) by root nodules and soil nitrogen (20-40%). However, the nitrogen supply to groundnut is very crucial and deficiency is observed during early growth stages till 45 days after emergence and at pod formation stages. During reproductive stage, N is mobilized continuously from leaves to the developing pods and hence deficiency occurs at this stage too. For these reasons N must be applied to groundnut at these stages.

In groundnut, as there is no early senescence of leaves, yellowing of older leaves which later on die with age of crop is the most common nitrogen deficiency symptom. Besides that the N deficient crop shows weak and prolonged stem. For optimum yield, the N, P and S ratio should be around 15:1:1 in the 4th leaves. The Spanish and Valencia bunch groundnut, because of lesser nitrogen-fixation and crop duration respond more to nitrogen than the Virginia. The Virginia cultivars accumulate N at faster rate and maintain higher N in leaves than the Spanish groundnut.

Nitrogen needs of groundnut can be met through any available N sources, however, urea, DAP and ammonium sulphate are the preferred sources. Ammonium sulphate is preferred, if deficiency is combined with S. In rainfed groundnut, N should be applied as basal in furrows, however split application is useful during good rain, if there is no drought. Nitrogen should be applied in 2-3 splits in irrigated groundnut.

The recommended doses of nitrogen are in between 20-40 kg ha^{-1} for rainfed and 30-50 kg ha^{-1} for irrigated crop. Application of 20 kg N ha–^{1} as basal dose followed by two topdressings of 10 kg N ha^{-1} each at 30 DAE (Days after emergence) and 70 DAE produce highest groundnut pod and increase the efficiency of N fertilizer. Higher amount of N leads to excessive growth and decreases harvest index and pod yield.

Yield data given in the above table 17.2 revealed that application of N @ 10 and 20 kg/ha proved highly significant over control and N @ 20 kg/ha also proved significantly superior over N @10 kg/ha. Response equation fitted for nitrogen data was Y = 766 + 9.10 X. Response to nitrogen was found to be 9.10 kg of dry pods per kg of N applied (Yadav, 1990).

Table 17.2 *Effect of nitrogen levels on dry pod yield (kg/ha) of groundnut.*

Nitrogen (kg/ha)	Pod yield (kg/ha)	Increase over control (kg/ha)
0	765	-
10	859	94
20	947	182

17.5 Phosphorus Management

Most of the Indian soils, where groundnut is grown are deficient in P due to its fixation and low availability. Phosphorus is a component of certain enzymes and proteins, and involved in various energy transfer reactions. Phosphorus limits N_2-fixation, plant growth, shelling percentage and oil content and it is very critical at flowering and pod formation stages.

The critical levels of soil available P, in surface soil, is 10 ppm for groundnut. Leaf P content, at flowering is the most standard plant part and age to determine the critical levels which is 0.22 % in Spanish and 0.25 % in Virginia cultivars.

The P fertilization at flowering and pod formation stages increases shelling percentage, oil yield and nodulation. The recommended doses are 30-40 kg P_2O_5 ha^{-1} during rainy season and 40-60 kg P_2O_5 ha^{-1} during post rainy season. Field application of 20 kg P_2O_5 ha^{-1} as basal and 20 kg P_2O_5 ha^{-1} at 30 DAE as single super phosphate (SSP) is best for maximum pod yield and fertilizer efficiency.

The Ca-P is the chief fraction involved in the P nutrition of groundnut. However, due to transformation of P into unavailable form in soil, the response of P to groundnut is not consistent for which following practices need to be adopted:

- Use of P-efficient genotypes in acid and calcareous soils.
- Broad-bed and furrow system of cultivation.
- Single superphosphate is better than triple superphosphate.
- Drilling of P below the seed in furrows is superior to broadcasting.
- Inoculation with efficient cultures of PSM and VAM.

Studies by (Yadav 1990) indicated that application of phosphorus @ 30 and 60 kg/ha proved to be highly beneficial. Application of phosphorus @ 60 kg/ha was also significantly superior over phosphorus applied at 30 kg/ha. Response equation fitted for phosphorous data was Y = 768 + 6.23 X which revealed that response to per kg of applied phosphorus was 6.23 kg of pods of groundnut (Table 17.3).

Table 17.3 Effect of Phosphorus levels on dry pod yield (kg/ha) of groundnut

Phosphorus (kg/ha)	Pod yield (kg/ha)	Increase over control (kg/ha)
0	765	-
30	962	187
60	1139	374

17.6 Potassium Management

Potassium is required for translocation of assimilates and involved in maintenance of water status of plant, the turgor pressure of cells and opening and closing of stomata. It increases the availability of metabolic energy for the synthesis of starch and proteins and helps in pod growth and filling. Most of the groundnut growing soils in western parts of India are rich in K, but the eastern parts are poor in K. However, the K deficiency is prominent in Tamil Nadu, Orissa, part of A.P., U.P., Gujarat and Maharashtra due to very high K requirement by groundnut, intensive cropping and very little or no potassium application.

As groundnut roots are highly efficient in obtaining soil K, the sufficiency level of K in soil is above 20 ppm K. The normal K content in leaf is 1.5 % and deficiency occurs below 0.5 %. The soil having less than 150 kg ha^{-1} exchangeable K is deficient in K and has to be fertilized with 50 kg ha^{-1} K_2O.

The recommended doses of K are 25-50 kg K_2O ha^{-1} for rainy season and 40-75 kg K_2O ha^{-1} for post rainy season irrigated crop. Large seeded groundnut requires more K than the small seeded one. It should be applied as basal in the furrows. However, top dressing also gives good results.

Application of 40 kg potassium/ha along with N and P @ 20 and 60 kg/ha respectively, proved significantly superior over application of the same without K (Table 17.4). (The balanced use of N and P in all the combinations proved highly significant over their application alone. Similarly balanced use of NPK proved highly significant over NP combinations alone (Yadav,1990).Thus it may be inferred that balanced use of N, P and K@ 20, 60 and 40 kg/ha respectively, proved highly desirable dose of boosting groundnut production under Chittorgarh conditions which increased the yield from 7.65 to 15.77 q/ha (Table 17.4).

Table 17.4 *Yield of dry pods of groundnut (q/ha) as affected by various treatments.*

Year			1978-79	1979-80	1980-81	1981-82	Pooled mean
No. of trials			14	13	14	11	52
N	P	K	4.27	4.16	10.68	12.23	7.65
0	0	0	4.70	4.90	15.16	14.43	9.62
0	30	0	5.64	5.54	18.74	16.26	11.39
0	60	0	4.87	4.79	12.16	13.28	8.59
10	0	0	5.52	5.69	17.21	15.88	10.90
10	30	0	6.45	6.41	20.57	18.24	12.74
10	60	0	5.70	5.49	13.16	14.26	9.47
20	0	0	7.05	6.59	20.12	18.04	12.78
20	30	0	7.76	7.39	23.16	20.08	14.54
20	60	0	8.28	7.91	25.99	21.57	15.77
20	60	40	6.02	5.89	17.74	16.43	11.34
SEd±			0.23	0.26	0.40	0.23	0.15
CD (P=0.05)			0.46	5.51	0.79	0.45	0.29

Hundred pod weight was significantly influenced by Long-term application of manure and fertilizers. The highest value was recorded with NPK + Gypsum + $ZnSO_4$. The next best treatments were T_4 (P alone), T_7 (NP) and T_{10} (NPK + Gypsum) during kharif , 2002 and T_2 (FYM alone) and , T_3 (N alone) and T_4 (P alone) during 2003 (Table 17.5). It was noticed that application of gypsum and $ZnSO_4$ alone with the NPK performed well. The treatment NPK + Gypsum + $ZnSO_4$ recorded the highest pod yield in both the years (Kishore Babu et al ., 2007).

Table 17.5 *Effect of long term application of manures and fertilizers on groundnut.*

Treatments	100 pod weight			Pod yield (kg/ha)		
	2002	2003	Mean	2002	2003	Mean
T_1- Control (no NPK/Manure)	63	68	65	830	1399	1115
T_2- FYM @ 5 t/ha	64	79	71	970	1763	1367
T_3-20 kg Nitrogen/ha	64	74	69	956	1677	1317
T_4-10 kg Phosphorus/ha	72	77	74	1020	1833	1427
T_5-25 kg Potassium/ha	63	68	66	946	1602	1273
T_6- 250 kg gypsum/ha	66	70	72	985	1827	1406
T_7-20 kg Nitrogen/ha + 10 kg Phosphorus/ha	70	73	71	1072	1855	1463
T_8-20 kg N/ha + 10 kg P/ha + 25 kg K/ha	68	73	71	1236	1860	1548
T_9-20 kg N/ha + 10 kg P/ha + 25 kg K/ha + 250 kg gypsum/ha	69	73	73	1272	1896	1584
T_{10}-20 kg N/ha + 10 kg P/ha + 25 kg K/ha + 100 kg lime/ha	74	73	77	1249	1886	1568
T_{11}-20 kg N/ha + 10 kg P/ha + 25 kg K/ha + 250 kg gypsum/ha + 25 kg $ZnSO_4$/ha	74	79	-	1286	2138	1712
SEm±	1.6	2.1	-	20.3	51.0	28.2
CD(P=0.05)	4.5	6.0		58.7	147.4	79.6

Field experiments conducted by (Rathore et al. 2014) at ICAR Research Complex, North East Hill region, Jharnapani Centre, Nagaland, on relative effectiveness of sulphate of potash (SOP) and schoenite (SCH) on groundnut revealed that schoenite as a source of potash significantly improved the growth, yield attributes and pod yield of groundnut. The maximum dry pod yield and oil yield were recorded with 60 Kg/ha potash through schoenite. Schoenite also helps in better nutrient uptake and their utilization in the plants. The indigenous potassic source was found to be highly effective in increasing the groundnut productivity compared to exotic sulphate of potash (Table 17.6).

***Table 17.6** Effect of potassic fertilizers on groundnut*

Treatments (kg K_2O /ha)	No. of pods / plant	Dry pod yield (kg/ha)	Oil yield (kg/ha)
Control	15.6	1626.5	475.7
20 SOP + SCH	17.8	1987.0	587.5
40 SCH	30.4	1911.5	659.8
40 SOP	22.2	2020.0	699.2
40SOP+SCH	30.3	2095.0	724.7
60 SCH	36.3	2725.0	965.6
60 SOP	24.5	2193.5	757.8
60 SOP+SCH	30.8	2602.0	875.0
SEm±	3.3	213.5	28.4
CD (P=0.05)	7.0	458.0	61.0

17.7 Sulphur Management

The sulphur is as important as phosphorus for groundnut as it constitutes methionine and cysteine, increases oil synthesis, nodulation and pod yield and reduces disease incidence. The groundnut grown on coarse-textured sandy soils generally suffers from S deficiency due to leaching of $S0_4^{2-}$ and is severe in Bihar, Gujarat, Punjab, M.P., U.P, Karnataka and Andhra Pradesh.

The S deficiency symptoms are like nitrogen but occur on young leaves and extend to middle showing pale yellow colour with vein yellow white. In calcareous soil the Fe and S deficiency symptoms appear together. The adequate concentration of S in leaf is in between 0.2-0.35% and S deficiency occurs when the concentration falls below 0.17%. The critical level of soil available S is 10 ppm (heat soluble) in the surface layer (0-15 cm) and most of the groundnut growing soils, in India, are deficient in S. Calcareous and red laterite soils are more S deficient and application of 20-30 kg S ha^{-1} is essential.

Generally 20 kg of S ha^{-1} is sufficient for groundnut which can be met through either of gypsum, elemental S, pyrite and phosphogypsum. Plant takes up S mainly as SO_4^{2-} and the sulphur added to soil is subjected to microbial transformation by *Thiobacillus sp.* Into SO_4^{2-} before the plants take it up. Application of 1 kg of nutrient S produces 13 kg pod and 4.3 kg oil. Application of 10 kg S ha^{-1} as basal and 10 kg S ha^{-1} at 40 days after sowing is best for maximum pod yield with high fertilizer efficiency. The natural deposits of gypsum and pyrites are found in abundance in India and are effective sources of S. Pyrite is a good and cheap sulphur source for calcareous and alkaline soils. The S also shows synergistic effect with K and P on yield and nutrient uptake.

(Patel et al. 2009) reported that both levels of sulphur (40 and 60 kg/ha) did not vary from each other in respect of number of pods and 100 kernel weight and showed superiority over lower levels on loamy sand soils at Dantiwada Agricultural University,

Gujarat. However, application of S irrespective of dose significantly increased shelling percentage over no sulphur application. A linear increase in pod and haulm yield was observed up to 40 kg S/ha. It increased pod yield over 0 and 20 kg/ha to the tune of 19.90 and 8.34% whereas for haulm yield the corresponding values were 13.49 and 6.45%, respectively (Table 17.7).

Table 17.7 *Effect of sulphur levels on yield attributes, yield and quality of groundnut*

Sulphur levels (kg/ha)	Pods/ plant	Shelling %	100 kernel weight	Pod yield (kg/ha)	Haulm yield (kg/ha)	Oil content (%)
0	27	74.29	33.05	3085	5161	47
20	31	75.29	35.98	3414	5502	48
40	33	75.57	38.17	3699	5857	49
60	33	75.43	38.04	3688	5843	49
CD(P=0.05)	3.2	0.42	0.49	89	117	0.3

(Dash et al. 2013) concluded that the pod yield was significantly increased up to 20 kg/ha, which was statistically at par with that obtained with 40 kg/ha application. The value of 100-seed weight was statistically differed due to application of graded doses of sulphur. The highest shelling percentage was observed with 40 kg/ha. The haulm yield was highest with application of 40 kg S /ha against the lowest value obtained with no sulphur (Table 17.8).

The seed yield due to application of graded doses of sulphur followed similar trend with that of pod yield. The mean oil content was the highest with 60 kg S/ha and the increase was 10.8% over the control.

The highest and the lowest values of sulphur use efficiency (SUE) were observed with 20 kg S/ha and 60 kg S/ha respectively. The SUE decreased with increasing dose of sulphur from 20 kg to 60 kg S/ha due to decrease in yield with higher dose of sulphur.

Table 17.8 *Effect of different levels of sulphur on yield attributes, yield, quality and sulphur use efficiency of groundnut*

Sulphur levels (kg/ha)	100 kernel weight (g)	Shelling %	Pod yield (t/ha)	Haulm yield (t/ha)	Seed Yield (t/ha)	Oil Content (%)	Sulphur use efficiency
No sulphur	48.1	70.4	1.52	2.72	1.08	43.23	-
20	49.5	72.2	1.8	3.01	1.31	46.60	15.42
40	49.1	72.6	1.79	3.06	1.31	47.77	6.67
60	48.8	71.0	1.66	2.91	1.18	47.90	3.62
SEm±	0.8	0.5	0.04	0.03	0.06	0.44	0.68
CD(P=0.05)	NS	1.4	0.11	0.09	0.16	1.28	1.94

17.8 Calcium Management

Calcium maintains cell integrity and membrane permeability, enhances pollen germination, activates number of enzymes for cell division and takes part in protein synthesis and carbohydrate transfer. In groundnut, Ca is more important as often lack of Ca reduces the yield and quality more than any other element. Early ovule abortion is prevented by adequate calcium supply. The groundnut grown on acid soils having low base saturation show prominent Ca-deficiency for which liming is must. Calcium is taken up directly from the soil by pods, and inadequate supply results in pods without seeds called "Pops" or blackened plumule inside the seed known as "Black heart". The plumule damage in seed is due to calcium deficiency. The critical levels of soil Ca is 250 ppm in root zone and 600 ppm in pod zone, however the soil Ca more than 2 meq 100 g of soil in root zone and more than 4 meq 100 g^{-1} in pod zone is ideal.

The runner genotypes disperse their pods more than bunch genotypes typically providing double the soil volume available for calcium exploitation to pods. The Ca accumulation in pod is also positively correlated with the pod surface area, days required to maturity, and negatively with the pod thickness.

Depending upon the soil Ca level, application of 500-1000 kg ha^{-1} lime and gypsum is sufficient to meet the requirement of groundnut, however, for effective use of Ca following practices are recommended:

- The lime-stones (90% $CaCO_3$), dolomites, gypsum, phospho-gypsum, calcium oxide, CAN, and Ca-rich waste products are useful Ca sources.
- In eastern and north-eastern states of India, farmers invariably apply lime or gypsum at 1 to 2 t ha^{-1} or higher depending upon the soil acidity and lime requirement.
- Calcium requirement is greater for pod filling than flowering, and it is greater for flowering than vegetative growth and shallow incorporation (in 5-10 cm soil) of 1000 kg ha^{-1} lime before planting in acid soil or 1000 kg ha^{-1} gypsum at 40 DAS increases the availability of Ca in the pegging and podding zone.
- In high rainfall areas, lime is better than gypsum as gypsum is leached below pegging zone due to heavy rain, but lime remains in place.
- Excess of Ca produce deficiency of Mg and K. The excess Ca as calcium carbonate increases the alkalinity and decreases the availability of Mn, Fe, Zn, Cu, B and P causing their deficiencies. Lime-induced iron-chlorosis is the major problem of calcareous soil.

Application of 60 kg calcium significantly increased the pod yield over control during both the seasons. Variation in yield due to 30 and 60 kg/ha was not significant (Table 17.9). As regards to effect of sulphur, the response was significant up to 60 and 30 kg/ha respectively during 1985 and 1986 (Gajendra Giri and Ganga Saran, 1989).

***Table 17.9** Pod yield of groundnut as influenced by calcium and sulphur*

Calcium (kg/ha)	Pod yield (q/ha)		
	1985	1986	Mean
0	10.8	7.0	8.9
30	13.0	8.0	10.5
60	14.1	8.5	11.3
CD at 5%	2.6	1.2	1.6
Sulphur (kg/ha)			
0	10.7	6.9	8.8
30	12.9	8.1	10.5
60	14.3	8.6	11.4
CD at 5%	2.6	1.2	1.6

17.9 Magnesium Management

Magnesium is a component of chlorophyll and it serves as a cofactor in most of the enzymes that activate phosphorylation processes. The deficiency of Mg is a problem of acid and sandy soils as under high rainfall it is leached out more easily. Like K the groundnuts are very efficient in extracting Mg from the soil, hence it rarely shows deficiency symptoms. However, the Mg deficiency in groundnut occurs in acid soils of Karnataka, Tamil-Nadu, part of A.P. and eastern and north eastern hill regions.

The Mg deficiency is conducive to the occurrence of "Tikka" disease. The adequate concentration of Mg in leaves, at 50-70 DAS, is 0.30-0.80% and the critical levels are 0.25% for Spanish and Valencia cultivars and 0.20% for Virginia cultivars. The sufficiency level of Mg in soil is 15 ppm for groundnut. In Mg-deficient soil, though clear cut Mg deficiency was not observed, the hunger sign of Mg deficiency caused up to 25 % yield losses of groundnut where application of 20 kg $MgSO_4$ ha^{-1} is helpful. This also reduces occurrence of tikka disease.

17.10 Conclusions

The desired yields only could be achieved with suitable fertilizer applied on soil test-crop response basis. This will enhance the fertilizer efficiency.

The average recommended doses of N, P and K are 20, 40 and 40 kg ha^{-1}, respectively for rainfed groundnut and 25, 50 and 50 kg ha^{-1}, respectively, for irrigated groundnut. However, due to extensive cropping, the deficiencies are becoming more and more. In the present context 30 kg N, 40 kg P_2O_5 and 40 kg K_2O per hectare for rainfed kharif and 40 Kg N, 50 kg P_2O_5 and 50 K_2O per hectare for irrigated rabi-summer must be applied for maximum yield. For secondary macronutrients 500-1000 kg ha^{-1} of gypsum, 20 kg ha^{-1} S as elemental sulphur and 10 kg ha^{-1} Mg as magnesium sulphate may be applied.

Among the sources, ammonium sulphate for N and S, SSP for P and S, gypsum for Ca and S.

Soil, foliar, seed dressing and drip application of fertilizers are the main application methods. After a product has been selected it has to be used at the right time and in the right way. Generally the split application is better than all as basal and application through drip is best.

References

1. *Gajendra Giri and Ganga Saran. 1989. Influence of calcium and sulphur on groundnut under rainfed conditions. Indian Journal of Agronomy. 34(1): 146-147.*
2. *Kishore Babu, G., Munaswamy, V., John, K and Padma raju, A. 2007. Effect of long-term application of manure and fertilizers on yield and yield attributes of rainfed groundnut, Arachis hypogaea L. Journal of Oilseeds Research. 24(2): 237-240.*
3. *Patel, G.N., Patel, P.H., Patel, D.M., Patel, D.K. and Patel, R.M.2009. Yield attributes, yield and quality and uptake of nutrient by summer groundnut, Arachis hypogaea L. as influenced by sources and levels of sulphur under varying irrigation schedules. Journal of Oilseeds Research. 26(2): 119-122.*
4. *Rathore, S.S., Chaudhary, D.R., Visya, L.K., Kapila Shekawat and Bhatt, B.P.2014. Schoenite and potassium sulphate: Indigenous potassic fertilizer for rainfed groundnut (Arachis hypogaea L.,). Indian Journal of Traditional Knowledge. 13(1): 222-226.*
5. *Singh,A.L and Basu,M.S.2005. Integrated nutrient management in groundnut.NRC on groundnut.*
6. *Yadava, G.L. 1990. Fertilizer requirement of groundnut (Arachis hypogaea L.) under dryland conditions on Cultivator's fields. Journal of Oilseeds Research. 7(1): 133-135.*

Chapter 18

Production Technology for Organic Groundnut

T. Giridhara Krishna

Associate Director of Research, RARS, Tirupati.

18.1 Introduction

Organic agriculture has grown out of the conscious efforts by inspired people to create the best possible relationship between the earth and men. Since its beginning the sphere surrounding organic agriculture has become considerably more complex. A major challenge today is certainly its entry into the policy making arena, its entry into anonymous global market and the transformation of organic products into commodities. During the last two decades, there has also been a significant sensitization of the global community towards environmental preservation and assuring of food quality. Ardent promoters of organic farming consider that it can meet both these demands and become the mean for complete development of rural areas. After almost a century of development organic agriculture is now being embraced by the mainstream and shows great promise commercially, socially and environmentally. While there is continuum of thought from earlier days to the present, the modern organic movement is radically different from its original form. It now has environmental sustainability at its core in addition to the founders concerns for healthy soil, healthy food and healthy people.

18.2 Concept of Organic Farming

Organic farming is very much native to this land. Whosoever tries to write a history of organic farming will have to refer India and China. The farmers of these two countries are farmers of 40 centuries and it is organic farming that sustained them. This concept of organic farming is based on following principles:

- Nature is the best role model for farming, since it does not use any inputs nor demand unreasonable quantities of water.

- The entire system is based on intimate understanding of nature's ways. The system does not believe in mining of the soil of its nutrients and do not degrade it in any way for today's needs.
- The soil in this system is a living entity
- The soil's living population of microbes and other organisms are significant contributors to its fertility on a sustained basis and must be protected and nurtured at all cost.
- The total environment of the soil, from soil structure to soil cover is more important.

In today's terminology it is a method of farming system which primarily aims at cultivating the land and raising crops in such a way, as to keep the soil alive and in good health by use of organic wastes (crop, animal and farm wastes, aquatic wastes) and other biological materials along with beneficial microbes (bio-fertilizers) to release nutrients to crops for increased sustainable production in an eco-friendly pollution free environment.

As per the definition of the **USDA** study team on organic farming "organic farming is a system which avoids or largely excludes the use of synthetic inputs (such as fertilizers, pesticides, hormones, feed additives etc) and to the maximum extent feasible rely upon crop rotations, crop residues, animal manures, off-farm organic waste, mineral grade rock additives and biological system of nutrient mobilization and plant protection".

In another definition **FAO** suggested that "Organic agriculture is a unique production management system which promotes and enhances agro-ecosystem health,including biodiversity, biological cycles and soil biological activity, and this is accomplished by using on-farm agronomic, biological and mechanical methods in exclusion of all synthetic off-farm inputs".

18.3 Growing Area Under Certified Organic Agriculture

- 35 million hectares of agricultural land are managed organically by almost 1.4 million producers.
- The regions with the largest areas of organically managed agricultural land are Oceania (12.1 million hectares), Europe (8.2 million hectares) and Latin America (8.1 million hectares). The countries with the most organic agricultural land are Australia, Argentina and China.
- The highest shares of organically managed agricultural land are in the Falkland Islands (36.9 percent), Liechtenstein (29.8 percent) and Austria (15.9 percent).
- The countries with the highest numbers of producers are India (3,40,000 producers), Uganda (1,80,000) and Mexico (1,30,000). More than one third of organic producers are in Africa.

- On a global level, the organic agricultural land area increased in all regions, in total by almost three million hectares, or nine percent, compared to the data from 2007.
- Twenty-six percent (or 1.65 million hectares) more land under organic management was reported for Latin America, mainly due to strong growth in Argentina. In Europe the organic land increased by more than half a million hectares, in Asia by 0.4 million.
- About one-third of the world's organically managed agricultural land – 12 million hectares is located in developing countries. Most of this land is in Latin America, with Asia and Africa in second and third place. The countries with the largest area under organic management are Argentina, China and Brazil.
- 31 million hectares are organic wild collection areas and land for bee keeping. The majority of this land is in developing countries – in stark contrast to agricultural land, of which two-thirds is in developed countries. Further organic areas include aquaculture areas (0.43 million hectares), forest (0.01 million hectares) and grazed non-agricultural land (0.32 million hectares). Almost two-thirds of the agricultural land under organic management is grassland (22 million hectares). The cropped area (arable land and permanent crops) constitutes 8.2 million hectares, (up 10.4 percent from 2007), which represents a quarter of the organic agricultural land.

18.4 Organic Agriculture in Asian Continent

The total organic agricultural area in Asia is nearly 3.3 million hectares. This constitutes nine percent of the world's organic agricultural land. 4,00,000 producers were reported. The leading countries by area are China (1.9 million hectares) and India (1 million hectares). Timor Leste has the most organic agricultural area as a proportion of total agricultural land (seven percent). Organic wild collection areas play a major role in India and China, while Aquaculture is important in China, Bangladesh and Thailand.

18.4.1 Organic Farming and Development Support

Both private and public development initiatives have contributed considerably in the last 25 years to the growth of the organic sector in many countries of the world. Activities have related to, for instance, building up the capacities of different stakeholder groups in the organic sector, developing domestic and international markets, and developing local standards and legislations. One of the new initiative is the proposed Organic Research Centers Alliance (ORCA), hosted by FAO, which intends to internationally network and strengthen existing institutions with scientific credentials and empower them to become centers of excellence in trans-disciplinary organic agriculture research. International trade, an engine for growth can substantially contribute to poverty reduction in developing countries. The Trade, Climate Change and Environment Programme of the

International Trade Centre (ITC) supports the organic sector through the provision of market information, training in standards compliance, and trade promotion; by supporting policies favorable to organic agriculture and trade; and by facilitating business contacts.

18.5 Organic Agriculture in India

18.5.1 Emergence

The growth of organic agriculture in India has three dimensions and is being adopted by farmers for different reasons. First category of organic farmers are those which are situated in no-input or low-input use zones, for them organic is a way of life and they are doing it as a tradition (may be under compulsion in the absence of resources needed for conventional high input intensive agriculture). Second category of farmers are those which have recently adopted the organic in the wake of ill effects of conventional agriculture, may be in the form of reduced soil fertility, food toxicity or increasing cost and diminishing returns. The third category comprised of farmers and enterprises which have systematically adopted the commercial organic agriculture to capture emerging market opportunities and premium prices. While majority of farmers in first category are traditional (or by default) organic they are not certified, second category farmers comprised of both certified and un-certified but majority of third category farmers are certified. These are the third category commercial farmers which are attracting most attention. The entire data available on organic agriculture today, relates to these commercial organic farmers.

18.5.2 Regulatory Mechanism

For quality assurance the country has internationally acclaimed certification process in place for export, import and domestic markets. National Programme on Organic Production (NPOP) defines the regulatory mechanism and is regulated under two different acts for export and domestic markets. NPOP notified under Foreign Trade Development and Regulation Act (FTDR) looks after the export requirement. The NPOP notified under this act has already been granted equivalence by European Union and Sweden. USDA has also accepted the conformity assessment system of NPOP. Due to this, the product certified by any Indian accredited certification agency under NPOP can be exported to Europe, Sweden and USA without the requirement of re-certification. To look after the requirement of import and domestic market the same NPOP has been notified under Agriculture Produce Grading, Marking and Certification Act (APGMC). Regulatory body of NPOP under FTDR act is Agricultural and Processed Foods Export Development Authority (APEDA) under Ministry of Commerce and of NPOP under APGMC act is Agricultural Marketing Advisor (AMA) under Ministry of Agriculture. Accreditation of Certification and Inspection Agencies is being granted by a common National Accreditation Body (NAB). 18 accredited certification agencies are looking after the requirement of certification process. Out of these 4 agencies are under public sector while remaining 14 are under private management.

18.6 Future Prospects

Although, commercial organic agriculture with its rigorous quality assurance system is a new market controlled, consumer-centric agriculture system world over, but it has grown almost 25-30% per year during last 10 years. In spite of recession fears the growth of organic is going unaffected. The movement started with developed world is gradually picking up in developing countries. But demand is still concentrated in developed and most affluent countries. Local demand for organic food is growing. India is poised for faster growth with growing domestic market. Success of organic movement in India depends upon the growth of its own domestic markets. India has traditionally been a country of organic agriculture, but the growth of modern scientific, input intensive agriculture has pushed it to wall. But with the increasing awareness about the safety and quality of foods, long term sustainability of the system and accumulating evidences of being equally productive, the organic farming has emerged as an alternative system of farming which not only address the quality and sustainability concerns, but also ensures a debt free, profitable livelihood option.

Principle of Care

Organic Agriculture should be managed in a precautionary and responsible manner to protect the health and well-being of current and future generations and the environment. Organic agriculture is a living and dynamic system that responds to internal and external demands and conditions. Practitioners of organic agriculture can enhance efficiency and increase productivity, but this should not be at the risk of jeopardizing health and well-being. Consequently, new technologies need to be assessed and existing methods reviewed. Given the incomplete understanding of ecosystems and agriculture, care must be taken. This principle states that precaution and responsibility are the key concerns in management, development and technology choices in organic agriculture. Science is necessary to ensure that organic agriculture is healthy, safe and ecologically sound. However, scientific knowledge alone is not sufficient. Practical experience, accumulated wisdom and traditional and indigenous knowledge offer valid solutions, tested by time. Organic agriculture should prevent significant risks by adopting appropriate technologies and rejecting unpredictable ones, such as genetic engineering. Decisions should reflect the values and needs of all who might be affected, through transparent and participatory processes. In totality organic agriculture aims at a sustainable production system based on natural processes. Key characteristics are that organic agriculture:

- Relies primarily on local, renewable resources;
- Makes efficient use of solar energy and the production potential of biological systems;
- Maintains the fertility of the soil;

- Maximises recycling of plant nutrients and organic matter;
- Does not use organisms or substances foreign to nature (e.g., GMOs, chemical fertilisers or pesticides);
- Maintains diversity in the production system as well as the agricultural landscape;
- Gives farm animals life conditions that correspond to their ecological role and allow them a natural behaviour.

Organic agriculture is also a sustainable and environmentally friendly production method, which has particular advantages for small-scale farmers. Available evidence indicates the appropriateness of organic agriculture for small farmers in developing countries like India. Organic agriculture contributes to poverty alleviation and food security by a combination of many features, such as;

- Increasing yields in low-input areas;
- Conserving bio-diversity and nature resources on the farm and in the surrounding area;
- Increasing income and/or reducing costs;
- Producing safe and varied food;
- Being sustainable in the long term.

The evaluations by IFAD in India and China (Giovannucci, 2005) reported that the income of participating farmers can increase substantially by adopting organic practices of farming. Certified production gives access to a premium market, or simply just better market access.

Organic production of groundnut relies on management techniques that replenish and maintain long-term soil fertility by optimizing the soil's biological activity. This is achieved through crop rotation, cover cropping or composting and using organically accepted fertilizers that feed the soil and provide nutrients. Use of either farmyard manure (FYM) and vermin compost along with other organic amendments like neem seed cake; and bio-fertilizers and bio-pesticides etc., in rain fed groundnut, found advantageous for sustainable crop production.

Groundnut being a legume leaves lot of residual fertility which in turn helps succeeding crop under rain fed farming situations. In groundnut, application of FYM at 10 to 15 t ha^{-1} increased the pod and haulm yields and improved the yield parameters like shelling percentage, 100 seed weight and sound mature kernel compared to the recommended dose of fertilizers (Subrahmaniyan et al. 2000).

The study indicates that the groundnut showed greater response to the application of organic manures (i.e., organic farming) under low rainfall situations with dry spell at the early growth stage (Malligawad, 2010).Organic farmers manage insects, diseases, weeds and other pests with an array of cultural, mechanical and biological options.

Studies conducted at RARS, Tirupati revealed under organic farming in Groundnut, the highest pod yield of 1690 kg/ha was recorded in Press Mud Cake treatment followed by FYM treatment (1630 kg/ha).RDF treatment recorded pod yield of 1180 kg/ha. There was slight buildup of Organic carbon in FYM treated plot (0.36%) with continuous application for six years over no application (0.29%) (Initial value 0.26%). Highest N uptake of 147.6 kg/ha, K uptake of 87.07 kg/ha and S uptake of 54.48 kg/ha in PMC treated plot followed by Poultry manure. Whereas highest P uptake was seen in Poultry manure plot followed by PMC (Annual Report, 2013).

18.6.1 Ecology and Fertility Requirement

Groundnut requires abundant sunshine and warm climate for normal development, but does not appear to be sensitive to day length, though it generally produces more flowers under long day conditions (Stalker, 1997).

Temperature also significantly effects development and growth rate of groundnut, the optimum temperature between 25-30 °C is good for vegetative and reproductive growth. Groundnut grows with an average rainfall from 500 mm to 1200 mm, but grows best when average rainfall is more than 500 mm throughout the crop season.

Organic production field of groundnut should have good tilth, high organic matter, good drainage and airflow for roots and nitrifying bacteria.

Groundnut grows well in slightly acidic soils with optimum pH ranged from 5.5–6.5. Organic groundnut growers need to get information of their soil fertility by obtaining a soil test report with recommendations for groundnuts.

Gypsum, a by-product of dry wall is not allowed; only mined sources of gypsum are allowed in organic production.

The importance of organic to groundnut plants was emphasized by (Ahmed et al. 1997) who stated that the highest dry matter accumulation, kernel yield and oil content were achieved by fertilization with farmyard manure.

Application of FYM @ 21.9 t ha^{-1} produced the highest DMP, pod yield and haulm yield, and gave higher net return and BCR (Chandrasekaran et al. 2007).

Application of panchgavya + neem leaf extract at branching and flowering is advantageous in increasing chlorophyll content, physiological growth, nutrient content and uptake, dry matter accumulation, yield and yield attributes and economics of groundnut (Kumawat et al. 2009).

Application of FYM @ 7.5 t ha^{-1} enriched with bio-fertilizers + bio-pesticide @ 2 kg ha-1 and seed treatment with bio-fertilizers + bio-pesticide + PGPR + Trichoderma @ 200 g ha-1 each + foliar application of panchgavya at 45 days was most productive and economic module for organic groundnut cultivation (Anonymous, 2012b).

Poultry manure is excellent organic manure, as it contains high nitrogen, phosphorus, potassium and other essential nutrients. Bio-fertilizers are the source of microbial inoculants, which are also important component in organic groundnut production.

18.6.2 Water Management

- Groundnut is a rainfed crop generally requires 500 mm rainfall throughout the crop season to raise a healthy groundnut crop.
- Excessive irrigation at different stages of crop growth may promote soil born diseases and iron chlorosis resulting losses in yield. If supplementary irrigation is available, it should be given at critical stages such as flowering, pegging and pod and seed development.
- Drought stress constitutes a serious threat for groundnut production because of the danger of aflatoxin. Aflatoxin is produced by two fungi such as Aspergillus flavus and Aspergillus parasiticus.
- Aflatoxin can cause financial losses for groundnut growers. Many countries and export markets are setting stricter tolerance levels for aflatoxin. Drought and hot weather conditions have been associated with increased aflatoxin contamination.
- In general field studies demonstrate that reduction of drought stress by irrigation reduces aflatoxin contamination. Pre-harvest aflatoxin contamination can occur when groundnut pods with insect or mechanical damage.
- Various approaches have been suggested for control of pre harvest aflatoxin contamination including the grow varieties with resistance to insects, resistance to plant stress (especially for tolerance to drought and high temperature).

18.6.3 Weeds Management

- It is essential to keep groundnut fields weed free for up to 45 days after crop emergence. Weeds plants may act as alternative hosts of plant parasites allowing the pathogen to survive between seasons and providing a source of inoculum in the new growing season.
- Weed control in organic systems, especially in groundnut production, relies heavily on crop rotations, cover crops and cultivation.
- For cultivation to be successful, a straight, well prepared bed, as well as straight seeding lines in a conventional diamond pattern is necessary for cultivating implements to remove the most weeds while leaving the crop undisturbed. Hoeing between plants eliminates weeds in the planting row.
- Flame weeding and other thermal devices can reduce broadleaf weeds, but grasses still need to be mechanically removed. Polythene mulching has been found effective in controlling weeds.

18.6.4 Diseases Management

Leaf spot: Rainfall or irrigation followed by high humidity during the growing season can contribute to early leaf spot and late leaf spot. Cultural practices such as resistant varieties, crop rotation, deep ploughing, and removal of volunteer groundnut plants may alleviate the leaf spot, as could foliar inoculation of micro organisms with compost teas or commercial products that include Bacillus subtilis bacteria, like Serenade.

The control of early leaf spot has been achieved by crop rotation with cotton, grain sorghum and corn. Biological control of early leaf spot in groundnut using Pseudomonas cepacia and Verticillium lecanii has been reported as a parasite on several groundnut pathogens in India (Subrahmanyam et al. 1990). Host range of this pathogen is very limited to only groundnut so that groundnut is rotated with any other crop for one or two or three years reduces disease inoculum in soil.

Rust: Temperatures of 20-25°C and high relative humidity favours the rust epidemics in groundnut. Rust disease also prevalent throughout the groundnut growing areas. Groundnut should not follow groundnut, as infected pods and foliage stay behind in the soil with plant residues. Pathogen may over-winter on these materials and provide inoculum in the following crop.

Crop rotation prevents the building up of pathogens in the soil. Planting of resistant cultivars, deep ploughing, removal of groundnut residue, elimination of volunteer groundnut plants, crop rotation and manipulation of planting dates and careful handling of pods during harvesting and shelling could prevent disease development.

Stem rot and collar rot: Stem rot (Sclerotium rolfsii Sacc.) and collar rot (Aspergillus niger van Tieghem) are prevalent in almost all groundnut producing areas especially in sandy loam and medium black soils. These diseases cause severe seedling mortality.

Soil solarization with polyethylene sheets for 3-4 weeks in the hot season before planting should prove useful in reducing stem rot and pod rot of groundnut.

Deep ploughing to bury all undecayed organic matter and plant debris to at least 9 cm can greatly reduce stem rot infection and increase yield.

Rotation of groundnut with wheat and corn is also effective against stem rot (Minton et al. 1991); this is a common practice in several parts of northern India and China.

Rotating groundnut with crops belonging to the Liliaceae family such as onion and garlic has proved effective against stem rot (Asghari and Mayee 1991).

Biological control has been proved to be a promising disease management technology especially against soil-borne plant pathogens. Trichoderma viride Tv1 was the most effective isolate against S. rolfsii with 69.40% growth inhibition followed by P. fluorescens resulting in 64.40% inhibition.

Among the organic amendments tested in greenhouse, mahua cake with T. viride each @ 5 g/kg of soil resulted in 3.75% stem rot incidence as against 39.98% in control (Karthikeyan et al. 2006). Seed treatment with Trichoderma viride or T. harzianum @ 10 g kg^{-1} seed also helps in managing seed and soil borne diseases.

18.6.5 Insect Pest Management

Groundnut is attacked by several insect pests, among them defoliators viz., groundnut leaf miner, red hairy caterpillars, tobacco caterpillar, gram pod borer and sucking pests like Aphids, leaf hoppers and thrips etc. are the economic important in India.

Biological and cultural methods can be used for controls of groundnut insect pest. Planting large expanses of a single, susceptible crop, or mono cropping, encourages pest problems. A diverse farming involving many types of plants and animals considerably diminishes the likelihood of severe insect pest outbreaks.

Farmers must create production methods that complement natural systems. The use of beneficial insect habitats with field borders increases the presence of beneficial insects. These habitats provide shelter, food and act as refuges that attract the natural enemies of pests. This would help in reducing pest populations.

Some botanical insecticides, including neem products such as Agroneem and Neemix, act as repellents, anti feedants and insect growth regulators should be used to management of insect pest in organic groundnut production.

18.7 Harvesting and Post-Harvest Management

Handling of organic groundnut is a complex issue. If organic groundnut is processed on farm in any way, a second certification as a handler or processor may be required. Following point should be considered for production of organic groundnut:

1. **Equipment:** Know what equipment is used for organic production, how to clean all parts of equipment, including planters, combines, wagons and trucks etc. Clean all the equipment's prior to use in organic fields and keep records of cleaning activities of the equipment.
2. **Storage unit**: All the storage units should be carefully inspected prior to use and thoroughly clean bins, dryers, cleaners and other storage units.
3. **Transport facility:** It should be making sure that transports units, including over-seas shipping containers are free of foreign matter. Keep records to document trucks, including clean transportation affidavits and bills of lading.
4. **Farm records**: Organic producers should maintain good records so that farmers will have a better chance of limiting the losses, identifying causes of problems and determining liability. The document with valid records of organic yields and sales may help to claims for losses due to contamination or by other means.

18.8 Markets for Organic Groundnut

- The marketing research on organic groundnut is very limited but demand of organic groundnut is increasing because of organic farmers enjoy high prices over conventionally grown groundnut.
- Some large organic food manufacturers seek dependable supplies of organic groundnut and willing to contract for organic production.
- The snack food industry, rather than the groundnut butter sector, may be the larger market for organic groundnut. This has implications for the type of groundnut desired and for the geographical area most suited to its production.
- However, shellers and manufacturers are the principal contractors for groundnut production, a much different processing and marketing infrastructure may need to be developed to assist organic marketing system.

18.9 Conclusion

Integrated agronomical management practices are to be more successful for production of organic groundnut. Cultural practices such as crop rotation, deep ploughing, and good seedbed preparation are the foundation of effective management of insect pest diseases of groundnut.

Biological control is likely to be an integral part of production of organic groundnut for disease management strategy in the future.

Development of pest and disease resistant varieties is crucial to any organic groundnut production on a scale comparable to that of conventional production.

References

1. *Ahmed MKA, Saad AOM, Thalooth AT and Kabesh MO. 1997. Utilization of biofertilizers in field crops production. 10-Yield response of groundnut to inorganic and biofertilizers.*
2. *Anonymous, 2012b. Annual report of AICRP on Groundnut (Kharif, 2012). Directorate of Groundnut Research, Junagadh.*
3. *Anonymous, 2013. Annual Report of Research, Dept. of Soil Science, Regional Agrl. Research Station, ANGRAU, Tirupati,Andhra Pradesh, India.*
4. *APEDA, National Program for Organic Production (NPOP). Present status in India. www.apeda.gov.in/organic_product.htm.*
5. *Asghari MA and Mayee CD. 1991. Comparative efficiency of management practices on stem and pod rots of groundnut. Indian Phytopathology.44: 328-332.*
6. *Chandrasekaran R, Somasundaram E, Mohamed A, Nalini K, Thirukkumaran K and Karthikeyan V, Sankaralingam A and Nakkeeran S. 2007. Biological control of*

Groundnut stem rot caused by Sclerotiumrolfsii (Sacc.). Archives of Phytopathology and Plant Protection, 39(3): 239 – 246.

7. *Kumawat RN, Mahajan SS and Mertia RS. 2009. Growth and development of groundnut (Arachis hypogaea) under foliar application of panchgavya and leaf extracts of endemic plants. Indian J. Agron. 54(3): 324-331.*
8. *Malligawad LH. 2010. Effect of organics on the productivity of groundnut and its residual effects on succeeding safflower under rainfed farming situations. 19th World Congress of Soil Science, Soil Solutions for a Changing World from 1 – 6 August 2010 at Brisbane, Australia.*
9. *Minton NA, Csinos AS, Lynch RF and Brenneman, TB. 1991. Effects of two cropping and two tillage systems and pesticides on peanut pest management. Peanut Science.18: 41-46.*
10. *Ramesh P, Panwar NR, Singh AB, Ramana S, Yadav SK, Shrivastava R and SubbaRao A. 2010.Status of organic farming in India.Current Science. 98(9):1190-1194.*
11. *Stalker HT. 1997. Peanut. Field Crop Research.53: 205-217.*
12. *Subrahmaniyan K, Kalaiselvan P and Arulmozhi N. 2000. Studies on the effect of nutrient spray and graded level of NPK fertilizers on the growth and yield of groundnut. Intern.J. Trop. Agric. 18(3): 287-290.*
13. *Subrahmanyam P, Reddy PM and Mcdonald D. 1990. Parasitism of rust, early and late leaf spot pathogens of peanut by Verticillium lecanii . Peanut Sci . 17:1-3.*

Chapter 19

Micronutrient Management in Groundnut

P.V.R.M. Reddy, V. Naga Madhuri, G. Krishna Reddy, P.M. Reddy and P. Bala Hussain Reddy

Scientist (Soil Science), IFT, RARS, Tirupati.

19.1 Introduction

The optimization of the mineral nutrition is the key to optimize the production of groundnut, as it has very high nutrient requirement and the recently released high yielding groundnut varieties remove still more nutrients from the soil. On contrary groundnut farmers, most part of the semi-arid region use very less nutrient fertilizer and sometime only one or two nutrients resulting in severe mineral nutrient deficiencies due to inadequate and imbalance use of nutrients is one of the major factors responsible for low yield in groundnut. India is the world's largest producer of groundnut where nutritional disorders cause yield reduction from 30 to 70% depending upon the soil types. Thus it is high time to look into the mineral nutrition aspects of groundnut for achieving high yield and advocate the suitable package of practices for optimization of yield (Singh, 2004).

19.2 Micronutrient Requirements of Crops

Crops require 16 essential elements to grow properly. The elements include carbon (C), hydrogen (H) and oxygen (02), which are derived from air and water. All the remaining nutrients used by plants come from soil in the form of inorganic salts. Legumes are an exception because they can also fix nitrogen from the air. The ***macronutrients*** obtained from the soil include nitrogen (N), phosphorus (P), potassium (K), calcium (Ca), magnesium (Mg), and sulphur (S). The remaining essential elements needed by plants are known as ***micronutrients*** because plants use them in relatively small amounts and they are equally essential for plants for proper growth, development and disease resistance. They include: boron (B), chlorine (Cl), copper (Cu), iron (Fe) manganese (Mn), molybdenum (Mo) and zinc (Zn). Carbon, hydrogen and oxygen comprise from 94.0 to 99.5 per cent of fresh plant tissue. The remaining nutrients, which come from the soil, make up the balance of the tissue. The term micronutrient refers to the relative quantity of a nutrient that is required for plant growth. It does not mean that they are less important to plants than other nutrients. Plant growth and development may be retarded if any of these elements is lacking in the soil or is not adequately balanced with other nutrients.

Continued use of only common fertilizers, which do not supply these micronutrients, is leading to the micronutrient deficiencies resulting in reduced yields of groundnut. Essential micronutrients required for groundnut, as in the case of other crops, are: Iron, Boron Zinc, Molybdenum, Manganese, Copper, Cobalt and Chlorine

These micronutrients are to be applied preferably based on soil testing. However, in a standing crop of groundnut, any micronutrient deficiencies can be corrected observing the deficiency symptoms in the crop. Application of organic manures like well decomposed farm yard manure (FYM) and compost, green manuring and green-leaf manuring practices provide most of the micronutrients to groundnut.

19.3 Sources of Micronutrients in Soils

Inorganic micronutrients occur naturally in soil minerals. The parent material from which the soil developed and soil forming processes determine what the micronutrient content of the soil will be. As minerals break down during soil formation, micronutrients are gradually released in a form that is available to plants. Two sources of readily available micronutrients exist in soil: nutrients that are adsorbed onto soil colloids (very small soil particles) and nutrients that are in the form of salts dissolved in the soil solution.

Organic matter is an important secondary source of some micronutrients. Most micronutrients are held tightly in complex organic compounds and may not be readily available to plants. However, they can be an important source of micronutrients when they are slowly released

In to a plant available form as organic matter decomposes.

19.4 Soil Factors that affect Micronutrient Availability

Physical and chemical characteristics of soil affect the availability and uptake of micronutrients

- Soils low in organic matter (less than 2.0 percent) may have lower micronutrient availability.
- Soils with higher amounts of clay (fine texture) are less likely to be low in plant available micronutrients. Sandy soils (course texture) are more likely to be low in micronutrients.
- Soil temperature and moisture are important factors. Cool, wet soils reduce the rate and amount of micronutrients that may be taken up by crops.
- As soil pH increases the availability of micronutrients decreases, with the exception of molybdenum.

19.5 Determining the Need for Micronutrients

Diagnosing a micronutrient deficiency can be a difficult and time consuming process. To identify a micronutrient deficiency follow these steps

1. Ensure that poor crop growth is not the result of a macronutrient deficiency, drought, salinity, disease or insect problem, herbicide injury or some physiological problem.
2. Find out if a micronutrient deficiency has been identified before in a particular crop or soil type in the area.
3. Examine the affected crop for specific micronutrient deficiency symptoms.
4. Take separate soil samples from both the affected and unaffected areas for complete analysis, including micronutrients.
5. Send plant tissue samples from both the affected and unaffected areas for complete analysis that includes tests for micronutrient levels.
6. If all indications point to a micronutrient deficiency, apply the micronutrient to a specific, clearly marked out affected area of land to observe results in subsequent seasons.

19.6 Micronutrient Deficiency Symptoms

Some micronutrients have characteristic deficiency symptoms. However, symptoms can be easily confused with other nutrient deficiencies, salinity, disease, drought, herbicide injury or other physiological problems. Visual symptoms alone are not a reliable method of determining a

Micronutrient problem, but they are useful indicators when used with other diagnostic tools.

Iron

Being relatively immobile, Fe deficiency appears first on the younger leaves of the plant.

In most species chlorosis is interveinal and a fine reticulate pattern can be observed in the newly formed leaves, the darker green veins contrasting markedly against a lighter green or yellow background. The youngest leaves may often be completely white and totally devoid of chlorophyll. In the leaves of cereals the deficiency is shown by alternate yellow and green strips along the length of the leaf. In fruit trees, iron deficiency is associated with high leaves of $CaCO_3$ in the soil, where it is called "lime induced iron chlorosis"

Manganese

Mottled chlorosis with veins green which appears first on younger leaves. Stem becomes yellowish green, often hard and woody. In groundnut leaves, the veins are surrounded by a definite zone of non chlorotic interveinal tissue. Characteristic symptoms of Manganese deficiency in certain crops have been given specific names. They are grey speck of oats, speckled yellow of sugar beet, marshy spot of peas, pahala blight of sugarcane, and frenching of tung tree. (Aleurites fordii).

Boron

B is immobile in most plant species and symptoms appear first at tops and roots. The youngest leaves are misshaped, wrinkled and are thicker and of a dark bluish green color. Shed prematurely. Plants dwarfed, stunted, flower development and seed production usually impaired.

In fruit trees, the bark of the stem may become rough and show splitting. Fruits often develop corky areas internally. Roots exhibits swollen root tips. Characteristic symptom of B deficiency in certain crops are Heart rot in sugar beet, Hollow stem in cauliflower, Brown heart of turnip, Corky pith of apples Hard fruit of citrus

Copper

Young leaves show chlorosis Terminal leaves, shoots are wilted and distorted, frequently followed by death resulting in development of several auxillary buds. Cereals show bushy growth with white twisted tips and reduced panicle formation. Poor or no heading in cabbage and lettuce. In citrus, Cu deficiency leads to exanthema or die-back.

Zinc

Little leaf and rosette are the common symptoms of Zn deficiency. In maize, Zn deficiency results in white or yellow emerging leaves, a condition called white bud. Leaves chlorotic and necrotic, young growth first affected, resulting, premature shedding. In monocots and particularly in maize, chlorotic bands form on either side of the midrib of the leaf.

Molybdenum

Symptoms of Mo deficiency are similar to those of N deficiency because in the absence of Mo, nitrate is not metabolized. Leaves show marginal scorching and rolling or cupping. In extreme deficiency, the leaf lamina is not formed and only the rib is present in cauliflower which is called whiptail disorder.

Chlorine

Chlorosis of younger leaves and an overall wilting of the plant. In some plant species, like tomato, leaves show chlorotic mottling, bronzing and tissue necrosis.

19.7 Soil Sampling and Testing for Micronutrients

Soil tests aid in determining whether a particular nutrient is responsible for poor production and provides the basis for deciding the type and amount of fertilizer needed to correct a nutrient shortage. A soil sample used for laboratory analysis must consist of a composite of a number of samples taken from the field. The DTPA (diethylenetriamine pentaacetic acid) method is used to extract the metal micronutrients. Hot water is used to extract boron. Either water or sodium nitrate is used to extract chlorine. Analytical results are usually given in parts per million (ppm). No soil test has proven particularly useful in predicting the availability of molybdenum. The micronutrients that are normally tested for are boron, copper, iron, manganese, and zinc.

Surface soil samples (0-15 cm) should be taken from both the affected area and an adjacent area of good crop growth for comparison. The general range levels used for determining when to add micronutrients for improving crop production are shown in Table 19.1. When a soil sample tests low in a micronutrient, a potential micronutrient deficiency may occur. Some soils with low micronutrient levels at the surface (0-15 cm) do not respond to fertilization because they have higher levels of the nutrient in the subsoil.

***Table 19.1** Range levels of micronutrients in soils*

	Deficient	Medium	Adequate
Boron (Hot Water Extractable-ppm)	0.0 - 0.4	0.5 - 1.2	> 1.2
Chlorino (Wator Extractable-ppm)	0.0 - 8.0a	-	-
Copper (DTPA Extractable-ppm)	0.0 - 0.2b	0.3 - 1.0	> 1.0
	0.0 - 0.5c	0.6 - 1.0	> 1.0
	0.0 - 2.5d	-	> 2.5
Iron (DTPA Extractable-ppm)	0.0 - 2.0	2.0 - 4.5	> 4.5

[a]Critical level Cl in cereals, [b]Dark Brown soil areas, [c]Black soil areas and [d]Organic soils

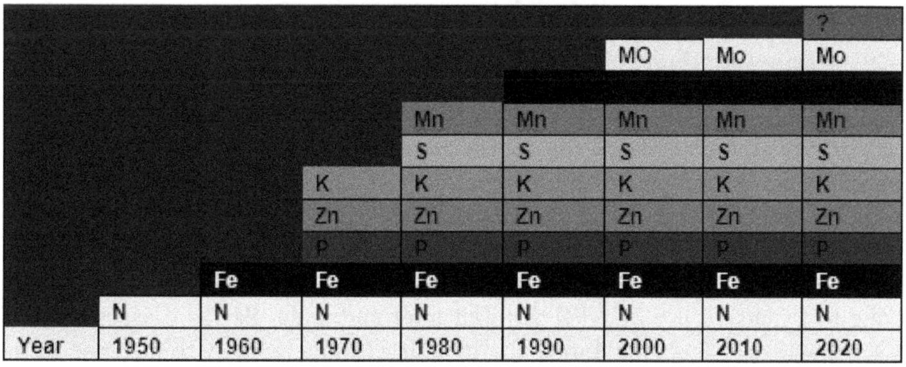

								?
						MO	Mo	Mo
				Mn	Mn	Mn	Mn	Mn
				S	S	S	S	S
			K	K	K	K	K	K
			Zn	Zn	Zn	Zn	Zn	Zn
			P	P	P	P	P	P
		Fe	Fe	Fe	Fe	Fe	Fe	Fe
	N	N	N	N	N	N	N	N
Year	1950	1960	1970	1980	1990	2000	2010	2020

***Fig. 19.1** Emerging multi-nutrient deficiencies in soils*

During the early years of Independence, when we started the fertilizer application only, nitrogen was the limiting nutrient. But, later with the advent of green revolution and with the introduction of High yielding Varieties, we have started dumping of fertilizers indiscriminately. This skewed fertilizer application ultimately resulted in deficiency of multiple nutrients as the time advanced Fig. 19.1. Presently along with NPK, other nutrients such as S, Fe, Zn, Mn, Mo are also being deficient in soils and limiting the crop growth and yield.

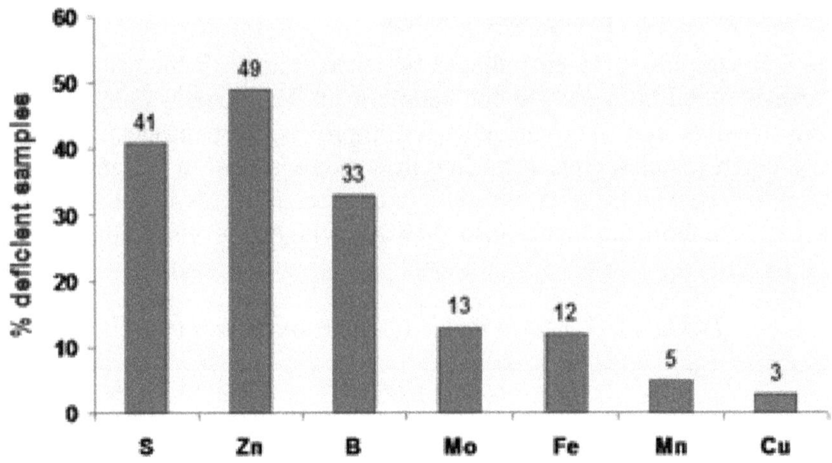

Fig. 19.2 Secondary and micro nutrient status of India

The graph in Fig. 19.2 shows that S, Zn, B, Mo, Fe, Mn and Cu are deficient to the tune of 41, 49, 33, 13, 12, 5 and 3 respectively.

19.8 Functions and Deficiency Symptoms and Corrective Measures in Groundnut

19.8.1 Zinc

In India, zinc is now considered the fourth most important yield-limiting nutrient after, nitrogen, phosphorus and potassium, respectively. Analysis of 2,56,000 soils and 25,000 plant samples from all over India showed that 48.5% of the soils and 44% of the plant samples were potentially zinc-deficient and that this was the most common micronutrient problem affecting crop yields in India. Deficiency of zinc has increased in Southern States due to extensive use of NPK without micronutrients. Periodic assessment of soil test data also suggests that zinc deficiency in soils of India is likely to increase from 49 to 63% by the year 2025 as most of the marginal soils brought under cultivation are showing zinc deficiency (Singh, 2006). The following graph (Fig. 19.3) shows the alarming situation of Zinc deficiency in different states of India.

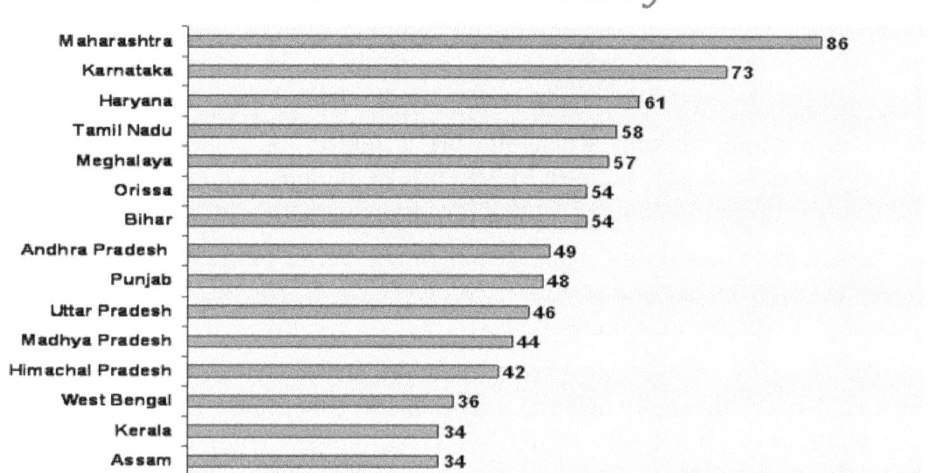

Fig. 19.3 Zinc deficiency in different states of India

Zn is taken up as the divalent cation Zn_{2+}. Early work suggested that Zn uptake was passive, but more recent work indicates that it is active (energy-dependent). Zn is required directly or indirectly by several enzymes systems, auxins and in protein synthesis, seed production and rate of maturity. Zn is believed to promote RNA synthesis, which in turn is needed for protein production. The mobility of Zn is low. The rate of Zn mobility to younger tissue is particularly depressed in Zn-deficient plants. Common symptoms of Zn deficiency are: stunted plant growth; poor tillering; development of light green, yellowish, bleached spots; chlorotic bands on either side of the midrib in monocots (particularly maize); brown rusty spots on leaves in some crops, which in acute Zn deficiency as in rice may cover the lower leaves; and in fruit trees the shoots may fail to extend and the small leaves may bunch together at the tip in a rosette-type cluster. Little-leaf condition is also a common symptom. Internodes are short. Flowering, fruiting and maturity can be delayed. Shoots may die off and leaves can fall prematurely. Deficiency symptoms are not the same in all plants.

Soil factors affecting availability of zinc

Sometimes zinc (Zn) may be present in the soil, but not available to plants. A high soil pH, or a calcareous soil means that zinc is less available (soluble). Crops under these soil conditions may suffer from zinc deficiency. In zinc deficient leaves wider chlorotic strips are found. These bands will usually be on the leaf portion nearest to the petiole. Zinc deficiency also results in stunted growth, while young leaves are smaller than normal. Zinc deficiency symptoms may occur concurrently with Iron deficiency symptoms and the two can easily be confused. However Iron chlorosis extends to full length of leaves where as Chlorosis due to zinc deficiency occur mainly on the lower half of the leaves. The following are some of the factors that affect the availability of Zinc

Soil pH In general, the soils where the pH is above 7.4, high calcium carbonate content in topsoil or in subsoil exposed by removal of the topsoil during field leveling or by erosion, coarse texture (sandy soil) with a low organic matter status

***Calcium carbonate*:** Calcareous soils with a high content of calcium carbonate (>15%) are typical soils of semi-arid and arid climates. Presence of calcium carbonate decreases the availability of zinc due to higher soil pH.

***Soil salinity*:** The main types of salt affected soils are the saline soils (Solonchaks), sodic soils (Solonetz) and both mainly occur in arid and semi-arid regions. Saline soils contain high concentrations of soluble salts which restrict the types of crops which can be grown and reduces the availability of zinc.

***Water logging*:** The poor availability of zinc caused by water logging can be due to a relatively high pH, zinc being present as the insoluble sulphide (ZnS) and elevated concentrations of ferrous, bicarbonate, and phosphate ions (Doberman and Fairhurst, 2000).

***Interactions between zinc and other plant nutrients*:** High available phosphate status, high bicarbonate or magnesium concentrations in soil or irrigation water and acid soil of low zinc status developed on highly weathered parent material shows deficiency of zinc. High soil phosphate levels are one of the most common causes of zinc deficiency in crops by cations added with phosphate salts can inhibit zinc absorption from solution, H+ ions generated by phosphate salts inhibit zinc absorption from solution and phosphorus enhances the adsorption of zinc into soil constituents. Nitrogen appears to affect the zinc status of crops by both promoting plant growth and by changing the pH of the root environment. In many soils, nitrogen is the chief factor limiting growth and yield and therefore, not surprisingly, improvements in yield have been found through positive interactions by applying nitrogen and zinc fertilizers. Several macronutrient elements, including calcium, magnesium, potassium and sodium are known to inhibit the absorption of zinc by plant.

Interactions of zinc with other micronutrients

Zinc interact with copper, iron, manganese and boron influence their concentration in plants by a) zinc-copper interactions occur due to copper and zinc sharing a common site for root absorption or copper nutrition affects the redistribution of zinc within plants, b) iron-zinc interactions study resulted increasing zinc supplies to plants have been observed to increase the iron status, to decrease it and to have no effect on it (Loneragan and Webb, 1993). In contrary to the previous report, Zn application had adverse effect on Fe concentration and Fe uptake in plants (Imtiaz et al., 2003). Mn and boron interaction resulted positive and negative response in availability in soil and uptake and distribution in the plant.

Zinc deficiency and its associated consequence in peanut

Zinc is needed for peanut as a tracer and aids in the use of other trace elements by the plants. The Zn deficiency in peanut caused irregular mottling and yellow-ivory interveinal chlorosis in the upper leaves. Under severe deficiency, the entire leaflets became chlorotic. The main symptoms of zinc deficiency are decreased internodal length and restricted development of new leaves. Deficient plants accumulate reddish pigment in stems, petioles and leaf veins (Alloway, 2008). In zinc deficient soils, fertilization of Zn increase the nodulation, chlorophyll content and pod yield. ILZRO (1975) reported that deficiency causes reduced pegging but no distinct leaf symptoms. The blade of the young fully expanded leaf is recommended for diagnosis of Zn deficiency in peanuts and 8-10 mg Zn/kg of dry matter is the critical value. When diagnosing the leaf sample and whole plant of peanut critical concentration is varied between 15 to 25 mg Zn kg^{-1} of dry matter (Bell et al., 1990).

Strategies to correct Zinc deficiency in groundnut

***Soil application*:** Zinc plays as an activator of many enzymes in plants and is directly involved in the biosynthesis of growth substances like auxin which produce more cells and dry matter that in turn will be stored in seeds as a sink. (Chitdeshwari and Poongothai 2003) reported that, the response of peanut to the soil application of zinc 5 kg ha-1 + boron 1 kg ha^{-1} + sulphur 40 kg ha^{-1} significantly increased the pod yield to the tune of 24.2% for TMV 7 and 14.8% for JL 24 over control. The micronutrient response of peanut studied in different States of India under rainfed conditions concluded that an application of Zn, B and S along with N+P was economical. Application of Zn through drip irrigation increased chlorophyll content, pod numbers and yield. It also increased fertilizer-use efficiency and kept the soil loose for peg penetration and pod development. The drip irrigation application was superior over the other soil and foliar Zn applications by precise application at appropriate times with desired concentration, uniform distribution, less damage to crop and soil and ultimately higher yield (Singh, 2007). The soil application of 10 kg ha^{-1} Fe and 5 kg ha^{-1} Zn was recommended by All India Coordinated Research Project on micronutrients (Singh, 2010). The requirement and response to zinc application varies with the soil types. Application of 40 kg Zn ha^{-1} recorded high pod yield, protein content and zinc uptake by peanut, whereas application of 20 kg Zn ha^{-1} recorded the highest oil content and sulphur uptake in vertisol (Tathe et al., 2008). An application of 10 kg of zinc in swell-shrink soils, 5 kg in zinc to alluvial red and lateritic soils per hectare was found optimum in ameliorating zinc deficiency (Singh, 2008). Basal application of 25 kg ha^{-1} of $ZnSO_4$ in alfisol increased the pod yield per plant in peanut varieties viz., TMV 7 from 19.2 to 21.4 g, TMV (Gn) 13 from 18.4 to 22.5 g and VRI (Gn) 6 from 35.7 to 38.6 g (Arunachalam et al., 2012).

Foliar application

Deficiency of Zn can also be effectively controlled with 2 to 4 spray of 0.5% zinc sulphate salt solution on standing crops (Singh, 2008). Foliar spraying of 1 g L-1 zinc recorded highest seed yield (2910 kg ha^{-1}) with increase in pod yield, plant height, 100 seed weight, seed length and seed width. The application of 80 kg ha-1 nitrogen along with zinc foliar application enhanced the seed yield to 3742 Kg ha-1 (Pendashtek et al., 2011).

Seed treatment and nano-coating of zinc

Seed treatment with Teprosyn Zn + P at the recommended level (8 ml kg-1 seed) increased the pod yield of peanut over NPK control. Seed treatment gave higher zinc use efficiency than soil application of zinc sulphate at the rate of 5.5 kg Zn ha^{-1} or two foliar spray of Zn at 30 and 45 days after flowering, also increased the peanut pod yield at similar magnitude (Singh et al., 2003). An attempt was made by the (Prasad et al. 2012) to study the effect of nanoscale zinc oxide on peanut. Peanut seeds treated with 1000 ppm ZnO showed better germination than the seeds treated with bulk $ZnSO_4$. Peanut seeds treated with nanoscale ZnO showed the maximum seedling vigour index at 1000 ppm. Nanoscale ZnO showed large root growth of seedling compared to bulk $ZnSO_4$ and control. Prasad and coworkers suggested that the growth promoting effect of nanoscale ZnO at optimum concentrations and inhibitory effect at high concentrations on root and shoot growth and pod yield in peanut. Seed treated with nanoscale ZnO enhanced the zinc level in seeds, which increased the germination, root growth, shoot growth, dry weight and pod yield. Significant zinc uptake by the leaf and kernel was observed with the foliar application of nanoscale ZnO compared to chelated zinc sulfate.

Zinc Fertilizers

There are different forms and source of zinc available as listed in Table 19.2. The soil and seed application of $ZnCl_2$ and $ZnSO_4$ showed a positive response with germination, pod number, pod yield and oil content.

***Table 19.2** Different forms and source of zinc*

Compound	Formula	Zn content (%)
Inorganic zinc fertilizers		
Zinc sulphate monohydrate	$ZnSO_4 . H_2O$	36
Zinc sulphate heptahydrate	$ZnSO_4.7H_2O$	22
Zinc oxysulphate	$ZnSO_4$ xZnO	20-50
Basic zinc sulphate	$ZnSO_4$.4Zn(OH)2	55
Ammoniated zinc sulphate	$Zn(NH_3)4SO_4$	10
Zinc oxide	ZnO	50-80

Table 19.2 ***Contd...***

Compound	Formula	Zn content (%)
Zinc carbonate	$ZnCO_3$	50-56
Zinc chloride	ZnC_{l2}	50
Zinc nitrate	Zn(NO3)2.3H2O	23
Chelated zinc fertilizers		
Disodium zinc EDTA	Na2Zn EDTA	8-14
Sodium zinc EDTA	NaZn EDTA	9-13
Sodium zinc HEDTA	NaZnH EDTA	6-10
Zinc polyflavonoid	-	5-10
Zinc lignosulphonate	-	5-8

Application FYM enriched Zn observed improvement in the available zinc status of deficient soils is probably the result of an increase in soluble, organically-complexed forms of zinc. Application of enriched $ZnSO_4$ as a basal was found to be economically viable and sustainable. In case of severe deficiency of zinc foliar application of ZnSO4 at 0.2 to 0.5% is recommended to temporarily arrest the deficiency.

19.8.2 Boron

Boron is essential for all plant growth. It aids in the transfer of sugars and nutrients from leaves to fruit, and increases pollination and seed development. Boron (B) is probably taken up by plants as the undissociated boric acid (H3BO3). It appears that much of the B uptake mainly follows water flow through roots. B in a plant is like the mortar in a brick wall, the bricks being the cells of growing parts such as tips (meristems). Key roles of B relate to: (i) membrane integrity and cell-wall development, which affect permeability, cell division and extension; and (ii) pollen tube growth, which affects seed/fruit set and, hence, yield. B is relatively immobile in plants and, frequently, the B content increases from the lower to the upper parts of plants.

B deficiency usually appears on the growing points of roots, shoots and youngest leaves. Young leaves are deformed and arranged in the form of a rosette. There may be cracking and cork formation in the stalks, stem and fruits; thickening of stem and leaves; shortened internodes, withering or dying of growing points and reduced bud, flower and seed production. Other symptoms are: premature seed drop or fruit drop; crown and heart rot in sugar beet; hen- and chicken-type bunches in grapes; barren cobs in maize; hollow heart in groundnut; unsatisfactory pollination; and poor translocation of assimilates. Death of the growing tip leads to sprouting of auxiliary meristem and a bushy broom-type growth. Roots become thick, slimy and have brownish necrotic spots. The following graph (Fig. 19.4) shows the alarming situation of boron deficiency in different states of India.

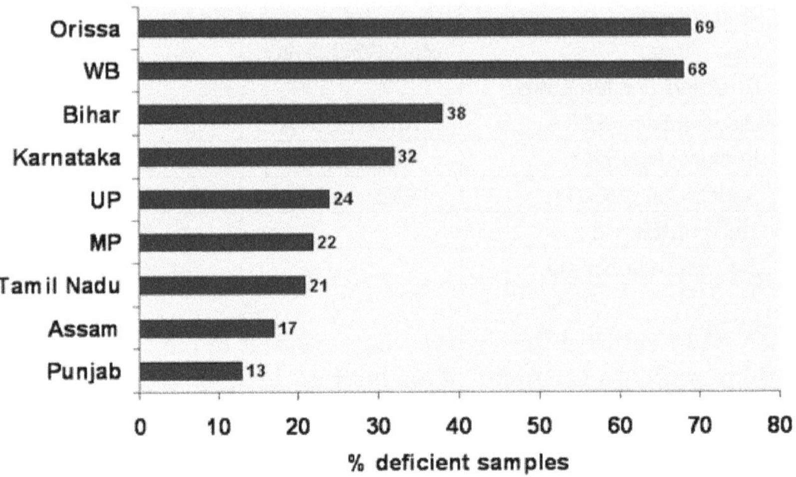

Fig. 19.4 Boron deficiency in different states of India

This deficiency results in stunted growth of young plants. They will be misshapen, thick, brittle and small. Because boron is not easily transferred from old to young leaves, older leaves usually remain green and appear healthy. Often dark brown, irregular lesions appear, followed by pale yellow chlorosis of young leaves. Stems are short and growing points may die.

The symptoms of a boron deficiency can be confused with a sulphur deficiency. Boron deficiency symptoms include death of the terminal bud, rosetting, yellowtop and poor flowering. When a boron deficiency is moderate, seed yield is often reduced without any evidence of severe deficiency symptoms during vegetative growth.

Boron in Groundnut

The nutrient requirements of peanuts are well known, and methods of fertilizer application and other cultural practices resulting in optimum production have been developed over the past 50 years. Boron has been universally recognized as the most important micronutrient for peanut production, but the B requirement for peanuts is not as high as that for some other leguminous species. While B is essential for all stages of plant growth, an available supply is most important during flowering and pod development. This is especially true with today's high-yielding varieties. Cell wall strength, cell division, fruit and seed development and sugar transport are plant functions related to B. Improved nut quality has been reported with B applications. While B requirements for optimum plant nutrition are low as compared with those of nitrogen, phosphorus and potassium, the need for B is especially significant in flowering and pod development.

The critical level of hot-water-soluble B for peanuts in most soils ranges from 0.2 to 0.5 ppm, depending on the soil pH, organic matter content and texture. Soils which are below the critical level generally will respond to applied B. The critical level of B in upper mature peanut leaves is about 25 ppm. Peanut plants with leaf B contents below the critical level should be sprayed one r more times with Solubor after flower initiation and during pod development.

Soil tests and plant analyses

Boron deficiencies may be suspected on coarse-textured soils where organic matter content is low, on recently limed soils, and where delayed maturity has been reported in peanuts with recommended N rates. Soil testing and plant analyses are both helpful in assessing the potential B-supplying capacity of the soil and the current B status of the growing plant. Boron deficiency is usually observed in deep black soils also. The availability of B in acid soils generally decreases when these soils are limed, so B is recommended on freshly limed soils. Response to applied B generally is greatest when there are adequate supplies of other nutrients, especially N.

Boron Deficiency Symptoms

Boron (B) deficiency has been reported from many groundnut-growing areas in India. Soil samples from parts of Tamil Nadu, UP, Maharashtra and Gujarat were found to be deficient in boron It occurs on deep, and highly weathered soils. The inner faces of the boron deficient groundnut cotyledons are depressed and discolored. This is classified as a form of internal damage and has been termed hollow heart.

Boron recommendations for Groundnut

Boron application need to be carefully done. Even little higher amount of application may result in boron toxicity which reduces groundnut yield. However, the amount of boron recommended on a soil test report improves good kernel development in groundnut.

Application of 10 kg per ha of borax as soil application at final ploughing is recommended for boron deficient soils. Foliar application of 0.25% borax results in overcoming the deficiency of boron in a standing crop. This should be done twice during the cropping season, once at the begining of flower developemnt and once at the pod set stage. Even, foliar application of boron as low as 0.1 ppm also increased the yield. Boron applications should be made every year for peanuts because soluble B can be easily leached from the root zone, especially in sandy soils in high rainfall regions, or with over-irrigation

B toxicity

B toxicity can arise under excessive B application, in arid or semi-arid areas, and where irrigation water is rich in B content (more than 1–2 ppm B). B-toxicity symptoms are yellowing of the leaf tip followed by gradual necrosis of the tip and leaf margins, which spreads towards the midrib (central vein). Leaves become scorched and may drop early.

For all types of applications, extreme care must be taken to apply the correct amount and to apply it uniformly to avoid toxicity

19.8.3 Iron

Iron is an essential element required by both the legume host and Rhizobium for a range of physiological and biochemical processes of particular importance in the iron nutrition of the symbiosis is the specific involvement of iron in several key proteins for nitrogen fixiation, e.g. nitrogenase, leghaemoglobin, ferrodoxin, ets., an adequate supply of iron is therefore required for the establishment and function of an effective symbiosis.The growth and metabolism of groundnut is affected by iron nutretion, apart from this the iron deficiency also limit the survival and growth of rhizobia in the soil and rhizosphere, nodule initiation and development and nodule function.

Functions and Deficiency symptoms of Iron in Groundnut

Fe is absorbed by plant roots as Fe2+, and to a lesser extent as Fe chelates. For efficient utilization of chelated Fe, separation between Fe and the organic ligand has to take place at the root surface, after the reduction of Fe3+ to Fe2+. Absorbed Fe is immobile in the phloem. Fe is generally the most abundant of the micronutrients with a dry-matter concentration of about 100 μg/g (ppm). It plays a role in the synthesis of chlorophyll, carbohydrate production, cell respiration, chemical reduction of nitrate and sulphate, and in N assimilation.

Fe deficiency begins to appear on younger leaves first. Otherwise, its deficiency symptoms are somewhat similar to those of Mn, as both Fe and Mn lead to failure in chlorophyll production. Yellowing of the interveinal areas of leaves (commonly referred to as iron chlorosis) occurs. In severe deficiency, leaves become almost pale white because of the loss of chlorophyll.

Chlorosis of the younger leaves characterizes an iron deficiency. The tissue between the veins gradually turns yellow, while the veins tend to stay green. The tips and margins of some leaves may turn brown and become dry and brittle.

- Groundnut is very susceptible to iron deficiency. Iron deficiency is seen in calcareous and alkaline soils with soil pH above 7.5 and in poorly drained soils.
- Plant growth is affected with chlorosis of younger leaves and results in reduced foliage yield
- Iron deficiency may not limit the number of nodules initiated as root infection of Rhizobia is not affected by iron content.
- Though the root nodule number was not affected by iron content, the nodule development i.e. root nodule mass will be seriously affected by the iron deficiency

Deficiency correction

- Soil application of 25 kg per ha of ferrous sulfate at the final ploughing is recommended in soils low in available iron.
- Foliar application of 0.5-1.0% ferrous sulfate solution was proved very successful for groundnut if iron deficiency is noticed in a standing crop.

The chlorosis or yellowing between the veins also results due to zinc deficiency. Iron chlorosis extends to full length of leaves where as Chlorosis due to zinc deficiency occur mainly on the lower half of the leaves.

19.8.4 Molybdenum

The plant needs some macronutrients, micronutrients for its normal growth. Some of these elements play an important role in the process of Rhizobium symbiosis: for example, molybdenum is a constituent of the nitrogenase enzyme and every bacterium, which fixes nitrogen, needs molybdenum during the fixation process. Molybdenum is a trace element found in the soil and is required for growth of most biological organisms including plants and animals. The availability of molybdenum for plant growth is strongly dependent on the soil pH, concentration of adsorbing oxides (e.g. Fe oxides), extent of water drainage and organic compounds found in the soil colloids

Molybdenum has a positive effect on yield, quality and nodules formation in legume crops. The functions of molybdenum in leguminous plants include nitrate reduction, fixation and general metabolism. Mo is absorbed as the molybdate anion MoO4 2- and its uptake is controlled metabolically. Mo is involved in several enzyme systems, particularly nitrate reductase, which is needed for the reduction of nitrate, and nitrogenase, which is involved in BNF. Thus, it is involved directly in protein synthesis and N fixation by legumes. Mo appears to be moderately mobile in the plant. This is suggested by the relatively high levels of Mo in seeds, and because deficiency symptoms appear in the middle and older leaves. The deficiency of molybdenum (Mo) is confined largely to acid soils with pH less than 5. Its deficiency is quite likely in high pH (more than 8) soils also.Among the micronutrients, only molybdenum availability decreases with decrease in the soil pH Mo deficiency in legumes can resemble N deficiency because of its role in N fixation. Mo deficiency can cause marginal scorching and rolling or cupping of leaves and yellowing and stunting in plants. Yellow spot disease in citrus and whip tail in cauliflower are commonly associated with Mo deficiency. Fodders containing more than 5 μg/g Mo in the dry matter are suspected to contain toxic levels of Mo for grazing animals (associated with the disease molybdenosis).

Mo resulted in enhanced nitrogenase and nitrate reductase activities in groundnut. Molybdenum had a significant promotive effect on nitrogenase activity of groundnut root nodules root nodules parameters and nitrogenase activity gave the indication for nitrogen fixing activity in nodules of groundnut was significantly influenced by molybdenum .

Functions of Molybdenum in Groundnut

***Plant growth*:** Molybdenum had a significant beneficial effect on all vegetative growth parameters such as plant height, number of branches and leaves per plant, leaf area, root length of groundnut plants. Nodule number and dry weight of groundnut were increased by applied molybdenum and its enhanced the rate of nitrogen fixation.

***Nodulation and Nitrogen Fixation*:** Molybdenum had a significant promotive effect on groundnut root nodules number per plant, fresh and dry weights of nodules with all Mo levels. Molybdenum at 12 ppm gave best figures of both number and weights of groundnut. Molybdenum at 12 ppm gave the highest activity of nitrogenase enzyme. Increasing Mo above than 12 ppm reduced the one. Molybdenum treatment significantly increases the nitrogenase activity of groundnut root nodules under different nitrogen levels

***Yield Characteristics*:** All yield parameters such as pods number and weight per plant, weight of 100 seeds and pods yield per feddan significantly increased as affected by different Mo doses compared with control plants. Molybdenum at 12 ppm gave the maximum yield parameters.

Molybdenum significantly improved the percentage of oil and oil-yield Seed Chemical

Constituents

Molybdenum application significantly increase groundnut seeds chemical content such as protein, total carbohydrates, total soluble sugars, total soluble solids, oil and total phenols percentages as compared with untreated plants

Molybdenum Deficiency Symptoms

The deficiency of molybdenum (Mo) is confined largely to acid soils (below 4.5 pH). Mo availability increase with the increase in pH. Its deficiency is quite likely in high pH soils (above 8.5 pH) also. In legumes, including groundnut, a shortage of molybdenum affects the nitrogen-fixing activities of soil microorganisms. This results in less availability of nitrogen to the plants. Hence, the symptoms of molybdenum deficiency are similar to those of ordinary nitrogen deficiency. Symptoms of molybdenum deficiency resemble those of nitrogen deficiency, that is, yellowing of leaves, stunted growth, and low yield.

Deficiency correction

- Soil application of 0.56 - 1.72 kg per ha of ammonium molybdate and/or foliar spray of 100-200 ppm ammonium molybdate before flowering was found to increase groundnut yield.
- Micronutrients like Manganese, Magnesium, Copper, Chloride and Cobalt are very rarely deficient in groundnut growing areas.
- Molybdenum at 12 ppm gave the best growth and yield parameters of groundnut.

19.8.5 Cobolt

Cobalt is considered to be a beneficial element for higher plants in spite of the absence of evidence for direct role in their metabolism. This is true in spite of essentiality for photosynthetic activities of lower plants such as euglena gracilis; it was frequently reported to be localized in various sub-cellular fractions as in chloroplasts

19.8.6 Manganese

Manganese (Mn) is taken up by plants as the divalent ion Mn2+. It is known to activate several enzymes and functions as an auto-catalyst. It is essential for splitting the water molecule during photosynthesis. It has certain properties similar to Mg. It is also important in N metabolism and in CO2 assimilation. Like Fe, it is generally immobile in the phloem.

Deficiency symptoms in groundnut

Mn-deficiency symptoms resemble those of Fe and Mg deficiency where interveinal chlorosis occurs in the leaves. However, Mn-deficiency symptoms are first visible on the younger leaves whereas in Mg deficiency, the older leaves are affected first. Mn deficiency in oats is characterized by "grey-speck" where the leaf blade develops grey lesions but the tip remains green, the base dies and the panicle may be empty. In dicots (e.g. legumes), the main symptom of manganese deficiency is chlorosis or yellowing between the veins of new leaves. There may also be a distinguishing characteristic of brown spots on the leaf margin. In legumes, deficiency symptoms include pale green young leaves and a pale yellow mottling develops in interveinal areas, while the veins remain green. Oats are an excellent indicator crop. Manganese is partly mobile in oats. White to grey flecks or specks first appear and become more severe on mature leaves about halfway up the shoot. If a deficiency persists, symptoms spread to old leaves then to the youngest leaves. The specked condition is referred to as "grey speck" and will appear in the interveinal area of the lower half of older leaves and extend toward the tip as symptoms develop Mn-toxicity symptoms lead to the development of brown spots, mainly on older leaves and uneven green colour. Some disorders caused by Mn toxicity are: crinkle leaf spot in cotton; stem streak; necrosis of potato; and internal bark necrosis of apple trees.

Deficiency correction

Low availability of manganese (Mn) is usually associated with alkaline soils (above 7.5 pH) and excess of Mn with acid soils (below 4.5 pH). In India since groundnut is mostly grown in the soils having pH more than 7.5 there is high incidence of Mn deficiency. Only limited information is available on manganese fertilization for groundnut. As a rule, broadcast applications are seldom effective. For cereals, a seed placed treatment of manganese sulphate should be most effective. Foliar application can also be used if deficiency symptoms develop during the growing season.

19.9 Foliar Spray of Nutrients on Groundnut

Foliar feeding is often the most effective and economical way to correct plant nutrient deficiencies. During the last decades, foliar feeding of nutrients has become an established procedure in crop production to increase yield and improve the quality of crop products (Roemheld and El-Fouly, 1999). Foliar application of nutrients could improve the nutrient utilization and lower environmental pollution through reducing the amounts of fertilizers added to soil (Abou-El-Nour 2002). Foliar feeding of a nutrient might have actually promoted root absorption of the same nutrient or other nutrients through improving root growth and increasing nutrients uptake (El-Fouly and El-Sayed, 1997). (Subrahmaniyan et al. 2000 reported that the application of combined nutrients (Diammonium phosphate 2.5 kg + ammonium sulphate 1.0 kg + borax 0.5 kg + NAA (naphthalene acetic acid) at 40 ppm) on 25th and 35th day after sowing (DAS), followed by foliar spray of NAA at 40 ppm alone on 45th and 55th DAS significantly gave the highest values of growth and yield attributes and pod yield of groundnut. (Ali 2003) observed that foliar spray of Zn (2%) improved the groundnut yield attributes and yield as well as quality. Beneficial bioactive compounds such as GA 3, IAA etc., and increased level of phosphorus and potassium and other essential nutrients present in composted pressmud will contributed to increased yield (Somasundaram, 2003). (Galindo et al., 2007) reported that the use of fermented, liquid organic fertilizers, effective microorganism (EM) and fermented plant extracts (FPE) as foliar fertilizers have been introduced to modern agriculture in recent years to produce food with good quality and safety.

19.10 Micronutrient Fertilizers

For the most current information on micronutrient recommendations, sources and methods of application for specific crops on problem soils, a regional soil or crop production specialist should be consulted. Micronutrient fertilizers are either in inorganic form (readily soluble inorganic salt) or organic form listed in table 19.3.

***Table 19.3** Lists some common inorganic micronutrient*

Nutrient	Form	%of nutrient in product
Boron	Borate	14-20
	Borax	11
Copper	Copper sulphate	25
Iron	Iron sulphate	19
Manganese	Manganese sulphate	26
Zinc	Zinc sulphate	18-23

References

1. Abou-El-Nour, E.A.A., 2002. Can supplemented potassium foliar feeding reduce the recommended soil potassium? Pak. J. Biol. Sci., 5(3): 259-262
2. Achim Dobermann and Thomas Fairhurst. Rice ecosystems. Nutrient management T.H. Fairhurst. Copyright © 2000 ... First edition2000. ISBN 981-04-2742-5.
3. Ali AAG, SAE Mowafy (2003). Effect of different levels of potassium and phosphorus fertilizers with the foliar application of zinc and boron on groundnut in sandy soils. Zagazig J. Agric. Res., 30: 335-358.
4. Alloway BJ (2008). Zinc in soils and crop nutrition, Second edition, published by IZA and IFA Brussels, Belgium and Paris, France ILZRO (1975). Zinc in Crop Nutrition, International Lead Zinc Research Organisation Inc, Research Triangle Park
5. Arunachalam P, Kannan P, Balasubramaniyan P, Prabukumar G, Prabhaharan J (2012). Response of groundnut (Arachis hypogaea L.) genotypes to biofortification through soil fertilization of micronutrients in Alfisol conditions. Annual Report of Dryland Agricultural Research Station, Chettinad, India.
6. Bell RW, Kirk G, Plackot, Lonoragan JF (1990). Diagnosis of zinc deficiency in peanut (Arachis hypogaea L.) by plant analysis. Comm. Soil Sci. Plant Anal. 21:273-285
7. Chitdeshwari T, Poongothai S (2003). Yield of groundnut and its nutrient uptake as influenced by Zinc, Boron and Sulphur. Agric. Sci. Dig. 23(4):263-266.
8. El-Fouly, M.M and A.A. El-Sayed, 1997. Foliar Fertilization: An environmentally friendly application of fertilizers. Dahlia Greidinger International symposium on "Fertilization and Environment" 24-27 March, Haifa, Israel, Ed. John, J. 346-357
9. Imtiaz M, Alloway BJ, Shah KH, Siddiqui, SH, Memon MY, Aslam M, Khan P (2003). Zinc nutrition of wheat II: Interaction of zinc with other trace elements. J. Asian Plant Sci. 2:156-160.
10. Loneragan JF, Webb MJ (1993). Interactions between zinc and other nutrients affecting the growth of plants. In: Robson, A.D. (ed.) Zinc in Soils and Plants. Kluwer Academic Publishers, Dordrecht, pp. 119-134.
11. Pendashtek M, Tarighi F, Doustan HZ (2011). Effect of foliar zinc spraying and nitrogen fertilization on seed yield and several attributesof groundnut (Arachis hypogaea L.). World Appl. Sci. J. 13(5):1209-1217.
12. Prasad TNV, Sudhakar KVP, Sreenivasulu Y, Latha P, Munaswamy V, Raja Reddy K, Sreeprasad TS, Sajanlal PR, Pradeep T (2012). Effect of nanoscale zinc oxide particles on the germination, growth and yield of peanut. J. Plant Nutr. 35(6):905-927
13. Roemheld, V. and M.M. El-Fouly, 1999. Foliar nutrient application: Challenge and limits in crop production. Proc. 2nd International Workshop on "Foliar Fertilization" April 4-10 Bangkok, Thailand: 1-32

14. Singh, A.L. 2004. Mineral nutrient requirement, their disorders and remedies in groundnut. Groundnut Research In India, National Research Centre for Groundnut, Junagadh, India. pp. 137-159.

15. Singh AL, Lal C (2007). Screening for high density groundnut genotypes in India. In:Proceeding of Zinc Crops 2007 Conference for improving crop production and human health, Istanbul, Turkey 24-26th May 2007.

16. Singh MV (2008). Micronutrient deficiencies in crops in India. In:Micronutrients in global crops. (ed. Alloway Brown) Springer, New York.

17. Singh MV, Patel KP, Ramani VP (2003). Crop responses to secondary and micronutrients in swell-shrink soils. Fert. News 48(4):63-66.

18. Singh MV (2010). Micronutrient deficiency in Indian soils and field usable practices for their correction. AICRP (micronutrient) Annual Report, Indian Institute of Soil Science, Bhopal, India

19. Singh, M.V. 2006. Micronutrients in crops and in soils of India. In Micronutrients for global crop production. ed. B.J. Alloway, Springer. Business.

20. Somasundaram E (2003). Evaluation of organic sources of nutrients and Panchagavya spray on the growth and productivity of maizesunflower-greengram system. Ph.D., Thesis, Tamil Nadu Agricultural University, Coimbatore.

21. Subrahmaniyan K, P Kalaiselvan, G Manickam, N Arulmozhi (2000). Spacing and fertilizer requirement for confectionery groundnut varieties. Crop Res., 19(2): 210-212.

22. Tathe AS, Patil GD, Khilari JM (2008). Effects of sulphur and zinc on groundnut in vertisols. Asian J. Soil Sci. 3(1):178-180.

Chapter 20

Long-Term Effect of Fertilizers and Manures on Groundnut Productivity and Soil Properties

K.V. Naga Madhuri*, P. V. R. M. Reddy, T. N. V. K. V. Prasad, and T. Giridhara Krishna

*Senior Scientist(Soil Science), RARS, Tirupati.

20.1 Introduction

Soil is a dynamic, living, natural body that is vital to the function of terrestrial ecosystems and represents a unique balance between physical, chemical and biological factors. Quality of soil as distinct from health is largely defined by the ability of soil to perform various intrinsic and extrinsic functions. Soil quality is represented by a suite of physical, chemical and biological properties that together; (1) provide a medium for plant growth; (2) regulate and partition water flow and storage in the environment; (3) and serve as an environmental buffer in the formation and destruction of environmentally hazardous compounds. Long term fertilizer experiments play an important role in understanding the complex interactions involving soils, plants, climate and management practices and their effects on crop productivity. It is well recognized that long term fertilizer experiments are repositories of valuable information regarding the sustainability of intensive agriculture. The LTFE serves as an important tool to understand the changes in soil properties due to intensive cropping and continuous fertilization. Consistent use of fertilizers and manures in soil alerts the physical, physico-chemical and chemical properties of the soil. Long term fertilizer experiments are normally intended to show how high a production can be reached under different conditions and how far the soils are able to with stand depletion and deterioration.

Rainfed farming is practiced on 80 Mha in India covering a wide range of soils. However Alfisols are the most predominant soils especially in southern India where 37.5 Mha (46.87 %) are cultivated by small land holders (Virmani et al.1991). The available water holding capacity of these soils is low because of coarse texture and shallow effective rooting depth. These soils are vulnerable to crusting because of the hard setting surface soil and structural instability. Most Alfisols are cultivated for a single rainy season crop with a low productivity of 0.7-0.8 Mg/ha. These soils are low in fertility. Therefore, crop and soil management practices used must ensure greater crop yield sustainability though adoption of appropriate cropping systems and techniques of nutrient management. In addition, the amount of rainfall and its distribution are important to harnessing the benefits of improved soil fertility. However, resource poor farmers in dry land regions apply meager quantity of nutrients and thus crops suffer from multi nutrient deficiencies (Srinivasarao et al. 2003). Soil and crop specific data on soil-plant nutrient balance are scanty for rainfed cropping systems in semi arid regions of India and elsewhere. Soil depletion of N in dry land agricultural systems is generally expected since its input is traditionally low despite a very rapid decomposition of organic residues because of the prevailing high temperatures. Much of the available research data from India on nutrient balance are based on experiments involving input of mineral fertilizers and output of nutrients removal by crop harvests (Yadav et al. 2000). Thus, elemental balances obtained from aggregated total nutrient application and removal does not account for nutrient losses by other processes. Traditionally, crop nutrient requirements for rain fed cropping systems are mostly met through inherent soil fertility rather than from off farm inputs. Therefore, nutrient balance studies based on long term cropping, fertilization and manuring can provide a realistic estimate of gain or loss of nutrients for a specific soil and a cropping system.

Applying an adequate amount of fertilizer is an important cultivation practice for the yield and quality of crops, environmental protection and soil sustainability. In addition it is important to apply organic matter to cultivated soil on the basis of sustainability because cultivation enhances the rate of soil degradation and decomposition of soil organic matter. Long term application of ammonium nitrogen (N) fertilizer and urea caused several soil problems, such as lowering soil organic carbon contents, acidifying soil and resulting in poor soil physical properties (Liu et al, 2010). These problems may partially be improved by applying organic amendments. Significant improvements in soil physical, chemical and biological properties have been reported in organic matter application systems. Application of different organic fertilizers in combination with chemical fertilizers to agricultural lands is a popular practice in crop production.

The term soil quality is related to chemical, physical and biological characteristics that enable soils to perform a wide range of functions. The biological properties of soils usually respond more rapidly to changing soil conditions than the chemical or physical properties. Soil enzymatic activity has been used as an indicator of soil quality, since it is the reflection of the effects of cultivation, soil properties and pedological amendments. However, agricultural management practices, particularly application of fertilizers, may have great impact on the size and structure of the soil microbial community. Research is needed to understand the diversity and function of the microbial community in agro ecosystems.

In this view an investigation has been initiated to assess the effects of long term application of fertilizers and manures on soil properties and productivity of groundnut monocropping under rainfed conditions in Alfisols.

20.2 Case Study

20.2.1 Experiment Site

A long term field experiment was initiated during the kharif season of 1981 at Regional Agricultural Research Station, Tirupati, A.P, India. Climate of the experiment site is semi arid tropical with mean annual temperature 28^{0}C and mean kharif rainfall of about 600 mm (June to October). The data in Table 3 show the monthly distribution of rainfall during the experiment period. Though the experiment was initiated in 1981, rainfall data is available since 1983. The soils of the experiment site are neutral in pH, normal in EC, low in organic carbon, medium in available phosphorus and potassium. Initial Nitrogen status was not available with the records. Soil are sandy loam in texture with CEC of 16.8 cmol (+)/kg.

20.2.2 Crop Management, Treatments, Soil Sampling and Analysis

Groundnut crop of popular varieties of this region and varieties released from the research station are grown viz., TMV 2, TAG-24, Abhaya, Narayani, Dharani etc during the kharif season. The land preparation involves one ploughing with cattle followed by leveling soon after receipt of monsoons. The experiment consisted of 11 treatments viz., T1 - Control (No fertilizer or manure), T2 - Farm yard Manure @ 5 t/ha once in 3 years, T3 - 20 kg N/ha,T4 - 10 kg P/ha, T5 - 25 kg k/ha, T6 - 250 kg Gypsum/ha (Top dressing), T7 - 20 kg N + 10 kg P/ ha,T8 – 20 kg N + 10 kg P + 25 kg K /ha, T9 - 20 kg N + 10 kg P + 25 kg K + 250 kg Gypsum/ha (top dressing),T10 - 20 kg N + 10 Kg P +25 kg K+ lime 100 kg/ha (Top dressing) ,T11 - 20 kg N + 10 kg P+25 kg K+ 250 kg/ha Gypsum (top dress) + 25 kg $ZnSO_4$ /ha once in 3 years. The experiment was laid in Randomized Block Design where in each treatment was replicated four times. Each treatment was imposed in individual plots with dimensions of 9.9 m x 4.8 m. NPK nutrients were applied as per treatments in the form of urea, SSP and MOP respectively. FYM manure was applied once in three years at the time of sowing. Gypsum and lime were applied at the time of flowering. Pre emergence application of pendimethalin and hand weeding was done for weed control. Data was recorded on yield and yield attributes at harvest of each crop. Soil samples were collected and air dried ground and passed through 2 mm sieve. Soil sample from each treatment was analyzed for pH, EC, OC, available N,P, K and S at harvest of each crop of every year since 19981 as per standard procedures prescribed by Jackson (1967).

20.2.3 Statistical Analysis

The data was analyzed statistically using the windows based Statistical Package for the Social Sciences (SPSS, 2001) program and determined the variance and statistical significance of the treatment effects.

20.3 Effect of Long Term Fertilization on Soil Properties

20.3.1 Soil Reaction (pH)

Long-term application of fertilizers and manures to soil is expected to affect changes in the soil reaction depending upon the nature of the soil and the composition of manure-fertilizer schedules adopted in the particular cropping system. Soil reaction of the surface layer in all the treatments was slightly acidic even in control and FYM applied treatments (Table 1). Lowest pH of 5.10 was recorded in T9 where NPK+ Gypsum was applied where as highest was recorded in control followed by FYM applied plot. In general, soil pH decreased with depth in all the treatments. Continuous cropping with or without inorganic fertilizers caused soil acidification. Continuous use of fertilizers increased soil acidity in plots where FYM was not applied compared with FYM applied treatment. Similarly, experiments conducted at Bengaluru in an Alfisol indicated that change in pH was minimum with FYM application (Sudhir et al, 1998).

When variation in soil pH over 34 years of application of fertilizers and manures in groundnut mono cropping was observed, highest variation was seen in T3 (100%N) followed by T9 & T11 (Fig.1). However, least variation in pH was seen in FYM applied plot followed by control. It was found that the continued cropping and fertilization over 34 years resulted in the decrease of soil pH by 0.1 to 0.2 units and the maximum fall in pH was due to the application of N alone in the form of urea and the acid producing nature of this was attributed as reason for the decrease in pH (Rajendra Prasad & Goswamy, 1992). Similar results of low pH due to fertilizer application to groundnut grown in sandy loam soils of humid and tropical climate of Meghalaya were attributed to oxidation of NH_4^+ and release of H^+ ions (Lolfakzuala et al, 2008).

20.3.2 Soil Salinity (EC)

The data showed that an increase in soil EC from 0.08 dS/m to 0.11 dS/m due to continuous application of fertilizers over 34 years. The minimum EC of soil was observed with T3 followed by control and T4 and the maximum was found in T9 (Table 1). The increase in soil EC with continued application of fertilizers was due to the addition of salts through fertilizers and solubilization of native minerals due to the reduction in the pH of the soil (Lohan & Dev, 1998). (Mairan et al 2005) concluded that there was decline in values of soil EC of Vertisol with crop residue incorporation over fertilizer application in LTFE on fixed plot with sorghum- sunflower sequence. Furthermore, the EC was lower in the surface layers compared with that in the subsurface layers.

Variation in soil EC over 34 years of application of fertilizers and manures revealed that least variation was seen in T9 followed by T2 however variation was high in T4 and T5 (Fig 2). The maximum salt concentration was seen in inorganic fertilizers applied plot compared with that in initial in the present study was due to the addition of soluble salts though fertilizers. Similar results were reported by (Srinivasarao et al 2013).

20.3.3 Soil Organic Carbon

Long term experiments are the primary source of information to determine the effect of cropping system, soil management, and fertilizer usage and residue utilization on changes in soil organic carbon (SOC) and to determine the agricultural sustainability. The soil organic carbon percent improved significantly in all fertilizer treatments and manure applied plots including control as compared to initial. Highest SOC was recorded in FYM applied plot followed by T7&T9. Whereas the least content was observed in T3 & control (Table 1). The SOC content depends largely on the annual turn over of root residues, rot exudates and stubbles (Srinivasarao et al, 2012). Thus the increase in SOC during the experiment in the plots treated with FYM and RDF over the control may be explained by the greater yields through fertilization and manuring associated greater amounts of root residues and stubbles and higher C input (Srinivasarao et al, 2012). Similar results were also been reported in other long term studies (Swarup and Wanjari, 2000).

The data on changes in variation in SOC over 34 years of application of fertilizers and manures showed that highest variation from initial was seen in FYM applied plot followed by NPK and NPK+ Gypsum + Zn applied treatment. However, the least change in variation in SOC was noticed in control followed by NPK + Lime applied treatment (Fig. 3). (Suresh et al 1999) found that the FYM had profound effect on the organic carbon content of soil. The organic matte content was 2.09 percent with FYM application, which was significantly higher than all other treatments. The addition of FYM helped in maintaining the organic matter status of 2.1% as observed when the experiment was initiated during 1984 (Singh et al, 1999).

(Alok Tiwari et al. 2002) reported from the experiment conducted on Typic Haplustalf; that the continued application of fertilizers alone also helped in increasing the organic carbon content of soil, which could be attributed to higher contribution of biomass to soil in the form of crop stubbles and residues. In Vertisols, the organic carbon content of both surface and sub-surface soils increased significantly with the incorporation of FYM either alone or in combination with recommended doses of fertilizers. This could be attributed to better root growth and more plant residues incorporation into the soil. The organic carbon and nitrogen contents of soil under 33 years of cropping decreased with increasing frequency of fallow than under continuous cultivation (Tolanur & Badanur, 2003). (Fan et al. 2005) observed that the combination of organic and inorganic fertilization enhanced the accumulation of soil organic carbon and maintained the highest productivity. Organic carbon content of soil was higher when higher rate of FYM was added (Rajashree et al, 1995). (Kuntal et al. 2007) stated that the balanced application of inorganic fertilizer and organic amendments greatly influenced the accumulation of organic matter in soil.

20.4 Effect of Long Term Fertilization on Soil Nutrient Status

The cultivation of crops without application of manures or fertilizers caused a drastic exhaustion of the native pool of nutrients from the soil under the present agro - climatic condition (dry land Vertisol). The combined application of organic source such as FYM (50 per cent of the requirement) with nitrogenous fertilizers N i.e., urea (50 percent of the requirement) + sustained the higher availability of N, P and K.

20.4.1 Available Nitrogen

Application of major nutrients especially N plays a key role in determining the yields of crops. Among the major nutrients, nitrogen is universally deficient and most vulnerable to losses especially under tropical climate. Its regular supply during crop growth period is possible through release of nitrogen from organic manures. The available nitrogen content differed widely between the control and other treatments. However, the highest Nitrogen was recorded in NPK+ Gypsum+ Zn treated plot and the lowest nitrogen content was recorded in T5 and Control plots. Long term application of fertilizers and manures for 34 years improved the available nitrogen status. However, the N status remained below the critical range showing no marked build up of N in these soils (Table 2). Despite continuous addition of organic and inorganic sources of plat nutrients for 20 years, all soil profiles were deficient in the available N (Tandon, 1993). Similar trends of N deficiencies were reported in soils cultivated for other food legumes with long term and continuous cropping (Srinivasarao et al, 2003).

Most crop land soils of India are deficient in N partly because of low concentration of SOC. Yet, maintaining available content of N at levels above the critical limit is essential to obtaining the high crop productivity. Therefore, application of 20 kg N/ha is recommended for legume crops assuming that remainder of the crop N requirement is met through biological N fixation (Rego et al, 2003). The fixation of N by Rhizobium is often adversely affected by drought stress and prevalence of high temperatures under semi arid conditions resulting in severe deficiency of N in most crops (Venkateswarlu and Rao, 1987). Thus, there is not enough residual N left in soil even after long term cultivation of leguminous crops under rainfed conditions (Srinivasarao et al, 2003). Higher input of nutrients for achieving the balanced fertilization of recommended doses of NPK and integrated use of organics and chemical fertilizer lead to better crop growth and higher biomass production compared to that of the control. However, nutrient mining is also affected by the rainfall deficits in most years ultimately resulting in net loss of nutrients.

Nitrogen accumulation was higher when manure or straw was added along with NP fertilizers. This might have been partially due to a slow release of N from manure and straw, resulting in reduced losses of N (Duraisami and Mani, 2000). Similar result was also reported by (Bhandari et al. 2002). In the red loam soils of the old and new permanent manurial experiments at Coimbatore, the available nitrogen content was found to be enhanced due to continuous incorporation of organic manures (Muthuvel et al, 1977). Continuous addition of potassium had decreased the available N status of the soil due to the blocking up of adsorbed NH_4 by added potassium preventing its release (Govindarajan and Gopala Rao, 1978). The results of the long-term fertilizer experiments were reviewed by (Nambiar 1985) and reported that there was a decline in the available N status of the soil in plots where N had not been applied for many years. Subramanian and (Kumaraswamy 1989) stated that the soil available N status in the long - term fertilizer experiment conducted on Inceptisol was the highest under100 per cent optimum NPK + FYM treatment while under cropped but unmanured treatment it was the lowest as against the initial status.

A comparison of available N status of the fallow plots with control clearly revealed that cultivation without any addition of fertilizers or manures drastically reduced the soil N availability. Increasing the NPK dose from 50 to 150 percent increased the available N content of the soils and the highest available N was noticed under FYM treated plots. The results of Long – Term Manurial Experiment showed that the N availability was not altered by the levels of phosphorus or potassium application, but their application without FYM caused depletion of available N (Rao and Dhakore, 1997). Any increase in the rate of FYM applied would increase the available N status of the soil (Mandal et al 1997). The available N content of the soil increased over the initial value under 100 per cent NPK + FYM treatment and was due to higher organic carbon content in Inceptisol. Continued cropping without fertilization depleted the available N status from the initial value in plots receiving only FYM. However, it increased significantly with regular dressing of N fertilizers. The increase in available N status due to organic manure application would be due to the multiplication of soil microbes leading to enhanced conversion of organically bound N into inorganic forms. Legume cropping increased the N status of the soil, which may be attributed to the litter fall and atmospheric N-fixation by legume crops. There was decline in availability of N in all the treatments when compared to the initial values in Vertisol at LTFE, MAU and Parbhani. Continuous manuring helped to increase the available nitrogen of soil (Mairan et al, 2005).

20.4.2 Available Phosphorus

A large depletion of available P was seen in all the treatments as compared with initial including control treatment after continuous application of fertilizers and manures over 34 years under rain fed conditions. The lowest available P was recorded in control plot followed by T5 and T6. However the highest available P was recorded in T11 followed by T10 and T8 (Table 2). It was observed depletion even in P applied plots continuously for 34 years. This could be due to insufficient application of P fertilizers than crop requirement. Increase in available P in soils with regular P additions through fertilizers and manures, is expected because groundnut crop utilizes a fraction of the applied P. Moreover, application of NPK at recommended dose added higher P inputs compared to organic plot. Probably because of this, higher content of available P was observed in the plots treated with 100% NPK compared to organic treatment.

The data on changes in variation in available P over 34 years of application of fertilizers and manures revealed that, highest variation in available P was seen in control plot from the initial where no nutrients were added since the experimentation followed by T3. The lowest variation was observed in T10 followed by T11 & T9 (Fig. 4).

A build up in the available P status of the soil over the initial value recorded in 1972, occurred in 100 per cent NPK + FYM and also in 150 per cent NPK treatments. The results of LTFEs all over the country have revealed that the use of N alone leads to accelerated depletion of soil P (Nambiar and Ghosh, 1984). Reviewing the results of long-term fertilizer experiments, (Nambiar 1985) reported a considerable build-up in the available P status of alluvial, red loam and submontane soils of Punjab, Ranchi and Palampur due to the application of increasing rates of P. Similar trends were also reported in the medium black soils of Coimbatore by (Subramanian and Kumaraswamy 1989) and (Muthusamy et al 1990). A marked rise in the available P status of the soil with the application of lime and FYM might be due to the inactivation of iron, aluminium and hydroxyl Al ions thereby reducing the P fixation in soil (Prasad and Singh, 1981). (Bahl and Avtar Singh 1997), stated that an increase in the available P status with the application of organics and the increased availability was a consequence of decrease in P adsorption and phosphate potential of soil. A balance sheet of P under different fertilizer treatments was worked out in LTFE of soil type Typic Ustrochrept. The mean uptake of P in unfertilized plots was the lowest (14.9 kg ha-1) followed by those receiving 100 per cent N alone (26.6 kg ha^{-1}) and 50 per cent NPK (28.0 kg ha^{-1}). The highest annual gain of 66.6 kg ha^{-1} was noted in 150 per cent optimum NPK treatment and the lowest (7.0) in suboptimal dose (Singh et al, 1986). Graded levels of P application were accompanied by a corresponding increase in the amount of available P owing to higher rates of P application in the amount exceeding crop uptake (Santhy et al, 2001). The available P status of soil after 10 years of rice - wheat cropping was minimum in control whereas the maximum available P was found in NPK + FYM + Zn treatment. The NPK + FYM + Zn and NPK + FYM treatments nearly maintained P availability as that of the initial value whereas in other treatments noticeable depletion in available P was seen (Singh et al, 1999).

(Jayasree Sankar et al, 2000) stated that the continued application of phosphatic fertilizer to a rice-wheat crop rotation over the years resulted in an increased available soil P status .The combined application of organic manure and inorganic fertilizer increased the available P status of the soil. The increased release of P may be due to the fact that organic anions compete with the phosphate ions for binding sites on soil particles thereby reducing the P fixation (Pannerselavam, 2000). A build-up of P in P- receiving plots and depletion in N and K treated plots was observed (Ghosh et al, 2000). In calcareous soil, CO_2 produced during the decomposition of organic matter plays an important role in increasing the phosphate availability. Organic matter enhanced the labile P in soil through complexation of cations like Ca and Mg when it is applied in combination with inorganic fertilizers (Tolanur and Badanur, 2003). Among the treatments, 100 per cent NPK + FYM recorded the highest level of P availability in the surface layer of soil in LTFE (Inceptisol). Also the addition of 150 percent of NPK had shown a beneficial effect. Phosphorus availability was found to increase with increase in the amounts of applied NPK in the 0-15 and 15-30 cm layers (Santhy et al, 2001). The availability of P was significantly increased in the treatments where 50 percent or 100 per cent of N was applied through organic source or where 150 percent NPK was applied (Mairan et al, 2005). Slope of the trend lines for soil available P indicated declines of about 0.20 mg kg-1 y-1 for the unfertilized and N only treatments, but increases of 0.19, 0.25, 0.38 and 0.67 mg kg-1 y-1 for the manure, NP, SNP and MNP treatments respectively (Fan et al, 2005).

20.4.3 Available Potassium

A large build up of Potassium was seen in all the treatments except in control where depletion of K was observed. The highest available K content was recorded in T5 followed by T6 and the lowest was seen in control followed by T9 (Table 1). The data on available potassium was ranged from 191 kg/ha to 401 kg/ha.

Decline in available K status was observed in the plot receiving only fertilizer N but the significant increase in available K content has been noted in the plot receiving either FYM or GM along with fertilizer N. This suggests that FYM and GM helped to maintain the supply of K by releasing the K from reserve source (Munishwar sigh et al, 2001). Intensive cropping with high yielding crop varieties led to depletion of K from soil which was evident from a number of long term field experiments conducted across the country under AICRP programme of ICAR (Anand, 2002).

The data on variation soil available K over 34 years of application fertilizers and manures showed that the highest variation in soil available K from the initial was seen in T11 followed by T10. However, the least variation was observed in control where depletion was observed followed by T4, T3 and T6 (Fig. 5). Mean potassium content in the soils have varied largely as significant differences were observed between the control and every other treatments. There was high depletion of potassium in the control plot while T3, 4 and 7 showed depletion of potassium, every other treatment had significantly higher potassium content when compared with the control treatment. T11 had the highest accumulated potassium.

20.4.4 Secondary Nutrients

Calcium, magnesium and sulphur are required in relatively larger amounts for good crop growth. Sulphur and magnesium are needed by plants in about the same quantities as phosphorus, and for many plant species, the calcium requirement is greater than that of phosphorus. In a multiple cropping system with higher fertilizer levels, use of high-analysis fertilizers like urea without secondary and micronutrients would decrease the availability of secondary nutrients like calcium, magnesium and sulphur in soil and cause imbalance of nutrients.

The treatments T6, 9, 10 and 11 are significantly higher in calcium when compared with the control treatment. However T10 appears to have retained higher mean (> 500%) calcium when compared to the rest of the treatments (Fig. 6). The data shows that all the treatments accumulated magnesium during the experimental period with highest accumulation observed in T2, T6, 7, 8, 9, 10. These treatments T2, 6, 7, 8, 9, 10 have retained significantly higher magnesium when compared to the control treatment (T1). Other treatments T3, 4, 5 and 11 were not different. Initial value was calculated based on average first year content (Fig.7).

In a multiple cropping system, higher fertilizer levels decreased the availability of calcium and magnesium while the application of FYM increased the levels of exchangeable Ca and Mg (Prasad and Singh, 1981). In the Old Permanent Manurial Experiment at the TNAU, the Ca content was lower in plots receiving continued application of nitrogen, which may be due to the increased uptake of Ca by crops, induced by nitrogen treatment (Muthuvel et al, 1977). A decrease in the exchangeable Ca and Mg contents of the soil was detected where ammonium sulphate had been used as the source of nitrogen. This decrease was more during the first seven years but the decreasing rate lessened subsequently due to the reduction in yield with less removal of nutrients. In the case of FYM application, there was first a decrease followed by an increase due to slow mineralization of organic manures (Suresh and Mathur, 1988). An increase in the Ca and Mg status of the soil was found due to the continued application of manure while the addition of N in soil tended to decrease it.

The exchangeable Ca and Mg status of continuously fertilized acid soil increased due to the application of lime and FYM (Prasad et al, 1996). The long-term Integrated Nutrient Management on soil fertility status leads to increased content of both Ca and Mg due to the incorporation of FYM along with fertilizers. The FYM has a positive effect on Ca and Mg content in soil because it is having high adsorptive capacity that might have adsorbed Ca and Mg which would otherwise be leached (Bellaki et al, 1998). The cultivated soils with the application of fertilizer + compost + lime contained a large amount of exchangeable Ca associated with fertilization and liming (Takeda et al, 2005).

Mean sulphur content was significantly higher in all treatments when compared with control. T9 had the highest mean sulphur content retained in the soils even after 34 years. There is depletion of Sulphur in control treatment when compared with all the other treatments which were significantly above controls. T9 had the highest accumulation while T3 and T2 recorded lowest accumulation (Fig. 8).

20.5 Effect of Long Term Fertilization on Pod Yield of Groundnut

20.5.1 Rainfall and Pod Yields

The amount of rainfall received during the cropping period at the location ranged from 394 mm in 1989 to 1060 mm in 1998 (Table 3). The nutrient management practices followed for 34 years had a significant effect on pod yields of groundnut (Table 4). The average pod yields over 34 years were higher in treatments receiving nutrients than control. Highest pod yields over 34 years were recorded in T11 treatment followed by T9. The FYM applied plot recorded pod yield of 1603 kg/ha which was on par with NPK treatment indicating the importance of organic manure application. FYM plot sustained the yields on par with inorganic fertilizer treatments on long term basis. Pod yield was significantly more in different fertilizer and manorial treatments than in the control.

Pod yield increased significantly with application of nutrient treatments over that of the control, yield gains were relatively more with better rainfall distribution during th growing season than with the improved nutrient management (Srinivasarao et al, 2013). Some studies have reported the beneficial effects of NPK+FYM on pod yield of groundnut over control on sandy loam soils (Lalfkzuala et al, 2008). The pod yields also differed significantly among years mainly due to variations in the rainfall distribution. The pod yield differed significantly because of the interaction between rainfall and nutrient treatments.

20.6 Conclusion

The importance of long term fertilizer experiment was felt in aspects of sustained crop and soil health. The limelight falls on sustainable maintenance of soil fertility in intensive cultivation of crops. The need of the hour is the production of more grain per unit area per unit time per unit input without detrimental effect on soil environment and soil health. There is a necessity for continuous monitoring of changes in nutrient dynamics. So, LTFE provide a resource of soil and plant material to further scientific research into soil and plant processes that control soil fertility, plant productivity, water, soil, and air quality.

Table 20.1 *Long term effect of fertilizers on soil physico - chemical properties, kharif, 2014*

	pH		EC (dS/m)		OC %	
	0-15 cm	15-30 cm	0-15 cm	15-30 cm	0-15 cm	15-30 cm
T1	5.76	5.30	0.079	0.051	0.293	0.260
T2	5.71	5.41	0.084	0.074	0.383	0.393
T3	5.50	5.11	0.073	0.178	0.243	0.333
T4	5.51	5.10	0.072	0.054	0.330	0.298
T5	5.56	5.10	0.074	0.046	0.298	0.320
T6	5.27	5.03	0.090	0.058	0.268	0.315
T7	5.31	5.13	0.084	0.063	0.360	0.228
T8	5.18	4.78	0.088	0.074	0.288	0.318
T9	5.10	4.89	0.111	0.090	0.355	0.300
T10	5.65	5.09	0.100	0.078	0.353	0.273
T11	5.14	4.81	0.101	0.092	0.320	0.230
Initial	6.7		0.08		0.18	

Table 20.2 *Long term effect of fertilizers on soil chemical properties, kharif, 2014*

	Avail.N (kg/ha)		Avail.P_2O_5 (kg/ha)		Avail.K_2O (kg/ha)	
	0-15 cm	15-30 cm	0-15 cm	15-30 cm	0-15 cm	15-30 cm
T1	144	144	24	28	191	84
T2	163	148	35	23	295	75
T3	154	154	31	27	295	60
T4	175	135	21	33	328	66
T5	144	163	35	28	312	105
T6	147	110	23	36	392	88
T7	147	132	30	31	401	83
T8	169	135	44	33	305	92
T9	147	145	40	31	235	92
T10	163	151	44	27	329	77
T11	188	166	45	33	362	97
Initial	NA		47.6		216	

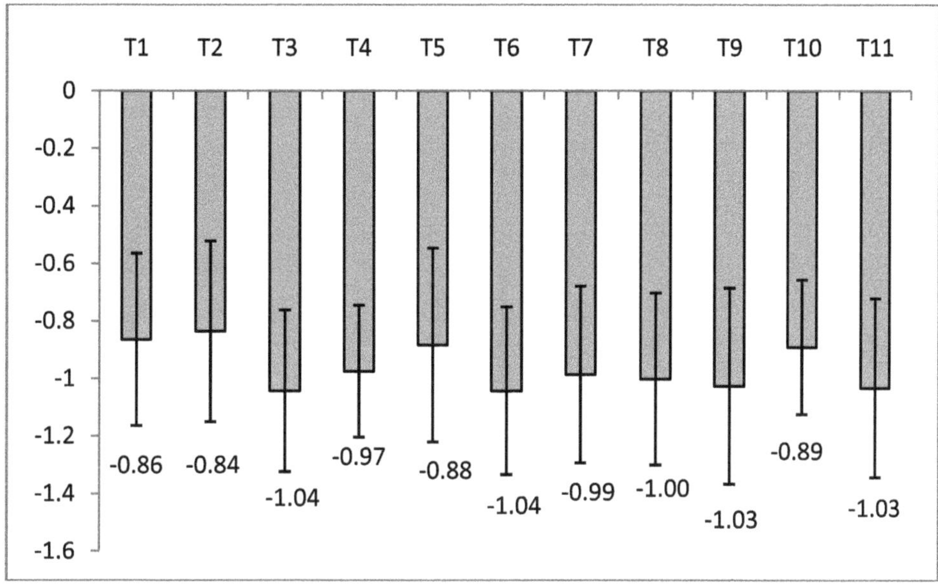

Fig 20.1 *Variation in soil pH over 34 years of application of fertilizers and manures in groundnut*

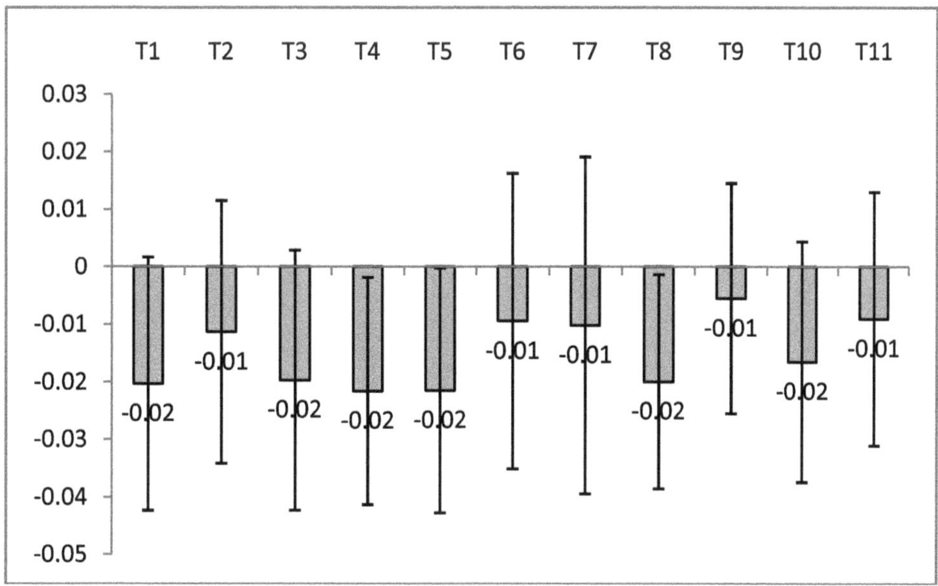

Fig 20.2 *Variation in soil EC over 34 years of application of fertilizers and manures in groundnut*

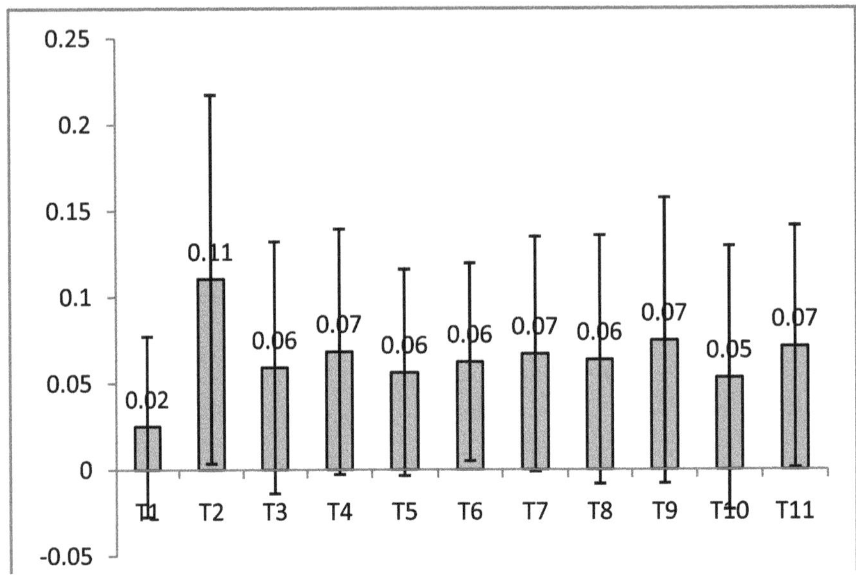

Fig 20.3 *Variation in soil OC over 34 years of application of fertilizers and manures in groundnut*

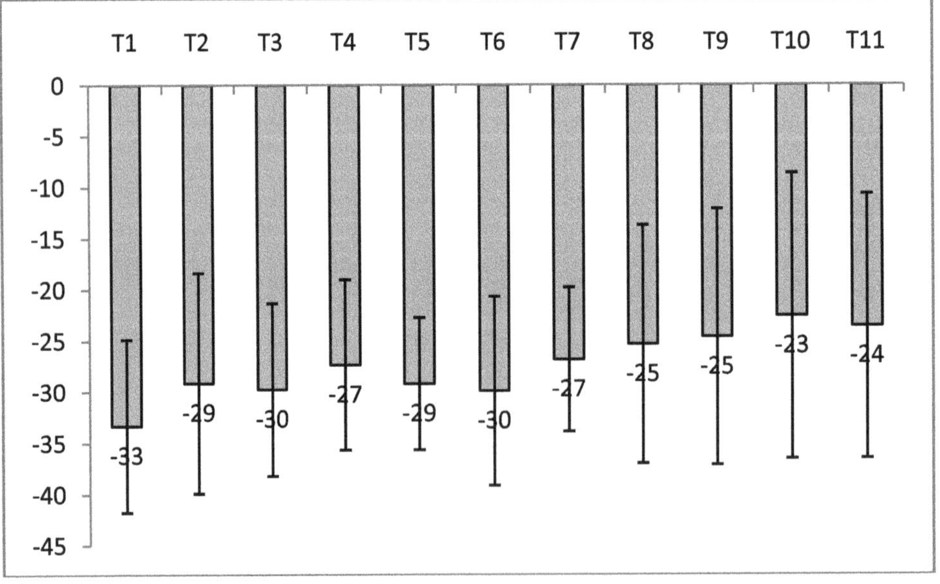

Fig 20.4 *Variation in soil P over 34 years of application of fertilizers and manures in groundnut*

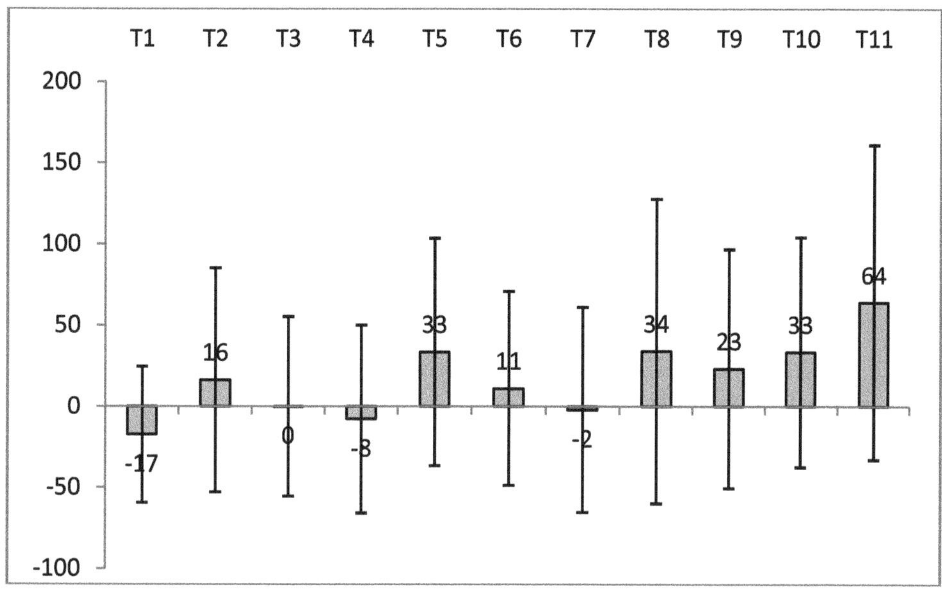

Fig 20.5 *Variation in soil K over 34 years of application of fertilizers and manures in groundnut*

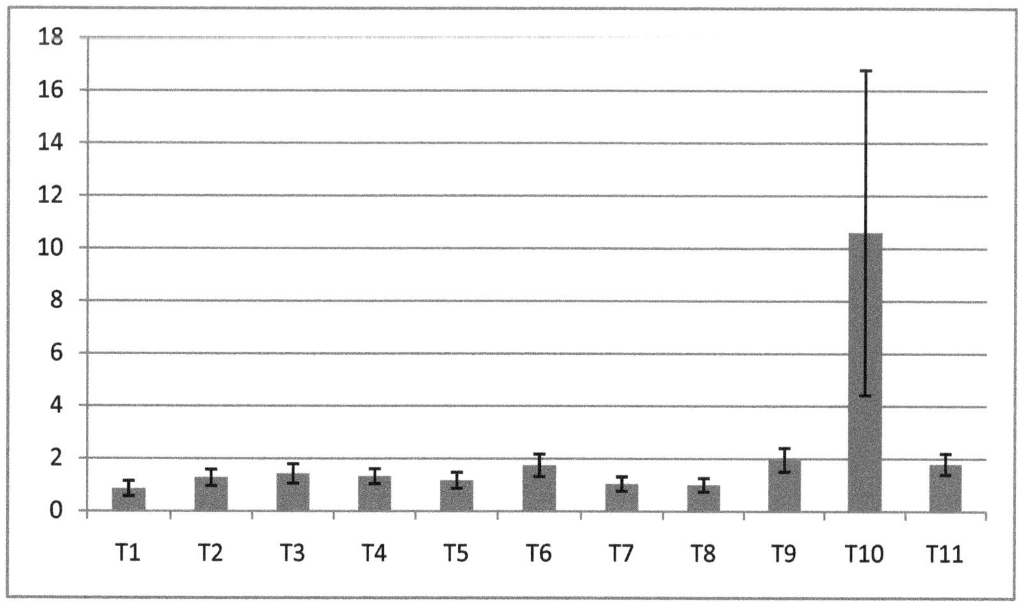

Fig 20.6 *Variation in soil Ca over 34 years of application of fertilizers and manures in groundnut*

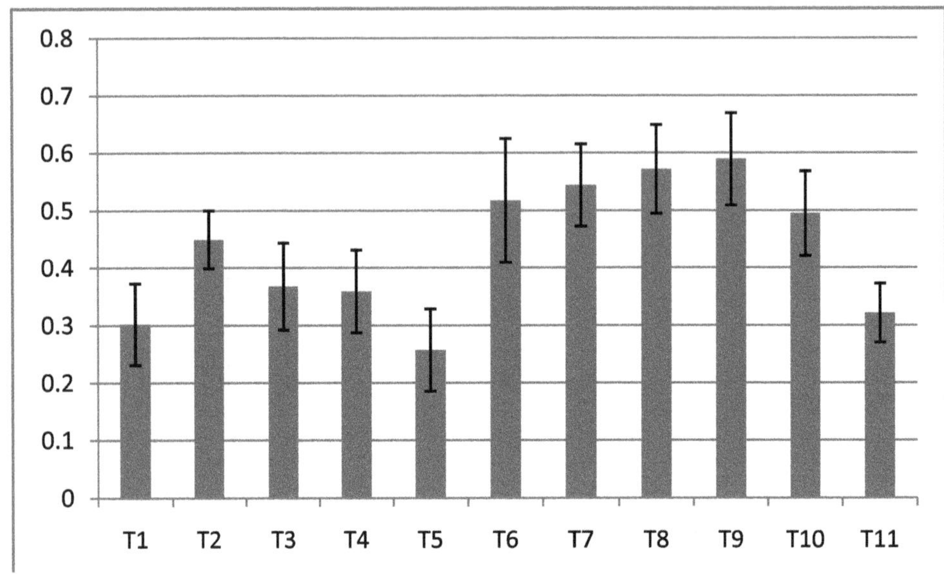

***Fig 20.7** Variation in soil Mg over 34 years of application of fertilizers and manures in groundnut*

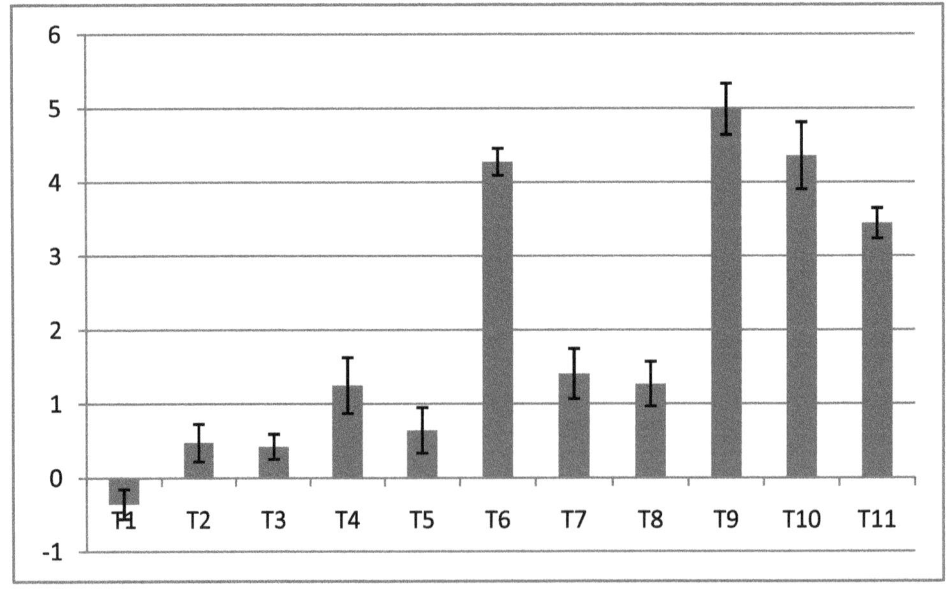

***Fig 20.8** Variation in soil S over 34 years of application of fertilizers and manures in groundnut*

Table 20.3 *Rainfall distribution over 34 years of cropping period*

Year	June	July	August	September	October	Total
1983	121	233.8	191.7	311.4	194.7	1052.6
1984	43.4	204.6	25.8	257.9	50.4	582.1
1985	74.4	126.6	59.6	134.6	175.6	570.8
1986	37.2	59.2	50.1	85.2	188.4	420.1
1987	29	47.2	165.2	91.6	89.1	422.1
1988	40.4	168.1	198	172.7	60	639.2
1989	77.8	155.1	2.8	139.5	18.6	393.8
1990	19.7	68.1	37.4	143.6	262.9	531.7
1991	216.3	153.1	257.2	58.7	363.9	1049.2
1992	41.4	70.5	31.7	114.2	209.7	467.5
1993	37.5	66	135.6	167.8	211	617.9
1994	7.6	182.4	127.5	79.5	169.5	566.5
1995	68.7	189.2	222.1	174	114.5	768.5
1996	256.1	89.1	186.4	238.3	290.2	1060.1
1997	90.8	65.6	58.5	104.7	111.8	431.4
1998	45	107	223.7	156	91	622.7
1999	55.3	59.6	8.8	49	157.2	329.9
2000	36	15.3	177.5	165.4	189.5	583.7
2001	-	-	-	354.0	280.2	634.2
2002	157.4	13.6	83.6	151.2	174.7	580.5
2003	136.9	295.4	106	156.1	197.5	891.9
2004	67.6	76.4	33	239.7	126.2	542.9
2005	41.6	157.6	69.6	65.2	320.8	654.8
2006	78.6	3.6	60	166.6	160.2	469
2007	131.4	37	298.3	84.3	272	823
2008	30.4	45.5	63.1	68.8	194.1	401.9
2009	48.2	27.8	167.2	56.2	23.1	322.5
2010	119.8	162	145.7	130.6	103.1	661.2
2011	44.4	112.6	121	110.8	115	503.8
2012	38	206.5	166	70	204.5	685
2013	23	66	110.2	300.1	194.5	693.8
2014	39	113	244	117	111.5	624.5

Table 20.4 *Effect of long term application of manures and fertilizers on pod yield (kg/ha) over 34 years*

	Mean of pod yield over 34 years				
Treatments	**1981-1990**	**1991-2000**	**2001-2010**	**2011-2014**	**Mean**
T1: Control	1129	1405	895	1536	**1241**
T2: Only FYM	1438	1703	1244	2026	**1603**
T3: Only N	1240	1513	1119	1795	**1417**
T4: Only P	1361	1595	1234	2015	**1551**

Table 20.4 Contd...

	Mean of pod yield over 34 years				
	1981-1990	1991-2000	2001-2010	2011-2014	Mean
T5: Only K	1239	1535	1136	1804	**1429**
T6: Only gypsum	1322	1614	1081	1920	**1484**
T7: N & P	1461	1685	1247	1937	**1583**
T8: NPK	1460	1807	1346	2077	**1673**
T9: NPK+ gypsum	1396	1749	1380	2204	**1682**
T 10: NPK+Lime	1467	1744	1331	2155	**1674**
T11: NPK+Gypum+ Zn	1466	1744	1458	2285	**1738**
Mean	**1362**	**1645**	**1225**	**1978**	**1552**

References

1. *Alok Tiwari, Dwivedi AK, Dikshit PR. Long-term influence of organic and inorganic fertilisation on soil fertility and productivity of soybean-wheat system in a Vertisol. J Indian Soc Soil Sci. 2002; 50: 319-321.*
2. *Anand Swarup. Lessons from Long Term Fertilzer Experiments in Improving Fertilizer Use Efficiency and Crop Yields. Fert News. 2002; 47 (12): 59 - 66 and 71 - 73.*
3. *Bahl GS, Avtar Singh. The inorganic soil phosphorus fractions and available P as influenced by greenmanuring, cropping and P fertilisation. J Indian Soc Soil Sci. 1997; 45: 19-23.*
4. *Bellaki MA, Badanur VP, Shetty RA. Effect of long-term Integrated Nutrient Management on important properties of Vertisol. J Indian Soc Soil Sci. 1998; 46: 176-180.*
5. *Bhandari AL, JK Ladha, H Pathak, AT Padre, D Dawe, RK Gupta. Yield and soil and nutrient changes in a long term rice-wheat rotation in India. Soil Sci Soc Am J. 2002; 66:162-170.*
6. *Duraisami VP, AK Mani. Effect of inorganic nitrogen, coir pith and Biofertilizer on availability and uptake of phosphorus and potassium under maize proceeded with sole sorghum. Madras Agric J. 2000;87(10-12):655-659.*
7. *Fan T, BA Stewart, William A. Payne, Wang Yong, Junjie Luo, Yufeng Gao. Long term Fertilizer and Water Availability Effects on Cereal yield and soil chemical properties in North West China. Soil. Sci. Soc. Am. J. 2005; 69: 842-855.*
8. *Ghosh BN, Ved Prakash, Singh RD. Phosphorus removal and P balance in soybean-wheat cropping sequence under long-term fertilizer experiment. J Interacad. 2000; 4: 260-263.*
9. *Govindarajan SV, HG Gopala Rao. Studies on soils of India. 1978, N fertilization and problems of N availability. Vikas Pub. House, New Delhi. pp. 276-290.*
10. *Jackson M.L. 1967. Soil Chemical Analysis. Prentice Hall of India Pvt. Ltd., New Delhi, India.*

11. Jayasree Sankar, Santhy P, Muthuvel P, Murugappan V. Land degradation due to P depletion under intensive cropping. Agropedology. 2000; 10: 34-39.
12. Kuntal M Hati, Anand Swarup AK, Dwivedi AK, Misra, KK Bandyopadhyay. Changes in soil physical properties and organic carbon status at the topsoil horizon of a vertisol of Central India after 18 years of continuous cropping, fertilization and manuring. Agric Ecosys Environm. 2007; 119:127-134.
13. Lalfakzuala L, Kayang H, Dhkar MS (2008). The effects of fertilizer on soil microbial components and chemical properties under leguminous cultivation. Am Eurasian J Agric Environ Sci 3:314-324.
14. Lohan S, Dev S. 1998. Effect of long-term use of different phosphatic fertilisers on some soil properties in semi-arid soil of Hariyana. In: National Seminar on Developments in Soil Science, 63rd Annual Convention. November 16-19 at Hariana Agricultural University, Hissar.
15. Mairan NR, SG Patil, KG Kachhave. Physico-chemical properties under sorghum-sunflower cropping sequence in Vertisols. J Soils Crops. 2005; 15(2): 352-355.
16. Mandal BC, Saha MN, Adhikari M. Yield of rice and nutrient status of soil due to continuous cropping and manuring in a jute-rice-wheat sequence on Indo-Gangetic alluvial soil. J Indian Soc Soil Sci. 1991; 39: 689-692.
17. Muneshwar Singh VP, Singh, K Sammi Reddy. Effect of integrated use of fertilizer nitrogen and farmyard manure or green manure on transformation of N, K and S and productivity of Rice-Wheat system on a Vertisol. J Indian Soc Soil Sci. 2001; 49(3): 430 435.
18. Muthusamy P, Santhy P, Ramanathan G. Long term use of fertilisers on soil fertility and yield of crops in irrigated Inceptisol. J Indian Soc Soil Sci. 1990; 38:541-542.
19. Muthuvel P, P Kandaswamy, KK Krishnamoorthy. Availability of NPK under long term fertilization. Madras Agric J. 1977; 64: 358-362.
20. Nambiar KKM. All India Coordinated Research Project on Long Term Fertiliser Experiments and its research achievements. Fert News. 1985; 30: 56-66.
21. Nambiar KKM, Ghosh AB. 1984, Highlights on Research of Long Term Fertilizer Experiment of India. LTFE Research Bulletein No. 1, ICAR, New Delhi.
22. Prasad B, Singh RP. Accumulation and decline of available nutrients with long term use of fertilisers, manure and lime on multiple cropped lands. Indian J Agric Sci. 1981; 51: 108-111.
23. Prasad J, Srivastava NC, Mathur BS. Available nutrient status of a continuously fertilised and cropped acid soil. J Indian Soc Soil Sci. 1996; 44: 171-173.
24. Rajendra Prasad, NN Goswami. Soil fertility restoration and management for sustainable agriculture in south Asia. Adv Soil Sci. 1992; 17: 37-71.
25. Rajshree MW, Maya MR, Swati VW, Bharti SB. Residual Effect of long-term fertilizer application of N, P, Zn and FYM on soil properties of Vertisols, Yield, Protein and oil content of Soy bean. J Soils Crops. 1995; 15(1) 155-159.

26. Rao J, RC Dakhore. Nutrient changes in a Vertisol under sorghum-wheat rotations for five years. J Soils Crops. 1997: 108-112.

27. Rego TJ, Rao VN, Seeling B, Pardhasaradhi G, Rao JVDK (2003) Nutrient balances a guide to improving sorghum and ground based dryland cropping systems in semi-arid topical India, Field Crop Res 81:53-68.

28. Rego TJ, Sahrawat KL, Wani SP, Pardhasaradhi G(2007) Widespread deficiencies of sulfur, boron, and zinc in Indian semi-arid tropical soils: on-farm crop responses , J Plant Nutr 30:1569-1583.

29. Santhy P, P Muthuvel, D Selvi. Status and Impact of Organic Matter Fractions on Yield, Uptake and Available Nutrients in a Long – Term Fertilizer Experiment. J Indian Soc Soil Sci. 2001;49(2): 281-285.

30. Singh NP, RS Sachan, PC Pandey, PS Bisht. Effect of decade long fertilizer and manure application on soil fertility and productivity of Rice-Wheat system in a Mollisol. J Indian Soc Soil Sci. 1999; 47(1): 72-80.

31. Singh HP, Sharma KD, Subba Reddy G, Sharma KL (2004) Dryland Agriculture in India. In: Challenges and strategies for dryland agriculture. Crop Science of America and American Society of Agronomy, Madison, USA. Publication No.32, pp 67-92.

32. SPSS (2001) Statistical package. Version 11.0 for Windows (SPSS Inc: Chicago, IL)

33. Srinivasarao Ch, Ganeshamurthy AN, Ali M (2003) Nutritional constraints in pulse production. Indian Institute of Pulse Research (IIPR). Publication No. 8, pp 1-34.

34. Subramanian KS, Kumaraswamy K. Effect of continuous cropping and fertilisation on chemical properties of soil. J Indian Soc Soil Sci. 1989; 37: 171-173.

35. Suresh Lal, Mathur BS. Effect of long term manuring, fertilisation and liming on crop yield and some physico-chemical properties of acid soil. J Indian Soc Soil Sci. 1988; 36: 113-119.

36. Suresh S, S Subramanian, T Chitdeshwari. Effect of Long Term Application of Fertilizers and Manures on Yield of Sorghum (Sorghum bicolor) — Cumbu (Pennisetum glacum) in Rotation on Vertisol under Dry farming and Soil properties. J Indian Soc Soil Sci. 1999;47(2): 272-276.

37. Srinivasarao Ch, Venkateswarlu B, Lal R, Singh AK, Kundu S, Vittal KPR, Balaguruvaiah G, Vijaya Shankar Babu M,Chary Gr, Prasadbabu MBB, Yellamanda Reddy T (2012) Soil carbon sequestration and agronomic productivity of an Alfisol for a groundnut-based system in a semiarid environment in southern India. Eur J Agron 43:40-48.

38. Sudhir K, Gowda SMM, Srikanth K (1998) Long term effects of continuous cropping and fertilizer use on sustenance of productivity and fertility of an Alfisol under irrigated finger millet-hybrid maize systems. In: Longterm soil fertility management through integrated plant nutrient supply. All India Coordinated Research Project on Long-term Fertilizer Experiments, Indian Institute of Soil Science, Bhopal, India, pp 139-145.

39. *Swarup A, Wanjari RH (2000) three decades of All India Coordinated Research Project on long-term fertilizer experiment to study changes in soil quality, crop productivity and sustainability. Indian Institute of Soil Science, Bhopal India.*
40. *Takeda A, Hirafumi Tsukada, Masami Nanzyo, Yuichi Takaku, Toyokazu Uemura, Shunichi Hisamatsu et al. Effect of Long Term Fertilizer Application on the Concentration and Solubility of Major and Trace Elements in a Cultivated Andisol. Soil Sci Plant Nutr. 2005; 51(2):251-260.*
41. *Tandon HLS (1993) Methods of analysis of soils, plants, waters and fertilizers, Fertilizer Development and Consultation Organization, New Delhi, p 150.*
42. *Tolanur SI, Badanur VP. Effect of integrated use of organic manure, green manure and fertiliser nitrogen on sustaining productivity of rabi sorghum- chickpea system and fertility of a Vertisol. J Indian Soc Soil Sci. 2003; 51: 41-44.*
43. *Tolanur SI, Badanur VP. Changes in organic carbon, available N, P and K under integrated use of organic manure, green manure and fertiliser on sustaining productivity of pearlmillet-pigeonpea system and fertility of an Inceptisol. J Indian Soc Soil Sci. 2003; 51: 37-41.*
44. *Venkateswarlu B, Rao AV(1987) Quantitative effects of field water deficits on nitrogen fixation in selected legumes of Indian desert, Biol Fert Soils 5:18-22.*
45. *Virmani SM Pathak P, Singh R (1991) Soil related constraints in dryland crop production in Vertisols Alfisols and Entisols of India. In: soil Related Constraints in Crop Production. Indian Society of Soil Science, New Delhi. Bulletin 15:80-95.*

Chapter 21

Dry Land Technologies for Enhancing Groundnut Productivity under Rainfed Conditions

B. Sahadeva Reddy and A. Malliswara Reddy

ARS Ananthapuramu, Andhra Pradesh.

21.1 Introduction

The growth in vegetable oil supplies is constrained due to their cultivation under shrinking resource base, input supplies and uncertain profitability, while the demand increases at an increasing rate due to rise in *per capita* income and improved standard of living. India occupies a prominent place in global oilseeds scenario with 12-15 percent of area, 6-7 percent of vegetable oil production, and 9-10 percent of the total edible oil consumption and 13.6 percent of vegetable oil imports. In India, oil seeds account for 13 percent of the gross cropped area, three percent of the gross national product and 10 percent of the value of all agricultural products. Nine annual oilseeds, which include seven edible oilseeds, *viz.,* groundnut, rapeseed-mustard, soybean, sunflower, sesame, safflower and niger and two non-edible crops, *viz.,* castor and linseed are grown in the country. Despite having the largest area under oilseeds in the world (28.53 m ha during 2013-14), India currently imports about 53 percent of total oil requirement (11.06 million tonnes) at an exchequer of Rs.56907 crores (2013-14). The proportion of import has increased from a meagre three percent in 1970-71 to almost 53 percent in 2013-14.

India grows oilseeds on an area of 27.25 million hectares, with production of 29.38 million tonnes and productivity of 1077 kg per hectare (quinquennium ending 2013-14). The average yields of most of oilseeds with castor as an exception *vis-à-vis* world are invariably low (Table 21.1) since they are confined to resource poor farmers in nutrient starved soils with small and fragmented land holdings primarily under rainfed ecosystem (72.8 percent) (Arvind Kumar, 2015).

***Table 21.1** Oilseeds productivity (kg/ha) in India vis-à-vis World (2012)*

Crop	India	World	Country with higher average productivity
Groundnut	996	1676	4699(USA)
Rapeseed-Mustard	1262	1873	3690 (Germany)
Soybean	1353	2374	2783 (Paraguay)
Sunflower	651	1482	2494 (China)
Sesame	402	518	1315 (Egypt)
Safflower	590	961	1489 (Mexico)
Castor	1592	1162	1455 (India)
Linseed	503	752	1358 (Canada)
Oil palm fruit	12380	14323	21901 (Malaysia)
			Source: FAO STAT, 2012

Groundnut *(Arachis hypogaea L.)* in an annual legume native to South American. It is now grown in most tropical, sub-tropical and warm temperate regions of the world between 40°S latitudes. Groundnut is unique among all the leguminous crops and designated as **"Wonder legume"** in that after flowering, fertilization and fruit set, the pegs (Gynophores) elongate and penetrate the soil (Positively geotropic) where the fruit enlarges and matures. Commercially, Groundnut is the world's fourth most important source of edible oil and third most important source of vegetable protein. Groundnut oil is considered as stable and nutritive as it contains right proportions of saturated fatty acid namely, Oleic acid (40-50%) and unsaturated fatty acid like linoleic acid (25-35%). Groundnut is one of the principal oilseed crops of the world. Groundnut is an important food legume of the world. It is grown in 26.4 m ha spread over 85 countries with the production of about 35.6 mt along with an average productivity of 1347 kg/ha. The major production are China, India, Indonesia, and the USA which together account for two-thirds of the world output. Other important producers are Nigeria, Senegal, Sudan and Argentina. India occupies 30% of global area (7.6 million ha) and 22% (7.8 million tons) of total groundnut production. Though India ranks first in the area and second, next to China in total production, its rank in productivity is tenth in the world. The low yield levels are attributed to cultivation of the crops mostly in rain fed areas and in marginal lands with low inputs, inadequate plant stand, imbalance fertilization, inadequate plant protection measures and improper post harvest handling. All these factors collectively contribute to low and unstable yields of groundnut in the country.

The major groundnut area (6.0 million ha) in India comprises marginal lands where the crop is grown under rainfed conditions. Eighty percent (6.0 million ha) of the total groundnut area (7.6 million ha) is confined to five states *viz.*, Gujarat, Andhra Pradesh, Tamil Nadu, Karnataka and Maharashtra, which account for 84% (6.5 million tones) of the total production (7.8 million tons).Groundnut is cultivated during *kharif* season (June-July to September-October) mostly under rainfed conditions with few protective irrigations. The crops is also grown during *Rabi* season (October-November to February-March) under residual moisture/minimal irrigation situation.

21.2 Major Production Constraints

The groundnut cultivation is unique in that it is being cultivated under four different production system/seasons like rainy (*Kharif*), winter/post rainy (*Rabi*), summer and spring fitted into different cropping pattern/sequences. Hence, the problems and constraints are multi-varied and multi-faceted according to the production system involved. However, the major production constraint of this crop has been the confinement mainly to dry or rain dependent areas (~ 6.0-6.5 million ha). Groundnut is energy rich C_3 crop, yielding about 50% of oil but grown under "energy starved conditions" in the dry and marginal lands.

The biotic and abiotic stresses also limit the production during the rainy season. Among the abiotic stress, soil moisture deficit stress at various stages of crop growth during rainy season and low temperature during germination and vegetative stage but high temperature during the pod filling and maturation stage during summer, hampers the productivity. In addition, salinity and acidity build-up and micronutrient deficiency in selected pockets of the country further limits the production. Among the biotic stresses, the foliar fugal (Early Leaf Spot, Late Leaf Spot, Rust, Alternaria), viral (Peanut Bud Necrosis, Stem Necrosis) soil borne (Stem rot, collar rot and pod rot complexes) diseases and the insect pests like defoliators (*Spodoptera, Helicoverpa,* Red hairy caterpillar and Leaf miner) and sucking pests (Jassids, Aphids, Thrips) are the major ones that limit groundnut production and productivity. In addition the pre and post harvest aflatoxin contamination in the kernels and meal also reduces the quality as well export values.

21.2.1 Approaches to Production Enhancement and Yield Stabilization in Groundnut

Necessarily, the challenges are enormous, not only to arrest the highly fluctuating production and productivity mainly in the rainfed eco-system, but also to make groundnut farming sustainable, competitive and remunerative. This calls for a concerted effort and focus to develop cost effective, sustainable, cropping system oriented production packages and development of input (mainly water and nutrients) use efficient cultivars with inbuilt resistance/tolerance to economic diseases and pests.

In India, since the *kharif* groundnut occupies a major share of total groundnut area and will continue to remain so especially when there are no viable alternative to match groundnut under the harsh environments, the primary target, should therefore, be to focus on a 250-300 kg yield increase per ha during *kharif* from the present level of 1 ton/ha. This very modest increase should be possible by developing broad based, location specific, and water use efficient cultivars and incorporation of traits associated with high water use efficiency into the proven stable cultivars.

The rainfed production system being basically risk prone and high cost associated with its cultivation, the crop management practices should be low input demanding and less risk prone. An important way to minimize the risk and enhance the resource use efficiency is to develop viable groundnut based cropping system involving potential and compatible field crops like Pigeon pea, Sesame and Castor under dry lands. The dynamics of nutrient and water uptake and use under different cropping systems should be worked out. Similarly, while targeting the breeding efforts to address economic pests through host resistance, integrated and cost effective approach (IPM) should go hand in hand in addition to the other pests management practices.

21.3 Seeds and Sowing

1. Time of sowing
2. Seed size
3. Method of sowing

21.3.1 Time of Sowing

Timely sowing helps in achieving optimum utilization of seasonal rainfall, reduces the incidence of pests and diseases, and escapes terminal drought. For groundnut, July first FN is the most appropriate time for sowing whereas for redgram, castor and cluster bean June month is the suitable time for sowing. August first FN is the optimum time to sow the sorghum under dryland *alfisols* of Anantapur district of Andhra Pradesh (Sahadeva Reddy *et al.*, 2011)

Time of sowing substantially influences the pod yield of groundnut. Experiments were conducted under irrigated conditions for 3 years at fortnightly interval throughout the year. The results revealed that pod yield of groundnut was high when sowings were taken up in June, July, November and December.

Similarly, several experiments were conducted for suitable time of sowing during *kharif* under rainfed condition. June, July months are best suited for sowing of groundnut throughout the country sowings are the best in Rayalaseema region of Andhra Pradesh (Table 21.2).

Table 21.2 *Pod yield (kg/ha) of groundnut as influenced by date of sowing*

Date of sowing	No. of pods / plant	Pod yield
July 11	8.4	1893
July 31	5.7	865
August 12	3.7	700

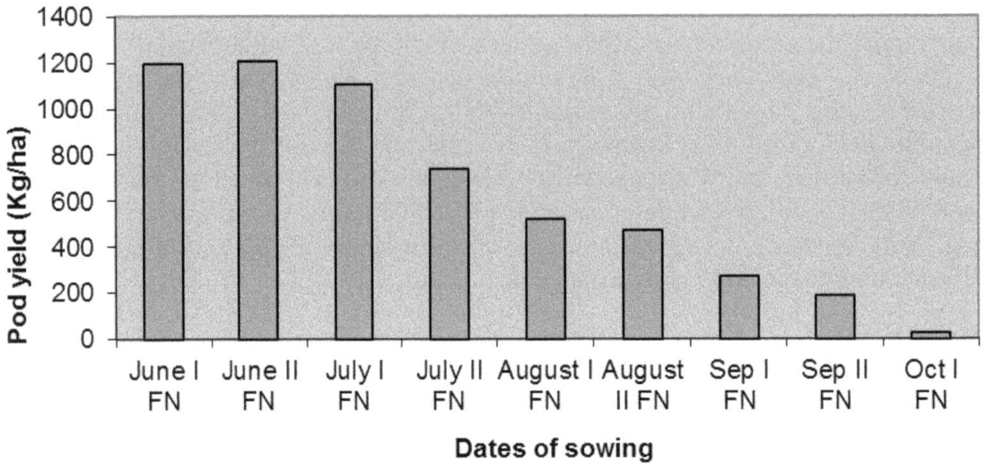

Fig. 21.1 **Groundnut pod yield vs dates of sowing**

21.3.2 Seed Size

Seed cost constitutes 36% of total cost of cultivation of groundnut. Hence, possibilities are explored to reduce cost of cultivation by reducing the quantity of seed. Experiments were conducted with different categories of seed viz., assort, bold, medium, small and partially shriveled seed with test weight of 38, 42, 28, 20 and 24 g of respectively. The results revealed that significant differences were not observed in root growth of seedling at 10 days after sowing (DAS). Emergence percentage at 10 DAS was significantly highest in medium, small and shriveled seed resulting in significantly high seedling vigour index in shriveled and small seed than bold seed (Table 21.3). Abnormal seedlings observed at 10 DAS were more in bold seed (34.9%) than small and shriveled seed (10.6%). Area of first and second leaves was significantly high in bold seed than small seed (Table 21.4). No difference was observed in number of pegs/plant, root length, root density/plant at harvest with different categories of seed. Pod yield and shelling percentage were not influenced by seed size (Table 21.5). No difference was observed in seed size distribution and emergence percentage in the progeny obtained from different sizes of seed used for sowing. Seed alone contributes from 14.6 (small seed) to 44.8 (bold seed) percent in total cost of cultivation. Pod yield did not differ significantly by using small seed compare to bold seed in groundnut. Higher net returns and benefit cost ratio were realized with small and medium compare to bold seed. Thus cost of cultivation was reduced by using small seed to sow groundnut crop Small and medium sized seeds germinated better and required less weight of seeds per unit area (Sahadeva Reddy *et al., 2009)*.

Table 21.3 *Seedling vigour attributes as influenced by different categories of seed in groundnut at 10 days after sowing*

Category of seed	Emergence percentage		Seedling vigour index		Abnormal seedlings percentage	
	2001	2002	2001	2002	2001	2002
Assorted	77.6	62.8	391	330	18.0	23.2
Bold	79.4	60.4	438	394	9.5	34.9
Medium	83.0	73.4	507	408	0.0	28.4
Small	80.0	78.6	453	456	4.2	12.4
Shriveled	80.0	75.4	385	477	4.2	10.6
S.Em±	2.9	2.25	44	20		
CD 5%	NS	6.7	NS	59		

Table 21.4 *Leaf area, number of leaves and variation in attainment of reproductive growth stages of groundnut as influenced by different categories of seed*

Category of seed	Leaf area (cm^2) First leaf		Root length (cm) Second leaf		No. of leaves/plant at 45 DAS		Coefficient of variation in age of plants (%)	
	2001	2002	2001	2002	2001	2002	2001	2002
Assorted	6.17	9.78	11.20	15.72	4.9	8.6	43.8	32.8
Bold	8.50	11.50	12.68	17.24	4.6	9.6	60.6	53.7
Medium	6.46	8.76	10.15	13.08	4.3	7.1	55.6	36.2
Small	6.25	7.75	9.19	12.68	3.9	8.0	58.3	40.6
Shriveled	6.37	8.67	10.04	14.28	4.2	8.4	57.5	37.9
S.Em±	0.32	0.52	0.51	0.80	0.2	0.7		
CD 5%	0.96	1.60	1.52	2.40	0.6	NS		

Table 21.5 *Effect of seed size on root length, root dry weight, pod and haulm yield of groundnut*

Category of seed	Pod yield (kg/ha)		Haulm yield (kg/ha)	
	2001	2002	2001	2001
Assorted	1097	626	2135	698
Bold	1180	552	2302	802
Medium	1169	568	2229	698
Small	1001	594	1801	687
Shriveled	987	626	1958	740
S.Em±	105	39	89	111
CD 5%	NS	NS	267	NS

***Fig 21.2** Different sizes of kernel in groundnut*

***Fig 21.3** Bold seeded kernel in groundnut (Farmers practice)*

21.3.3 Method of Sowing

21.3.3.1 Ananta Groundnut Planter

Small and marginal farmers are unable to maintain cattle pair due to shortage of fodder. With the limited amount of soil moisture available for sowing, the sowing operation has to be completed within short period. The seed has to be placed in the moist zone for proper germination. Tractor drawn Anantha groundnut planter (8 row) was developed and popularized to mechanize the groundnut sowing for timely operation with mechanical advantage.

Sowing of groundnut within short period of time and without loss of soil moisture can be done by establishing custom hiring centers.

***Fig 21. 4** Tractor drawn Anantha Planter*

***Fig 21.5** Groundnut crop sown with Ananta planter*

21.3.3.2 Crop Establishment Techniques in Castor, Pigeon Pea and Groundnut using Aqua Seed Drill

Timely sowing of crops such as castor and Redgram can be done in wide spaced dryland crops by way of pot watering @ 500 ml per hill. Groundnut crop can established with aqua seed drill. Timely sowing can be done with aqua planter developed by ARS, Anantapur by using 1600 to 3200 lt per acre depending up on the moisture content of the soil.

***Fig 21.6** Aqua planter*

***Fig 21.7** Groundnut sown with aqua planter*

21.3.3.3 Tractor Drawn Planter Cum Herbicide Boom Sprayer

It is used for simultaneous sowing of groundnut and herbicide spraying for control of weeds in initial stages. It is provided mainly with piston type pump for developing the sufficient pressure, six flat pattern nozzles arranged on the boom to develop spray, two drums of each 200 liters capacity, seed hopper of 40 kg capacity and inclined plate metering mechanism for sowing of seed. The pump gets the power from PTO shaft of tractor for pumping of the chemical from tank into nozzles through pipes. It can sow seed in 8 rows at a time and spray the herbicide at the application rate of 200 liters per acre. The field is 5 – 6 ha/day and maintains required seed rate of 100 kg/ha, seed to seed distance 10 cm in a row and depth of seed placement 4 – 5 cm. Its cost is about Rs. 65,000/-.

***Fig 21.8** Tractor drawn planter cum herbicide boom sprayer*

21.4 Cropping Systems

21.4.1 Crops and Varietal Choice

The crops in an intercropping system should together utilize the available resources synergistically and efficiently. In a two crop intercropping system, a tall crop matches with short growing, and a deep-rooted crop with a shallow rooted one, so that both above and below ground resources could be exploited. There should be a difference of about 25% in crop duration among the intercrops in a system for effective complementarity.

21.4.2 Plant Population and Row Arrangement

Intercropping gives higher yield advantages when the total population in the system is higher than that of the individual sole crops. The optimum plant densities and row ratios recommended for various intercropping systems are given in the table.

21.4.3 Time of Planting

Under rain fed conditions, it is desirable to plant crops at earliest opportunity to allow maximum growth during the most favourable periods. In general, all the component crops may be planted simultaneously to achieve higher yields. Staggered planting of component crops is found to be advantageous to safe guard drought against uncertain.

Rainfall regimes to utilize late season rainfall and also to reduce competition between the component crops. Staggered planting of pigeon pea, castor, sesame and sunflower (30 days after groundnut sowing) with groundnut has been recommended for Saurashtra region of Gujarat. This practice allows operation of bullock drawn implements for inter cultivation in groundnut. However, staggered planting has limited application in semi-arid tropics because rainfall is uncertain in rain fed areas. Besides, the opportunity to establish a second component crop is risky; it is very difficult to establish a second component crop with mechanical equipment in an established crop Simultaneous sowing is common practice in intercropping of cereals (Pearlmillet, sorghum and maize), sesame and short duration pules (green gram and blackgram) with groundnut. In case of cotton + groundnut, planting of groundnut is done 25-30 days after cotton sowing.

The main cropping system in rainfed areas is mono cropping of groundnut and intercropping of groundnut with redgram or castor. A variety of intercropping systems in India are given table 21.6.

Table 21.6 *Different intercropping systems being followed in India*

State	Region	Intercropping	Row ratio of base crop (g'nut) and intercrop
Andhra Pradesh	Telangana	Groundnut + Pigeonpea	6 : 1
		Groundnut + Pearlmillet	3 : 1
		Groundnut + Cowpea	6 : 1
	Rayalaseema	Groundnut + Pigeonpea	11 : 1
		Groundnut + Castor	11 : 1
Jarkhand		Groundnut + Maize	6 : 1
		Groundnut + Sesame	4 : 1
Gujarat	Saurashtra	Groundnut + Castor	3 : 1
		Groundnut + Pigeonpea	3 : 1
		Groundnut + Pearlmillet	1 : 1 or 2 : 1
		Groundnut + Sesame	1 : 1
		Groundnut + Sesame	1 : 1
	Rajkot	Groundnut + Castor	3 : 1
		Groundnut + Sunflower	1 : 1
Karnataka		Groundnut + Pigeonpea	4 : 1
		Groundnut + Cotton	3 : 1
		Groundnut + Sunflower	4 : 2
		Groundnut + Sorghum	6 : 1
		Groundnut + Chilli	4 : 2
Kerala		Groundnut + Tapioca	3 : 1
Maharashtra		Groundnut + Pigeonpea	6 : 1
		Groundnut + Pearlmillet	4 : 1
		Groundnut + Sorghum	4 : 1
		Groundnut + Sunflower	3 : 1
Madhya Pradesh	Rewa region	Groundnut + Maize	3 : 1
		Groundnut + Pigeonpea	3 : 1
		Groundnut + Pearlmillet	4 : 1
		Groundnut + Sesame	4 : 1
Tamilnadu		Groundnut + Pigeonpea	6 : 1
		Groundnut + Cotton	5 : 1
		Groundnut + Sorghum/ Pearlmillet	6 : 1
		Groundnut + Sesame	4 : 1
		Groundnut + Blackgram	6 : 1
		Groundnut + Rice	2 : 2
Uttar Pradesh		Groundnut + Pearlmillet	4 : 1
		Groundnut + Sesame	4 : 1
		Groundnut + Pigeonpea	6 : 1
		Groundnut + Rice	2 : 2
Orissa and West Bengal, NEH region		Groundnut +Maize	3 : 1
		Groundnut + Rice	3 : 2
		Groundnut + Soybean	2 : 1

21.5 Nutrient Management

N, P, K, Ca and S nutrition are important in groundnut cultivation. N, P, K status soils during 1980 and 2000 years revealed that there was no change in organic matter content of soil, but build up of phosphorus and deletion potassium was observed in *alfisols* of Andhra Pradesh.

The results from long term experiments conducted at ARS, Anantapur under rainfed condition revealed following:

1. There was no response to application of farm yard manure in many years.
2. There was no response to application of recommended dose fertilizer in 11 years out of 18 years.

In several of experiments conducted in groundnut further revealed that the response to rhizobium is erratic and response is very less. Response to gypsum was observed where the sulphur is deficient and yield levels are over and above 1500 kg/ha. The recommended fertilizer doses in different states are given Table 21.7. The recommended N ranges from 8 to 40 kg/ha, phosphorus from 16 to 80 kg/ha and potassium from 0 to 60 kg/ha.

***Table 21.7** Fertilizer doses (kg/ha) recommended for major groundnut growing states during kharif*

States	N	P_2O_5	K_2O	Gypsum
Andhra Pradesh	20	40	20	---
Jharkhand	25	50	20	---
Gujarat	12.5	25	0	---
Karnataka	15	30	25	250
Maharashtra	20	40	0	---
Madhya Pradesh	20	40	20	---
Orissa	8	16	15	---
Punjab	15	40	25	125
Rajasthan	20	60	0	---
Tamilnadu				
a) Aliyarnagar-Polachi tract	11	22	33	250
b) Tindivanam	10	20	45	250
Uttar Pradesh	15	30	45	200
West Bengal				
a) Alluvial soil	20	40	00	250
b) Red soils	40	80	60	250

- **Soil test based fertilizer application**

  Soil test based fertilizer application for groundnut crop was suggested to reduce cost of cultivation. Phosphorus and potassium fertilizers were applied in full dose if soil tests values was low, half of the recommended fertilizers are sufficient when soil test values are medium and no fertilizer if the soils are high in P_2O_5 and

K_2O. Pod yield of groundnut was increased by 5 to 22% due to soil test based fertilizer application in different on- farm demonstrations besides saving of Rs. 555 to 995/ha.

Fertilizer doses are recommended based on soil test values as indicated Table 21.8.

***Table 21.8** Recommended doses (RD) based on soil test values*

Soil test value	N	P_2O_5	K_2O
Low	Full RD	Full RD	Full RD
Medium	Half RD	Half RD	Half RD
High	No application	No application	No application
			RD: Recommended dose

- **Sheep penning for increasing the productivity of groundnut under rainfed situations**

  Four years (2009 - 10 to 2012 - 13) of study on sheep penning ($2m^2$/animal) overnight on the field resulted that 13% increase in pod yield and 41% increase in haulm yield was recorded compared to contol. Available Potassium in soil was significantly increased in sheep penning treatment compared to control. Available micronutrients in soil such as Fe and Zn also increased in sheep penning treatment compared to control.

21.6 Rain Water Management

Water is a prime natural resources and an important input for assured agricultural production. Recognizing the vital role of water as critical input in agricultural production huge.

***Table 21.9** Water requirement of groundnut at various states of India*

Place/state	Soil type	Water requirement (mm)	References
Kanpur, UP	Loamy soil	550	Mehrotra et al. (1967)
Bhavanisagar, Tamil Nadu	Sandy Loam	600	Chandra Mohan (1970)
Rajendranagar, AP	Medium black clay	655	AICSRWMSS (1972)
	Sandy loam	576	AICSRWMSS (1972)
Chiplima, Orissa	Sandy loam	470	Taha and Gulati (2001)
Bhubaneswar, Orissa	Sandy loam	322	Satpathy and Patro, (1992)
Ludhiana, Punjab	Loamy sand	510	Cheema et al. (1977)
Tirupati, AP	Sandy Loam	420	NARP-T (1983)
Junagadh, Gujarat	Medium black clay	356-363 (Kharif)	Singh and Devi Dayal (2002)
Bangalore, Karnataka		597-696	Ramachandrappa and Kulkarni (1993)

21.6.1 Water Requirement

Water amount of water used by the crop is determined by the potential evapo-transpiration during the crop period and the degree of soil cover. The water requirement reaches a maximum during flowering and continues up to pod formation. On an average, groundnut crop requires about 550-650 mm of water depending upon the soil types, season and groundnut cultivars. Water requirement of groundnut worked out by different scientists at various places in the country is given in table 21.9.

21.6.2 Critical Period for Moisture Stress

In general, groundnut plant is considered to be more drought tolerant than other legume crops. However, some of the phenophases of crop growth are more sensitive to moisture stress than other growth stages. In general, water deficit during the vegetative period causes delayed flowering and maturity and reduce growth and yield. Water deficit during flowering cause flower drop or impaired pollination and during pod setting reduce pod weight. The early part of the pod setting is particularly sensitive to water deficit.

Considerable work has been done to find out the most sensitive growth stage to moisture stress. (Sankara Reddy 1976) reported that flowering, peg penetration and early pod formation are the moisture sensitive period for groundnut since most of the dry matter is accumulated during these periods where as (Subhash Babu *et al.* 1977) and (Raju *et al.* 1981) found pegging to pod formation stage as the most sensitive period. In Gujarat, Khatri and (Patel, 1983) while working on a model to predict moisture availability during the crop growth period, observed that germination, vegetative development and full pegging and pod development were the critical period for moisture availability. However, many workers observed that peg initiation to pod development was the most sensitive crop stage for moisture deficit in the soil (Patel and Golakia, 1991; Sakarvadia and Yadav, 1994). In a lysimeter study, (Golakiya and Patel 1992) reported that the growth of groundnut crop was curtailed maximum with moisture stress at flowering, but yield reduction was the maximum at pod development stage due to reduced crop reproductive efficiency. (Singh and Devi Dayal 2003) reported that the most sensitive crop stage for moisture stress is pod development followed by peg initiation and peak flowering. (Nautiyal *et al* 2002) observed that moisture stress at vegetative phase was even advantageous and imposing moisture stress about 20 days enhanced pod yield Moisture stress at vegetative stage followed by two frequent irrigation at 5 days interval creates favourable condition for early flush of synchronous flowering, more peg to pod ratio resulting in enhanced yield. Higher pod yield with irrigation at 10 day interval than irrigating at critical growth stages was also reported from Khargone, Madhya Pradesh (Mishra, 1997). Four irrigations at vegetative, flowering, pegging and pod development stages in Kharif groundnut at Sriganganagar, Rajasthan gave the highest mean pod yield of 1.60 t/ha along with the maximum consumptive use of water and WUE (Sing *et al.*, 1994). In Punjab, providing three irrigations, each one at flowering, pegging and pod formation was considered to be the optimum for Kharif groundnut. However, if only one irrigation has to be provided, and then apply only at pod formation in the June sown crop and at peg formation in July sown crop (Sing *et al.*, 1986). Drought is the main constraint in reducing the yield of groundnut from 5 t/ha to 0.5 t/ha. Drought reduces leaf area expansion and photosynthesis resulting in poor growth. The crop growth stages are also delayed by drought. Groundnut can tolerate drought up to 55 days in the early stages.

Some of the drought management practices are:

1. Contour cultivation.
2. Application of groundnut shells as mulch @ 5 t/ha at 10 days after sowing.
3. Application of sand @ 40 t/ha during summer once in 4 years.
4. Water harvesting and supplemental irrigation.

Table 21.10 *Influence of sand on infiltration of rain water and pod yield of groundnut*

Sand (Tonnes/ha)	Infiltration rate (mm/h)	Pod yield (kg/ha)	
		1987	1988
O	26.1	1661	2337
10	29.6	1744	2297
20	31.3	1775	2546
30	46.4	1759	2957
40	38.7	1848	2756
CD (P=0.05)		NS	432

Table 21.11 *Yield of groundnut (kg/ha) as influenced by drought management practices*

Treatment	Early stress		Late stress	
	Pod	Haulm	Pod	Haulm
Control	1269	1029	1462	1198
Mulching with groundnut shells (5 t/ha)	1484	1016	1683	1289
Kaolin spray 5% during stress	1445	1172	1693	1159
Lime spray (1%) during stress	1369	1107	1687	1198
Urea spray (2%) after stress	1350	1016	1738	1211
Soil application of urea after stress (20 kg N/ha)	1367	1050	1824	1368
Supplemental irrigation (5 cm)	1413	1094	1863	1485
Mulching, kaolin (5%) during stress and urea spray (2%) after stress	1415	1040	1783	1250
C.D @ (5%)	NS	NS	195	196

21.6.3 In-Situ Moisture Conservation

The soils of Anantapur district are predominantly red sandy loams occupying 78 percent of the total area. The remaining are shallow black soils. The red soils arc shallow in depth (10 – 30 cm deep) with compact sub soil that restricts the root growth. The topography is undulated with an average slope of 2 percent. The water holding capacity of the soils is low (40 mm/30 cm of soil depth). On an average five runoff events occur inspite of low rainfall resulting in loss of soil. Insitu soil and moisture technologies conservation are to be developed to overcome the problem. In this direction, experiments planned were technologically relevant to the situation.

Field experiments conducted at Agricultural Research Station, Anantapur on the effect of in-situ moisture conservation methods in inter bunded area on run off and groundnut yields revealed that it could be possible to reduce soil and water losses by adopting insitu conservation practices like dead furrows at 3.6 m interval across the slope (25-30 days after sowing) with the receipt of rains, coupled with compartmental bunding with 20 m length and 10 m width before germination of crop.The soil moisture availability increased considerably by adopting these practices, which lead to the increased groundnut and yields.

When the technology on *in situ* moisture conservation practices was verified over farmers' fields (1995-98), contour cultivation + compartmental bunding resulted in 13% increased yield over farmers' practice of sowing across the slope only.

Table 21.12 *Groundnut pod yield (kg/ha) as influenced by in situ moisture conservation practices*

Treatment	1995	1996	1997	1998	Mean
Contour cultivation + compartmental bunding	1610	885	439	2143	1269
Farmers' practice (sowing across the slope only)	1415	863	188	1940	1102

These practices are quite simple, require no input and can be implementable.

Table 21.13 *Groundnut yields as influenced by water harvesting practices (in situ)*

Treatments	Yield q/ha					
	1978-79	1979-80	1980-81	1981-82	1982-83	1983-84
1. Sowing across the slope (check)	8.1	5.65	2.19	Drought year	5.10	13.0
2. Dead furrow at 1.8 m interval	---	5.43	2.33		5.09	13.6
3. Dead furrow at 2.4 m interval	8.9	---	2.70		5.44	13.8
4. Dead furrow at 3.6 m interval	---	5.97	2.35		6.65	14.6
5. Dead furrow at 4.8 m interval	8.0	---	---		---	---
	0.3	---	---		0.35	0.33
	0.8	NS	NS		1.06	0.6

Forming dead furrows at 3.6 m intervals gave higher yields of groundnut compared to sowing of groundnut across the slope (check).

Fig. 21.9 Dead furrows in groundnut with 3.6 m interval

Fig. 21.10 Contour bund

Intercropping of groundnut with redgram + Mixed pulses for moisture conservation and higher yields

In Anantapur district, though the annual rainfall is low (540 mm), the intensity is high, causing severe soil erosion and runoff losses. On an average five runoff incidents are recorded during the season. The lands are sloppy and shallow in depth with moisture storage capacity of only 40 mm/30 cm soil depth with no carrying over capacity. The available soil moisture can be increased by reducing runoff losses and allowing more opportune time for rainwater to infiltrate.

In Anantapur district there is a practice of growing groundnut intercropped with redgram in Anantapur district. However, the system does not prevent the runoff losses. By improving the existing intercropping system with the innovative practice of mixing redgram with other pulses or by practicing intercropping with cowpea alone reduces the runoff to a greater extent as this acts as a vegetative barrier. This practice does not involve any additional expenditure. Besides this, this system increases the total returns of the farmer and reduces runoff and soil losses. In this direction, the experiments carried were significantly relevant and technologically feasible to the existing climatic conditions and poor economic status of farmer as the developed practice is low input technology. This intercropping system has gone as a recommendation from the university.

Results of the experiments (1996 and 1997) revealed that intercropping of groundnut either with cowpea or mixed pulses (redgram + cowpea + horsegram) in 15 : 1 row ratio not only arrested the runoff, but also increased the total returns by 17.8 and 8.5 percent respectively over sole crop of groundnut. Results of the experiments under AICAR Project (1998–2001) also revealed that intercropping of groundnut with mixed pulses reduced the runoff and soil losses, thereby conserving more soil moisture .Groundnut pod equivalents were also high with the same system as compared to sole groundnut.

***Table 21.14** Impact of different inter crops with groundnut on unoff and crop yields*

Intercrop	Mean Runoff (%)	Pod yield (Kg/ha)		Total Returns (Rs/ha)		% Increase Over control
		1996	1997	1996	1997	
Horsegram	23.4	532	193	11156	5327	0.6
Redgram	61.2	531	210	11067	5655	3.2
Cowpea	0.0	615	270	12683	6425	17.8
Mixed pulses	6.3	588	272	11909	5808	8.5
Pure crop	31.3	559	243	10857	5413	---

***Fig. 21.11** Groundnut with mixed pulses (pigeonpea + cowpea) intercropping*

Selection of suitable bund for prevailing soil and climate conditions is most important hence contour bunds and graded bunds with different grades are tested. It was observed that 0.2% and above grades caused erosion at the discharged ends. Among various types of bunds tested, contour bund proved most effective conserving soil and water and for increased yields (Table 21.15).

***Table 21.15** Effect of different types of bunds on yield of crops*

Treatment	Groundnut pod yield (kg/ha)	Setaria grain yield (kg/ha)
Coconut bund	905	895
0.1% graded bund	840	840
0.2% graded bund	805	840

In an effort to identify a suitable cross section for a contour bund. It was found that a cross section of 0.63 sq.m. was superior than 0.45 sq.m. or less because the latter is subjected to breaching even with low intensity rain.

Evaluation of different horizontal spacing's (25m to 125 m) between contour bunds indicated that closure the spacing, higher was the yield. Considering the operational convenience. 50 m horizontal spacing was recommended.

Table 21.16 *Mulching and in-situ Rain Water Harvesting (RWH) through micro catchments on soil water dynamics and groundnut yields (2007-10)*

S.No.	Treatment	Pod Yield (kg ha^{-1})					
		2007-08	2008-09	2009-10	Mean	Mean BC ratio	Mean RWUE (kg/ha-mm)
			No Mulch				
1	Control	944	1202	662	936	1.85	2.16
2	30cm apart 4 lines	949	1176	694	939	1.87	2.18
3	30cm apart 6 lines	983	1444	677	1034	2.09	2.39
4	45cm apart 4 lines	998	1204	706	969	1.91	2.24
5	45cm apart 6 lines	1125	1205	709	1013	2.04	2.31
			Mulch				
6	Control	1096	1286	587	989	1.99	2.23
7	30cm apart 4 lines	1175	1415	724	1104	2.23	2.52
8	30cm apart 6 lines	1127	959	724	936	1.86	2.13
9	45cm apart 4 lines	1062	1044	742	949	1.88	2.18
10	45cm apart 6 lines	1081	1097	655	944	1.84	2.14
	CD (Main)	NS	NS	NS	NS	--	--
	CD (Sub)	NS	NS	NS	NS	--	--
	CD (M x S)	NS	306.5	NS	NS	--	--

In pooled data of three years of experiment on *in-situ* moisture conservation indicated no significant differences in pod and haulm yields among all the treatments. However, higher B:C ratio was observed in micro catchments arranged after every 4 rows of groundnut and 30 cm apart in plots under mulch treatment, whereas in without mulch treatment higher pod yield was obtained in micro-catchments arranged after every six rows of groundnut with 30 cm inter catchment distance.

Fig. 21.12 Micro catchments for every 4 rows

Fig. 21.13 Mulching on micro catchments

Rain water harvesting and supplemental irrigation to groundnut

During 2003, the supplemental irrigation through sprinkler at 2 spells, 10 mm each has increased the pod yields by 40% and haulm yield by 33% with supplemental irrigation. During 2004, pod yield of rainfed crop is 315 kg/ha compared to 698 kg/ha with supplemental irrigation. Similarly haulm yield was higher (1843) with supplemental irrigation than with rainfed crop (1250 kg/ha). During 2005, by giving two supplemental irrigation on (26 DAS) flowering and (63 DAS) pod initiation stages of groundnut crop, the pod yield increase was 60% and haulm yield increase was 93% over rainfed crop.

Table 21.17 *Effect of Dikes in alfisols for increasing rain water productivity under groundnut based farming system*

Treatment	Kharif 2009	Kharif 2010	Kharif 2011	Mean
Control (normal sowing)	288	1948	546	927
4 Rows 1 dike	334	2046	628	1002
6 Rows 1 dike	342	1974	527	947
4 Rows 1 dike with mulching	301	1792	546	879
6 rows 1 dike with mulching	367	2034	556	985
Paired row	757	1448	559	921
Paired row with earthing up	781	1428	539	916
Paired row with mulch	767	1601	463	944
CD (0.05)	58.4	NS	124	NS

Dikes has no effect on soil and water conservation in groundnut.

21.6.4 Deep Ploughing

Summer deep ploughing is one of the most important low cost and traditional practices in rainfed areas for *in situ* harvesting of rain-water. The practice is generally followed in

April-May when fields are ploughed deep across the slope and also through criss-cross ploughing. It has been observed that summer ploughing is very effective for *in situ* rain water harvesting during initial storms of monsoon with low to moderate intensity specially in light porus soils. Deep ploughing conserves the rain water into the deeper soil layer and ensures better growth of *rabi* pulses. Apart from moisture conservation, summer ploughing also reduces weed intensity and insect pest disease incidences in the following *rabi* pulses in Bundelkhand region, summer ploughing helps in successful cultivation of *rabi* pulses like lentil and chickpea. Experiments conducted under National Agricultural Technology Project, ICAR revealed that in shallow soils, higher yields were recorded with deep ploughing, residue incorporation and conservation furrow as compared to farmer's practice of criss-cross ploughing with wooden *desi* plough. At Varanasi, summer tillage with Mould Board plough following by line sowing of pigeonpea and sesame resulted in doubling of pigeonpea equivalent yields over farmers' practice of summer criss-cross ploughing with country plough and broadcast sowing. Ploughing with tractor drawn mould board plough resulted in 12% additional moisture storage in the profile 105 reduction in run off from treated fields.

Deep ploughing significantly increased the grain yields of castor, bajra during 1972-73. In 1973-74 and 1974-75 kharif there was significant increase in the grain yield of castor, redgram, bajra and groundnut. Setaria in 1972-73 and sunflower in 1973-74 did not give any significant increase in grain yield.

Fig. 21.14 Sub-soiler

Fig. 21.15 Mould board plough

21.6.5 Ex-Situ Water Harvesting

21.7 FARM Pond Designs and Dimensions

A farm pond of 150 m^3 capacity, with side slopes of 1.5:1 is considered sufficient for one hectare of catchment area in black soils with provision of emptying the pond after it is filled up to accommodate subsequent events of runoff. While, deciding the size, i.e., storage capacity of pond meant for collecting runoff for supplemental irrigation, factors such as area to be irrigated, the estimated soil moisture deficiency which is required to be replenished, the storage losses in pond and the efficiencies of conveyance and application system will have to be considered.

Although circular ponds have the geometrical advantage in that they offer highest storage capacity, square or rectangular ponds are preferred. Stepping method of construction of farm pond has several advantages. A silt trap of suitable dimensions should be formed in water course just near the entrance to the inlet in order to have a check on bed load entering the pond. The length of a silt trap should be slightly greater than the width of the water course and depth may be about 0.5 to 1 m with side slopes of 1.5:1.

In an experiment conducted at Agricultural Research Station, AICRPDA, Anantapur, it was found that, a farm pond of dimensions 10 m length, 10 m width and 2.5 m depth with side slope of 1.5:1 is sufficient for the catchment of 5 ha of field. Once it was filled with runoff, 2,50,000 liters of water will be available for supplemental irrigation.

Farm pond of 250 m^3 capacity (size of 10 m × 10 m with 2.5 m depth) with side slopes of 1.5:1 is sufficient for catchment area of 10 ha. Soil + Cement lining with 6:1 ratio was found very effective in reducing seepage losses.

21.16 Lined farm pond

21.7.1 Storage and Lining Requirements and Materials

In *alfisols* higher infiltration rate is a major problem for storing water in ponds. Preliminary investigations in small ponds revealed that it was cheap and practicable to reduce the seepage loss effectively by plastering the pond with alkali soil (5 cm thick). These results were also confirmed on a larger farm pond of 23 m x 16 m (552 m^3). The water stored in the pond was used for retting mesta sticks harvested from one hectare area.

Water loss due to seepage and percolation in red sandy loam soils of Anantapur is 158.6 $l/m^2/day$. A 7.5 cm sodic soil layer on sides and bottom initially reduced the losses to 107.1/1/m^2/day. But sodic soil on sides creeped down on filling with water, bringing the loss to normal rate of 157.21/m^2/day. The 7.5 cm concrete to the bottom and cemented sides and 7.5 cm soil + cement layer reduced the loss to 72.3 L/m^2/day initially. After application of cement water on sides, the loss was finally reduced to 20.4 and 5.29 L/m^2/day, while lining of farm ponds with Cuddapah slabs reduced the loss of water to 1.64 L/m^2/day. Cuddapah slabs was found better in lowering (1:3:6) and also with supplemental irrigation.

In an experiment to study the different lining materials indicated that , soil + cement lining with 6:1 ratio was found very effective in reducing seepage loses in sandy loamy soils at Anantapur (Andhra Pradesh)

In alluvial sandy loam to loamy sand soils of Vasad (Gujarat) and red sandy loam soils of Bangalore (Karnataka), a soil + cement (8:1) mixture proved to be the best lining material (Radder et al.,1995). In laterite silty clay loam soils of Ooty (Tamil Nadu) and medium black soil of Kota (Rajasthan), bitumen and high density polyethylene sheet, respectively reduced seepage losses considerably. Cement + bricks reduced seepage of 10.0 cm day^{-1} to 2.0 cm day^{-1} at Dehradun (Uttar Pradesh) having silty loam to silty clay loam soils. At Solapur (medium deep black soils) lining with saline sodic soil gave the least losses per day (362 L day^{-1}), as against a loss of 818 L day^{-1} with soil + cow dung + straw treatment.

Fig. 21.17 *Cement + soil 1:6*

Fig. 21.18 *Cement + soil 1:8*

Fig. 21.19 *Kadapa slabs*

Fig. 21.20 *Plastic sheet*

21.7.2 Efficient use of Water through Supplemental Irrigation

In semi-arid areas, though the rainfall is low, its intensity is high, causing erosion and 20-40% of rainfall is lost through runoff. Water requirement of rainfed crops like groundnut, sunflower, chickpea, cotton, sorghum, pearl millet are in the range of 300 to 550 mm. The rainfalls in drought prone districts are also in the same range and more. However, crop failure occurs due to prolonged dry spells. However, rainfall also occurs with high intensity resulting in runoff. Hence, water harvesting, storage and supplemental irrigation during dry spells helps in reducing adverse effects of drought in rainfed crops. One supplemental irrigation of at critical stages of crop growth increases the yield considerably. Introduction of high value crops such as hybrid cotton under protective irrigation further helps in enhancing the income of dryland farmers.

Table 21.18 *Yield and yield attributes of groundnut as influenced by supplemental irrigation*

Parameters	Rainfed	Supplemental Irrigation	% increase
Pod yield (kg/ha)	515	686	33.2
Haulm yield (kg/ha)	727	824	13.34
Shelling percentage	59.5	66.0	10.92

Application of 1.0 cm irrigation to groundnut during stress at pod development stage gave 1023 kg/ha compared to control of 820 kg/ha. There is an yield increase of 24.8% with the supplemental irrigation. The capital (Rs. 50,000/-) recovery can be made over a period of 4 to 5 years by adoption of this technology with a groundnut pod yield increase by 20 to 30 percent annually.

At Anantapur yield of groundnut can be increased by 33% with one supplemental irrigation of 10 mm during pod filling stage (Table 21.18). In Maharashtra, many workers reported flowering and pod development stages as the most critical crop growth stages for moisture stress (Patil and Gangawane, 1990; Patil and More 1993) while, pod initiation to pod development (70 days after sowing to harvest) was reported to be the most sensitive growth stage from Karnataka (Ramachandrappa *et al.*, 1992).

21.7.3 Time of Irrigation

In groundnut, it was found that one life saving irrigation at 45th day (peg penetration) resulted in 28 percent increase in yield over unirrigated control. Irrigating at this stage was found superior to the treatment which received one protective irrigation (2 cm depth) when the symptoms of wilting were seen. In groundnut, supplemental irrigation during pegging to pod development stage is beneficial. Flowering and pod formation stages were reported as the most sensitive growth period by (Sarma and Sivakumar 1989), (Reddy and Reddy 1993) and (Reddy *et al.* 1996). Proving one or two protective irrigations during moisture stress in rain fed conditions in Tamil Nadu (Gnanamurthy *et al.*, 1992);

and Subramanian *et al.,* 1999) A.P. (Reddy, 1994), Maharashtra (Meher *et al.,* 1992; Uke *et al.,* 1995) and Punjab (Kachroo and Walia, 1991) enhances the productivity and the sustainability of groundnut production.

21.7.4 Depth of Irrigation

The quantity of water given for supplemental irrigation is very important. The depth of irrigation has a direct bearing on the yield of crops, depending on the time of irrigation, available soil moisture in the root zone, and critical stage of crop growth. As the amount of water available in farm ponds is limited, one should work out the minimum quantity of water required where the crop highly responds. For example, it was studied at Anantapur; there was no significant difference in pod yield of groundnut between 10, 15 and 20 mm depth of irrigation water. Hence, 10 mm depth of water is to be given.

Table 21.19 *Effect of depth of irrigation through sprinkler irrigation in groundnut*

Depth of irrigation (mm)	Pod yield (kg/ha)	
	2001	2002
0	1146	493
5	1198	530
10	1395	650
15	1400	695
20	1453	--
CD 5%	157	133

21.7.5 Method of Irrigation

The method of irrigation plays a vital role in increasing the efficiency of water use. For a given depth of water, drip method of irrigation saved 50% of pond water compared to modified sprinkler and surface method of application. In groundnut 10 mm supplemental irrigation through sprinkler irrigation is more efficient compared to surface irrigation.

Fig. 21.21 *Sprinkler irrigation to groundnut*

Fig. 21.22 *Drip irrigation to groundnut*

21.8 Mobile Irrigation Units

Lorry Tanker (12000 l) fitted with motor both for lifting water and supplemental irrigation to crop can be used. The water available at one placed will be filled by the motor and can be transported where supplemental irrigation by sprinklers with motor of the tanker. Even tractor tanker water with 4000 l capacity can be used for either drip or sprinkler irrigation.

21.9 Soil Amendments

21.9.1 Sand Application to Increase Infiltration of Rainwater

In the district, though the annual rainfall is low, the intensity is high, causing severe soil erosion and runoff losses. On an average five runoff incidents are recorded during the season. The lands are sloppy and shallow in depth with moisture storage capacity of only 40 mm/30 cm soil depth with no carrying over of moisture. The available soil moisture can be improved either by allowing more opportune time for rain water to infiltrate or by arresting runoff losses. **This can be achieved, by simple agronomic manipulations like application of sand so as to increase infiltration**. In this direction, the experiments carried out on sand application are apt and scientifically relevant and technologically feasible to the existing situation.

Besides this hard setting sandy loam soils form the major chunk of groundnut growing soils in Anantapur district. Surface crusting is one of the important problems of the soils, limiting rain water infiltration, delayed and poor emergence of the crop. Pegging and pod development are also affected due to soil crustation, leading to reduced groundnut pod yields. Application of easily and cheaply available sand is an effective way of improving soil physical conditions, in turn leading to reduced runoff losses and higher soil moisture availability.

Results of the field experiments (1997–2000) revealed that the soil water in the profile was high during critical dry spells when sand was applied @ 40 t/ha once in four years, due to more infiltration of rain water. Crust strength and soil resistance were low with the application of sand. Besides improvement in soil physical condition, groundnut pod yields were found to increase upto 150 kg/ha.

21.10 Other Practices

21.10.1 Antitranspirants

Spraying urea (2%) after prolonged dryspell showed 48% an 10% increase in groundnut pod and haulm yields over control in spray. In another study spraying di ammonium phosphate (2%) on groundnut immediately after the dryspell recorded better yields than water spray.

Among different drought management practices, pod and haulm yields of groundnut were higher with application of groundnut shells (5t/ha) at 10 days after sowing. The other drought management practices which influenced the pod yield were sand application @ 40 t/ha before sowing, lime spray (1%) during late stress, spraying of urea (2%) after late stress. In late planted sunflower, kaolin spray (5%) during stress increased yield over control. In Pomegranate the number and the average weight of fruit was higher with kaolin spray (9%) followed by liquid paratin (0.005%).

21.11 Weed Management

Weed management especially under rainfed conditions is well maintained by farmers. Indigenous implements are available with farmer both in Andhra Pradesh and Gujarat. Intercultivation is generally done from 25 to 30 days after sowing and second intercultivation is being done at 40 to 45 DAS. Inter cultivation followed by line weeding is generally practiced. Weeding becomes a problem once in 4 to 5 years when continuous rains are received in early stages of the crop.

Chemical weed control is possible under irrigated conditions. The most effective herbicides are pendimelhalin @ 1 kg *ai*/ha, Butachlor @ 1 kg *ai*/ha. Pre emergence application of these herbicides controls weeds effectively for 30 days.

21.12 Harvesting

Harvesting is done when 75% of pods are matured. Duration of most of the groundnut varieties are from 95 to 120 days. Delay in maturity occurs with severe drought. The duration of the TMV-2 variety is 105 days but crop was harvested from 90 to 145 days depending on the distribution of rainfall. If rains are received around 90 to 95 days, farmers pull the plants, even if it is 10 days early since pulling of plants become difficult if soil dries. Spraying of fungicides for the control of late leaf spot helps in harvesting at correct time and even if it is delayed for a week to ten days, there will be no yield loss.

***Table 21.20** Gross and net returns as influenced by late leaf spot management and extending pod growth period*

Treatment	Gross returns (Rs./ha)	
	1998	1999
T1 : Control	16218	15338
T2 : Fungicidal spray	17523	15850
T3 : Spray + harvest extended by 10 days	18233	12348
T4 : Spray + harvest extended by 20 days	13642	12655
C.D @ 5%	2674	1493

21.12.1 Dry Pod Thresher

The farmers are accustomed to hand stripping of groundnut pods from the plants at a later date after harvesting. The harvested crop is transported and heaped in the farm houses and stripping will be done as and when the labour is available. The stripping capacity of one labour varies from 40 to 50 kg/day and stripping cost comes to Rs. 200 per 100 kg of pods.

The groundnut threshers are introduced in the region to make the stripping operation easy and economical. The threshers are having high stripping capacity depending upon the crop yield. The threshers can be used immediately 5 days after harvesting without transportation and heaping. The operation is simple and repairs if any can be done by local artisans. Due to early threshing of groundnut crop, the haulms are better preferred by the animals.

The groundnut threshers are having high stripping capacity of 200 to 300 kg/hr. The cost of operation comes to Rs. 33 per 100 kg of pods. The thresher can be moved from one field to another on pneumatic tyres with a pair of bullock or by a tractor. By introduction of this new technology, it is estimated that Rs. 5.25 crore is saved to the farmers of Anantapur region. The quality produce free of aflatoxin and better price can be achieved by the farmers due to timely an early threshing.

Fig. 21.23 Groundnut dry pod thresher

21.12.2 Wet Pod Thresher

The farmers are accustomed to hand stripping or dry pod threshing with thresher of groundnut pods from the plants at a later date after harvesting. The harvested crop is transported and heaped in the farm houses and stripping will be done as and when the labour is available by dry pod thresher or hand stripping. Immediately after harvest of groundnut (October to November) coincides with cyclonic rains resulting heavy losses to the farmers. Even after the rains the pods were rotten with aflatoxin. With the help of wet/ fresh pod thresher groundnut crop can be threshed immediately after harvest of groundnut.

It is used for separating pods form plants immediately after harvesting of groundnut crop. The farmer no needs to wait for drying of crop after harvesting for threshing. It is operated with 540 RPM PTO shaft of 35 – 45 H.P tractors. It consists of frame, feed hopper, drum type threshing cylinder, concave, oscillating sieves, blower and pod outlets. The groundnut pods are separated from plants between cylinder and fixed concave by rubbing and *impact forces and fall down through cleaning sieves. Blower separates plant stalk and leaves from pods. Cleaned pods come from main outlet. It has the output capacity of 300 kg/h with 96 % threshing efficiency. It is very useful where the influence of north east monsoon will be more and continuous rains occur at harvesting time.*

Fig. 21.24 *Groundnut wet pod thresher*

21.13 Integrated Farming Systems

The farming systems in rainfed areas are quite diverse with a variety of crops, cropping systems, agro forestry and horticulture and livestock production. The integration of various crops and animals enables synergistic interaction, which has a great total contribution than the sum of their individual effects (Edward *et al.,* 1988). As only mono cropping of groundnut is being followed in the district due to low and erratic rainfall situations and with only 14-17 weeks of rainy period, the employment opportunity is only a short period of 3-4 months. On farm studies also indicates that under integrated farming systems, farmers benefitted by getting additional income of Rs. 200 per sheep per month after harvest of groundnut.

Sheep fed with available dry fodder such as groundnut haulms and horsegram *bhusa* for 5 months recorded higher body weight and profitable for sheep rearing after harvest of rainfed groundnut. So, rainfed groundnut and sheep rearing is profitable farming system compared to groundnut crop alone for scarce rainfall zone of Andhra Pradesh. (Sahadeva Reddy et al., 2011)

Fig. 21.25 Stall feeding of sheep with green and dry fodder

References

1. AICSRWMSS 1972. Annual Report 1971-72. All India Coordinated Scheme for Research in Water management and Soil salinity, Chiplima center, ICAR, New Delhi.
2. Arvind Kumar 2015. Dr.M.V.Rao Lecture on Stakeholders Access to Oilseed Technologies towards Self Sufficiency, Feb.19th 2015 at Directorate of Rapeseed Mustard Research, Rajastan
3. Chandramohan 1970. Efficient water management for groundnut. Indian Farm. 19 (10): 9-11.
4. Cheema S S, Minhas K S, Tripathi H P and Kundra H C 1977. The effect of applying phosphorous and nitrogen to groundnut under different regimes of soil moisture. J.Research Punjab Agril. Univ.14 (1):9-14.
5. Gnanamurthy P, Hendry X and Balasubramanian P 1992. Influence of protective irrigation, recommended method of weed control, fertilizer application and plant protection measures on the yield of rainfed groundnut. Indian J. Agron. 37(4): 855-856.
6. Golakiya B A and Patel M S 1992. Growth dynamics and reproductive efficiency of groundnut under water stress at different phenophases. Indian J.Agric.Res.26 (4): 179-186.
7. Kachroo D and Walia AS 1991.Effect of pre and post monsoon schedule of irrigation to groundnut crop sown on different dates. Indian J.Agron.36 (Supplement), 304-305.
8. Kar D N, Ray M and Garnayak L M 2001. Determination of crop coefficient of groundnut for modified penman method. Environment and Ecology 19 (2) : 474-476
9. Khatri T J and Patel R M 1983. Use of yield prediction model for scheduling irrigation in groundnut. Indian J. Agric. Sci. 53(9): 831-832.

10. Meher A L, Ingle V N,Lanjewar M M,Bambal S M and Panchabhai S S 1992. Response of pre-monsoon sown groundnut varities to irrigation levels.J.Soils and Crops 2(1): 39-41

11. Mehrotra C M, Mathur A K and Ali A 1967. Irrigation studies in groundnut. Proc. Symp. on water management, 40th ANNUAL Session, Central board of irrigation and Power, New Delhi.

12. Mishra C M 1997. Effect of irrigation and phosphorous on the yield of summer Ground nut. Madras Agric. J. 84(2), 113-114.

13. Nautiyal P C, Joshi Y C and Devi Dayal 2002.Response of groundnut to deficit irrigation during vegetative growth .Deficit irrigation practices. Water Report 22, Food and Agricultural Organization of the United Nations, Rome: pp 39-46.

14. Padmalatha Y, Ravindranatha Reddy B, Sahadeva Reddy B, Bhargavi K and Narayana Swamy G 2009.Technical Bulletin on Technologies for Dryland Agriculture in Scarce Rainfall Zone of Andhra Pradesh, ARS, Anantapur

15. Patel M S and Golakiya B A 1991. Effect of water stress on yield and yield attributes of groundnut (Arachis hypogaea L.).Madras Agric.J.78 (5-8): 178-181

16. Patil B P and Gangawane S B 1990. Effect of water stress imposed at various growth stages on yield of groundnut and sunflower. J. Maharashtra Agric.Univ. 15 93): 322-324.

17. Patil S M and Morey D K 1993. Growth and yield of summer groundnut as influenced byantitranspirants and water management practices. Journal of Soils and Crops, 3(1):18-22.

18. Raju A P, Sankara Reddi GH and Bapi Reddy T 1981. Studies on productivity of rainfed groundnut under moisture stress using neutron moisture probe. J. Nuclear Agriculture and Biology 10 (4):131-132.

19. Ramachandrappa B K, Kulkarni K R and Nanjappa H V 1992. Stress day index for scheduling irrigation in summer groundnut. Indian J.Agron.37(2): 276-279.

20. Ramachandrappa B K and Kulkarni K R 1993A. Flowering pattern,production and pod yield in summer groundnut as influenced stress at different stages.Current Research (Univ. Agric. Sci. Banglore) 22 (1):11-13.

21. Reddy C R and Reddy S R 1993. Scheduling irrigation for peanuts with variable amounts of available water. Agricultural Water Management, 23(1): 1-9.

22. Reddy P R R, Reddy B B , Rao V P and Sarma P S 1996. Effect of water deficit at different crop growth periods on growth and yield of groundnut. J. Research APAU, 24(1-2):147-149.

23. Reddy T Y 1994. Management practices for amelioration of moisture stress in groundnut in arid tropics. Indian J.Agron.39 (3): 489-490.

24. Sahadeva Reddy B, Malliswara Reddy A, Padmalatha Y and Shakuntala P 2009. Influence of seed size in different varieties of groundnut on its productivity. Legume Research 32(4): 298-300.

25. Sahadeva Reddy B, Malliswara Reddy A and Ravindranatha Reddy B 2011. Effect of sowing time on productivity and economics of different kharif crops in scarce rainfall zone of Andhra Pradesh. Indian Journal of Dryland Agriculture and Development, 25 (2): 68-72

26. Sahadeva Reddy B, Malliswara Reddy A, Jalajakshi K, Ravindranatha Reddy B and Vijaya Sai Reddy M 2010. Studies on groundnut based farming systems for scarce rainfall zone of Andhra Pradesh. Progressive Research 6(2): 177-179.

27. Sahadeva Reddy B, Vijayasai Reddy M, Vijayasankar Babu M, Madhusudhan Reddy K, Radha Kumari C, Padmalatha Y, Ravindra Chari G and Srinivasa Rao Ch 2014. Achievements of Four Decades (1971-2011) of Dryland Agriculture Research, AICRPDA, ARS, Anantapuramu

28. Sakarvadia H L and Yadav B S 1994. Effect of water stress at different growth stages of groundnut on yield and nutrient absorption.L.Indian Soc. Soil Sci. 42 (1): 147-149

29. Sankara Reddy G H 1976. Management of irrigation water. Technical Bulletin No. 33, Andhra Pradesh Agricultural University, Rajendranagar, Hyderabad, pp.175.

30. Sarma P S and Sivakumar M V K 1989. Response of groundnut to drought stress in different growth stages. Agric. Water Manag.15 93):301-310.

31. Satpathy M R and Patro N 1992. Moisture utilization pattern in rabi groundnut at different irrigations and phosphorous levels. Environ. ECOL. 10(1):194-196.

32. Sing Ved, Sharma S K, Verma B L, Singh R K and Singh V 1994. Effect of irrigation schedule and phosphorous fertilization on yield, water use and phosphorous uptake by groundnut in north-west Rajasthan. Indian J. Agron. 39(1):58-61.

33. Singh Y V and Devi Dayal 2002. In situ water harvesting for sustainable production of rainfed groundnut in semi arid region. Extended summary, Vol 2, 2nd Intl. Agron. Cong. November 26-30, 2002, New Delhi,pp 1312-1313.

34. Subhash Babu P, Sankara Reddi G H and Nageswara Reddy M 1977. Studies on irrigation scheduling of groundnut based on cumulative evaporation from free water surface. Current Agric. 1: 5-18.

35. Subramanian K, Arulmozhi N, Kalaiselvan P and Kalaiselvan P 1999. Effect of different irrigation layout irrigation and fertilizer levels on the yield of rainfed groundnut. Crop Res. 18 (1): 22-24

36. Taha M and Gulati JML 2001. Influence of irrigation on yield and moisture utilization of groundnut. Indian J. Agron. 46 (3): 523-527.

37. Uke P C, Ingle V N, Ingole P G, Malvi S D and Peshne Y P 1995. Effect of advance sowing date and protective irrigation on growth, yield and quality of kharif groundnut variety JL-24.PKV Research Journal 19 (2):169-170.

Chapter 22

Water Management Practices for Groundnut

Reddy S. R. and Deepthi Kiran Y.

Department of Agronomy, ANGRAU, S.V. Agricultural College Campus, Tirupati, Andhra Pradesh.

Globally, India ranks first in area and second in production of groundnut. India accounts for about 31 percent of the total groundnut area (Table 22.1) in the world (24.6 M ha) and 22 percent of the total production (35.7 M t.) In India, it is largely grown in Andhra Pradesh, Gujarat, Maharashtra, Tamil Nadu and Karnataka. In Andhra Pradesh it is largely cultivated in Rayalaseema districts (Anantapur, Kadapa, Kurnool and Chittoor districts).

***Table 22.1** Area, production and productivity of groundnut.*

Particulars	Area (M ha)	Production (M t)	Productivity (kg ha^{-1})
World	24.6	35.7	1451
India	7.6	7.8	1026
AP	1.92	1.71	924
AP (rabi)	*0.3*	*0.57*	*1919*

Groundnut can be cultivated in *kharif, rabi* and summer seasons. In *kharif,* it is mainly grown as rainfed crop. In *rabi* and summer, it is grown with irrigation. Average productivity of groundnut is relatively low as it is mostly grown rainfed during *kharif.* Productivity of *rabi* and summer crop is 2 to 3 times higher than that of *kharif* rainfed crop.

Challenge for the coming decades will be the task of increasing food production with less water, particularly in countries with limited water and land resources. Crop water management is a key area to optimise crop production with limited and dwindling water supplies. Many factors of climate, soil, crop and management practices influence crop water requirements and the decision making process of when to irrigate, how much quantity of irrigation water to apply and which method of irrigation is appropriate to satisfy the crop water needs.

Sound irrigation water management programme requires dependable information on crop water requirement, irrigation requirement, irrigation scheduling and methods of irrigation to crops in order to ensure that production proposed and yields predicted can be achieved with maximum water use efficiency. Fig 22.1 depicts planning for efficient water management programme leading to high water use efficiency and enhanced crop productivity.

22.1 Determination of Crop Water Requirements

Determination of crop water requirements (CU/ETc) is one of the basic needs for planning of any irrigation project. Methods for determining crop water requirements depends on the desired level of accuracy, availability of equipment and technical know how. Commonly mentioned methods for determining crop water requirements in the literature are: Direct measurement of crop water requirements (CU/ETc) and Computation of ETc from climatic data (empirical methods).

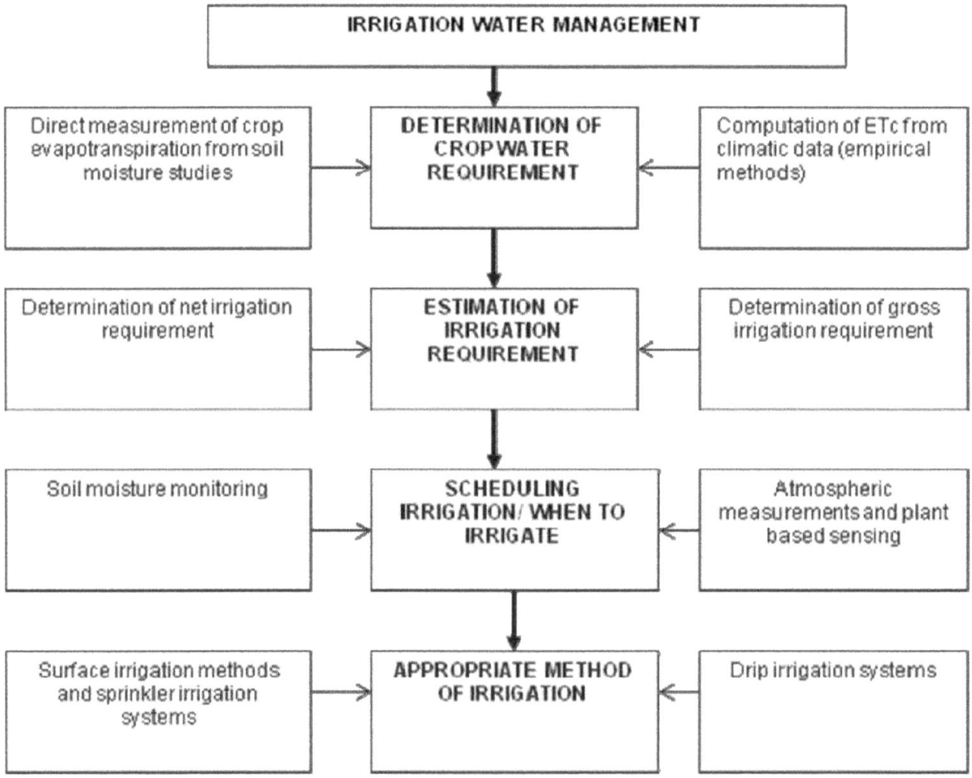

Fig. 22.1 Planning for efficient use of irrigation water

Direct measurement of crop water requirements (CU/ETc)

- Transpiration ratio method
- Depth-interval-yield approach
- Lysimeter experiments
- Soil moisture depletion studies
- Field experimentation
- Soil-water-balance method.

Computation of ETc from climatic data (empirical methods)

- Blaney-Criddle method
- Radiation method
- Pan evaporation method
- Modified Penman method
- Penman- Monteith method.

22.1.1 Direct Measurement of Crop Water Requirements (Cu/Etc)

The amount of water required to produce a unit quantity of dry matter is called **transpiration ratio**. Transpiration ratio measured in green house with pot culture studies differ under natural conditions.

In **depth-interval-yield approach**, irrigations are scheduled with different depths of water at different intervals between irrigations. Total amount of water used for highest yield is taken as crop water requirement. It may give an idea of crop water needs, but not reliable estimate due to differences in climate in different seasons.

22.1.1.1 Lysimeter Experiments

Different terms in the soil-water balance equation can be determined with greater accuracy using lysimeters, where the crop grows in isolated tanks filled with either disturbed or undisturbed soil.

Lysimeters are either weighable or non-weighable. Weighable lysimeters provide information about the change of water storage for any time period; non-weighable lysimeters collect only the water percolating from the soil column.

In precision weighing lysimeters, where the water loss is directly measured by the change of mass, evapotranspiration can be obtained with an accuracy of a few hundredths of a millimeter and small time periods such as an hour can be considered. In non-weighing lysimeters the evapotranspiration for a given time period is determined by deducting the drainage water, collected at the bottom of the lysimeters, from the total water input.

A requirement of lysimeters is that the vegetation both inside and immediately outside of the lysimeter be perfectly matched (same height and leaf area index). This requirement has historically not been closely adhered to in a majority of lysimeter studies and has resulted in severely erroneous and unrepresentative ETc and Kc data. As lysimeters are difficult and expensive to construct and as their operation and maintenance require special care, their use is limited to specific research purposes.

Evaporation measurements are obtained from weight change of the lysimeter, using the water balance equation:

$$ET = \text{Weight change} + \text{Water added} - \text{Percolation}$$

22.1.1.2 Soil Moisture Depletion Studies

Consumptive use (CU) of water (ETc) can be determined by estimating changes in soil moisture from root zone depth at particular intervals during the crop season. It is usually employed in fairly uniform soils, where the groundwater table is deeper enough to influence soil moisture in the crop root zone depth.

$$CU = \sum_{i=1}^{n} \frac{M1i - M2i}{100} \text{ X BDi X Di}$$

where,

CU = Water use from root zone two successive sampling periods (mm)

n = Number of soil layers considered in root zone depth D

M1i = Soil moisture percentage at first sampling in ith layer

M2i = Soil moisture percentage at second sampling in ith layer

BDi = Bulk density of ith layer (g cm-3)

Di = depth of ith soil layer (cm).

Effective precipitation, groundwater contribution and deep percolation losses of water, if any are to be considered for reliable estimates of crop water requirements. Seasonal consumptive use can be obtained by adding CU values of each sampling interval.

22.1.1.3 Field Experimentation

Crops are grown with recommended management practices, including an ideal irrigation schedule for optimum yields. Irrigation water applied, effective precipitation and soil profile moisture contribution are measured. Seasonal crop water requirements can be computed by adding these components.

$$ETc = IR + ER + \left[\sum_{i=1}^{n} \frac{Mbi - Mei}{100}\right] \text{ X BDi X Di}$$

where,

ETc = Water requirement of the crop (mm)

IR = Amount of irrigation water applied (mm)

ER = Effective rainfall during the season (mm)

n = Number of soil layers considered in root zone depth D

Mbi = Soil moisture percentage at the start of the season in ith layer

Mei = Soil moisture percentage at the end of the season in ith layer

BDi = Bulk density of ith layer (g cm-3)

Di = depth of ith soil layer (cm).

22.1.1.4 Soil-Water Balance Method

Soil-water balance, like a financial statement of income and expenditure, is an account of all quantities of water added, removed or stored in a given volume of soil during a given period of time. Soil-water balance equation in its simplest form of expression is:

Change in soil-water = Inputs of water – Losses of water

Inputs (addition) of water to the soil: Water is usually added to the soil in three measurable ways - precipitation (P), irrigation (I) and contribution from the groundwater table (C). Contribution from the groundwater will be significant only if the groundwater table is near the surface. Inputs of water can be presented as:

Water inputs = P + I + C

Loss (removal) of water: Water is removed from the soil through evapotranspiration (ET) and deep drainage (D). Further, a part of the rainwater received at the soil surface may be lost as surface runoff (RO). Losses of water from soil can then be represented as:

Water losses = ET + D + RO

Soil-water balance: Principle of water balance method is shown in Fig 22.2 Change in the soil-water content which is the difference between the water added and water withdrawn will be:

Change in soil-water = (P + I + C) – (ET + D + RO)

Soil-water refers to the amount of water held in the root zone at a given time. The change in soil-water from one measurement to another depends on the contribution of components in the equation. Suppose the amount of water in the root zone at the beginning is M1 mm and at the end of a given period is M2 mm, thus the equation is expressed as:

$$M1 - M2 = (P + I + C) - (ET + D + RO) \text{ or}$$

$$M1 + ET + D + RO = M2 + P + I + E$$

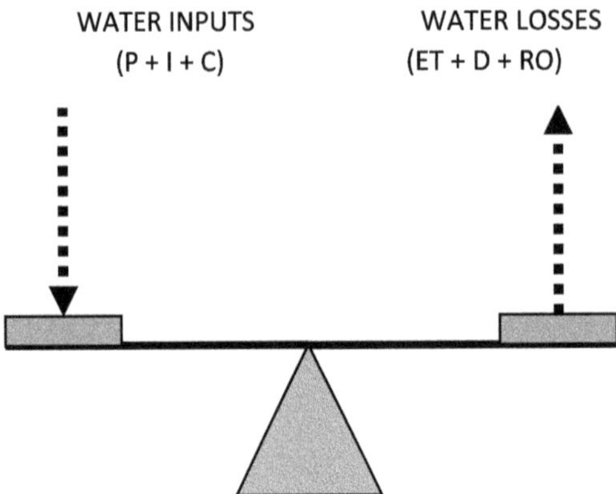

Fig. 22.2 Principle of soil-water balance method

With the help of this equation one can compute any one unknown parameter in the equation if all others are known.

Quantitative data on rainfall (P) evapotranspiration (ET), deep drainage (D) and soil moisture at a given time (M1 or M2) for different locations are necessary for estimating reliable crop water requirements.

Soil-water balance computation: An example using the soil-water balance equation to appreciate the usefulness of this model is presented below:

Given:

Soil = Vertisol, crop = sorghum, period = 01 to 31 Aug, area = 2 ha

Soil moisture in the profile on Aug 01 (M1) = 300 mm

Precipitation or rainfall (P) = 70 mm

Irrigation (I) = nil

Contribution from groundwater (C) = nil

Runoff of 200 cubic m from 2 ha field (R) = 10 mm

Deep drainage (D) = nil

Soil-moisture in the profile on Aug 31 (M2) = 250 mm.

Estimate:

Evapotranspiration (ET) from the field during 01 to 31 Aug?

Solution:

$$M1 + ET + D + RO = M2 + P + I + E$$

$$300 + ET + 0 + 10 = 250 + 70 + 0 + 0$$

$$ET = 320 - 310 = 10 \text{ mm.}$$

Crop water requirement for a specified period can be obtained as the product of:

$$ET \times K_{pan}$$

where,

ET = Evapotranspiration (mm)

K_{pan} = Pan coefficient (usually 0.7).

22.1.2 Computation of Etc from Climatic Data

It is necessary to conduct field experiments for precise data on crop water requirements. In view of difficulties associated with direct measurement of crop water requirements, empirical methodologies have been developed to predict the crop water requirements, primarily, based on climatological data and crop factors. The FAO group of scientists screened 31 empirical formulae for predicting ET and recommended four formulae for use under different climatic conditions through FAO IDP 24 (Doorenbos and Pruitt 1975).

Advances in crop water research and more accurate procedures in determining crop water requirements revealed need for revision of FAO IDP 24, particularly in relation to the estimation of reference evapotranspiration (ETo). An expert consultation was held in May 1990 in Rome to advice on options to revise the procedures introduced in FAO IDP 24 (Kassam and Smith 2001). The expert consultation reached unanimous agreement in recommending the Penman-Monteith approach as the best performing method to estimate evapotranspiration of a reference crop ETo and published Irrigation and Drainage Paper No 56 (Allen et al 1998) giving the detailed procedures for the determination of reference crop evapotranspiration (ETo).

22.1.2.1 Reference Evapotranspiration (ET_o)

ETO can be defined as the rate of evapotranspiration of an extended surface of an 8 to 15 cm tall, green cover, actively growing, completely shading the ground and not short of water.

Selection of a method for estimation of ETO depends on availability of meteorological data and amount of accuracy needed. Among the four methods for estimation of ETO, Blaney-Criddle method is simple, easy to calculate and requires data on sunshine (SS) hours, wind velocity (WV), relative humidity (RH) in addition to temperature (T).

Blaney-Criddle Method

$$ETO = C\,[P\,(0.46\,T + 8)] \text{ mm day-1}$$

where

ETO = Reference crop ET in mm day-1 for the month considered

T = Mean daily temperature in oC over the month considered

P = Mean daily percentage of total annual day time hours of a given month and latitude (from standard table)

C = Adjustment factor depends on minimum RH, sunshine hours and day time wind estimates.

Radiation Method

$$ETo = C\ (W \times Rs)$$

where,

Rs = Measured mean incoming shortwave radiation (mm day^{-1})

W = Temperature and altitude dependent weighing factor

C = Adjustment factor made graphically on W X Rs using estimated values of RH_{mean} and U day time.

Pan Evaporation Method

$$ETO = K_{pan} \times E_{pan}$$

where,

K_{pan} = Pan coefficient

E_{pan} = mean pan evaporation (E_{pan} is the pan evaporation).

Modified Penman Method

$$ETO = C\ [W.\ Rn + (1\text{-}W).\ f\ (U).\ (ea - ed)]$$

where,

Rn = Net radiation in equivalent evaporation expressed as mm day^{-1}

W = Temperature of altitude related factor

F (U) = Wind related function

ea – ed = Vapour pressure deficit (milli bar)

C = Adjustment factor (ratio of Uday to Unight)

Rn = (0.75-Rns)

ea = Saturated vapour pressure (milli bar)

ed = Mean actual vapour pressure of the air (milli bar).

FAO Penman-Monteith method

The FAO panel of experts, recently, recommended the adoption of Penman-Monteith combination method as new standard for reference evapotranspiration and advised on procedures for calculation of the various parameters. The following Penman-Monteith equation has been developed:

$$ETo=\frac{0.408D(R_n-G)+g\left[\frac{900}{T+273}\right]u_2(e_s-e_a)}{D+g(1+0.34u_2)}$$

where,

ETo = Reference evapotranspiration (mm day-1)

Rn = Net radiation at crop surface (MJ m-2 day-1)

G = Soil heat flux density (MJ m-2 day-1)

T = Mean daily air temperature at 2 m height (0C)

u2 = Wind speed at 2 m height (m s-1)

es = Saturated vapour pressure (kPa)

ea = Actual vapour pressure (kPa)

es – ea = Saturated vapour pressure deficit (kPa)

Δ = Slope vapour pressure curve (kPa 0C-1)

γ = Psychrometric constant (kPa 0C-1).

Full details on this aspect can be found in FAO IDP 24: Guidelines for predicting crop water requirements by (Doorenbos and Pruitt 1977) and FAO IDP 56: Crop evapotranspiration – Guidelines for computing crop water requirements by (Allen et al 1998).

Crop water requirement of any crop depends on crop factors such as variety, growth stage, duration of the crop, plant population and growing season. Climatic factors such as temperature, relative humidity, wind velocity and crop management practices such as tillage, fertilisation, weeding etc. Water requirement of crops vary from area to area and even field to field in a farm depending on the above mentioned factors.

22.1.2.2 Calculation of Crop Water Requirement

Crop water requirement is the amount of water required by crop to meet the evapotranspiration losses. Therefore, measurement of evapotranspiration rate is the first step to to estimate the crop water requirement. Two methods commonly used to measure or predict the evapotranspiration rate of crops are: evaporation pans and empirical equations that predict the evapotranspiration rate based on climatic parameters.

Evaporation Pan

Loss of water from the pan (USWB Class A Pan) is measured in mm per day (Fig 22.3). Evaporation from evaporation pan takes into account the effect of wind, temperature, radiation and humidity, which are the same factors that affect crop transpiration rate.

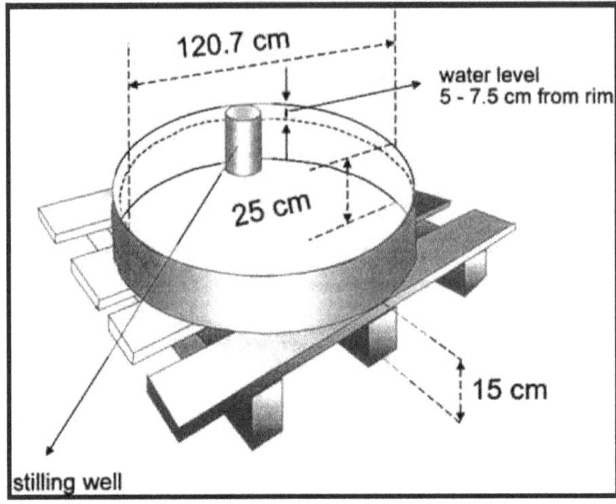

Fig. 22.3 Evaporation pan

However, there are a few factors that prevent this recording from being entirely accurate. Solar radiation results in heat storage in the pan. This can lead to increased reading of the evaporation rates at night, when transpiration usually does not occur. In addition, temperature and humidity levels above the pan surface will vary from what would naturally occur. To account for these variations, a pan coefficient (Kpan) is used. Crop water requirement or reference crop evapotranspiration (ETo) is estimated as the product of Epan and Kpan:

$$ETo = E_{pan} \times K_{pan}$$

Reference crop evapotranspiration (ETo) represents the maximum or potential evapotranspiration rate that can occur. However, crop water requirement is usually less than ETo due to several crop characteristics.

These crop characteristics include crop growth stage, the leaf coverage that provides shade to the ground, crop management factors etc. To account for these variations, ETo is converted into crop evapotranspiration (ETc), using crop-coefficient (Kc).

$$Kc = \frac{ETc}{ETo}$$

$$ETc = ETo \text{ X } Kc.$$

The ETc represents the evapotranspiration rate of the crop under standard conditions (no stress conditions). When calculating ETc, one must identify the crop growth stages, their duration and select the proper Kc that need to be used (Fig 22.4 and Table 22.2).

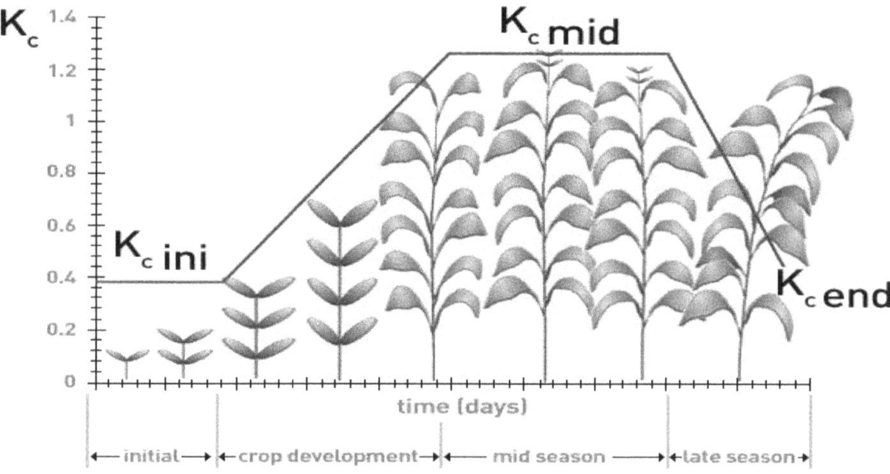

Fig. 22.4 Crop growth states and their duration

Table 22.2 Crop growth stages, duration and crop coefficients.

Growth stage	Duration of the stage (days)	Crop coefficient (Kc)
Initial	Seeding to 20 DAS (20)	0.20
Crop development	21 to 55 (34)	0.60
Mid-season	56 to 85 (29)	0.80
Late season	86 to 105 (22)	0.20
Total	105	0.7
Flowering continues during part of the yield formation but the pods from the late-formed flowers do not reach maturity.		

Example

- Crop: Groundnut
- Growth stage: Crop development
- Kc crop development: 0.45
- ET_o (from a local meteorological observatory): 9 mm day^{-1}

Crop water requirement (ETc) = Kc × ETo = 0.6 × 9 = 5.4 mm day^{-1}

22.2 Irrigation Requirement

When once we know the crop water requirements, the most important step is to apply the right quantity of water (irrigation requirement) at the right time (scheduling irrigation) through an appropriate method of irrigation to satisfy the crop water needs.

Accurate information on irrigation requirement is essential in irrigated agriculture. Such information aids in efficient use of scarce irrigation water.

22.2.1 Irrigation Requirement

Amount of water required to compensate the evapotranspiration losses from the cropped field is defined as crop water requirement. Although the values for crop water evapotranspiration and crop water requirement are identical, crop water requirements refers to the amount of water that is lost through evapotranspiration. Irrigation water basically represents the difference between the crop water requirement and the effective precipitation. Irrigation requirement also includes additional water for leaching of salts and to compensate for non-uniformity of water application.

The usual way of expressing crop water requirement is: It is the quantity of water, regardless of its sources, required by a crop or diversified pattern of crops in a given period of time for their normal growth under field conditions. It includes evapotranspiration and other economically unavoidable losses. When the water requirement is supplied entirely by irrigation, irrigation requirement and water requirement will be same.

22.2.2 Consumptive Use/Consumptive Water Use (cu)

It is the sum of the volumes of water used by crop over a given area in producing plant tissue, in transpiration (T), plus that evaporated (E) from adjacent soil or from moisture intercepted on plant foliage. Since the volume of water used in producing plant tissue is negligible (<1 percent) compared with the volumes used in E and T, the CU can be taken to be approximately equal to evaporation plus transpiration or evapotranspiration (ET).

$$WR \approx CU \approx ET$$

It is not always possible to apply exactly the required quantity of water to bring the root zone depth to field capacity. Some losses (deep percolation, runoff, seepage etc) are unavoidable. Such losses are known as application losses (AL). Water is also needed for special operations like leaching of excess soluble salts, land preparation, pre-sowing irrigation etc. These are collectively known as water for special operations (WSO). Water requirements (WR), therefore, include all these components.

$$WR = ET + AL + WSO$$

Water requirement can be expressed in terms of source of water.

$$WR = IR + EP + S$$

where,

IR = Irrigation water; EP = Effective precipitation

S = Soil profile moisture contribution as stored soil moisture or shallow groundwater table.

22.2.3 Leaching Requirement

To prevent excessive salt accumulation in the soil, it is necessary to remove salts periodically by application of water in excess of the consumptive use. The excess water applied will remove salts from the root zone provided the soil has adequate internal drainage.

This concept is quantified in the term leaching requirement (LR). By definition, leaching requirement (LR) is the fraction of total water applied that must drain below the root zone to restrict salinity to a specified level according to the level of tolerance of the crop.

$$LR = \frac{Ddw}{Diw}$$

where,

- D = Depth of water
- dw and iw = drainage and irrigation water.

Assuming strict balance conditions in soil-water system:

$$Diw \times Ciw = Ddw \times Cdw$$

$$LR = \frac{Ciw}{Cdw} \text{ or } \frac{Eciw}{Ecdw}$$

where,

- C = Concentration of salt.

Therefore,

The value of Ecdw represents the maximum salinity tolerated by the crop. For a crop, if the value of EC dw is 8 dSm m-1, the leeching requirement will be ECdw/8.

Thus for irrigation water with conductivities 1, 2 and 4 dSm m-1, respectively, the leching requirements will be 12, 25 and 50 percent respectively.

22.2.4 Irrigation Requirement (Ir)

It is the total amount of water applied to a cropped field for supplementing precipitation and soil profile moisture contribution to meet crop water needs for optimum growth and yield. Irrigation requirement is only part of water requirement.

$$IR = WR - (EP + S)$$

22.2.5 Net Irrigation Requirement (NIR)

It is the amount of water, exclusive of precipitation and profile soil moisture contribution, required for optimum crop production. In other words, it is the amount of irrigation water that must be stored in the crop root zone to meet the consumptive use requirement of a crop. Simply, it is the amount of irrigation water required to bring the soil moisture level in the effective root zone to field capacity. Thus, it is the difference in depth or percentage soil moisture between field capacity and the soil moisture content in the root zone before starting irrigation. In terms of depth, it can be expressed as:

$$NIR = \sum_{i=1}^{n} \frac{M1_i - M2_i}{100} \times BD_i \times D_i$$

where,

NIR = Net amount of irrigation water to be applied at each irrigation (mm or cm)

n = Number of soil layers considered in root zone depth D

M1i = Soil moisture percentage at first sampling in ith layer

M2i = Soil moisture percentage at second sampling in ith layer

BDi = Bulk density of ith soil layer (g cm-3)

Di = Depth of ith soil layer (mm or cm).

Effective precipitation, groundwater contribution and deep percolation losses, if any are to be considered for reliable estimates of crop water requirements.

$$NIR = (CU - EP) + AL$$

Seasonal irrigation requirement can be obtained by adding NIR values at each irrigation.

22.2.6 Gross Irrigation Requirement (GIR)

It is the net irrigation requirement and losses in conveyance, distribution and application of water in operating the system.

$$GIR = \frac{NIR}{\text{Irrigation efficiency}} \quad \text{(or)} \quad GIR = \sum_{i}^{n} \frac{D}{E_a}$$

where,

D = Net depth of irrigation water to be applied at each irrigation

Ea = Water application efficiency.

Calculation of Irrigation Requirement

Problem

With the information provided below, find out the net and gross amounts of irrigation water requirements if the irrigation efficiency is 60 per cent?

Soil depth (cm)	Soil moisture at FC (%)	Actual soil moisture (%)	Bulk density (g cm^{-1})
0 to 15	20	10	1.4
15 to 30	18	12	1.5
30 to 45	18	14	1.6
45 to 60	19	16	1.6

Solution:

Soil moisture deficit in the four soil layers:

$$0 \text{ to } 15 \text{ cm} = \frac{20-10}{100} \times 1.4 \times 15 = 2.10 \text{ cm}$$

$$15 \text{ to } 30 \text{ cm} = \frac{18-12}{100} \times 1.5 \times 15 = 1.35 \text{ cm}$$

$$30 \text{ to } 45 = \frac{18-14}{100} \times 1.6 \times 15 = 0.96 \text{ cm}$$

$$45 \text{ to } 60 = \frac{19-16}{100} \times 1.6 \times 15 = 0.72 \text{ cm}$$

Net depth of application from 0 to 60 cm = 5.13 cm

Gross amount to be applied to 60 per cent irrigation efficiency = $\frac{5.13}{60} \times 100 = 8.55$ cm

22.2.7 Irrigation Frequency

It refers to the number of days between two successive irrigations (irrigation interval) during the periods without precipitation. It depends on consumptive use rate of a crop and on the amount of available soil moisture in the crop root zone. Sandy soil requires more number of frequent irrigations than fine textured soils. In general, irrigation should start when around 50 per cent of the available soil moisture has been depleted from the root zone depth.

In designing irrigation systems, the irrigation frequency used is the time (days) between two irrigations in the period for highest consumptive use of the crop. The average consumptive use rate during this period is the rate to be considered in designing and planning irrigation system and this is known as design frequency.

$$\text{Design frequency} = \frac{\text{Net amount of moisture between FC and starting level}}{\text{Peak period moisture use rate}}$$

$$= \frac{\text{Net depth of application}}{\text{Peak period moisture use rate}}$$

Problem

Given the following, determine the irrigation frequency?

Field capacity of the soil = 30%

Permanent wilting point = 11%

Soil bulk density = 1.3 g cm-3

Effective root zone depth = 700 mm

Level of DASM for irrigation = 25%.

Solution:

ASM = 30 – 11 = 19%

25% of ASM = 0.25 X 19 = 4.75%

Total soil moisture level for starting irrigation = DASM + PWP

= 4.75 + 11.0 = 15.75%

Water depth in root zone depth = (0.3 – 0.1575) = 0.1297 = 12.97 cm

$$\text{Irrigation interval} = \frac{12.97}{12} \times 10 = 10.8 \text{ days or } 11 \text{ days.}$$

22.2.8 Irrigation Period

The number of days that can be allowed for applying one irrigation to a given design area during the peak period consumptive use of the crop is known as irrigation period. It is the basis for capacity and equipment design. Irrigation systems are to be designed in such a way that the irrigation period is not greater than irrigation frequency.

$$\text{Irrigation period} = \frac{\text{Net amount of moisture between starting of irrigation and lower limit of depletion}}{\text{Peak period moisture use rate of the crop}}$$

Problem

Given the following:

Soil moisture content at FC = 30%

Soil moisture content at PWP = 15%

Soil bulk density 1.5 g cm-3

Effective root zone depth = 75 cm

Irrigation efficiency = 80%

ET per day = 0.6 cm

Compute the following?

Net depth of irrigation water application

Gross depth of water application

Irrigation period.

Solution:

$$\text{ASM in the root zone depth} = \frac{(30-15)}{100} \times 1.5 \times 75 = 16.88 \text{ cm}$$

$$\text{Net depth to be applied to the root zone} = \frac{75 \times 16.88}{100} = 12.66 \text{ cm}$$

$$\text{Gross depth of application with 80\% efficiency} = \frac{12.66}{80} \times 100 = 15.83 \text{ cm}$$

$$\text{Irrigation period} = \frac{12.66}{0.6} = 21 \text{ days}$$

22.3 Irrigation Scheduling (when to Irrigate)

When once we know the right quantity of water (irrigation requirement), next important step is to apply the right quantity of irrigation water at the right time (scheduling irrigation) through an appropriate method of irrigation to satisfy the crop water needs. Accurate information on the quantity of irrigation requirement is essential for efficient use of scarce irrigation water.

Maintenance of soil moisture around field capacity (-0.33 bar) is ideal for many crops. As the soil moisture content drops from FC due to evapotranspiration and other losses, soil moisture tension increase and crops can't extract needed moisture from soil for optimum growth. By knowing the available soil moisture (ASM) in the crop root zone and the evapotranspiration demands of the crop and atmosphere, it is easy to determine when irrigation is needed. There are several approaches to decide 'when to irrigate' based on soil, plant and atmospheric parameters as indicated in the Fig 22.5.

22.3.1 Soil Moisture Monitoring Approaches

Both direct and indirect methods are used for soil moisture monitoring for determining crop water requirements. There are several direct methods for soil moisture measurement. Methods that are commonly in use only are briefly presented.

22.3.1.1 Direct Methods

These methods involve determination of soil moisture content and finding the deficit level in available soil moisture. Soil-water content can be determined either by direct measurement (gravimetric method) or inference from measurements of other methods (tensiometer method, electrical resistance method etc). Based on pre-determined minimum water content, irrigation is given to bring the soil to field capacity.

Gravimetric Method: It is the most widely used standard oven dry method for the calibration of all other soil moisture determination techniques.

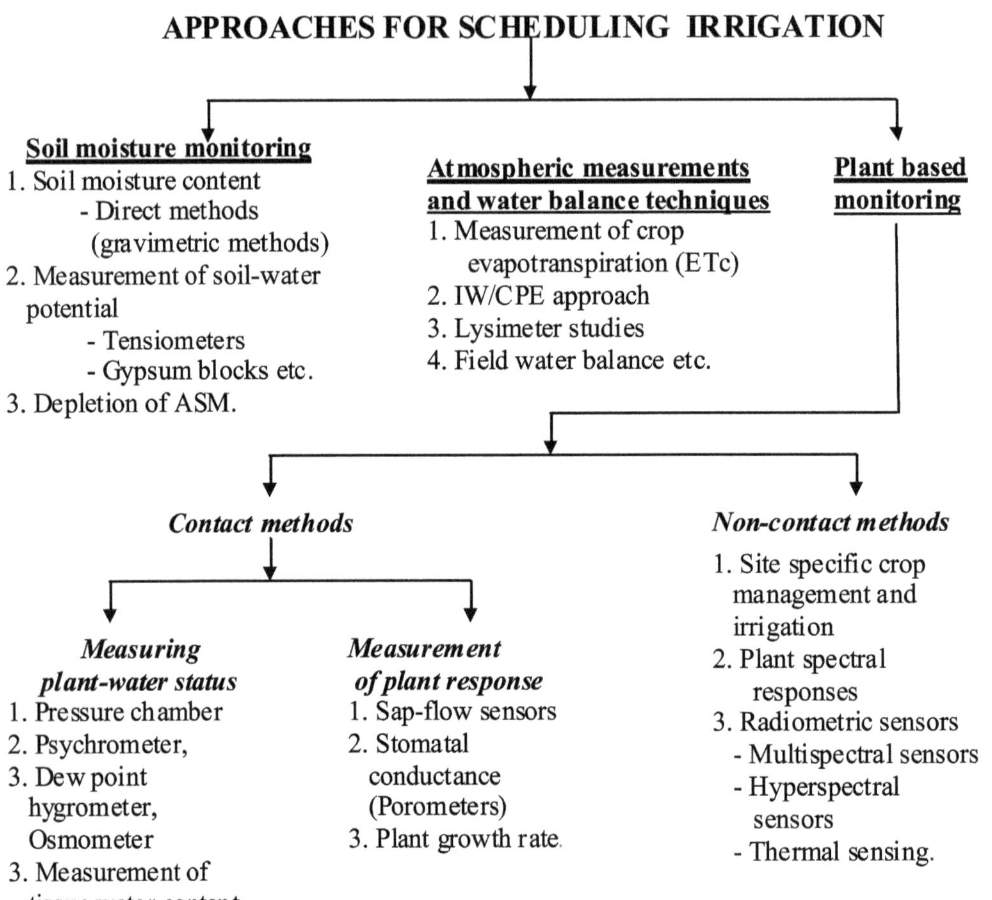

Fig. 22.5 Approaches for scheduling irrigation

Drying the moist soil to a constant weight in a drying oven is controlled at 110° ± 5°C. Samples should be drried for at least 24 hrs. Weight of soil remaining after oven drying is used as the weight of soil solids. Moisture content expressed as a per cent is equal to the weight of water divided by weight of soil solids multiplied by 100.

Soil moisture content by weight (per cent) can be obtained as indicated below:

$$\text{Moisture content (per cent)} = \frac{\text{Wt of wet soil} - \text{Wt of dry soil}}{\text{Wt of dry soil}} \times 100$$

Volumetric water content (per cent) can be calculated if soil bulk density (BD) is known.

$$\text{Volumetric water content (\%)} = \frac{\text{Wt of wet soil} - \text{Wt of dry soil}}{\text{Wt of dry soil}} \times 100 \times \text{BD}$$

Depth of water (cm) per unit depth of soil can be expressed if soil depth is known.

Depth of water per unit soil depth = Volumetric water content x Soil depth.

Merits

- Oven drying is considered the standard method for obtaining soil moisture content
- Very simple equipment requirement
- No specific site calibrations are required
- Inexpensive compared to other methods
- No health risks associated with this method.

Drawbacks

- Sampling is very tedious and time consuming
- Time to dry sample is approximately 24 hrs, which will drastically reduce sample rate
- Obtaining representative soil moisture values in a heterogeneous soil profile is difficult
- If sampling is required over long periods, it is very destructive to the site
- It is difficult to determine moisture content at specific depths.

Example

Given the following:

Weight of empty can = 40 g

Weight of moist soil + can = 150 g

Weight of dry soil + can = 140 g

Soil bulk density = 1.5 g cm^{-3}

Sampling soil depth = 30 cm.

Answer the following questions?

Soil moisture content by weight (per cent)

Volumetric water content (per cent)

Depth of water in sampled soil depth (cm).

Solution:

Soil moisture content by weight (per cent)

Weight of moist soil = 150 – 40 = 110 g

Weight of dry soil = 140 – 40 = 100 g

$$= \frac{\text{Wet weight - Dry weight}}{\text{Dry weight}} \times 100$$

$$= \frac{110-100}{100} \times 100 = 10\%$$

Volumetric water content (per cent)

$$= \frac{\text{Wet weight - Dry weight}}{\text{Dry weight}} \times 100 \times \text{BD}$$

$$= \frac{110-100}{100} \times 100 \times 1.5 = 15\ \%$$

Depth of water in sampled soil depth (cm)

= Soil moisture content by weight x BD x Soil depth

$$= \frac{10}{100} \times 1.5 \times 30 = 4.5\ \text{cm}$$

OR = Volumetric water content x Soil depth

$$= \frac{15}{100} \times 30 = 4.5\ \text{cm}$$

Feel and Appearance Method

With experience, farmer can judge soil-water content by feel and appearance of the soil. Soil samples are taken from each quarter of the root zone depth, formed into a ball with hand, tossed into air and caught in one hand. From the description given in Table 22.3, ASM can be assessed. Considerable experience and judgment are necessary to estimate ASM content in the sample within reasonable accuracy.

22.3.1.2 Indirect Methods

These methods are relatively quick and accurate. However, they require technical skill to operate and are relatively expensive. It is beyond the scope of this book to discuss all the indirect methods of soil moisture measurement. Discussion is limited to three methods only.

Gypsum Blocks (porous/electrical resistance blocks)

Gypsum blocks use the principle of resistivity measured indirectly from a material in equilibrium with the soil. When these blocks reach equilibrium with the surrounding soil, their electric or thermal properties are often regarded as an index of soil-water content.

Gypsum blocks (Fig 22.6) are most widely used for moisture measurement and are available in a variety of porous materials ranging from nylon cloth and fiberglass to casting plasters. Each block is calibrated in soil and soil density typical of the site, the block is to be used in. Electrical resistance of the block is measured by a meter based on the principle of Whetstone Bridge. Resistance blocks read low resistance (400-600 ohms) at FC and high resistance (50,000-75,000 ohms) at wilting point. Interpretation of readings on latest meters is given in Table 22.4.

Table 22.3 Guide for judging the amount of ASM in soil

Available soil moisture range	Coarse texture (loamy sand)	Moderately coarse texture (sandy loamy)	Medium texture (loamy and silt loamy)	Fine texture (clay loamy and silty clay loamy)
Field capacity (100%)	On squeezing, no free water appears on soil, but wet outline is left on hand	Similar symptoms		
75 to 100%	Tends to stick together slightly, sometimes forms a very weak ball under pressure	Forms weak ball, breaks easily, don't slick	Forms a ball, very pliable, slicks readily	Easily ribbons out between fingers, has slick feeling
50 to 75%	Appears to be dry don't form a ball with pressure	Tends to form a ball under pressure but seldom holds together	Forms a ball somewhat plastic, some-times slick slightly with pressure	Forms a ball, ribbons out between thumb and fore-finger
25 to 50%	As above, but ball is formed by squeezing very firmly	Appears to be dry, don't form a ball unless squeezed very firmly	Some what crumbly but holds together with pressure	Somewhat pliable, forms a ball under pressure
0 to 25%	Dry, loose, single grained flows through fingers	Dry, loose, flows through fingers	Powdery dry, sometimes slightly crusted but easily broken down into powdery conditions.	Hard, baked, cracked, sometimes has loose crumbs on surface.

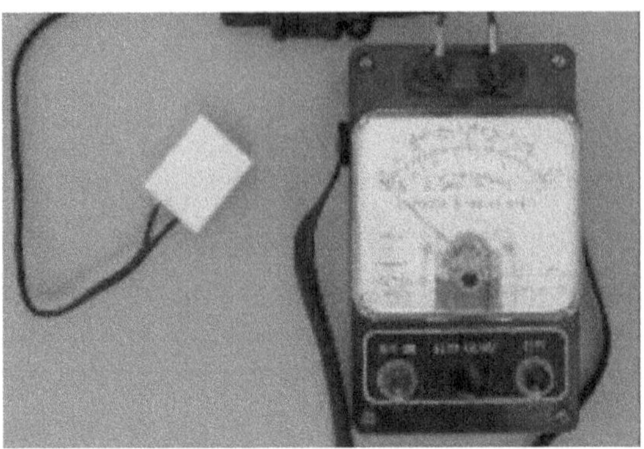

Fig. 22.6 Gypsum block

Table 22.4 Interpretation of readings on typical electrical resistance meter.

Soil-water condition	Meter readings (0 to 200 scale)	Interpretation
Nearly saturated	180 to 200	Near saturated soil for few hours from irrigation. Danger of waterlogged soil, high water table or poor soil aeration if readings persist for several days.
Field capacity	170 to 180	Excess water has mostly drained out. No need to irrigate. Any irrigation leads to leaching losses of nutrients.
Irrigation range	80 to 120	Usual range for starting irrigation. Irrigation at this range ensures optimum available soil moisture for normal crop growth.
Dry	< 80	This is the stress range. Yield may not necessarily reduce. Some soil moisture is available for crop use, but is getting dangerously low.

Water content of a porous block coming to water equilibrium with a soil depends on the energy status of the water rather than the water content of the soil. Soils with fine pores will contain more water than soil with coarse pores at equal matric potential. Hence, if porous blocks are to provide an indirect measure of soil-water content, calibration is necessary. Since the equilibrium is a matric potential and not a water content equilibrium, associated soil-water content must be obtained from a calibration curve.

Neutron Scattering

This method consists of americium and beryllium (241Am/9Be) as neutron source and boron trifluoride (BF_3) gas as detector (Fig 22.7). The instrument works on the principle that when fast moving neutrons are emitted into the soil, they undergo elastic collisions with hydrogen having proton mass equal to mass of neutron and get scattered. Their energy is reduced and speed is slowed down.

The slowed neutrons are absorbed by BF_3 gas, which can be counted with a scalar, rate meter or electronic microprocessor aided circuitry, which directly displays the moisture content. The density cloud of neutrons so formed around the source (probe) is proportional to the concentration of hydrogen in the soil. Since water is the main source of hydrogen in soil, the density of slowed neutrons is proportional to the volume fraction of water present in the soil. The count rates are then calibrated against volumetric determination of soil moisture content.

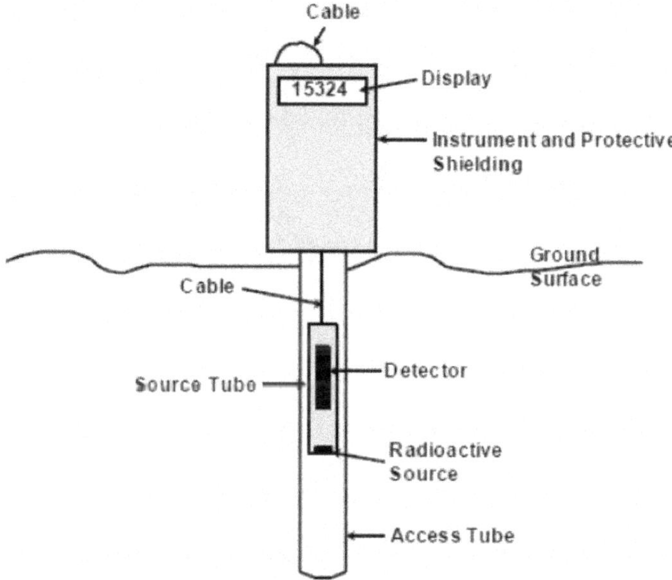

Fig. 22.7 Components of neutron probe

Merits

- Quick result
- Once the tube is inserted, the same location can be used repeatedly
- Non destructive and less laborious
- Fairly repeatable.

Drawbacks

- Expensive
- Probe must be calibrated for each soil/access tube type
- Top 15 cm soil depth is difficult to measure due to escape of neutrons into atmosphere

- Radiation exposure danger
- Difficulty with maintenance, repair or parts in developing countries.

Tensiometer

The working principle of tensiometer is that when a sealed water-filled tube is placed in contact with the soil through a permeable and saturated porous material, water (inside the tube) comes into equilibrium with the soil solution (it is at the same pressure potential as the water held in the soil matrix). Hence, the soil-water matric potential is equivalent to the vacuum or suction created inside the tube.

The tensiometer consists of a sealed water-filled plastic tube with a ceramic cup at one end and a negative pressure gauge at the other (Fig 22.8). The shape and size of the ceramic cup can be variable and the accuracy depends on the gauge or transducer used (about 0.01 bar). Typically, the measurement range is 0 to 0.80 bar, although there are low tension versions (0 to 0.40 bar) designed for coarse soils.

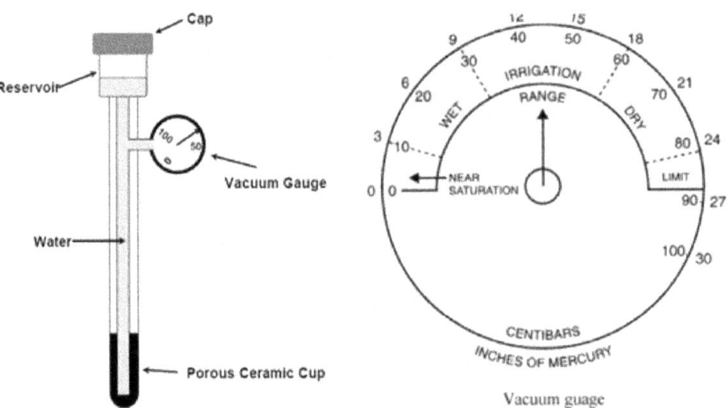

Fig. 22.8 Tensiometer

Soil-water potential or soil moisture tension is a measure of the tenacity with which water is retained in the soil and shows the force per unit area that must be exerted to remove water from the soil. It is measured in terms of the potential energy of water in the soil measured, usually, with respect to free water. It is usually expressed in atmospheres, the average air pressure at sea level.

22.3.2 Atmospheric Measurements/ Climatological Approaches

An integrated approach to soil-plant-atmosphere continuum (SPAC) is essential for an ideal irrigation schedule. Climatological approach for scheduling irrigation is also based on two assumptions: crop evapotranspiration (ETc) is, generally, a function or at the most equal to evaporation from free water surface and ETc is considered desirable for optimum crop yield. Climatological approaches for scheduling irrigation include:

- Measurement of crop evapotranspiration (ETc)
- IW/CPE approach
- Lysimeter studies
- Field water balance etc.

22.3.2.1 Measurement of Crop Evapotranspiration (Etc)

Crop evapotranspiraton (ETc) can be estimated by estimating reference evapotranspiration (ETo) with any of the empirical methods, preferably with Penman-Monteith method, elaborated under 1.2.1 and using the crop coefficient (Kc).

Actual crop water requirement, in addition to climate, include the effect of crop characteristics like variety, its duration, stage of crop growth, season, tillage, irrigation, plant protection etc. Crop coefficient (Kc) is used to account for all these variations.

Daily crop water use (crop evapotranspiration (ETc) is calculated using a crop coefficient (Kc) and a reference evapotranspiration (ETo) obtained from atmospheric measurements as: ETc = Kc x ETo

To sum up, irrigation schedules can be determined as indicated below:

Step 1: Determine the reference crop evapotranspiration: ETo

Step 2: Determine the crop factors: Kc

Step 3: Calculate the crop water need: ETcrop = ETo × Kc

Step 4: Determine the effective rainfall: ER

Step 5: Calculate the irrigation requirement: IR = ETcrop – ER.

Irrigation can be scheduled if the level of depletion of available soil moisture (DASM) in the root zone and evaporation during the crop period is known. Full details on this aspect can be found in FAO IDP 24: Guidelines for predicting crop water requirements by (Doorenbos and Pruitt 1977) and FAO IDP 56: Crop evapotranspiration – Guidelines for computing crop water requirements by (Allen et al 1998). Lysimeter studies and field water balance have been discussed under 1.1.4.

22.3.2.2 IW/CPE Approach

Pan evaporation (E_{pan}) can be used for scheduling irrigations. (Prihar and Sandhu 1987) suggested relatively practical approach of **IW/CPE ratio** (ratio between a fixed amount of irrigation water and cumulative pan evaporation) as the basis for scheduling irrigations.

In this approach, a known quantity of irrigation water (IW) is applied when cumulative pan evaporation (CPE) reaches a predetermined level. The amount of water given in each irrigation ranges from 4 to 6 cm, the most common being 5 cm of irrigation. Scheduling irrigation at an IW/CPE ratio of 1.0 with 5 cm of irrigation water is applied when the CPE reaches 5 cm. Generally, irrigation is scheduled at 0.75 to 0.8 ratio with 5 cm of irrigation water. In IW/CPE ratio approach, irrigation can also be scheduled at fixed level of CPE by varying amount of irrigation water. However, the equipment to measure CPE and IW are not easily available with the farmers.

Problem

Calculate cumulative evaporation required for scheduling irrigation at 0.5, 0.6, 0.75 and 0.8 with 5 cm of irrigation water ?

Solution:

Cumulative pan evaporation at IW/CPE ratio of 0.5 = IW/CPE = 0.5

Irrigation of 5 cm is given when CPE is 10 cm

CPE at 0.6 ratio = 5/0.6 = 8.33 cm

CPE at 0.75 ratio = 5/0.75 = 6.66 cm

CPE at 0.8 ratio = 5/0.8 = 6.25 cm

In IW/CPE ratio approach, irrigation can also be scheduled at fixed level of CPE by varying amount of irrigation water.

Problem

Calculate the amount of water for each irrigation for scheduling irrigation at 0.5 and 0.8 IW/CPE with 10 cm of CPE?

Solution:

1. Amount of water to be given at IW/CPE ratio of 0.5 = IW/10 = 0.5

   IW = 0.5 × 10= 5 cm.

2. Amount of water to be given at IW/CPE ratio of 0.8 = IW/10 = 0.8,

   IW = 10 × 0.8 = 8 cm.

Lysimeter studies and Field water balance have been discussed under subheads 22.1.1.1 and 22.1.1.4, respectively. Irrigation can be scheduled if the level of depletion of available soil moisture (DASM) in the root zone and evaporation during the crop period is known.

22.3.2.3 Yield-Moisture Stress Relationship

A simple, linear crop-water production function was introduced in the FAO Irrigation and Drainage Paper No 33 (FAO IDP 33) to predict the reduction in crop yield due to stress (shortage of soil-water):

The upper limit for yield is set by climate, soil fertility and management practices. Where all of these are optimal, yield reaches the maximum value as does evapotranspiration. Any significant decrease in soil-water storage has an impact on water availability for a crop and, subsequently, on actual yield and actual evapotranspiration. A standard formulation relates these four parameters to a fifth: the **yield response factor Ky)**, which links relative yield decrease to relative evapotranspiration deficit, as follows:

$$1-\frac{Ya}{Ym}=K_y X\left(1-\frac{ET_a}{ET_m}\right) \quad .....(1)$$

where,

Y_a = actual yield (kg ha^{-1})

Y_m = maximum yield (kg ha^{-1})

ET_a = actual evotranspiration (mm)

ET_m = maximum evapotranspiration (mm)

K_y = yield response factor.

Calculations of Y_m, ET_m and ET_a are well documented (FAO IDP 1977 and FAO IDP 1998) and the literature has provided values for K_y (FAO 1979). From these four parameters, it is possible to calculate Y_a where the available water supply does not meet the full moisture requirements of the crop. Where water deficit occurs during a specific growth stage, the yield response will depend on crop sensitivity during that period. Therefore, the timing of the deficit is a tool for scheduling the use of a limited water supply and in setting priorities among several irrigated crops.

Yield Response Factor (ky)

The Ky values are crop specific and vary over the growing season according to growth stages with:

Ky > 1: Crop response is very sensitive to water deficit with proportional larger yield reductions when water use is reduced because of stress

Ky < 1: Crop is more tolerant to water deficit and recovers partially from stress, exhibiting less than proportional reductions in yield with reduced water use

Ky = 1: Yield reduction is directly proportional to reduced water use.

Based on the analysis of an extensive available literature on crop-yield and water relationships and deficit irrigation, Ky values were derived for several crops.

Analysis of deficit irrigation studies also allowed, for a majority of crops, the development of crop response functions when water deficits occur at different crop stages. As illustrated for groundnut (Figs.9 and 10), yield response will differ largely depending on the stage the water stress occurs. Typically, flowering and yield formation stages are sensitive to stress, while stress during the ripening phases has a limited impact, as in the vegetative phase, provided the crop is able to recover from stress in subsequent stages.

Vegetative and reproductive growth shows a definite response to water supply. Flowering period (2) is most sensitive to water deficit, followed by the yield formation period (3). In general, water deficits during the vegetative period (1) cause delayed flowering and harvest and reduced growth and yield. Water deficits during flowering (2) cause flower drop or impaired pollination, whereas water deficits during the yield formation period (3) give reduced pod weight. Early part of the yield formation period (pod setting) is, particularly, sensitive to water deficit. In the case of limited water supply, water savings should be made during the periods other than flowering (2) and early yield formation (3). Higher total production can be obtained by increasing the cultivated area and partially meeting crop water requirements rather than by meeting full crop water requirements over a limited area.

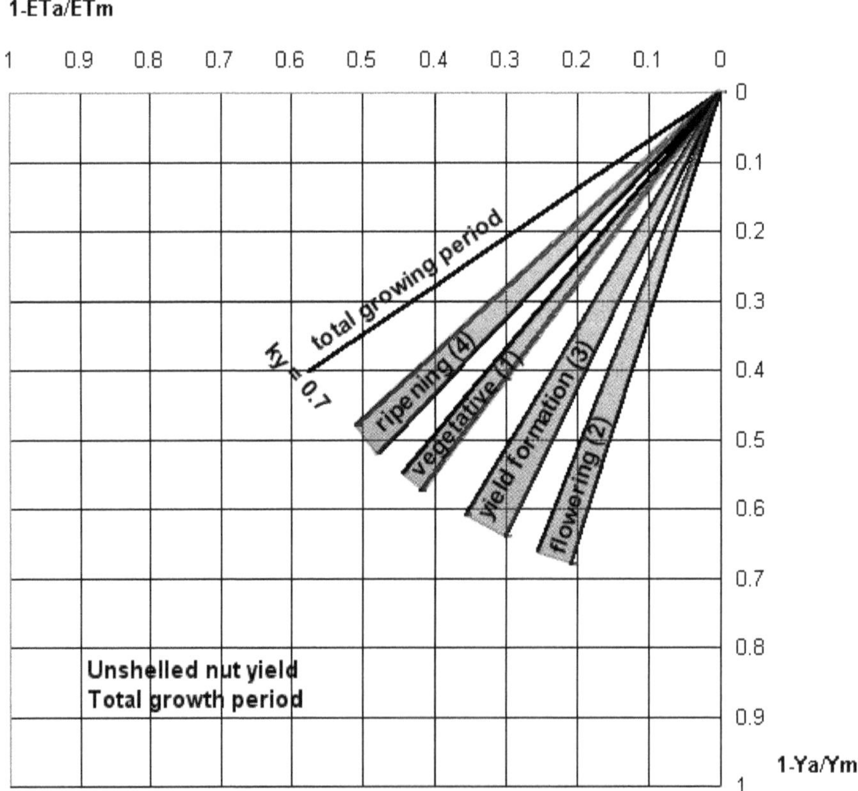

Fig. 22.9 Relationships between relative yield decrease (1-Ya/Ym) and relative evapotranspiration deficit (1-ETa/ETm) for the individual growth periods

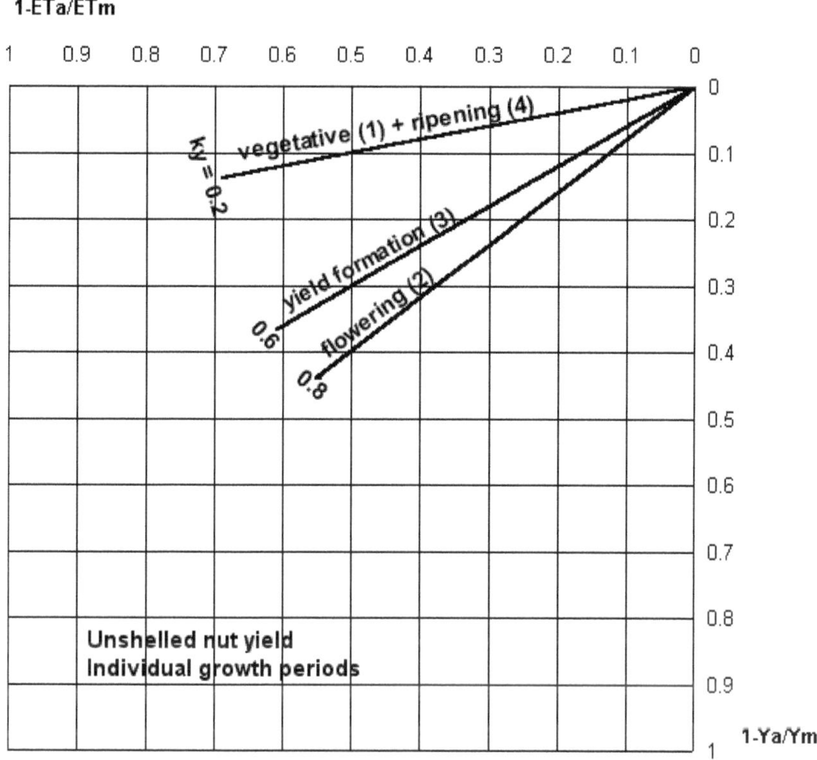

Fig. 22.10 Relationships between relative yield decrease (1-Ya/Ym) and relative evapotranspiration deficit (1-ETa/ETm) for the individual growth periods

Depending on the level of ET and water retentive capacity of the soil, irrigation intervals vary from 6 to 14 days up to 21 days from loamy soils, with shorter intervals during flowering when DASM should not exceed 40 per cent. In the case of supplemental irrigation, best results are obtained when water is applied during flowering.

Calculation Procedure

Calculation procedure for Equation (1) to determine actual yield (Ya) has four steps:

1. Estimate maximum yield (Ym) of an adapted crop variety, as determined by its genetic makeup and climate, assuming agronomic factors (water, fertilisers, pest and diseases) are not limiting
2. Calculate maximum evapotranspiration (ETm) according to established methodologies and considering that crop-water requirements are fully met
3. Determine actual crop evapotranspiration (ETa) under the specific situation, as determined by the available water supply to the crop

4. Evaluate actual yield (Ya) through the proper selection of the response factor (Ky) for the full growing season or over the different growing stages.

Maximum yield (Ym)

The FAO IDP No 33 recommended procedures for estimating maximum yield either from available local data for maximum crop yields or based on the calculation of maximum biomass and a corresponding harvest index, following two different procedures: Wageningen procedure (De Wit 1968) and Ecological zone approach (Kassam 1977)

Maximum Crop Evapotranspiration (ETm)

Procedures for determining ETm were based on FAO guidelines for crop-water requirements (ETc) and the ETm component of Equation 1, which is equal to ETc, was determined through the product of the reference-crop evapotranspiration (ETo) times the crop coefficient (Kc), i.e

$$ETm = Kc \times ETo \qquad .....(2)$$

Original procedures for determining ETo are described in FAO IDP 24 (Doorenbos and Pruitt 1977), offering different equations for its calculation according to the available climate data. Kc values were provided for a large number of crops and procedures to determine ETc over the growing season. Subsequently, revised procedures for calculating ETo were introduced in FAO IDP No 56 (Allen et al 1998), according to the FAO Penman-Monteith equation, which has now become the standard for estimating reference crop evapotranspiration.

Actual Crop Evapotranspiration (ETa)

FAO IDP No 33 provided tables from which ETa could be estimated from data on evapotranspiration rate, available soil-water and wetting intervals. The tables however proved cumbersome and later were replaced by more accurate ETa calculations based on daily water balance calculations and digital computation methods.

Water balance calculations allow the level of available soil-water in the root zone to be determined on a daily basis. As long as soil-water is readily available for the crop, then ETa = ETm. When a critical soil moisture level is reached, defined as a fraction of the total available soil-water content (p), transpiration is reduced because the stomata close and thus $ETa < ETm$, until the level of soil-water in the root zone reaches the permanent wilting point, when ETa is assumed to be zero. This critical soil-water content is estimated from soil, crop and rooting characteristics and from the ETo rate. Depletion of soil-water content between p and the permanent wilting point will result in a proportional reduction of ETa.

FAO IDP No 56 provides detailed procedures to assess the impact of stress on reduced evapotranspiration based on the water balance calculations with parameters on critical soil-water content values and rooting depth.

Actual Crop Yield (Ya) and Yield reduction

Based on the estimated Ym and the calculated ETm and ETa, actual yield (Ya) may be determined using Equation (1).

$$1-\frac{Ya}{Ym}=K_y \text{ X }\left(1-\frac{ET_a}{ET_m}\right)$$

Solving for Ya/Ym:

$$\frac{Ya}{Ym}=\left[1-Ky\left(1-\frac{ETa}{ETm}\right)\right] \quad .....3$$

However, in many planning and management studies requiring the estimation of yield in relation to the water availability, the yield reduction is expressed in relative terms, e.g. as a fraction or percentage $\left(1-\frac{Y_a}{Y_m}\right)$ rather than absolute (Ya).

22.3.3 Rough Methods for Farmers

Simple methods are suggested to the farmers to find out when to start irrigation and how much water to apply.

22.3.3.1 Can Evaporation Method

Small can of one liter capacity (14.3 cm height and 10 cm diameter) can be used to indicate evaporation from the cropped field. The cans are white painted and covered with 6/20 size mesh. A pointed indicator is fixed at 1.5 cm below the brim of can. When irrigation is given (bringing the soil to field capacity), the can is filled up with water to pointer level and kept to the crop height. Evaporation from the can is directly related to the crop evaporation.

Irrigation is scheduled when water level in the can falls to a predetermined level (equal to the amount of water to be applied at each irrigation) and can is filled again to pointer level.

22.3.3.2 Soil-Cum-Mini-Plot Technique

In this method, 1 × 1 × 1 m size pit is dug in the middle of the field. About 5 percent of sand (by volume) is added to the pit, mixed well with soil and the pit is filled up in natural order. Crops are grown normally in all areas including pit area. Plants in the pit show wilting symptoms earlier than the other areas. Irrigation is scheduled as soon as wilting symptoms appear on the plants in the pit.

22.3.3.3 Sowing High Seed Rate

In an elevated area, one square meter plot is selected and crop is grown with four times thicker than the normal seed rate. Because of high plant density, plants show wilting symptoms earlier than in the rest of the crop area indicating the need of scheduling of irrigation.

22.3.3.4 Critical Stage Approach

In each crop, there are certain growth stages at which moisture stress leads to irrevocable yield losses. These stages are called as critical periods or moisture sensitive periods. Hence, irrigation must be given to these stages to avoid yield losses. In general, tillering/branching, panicle initiation, flowering and grain development stages are considered soil moisture sensitive stages for soil moisture in field crops.

22.3.4 Plant Based Irrigation Schedules

The primary objective of irrigation is water supply to meet the crop needs. Monitoring crop is the most direct method of irrigation scheduling. Plant parameters have to be related to soil-water content to determine irrigation scheduling.

22.3.4.1 Appearance and Growth (Visual Plant Symptoms)

Deliberate visual indicators to assess water needs of the crops are leaf and shoot wilting, leaf colour, drooping of leaves, rolling of leaves etc.

However, visual plant symptoms are not often effective parameters for deciding irrigation scheduling, as plants exhibit visible symptoms of soil-water deficiency long after they experience moisture stress.

22.3.4.2 Plant Growth Rate

Plant growth rate sensors suitable for irrigation scheduling include auxanometers and stem or fruit diameter sensors (dendrometers).

Auxanometers are used to measure plant height and the extinction in plant height over time. Dendrometers can also be used to measure plant growth rate. Major limitation is their point based measurement and the need for adequate representation of the field.

22.3.4.3 Other Plant Based Methods

Leaf water potential (pressure chamber), stomatal resistance (leaf diffusion porometer), canopy temperature (infrared thermometer), stem diameter (dendrometer), relative water content etc can be used for scheduling irrigation. Readers are advised to refer the book

Irrigation Agronomy (Third edition) by (Reddy and Reddy 2015) for further details on plant based irrigation schedules.

22.4 Methods of Irrigation

The manner in which irrigation water is applied to the land is referred to as **method of irrigation**. Several irrigation methods are adapted towards this end under different circumstance. Selection of appropriate irrigation method for any given combination of physical and socioeconomic conditions involves numerous complex and often conflicting considerations. Since the economic considerations along with physical conditions and cropping patterns are necessarily specific to each location, an irrigation system optimal in a particular region may not be so in another.

22.4.1 Choosing an Irrigation Method

To choose an irrigation method, the farmer must know the advantages and disadvantages of the various methods. Farmer must know which method suits the local conditions best. Unfortunately, in many cases there is no single best solution: all methods have their advantages and disadvantages. Irrigation systems are often designed to maximise efficiencies and minimise labour and capital requirements. Most effective management practices are dependent on the type of irrigation system and its design. A large number of considerations must be taken into account in the selection of an irrigation system. These will vary from location to location, crop to crop, year to year and farmer to farmer. Broad guidance and several important criteria in the selection of suitable irrigation method are discussed.

In general, these considerations will include the compatibility of the system with other farm operations, economic feasibility, topographic and soil properties, crop characteristics and social constraints.

22.4.1.1 Surface, Sprinkler or Drip Irrigation

Suitability of surface, sprinkler or drip irrigation depends mainly on the following factors:

- Compatibility
- Natural conditions
- Type of crop
- Type of technology
- Previous experience with irrigation
- Required labour inputs
- Costs and benefits.

Compatibility

Irrigation system for a field or a farm must function alongside other farm operations such as land preparation, cultivation and harvesting. Use of large mechanised equipment requires longer and wider fields. Irrigation systems must not interfere with these operations and may need to be portable or function primarily outside the crop boundaries (surface irrigation systems). Smaller equipment or animal powered cultivating equipment is more suitable for small fields and more permanent irrigation facilities.

Natural Conditions

Natural conditions such as soil type, slope, climate, water quality and availability, have the following impact on the choice of an irrigation method:

Soil type: Sandy soils with low water storage capacity and high infiltration rate need frequent but small irrigation applications, in particular when they are also shallow. Under these circumstances, sprinkler or drip irrigation systems are more suitable than surface irrigation. On loam or clay soils, all three irrigation methods can be used, but surface irrigation is more commonly found. Clay soils with low infiltration rates are ideally suited to surface irrigation.

When different soil types are found within one irrigation scheme, sprinkler or drip irrigation are recommended as they will ensure a more even water distribution.

Slope: Sprinkler or drip irrigation are preferred over surface irrigation on steeper or unevenly sloping lands as they require little or no land levelling. An exception is rice grown on terraces of sloping lands.

Climate: Strong wind can disturb spraying of water from sprinklers. Under very windy conditions, drip or surface irrigation methods are preferred. In areas of supplementary irrigation, sprinkler or drip irrigation may be more suitable than surface irrigation because of their flexibility and adaptability to varying irrigation demands on the farm.

Water availability: Due to higher water application efficiency with sprinkler and drip irrigation than surface irrigation, these methods are preferred when water is in short supply.

Water quality: Surface irrigation is preferred if the irrigation water contains much sediment, to prevent clogging of drip or sprinkler irrigation systems.

If the irrigation water contains dissolved salts, drip irrigation is particularly suitable, as less water is applied to the soil than with surface methods.

Type of Crop

Surface irrigation can be used for all types of crops. Sprinkler and drip irrigation, because of their high capital investment, are mostly used for high value cash crops, such as vegetables and fruit trees. Drip irrigation is ideal for irrigating individual plants or trees or row crops such as vegetables and sugarcane. It is not suitable for close growing crops like wheat, rice, sorghum, groundnut, pulse crops etc.

Type of Technology

In general, drip and sprinkler irrigation are technically more complicated methods. Purchase of equipment requires high capital investment. Equipment maintenance requires high level of know-how. Purchase of spare parts may require foreign currency.

Surface irrigation systems, in particular small scale schemes, usually require less sophisticated equipment for both construction and maintenance (unless pumps are used). Equipment needed is often easier to maintain and less dependent on the availability of foreign currency.

Previous Experience with Irrigation

Choice of an irrigation method also depends on the irrigation tradition within the region or country. Introducing a previously unknown method may lead to unexpected complications. It is not certain that the farmers will accept the new method. The servicing of the equipment may be problematic and the costs may be high compared to the benefits.

It will be easier to improve the traditional irrigation method than to introduce a totally new method.

Required Labour Inputs

Surface irrigation often requires a much higher labour input - for construction, operation and maintenance - than sprinkler or drip irrigation. Surface irrigation requires accurate land levelling and regular maintenance. Sprinkler and drip irrigation require little land levelling; system operation and maintenance are less labour intensive.

Costs and Benefits

Before choosing an irrigation method, an estimate must be made of the costs and benefits of the available options. On the cost side, not only the construction and installation, but also the operation and maintenance should be taken into account. These costs should then be compared with the expected benefits (yields). It is obvious that farmers will only be interested in implementing a certain method if they consider this economically attractive.

Surface irrigation is by far the most widespread irrigation method. It is normally used when conditions are favourable: mild and regular slopes, soil type with medium to low infiltration rate and sufficient supply of surface or groundwater. In the case of steep or irregular slopes, soils with a very high infiltration rate or scarcity of water, sprinkler and drip irrigation may be more appropriate.

22.4.1.2 Basin, Furrow or Border Irrigation

Important factors determining which surface irrigation method (basin, furrow or border (strip) irrigation) is most suitable are briefly discussed. It is not possible to give specific guidelines leading to a single best solution; each option has its advantages and disadvantages. Factors to be taken into account include:

- Natural circumstances (slope, soil type)
- Type of crop
- Required depth of irrigation application
- Level of technology
- Previous experience with irrigation
- Required labour inputs.

Natural Circumstances

Flat lands, with a slope of 0.1 percent or less, are best suited for basin irrigation: little land levelling will be required. If the slope is more than 1.0 per cent, terraces can be constructed with considerable land levelling.

Furrow irrigation (short, near horizontal furrows) can be used on flat land and on mildly sloping land with a maximum slope of 0.5 percent. On steeper sloping land, contour furrows can be used up to a maximum land slope of 3.0 percent. A minimum slope of 0.05 percent is recommended to assist drainage.

Border irrigation can be used on sloping land up to 2.0 percent on sandy soil and 5.0 per cent on clay soil. A minimum slope of 0.05 percent is recommended to ensure adequate drainage.

Surface irrigation may be difficult to use on irregular slopes as considerable land levelling may be required to achieve the required land gradients.

All soil types, except coarse sand with an infiltration rate of more than 30 mm per hour, can be used for surface irrigation. If the infiltration rate is higher than 30 mm per hour, sprinkler or drip irrigation should be used.

Type of Crop

Many crops (maize, sorghum, trees etc) can be grown in basins/check basins. Those that cannot stand a very wet soil for more than 12-24 hrs should not be grown in basins/check basins.

Furrow irrigation is best used for irrigating row crops such as maize, cotton, vegetables etc. Border irrigation is, particularly, suitable for close growing crops such as groundnut, fingermillet etc, but it can also be used for several other row crops.

Required Depth of Irrigation Application

When the irrigation schedule has been determined, it is known how much water has to be given per irrigation. It must be checked that this amount can indeed be given, with the irrigation method under consideration.

Field experience has shown that most water can be applied per irrigation application when using basin irrigation, less with border irrigation and least with furrow irrigation. In practice, in small scale irrigation projects, usually 40-70 mm of water are applied in basin irrigation, 30-60 mm in border irrigation and 20-50 mm in furrow irrigation.

This means that if only little water is to be applied per application, as for on sandy soils and a shallow rooting crop, furrow irrigation would be most appropriate. None of the surface irrigation methods can be used if the sand is very coarse (if the infiltration rate is more than 30 mm hr^{-1}).

If, on the other hand, a large amount of irrigation water is to be applied per application, as on a clay soil and with a deep rooting crop, border or basin irrigation would be more appropriate.

Above considerations have been summarised in Table 22.5. Net irrigation application values used are only a rough guide. They result from a combination of soil type and rooting depth. For example: if the soil is sandy and the rooting depth of the crop is medium, it is estimated that the net depth of each irrigation application will be in the order of 35 mm. Last column indicates which irrigation method is most suitable. In this case, medium furrows or short borders.

Table 22.5 *Selection of an irrigation method based on depth of net irrigation application*

Soil type	Rooting depth of the crop	Net irrigation depth per application (mm)	Irrigation method
Sand	Shallow	20-30	Short furrows
	Medium	30-40	Medium furrows, short borders
	Deep	40-50	Long furrows, medium borders, small basins
Loam	Shallow	30-40	Medium furrows, short borders
	Medium	40-50	Long furrows, medium borders, small basins
	Deep	50-60	Long borders, medium basins
Clay	Shallow	40-50	Long furrows, medium borders, small basins
	Medium	50-60	Long borders, medium basins
	Deep	60-70	Large basins

Level of Technology

Basin irrigation is the simplest of the surface irrigation methods. If the basins are small, they can be constructed by hand or animal traction. Their operation and maintenance is simple.

Furrow irrigation - with the possible exception of short, level furrows - requires accurate field grading. This is often done by machines (tractor drawn implements). The maintenance - ploughing and furrowing - is also often done by tractors.

Short, level furrows - also called furrow basins - can, like basins, be constructed and maintained by hand.

Borders require the highest level of sophistication. They are constructed and maintained by machines. The grading needs to be accurate. Machine operation requires a high level of skill, organisation and usually foreign currency.

Previous Experience with Irrigation

If there is no tradition in irrigation, the most simple irrigation method to introduce is basin/check basin irrigation. The smaller the basins, the easier their construction, operation and maintenance.

If irrigation is used traditionally, it is usually simpler to improve the traditional irrigation method than it is to introduce a previously unknown method.

Required Labour Inputs

Required labour inputs for construction and maintenance depend heavily on the extent to which machinery is used.

In general, it can be stated that to operate the system, basin irrigation requires the least labour and the least skill. For the operation of furrow and border irrigation systems, more labour is required combined with more skill.

22.5 Water Management for Groundnut

On an average, water requirement of groundnut varies from 450-600 mm depending upon the soil and climate. Estimated water requirements of groundnut are given in Table 22.6.

***Table 22.6** Water requirements of groundnut*

Place	Soil type	Water requirement (mm)
Israel	Light	600 to 700
Israel	Heavy	500 to 600
	India	
--Kanpur	Loamy	550
--Bhavanisagar	Sandy loam	600
--Rajendranagar	Medium-black clay, sandy - loam	550 to 650
--Chiplima	Sandy loam	650 to 750
--Punjab	Loamy sand	500 to 550
--Tirupati	Sandy loam	450 to 500

Soil moisture status invariably affects growth and development since 80 to 90 percent of plant fresh weight is water. Reduction in plant-water status much below this level causes visible wilting. Moisture stress in groundnut crop leads to:

- Reduction in leaf area leading to reduced photosynthetic area and crop growth rate
- Reduced plant height, fewer branches and flowers
- Reduction in yield up to 50 percent if the moisture stress is around 50 to 80 days after sowing (DAS)
- More number of ill-filled pods and reduction in sound matured kernels.

Critical stages of irrigation in groundnut are flowering, pegging and pod development. Soil moisture stress during initial stage (till flower initiation) results in optimum root growth and profused basal flowering leading to higher yields. Gynophore production seems relatively insensitive to early drought. Hence for fruiting to occur, gynophores must enter the soil for which soil moisture is important. Gynophores can exert a force equivalent to 3-4 grams only on the soil and mechanical resistance of a dry soil during pegging stage is a problem. As such, care should be taken to maintain optimum soil moisture for easy peg penetration during the pegging period. Similarly soil moisture stress during pod development results in more number of ill-filled pods, which drastically reduces the kernel yield of groundnut. Hence adequate soil moisture should be maintained during the pod development period to avoid water deficits.

22.5.1 Water Management Practices for Rabi Groundnut

Groundnut crop from second fortnight of November – second fortnight December to the end of March is known as *rabi* groundnut. In southern part of the country, where irrigation water is inadequate for a second crop of *rabi* rice after *kharif* rice, groundnut crop is the usual choice under well-irrigation.

22.5.1.1 Irrigation Schedules Based on Depletion of Available Soil Moisture (Dasm)

Inspite of advances in irrigation water management technology for crop production, cumbersome depletion of available soil moisture (DASM) approach is still the major irrigation scheduling approach to crops including groundnut.

On sandy loam soils of Tirupati, pod yield of *rabi* groundnut did not differ significantly due to irrigation schedules at 25 and 50 per cent DASM. Scheduling irrigations at 25, 50 and 75 per cent DASM required 600, 410 and 360 mm irrigation water, respectively (Ramesh Babu 1984).

Scheduling irrigation at 25 and 50 percent DASM during flowering and yield formation stages and at 75 percent DASM during maturity phase appears ideal for groundnut on sandy loam soils of Tirupati (Table 22.7). Soil moisture stress at flowering and yield formation stages drastically reduced the pod yield of groundnut (Srilatha 1992).

***Table 22.7** Pod yield as influenced by water management practices*

Treatments	DASM level (%) considered for irrigation				Pod yield (kg ha^{-1})
	Stages				
	Vegetative	Flowering	Yield formation	Maturity	
T_1	Irrigating the crop once in 10 days				3090^{c}
T_2	X	25	25	50	3631^{a}
T_3	X	25	25	75	3514^{a}
T_4	X	25	50	75	3406^{ab}
T_5	X	50	50	75	3194^{b}
T_6	X	50	75	75	2999^{cd}

22.5.1.2 Irrigation Schedules Based on IW/CPE RATIO

On sandy soils of Tirupati, application of 5 cm depth of irrigation water at each irrigation resulted in highest pod yield (25.5 q ha^{-1}) closely followed by 6 cm depth (22.6 q ha^{-1}). Lowest yield (19.2 q ha^{-1}) was due to 7 cm depth of water at each irrigation (Table 22.8). Among IW/CPE ratios, scheduling irrigation at 1.0 IW/CPE gave significantly highest pod yield (28.7 q ha^{-1}). The next best treatment was IW/CPE of 0.75 which was superior to IW/CPE of 0.5. Among interactions, scheduling irrigation at IW/CPE ratio of 1.0 with a depth of 5 cm of water at each irrigation resulted in higher pod yield (32.1 q ha^{-1}).

***Table 22.8** Pod yield, no. of irrigations, water requirements and consumptive use as influenced by irrigation depths and IW/CPE ratios.*

Treatments	1.0 IW/CPE	0.75 IW/CPE	0.5 IW/CPE	Mean
Pod yield (q ha^{-1})				
1. (5 cm)	32.1	24.9	19.5	25.5a
2. (6 cm)	26.7	24.8	16.3	22.6a
3. (7 cm)	27.2	23.1	7.4	19.2b
Mean	28.67a	24.27b	14.4c	
Consumptive use (mm)				
1. (5 cm)	333	263	134	243
2. (6 cm)	341	185	92	206
3. (7 cm)	241	148	60	150
Mean	305	199	95	

High consumptive use of water was with 5 cm depth (243 mm) and low consumptive use with 7 cm depth (150 mm). Consumptive use also differed among the three ratios with 1.0 IW/CPE resulting in highest consumptive use (305 mm) and the lowest with 0.5 IW/CPE ratio (Seshasailasree 1994).

At Anantapur (AP), scheduling irrigation at IW/CPE ratio of 1.0 resulted in significantly highest yield of (2117 kg ha^{-1}) (Table 22.9). Pod yields due to 0.8 (1800 kg ha^{-1}) and 0.6 (1500 kg ha^{-1}) did not differ significantly. Increase in pod yield (pooled) due to IW/CPE ratio of 1.0 was 41.1 and 17.6 per cent over IW/CPE ratio of 0.6 and 0.8 respectively (Padmalatha et al 2001).

***Table 22.9** Pod yield (kg ha^{-1}) as influenced by sowing dates and irrigation schedules*

Dates of sowing	Pooled over two years			
	IW/CPE ratio			
	1.0	0.8	0.6	Mean
November 01	2026	1779	1570	1792^{a}
November 16	2253	1972	1570	1931^{a}
December 01	2480	2035	1556	2024^{a}
December 16	2336	2038	1766	2047^{a}
December 31	2142	1915	1523	1860^{a}
January 15	2009	1670	1505	1728^{a}
January 30	1573	1194	1007	1258^{b}
Mean	2117^{a}	1800^{b}	1500^{b}	
CD (P = 0.05)	364.30	308.42		NS

Irrigation requirement increased with increasing IW/CPE ratio from 0.6 to 1.0 (Table 22.10). Irrigating the crop at IW/CPE ratio of 1.0 had highest total irrigation requirements (705.8 and 737.6 mm) and it was higher by 15.2 and 14.4 percent than 0.8 IW/CPE ratio and 35.7 and 33.6 percent than IW/CPE ratio of 0.6 during 1994-95 and 1995-96 respectively. Latest sowings irrigated at 1.0 IW/CPE ratio had highest total irrigation requirement in both the years (Padmalatha 2001).

Studies on the effect of irrigation regimes at different phenophases of *rabi* groundnut at Ratnagiri. (MS) indicated that scheduling at 0.6 – 1.0 – 0.8 IW/CPE ratios (IW/CPE of 0.6 up to flower initiation, 1.0 from flowering to early pod formation and 0.8 from pod formation to maturity) recorded significantly higher pod yield over all other irrigation treatments (Table 22.11). Stress at vegetative phase, no stress at reproductive phase and moderate stress at maturity phase appears ideal for saving irrigation water leading to high water use efficiency (Zagade and Chavan 2006).

***Table 22.10** Total irrigation requirement (mm) of groundnut as influenced by sowing dates and irrigation schedules*

Dates of sowing	1994-95				1995-96			
	IW/CPE ratio				IW/CPE ratio			
	1.0	0.8	0.6	Mean	1.0	0.8	0.6	Mean
November 01	600.0	500.0	450.0	516.7	650.0	550.0	450.0	550.0
November 16	650.0	550.0	450.0	550.0	650.0	600.0	550.0	600.0
December 01	600.0	500.0	450.0	516.7	650.0	550.0	500.0	566.7
December 16	700.0	600.0	500.0	600.0	745.4	645.4	545.4	645.4
December 31	775.8	675.8	575.8	675.8	820.2	720.2	570.2	703.5
January 15	807.4	757.4	607.4	724.1	770.2	720.2	620.2	703.5
January 30	807.4	707.4	607.4	707.4	877.2	727.2	627.2	743.9
Mean	705.8	612.9	520.1		737.6	644.7	551.9	

***Table 22.11** Effect of varying IW/CPE ratios at different phenophases on yield attributes and yield of groundnut (Pooled data).*

Treatments	100 kernel weight (g)	Shelling percentage	Pod yield (kg ha-1)
I1- 0.8-0.8-0.8	43.00	71.65	4060
I2-0.6-0.8-0.8	42.77	71.50	3990
I3- 0.6-0.8-1.0	42.71	71.25	3960
I4- 0.6-1.0-0.8	43.30	73.80	4380
I5- 0.6-0.8-0.6	40.60	62.40	2940
CD (P = 0.05)	1.20	0.67	120
(Sowing to 40 DAS:flowering initiation, 40 to 80 DAS:flowering to early pod formation and 80 DAs to maturity:pod formation to maturity)			

22.5.1.3 Irrigation Schedules based on Meteorological Parameters

Maintenance of adequate soil moisture (ETa = ETm) all through crop period gave significantly highest yield (2777 kg ha^{-1} in 1984 and 3208 kg ha^{-1} in 1985). The next best treatments, which produced higher yields are those where irrigations were scheduled to supply adequate soil moisture at flowering and yield formation stages, irrespective of moderate and severe stress at vegetative and maturity stages (Table 22.12). Moderate stress (60 percent DASM) all through the crop period gave significantly lesser pod yield compared with treatments which received adequate moisture only during flowering and yield formation stages. Severe stress either at flowering and yield formation stages or although the crop period produced 1496 and 1395 kg ha^{-1} pods respectively in 1984 and 1920 and 1810 kg ha^{-1} respectively in 1985 as against 2777 in 1984 and 3208 in 1985 when adequate moisture was provided all through the crop period (Raghava Reddy 1998).

Table 22.12 Pod yield (kg ha^{-1}) as influenced by irrigation treatments.

Treatments	Moisture depletion level (per cent) considered for irrigation in stages				1984	1985	Mean
	Vegetative (seeding to 20 DAS)	Flowering (21 to 55 DAS)	Yield formation (56-85)	Maturity (86 DAS to maturity)			
T_1	ETa = ETm all through the crop period				2777^a	3208^a	2993
T_2	80 DASM all through the crop period				1395^e	1810^e	1603
T_3	60 DASM all through the crop period				2179^c	2866^c	2523
T_4	80	ETa=ETm	ETa=ETm	80	2440^b	3024^b	2732
T_5	80	ETa=ETm	ETa=ETm	60	2536^b	3083^b	2810
T_6	60	ETa=ETm	ETa=ETm	60	2548^b	3098^{ab}	2823
T_7	60	ETa=ETm	ETa=ETm	80	2435^b	3009^b	2722
T_8	80	60	60	80	2047^d	2700^d	2374
T_9	60	80	80	60	1496^e	1920^e	1708
CD (P =0.05)					115	123	

Yield Response Factor (ky)

Response of yield to water supply can be quantified through the yield response factor (Ky) which relates relative yield decrease to relative evapotranspiration deficit as discussed under 3.2.3 above. Slope of the curve is the measure of the crop sensitivity to degree of stress. The Ky values were lowest (0.23) with the crop irrigated at ETa = ETm during flowering and yield formation stages in combination with 60 per cent DASM at vegetative and maturity stages (T_6) compared with T_7, T_5 and T_4 treatments (Fig 22.11). Corresponding values in these treatments varied from 0.40 to 0.43. This indicate that irrigating the crop with ETa = ETm at flowering and yield formation stages along with irrigations at 60 per cent DASM at vegetative and maturity stages appears most favourable for groundnut. The Ky values were higher in the crop irrigated at 80 per cent DASM (1.57) revealing that groundnut crop is sensitive to moisture stress and the yield decrease is severe with higher ET deficits (Raghava Reddy 1998).

22.5.1.4 Irrigation Frequency

Ten irrigations gave the highest pod yield of 2569 kg ha^{-1} which was comparable with that due to 7, 8 and 9 irrigations (Table 22.13). When irrigations were not scheduled beyond 70 DAS, yield reduced considerably relative to irrigation schedules up to 90 DAS as in the case of T_8 and T_9. Pod yield reduced drastically, when irrigations were not scheduled beyond 30, 45 and 56 DAS (Srinivasa Suman 1983).

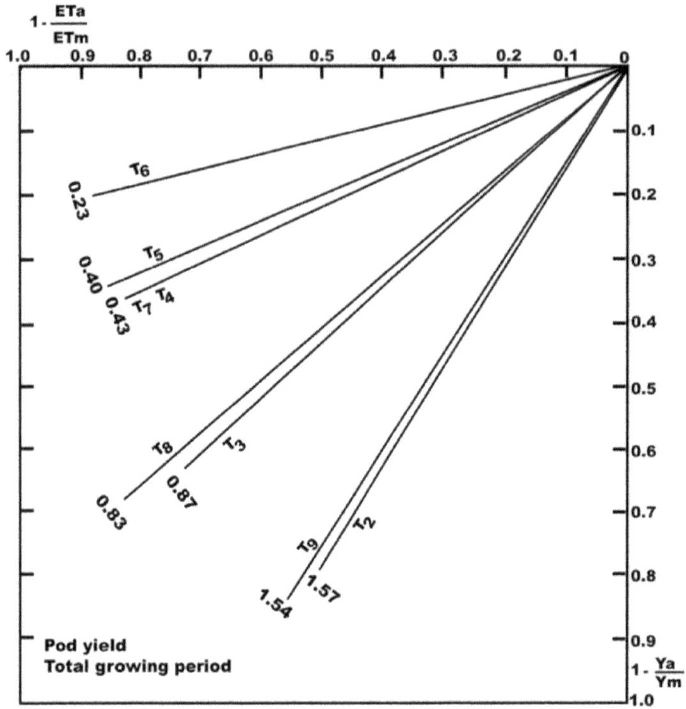

Fig. 22.11 *Relationship between relative yield decrease (1 - Ya/Ym) and relative evapotranspiration deficit (1 - Eta/ETm)*

Table 22.13 *Pod yield of groundnut as influenced by irrigation schedules*

Treatments	Pod yield (kg ha^{-1})	CU (mm)	WUE (kg ha-mm^{-1})
T_1:No irrigation after sowing	250[f]	78	3.20
T_2:Irrigatin at 0,30 DAS (2)	666[e]	119	5.59
T_3:Irrigatin at 0,25,45 DAS (3)	954[d]	182	5.24
T_4:Irrigatin at 0,25,34,45 DAS (4)	1007[d]	195	5.16
T_5:Irrigatin at 0,25,34,45,56 DAS (5)	1260[c]	229	5.50
T_6:Irrigatin at 0,10,20,34,45,56 (6)	1268[c]	289	4.38
T_7:Irrigatin at 0,10,20,34,45,56,70 DAS (7)	2152[b]	333	6.46
T_8:Irrigatin at 0,25,34,45,56,70 DAS (6)	2220[b]	330	6.72
T_9:Irrigatin at 0,25,34,45,56,70,90 DAS (7)	2463[a]	355	6.93
T_{10}:Irrigatin at 0,10,20,34,45,56,70,90 DAS (8)	2526[a]	398	6.34
T_{11}:Irrigatin at 0,20,30,37,44,51,61,75,90 DAS (9)	2570[a]	402	6.41
T_{12}:Irrigatin at 0,10,20,30,37,44,51,61,75,90 DAS (10 irrigations)	2596[a]	441	5.88
CD (P = 0.05)	159		

Highest consumptive use (441 mm) was due to 10 irrigations followed by 9 and 8 with a consumptive use of 402 and 398 mm respectively (Table 22.13). Maximum water use efficiency (6.93 kg ha-mm^{-1}) was due to 7 irrigations followed by that due to 6 irrigations (T_8) with 6.72 kg ha-mm^{-1}. Application of 10 irrigations resulted in WUE of 5.88 kg ha-mm^{-1}.

High frequency of irrigating the crop once in two or thee days, resulted in highest pod yield (3574 kg ha^{-1}). Pod yield decreased significantly with decrease in irrigation frequency from 2-3 days to 8-12 days (Table 22.14). Increasing the IW/CPE ratio from 0.50 to 1.00 significantly increased the pod yield of groundnut. Scheduling irrigation at IW/CPE ratio of 1.0 resulted in highest pod yield of 3551 kg ha^{-1} followed by 3219 kg ha^{-1} with IW/CPE ratio of 0.75 and 2773 kg ha^{-1} with IW/CPE ratio of 0.5 (Bhaskara Reddy 1979).

On deep black soils of Dharwad (Karnataka) also, high frequency irrigation resulted in highest pod yield (Hosamani and janawade 2007). Pod yield of *rabi* groundnut was significantly high (2210 kg ha^{-1}) due to five irrigations at critical stages compared with three irrigations at critical stages and irrigation at fortnightly interval all through the crop period (1790 kg ha^{-1}).

***Table 22.14** Pod yield (kg ha^{-1}) as influenced by frequency and depth of irrigation (pooled mean 2 years).*

Irrigation frequency	Depth of irrigation (pooled data)			
	IW/CPE = 0.5	IW/CPE = 0.75	IW/CPE = 1.0	Mean
F_1: Irrigation when cumulative evaporation from free water surface (E_0) is 2 cm (irrigation once in 2-3 days)	3161	3551	4011	3574a
F_2: Irrigation when cumulative evaporation from free water surface (E_0) is 4 cm (irrigation once in 4-6 days)	2950	3468	3677	3365b
F_3: Irrigation when cumulative evaporation from free water surface (E_0) is 6 cm (irrigation once in 6-9 days)	2585	3022	3336	2981c
F_4: Irrigation when cumulative evaporation from free water surface (E_0) is 8 cm (irrigation once in 8-12 days)	2398	2834	3182	2805d
Mean	2773c	3219b	3551a	

22.5.2 Water Management Practices for Summer Groundnut

Groundnut crop cultivated from second fortnight of February to the end of June is, generally, known as *rabi* groundnut. Since the irrigation water is scarce during summer, the extent of area under groundnut is limited. However, productivity of groundnut is highest in summer season.

At Sardar Krushinagar, Gujarat, scheduling irrigation at 40 mm CPE (high frequency irrigation) resulted in significantly higher pod yield (3.79 t ha^{-1}) over 50 (3.48 t ha^{-1}) and 60 mm (3.14 t ha^{-1}) CPE. Consumptive use of water increased with decrease in CPE. Highest mean CU (795.77 mm) was due to high frequency irrigation of scheduling irrigation at 40 mm CPE (Table 22.15). The WUE increased with decrease in irrigation interval. Highest mean WUE (4.76 kg ha^{-1} mm^{-1}) was with scheduling irrigation at 40 mm CPE while the lowest was with 60 mm CPE (Patel et al 2006).

***Table 22.15** Consumptive use of water, water use efficiency and pod yield of summer groundnut as influenced by irrigation schedules*

Irrigation schedules	Consumptive use (mm)			Water use efficiency (kg ha-mm^{-1})			Pod yield (t ha^{-1})		
	2003	2004	Mean	2003	2004	Mean	2003	2004	Mean
40 mm CPU	809.87 (985)	781.66 (935)	795.77 (960)	5.25	4.27	4.76	4.24	3.33	3.79
50 mm CPU	760.64 (875)	924.71 (825)	742.68 (850)	5.13	4.22	4.68	3.90	3.05	3.48
60 mm CPU	735.79 (765)	697.07 (770)	716.43 (768)	4.83	3.96	4.40	3.51	2.76	3.14
CD (P = 0.05)	-	-	-	-	-	-	0.38	0.31	0.23

Figures in parentheses indicate total amount of water applied.

Summer groundnut requires high frequency irrigation from pegging to pod development. Highest pod yield and water use efficiency was due to irrigations at IW/CPE ratio of 0.5 from seeding to flowering, 0.9 from pegging to pod formation and 0.7 from pod development to maturity. Pod yield due to this and that due to irrigation schedules at IW/CPE ratio of 0.9 from sowing to maturity did not differ significantly. High frequency irrigation, however, increased the crop water requirement (Pahalwan and Tripathi, 1984).

At Akola, summer groundnuts on sandy loam soils require 15 to 17 irrigations depending on climate, soil and management practices. Irrigation interval should be 10-12 days in February, 8-10 days in March, 6-8 days in April and 4-6 days in May. Withholding irrigation up to 25 days, depending upon soil type, can improve the yield by around 20 per cent.

A Simple Irrigation Scheduling Method

A simple irrigation scheduling method, indicating irrigation quantity and irrigation intervals, is the most basic approach for scheduling irrigation. By assessing the average temperature in the location of the farm, the crop, the soil texture and the use of some data tables given by FAO, the farmer will be able to know how much water to apply during each irrigation and periodicity of irrigation.

This method assumes that the crop is grown during the period of peak water demand, little or no rainfall occurs during the growing season and that the maximum possible net application depth is 70 mm. Also it considers only three different soil types (sandy, loam and clay) and three different climates (mean daily temperature <15°C, 15-25°C; >25°C). This method considers three steps as dictated below:

Step 1: Determine the type of soil between sand, loam and clay, which have, respectively, low, medium and high available water content (Table 22.16).

Table 22.16 *Type of soil and its irrigation characteristics*

Type of the soil	Soil characteristics
Shallow and/or sand	In sandy soil or shallow soil (with a hard pan or impermeable layer close to the soil surface), little water can be stored; irrigation will thus have to take place frequently but little water is given per application.
Loam	In loamy soil more water can be stored than in a sandy or shallow soil. Irrigation water is applied less frequently and more water is given per application.
Clay	In clay soil even more water can be stored than in a medium loamy soil. Irrigation water is applied even less frequently and again more water is given per application.

***Step 2*:** Determine climate of the farm location. This method distinguishes between three different climates; climate 1, climate 2 and climate 3 as described in the Table 22.17.

Table 22.17 *Types of climate*

Type of climate	Reference crop ET (ETo)	Climatic zone and mean daily temperature
Climate 1	4 - 5 mm day^{-1}	Desert/arid, semi arid and sub-humid with low temperatures (less than 15°C).
Climate 2	6 - 7 mm day^{-1}	Desert/arid, semi arid and sub-humid with medium temperatures (15-25°C).
Climate 3	8 - 9 mm day^{-1}	Desert/arid, semi arid and sub-humid with high temperatures (more than 25°C).

Step 3: Determine the recommended interval and irrigation depth.

Table 22.18 provides net irrigation depth for each type of soil and climate and also the interval of days to apply irrigation for different type of soil in brackets.

Table 22.18 *Interval and net irrigation depths for some crops grow in shallow, loam and clays soil in different types of climates.*

Type of the soil	Shallow and/or sand Interval (net irrigation depth mm)			Loam soil Interval (net irrigation depth mm)			Clay soil Interval (net irrigation depth mm)		
Type of climate	1	2	3	1	2	3	1	2	3
Crop: groundnut	6	4 (30)	3	8	6 (40)	4	10	7 (50)	5

Groundnut grown in loamy soil, in a semi-arid area with high temperatures >25°C (Climate 3 according to table above) require 40 mm of net irrigation depth every 4 days.

22.5.3 Rainwater Management for Kharif Groundnut

The rainfed groundnut is usually exposed to mid or late season drought (August) leading to yield reduction by more than 50 per cent. Results of experiments clearly demonstrated need for one or two protective irrigations to *kharif* groundnut during the dry spell. Irrigations should be given at an interval of 12 to 15 days and 12-15 days after last rain.

A large number of simple *in situ* soil and moisture conservation practices such as broad bed and furrows, compartmental bunding, dead furrows, tied ridges, vegetative barriers etc have been tested, with varying success, for minimising the vagaries of rainfall on *kharif* groundnut. This topic will be dealt elaborately under dryland technology for groundnut production.

22.5.3.1 Rain Water Harvesting and Supplemental Irrigation

Rainwater is dissipated as runoff, evaporation, transpiration and deep percolation. Techniques have been developed to harvest runoff, at times of excess rainfall, for efficient use incrop production.

In semiarid areas, high intensity rains leads to erosion and 20-40 per cent of rainfall is lost through runoff. Prolonged dryspells during the crop season drastically reduce the productivity of rainfed groundnut. The extent of yield reduction depends upon the time and duration of the dryspell during crop period.

Results of experiments at Agricultural Research Station, Anantapur (AP) indicated scope for saving the rainfed groundnut if supplemental irrigation is given at times of prolonged dryspells. Hence, feasibility of harvesting the rainwater and its use for groundnut as supplemental irrigation at times of dryspells, using micro-irrigation methods were takenup. Farm ponds were developed and supplemental irrigation, were given through sprinklers.

One supplemental irrigation of 10 mm from farm pond through sprinkler irrigation for a 20 days dryspell at pod development stage increased the pod yield by 30-35 per cent (Table 22.19).

Table 22.19 *Effect of supplemental irrigation on groundnut yield.*

S. No	Treatment	Pod yield (kg ha^{-1})						RWUE (kg ha-mm^{-1})
		2005-06	2006-07	2007-08	2008-09	2009-10	Mean	
1.	Rainfed	952	1095	1055	-	332	859	2.1
2.	Irrigated	1428	1478	1250	-	374	1132	2.7

Farm pond of 250 m^3 capacity (size of 10 × 10 m with 2.5 m depth) with side slopes of 1.5:1 will be adequate for catchment area of 5 ha. Soil + cement lining with 6:1 ratio appears to be effective in reducing seepage losses. A supplemental irrigation of 10 mm through sprinkler at 45 days after sowing (pod formation stage) increased the pod yield of groundnut significantly.

Depth and method of irrigation: Normal depth of water, at each irrigation, for irrigated crops is around 50 mm. Since water available in farm ponds is limited, available water has to be effectively utilised. At Anantapur; there was no significant difference in pod yield of groundnut due to 10, 15 and 20 mm depth of irrigation water. As such, 10 mm depth of water was given. However, for Vertisols 40 mm depth of irrigation appears necessary for optimum yield. Thus, supplemental irrigation can be given to about one ha of land, using water stored in a farm pond of size 250 – 300 m^3 through sprinkler irrigation. However, instead of traditional methods, micro-irrigation methods like drip or sprinkler systems are to be used for effective conveyance of harvested rainwater (Fig 22.12).

Fig. 22.12 *Sprinkler and drip irrigation systems for supplemental irrigation*

22.5.3.2 Mobile Irrigation

Under irrigated conditions as in the case of well-irrigation, the source of irrigation water (well) will be within the cropped field. Water will be directly let in to the cropped field through field channels. There is no necessity for transporting irrigation water from well to the field. If the distance between the source of irrigation water and the cropped field is more, water conveyance to the field is limited by costs associated with conveyance as in the case of supplemental irrigation to dryland crops. Under such conditions, irrigation water has to be transported through automobiles from a source far away from the cropped field. Mobile irrigation is a system of irrigation in which crops are irrigated by transporting water from far away places to the cropped fields. Mobile irrigation is practiced for high value crops only. It is similar to transporting water from distant places to the residential areas through tractors (Fig.22.13).

Fig. 22.13 Mobile irrigation units

Lorry tanker (12000 liters) fitted with motor both for water lifting and supplemental irrigation appears to be ideal. Water from a source will be filled by the motor and transported to the field for giving supplemental irrigation through sprinklers with motor of the tanker. Even tractor water tanker of (4000 liters) capacity can be used for drip or sprinkler irrigation.

22.6 Appropriate Methods of Irrigation for Groundnut

Scientific irrigation aims at efficient use of irrigation water in conjunction with other essential inputs to sustain crop production. The manner in which irrigation water is applied to the land is commonly referred to a method of irrigation. Selection of appropriate method of irrigation has been discussed under 4 Methods of Irrigation.

22.6.1 Surface, Sprinkler or Drip Irrigation

Effect of micro-sprinkler irrigation with fertigation under different land configurations on groundnut yield and water use efficiency was studied at Madurai (TN), during June-September 2002 and January-April 2003 (Vijayalakshmi et al 2006). Micro-sprinkler irrigation gave higher pod yield (Table 22.20) compared to surface irrigation. Increase in pod yield with micro-sprinkler irrigation was due to high frequency irrigation which maintained soil moisture content in the active root zone at adequate level throughout the crop period. Within the micro-sprinkler irrigation treatments, pod yield was higher due to irrigation scheduling at 100 per cent ETc over micro-sprinkler irrigation scheduling at 75 per cent ETc. Fertigation through micro-sprinkler irrigation resulted in significantly higher pod yield than soil application under micro-sprinkler irrigation or surface irrigation irrespective of the effects of land configuration.

***Table 22.20** Effect of land configuration, micro-sprinkler irrigation and fertigation on pod yield and water use efficiency of groundnut*

Treatments	Pod yield (kg ha^{-1})		Water use efficiency (kg ha-mm^{-1})	
	2002	2003	2002	2003
T1. CB + SI + SAF	1563	1511	2.64	2.48
T2. BBF + SI + SAF	1909	1867	3.22	3.07
T3. CB + MSI 100 % ETc + SAF	2171	2051	5.35	4.65
T4. CB + MSI 100 % ETc + F	3430	3387	8.45	7.69
T5. CB + MSI 75 % ETc + SAF	2067	2009	5.15	5.34
T6. CB + MSI 75 % ETc + F	3053	2899	7.61	7.71
T7. BBF + MSI 100 % ETc + SAF	2367	2255	5.83	5.12
T8. BBF + MSI 100 % ETc + F	3844	3776	9.47	8.57
T9. BBF + MSI 75 % ETc + SAF	2113	2047	5.27	5.44
T10. BBF + MSI 75 ETc + F	3202	3128	7.98	8.32
CD (P = 0.05)	225	234	0.53	0.55

(CB-check basin, BBF- broad bed and furrow, SI-surface irrigation, SAF-surface application of fertiliser, MSI-micro-sprinkler irrigation, F- fertigation)

Micro-sprinkler irrigation improved water use efficiency relative to surface irrigation (Table 22.21) mainly due to higher pod yield and saving in irrigation water. Micro-sprinkler irrigation at 100 percent ETc gave higher water use efficiency than 75 percent ETc in Crop I and vice versa in Crop II. Broad bed and furrow results in higher water use efficiency relative to check basin.

Table 22.21 *Effect of irrigation methods and irrigation schedules on consumptive use (CU).*

Treatments	Irrigation water (mm)		Water saving (%)		Consumptive use (mm)		Total water saving (%)	
	2002	2003	2002	2003	2002	2003	2002	2003
Surface irrigation at 0.8 IW/CPE	480.00	600.00	-	-	592.20	608.50	-	-
Micro-sprinkler irrigation at 100% ETc	332.65	433.13	31	28	405.82	440.73	31	28
Micro-sprinkler irrigation at 75% ETc	279.49	368.46	42	39	401.11	376.06	32	38

Surface irrigation required higher quantity of irrigation water relative to micro-sprinkler irrigation (Table 22.21). Micro-sprinkler irrigation scheduling at 100 per cent ETc improved water use relative to that due to irrigation schedule at 75 percent ETc. Consequently, micro-sprinkler irrigation at 100 percent ETc saved irrigation water up to 31 percent in Crop I and 28 percent in Crop II as compared to surface irrigation at 0.80 IW/CPE ratios with 6 cm depth. Micro-sprinkler irrigation at 75 per cent ETc saved more irrigation water up to 42 percent and 39 percent in Crop I and Crop II respectively.

Consumptive use was higher due to surface irrigation (592.20 mm – 608.50 mm) compared to micro-sprinkler irrigation. Micro-sprinkler irrigation at 100 percent ETc resulted in higher consumptive use (405.82 mm – 440.73 mm) than that at 75 percent ETc (376.06 mm – 401.11 mm) in both crops. Micro-sprinkler irrigation at 100 ETc resulted in 31 percent water saving during 2002 and 28 percent during 2003. Water saving due to micro-sprinkler irrigation at 75 ETc was 32 percent in 2002 and 38 percent in 2003 (Vijayalakshmi et al 2006).

Results of the investigations lead to formulation of the following best package of management practices for realising high yield and income from groundnut. Individually, each of the following treatments proved to be the most effective.

- Micro-sprinkler irrigation at 100 percent ETc
- Fertigation of 100 percent RDF
- Fertigation in 9 equal splits at 7 days interval from 14 DAS to 70 DAS
- Land configuration of broad bed and furrow formation.

On sandy loam soils of Tirupati, *rabi* groundnut gave highest pod yield (Table 22.22) with drip irrigation (RARS 2006).

Table 22.22 Relative efficiency of irrigation methods on yield, irrigation requirement and water use efficiency of groundnut

Methods of irrigation	2005			2006		
	Pod yield (kg ha^{-1})	Irrigation requirement (mm)	WUE (kg ha-mm^{-1})	Pod yield (kg ha^{-1})	Irrigation requirement (mm)	WUE (kg ha-mm^{-1})
Drip	2880	433	6.65	2267	362	6.15
Sprinkler	2380	487	4.89	2469	365	5.15
Flat bed	2250	541	4.16	1831	420	4.35

Irrigation requirement was least with drip system followed by sprinkler and drip systems in both the years. High water use efficiency was with drip irrigation followed by sprinkler and drip systems.

22.6.2 Supplemental Drip Irrigation

Drip irrigation is the process of slow application of water in the form of discrete, continuous drops or minute sprays through mechanical devices called emitters or applicators located at selected points along water delivery lines. Fertilisers and other agrochemicals can be efficiently applied to crops using drip irrigation.

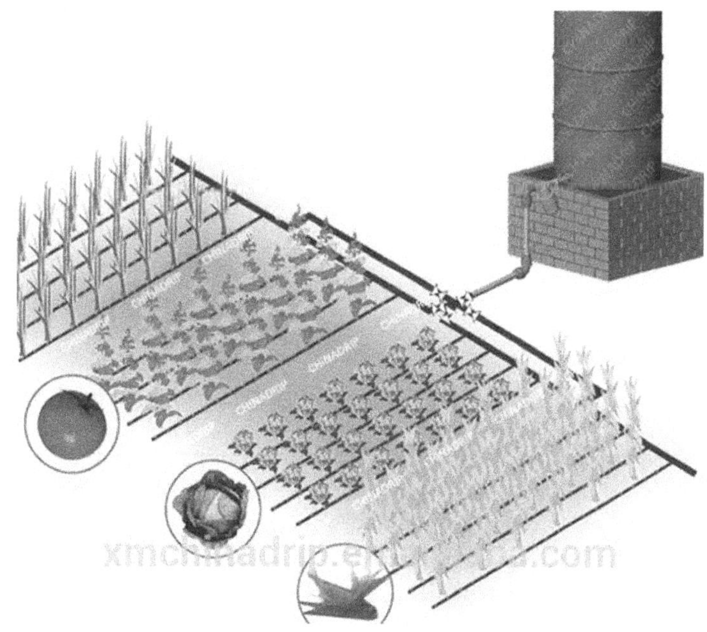

Fig. 22.14 Drip system with mobile irrigation water

Drip irrigation can save water by reducing the portion of the soil surface that is wetted, thus decreasing the amount of direct evaporation. A container filled with water can be the source of irrigation water to which drip system can be connected (Fig 22.14). The container can be filled with mobile water service for irrigating the crop. Mobile supplemental drip irrigation system appears ideal for dryland groundnut, especially at times of prolonged drought period.

22.6.3 Planting Technique for Saving Irrigation Water

Paired row planting of groundnut at 45 × 15 cm spacing (paired row at 15 cm on raised beds with 45 cm furrow spacing) can save irrigation water over normal sowing.

Paired row planting (Fig 22.15) on raised beds gave pod yield advantage by 13-20 percent over flat bed planting of groundnut, irrigation water saving to the extent of 27-41 percent and enhancement of crop WUE by 40-45 percent. Improved planting technique has the potential advantage of saving irrigation water. This technique would bring more areas under irrigation for post-rainy season crops (Rama Krishna and Rao 2006).

Fig. 22.15 Groundnut crop with paired-row planting

Adequate available soil moisture in the upper 60 cm soil depth is important for high yield of groundnut. Most soils at field capacity hold about 3.1 cm of water at 30 cm depth. Highest groundnut yields are observed when available soil moisture is maintained above 50 per cent of field capacity. Therefore, sprinkler irrigation is recommended when the moisture has been depleted to 50 per cent of field capacity in the top 60 cm. During peak water use periods (0.6-0.7 cm day^{-1}), the crop may require 3.0 to 3.5 cm of water every 5 days. Irrigation intervals can be prolonged for 7-10 days during periods of low daily water requirements.

If the soil-water potential is measured, irrigation should maintain the soil-water potential above –0.6 bars. When long, dry, hot periods occur at the sensitive growth stages, such as pegging, pod formation and early pod filling, sprinkler irrigations are necessary to maintain the soil-water potential above – 0.25 to – 0.50 bars. This means that the upper 30 cm of soil should appear and feel moist and the plants should not wilt from lack of moisture during the afternoon.

Summary of Water Management Practices for Groundnut

From the results of the experiments at several situations (climate, soils and management practices) in the country, the in following are the recommended water management practices for *kharif*, *rabi* and summer groundnut:

Rainfed (*kharif*) groundnut

- Pronged breaks (dry spells) during the monsoon, especially during August, are the major factors limiting groundnut productivity during *kharif.*
- Deep summer tillage increases the water intake and storage capacity of the soil leading to sustaining the crop for longer periods of drought than due to normal tillage.
- *In situ* soil moisture conservation measures (broad beds and furrows, dead furrows, compartmental bunding, mulching etc) can sustain the crop for longer periods at times of breaks in the monsoon.
- For a crop on shallow soils, soil moisture stress up to 20 DAS may not have any adverse effect on final pod yield. Soil moisture stress up to 25 DAS appears to be beneficial for a crop on medium to deep soils.
- In the case of supplemental irrigation (farm ponds or mobile irrigation), best results are obtained when water is applied during the flowering period.
- If there is scope for two supplemental irrigations, they should be given at flowering and pod development stages for best results.
- Supplemental irrigation through sprinkler or drip system aids in efficient uses of scarce water for irrigation. Larger area can be irrigated with these systems of irrigation.
- Perfect weed management can sustain crop for longer period of drought relative to weedy crop.

Irrigated (*rabi)* groundnut

Water management for groundnut crop is complicated by due to its prolonged flowering period with two or three flushes of flowering. It is not uncommon to observe the crop with flowers, pegs, developing the pods and mature pods at a time, especially, when the crop is subjected to water deficits at first or second flush of flowering.

Frequency of irrigation

Groundnut crop requires a minimum of 7 to 10 irrigations depending on the soil type. Light textured soils require more frequent irrigations relative to heavy soils. General irrigation schedule for bunch groundnut with duration of 105 days is as follows:

- I irrigation: Immediately after sowing (not necessary if the soil contains optimum soil moisture for seed germination and establishment).
- II irrigation: 5 to 7 DAS, if necessary on light sandy loam soils (vegetative growth).
- III irrigation: 25 DAS (flowering).
- IV irrigation: 32 DAS (flowering).
- V Irrigation: 39 DAS (flowering and peg formation).
- VI irrigation: 46 DAS: (pegging).
- VII irrigation: 56 DAS (pegging).
- VIII irrigation: 66 DAS (pod filling).
- IX irrigation: 78 DAS (pod maturity).
- X irrigation: 88 DAS (final).

Generally, first two irrigations are not necessary for groundnut. As such, eight irrigations appear ideal for a crop of 105 days duration.

The crop needs no irrigation from 88 DAS to harvest. If the crop has to be harvested by hand pulling, one light irrigation may be given one or two days before manual harvest.

Irrigation scheduling based on critical stages approach

- Pre-sowing irrigation (stand establishment), if necessary. No necessity there after up to 20 DAS.
- Starting from 20 DAS (flowering) to pod formation, 6 irrigations at 8 days interval on light soils and 5 irrigations at 10 days interval on medium soils.
- Last two irrigations at 15 days interval.

Irrigation scheduling based on DASM approach

- No need for irrigating relatively heavy soils up to 25 DAS.
- Maintenance of around 80 per cent DASM appears ideal for a crop on light soils up to 20 DAS.
- Starting from 20 DAS (flowering) to pod formation, DASM should not exceed 40 per cent
- From pod development to maturity, scheduling irrigation at 80 per cent DASM is adequate.

Irrigation scheduling based on IW/CPE approach

- No necessity for irrigation a crop on relatively heavy soil up to three weeks after seeding.
- Crop on light soil require irrigation schedules at an IW/CPE ratio of 0.5.

- Irrigate the crop at an IW/CPE ration of 1.0 from flowering to pod development.
- From pod development to maturity, irrigations are to be scheduling at an IW/ CPE ratio of 0.8

Sprinkler and drip irrigation systems with an irrigation efficiency around 80 per cent leads to high water use efficiency. Among the surface irrigation methods, border strip irrigation appears ideal under several situations.

Summer groundnut

Summer groundnut area is limited by the scarcity of irrigation water in arid and semiarid regions. Its water requirements are also higher than *rabi* groundnut due to hot weather during the summer. Water management practices for summer groundnut crop are almost similar to that of *rabi* crop except that the summer crop requires more frequent irrigations relative to rainfed and *kharif* crops.

- Irrigation at 75 DASM appears adequate from sowing to flowering.
- From flowering to pod development, irrigations are to be scheduled at 40 per cent DASM.
- From pod development to maturity, irrigations can be scheduled at 60 per cent DASM in semiarid regions.
- Scheduling irrigation at IW/CEP ratio of 0.6 is optimum from seeding to flowering under several situations.
- From flowering to pod development, irrigation may be scheduled at an IW/CPE ratio of 1.2.
- From pod formation to maturity, scheduling irrigation at IW/CPE ratio of 0.8 appears idea.

References

1. *Allen RG, Pereira LS, Raes D and Smith M 1998. Crop evapotranspiration: Guidelines for computing crop water requirements. In: FAO IDP No 56, Rome.*
2. *Bhaskara Reddy G. 1979. Studies on frequency and depth of irrigation for groundnut. M Sc (Ag) Thesis submitted to ANGRAU, Rajendranagar, Hyderabad, Andhra Pradesh.*
3. *De Wit CT 1968. Plant production. In: Miscellaneous Papers Landbouw Hogeschool, Wageningen No. 3: 25-50.*
4. *Doorenbos J and Kassam AH 1979. Yield response to water. FAO IDP 33, Rome.*
5. *Doorenbos J and Pruitt WD 1977. Guidelines for predicting crop water requirements. FAO IDP 24, Rome.*

6. *Hosamani MH and Janawade AD 2007. Response of rabi groundnut to irrigation schedules and integrated nutrient management in deep black soils of upper Krishna command. Karnataka J Agric Sci 20 (3): 453-456.*

7. *Kassam AH 1977. Net biomass production and yield of crops. Present and potential land use by agroecological zones project. Rome, FAO.*

8. *Padmalatha Y, Reddy SR and Reddy TY 2001. Soil moisture studies on rabi groundnut under dates of sowing and irrigation schedules in low rainfall zone of Andhra Pradesh. J Oilseeds Research Vol 18 (2): 187-192.*

9. *Pahalwan DK and Tripathi RS 1984. Irrigation scheduling based on evaporation and crop water requirement for summer peanuts.* **www.peanut** science.com.

10. *Patel GN, Patel PT and Patel PH 2006. Yield, water use efficiency and moisture extraction pattern of summer groundnut as influenced by irrigation schedules, sulfur levels and sources. ejournal.icrisat.org/Volume6/Groundnut/GN Patel.pdf.*

11. *Prihar S and Sandhu BS 1987. A critical appraisal of research on irrigation scheduling to crops. Pro Sec World Cong on Water Resources. International Water Resource Association, New Delhi.*

12. *Raghava Reddy C 1988. Studies on yield response to water in groundnut. Ph D Thesis submitted to ANGRAU, Rajendranagar, Hyderabad, Andhra Pradesh. Ramakrishna YS and Rao GGSN 2006. Water management - Water use in rainfed regions of India.* ***www.wamis.org/agm/meetings/worisk06/S5*** *Rama-krishna_WORISK.pdf.*

13. *Ramesh Babu N, Reddy SR, Reddi GHS and Reddy DS 1984. Effect of fertilisers, irrigation and plant density on groundnut. Fertiliser Research, Vol 5 (**3**): 273-277.*

14. *RARS 2006. Annual report 2005-06, Regional Agricultural Research Station, Tirupati, Andhra Pradesh.*

15. *Sesha Saila Sree P 1994. Water requirements of groundnut at different IW/CPE ratios. M.Sc (Ag) Thesis submitted to ANGRAU, Rajendranagar, Hyderabad, Andhra Pradesh.*

16. *Simhachalam P 1981. Studies on high frequency irrigation and nitrogen levels on groundnut. M Sc (Ag) Thesis submitted to ANGRAU, Rajendranagar, Hyderabad, Andhra Pradesh.*

17. *Srilatha D 1992. Water management practices for groundnut at times of deficit water supplies. M Sc (Ag) Thesis submitted to ANGRAU, Rajendranagar, Hyderabad, Andhra Pradesh.*

18. *Srinivasa Rao M 1989. Effect of sources of phosphorus and irrigation frequencies on growth and yield of groundnut. M Sc (Ag) submitted to ANGRAU, Rajendranagar, Hyderabad, Andhra Pradesh.*

19. *Srinivasa Suman C 1983. Studies on scheduling of irrigation to summer groundnut. M Sc (Ag) Thesis submitted to ANGRAU, Rajendranagar, Hyderabad, Andhra Pradesh.*

20. *Vijayalakshmi R, Veerabhadran V, Shanmugasundaram K and Kumar V 2006. Micro-sprinkler irrigation and fertigation and land configuration as a best management technology package for groundnut. irncid.org/Get File Articles.aspx? FilePrm = 6878_85877.pdf.w*

21. *Zagade MV and Chavan SA 2006. Growth and yield of* rabi *groundnut as influenced by moisture regimes, polythene mulch and plant densities. Ann Agric Res New Series Vol 27 (4): 355-359.*

Chapter 23

Soil Profile Moisture Study – Hands on Experience

G. Krishna Reddy
PS (Agronomy), RARS, Tirupati.

23.1 Introduction

The NMM is available as both a surface meter, which lies flat on the soil surface, and as a profiling meter. The latter consists of a cylindrical probe which is connected by a cable to a case containing the power supply, display, keypad and microprocessor. During use, the probe is lowered into an access tube in the soil for readings, while the case remains at the surface. When not in use, the probe is locked inside the case, which contains a high density plastic shield. The surface meter has not proved to be useful for agricultural and environmental uses. Soil moisture can be estimated quickly and continuously with neutron moisture meter without disturbing the soil. Another advantage is that soil moisture can be estimated from large volume of soil. This meter scans the soil about 15 cm diameters around the neutron probe in wet soil and 50 cm in dry soil. It consists of a probe and a scalar or rate meter. This contains a fast neutron source which may be a mixture of radium and beryllium or americium and beryllium. Access tubes are aluminum tubes of 50-100 cm length and are placed in the field when the moisture has to be estimated. Neutron probe is lowered in to access tube to a desired depth. This is an indirect method of soil moisture determination which uses the interaction of neutron with water molecules as a basis of monitoring soil water content in situ.

23.2 Neutron Probe

Neutron meter (neutron scattering method)

Soil moisture can be estimated quickly and continuously with neutron moisture meter without disturbing the soil. It is one of the most accurate devices available for measuring the amount of moisture in the soil profile. Another advantage is that soil moisture can be estimated from large volume of soil. This meter scans the soil to about 15 cm. diameter around the neutron probe in wet soil and 50 cm in dry soil. Access tubes are aluminum tubes of 50 to 100 cm length and are placed in the field where moisture is to be estimated. Aluminum is practically transparent to neutrons and therefore, is a desirable material for access tubes

Fast neutron: A high-energy neutron, which travels at a speed of about 6,000 miles per second.

Flux: The number of particles crossing a unit of area per unit of time.

Half-life: The amount of time required for a radioactive element to decay to one-half of its original strength.

Neutron: Elementary nuclear particle with a mass of one and a charge of zero.

Slow neutron: A low-energy neutron, which travels at a speed of about 1.7 miles per second.

Source: A small sealed capsule which contains a mixture of radioactive material americium and beryllium.

23.2.1 Access Holes and Tube

An access hole and tube must be provided at each measurement site to allow the probe to be lowered into the soil. Tubes should be installed in snug-fitting holes pre-drilled with a hand or machine operated soil core barrel or auger. Placing the access tube in an oversized hole and back filling is not recommended. Access holes should be lined with a durable tubing which will hold the dimensions of the hole at a constant size and exclude free water from moving into the hole. The access tube should be as small as possible, yet allow easy movement of the probe, have minimum wall thickness, and be made of a material that does not hinder the movement of neutrons.

Conditions met while placement of access tubes

1. The site should be representative of the field to be monitored.
2. No void spaces should exist between the access tube and soil.
3. Mixing of soils adjacent to access tubes should be avoided during placement.
4. No vertical flow of water should occur along the access tube from the surface.
5. The top of the tube must be plugged to exclude water either from irrigation or rainfall.
6. Permanent tubes are buried below plow depth and covered with soil. The soil cover is removed when measurements are to be made.
7. When tubes are to be monitored more often, they should be installed in the crop row after seedling emergence with the top extending above ground.

23.2.2 Method

The probe is lowered into the soil down an access tube where neutrons are emitted and interact with soil water. The density of the neutron flux is dependent upon the amount of water present in the surrounding soil. It consists of a probe and a scalar or rate meter (gauge). The probe contains **fast neutron source,** which may be a mixture ***of radium and beryllium or Americium and beryllium*** and the gauge monitors the flux of slow neutrons scattered by the soil. In using the neutron meter, a cased hole in the ground is necessary for lowering the probe to obtain readings.

23.2.3 The Source

The source of fast neutrons is generally obtained by mixing alpha particles from a radioactive emitter with beryllium. Beryllium atoms contain a loosely held neutron in their nucleus. When bombarded with alpha particles, beryllium is converted to carbon, and a high-energy neutron is released. The source materials are chosen for their longevity. They can be used for a number of years without an appreciable change in radiation flux. Neutrons emitted by the source have an average speed of about 6,000 miles per second and are called "fast neutrons."

23.2.4 How it works

The fast neutrons are emitted in all directions, where they collide with atoms in the soil through repeated collisions, they are scattered and some lose part of their energy. As the energy is lost, the neutrons are slowed to a speed of about 1.7 miles per second. Neutrons that have been slowed are said to be thermalized" and are called "slow neutrons." The neutron detector in the neutron moisture meter is designed to ignore fast neutrons and count only ***thermalized*** or *slow neutrons.* The average energy in a collision with another atom is much greater with atoms of low atomic weight than in collisions involving heavier atoms. Hydrogen is the only element with a low atomic weight that is found in significant amounts in typical soils. Hydrogen found in soils is almost entirely in the form of water. The occurrence of hydrogen in water allows for the use of the neutron probe to measure soil water content.

23.3 Principle

When a source is lowered into the soil and fast neutrons are emitted, the soil causes the neutron to slow and scatter. The result is that the probe becomes surrounded by a cloud of slow neutrons. The density of the neutron cloud around the probe does not increase indefinitely, but reaches an equilibrium determined by the rate neutrons are absorbed into the soil. *If the soil is dry*, the cloud of neutrons will be less dense and extend further from the probe. *If the soil is wet*, the neutron cloud will be denser and extend a shorter distance. The density of the neutron cloud is measured by the detector and the resultant signals are processed electronically and displayed as a number on the front panel of the gauge. The gauge reading is an index of volumetric soil moisture content (%).

23.3.1 Calibration

Initially, standard counts are taken with the probe inside the field. The shield acts as a standard having moisture 50 percent by weight. Counts are generally taken for 1 minute intervals but for precision, they should be taken for greater intervals. Observations are taken successively at 30 cm intervals after lowering the probe in the access tube. Count ratio is then computed by dividing the observed counts (Nm) by the standard counts (Ns). The Calibration curve is obtained by taking series of counts and soil samples for gravimetric estimation simultaneously at different depths.

23.3.2 Merits

1. It is very quick, non-destructive method.
2. It measures the soil water content on volume basis, in the whole range from saturation to wilting point.
3. It is not influenced by hysteresis or salts present in soil.
4. No cables are required in the property.
5. Soil moisture is estimated in millimetres, which is useful in deciding when and how much water to apply.
6. Reasonably well suited to overhead irrigation as 2 or 3 tubes are used at a site and averaged, overcoming possible variations in water distribution uniformity.
7. Soil moisture can be measured regardless of its physical state.
8. Probes can be interfaced for automatic downloading of stored data.
9. Measurements can be made repeatedly and nondestructively at the same site.

23.3.3 Limitations

The three drawbacks of the instruments are that It is expensive and *moisture content from shallow top layers (top 15cm) cannot be estimated* because of the escape of fast neutrons through the soil surface which is relatively dry.

1. The fast neutrons are also slowed down by other source of hydrogen (present in the organic matter). Other atoms such as **chlorine, boron, cadmium and iron also** have a high absorption capacity for slow neutrons, thus *overestimating the soil moisture content.*
2. Other forms of soil hydrogen, such as humus, plant organic matter, and water bound in the crystalline structure of calcium carbonate and gypsum cause inaccurate readings.

Therefore, measurements taken with a neutron moisture meter must be calibrated to account for these factors. Neutron moisture meter measurements are correlated to soil types. The neutron moisture meter measures the total water content of a soil. Only a part of the water in a soil is available for plant use.

The amount of water depleted from a soil (soil water deficit) is determined by subtracting the corrected neutron meter measurement from the total water-holding capacity of the soil. **A special surface probe** is used to measure the moisture in the top layer of the soil.

23.4 Safe Handling of the Equipment

Common procedures for handling the neutron moisture meter are necessary.

Two objectives which must always be kept in mind are:

1. Keep radiation exposure to a minimum, since any radioactive substance can be a potential health hazard no matter how small the quantity
2. Avoid physical damage to the instrument.
3. Shielding used with commercially available neutron moisture probes is a compromise between complete shielding and ease of handling.
4. Understanding the hazards and careful handling and use of moisture probes result in radiation exposures within permissible limits.

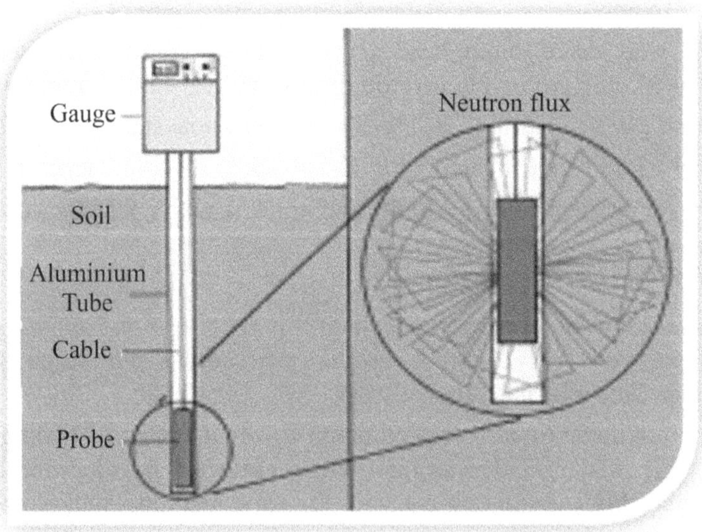

***Fig. 23.1** The Neutron probe*

References

1. *Mike Risinger, USDA-NRCS and Ken Carver, High Plains Underground Water Conservation District.*
2. *Hanson, B.R. and Orloff, S. 2007 Monitoring soil moisture for irrigation water management. UC ANR Pub 21635.*

Chapter 24

Systems of Irrigation for the Enhanced Productivity in Groundnut

K. Madhusudhana Reddy[1], Y. Pavan Kumar Reddy[2] and V. Siva Jyothi[3]

[1]Scientist (Agri.Engg),
[2]Scientist (Agronomy), Agricultural Research Station,
[3]Research Scholar. Acharya N.G.Ranga Agricultural University, Anantapuramu, Andhra Pradesh.

24.1 Introduction

Indian agriculture accounts 25 percent of the nation's Gross Domestic Product (GDP), 15 percent of exports and 60 percent of the employment continues to be the monitoring of the economy. Having achievable land able success in agriculture production in the last 50 years (50 mt to 210 mt) India has transformed herself from a deficit country to a food surplus country. But still there are many challenges which Indian agriculture being in a fast challenging technical and socio economic scenario. Relating to the natural resources and production base water has emerged as the most crucial factor for the sustaining the agriculture.

India accounts for 16 percent of the world's human populations and nearly 30% of the cattle with only 2.4 % of the land and 4% of the water resources. Even if the full irrigation potential is exploited about 50% of the country's cultivated area still remains unirrigated. Particularly with the current level of irrigation efficiently. The share of the water for agriculture would reduce for their increasing demand from other sectors.

But the water demand for the agricultural purpose is estimated to increase in order to produce more cereals, horticultural produce and raw material for a fast expanding food industry. Efficient water management of water is therefore, key to future growth of Indian agriculture.

The requirement of water by all sectors in 2025 is estimated to be 105 mha.m. The share of the water for agriculture is expected to get reduced from present level 85 to 69 percent by 2025 on the other hand, the actual agricultural water demand is estimated to increase from 470 billion cubic meters (BCM) in 1985 to 740 BCM in 2025.

During the same period the demand for non-agricultural sector will multiply four folds from 70 BCM to 280 BCM. The water use efficiency (WUE) in Indian agriculture at about 30-40% is one of the lowest in the world against 53% in china. The vulnerability of Indian agriculture is bound to be sure if the present trend of water use and management efficiency is not changed.

The international water management institute forecasts that by 2025, 33 percent of the Indian population will live under absolute water scarcity condition. Presently the problem facing the country is not the development of water resources but management of the developed water resources in a sustainable manner. By adopting efficient water management practices the bulk of India's agricultural lands could be brought under irrigation. Micro irrigation is once such practice.

Ramaprabhanalini (1999) stated that micro sprinkler irrigation was more efficient and economically viable method as compared to surface irrigation for the groundnut. Mikamadfierake (2002) substantiated that micro irrigation systems have turned and to better than surface irrigation in terms pod higher pod yield and field water use efficiency (FWUE). However (Krishnamurthy *et al.* 2003) reported that irrigating groundnut at 80 per cent I/PE at once in two days instead through micro sprinkler recorded higher pod yield and was saving other than other methods of irrigation.

24.2 Irrigation Present Method

Gravity irrigation is the dominant method of irrigation in India and it is going to be in the coming years also. Generally under open canals conveyance and surface irrigation methods less than one half of the water alone released reaches the plants. In major irrigation projects the overall efficiency ranges from 30 to 35 percent.

Experiments were conducted in various research institutes in India, Israel, Spain, USA and Australia using drip irrigation. The ranges have indicated that in this method the water saving was 40-70 percent and the crop yield for all crops was increased by 10-100 per cent. Farmers usually follows common surface methods of irrigation *i.e.* border strip and check basin method for the groundnut. Generally under open ditch conveyance and surface irrigation methods less than half of the water released reaches the crop plants.

Thus the water requirement of groundnut normally ranges from 400 to 720 mm under surface methods. But in micro irrigation systems water is applied once in 3-6 days interval which in turn reduces the moisture and results in higher pod yields and water use efficiency (WUE) in sandy loamy soils and in areas of water scarcity. Under optimum conditions about 30 to 40 percent irrigation water is reported to be saved with sprinkler irrigation system compared to surface irrigation. Frequent application of water through sprinkler maintains higher moisture status in the soils.

The research was carried out at Regional Agricultural Research Station Tirupati, Andhra Pradesh during *kharif* 2006 and *Rabi* 2007 revealed that the pod yield increased from 21 to 26 percent over flood irrigation.

24.3 Development of Micro Irrigation Systems

Micro irrigation is the fundamental concept of irrigating only the root zone of the crop rather than the entire land surface and maintaining the water content of the root zone at near optimum levels. It is one of the latest methods of irrigation, which is becoming popular in areas with water scarcity and salt problems. Economic considerations usually limit the use of drip irrigation system to orchards and vegetables in water scarcity areas.

The concept of drip irrigation goes back at least to the ancient Egyptians who placed porous pots in the soil and filled them with water. Use of subsurface claypipes for irrigation in Germany led to doubling of crop yields in 1860 and perhaps that was the beginning of the concept of applying irrigation water directly to the root zone. Nehemiah Clark in 1873, obtained the first known US patent for a water emitting device as dripper (a simple hole). Almost at the same time experiments of irrigating with porous pipes began. Drip Irrigation in principle evolved from subsurface irrigation when in 1920, the drainage tubes (with open joints) of subsurface irrigation were replaced experimentally by porous pipes for the purpose of irrigation. The perforated pipe was used in Germany, which made the concept feasible, and various experiments were then centered on the development of drip system using perforated pipe made of various materials.

During early 1940, Symchablass, an Israeli engineer observed that good flourished tree near a leaking faucet exhibited a more vigorous growth than the other tree in the area, which were not reached by the water from the faucet. This example of leaking faucet led him to the concept of an irrigation system that would apply water in small amounts. In India the farmers are conversant with micro irrigation techniques since 1970s. However, micro irrigation systems were initially not popular due to poor quality of components and lack of services from the manufacturers/traders. It took more than ten years before Indian business started showing interest in the systems, after getting convinced that the system has potential in the country. By the time researchers in different ICAR institutions and State Agricultural universities had provided adequate data to prove that the systems not only save water but also help in improving crop yields and the quality of produce. (Table 24.1).

***Table 24.1** Theoretical potential area for the drip and Sprinkler irrigation in India*

S.No.	Crop	Area (Million ha)		
		Drip	Sprinkler	Total
1	Cereals	-	27.6	27.6
2	Pulses	-	7.6	7.6
3	Oil seeds	3.8	1.1	4.9
4	Cotton	7.0	1.8	8.8
5	vegetablos	3.6	2.4	6.0

Table 24.1 ***Contd...***

S.No.	Crop	Area (Million ha)		
		Drip	Sprinkler	Total
6	Spices and condiments	1.4	1.0	2.4
7	Flowers, Medicinal, aromatic plants	-	1.0	1.0
8	Sugarcane	4.3	-	4.3
9	Fruits	3.9	-	3.9
10	Coconut, plantation and oil palm	3.0	-	3.0
	Total	**27.0**	**42.5**	**69.5**

Based on the recommendations of the task force, Government of India had initiated a centrally sponsored micro irrigation scheme in the country with 50% financial support to farmers from central as well as state Government budgets in the ratio of 4:1 National committee on Plasticulture Applications in Horticulture was made its coordinating agency. The results of the efforts of the scheme may be seen from table 24.2.

Table 24.2 *Growth rate of drip irrigation in India*

S.No.	Year	Micro irrigation area in ha
1	1985	1,000
2	1991	55,000
3	1999	2,54,000
4	2001	3,10,000
5	2003	5,00,000
6	2010	19,00,000

Table 24.3 *Comparison of micro irrigation over conventional system*

Performance indicator	Conventional irrigation method	Micro irrigation method
Water saving	Wasteful of water, losses occur due to percolation, runoff and evaporation	40-100% of water can be saved over flood method. Runoff and deep percolation losses are nil or negligible
Water Use efficiency	30-50% because losses are very high	90-95 %
Saving in labour	Labour engaged per irrigation is higher than Micro	Labour required only to start or stop the system
Reduced weeds problem	Weeds infestation is very high	Less wetting of soil, weed infestation is very less or almost nil
Use of saline water	Concentration of salts increases and adversely affects the plant growth. Saline water ca not be used for irrigation	Frequent irrigation keeps the salt concentration within root zone soil below harmful level

Table 24.3 *Contd...*

Performance indicator	Conventional irrigation method	Micro irrigation method
Diseases and Pest problems	High	Relatively less because of less atmospheric humidity
Suitability under physical soil constraints	Deep percolation is more in light soil and with limited soil depths. Runoff is more in heavy soil.	Suitable under various soil physical constraints as flow rate can be controlled.
Water control	Inadequate	Very precise, high and easy
Efficiency of fertilizer use	Efficiency is high because of heavy losses due to leaching and runoff	Very high due to reduced loss of nutrients through leaching and runoff water
Soil erosion	Soil erosion is high because oflarge stream sizes used for irrigation	Partial and control the wetting of soil surface eliminates any possibility of soil erosion
Increase in crop yield	Non Uniformity of available moisture reduced the crop yield	Frequent watering eliminates moisture stress and yield can be increased up to 20-100% as compared to flood

24.4 Different Methods of Irrigations

24.4.1 Drip Irrigation

Drip irrigation is an efficient method of providing water directly into the root zone of the plants to keep the soil moisture optimum for the desired range of the plant growth. The drip system is also used for applying fertilizers and pesticides as well.

***Table 24.4** The main components of drip irrigation System can be stated as under*

Safety Equipment	Non Return Valve , Pressure Relief Valve, Pressure Regulators, Air Release and vacuum Breaker Valve
Filter unit	A) Primary Filter-Gravel Filter, hydro Cyclone Filter B) Secondary Filter-Disc Filter , screen Filter
Fertigation Equipment	Ventury,Fertilizer Tank and Fertilizer Pump
Automation Equipment	Solenoid valves and controllers
Main and Sub main line	PVC or HDPE Pipes
Control Valves	Ball valve or Gate valve
Lateral (LLDPE) Pipe	20,16 and 12 mm
Emitting system	Integrated drip line , Online dripper (Pressure Compensating and Non Pressure compensating)

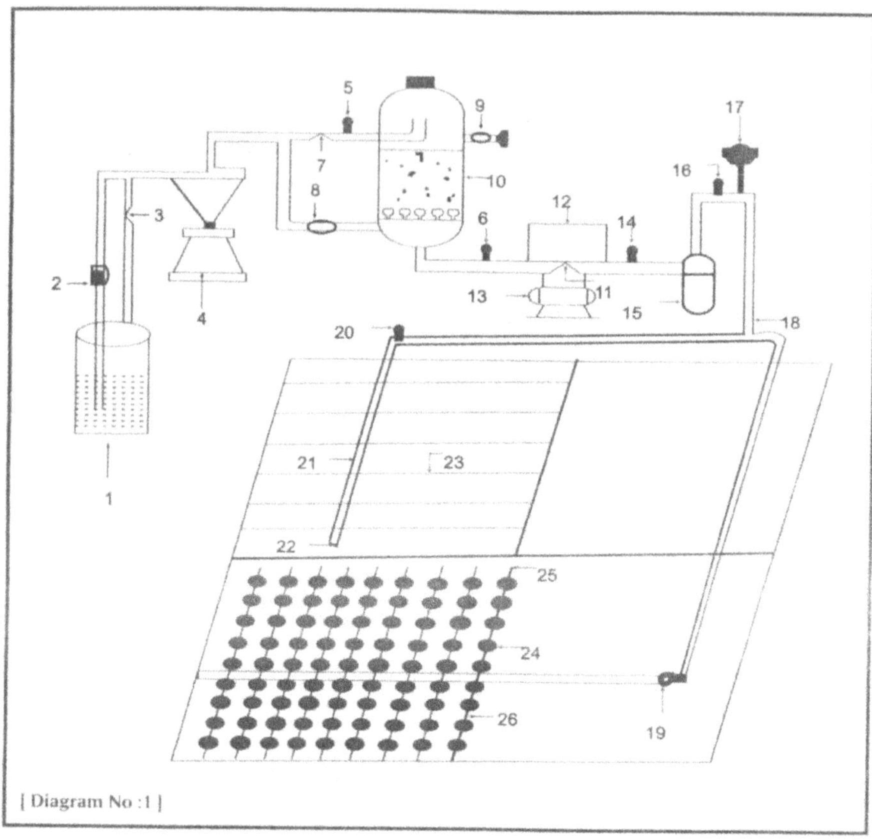

1. Well / Tank
2. Pump
3. Bypass / Pressure Relief Valve
4. Hydro cyclone Filter
5. Pressure Gauge No. 1
6. Pressure Gauge No. 2
7. Inlet Valve
8. Back flush Valve No. 1
9. Back flush Valve No. 2
10. Gravel Filter
11. Head Unit Control Valve
12. Fertilizer Pump (optional)
13. Fertilizer Tank (optional)
14. Pressure Gauge No. 3
15. Disc Filter
16. Pressure Gauge No. 4
17. Air Release Valve
18. Supply Mains
19. P. P. Ball Valve
20. Pressure Regulator
21. Sub Main
22. Sub Main Flush Valve
23. Laterals
24. Drippers
25. End Plug
26. Dripline

***Fig. 24.1** Components of the drip irrigation system*

***Courtesy:** Netafim Irrigation Manual*

24.4.2 Sprinkler Irrigation

The sprinkler or overhead irrigation, means the method of application of water similar to rainfall through a set of pipes normally by pumping the one or more control locations within the field and distributed by over head high-pressure sprinklers nozzles or rain guns. The sprinkler nozzles or rain guns spray the water in the air in form of small water droplets which fall on the ground and provide efficient coverage.

Types of Sprinklers

(a) Based on their Discharge sprinklers can be classifies as

1. Mini Sprinklers (40-600 lph)
2. Medium Sprinklers (1200-2000 lph)
3. Rain Gun sprinklers (>10000 lph).

(b) Based on portability sprinkler systems are classified as follows

1. ***Portable system:*** A portable system has portable mainlines, and laterals and portable pumping unit. It is designed to be moved from field to field or to different pump sites in the same field.
2. ***Semi-Portable system:*** A semi-portable system is similar to a fully portable system except that the location of water source and pumping unit are fixed. Such a system may be used on more than one field where there is an extended mainline, but may not be used on more than one farm unless there are additional pumping units.
3. ***Semi-Permanent system:*** A semi-permanent system has portable lateral lines, permanent main lines and sub mains, and a stationary water source and pumping plant. The main lines and sub mainlines are usually buried, with risers for nozzles located at suitable intervals.
4. ***Solid-set system:*** A solid-set system has enough laterals to eliminate their movement. The laterals are positioned in the field early in the crop season and remain for the season. The system is used for crops frequent irrigations.
5. ***Permanent system:*** A fully permanent system consists of permanently laid main lines, sub mains and laterals and a stationary water source and pumping unit.

For satisfactory sprinkler irrigation system

1. Components used should be good quality. This has to be taken care of as a large number of parts are going in to this system.
2. Systems has to design in such a way that all the components together perform the desired function. Otherwise the system will not work successfully through the components used are of good quality.
3. The third and most important aspect in this system has to be installed and operated properly and as specified. Otherwise the system will not give good moisture through the components are of good quality and the system has to be designed carefully.

Advantage of sprinkler irrigation systems

1. This saves considerable amount of the water in comparison with any surface method of irrigation.

2. Where land is sloppy/undulating sprinkler irrigation systems does not need any levelling.
3. Infact land levelling takes away the top soils which has all the fertility.
4. Once this top soils is removed it takes a long time and effort to restore the fertility.
5. Sprinkler system if properly designed can provide very high uniformity and irrigation efficiencies than any other surface method of irrigation.
6. It is a controlled method of irrigation whereby we can have a very good control over the amount of irrigation water to be given depending upon the stage of growth of crop. This control is not easily available in any surface method of irrigation.
7. No land is wasted as bunds, ditches, channels and consequently putting the land to optimum productivity of crops.
8. In most the surface methods of irrigation, water logging is a problem due to over irrigation, which creates a serious drainage problem, particularly in heavy soil. This problem does not exist in the sprinkler system.
9. Sprinklers are better to irrigate light soils than the surface method as frequent and small applications of water can be applied readily by sprinkler systems.
10. Compared to any methods of irrigation, sprinkler irrigation system gives a far better germination, there by assuming a higher output of crop.

24.4.3 Micro Sprinkler

The spraying devices throw the water in air, usually in predetermined pattern. Based on the discharge rate and pattern of water throw, the micro –sprinklers may be in the following types:

1. Mini-micro sprinklers
2. Micro-sprinklers
3. Jets or spinners

Fig. 24.2 Micro Sprinkler Sprinkler

Micro sprinkler heads are connected to the lateral pipe through micro tube. In field, the sprinkler heads are mounted on stake. The flow rates of this system vary from 20 to 600 lph depending on the size of orifice and operating pressure. Spray system wet greater soil volume than the drip system and decrease the labour requirement to shift sprinkler system. Micro sprinkler supplies less amount of water per unit area than the drip emitters because they spread water over a large area. The disadvantage of micro sprinkler system are greater loss of water due to evaporation and wind drift. In groundnut cultivation micro sprinkler system useful in coastal sandy soils where the infiltration rate is high when compared to other soils.

24.4.4 Rain Gun Irrigation

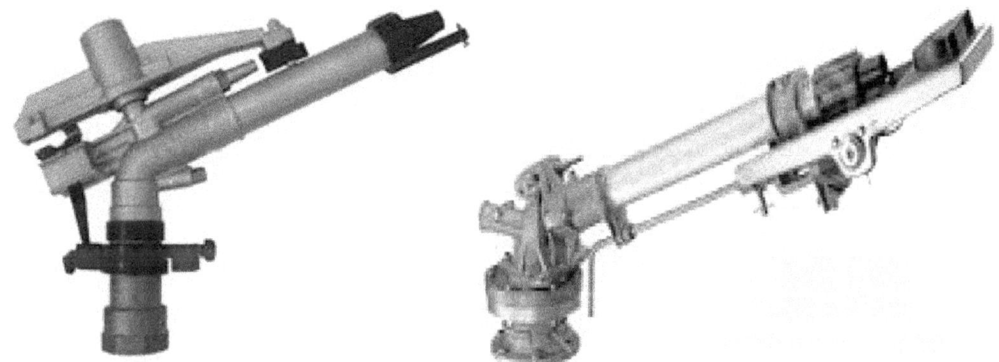

Fig. 24.3 Rain gun sprinklers

The term raingun denotes the meaning of large size sprinkler. It throws the water droplets to a large distance, and covers large area as compared to other types of sprinkling system. At once setting this system is capable to irrigate the area up to 4 ha. Its application rate varies from 5 to 35 mm per hour.

Hose reel (traveler) irrigator

To cater to the need of uniform application of water to medium to large farms Hose reel (Traveler). Gun sprinkler system is useful. In this system gun sprinklers are mounted on a moving trolley connected to the water source by a hose. Which would on reel during irrigation. As the hose is wound on the drum powered by the irrigation water, the sprinkler is pulled across the field. The size of the hose decided on the basis of size of the fields and total area to be irrigated. The rate of application of water is controlled by adjusting the speed of the reel vis-à-vis the water carrying hose.

24.5 Fertigation

Fertigation is defined as the application of fertilizers along with irrigation water using irrigation system. Drip irrigation and fertigation should go hand in hand in order to improve the efficiency of water and nutrients in crop production. The traditional methods of application of fertilizers to the soil followed by drip or sprinkler irrigation systems are not that effective and have resulted in reduced availability of nutrients to the plants because of limited root zone and reduced amount of mineralization in the restricted wetted zone.

Fertigation overcomes these aspects and is currently recognized as the best means of supplying fertilizers and making use of the scarce resource of water very economically, especially in green houses, orchards, vegetables, ornamental nurseries, high value plantations and even field crops and other crops such as capsicum, maize, sunflower, cotton, coconut, banana, sugarcane and mulberry. Drip irrigation permits application of fertilizers through irrigation water directly at the site of high concentration of root activity and cause for improving the fertilizer use efficiency in crop production.

Need for fertigation

- Uneven growth in fertilizer consumption resulting in state -wise crop–wise variation in consumption.
- Mining of nutrients from the soil at alarming rate.
- Decline in crop response to fertilizer.
- Stagnation in fertilizer production.
- Weakening relationship between fertilizer use and food grain production.
- Increasing dependence on fertilizer imports.

Advantages of fertigation

(i) Improved efficiency in fertilizer use generally 60 to 80 percent of the recommended dose of fertilizers.

(ii) Better use of fertilizers through fertigation due to:
  - Least loss of nutrients through leaching to around 10 percent as compared to 40 to 55 percent in the traditional system.
  - Optimization of nutrient balance of N: P_2O_5: K_2O by supplying the desired and required quantities of nutrients directly to the root zone in available forms.
  - Fertigation allows for easy application of water soluble fertilizers of different grades and combinations within the rhizosphere so as to match the physiological needs of the crop at different stages of crop growth.

(iii) High nutrient availability due to maintenance of soil moisture near field capacity under drip irrigation.

(iv) No damage to root system while top dressing of fertilizers.

(v) Considerable savings in the application cost of fertilizers as in plough sole application, drilling or band placement of fertilizers.

(vi) About 25 to 50 percent reduction in the quantity of fertilizer applied is possible through fertigation without affecting the growth and yield.

(vii) Higher WUE and 30 to 40 percent economy in the use of irrigation due to ferti-drip in crop production.

Principles of Fertigation

(i) ***Method of planting:*** It is an important factor deciding the economic feasibility of drip/trickle or sprinkler irrigation method for fertigation. Drip/trickle or sprinkler irrigation methods are normally well suited for feeding the fertilizers to root zone or zone of immediate utility. Since these two methods involve more initial investment there is a need for economizing cost on drip irrigation method by suitably altering the planting method. The more convenient planting methods to reduce the cost on laterals are paired row and four row pairing systems. Eg: In widely spaced crops viz., Maize, Sunflower, Bell pepper, Sugarcane, tomato and chilli two row paired system would reduce the cost on lateral lines.

(ii) ***Scheduling of irrigation:*** Which is based on percentage replenishment of daily evaporation based on the observation of USWB class Aopen pan evaporimeter. Optimum depth vary depending on the crop, soil type, season and region.

(iii) ***Sources of fertilizers:*** Fertilizers supplying different primary, secondary and micronutrients are mixed with irrigation water and applied to zone of immediate utilization. It is water soluble which do not leave suspended particles in the irrigation water which are non-corrosive and do not damage/ block the components of fertigation unit are preferable for fertigation.

(iv) ***Dose of fertilizers:*** The quantity of different nutrients required for fertigation is less than soil application. However, depending on the nature of crop, variety and yielding ability.

(v) ***Method of fertigation:*** It should be targeted for minimizing the losses of nutrients or fixation and enhancing the efficiency of applied fertilizer solution. The options available are surface or subsurface fertigation. Under surface fertigation along with water, nutrients are also released on the soil surface near the root zone. While subsurface fertigation ensures application of fertilizer solution beneath the soil near the rhizosphere.

(vi) ***Frequency of fertigation:*** It is governed by the nature of crop, duration, growth habit and yielding ability. Frequency of fertigation can be considered depending on the crop response.

Characteristics of fertilizer suitable for fertigation

- High nutrient content readily available to plants.
- Fully water soluble at field temperature conditions.

- Fast dissolution in irrigation water.
- No clogging of filters and emitters.
- Low content of insoluble (< 0.02%).
- Minimum content of conditioning agents.
- Compatible with other fertilizers.
- Minimal interaction with irrigation water.
- No drastic changes of water pH.
- Low corrosive for control head and system.

24.6 Types of Fertilizers

Nitrogen

The different sources of fertilizers supplying N are anhydrous ammonia, ammonium phosphate, urea, ammonium nitrate and calcium nitrate. Some of the N sources particularly anhydrous ammonia will increase pH, leading to precipitation with calcium magnesium resulting in clogging of the drip system. Calcium nitrate is relatively soluble and does not cause pH shift.

***Table 24.5** Solubility of Nitrogenous Fertilizers*

Type of Fertilizer	Nitrogen Content (%)	Solubility (g/liter)
Ammonium Sulphate	21	750
Urea	46	1100
Ammonium nitrate	34	1920
Calcium nitrate	15.5	1290

Phosphorus

Generally the application of phosphotic fertilizers through the drip irrigation system is not recommended because

(i) In majority of the cases basal application of phosphorus satisfies the plant P needs.
(ii) Phosphorus moves slowly from the point of placement.
(iii) It has high rate of fixation with soil.
(iv) Applied phosphorus creates chemical and physical precipitation leading to clogging problems.

Phosphoric acid being soluble with low pH, no clogging occurs with orthophosphate. Glycerophosphate and inorganic phosphorus like MAP and DAP are the other sources of phosphorus for fertigation.

Potassium

Potassium is easily soluble in water, easily leached in sandy soils and can be fed through drip irrigation to maintain a proper N: K ratio for crop production. It can be applied as Potassium sulphate, Potassium chloride, Potassium nitrate, and Mono potassium phosphate.

Table 24.6 *Solubility of Potassic Fertilizer*

Type of Fertilizer	K Content (%)	Solubility (g/liter)
Potassium sulphate	50	110
Potassium chloride	60	340
Potassium nitrate	44	133

Micro nutrients

They are generally applied separately to plants in most crops and in most soils as their application through fertigation would react with salts in the irrigation water and cause precipitation and clogging. Chelated micronutrients are highly water soluble and can be used for fertigation since they cause very little clogging or precipitation.

Table 24.7 *Solubility of Micronutrient Fertilizer*

Fertilizers	Content (%)	Solubility (g/l)
Solubor	20 % B	220
Copper sulphate	25 % Cu	320
Iron sulphate	20 % Fe	160
Magnesium sulphate	10 % Mg	710
Ammonium molybdate	54 % Mo	430
Zinc sulphate	36 % Zn	965
Manganese sulphate	27 % Mn	1050

Liquid Fertilizers

The liquid fertilizers are pure and do not precipitate. Normally the liquid fertilizers are acidic (pH 5.5-6.5) and help in correcting the soil pH to some extent and also help in the prevention of clogging of emitters. For acidic soils, liquid fertilizers with neutral pH or even higher pH could be used.

Table 24.8 *Fertilizers for fertigation*

Speciality water soluble fertilizers
Polyfeed (19-19-19)
MAP (12 -61-0)
Multi -K (13-0-46)
MKP (0-52-34)
SOP (0-0-50)
Normal Fertilizers
Urea (46-0-0)
Potash (0-0-60)

***Table 24.9** Compatibility chart for different water soluble fertilizers*

Fertilizers	Urea	Ammonium nitrate	Ammonium sulphate	Calcium nitrate	MAP	MPP	Potassium nitrate
Urea		C	C	C	C	C	C
Ammonium nitrate	C		C	C	C	C	C
Ammonium sulphate	C	C		C	C	C	C
Calcium nitrate	C	C	LC		NC	NC	C
MAP	C	C	C	NC		C	C
MPP	C	C	C	NC	C		C
Potassium nitrate	C	C	L	C	C	C	
C – Compatible, NC – Not Compatible, LC- Limited Compatible							

Fertilizer Solubility and Compatibility Fertilizers

Must be dissolved (not suspended) – dissolved means that it cannot be filtered out.

(a) Do a "Jar Test" for fertilizer compatibility.
- Get a sample of the water and mix the nutrients in proportion.
- Wait and watch for precipitation.

(b) Solid fertilizers vary in their dissolution rate and the amount that can be dissolved in water.

(c) Fertilizer solubility increases at higher water temperature.

Nutrient requirement of crop plants

The nutrient requirement of crop plants varies with type of crop, variety and stage of the crop. In drip fertigation nutrient is supplied according to the physiological stage of the crop and the site – specific requirements (crop type and cultivar, growing system, yield target, soil type and climatic conditions). Nutrients status should be monitored regularly in the plant tissue (Leaf and petiole analysis), in the soil and in the irrigation and leaching solutions.

Fertigation equipment

1. ***Veutury Injector:*** This is very simple device. A partial vacuum is created in the system which allows suction of the fertilizers into the irrigation system through ventury action. The vacuum is created by diverting a percentage of water flow from the main and pass it through a constriction which increases the velocity of flow thus creating a drop in pressure. When the pressure drops the

fertilizer solution is sucked into the ventury through a suction pipe from the tank and from there enters into irrigation stream. The suction rate of ventury is 30 to 120 lit per hour.

2. ***Fertilizer Tank:*** In this system part of irrigation water diverted from the main line of flow through a tank containing fertilizer in a fluid or solid form, before returning to the main line, the pressure in the tank and the main line is the same but a slight drop in pressure is created between the off take and return pipes for the tank by means of a pressure reducing valve. This causes water from main line to flow through the tank causing dilution and flow of the diluted fertilizer into the irrigation stream. With this system the concentration of fertilizer entering the irrigation water charges continuously with the time, starting at high concentration. Fertilizer tanks are available in 90, 120, 160 liters capacity.

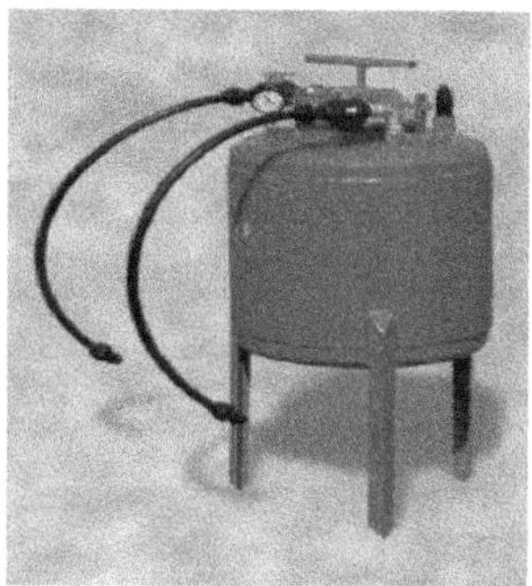

3. ***Fertilizer Injector Pump:*** These are piston or diaphragm pumps which are driven by the water pressure of the irrigation system and such as the injection rate is proportional to the flow of water in the system. A high degree of control over the fertilizer injection rate is possible, no serious head losses are incurred and operating costs are low. Another advantage is that if the flow of water stops. Fertilizer injection also automatically stops. This is perfect equipment for accurate fertigation. Suction rates of pumps vary from 40 to 160 liters per hour.

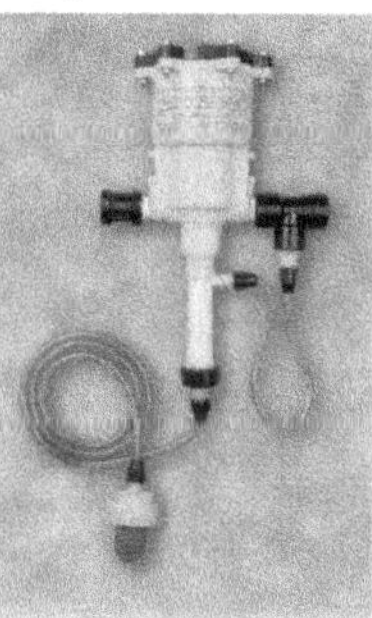

Fertigation Program

Nutrient rates

- When a crop is first seeded/ transplanted, begin with small amounts of fertilizer as the crop is growing slowly.
- Increase the amounts as the growth rate increases.
- As the crop matures and growth slows, reduce the fertilizer rates accordingly.
- For most crops, is sufficient to program no more than four or five rate changes during the production season.

Nutrient ratio

- Early in the crop's season, apply P and K in small doses for best rooting and plant establishment.
- When plants are vegetative, apply adequate quantities of N for best plant growth and development.

- As fruit load grow, increase K for best fruit setting, development and quality. High absorption of K is noticed during heavy fruit loads of vegetable crops and a massive stalk development of flower crops.
- Reduce N doses to avoid lush growth, soft fruit and pest problems.

Requirements for fertigation

1. Test the native soil fertility status
2. Fix the correct fertilizer dose for selected crop
3. Develop appropriate fertigation schedule
4. Selected suitable fertilizer grade according crop stage
5. Install appropriate fertigation device along the main line

Steps in fertigation

1. Calculate the required fertilizer quantity for the actual cropped area
2. Prepare the nutrient stock solution (Dissolve the solid fertilizer with water at 1:5 ratio)
3. Operate the drip system for 10-20 mts for wetting (1st process)
4. Regulate the valves and initiate fertigation (2nd process) at 95 or 186 lph injection rate as per the fertigation device selected.
5. Complete the fertigation and finally flushing (3rd process) for 10-15 mts for removal of left out fertilizer in pipe network.

Points to remember for adopting fertigation

(i) Fertilizer tank or injection pump may be used to inject the fertilizer.

(ii) For precise placement of both water and fertilizer, it is necessary to use pressure compensating drippers or online drippers instead of micro tubes.

(iii) Daily feeding is most desirable, if not possible feed alternate day.

(iv) Urea can be combined with WSF for meeting N requirements.

(v) Fertigation should be done during at the end of irrigation period, run the drip system for 5-6 minutes after the completion of fertigation.

(vi) It is not advisable to use a very concentrated stock solution (generally not more than 10%)

(vii) The fertilizer solution should be compatible with the quality of water into which it is being injected.

(viii) Do not inject fertilizers in combination with pesticides or chlorine.

(ix) The injection point must be upstream of the filter system so that the filter will remove any un-dissolved fertilizer or precipitates that occur.

(x) Select fertilizer solutions to help adjust water pH if necessary.

(xi) The time needed to distribute fertilizer should be less than the time needed to supply enough water to the field.

(xii) Do not over irrigate because this will leach some of the fertilizer out of the root zone.

Table 24.10 *Fertilizer Use Efficiency*

Nutrient	FUE		
	Soil application	**Drip**	**Drip fertigation**
Nitrogen	30-50	65	95
Phosphorus	20	30	45
Potassium	50	60	80

Limitations and precautions of fertigation

(i) Fertilizer Suitability–Only completely water soluble fertilizers interactions of injected chemicals with irrigation water resulting in formation of precipitates which clog the emitters.

(ii) Incorrect application may cause damage to crop, leaching of nutrients and contamination of groundwater.

(iii) Knowledge of fertilizer compatibility while mixing is necessary.

(iv) Correct operation of irrigation system is required.

(v) Corrosion of irrigation and injection system components.

(vi) Toxicity of water when water source is linked to potable water supply.

References

1. *Asokaraja, N, 2002. Performance evaluation of drip fertigation system with liquid and water soluble fertilizers for increasing the yield and quality of vegetable crop (Tomato), final report of ICAR Adhoc Project 1999 – 2002, WTC, TNAU, Coimbatore.*
2. *Asokaraja, N, 2002. Maximizing the productivity and quality of banana and sugarcane with water soluble fertilizer through drip fertigation. Annual Report.2001 - 2002, WTC, TNAU, Coimbatore.*
3. *Rajput T.B.S and Neelampatel, 2009. Micro irrigation. Proceedings of the winter school on micro irrigation, March 4-24, 2009, Water Technology Centre, IARI, New Delhi.*
4. *Rajput T.B.S, 2010. Micro irrigation technologies of renhancing water productivity. paper presented at National Seminar on Engineering Agriculture for evergreen Revolution September, 24-25, 2010, Titupathi pp 34-41.*
5. *Ramachandrappa, B.K, and Nanjappa, H.V, 2008. Fertigation Technology. Agrobios. India, pp:161-233.*
6. *Suresh, R, 2010. Micro irrigation Theory and Practice. Netafim irrigation indiaPvt Ltd. Drip irrigation operation and Maintenance Manual.*

Chapter 25

Farm Pond Technology

T. Yellamanda Reddy

Chief, Sustainable Agriculture, A. F. Ecology Centre, Anantapuramu, Former Dean of Agriculture, Acharya N. G. Ranga Agricultural University, Hyderabad.

25.1 Farm Ponds

In arid and semi-arid regions, rains are sometimes received in heavy down pours resulting in runoff (Singh, 1983). The percentage of runoff ranges from 10 to 30% of total rainfall. Alfisols have high runoff generating potential than vertisols with deep cracks at the commencement of the monsoons. Runoff starts earlier and more frequently during rainy season in alfisols compared to vertisols. On alfisols even with contour bunds, there is at least 20 to 30 percent runoff. Simple treatment of the land such as shaping, removing obstructions etc. enhance the harvesting efficiency of runoff water.

25.1.1 Size of Farm Pond

Small farm ponds of size 100-300 m^3 can be dug for storing runoff water. The size of the farm pond depends on the rainfall, slope of the soil and catchment area. The dimensions may be in the range of 10m x 10m x 2.5 m to 20 m x 20 m x 3.5 m (Yellamanda Reddy and SankaraReddi, 2010). The side slope 1.5:1 is considered sufficient.

(Samindre and More 2012) reported that the maximum depth of water impounded and storage volume in pond was 3 m and 1079 m^3, respectively. Total evaporation and seepage loss through farm pond for the period 1st July 2010 to 31st December 2010 was 229 m^3 and 2776 m^3, respectively. The weight of silt deposited in the farm pond for the monsoon season 2010 was recorded as 11.95 t. Series of field trials were conducted during 1987-2010 at Central Institute for Cotton Research, Nagpur indicated that in shallow, medium and deep soils, it was estimated to collect 150, 300 and 350 m^3 of runoff water/ha in farm ponds by September end, and 56-82 percent of the pond capacity was stored. The pond size of 200 m^3 was not accepted by the farmers as it was economically non-viable at field scale, but it can be helpful in improving the recharge of existing gravitational wells (Raju et al., 2011). Catchment to command area ratio should be larger than 7.4 in order to provide sufficient amounts of runoff water to the terrace for the cultivation of olive trees for an average annual rainfall amount of 235 mm (Gabriels et al., 2005).

25.1.2 Lining of Farm Ponds

The problem associated with farm ponds is high seepage loss. This can be reduced by lining walls. Some of the traditional methods for seepage control are: use of bentonite, soil dispersants and soil-cement mixture (Maheswari and Turner, 1986). Bentonite has excellent sealing properties if kept continuously wet, but cracks develop when dried. Soil-cement mixture can be used, but surface cracking develops when exposed to sun drying. A soil-cement lining of 100 mm thickness reduces seepage losses up to 100 percent. The pit lined continuously develops cracks but no cracks develop when applied in blocks (AICRPDA.1986). The other alternative sealant for alfisols is a mixture of red soil and black soil in the ratio of 1: 2.

The different types of lining materials are used depending on the availability, cost and soil type. They are: soil-cement, red and black soils, cement-concrete, bricks, Cuddapah slabs, stone pitching, polythene sheet etc. In alluvial sandy loam to loamy sand soils of Gujarat and red sandy loams soils of Bangalore, a soil + cement (8:1) mixture is the best lining material. At Anantapur in Andhra Pradesh, soil without sieving and cement in 6:1 ratio is very effective and cheap lining materials for red sandy loam soils (Yellamanda Reddy et al, 2005). In laterite silty clay loam soils of Ooty, medium black soils of Kota, bitumen was effective. Cement and bricks is found useful for silty loam to silty clay loam soils at Dahradun. At Solapur, lining with sodic soils is better than lining with soil + cow dung + straw in medium deep soil. Water can be stored for two months in deep heavy soils without lining at Nandyal in Andhra Pradesh, India. Clay soils linings are generally the most economical.

(Mathew et al., 2008_ reported that it is essential to design, construct and manage small water harvesting structure, a farm pond of 1.5 lakh litres of water, in porous lateritic formations. A suitable lining material to reduce percolation losses are soil cement mixture with 10% cement content; and HDPE lining at bottom and sides and both are equally effective. Since the HDPE lining was cheaper than the soil cement mixture the pond was lined with HDPE film both at the sides and bottom. It is better to use the stored water in conjunction with the available ground/surface water. The harvested water is used for irrigating arecanut during summer months.

Geosynthetic is a new kind of geotechnical material made of polymer, which can be used in engineering for the purpose of seepage prevention, reinforcement, protection and dewatering. In the pilot project, such geosynthetic was used by (Su and Zuo 2006) for waterproofing of water storage tanks of rainwater harvest project. The materials and technologies have been widely used in western Henan, northern Henan and other areas, creating remarkable social, economic and ecological benefits.

Adilabad is one of the backward districts of Andhra Pradesh, is blessed with high rainfall (1159 mm) but uneven rainfall distribution, undulating topography, black cotton soils (80%), less per cent of irrigated area (11.7%), more forest area (43%). The district receives larger part of the rains within four months (June to September) which is very irregular. Though rainfall is high in the district, infiltration is very low due to uneven topography, thus major portion of rainwater goes away unused as runoff. Community farm pond technologies were adopted in the villages. They made attempts to revise old concept of small individual farm ponds (10 × 10 × 2.5 m) to large community farm ponds (20 × 20 × 4.5m) by increasing the dimensions as well as involving the group of adjacent farmers. Utilization of harvested water for life saving irrigation was found to be highly productivein the rainfed areas (Rmesh et al., 2012).

The tribal farmers of Andhra Pradesh in Adilabad and agency tracts of north-east part, use five per cent of their land holding for digging pond for rain water harvesting. They use collected water for irrigating the high value vegetable crops, cotton and sometimes for Paddy (Venkateswarlu, 2010). Harvested water (farm pond) is precious than surface and ground water. Hence, need to be utilized efficiently. If the harvested is used for irrigating the crops during dry spell especially at critical stages, through flood irrigation, it requires more water thus less cropped area will be irrigated. Further, water use efficiency will also be less. Hence, the harvested water if utilized through drip and sprinkler irrigation methods, more cropped area can be irrigated with lifesaving irrigation besides enhancing water use efficiency.

Palem in Mahabubnagar district of Andhra Pradesh receives high intensity rainfall during summer (April or May), starting (June) or ending (September to October) of *kharif* season. It leads to loss of precious water and top fertile soil due to heavy runoff. A farm pond of 25m × 25m × 2m dimensions was dug and lined with with black polythene sheet lining. The harvested water was utilized for establishing and scheduling lifesaving irrigation to cotton, redgram and castor which are the predominant long duration crops grown in Mahabubnagar district. Under the conditions where monsoon is delayed by 15-30 days, late planting of long duration crops hampers the productivity. A technique called pot wateringwith harvested water from farm pond was employed.Pot wateringtechnique was found promising for cotton where 50% higher yield has been observed as compared to control. This technique has become popular in Bt cotton growing tracts in Andhra Pradesh as it is a high value cash crop and fetches good monetary returns to the farmers. When long dry spell occurs during crop growth stages, lifesaving irrigation was given to save the crop and enhance the yield. Very good response has been observed for drip irrigation. Many dry land farmers in the state have dug farm ponds of 15-25m × 15-25m × 3m size under watershed programme or National Rural Employment Guarantee Programme (NREGP).

Calcareous soils are characterized by extremely low rate of infiltration due to presence of high content of $CaCO_3$. Dugout pond constructed with lining of calcareous soil recorded minimum seepage rate from 1.05 to 1.08 cm/m^2/day at different heads over without calcareous soil lined pond. The reduction of seepage rate was varied from 61.4 to 62% over control. The seepage rate decreased with increase (%) of calcareous soil in the dugout farm pond. Increased calcareous soils in the farm pond slightly changed the water quality but it was within the suitability range of irrigation. Seepage rate and water head in the pond are positively correlated. Lining of farm pond can be done with calcareous soil either obtained from *in situ* or transported from outside. It has been found that calcareous soil lining is cost effective technique for reduction of seepage loss from the ponds (Jat et al., 2011).

25.1.3 Evaporation Losses

Evaporation losses can be reduced in farm ponds especially in arid regions by rubber or plastic floats. White plastic sheet is economical and easily available. Neem oil spray reduces the evaporation from the farm pond.

25.1.4 Economic Viability

Farm pond technology is economically viable. Studies undertaken in the Jhanwar model watershed in Rajasthan showed that water harvesting in a farm pond of size 271 m^3 (2,71,000) and utilizing the water for supplemental irrigation is economically viable (Goyal et al., 1995). The cost benefit ratio was 1: 1.7. Economic analysis of irrigation from farm pond at Diara (Rajasthan) recorded a cost – benefit ratio of 1: 2.5. Before water harvesting was introduced, onions were not grown in the area due to lack of seedlings. Onion seedlings were grown on 100 m^2 plots using water from the ponds in the dry season, then sold or planted under rainfed conditions during the rainy season. The results obtained show that the average net income from onion seedlings and from bulb onions grown rainfed in the field was higher than from rainfed wheat (Akalu et al., 2010).

Despite its weather-risk reducing advantage, the average adoption rate of water harvesting technology was 58% between 2005 and 2010 in Ethiopia. The adoption rate can be improved by increasing availability of plastic sheets, water pumps, easier market and credit access, and the cultivation of perennials (Wakeyo, 2012).

As the availability of water is limited than land, productivity water can be increased by, timely sowing, use of improved varieties, soil test based fertilizer application, timely weeding and efficient irrigation system. Water harvesting and supplemental irrigation is risk reducing investment. Economic study conduct in sub–Saharan Africa revealed that water harvesting and supplemental irrigation increased net profit of 151 to 626 USD/ year compared to – 83 to 15 USD/year for the existing farming practices in the Sahel (Fox et al, 2005). Water harvesting and use of water for pre-sowing and supplemental irrigation increase the productivity of wheat and onion in mountainous watershed in Himachal Pradesh (Goel and Kumar, 2005). The benefit – cost ratio ranges from 0.41 to 1.33 for water harvesting structures of different sizes with an estimated life of 25 and 40 years respectively.

25.2 Supplemental Irrigation

In dry areas, water, not land is the most limiting resource for crop production. Productivity of harvested water can be increased by applying small quantity of water to large areas than heavy irrigation to small area. If rains occur immediately after irrigation, there will be no impact of irrigation and in black soil, it may reduce yield. Maximizing the water productivity but not the yield per unit land is the better strategy for dry farming areas (Yellamanda Reddy and SankaraReddi, 2010). Supplemental irrigation is a highly efficient practice for increasing productively of crops in arid regions (Fox and Rockstrom, 2003).

25.2.1 Response to Irrigation

The yields of rainfed crops are less than half of their yields under irrigated conditions. Yield of groundnut under irrigated conditions in shallow alfisols in arid tropic conditions of Anantapur district in Andhra Pradesh is 2500 kg/ha. Groundnut yield ranges from 100 to 1500 kg/ha under rainfed condition. Similarly, maize yields in arid Sub–Sahara Africa are less than half of the potential yield under rainfed conditions (Jennie and Okwach, 2005). Providing two protective irrigations to finger millet during dry spells gave the highest grain and straw yields as compared to normal cultivation practices alone (Nagaraju and Mohankumar, 2010).Yield of castor, red gram and cotton increased with protective irrigation in light soils even with small quantity of water.

***Table 25.1** Effect of pot watering and drip irrigation on the productivity of dry land crops*

Treatments	Seed yield of castor (kg ha-1)	Seed yield of red gram (kg ha-1)	Yield of cotton (kg ha-1)
No irrigation	869	511	708
10 mm drip	937	601	960
15 mm drip	1179	759	1233
20 mm drip	1400	1017	1320

For early withdrawal of rainfall around 45 DAS, two supplemental irrigations given at tasseling and silking stages through run off recycling from farm pond increased growth, yield parameters, grain yield by 134 percent and net returns by Rs.3389 ha^{-1} over rainfed maize. For moisture stress at tasseling and soft dough stages, two supplemental irrigation increased growth, yield attributes, grain yield by 88 percent and net return by Rs. 3305 ha-1 over rainfed maize (Lakshmi et al 2009).

Yields of crops can be increased by judicious application of small amounts of water through alternate furrows to wet the root zone during a stress period. Supplemental irrigation is beneficial in both sub-normal and normal rainfall conditions during the cropping period. With only 20 mm of irrigation in a intercropping system of pigeon pea + pearl millet, gross yields increases significantly. Vegetable yields increases significantly with irrigation (Vijayalakhsmi et al, 1989).

Crops differ in responding to amount of irrigation water to supplemented irrigation during dry spell. Groundnut responds to 10 mm of irrigation through sprinkler on alfisols during pod development stage (Yellamanda Reddy and Sulochanamma, 2008). The benefit of supplemental irrigation lost for one week (Yellamanda Reddy and Sulochanamma, 2003). Cotton needs a minimum of 30 mm of water to respond to irrigation applied either sprinkler or drip irrigation system on vertisols. Chickpea similarly need 30 to 40 mm of supplemental irrigation applied as drip or sprinkler irrigation during flowering. Pigeonpea responds to 20 mm irrigation water applied at pod development state with drip irrigation to 20% of the cropped area. Irrigation can be provided near the row, covering about 20% of the cropped area, leaving 80% of interrow zone.

Atriplex, a desert fodder shrub that can be grown on saline soils, can be established with harvested water, in Jordan with an annual rainfall of 100 – 200 mm. Application of harvested water from macro-catchments has substantive effect on shrub survival, biomass production and water use efficiency. The survival of the transplants averaged 67% under rainfed conditions compared to 95% when additional harvested water of 39 mm was applied. The production of grazable browse of 3 year old shrubs is 380 and 1151 kg/ha with and without supplemental irrigation. Yield of watermelon increased by 36% when 45 mm irrigation was given through drip irrigation during which the seasonal rainfall was 126 mm (Abuanat et al., 2004).

25.2.2 Time of Irrigation

Unlike in irrigated agriculture, the critical stage concept does not suit well, as dry spell may reduce the growth and yield of crop at any stage. Vegetative stage is considered as non-critical stage in irrigated agriculture but in arid regions, dry spell during vegetative state prolongs the crop duration which may ultimate result in crop failure due to end season drought. Death of seedlings also cause reduction in yield due to dry spell in vegetative stage, therefore, the strategy for getting successful crop is providing small quality of water, if available, at any stage if the dry spell is more than 10 days in light soils and 15 days in heavy soils (Yellamanda Reddy and SankaraReddi, 2010).

Supplemental irrigation to mustard, taramira and Bengal gram at 45 DAS is more profitable than irrigation at 60 DAS and no irrigation. The in-situ moisture conservation practices and recycling of harvested rainwater techniques were found to very effective for drought mitigation (Jat et al., 2008).

Use of farm pond water through supplemental irrigation increased seed cotton yield by 20, 28 and 26 percent over rain fed cotton in shallow, medium and deep soils respectively. It is difficult to provide lifesaving irrigations from farm ponds in first fortnight of July. Terminal droughts of August can overcome by supplemental irrigations from farm ponds of 10 ha catchment with a breakeven of 12 years. Bt cotton responded for two supplemental irrigations from harvested runoff water at seedling/terminal droughts in July or in mid-September in vertisols (Raju et al., 2011).

25.2.3 Method of Irrigation

Surface methods of irrigation like check basin, basin, and furrow method irrigation are not suitable for supplemental irrigation, mainly for three reasons: the rainfed lands are uneven, conveyance losses may go up to 30% and limited amount of water available for irrigation. Drip and sprinkler irrigations are more suitable because small amount of water can be delivered, even on uneven soils with no conveyances losses. Subsurface drip irrigation is very efficient for providing supplemental irrigation. The main drawback of micro-irrigation system is high initial cost of the system. Pot watering is another efficient method being used by the farmers for transplanting crops like tobacco, chilli, and tomato (Yellamanda Reddy and SankaraReddi, 2010).

25.2.4 Quantity of Irrigation

Crops differ in responding to amount of irrigation water to supplemented irrigation during dry spell. Groundnut responds to 10 mm of irrigation through sprinkler on alfisols during pod development stage. The benefit of supplemental irrigation lost for one week. Cotton needs a minimum of 30 mm of water to respond to irrigation applied either sprinkler or drip irrigation system on vertisols. Chickpea, similarly need 30 to 40 mm of supplemental irrigation applied as drip or sprinkler irrigation during flowing pigeon pea responds to 20 mm irrigation water applied at pod development stage with drip irrigation to 20% of the cropped area.

Irrigation can be provided near the row, covering above 20% of the cropped area, leaving 80% of inter-row zone. Pot watering, applying small quantity of water (around 250 ml) manually to each plant either to for crop establishment or for sowing or for transplanting in widely spaced crops like cotton, red gram castor, tomato, tobacco etc. Similarly, pot watering to protect the seedlings during early crop growth state is highly useful. The amount of water, if calculated over the entire area, works out less than 5 mm. For example, pot watering cotton seedlings at 250 ml/ plant work out 5000 l/ ha which works out to 5 mm.

Rainfall amount received does not meet crop water requirements, hence requires supplementary irrigation water to mitigate dry spells. Crop yields in fields that received supplemental irrigation increased by more than 120% as compared to fields that received rainfall only. The result also showed that the amount received in the field is not the only factor that can contribute to the water use efficiency but also depends on other management practices (Mzirai and Tumbo, 2010).

25.2.5 Quality of Irrigation Water

The quality of water harvestedin farm ponds is generally better than ground water. The quality of rainwater was higher than that of the groundwater. The crop yield (yellow melon) was by 9.2% higher when irrigated with rainwater, compared with groundwater (Nicolescu et al., 2010).

25.2.6 Water Productivity

As the availability of water is limited than to land, productivity water can be increased by, timely sowing, use of improved varieties, soil test based fertilizer application, timely weeding and efficient irrigation system.

References

1. *Abuanat, M.W., Ruyle, G.B., Adel Hamid, N.F., 2004. Increasing range production from fodder shrubs in low rainfall areas. Journal of Arid Environments 59(2), 205-216.*
2. *Fox, P. and Rockshrom, J. 2003. Supplemental irrigation for dry spoll mitigation of `rainfed agricultural in the Sahel. Agricultural Water Management, 62 (1): 29 – 41.*
3. *Goel, A.K. and Kumar, R.2005. Economic analysis of water harvesting in a mountainous watershed in India. Agricultural Water Management. 71 (3): 257-266.*
4. *Jat, M. L. Jain, P. M. Sharma, S. K. Kothari, A. K. Jain, L. K. Balyan, J. K. Sharma.*
5. *R. K. 2008. Drought management through in situ moisture conservation practices and rainwater harvesting strategies based on rainfall analysis in semi arid tropics. Indian Journal of Dryland Agricultural Research and Development. 23: 2, 48-60*
6. *Lakshmi, K. V. Balasubramanian, A. Sankaran, N. 2009. Supplemental irrigation, green manuring and nitrogen levels on growth, yield and economics of dry land maize. Madras Agricultural Journal; 96(1/6):129-134.*
7. *Nicolescu, C. Cimpeanu, S. Bucur, R. D. Burnichi, F. Sovaiala, G. 2010. System of rainwater collection, storage and pumping. Journal of Food, Agriculture & Environment. 8: 3/4, 1326-1328.*
8. *Samindre, M. S., More, M. R. 2012. Assessment of farm pond with respect to water harvesting and recycling. International Journal of Agricultural Engineering; 2012. 5: 2, 198-201.*
9. *Wakeyo, M. B. 2012. Economic analysis of water harvesting technologies in Ethiopia. Economic analysis of water harvesting technologies in Ethiopia; 2012. Thesis, 150 pp.*
10. *Yellamanda Reddy T and Sankara ReddiGH. 2010. Principles of Agronomy, Kalyani Publishers, New Delhi.*

Chapter 26

Drought Mitigation Strategies for Dryland Agriculture with Special Reference to Groundnut

P. Sudhakar, P. Latha and K. Sujatha

Professor (Crop Physiology), S. V. Ag College, Tirupati

26.1 Introduction

Rainfed farming is practiced in nearly two third of the arable land (96 mha) in India. So far much of the agricultural growth achieved in past decades occurred in irrigated areas. The potential for additional production gains in these areas may lessen with time from inherent problems. It is in the rainfed belt where cultivation of coarse cereals (91%), pulses (91%), oilseeds (80%) and cotton (65%) predominates. About (44%) of the total production is contributed by rainfed region. Rainfed agriculture supports (40%) of country's population. The rainfed areas are increasingly being warranted to help meet the raising demand for food, pulses, oilseeds, feed, fuel, fruits, vegetables etc. Thus, the country's economy depends on a sustained increase in the productivity from drylands. However productivity in dryland crops affected by recurrent droughts. Hence development of drought mitigation strategies to augment productivity of dryland crops is need of the day.

Groundnut (*Arachis hypogaea* L.) is an important summer oil seed crop and food grain legume, it contains about 50% oil, 25-30% proteins, 20% carbohydrates, 5% fibre and ash which make a substantial contribution to human nutrition. Groundnut is grown largely in warm and hot regions of the world. In India the crop is traditionally grown by small-holding farmers, on less fertile soil, in rainfed conditions, with low inputs. Hence, there is a low productivity levels of groundnut across the country when compared to counties like china, USA where the crop is grown on large farms with assured inputs, with high productivity levels.

Moisture stress during crop growth has been reported to adversely influence water relations and thereby growth and yield of groundnut. Flowering and pegging stages were considered as most sensitive ones and the period of maximum sensitivity to drought occurs between 50-80 days after sowing (Reddy .1988) during which the demand of photosynthetic products for active sinks (pods) is higher.

26.2 Drought Effects in Groundnut

26.2.1 Flowering

The rate of flower production is reduced by drought stress during flowering, but the total number of flowers per plant is not affected due to an increase in the duration of flowering. A significant burst in flowering on alleviation of stress is a unique feature in the pattern of flowering under moisture stress, particularly when drought is imposed just prior to re-productive development (Janamatti et al., 1986). When stress is imposed during 30-45 days after sowing the first flush of flowers produced up to 45 days do not form pegs during that time, however, flowers produced after re-watering compensated for this loss (Gowda & Hegde, 1986).

26.2.2 Pegging and Pod formation

Groundnut plants may experience water stress during pegging and pod development and then may have adequate amount of water. This would result in a drastic reduction of crop yield, and the magnitude of reduction would depend on groundnut cultivars. Peg elongation, which is turgor dependent, is delayed due to drought stress (Boote & Ketring, 1990). Pegs fail to penetrate effectively into air-dry soil, especially in crusted soils. Often, within 4 days of withholding water, the soil surface becomes too dry for peg penetration. Once pegs are in the soil, adequate moisture and darkness are needed for pod development. Adequate pod zone moisture is critical for development of pegs into pods. Peg initiation and pod growth during drought stress demonstrated ability to suspend development during the period of soil water deficit and to re-initiate pod development after the drought stress was relieved (Sexton et al., 1988). Under water stress, pegging and seed set responses of various groundnut cultivars varied substantially, this leads to a large reduction in pod yield. (Nageswara Rao et al., 1989).

26.3 Strategies to Mitigate Drought Effects

Climate change is projected to intensify drought stress in groundnut crop in rainfed regions. This will require identification of reliable drought management strategies and developing high yielding groundnut cultivars to mitigate drought stress effects and sustainable yields

A. Drought management strategies

B. Crop improvement planning

26.3.1 Drought Management Strategies

I. Pre – drought strategies

II. Post – drought strategies

26.3.1.1 Pre - Drought Management Strategies

26.3.1.1.1 Crop diversification

Crop diversification is essential in areas prone for frequent droughts and where mono cropping is predominantly exists. Choice of crop diversification has to be done based on rain fall pattern of the area and drought tolerant nature of the crops selected.

Crop tolerant to drought (moisture stress)

High tolerance	-	minor crops millets, seasamum, safflower, guar
Medium tolerance	-	Sunflower, Red gram, Bengal gram, Black gram, Green gram, Cotton, Jowar, Sugarcane, Ground nut.
Low tolerance	-	Maize
Susceptible	-	Rice, Vegetables

In parts of Rayalaseema where drought frequently causes crop loss, and groundnut is mono crop. Options for diversification are guar, castor for kharif groundnut and sunflower, Bengal gram for rabi groundnut.

26.3.1.1.2 Adjusting sowing time according to rainfall distribution

In drought prone areas sowing dates have to be adjusted both in kharif and rabi based on the mean rainfall distribution over past 10 years. In rayalaseema the areas benefited by both S.W & M. E monsoons July is best for Kharif season, where as in other parts where S.W monsoon is the source, June sowing is recommended.

26.3.1.1.3 Cropping patterns

Cropping patterns are designed based on the intensity and nature of the drought. Intercropping is one potential option where compensatory yields benefit the farmer. The following are several cropping pattern options for adaption.

Groundnut	+	Red gram	-	7:1
Groundnut	+	Castor	-	7:1
Groundnut	+	Sorghum	-	6:2
Groundnut	+	Pearlmillet	-	6:2
Groundnut	+	Fieldbean	-	7:1

Identification of location specific contingency crop plans against delayed monsoons is essential to compensate the monitory loss of farmers. Several contingent crops were identified against kharif groundnut across the country. The following are contingent crops identified for four major districts of Andhra Pradesh, where groundnut is predominantly kharif crop.

District	August	September
Chittor	Redgram, Jowar, Greengram,	Blackgram,Greengram, Redgram, Jowar, Horsegram
Ananatapur	Jowar, castor, Horsegram, Groundnut	Horse gram, Greengram, Bajra & Fodder jowar
Kadapa	Redgram, Castor, Cowpea, Fieldbean	Cowpea, Fieldbean Horsegram
Kurnool	Sunflower, Castor	Sunflower

26.3.1.1.4 Soil management technique

Generally, 95% of water received through rain is lost through run-off. There is every need to hold & conserve the every drop in drought prone areas for maintaining high soil moisture. The following are some of the premonsoon methods with which the dynamic rain water can be conserved.

- Deep ploughing (80 cm) once in three years
- Ploughing and sowing across slope
- Formation of contour bonding for every 50 mt (in 2% slope)
- Soils prone to sub surface crust plough with "subsoiler" where groundnut can be cultivated with better moisture retention, root growth and yield in rainfed areas.

26.3.1.1.5 Water management technique

Among the water management technologies, maintaining a 'farm pond' is potential option for providing supplemental irrigation (sprinkler or drip) at critical growth stages during drought. A farm pond of size 14 × 14 × 3.5 mt can harvest 6 lakh liters of rain water which is sufficient to give two supplemental irrigation to 7.5 acres.

Use of micro irrigation methods viz. drip and sprinker compared to traditional methods is another option in drought prone areas to reap maximum water use efficiency.

Methods of irrigation	Efficiency of water use	Water saving
Traditional	30-45%	-
Sprinkler	55-70%	15-25%
Drip	90-95%	21-50%

26.3.1.1.6 Crop insurance

Crop insurance is a boon to rainfed farmer as it insures guaranteed amount in severe drought conditions. Now crop insurance is being adopted to most of the crops including groundnut. Rainfed farmers can be motivated to insure groundnut crop by paying prescribed premium amount.

26.3.1.2 Post - Drought Management Strategies

26.3.1.2.1 Early season drought

Non recieval of rainfall after sowing for a period upto grand growth stage.

Effects

1. Extension of crop duration
2. Poor germination and establishment
3. Poor root growth
4. Low plant stand (if drought prolongs)

Management

Spraying 2% urea or 0.5% KNO_3 to groundnut crop for early recovery after recieval of rain.

26.3.1.2.2 Mid season drought

No rain (Moisture stress) from flowering stage to pod formation

Effects

1. Low flowering
2. Low flowering to peg formation
3. Plant growth and no. of secondary branches restricted
4. Poor seed size.
5. Low pod yields.

Management of mid season drought

1. Supplementary irrigation through farm pond.
2. Spraying 2% urea or 0.5 % KNO_3 to groundnut crop for early recovery after recieval of rain.

26.3.1.2.3 Terminal drought

Early cessation of rains. It is the most vulnerable growth stage of groundnut crop

Effects

1. Crop duration reduces (specially reproductive growth phase)
2. Poor grain filling
3. Low shelling percentage
4. Yield and yield components are affected drastically

Management

1. Life saving irrigation using farm pond
2. Spraying 2% urea or 0.5 % KNO_3.
3. Sowing of short duration varieties for escaping terminal drought.

26.3.2 Crop Improvement Strategies

Development of drought tolerant varieties is the potential option to sustain yields in drought prone areas. The following are the strategies for development of drought varieties.

1. ***Characterization of stress:*** Identification of specific abiotic stress viz. moisture stress, heat stress, salinity stress etc., which frequently affect the crop yields in the drought prone area.
2. ***Identification of relevant drought traits:*** Reliable drought traits like WUE, root growth, relative water content, Relative membrane injury, Omolyte accumulation etc., are to identified. The identified traits are to be used in screening parents as well as segregating populations. The identified genotypes can be used as donor parents for transferring these traits into a locally adopted and high yielding groundnut genotypes either through traditional and molecular breeding.
3. Appropriate methodologies have to be designed for rigorous evolution of stable lines specially testing for its ability to tolerant midseason and end of session moisture stress.
4. Thus identified genotypes can be tested properly across the various drought prone locations before to farmers.

In developing drought tolerant groundnut varieties ANGRAU has made substantial progress and released drought tolerant varieties DHARANI, ABHAYA from RARS, Tiruapti and K-9 from ARS, Kadiri. Dharani is a Promising throughout A.P and Telangana States due to its yielding ability and high shelling percentage under rainfed conditions.

To conclude, there are limited available technologies to save the crop if prone to midseason and thermal drought conditions. Among them selection of drought tolerant groundnut variety, supplemental irrigations by adopting farm pond technology and crop insurance are prime options to sustain dry land farm productivity and to save the dry land farmers.

26.4 Aflatoxin Management in Groundnut

Aflatoxins are highly carcinogenic, immunosuppressive agents, highly toxic and fatal to humans and particularly affecting liver and digestive tract, (Verma and Mehtha 1998),

(Katiyar et al., 2000), reported the risk of aflatoxins with hepatitis-B infection to human and livestock population in India. It is currently known that there are synergistic effects between aflatoxin and Hepatitis-B infection causing liver cancer. Children exposed to aflatoxin suffer from poor growth and immune suppression, making them susceptible to HIV and malaria.

Agricultural products are often invaded by fungi that can produce poisonous substances called Mycotoxins. Among mycotoxins, aflatoxins, produced by Aspergillus flavus and A. parasiticus, occur globally. Groundnut, maize, sorghum, pearl millet, chilies, cassava and other food products are contaminated by aflatoxin each year. Contamination can occur at any stage from pre-harvest to storage. Studies have shown that the fungi causing this poisoning are present in air and soil, affecting between 20 and 80% of crops at different levels in warm humid environments.

More than five billion people in developing countries are constantly exposed to aflatoxins by unknowingly consuming contaminated foods. A World Bank study estimates that the EU's standards for allowable aflatoxin in imported grains and nuts (4 ppb) costs Africa alone US$750 million every year. The Food and Agriculture Organization attributes a quarter of food spoilage due to mycotoxins, with losses of up to US$26 million recorded in the USA from groundnut aflatoxin contamination alone.

- 76 Countries: Total aflatoxins 0-35 ppb
- 61 Countries: aflatoxin B_1 only 1-20 ppb
- INDIA: Groundnut food: 20 ppb/kg , Groundnut feed: 100 ppb/kg

Aflatoxin is one of the major problems in groundnut, which hinders not only the domestic consumption and the export of groundnut since, the international regulation for minimum standards for aflatoxin contamination is becoming stringent (4 ppb/kg). Aflatoxins are major impediment in the export of groundnut. In groundnut aflatoxins contaminate pods both at pre harvest stage as well as at harvest stage. When groundnut is exposed to end season moisture stress (After 70 DAS) aflotoxin production in kernels is most likely occur. Hence compared to rainfed kharif produles, rabi produced groundnuts are safer in terms of aflatoxins (because rabi is grown under irrigated conditions). As development of resistant groundnut varieties against aflatoxin risk was not successful, agronomic management of aflatoxin is still a potential option. (Choudhary et al.2010). The fallowing are findings of research work done at RARS, Tirupati and work elsewhere related to aflatoxin problem and management options in groundnut.

- Group of toxic compounds produced by fungi Aspergillus flavus , Aspergillus parasiticus
- 18 types of aflatoxins identified and major ones - B1, B2, G1, G2 and M1, M2 (Milk). Af B1 is most abundant and highly carcinogenic.

The fungus, Aspergillus flavus

- Grows over a wide range of temperatures →10-40°C
- Thrives at high relative humidity and kernel moisture contents of 10-30%.
- Optimum temperature for growth and aflatoxin production about → 25-35°

Groundnut Products – Vulnerable to Aflatoxin contamination

- Physically damaged or shriveled pods and/or kernels
- Left over pods after harvest
- De-oiled groundnut cake
- Fodder left with contaminated immature pods
- Groundnut butter and other products of groundnut
- Filtered groundnut oil

Strategies for minimizing aflatoxin contamination in groundnut

1. **Genetic Tolerance in Groundnut**
    - No genotype identified with complete resistance
    - Long Duration varieties are susceptible to A.flavus invasion
2. **Cultural Methods**

    (a) ***Pre-Sowing stages***

    - Remove stubbles of previous crops/weed flora and keep the field clean
    - Deep ploughing
    - Application of Trichoderma mixe with neem cake
    - Application of FYM @ 5-10 t/ha
    - Adopting paired row method of planting

    (b) ***Sowing and Post-Sowing operations***:

    - Select short/medium duration variety to escape end-of-season drought
    - Sowing only sound seeds and treat the seeds with Bavistin @ 2g/kg
    - Apply gypsum @ 400-500 kg/ha at the time of flowering
    - Control of foliar diseases like rust and leaf spots
    - Avoid end-of-season drought
    - Harvest the crop at right maturity

    (c) ***Harvest and Post-harvest operations***

    - Avoid mechanical damage to the pods during harvesting
    - Rapid drying of the uprooted plants along with the pods in small heaps

- Isolation of immature pods
- Discard mechanical or insect damaged pods
- Thoroughly drying of the pods to safe moisture level of 8%

3. Precautions for storage

- Store the produce in a clean and well aerated storage conditions
- Prevent the insect damage to the pods in storage
- Drying the pods to 7-8% moisture level
- Sorting our damaged and discolored pods
- Storing at low temperature and low humidity levels
- Storing in clean polythene lined gunny bags

References

1. A.K. Choudhary, P. Kumara, Management of mycotoxin contamination in preharvest and postharvest crops: Present status and future prospects, J. Phytol. 2 (2010) 37-52.
2. Boote, K.J. & Ketring, D.L. (1990). Peanut. In: Stewart B.A. And Nielson D.R. (Eds), Irrigation of Agricultural Crops. Asa- Groundnut - A Global Perspective. International Crops Research CSSA-SSSA, Madison.
3. Gowda, A. & Hegde, B.R. (1986). Moisture Stress and Hormonal Influence on The Flowering Behavior And Yield Of Groundnut, Plant Physiology, Vol.66, pp. 835-837.
4. Janamatti, V.S.; Sashidharv, R., Prasad, I.G., Sastry, K.S.K. (1986). Effect Of Cycles Of Moisture Stress On Flowering Pattern, Flower Production, Gynophore Length and Their Relationship To Pod Yield In Bunch Types Of Groundnut, Journal of Agricultural Research, Vol.1, No.2, pp. 136-142.
5. Katiyar, S., Dash, B.C., Thakur, V., Guptan, R.C., Sarin, S.K. and Das, B.C. (2000) tumor suppressor gene mutations in hepatocellular carcinoma patients in India. Cancer, 88, 1565–1573.
6. Nageswara Rao, R.C.; Williams, J.H., Sivakumar, M.V.K. & Wadia, K.R.D. (1998). Effect of Water Deficit At Different Growth Phases Of Peanut. Ii. Response to Drought during Pre-Flowering Phase, Agronomy Journal, Vol.80, pp.431-438.
7. Reddy C R 1988 Studies on yield response to water in groundnut (Arachis hypogaea L.) Ph. D Thesis, Andhra Pradesh Agricultural University, Hyderabad.
8. Verma RJ and Mehta DN. 1998. Occurance of hemolytic anemia during aflatoxicosis. Indian Journal of Environment. & Toxicology, 1998, 8(1):5-7.

Chapter 27

Integrated Weed Management in Groundnut

A. S. Rao

Nodal Officer, ANGRAU, Rajendranagar, Hyderabad

27.1 Introduction

Groundnut is an important oilseed crop of India which is cultivated in nearly 6.0 million ha area with a production of 7.5 million tonnes and average productivity of 1.27 t/ha. Though, India ranks first in the world under groundnut area, productivity is less compared to other countries. One of the major factors responsible for low productivity of groundnut is the severe weed infestation. Being a short statured plant and initial slow growth, weeds effectively compete with the groundnut crop for light, moisture, nutrients and space and reduce the yields ranging from 13-80% (Ghosh *et al.*, 2000). Further, the additional losses caused by weeds in groundnut crop compared to other crops are:

1. Weeds hinder pegging
2. Harbours pests Eg. *Portulaca oleracea* act as alternate host for *Helicoverpa*
3. Harbours diseases Eg. *Solanum nigrum* harbours *Verticillium dahli(wilt) Parthenium harbours*, Peanut bud necrosis virus
4. Interfere with Harvesting
5. Competition for Under Ground Space

During *kharif* groundnut is mostly grown as rain-fed, while in *rabi*, it is grown with the residual moisture of the previous crops or with limited irrigation facilities. The weed problem in *kharif* groundnut is more intense due to dominant role of grasses, besides sedges and broad leaved weeds. The hot, humid climate and plentiful rainfall favours profuse growth of weeds and crops suffer adversely. The weed problem in *rabi* reason is lesser magnitude and easy to manage.

27.2 Major Weed Flora of Groundnut

The major weeds that cause severe yield loss in groundnut crop are as under

***Table1 27.1** Major Weed Flora of groundnut Crop*

S. No.	Common Name	Scientific Name
A	**Grassy weeds**	
1	Bermuda grass	*Cynodon dactylon*
2	Sawa millet, Jungle rice	*Echinochloa colonum*
3	Indian goosegrass,	*Eleusine indica*
4	Swollen fingergrass	*Chloris barbata*
5	Crowfoot grass	*Dactyloctenium aegyptium*
6	Torpedograss,	*Panicum repense*
7	Common crabgrass,	*Digitaria sanguinalis*
B	**Sedges**	
1	Purple nut sedge	*Cyperus rotundus*
2	Yellow nutsedge, tiger nut sedge,	*Cyperus esculentus*
3	Globe fringe rush and hoorahgrass	*Fimbrystylis miliacea*
C	**Broad leaf weeds**	
1	Cock's comb	*Celosia argentea*
2	Wild mustard	*Cleome viscosa*
3	Slender amaranth or green amaranth.	*Amaranthus viridis*
4	Congress weed	*Parthenium hysterophorus*
5	Leucas	*Leucas aspera*
6	Stonebreaker or seed-under-leaf.	*Phyllanthus niruri*
7	Desert horse purslane, black pigweed, and giant pigweed	*Trianthema portulacastrum*
8	Garden spurge	*Euphorbia hirta*
9	Spiderling, spreading hogweed	*Boerhavia diffusa*
10	False amaranth	*Digera arvensis*
11	Coat buttons	*Tridax procumbens*
12	Common purslane	*Portulaca oleracea*
13	Spurge	*Euphorbia spp.*
14	Benghal Dayflower, Tropical spider wort	*Commelina benghalensis*
	(Harpreet Singhand Surjit Singh, 2009; Kalhapare *et al.,2013;* Majumder *et al.,* 2009)	

27.3 Critical Period of Crop Weed Competition and Extent of Crop Losses Due to Weeds

The critical period of crop weed competition (CPWC) is a key component of an IWM programme. This is the period in the life cycle of a crop, when it must be kept weed free in order to prevent a specific level of yield loss (Van Aeker *et al.*,1993).Weed presence before and after CPWC should not significantly reduce yields. If the weeds are managed up to a certain stage of crop i.e one third of the crop duration, the yield of crop can be increased. The critical period of crop weed competition in groundnut crop vary from 15-50 days after sowing (Majumder *et al.*, 2009) depending on cultivar, weed species, cultural practices and environmental conditions etc. The critical period of crop weed competition, extent of yield losses and nutrient depletion due to weeds reported by some researchers are given below.

***Table 27.2** Critical period of crop- weed competition, nutrient removal and yield losses in oilseeds due to weeds*

S. No.	Critical period of crop-weed competition (DAS)	Average yield loss (%)	Nutrient losses NPK in kg/ha			Reference
			N	P	K	
1	-	-	38.8	9.2	23.3	Naidu *et al.*, 1982
2	30-60	30-75	39.0	9.1	23.5	Sasikala *et al.*, 2004
3	15-50		-	-	-	Majumder *et al.*,2009
4	28-56	62	58.0	6.0	45.0	Bhale *et al.*, 2012

To overcome this hurdle and for getting higher production as well as productivity of groundnut, an integrated effort will be needed to minimize the weed competition.

27.4 Integrated Weed Management (IWM)

IWM can be defined as "a system which brings together all feasible methods of weed control, harmonizing them into a single unified co-ordinated system designed to maintain weeds at below those at which they cause economic loss, a system which takes into consideration all weeds and all classes of weeds in a plant eco system and which rests on firm ecological and economic principles and approaches". This approach is gaining importance as herbicides at higher doses cause hazards and the labour availability becomes problem and at time weather conditions do not permit inter cultivation.

27.4.1 Need for IWM

1. One method of weed control may be effective and economical in a situation and it may not be so in other situation.
2. No single herbicide is effective in controlling wide range of weed flora.
3. Continuous use of same herbicide leads to development of resistance/shifts.
4. Continuous use of only one practice may result in some undesirable effects.
5. Indiscriminative herbicide use and its effects on environment and human health.

27.4.2 IWM should be

1. Flexible enough.
2. To incorporate innovation and practical experience of local farmers.
3. Developed for the whole farm/cropping system
4. Economically viable and practically feasible.

IWM programmes are designed to address one or more of the following

1. Suppression of weed growth particulary during critical period of weed control.
2. Prevention or suppression of weed seed production.
3. Reduction of weed seed reserves in soil.
4. Prevention or reduction of weed growth.

27.5 IWM Strategies and Components

27.5.1 Strategies

Attention must focus on managing threatened resources (crop, live stock, natural areas, landscape etc) to maintain their competitive ability rather than on eradication of weeds once they invade. Control of undesirable plants without improving management is usually futile. A balanced approach to successfully managing the resources must includes the following processes

1. Managing the resources to prevent weeds from invading
2. Proper identification and knowledge of invasive weed species
3. Inventory, mapping and maintaining of weed population and damage.
4. Making control decision based on knowledge of potential damage, cost of control methods and environmental impact of weed and weed control decision. Using control strategies that may include a combination of methods to reduce weed population to an acceptable level.
5. Evaluating the effectiveness and effect of IWM.

27.5.2 Components

The aim of weed management practices is to shift the crop weed competition balance in favour of crop growth. Several weed control measures such as physical/mechanical, biological, cultural and chemical methods should be used alone or in combination with each other depending upon situations

27.5.2.1 Preventive Measures

- Selection of weed free seeds
- Do not feed animals with matured weeds containing seeds
- Do not allow animal from weed infested fields to weed free fields
- Clean the tillage and other implements before entering from one to other
- Use well rotten FYM only
- Do not transport soil from weed infested field to others
- Control the weeds on field bunds and channels

- Ensure no weeds on vacant lands, railway tracks and roads
- Try to control the weeds before seed setting
- Take all possible precautions to prevent spread of weeds to new places

27.5.2.2 Control Measures

27.5.2.2.1 Physical/mechanical methods

These are age old practices in groundnut. A major portion of the weed can be managed by hand weeding, hoeing, and cutting of weeds by sickles, mechanical weeding with tools and implements, burning, soil mulching, soil solarisation etc. In general weed control by hand weeding two times at 20 and 40 DAS during critical period of crop weed competition is required to control the weeds in groundnut crop in order to get the higher yields. Weed control by hand weeding is not only laborious, inefficient, time consuming but also expensive.

27.5.2.2.2 Cultural methods

A major portion of the weeds can be managed by preparatory cultivation, use of optimum seed rate, adjustment in time of sowing, planting methods, selection of proper variety, time, rate and methods of fertilizer application, inter-cultivation, intercropping, crop rotation etc . Cultural methods of weed control are most effective, as it pulverizes the soil and thus provide better aeration for root penetration, nodulation, pegging and pod development. Experimental results indicated that two intercultivations at 20 and 40DAS were found to be effective in reducing the weed growth. However, sometimes, wet soil conditions due to frequent premonsoon//monsoon rains do not permit intercultivation when the crop is due for this operation.

27.5.2.2.3 Biological methods

Biological control of weeds involves utilizing living organisms to attack a weed population such that it is maintained at or below desirable low level without significantly injuring desirable plants. Biological weed control is found to be successful mostly in non-crop situations. Unfortunately biological weed control has not developed to the point that it has had any appreciable impact on the production of agronomic crops in general and groundnut in particular.

27.5.2.2.4 Chemical methods

Use of herbicide (or) herbicidal approach of weed management in groundnut found to be efficient and economical due to increased cost of labour wages and non availability of labour in time for weeding operations besides unfavorable soil and weather conditions. In fact chemical weed control is cheaper than manual weeding in all most all parts of the country. Therefore the farmers are resorting to chemical weed control in groundnut crop also. Even in IWM systems chemical weed control is often necessary at decreased application rates. Some of the herbicides recommended in groundnut crop are given below.

Table 27.3 *List of some herbicides with their recommended rates and time of application in groundnut crop*

Time of application	Herbicide	Rate (kg ai/ha)	Weeds controlled	Remarks
Preemergence	Alachlor	1.0-1.5	G	
(within 2 DAS)	Butachlor	1.0-1.25	G, BLW	
	Oxyfluorfen	0.1-0.15	G, BLW	
	Pendimethalin	1.0-1.5	G, BLW	
Postemergence	Fenoxaprop ethyl	0.056	G	
(15-25 DAS)	Quizalofopethyl	0.05	G	
	Fluazifopbutyl	0.25	G	
	Imazethapyr	0.05	G, S, BLW	
				Causes slight injury
(Samant and Mishra, 2014; Vaghasia *et al*, 2014; Malunjkar *et al*, 2012;Malligawada *et al*, 2000) DAS: Days after sowing G: Grasses S: Sedges BLW: Broad Leaf Weeds				

27.6 IWM in Groundnut

The control of weeds by integrating mechanical, cultural and chemical weed control is more efficient than their isolated control measures. The burden of chemicals on non target organisms and environment can be made as low as possible through IWM. This approach is gaining importance as herbicides at higher doses cause hazards and the labour availability becomes problem and at time weather conditions do not permit intercultivation. Some of the IWM approaches which proved very effective in groundnut are given below (Samanth and Mishra, 2014; Vaghasia *et* al., 2014; Kalhapare *et al.*, 2013; Sasikala *et al.*,2013; Malligawada *et al.*, 2000).

1. Summer deep ploughing of fields for destroying the underground vegetative parts of perennial weeds
2. Chemical/cultural Pre emergence application of pendimethalin 0.75-1.0 kg/ha *fb* inter-cultivation at 25-30 DAS.
3. Chemical/mechanical Pre emergence application of pendimethalin 0.75-1.0 kg/ha or alachlor 1-1.5kg/ha *fb* hand weeding/mechanical at 25-30 DAS.
4. Chemical/chemical Pre emergence application of pendimethalin 0.75-1.0 kg/ha or alachlor 1-1.5kg/ha *fb* post emergence application of imazethapyr 50g/ha at 25-30 DAS.

5. Chemical/mechanical/chemical Pre emergence application of pendimethalin 0.75-1.0 kg/ha or alachlor 1-1.5kg/ha *fb* hand weeding/mechanical at 20-25 DAS. Followed by post emergence application of imazethapyr 50g/ha at 35-40 DAS
6. Chemical/cultural/chemical Pre emergence application of pendimethalin 0.75-1.0 kg/ha or alachlor1-1.5kg/ha *fb* intercultivation at 20-25DAS. Followed by post emergence application of imazethapyr 50g/ha at 35-40 DAS. Followed by post emergence application of imazethapyr 50g/ha or quizalofopethyl 50g/ha at 35-40 DAS.

27.7 Weed Management in Groundnut based Inter-Cropping Systems

Weed management is a big problem in inter-cropping system than in sole cropping system. In such situations the herbicides selected or used should be selective to both the crops.

27.7.1 Groundnut+ Sunflower Inter-cropping System

Pre-emergence application of alachlor 1.5 kg/ha or pendimethalin 1.0 kg/ha are found to be effective in controlling grassy as well as broad leaf weeds.

Post emergenc application of fenoxaprop ethyl 56 g/ha or quizalofopethyl 50g/ha at 20-25 DAS are found to be effective in controlling grassy weeds without any injury to both crops

27.7.2 Groundnut+ Red Gram Inter-Cropping System

Pre-emergence application of alachl or 1.75kg/ha or pendimethalin 1.0 kg/ha are found to be effective in controlling grassy as well as broad leaf weeds.

Post emergenc application of imazethapyr 50g/ha at 20-25 DAS found to be effective in controlling grasses, sedges and broad leaf weeds without any injury to both crops.

27.8 Advantages of IWM

- It shifts the crop weed competition in favor of crop.
- Prevents weed shifts towards perennial nature.
- Prevent resistance in weeds to herbicide.
- Potential to reduce herbicide application by 10-50%.
- Reduction in weed control costs.
- Less environmental problems when herbicides are applied in limited quantities.
- Reduction in cost of cultivation.

- Increased yields /net profits.
- No danger of herbicide residues in soil or plants.
- Reduced water contamination.
- Reduced health risk to the public.

27.9 Bottlenecks of IWM

- Main bottle neck is only few efficient and economic herbicides are available for groundnut crop.
- Non availability of labour/cattle pair.
- Higher cost of labour.
- Survival of problematic weeds.
- Some systems require more initial investment.

27.10 Socio - Economic Aspects

- Only a small portion of literature addresses cost –benefit considerations of existing or prospective control strategies.
- Farmers compare the cost of weed control with the value of yield increase resulting from the control, which means choosing control inputs that provide the desired degree of control, at the lowest cost.
- The final choice of control input then depends not only on its efficacy but also on its cost, in addition to
  - Farmer's perception of risks involved.
  - Weed economic threshold level or severity of infestation.
  - Crop potential productivity.
  - Grower's Economic resources and technical knowhow.
  - Availability of IWM technology.

27.11 Summary and Conclusion

- Weeds are and will continue to be a major challenge to groundnut production.
- Dependence on any one type of weed control method alone will lead to unforeseen problems.
- Majority of the literature on IWM indicates that use of pre-emergence herbicide *fb* hand weeding/inter-cultivation.

- The successful integration of soil, crop, and weed management will require active collaboration among microbiologists, biochemists, agronomists, weed ecologists, agricultural engineers, and members of other disciplines.
- A better understanding of weed dynamics and soil-related processes on organic and low-external-input farms, which often use diversified cropping systems, crop residues, composts, and manures to maintain soil productivity, will also further the development of integrated weed management systems.
- IWM has potential economic and environmental benefits, however the techniques need to be modified for local cropping situations.
- IWM studies on cost-effective control methods are few despite the fact the farmers aim for maximum profits in contrast to researcher's quest for excellent weed control.
- To date, economic and technical limitations for IWM have not favoured its wide spread adoption.

27.12 Priority Areas for Research/Extension

- Survey of weed flora and preparation of weed maps/atlas in different agro-climatic zones to have clear understanding of the weed problems.
- In -depth studies on weed biology of important weeds associated with groundnut crop and cropping systems to find out vulnerable stages for weed management.
- Monitoring weed flora shifts due to cultural practices, herbicide uses etc to formulate strategic research for managing them.
- Breeding of weed competitive cultivars particularly for weeds which are difficult to control.
- Investigations on the effect of herbicides on soil micro flora, earthworms etc.
- Identification of new herbicide formulations for broad spectrum weed control with better crop safety.
- Intensive work on eco-friendly methods like biological control, development of bio-herbicides etc.
- Development of herbicide resistant varieties so as to facilitate use of non selective herbicides.
- Studies on integrated approaches for weed control.
- Evaluation of herbicide residues in soil, water and crop.
- Technology verification in farmers' field and motivating the farmers to adopt integrated weed management.
- Training to the extension personnel and farmers on IWM technologies.

References

1. Bhale, M. Vilas, Jayashri, V. Karmore, Yuvaraj Patil, R. 2012. Integrated weed management in groundnut (Arachis hypogaea). Pakistan Journal of Weed Science.18(special issue):733-739.
2. Ghosh, P.K., Mandal, K.G and Kuntal, M.H.2000. Allelopathic effects of weeds on groundnut(Arachis hypogaea) in India-A review. Agricultural Reviews 21(1):66-69.
3. Kalhapare, A.H., Shete, B.T. and Bodake, P.S.2013. Integration of chemical and cultural methods for weed management in groundnut. Indian Journal of Weed Science 45(2).116-119.
4. Majumder, A., Sahu, A.K., Karan, A.K, Kundu. C.K, Nath,. R. and Bera, P.S.2009. Critical period of weed control in summer groundnut (Arachis hypogaea) in gangetic alluvials in West Bengal. Journal of Crop and Weed. 5(1):243-247.
5. Malligawada,L.H, Kannur, V.S and Giriraj, K. 2000.Integrated weed control in kharif groundnut. Karnataka Journal of Agricultural Science.13 (2):288-294.
6. Malunjkar, B.D., Malik, B.B., and Patil, S.C.2012. Evaluation of post emergence herbicides in rainy season groundnut. IndianJournal of Weed Science.44 (2): 95-97.
7. Naidu, L.G.K., Reddy, G.H.S. and Rajan, M.S.S.1982. Nutrient uptake as influenced by crop weed competition in groundnut. Indian Journal of Weed Science. 14(2): 137-140.
8. Samant, T.K. and Mishra, K.N.2014. Efficacy of post emergence herbicide quizalofp ethyl for controlling grassy weeds in groundnut. Indian Journal of Agricultural Research 48(6): 488-492.
9. Sasikala, B., Reddy, Y. and Rahava Reddy.C.2004. Pre and post emergence herbicides on weed control and yield of groundnut (Arachis hypogae).Indian Journal of Dryland Agriculture Research and Development 19(1):78-80.
10. Singh, Harpreet and Surjit Singh. 2009. weed management and soil microorganisms studies in irrigated summer groundnut (Arachis hypogaea). Indian Journal of Weed Science. 41(1&2):103-107.
11. Vaghasia, P.M., Bhalu, V.B., and Kavani, R.H. 2014. Influenc of pre and post emergence herbicides in groundnut (Arachis hypogaea). Journal of oilseed Research. 31(1):37-40.
12. Van Acker, R., Swaton, C.J. and Wise, S.E.1993. The critical period of weed control in soyben (Glycine max(L).Weed Science.41:194-200.

Chapter 28

Soil and Water Conservation Measures in Rainfed Agriculture

A. Pratap Kumar Reddy and S. Hemalatha
Farmers Call Centre, ANGRAU.

28.1 Introduction

A significant cause of low crop production and crop failure in rainfed agriculture in the tropics is low and erratic rainfall. However, in many areas crop and land management do not optimize water flow along the rooting zone of the crop. Thus, poor yields are related to an insufficiency of soil moisture rather than to an insufficiency of rainfall. Tropical and subtropical rainfed agriculture depends on an adequate supply of water in the rooting zone of the soil. Soil water limits crop production in dry and semiarid tropics and subtropics with restricted, irregular or markedly seasonal rainfall with annual precipitation from 400 to 1000 mm (Rao et al., 1998). But also in areas, where seasonal shortages of rainfall can limit crop productivity, it is necessary to give more attention on the efficient capture of precipitation in situ and improve the soil moisture content of the rooting zone.

28.2 Strategies of Water Conservation and Management in Arable Lands

1. Reducing runoff amount through:
   (a) Increased infiltration
   (b) Increased time for infiltration
   (c) Reducing rain drop impact and intensity
2. Decreasing runoff velocity by:
   (a) Reducing slope gradient
   (b) Decreasing slope length
   (c) Increasing roughness

3. Adopting sustainable farming systems like:
   (a) Low cropping intensity
   (b) Use of cover crops
   (c) *In situ* water conservation

28.3 Measures that can be Adopted for Water Conservation in Arable Lands

1. Through Crop management
   (a) Early planting
   (b) Adequate stand
   (c) Crop rotation
   (d) Multiple cropping
   (e) Strip cropping
   (f) Balanced fertilizer application
   (g) Use of organic farming
   (h) Agro forestry
2. Through soil management
   (a) Mulch farming
   (b) Conservation tillage
   (c) Rough seed bed
   (d) Contour cultivation
   (e) Ridge furrow system of planting
   (f) Formation of tie ridges
   (g) Soil Conservation systems
3. Through slope management
   (a) Terracing
   (b) Contour bunds
4. Through Runoff management
   (a) Surplus water disposal structures
   (b) Grassed Waterways
   (c) Soil Conservation structures
   (d) Rainwater harvesting

28.4 Soil and Water Conservation Measures

Soil and water conservation measures consist of agronomical and mechanical methods. Agronomic methods are supported with mechanical measures where land slope exceeds permissible limits and runoff gains erosive velocities.

28.4.1 Agronomic measures

28.4.1.1 Contour Tillage

All agricultural operations such as ridging, ploughing, harrowing, sowing, trenching, etc., are recommended to be done on the contour wherever possible or at least generally across the direction of the slope where holdings are very small. Even though the operation is very simple, it plays a major role in retarding the process of soil erosion through runoff. It also conserves soil, and due to increased time of concentration, more rainwater seeps through the soil profile to recharge ground water. Summer ploughing leaves the soil highly absorbent of initial rains.

28.4.1.2 Dead Furrows

When all tillage operations are complete, it is advisable to leave a deep dead furrow at every 10 m interval. This should remain in position until the groundnut crop is harvested. Dead furrows aid in reducing the runoff velocity and they also conserve water.

28.4.1.3 Organic Matter

Indian soils are very poor in organic matter, especially in drought-prone areas. This can be improved by leaving the crop residue in situ (on the fields). Adding organic manures such as farmyard manure and compost every year as basal application to the soil improves the physical condition of the soil considerably. Soil – Air, Soil – Temperature, and Soil – Moisture relationships are well balanced with the presence of organic matter. Organic matter improves the activities of soil microorganisms and also provides the much needed micro plant nutrients of all kinds, besides nitrogen, phosphorus and potash. Addition of large amounts of chemical fertilizers to dry land crops should be discouraged as it damages the soil due to:

- excessive depletion of scarce soil moisture for its own transformation;
- reduction in all soil microorganism activity; and
- destruction of soil structure

28.4.1.4 Selection of Drought-Resistant Varieties

It is important that varieties which have proven genetic character to withstand longer periods of drought are chosen so that the crops can do well even in situations where the intervals between rainy days are long.

28.4.1.5 Early Maturing Varieties

Where the distribution and the amount of rain is unpredictable, it is important to select varieties which have a shorter duration life cycle (seed to seed) to cut down the water requirements of the crop. In drought-prone areas, the success rate of short duration crops is greater than long-duration crops.

28.4.1.6 Strip-Cropping

Raising Erosion Permitting Crops (EPC) with Erosion Resistant Crops (ERC) having abundant adventitious root system and providing high percentage of canopy in strips in a ratio of 2:1 or 3:1 (i.e., 20 to 10 rows or 30 to 10 rows) helps in trapping soil from EPC strips to ERC strips. The increased resistance to runoff in ERC results in higher volume of water percolating through soil profile, due to increased time of (on-ground) concentration. The close-growing ERC strips are generally legumes which fix nitrogen in the soil and enrich it. The canopy of the ERC also protects the soil from beating action of rain drops. Strip cropping also helps in stabilizing crop production.

28.4.1.7 Wind Erosion

Wind Erosion is as important as water erosion, the same system of cropping is recommended to be adopted, generally across the wind direction. Farmers can also raise wind belts across the wind direction. Alley cropping with rows of trees at 30 to 45m intervals also controls the desiccating effects of hot and dry winds.

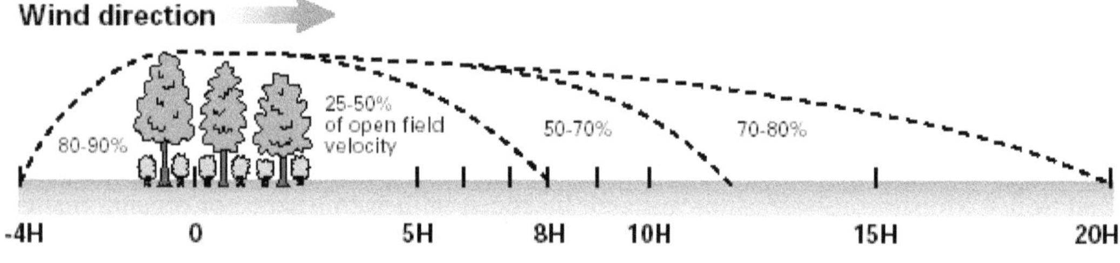

Fig. 28.1 Windbreak

28.4.1.8 Agro-Horticulture

Marginal lands do not produce good annual crop returns even in normal season. These kinds of soils are best used for raising trees of economic value and creating permanent assets. Some of these lands are also very good for raising dry land horticultural crops such as mango, ber, pomegranate, tamarind etc. A part of the land could be earmarked specially for planting mixed tree species known in the area for providing fuel, fodder and timber for household needs and agricultural implements. Trees provide stable and sustained income every year, especially in drought years.

28.4.1.9 Role of Vegetation

Vegetation is crucial in preserving productive soil and conserving rain water for sustaining life. Soil and water need to be preserved for crop production (both annual and plantation crops) as well as plants that give fuel, fodder, fruits, industrial raw materials, medicinal and aromatic plants and the like. Minor vegetation such as creepers, shrubs, various kind of grasses, legumes and plants like Agave, which yield fiber have an important role when grown in gullies of various kinds and also on upstream and downstream sides of mechanical structures such as gully checks, water harvesting structures, etc. These provide reinforcement and extend the life span of the structures by binding the soil through the network of their root systems. These plants also provide fodder to animals in the area.

28.4.1.10 Seed Rate

Normal season Sowing is done with the normal seed rate. However, if there is a drought during the plant's growth period and wilting is likely to occur, selective thinning is recommended to reduce the plant population to effectively use the scarce soil moisture among fewer plants. Late season: Where the monsoon is moderately delayed, normal cropping with reduced seed rate is advised.

28.4.1.11 Line Sowing

Line-sowing on contours is essential. It arrests runoff and conserves soil being eroded. It helps in the use of labour-efficient implements in weeding (i.e., removal of unwanted vegetation through use of different sizes of blade harrows between the rows).

28.4.1.12 Wider Spacing

In all drought-prone areas, the most important objective is to raise a successful crop under scarce soil moisture conditions. One of the recommended practices is wider spacing between rows and between plants within the row. This reduces plant population and competition between plants for scarce soil moisture. Fewer plants have greater access to limited available soil moisture.

28.4.1.13 Weeding

Frequent weeding is an important part of dry land agriculture. Line sowing and mechanical weeding, with appropriate size of blade harrows, removes unwanted vegetation which competes with the main crop. It is not uncommon to see the dry land farmer hitching several blade harrows to one yoke and a pair of bullocks. Weeding within rows can be done using hand hoes. Removal of unwanted vegetation helps the main crop obtain greater accessibility to soil moisture and plant nutrients for its own growth.

28.4.1.14 Mixed/Inter Cropping

Mixed cropping of different crops along with the main crops, such as millets and different legumes, is an insurance against the vagaries of the monsoon. The different root systems of mixed crop feed at different depths of the soil. Moreover, mixed cropping provides small quantities of grain of different kinds for home consumption at different times.

28.4.1.15 Mulching

Mulches are ground covers that prevent the soil from being washed away, reduce evaporation, increase infiltration, and control growth of unwanted weeds. Mulch can be organic crop residue, pebbles, or materials such as polythene sheets. Mulching prevents the formation of hard crust after each rain. Organic mulches add plant nutrients to soil upon decomposition. Use of blade harrows between rows also creates "dust mulch" by breaking the continuity of capillary tubes of soil moisture.

28.4.1.16 Contingent Planning

With every care taken to undertake timely agricultural operations, it is still possible that the whole operation becomes a gamble due to unpredictable monsoons. The main crop could fail in the early part of its life cycle. In such cases, the farmer should come up with an alternate crop that can mature in a very short time and under hard conditions to take advantage of what is left of the rainy season. Contingent planning helps catch and make the best use of late rains. Advance planning is necessary in selecting a contingent crop. And all the requisites for its sowing should be ready within the main season itself. Credit for farmers must be made available at the right time.

28.4.1.17 Buffer Strips

In steep slopes and highly-eroded soils, raising permanent vegetation in small strips at appropriate intervals along the contour helps in arresting soil erosion and conservation of rain water. Raising trees in buffer strips is an asset to counter unpredictable monsoons.\

28.4.2 Mechanical Methods

Where the slope of the soil is more than permissible, mechanical measures such as bunding, terracing and trenching are recommended in addition to agronomic methods. Agronomic methods are used in inter-bunded areas and mechanical practices complement to help boost crop yields in rainfed dry lands.

28.4.2.1 Soil and Water Management Practices

Earlier in soil and water conservation programs, efforts were concentrated on construction of various types of bunds across the slope and community water harvesting structures. This helped in controlling erosion and reducing soil loss rather than increasing crop yields through additional moisture conservation. Current emphasis is more on improving moisture through various field- and community-based *in-situ* moisture conservation practices. Some of the *in-situ* soil and water conservation interventions that were found promising for improving moisture regime, productivity and reducing land degradation are discussed here.

28.4.2.2 Establishment of Field Drains, Land Preparation and Tillage

Excess runoff frequently occurs often in medium – high average rain fall areas. Therefore provision must be made for safe disposal of excess runoff. It is possible to improve the natural drains by clearly delineating and straightening them and adjusting their longitudinal slopes so that they will be stable after the development of grass cover. Land smoothing is most efficiently done if executed in the direction of envisaged cultivation by first plowing utilizing small animal-drawn scrapers. It can be implemented at a much lower cost compared to other form of land leveling/terracing. Although it is ideal to initiate these activities immediately after the previous crop, dry-season showers are often necessary to make plowing with animals feasible.

Land preparation is crucial since it provides suitable environment for the plant growth.Tillage ensures the proper establishment of crops as well as it helps for the retention andmovement of water and nutrients to the roots. Moreover, tillage is also practiced to control the weeds. Tillage can be classified in two classes viz. pre-planting tillage and post-emergence tillage. The objective of pre planting tillage is to loosen the soil up to the rootingdepth for solarization, rainwater conservation as well as for easy penetration of the roots. These operations performed preferably immediately after harvesting the previous season's crop, when the soil has adequate moisture.

28.5 Moisture Conservation through Tillage and Organic Residues Application

(i) Deep tillage/off-season (summer) tillage (20-25 cm deep), once in three years in black soils and once a year in red soils with under lying hard pans

(ii) Shallow cultivation (about 10 cm deep) for entire semi-arid soils, particularly crusting nature of soils.

(iii) Application of organic residue/ manure @ 3-4 t/ha and incorporation through shallow cultivation with first rains in April-May along with heptachlor of 15 kg/ha. The residue should be cut in to small pieces.

(iv) Application of clay rich tank silt on light soils

(v) Mulching

Benefits:

(i) Improves moisture conservation

(ii) Better root proliferation

(iii) Improve soil quality

28.5.1 Conservation Furrow on Contour

Conservation furrow on contour system is a promising land and water management practice for *in-situ* soil and water conservation in semi arid red soils.

Specification: Series of furrows are opened across the slope at 3-5 m apart. The size of the furrows is about 20 cm wide and 15 cm deep.

Area of applicability and scope: This practice is quite suitable in semi arid red soils having gentle to moderate slope with an average annual rainfall of 500-600 mm. Surface runoff from small storms moving down the slope is intercepted by furrows, the velocity is slowed down and runoff water is collected in these furrows.

28.5.2 Broad Bed and Furrow Land and Water Management System

There is a need for an improved *in-situ* soil and water conservation and proper drainage technology particularly in deep black soils that can protect the soil from erosion throughoutthe season and provide control at the place where the rain falls. Broad bed and furrow (BBF) system has been found to satisfactorily attain these goals on deep black soils, especially under conditions of high and assured rainfall by facilitating earlier and more effective crop cover.

Specification: The BBF consists of a relatively flat bed or ridge approximately 100 cm wide and shallow furrow about 50 cm wide and 15 cm deep. The BBF system is laid out on a grade of 0.4 – 0.8% for optimum performance (Fig. 1).

Area of applicability and scope: Semiarid tropics with deep black soils and for groundnut crop in red soils with a reduced gradient along the BBF (0.2-0.3%) and deep black soils with an average rainfall of 600-800 mm.

Benefits:

1. The raised bed portion acts as an *in-situ* 'bund' to conserve more moisture and ensures soil stability; the shallow furrows provides good surface drainage to promote aeration in the seed and root zone; prevents water logging of crops growingon the bed (Velayudham et al, 1994)
2. The BBF design is quite flexible for accommodating crops and cropping systems with widely differing row spacing requirements.

3. Precision operations such as seed and fertilizer placement and mechanical weeding are facilitated by the defined traffic zone (furrows), which saves energy, time, cost operation and inputs.
4. It minimizes loss of fertilizer nutrients as runoff water does not come in contact withfertilizers as they are applied on beds near the plant.
5. Provides more opportune time for rainwater on soil increasing infiltration during low rainfall events and during excess rainfall event takes out extra water safely with minimum soil loss. Reduces runoff and soil loss and improves soil properties over the years.
6. Facilitates double cropping
7. Improves crop yield.

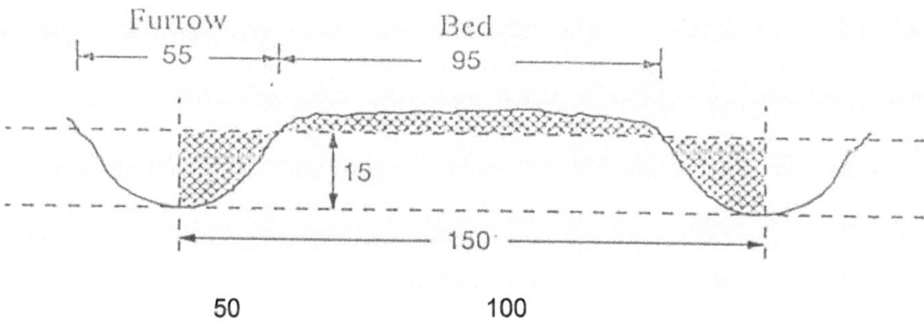

Fig. 28.2 Broad bed and furrow system (all dimension in cms.)

28.5.3 Contour Cultivation Across the Slope

The common method of cultivation on sloping lands is up and down the slope. This is one of the causes of poor rainfall infiltration and accelerated soil erosion. Contour cultivation or cultivation across the slope is simple methods of cultivations which can effectively reduce the runoff and soil loss on gentle sloping lands (Shashekala et al, 2012).

Recommended agro-ecology: ***Soil:*** All most all soil types

Rainfall: Up to 1000 mm ***Slope:*** 1.5 –4.0%

Benefits:

- Reduce runoff and soil losses
- Increase in crop yields
- Simple, low cost and technically feasible even for small farmers

28.5.4 Graded Border Strips

Series of graded border strips are constructed across the slope for *in-situ* moisture conservation. The normal size of the strips could be 5-10 m width and length not exceeding 100 m with 0.2-0.3 % gradient along the strip and avoid cross-slope.

Area of applicability and scope: This practice is quite suitable in semi arid red soils having gentle to moderate slope with an average annual rainfall of < 500 mm. It also facilities good distribution of water where ever the irrigation facility is available. (Sandhya et al, 1994) Border strips can be constructed by using animal drawn bund former andleveller.

28.5.5 Vegetative Barriers

Vegetative barriers or vegetative hedges or live bunds are effective in reducing soil erosion and conserving moisture. In several situations the vegetative barriers are more effective and economical than the mechanical measures viz. bunding. Vegetative barriers can be established either on contour or on moderate slope of to 0.8%. In this system, the vegetative hedges acts as a barrier to runoff flow, which slow down the runoff velocity resulting in the deposition of eroded sediments and increased rainwater infiltration. It is advisable to establish the vegetative hedges on small bund. This increases its effectiveness particularly during the first few years when the vegetative hedgesare not so well established (Poonia and Singh, 2005).

If the main purpose of the vegetative barrier is to actas a filter to trap the eroded sediments and reduce the velocity of runoff then the grass species such as vetiver, sewan (*Lasiurus sindicus*), sania (*Crotolaria burhia*) and kair (*Capparis aphylla*) could be used. But if the purpose of vegetative hedges is to stabilize the bund then the plants such as *Gliricidia* or others could be effectively used. The *Gliricidia* plants grown on bunds not only strengthen the bunds while preventing soil erosion, but also provide N-rich green biomass, fodder and fuel. The cross section of earthen bund can also be reduced.

Recommended agro-ecology: ***Soil:*** Alfisols, Vertisols, Vertic-Inceptisols & associated soils ***Rainfall:*** 400-2500 mm Slope: More than 2.5%.

Benefits:

- Once properly established the system is self sustaining and almost maintenance free.
- Land under the hedge is used for multipurpose viz. N-rich biomass, fodder and fuel.
- Can be successfully used under wide range of rainfall (400-2500 mm) and topography.
- Economical and often more effective than other erosion control measures.

28.6 Conclusion

Areas which urgently require conservation methods should be prioritized based on a five-year perspective and resources have to be mobilized for both hardware and software components. Convergence of different programs related to agriculture, horticulture and soil and water conservation is required, besides involving different institutions and agencies both private and public. A master plan is required which takes into consideration both the micro and macro requirements, but local planning and implementation should be based on active participation of the stakeholders.

References

1. *Poonia, T.C and Singh G.D. 2005. Impact of vegetative barriers on soil and water conservation and production of kharif crops on marginal lands affected by wind erosion. Annals of Arid zone, 44(2):199-201*
2. *Rao, K.P.C, Steenhuis, T.S., Cogle, A.L, Srinivasan, S.T, Yule, D F and Smith G D (1998) Rainfall infiltration and runoff from an Alfisol in semiarid tropical India, 1. No-till systems."Soil Tillage Res. 48:51-59*
3. *Sandhya N.S., Reddy, S. R., Reddy, A.P.K. and Reddy M.K (1994). Relative efficiency of soil and water management practices on productivity of rainfed groundnut (Arachis hypogeae) Indian J. Agron 39:62.*
4. *Shashekala, S.G., Govinda Gowda, V., Ravikumar P. (2012) Adoption of soil and water conservation practices by groundnut farmers. I.J.A.B.R 2(1):166-167*
5. *Velayudham, K, Ali, A H, Arunachalam A A, Velayudham, A and Seeganpal, (1994).Effect of in situ moisture conservation practices on yield of rainfed crops. IndianJ. Agron. 39:160*

Chapter 29

Types of Herbicide Formulations and improvement in Application Techniques

P. Maheswara Reddy, G. Krishna Reddy and P. V. R. M. Reddy

Scientist (Agronomy), RARS, Tirupati.

29.1 Herbicide Formulations

The active ingredient in a herbicide is the chemical that controls the target weed. The herbicide product we purchase is rarely made up only of active ingredients. Often the herbicide is diluted in water or a petroleum solvent, and other chemicals are added before the product is offered for sale. These other chemicals may include wetting agents, spreaders, stickers, extenders or diluents. They usually make the product easier to apply and more convenient to handle. This mixture of active and inert ingredients (inactive) is called a formulation. Herbicide efficiency depends upon the type of formulation it is made into. The development of a formulation depends on the physicochemical properties of the active ingredient to be formulated and of the inert materials to be used in the formulation. A single active ingredient often is sold in several different kinds of formulations. we must choose the formulation that will be best for each use. In making our choice, we have to consider: application equipment available and best suited for the job, hazard of drift and runoff (nearness to sensitive areas, likelihood of wind or rain), safety to applicator, growth patterns of the plant (granular vs. foliar spray) and cost. So, the Formulated product should possess (a) required chemical stability (b) retain desirable physical properties like suspendability; (c) stability of emulsifiable concentrates (no separation of contaminants present in the colloidal system), (d) stable particle size .

29.1.1 Types of Formulations

The two major objectives of formulating herbicides are to ensure their (a) ease of handling and (b) high controlled activity on the target plants. Different formulations of herbicide are:

1. Wettable powders (WP)
2. Soluble powders (SP)

3. Emulsifiable concentrate (EC).
4. Soluble Concentrates (SC)/Water soluble concentrates (WSC)/Soluble Liquids (SL).
5. Dry flowables (DF).
6. Flowable liquid (FL).
7. Granules (G).
8. Others – Capsules, wax bars, soluble mulches, foam pieces, aerosols etc.

The most important types of formulations are wettable powder, soluble powder/liquid, emulsifiable concentrate, water-soluble concentrate, granules and fumigants.

29.1.1.1 Wettable Powder (WP)

Wettable powders are dry, finely ground formulations in which the active ingredient is combined with a finely ground carrier (usually mineral clay) along with other ingredients, to enhance the ability of the active ingredient plus carrier to suspend in water. The powder is mixed with water for application. Wettable powders are one of the most widely used herbicide formulations and offer low cost and ease of storage, transport, and handling; lower phytotoxicity potential than ECs and other liquid formulations; and less skin and eye absorption hazard than ECs and other liquid formulations. Some disadvantages are that they require constant and thorough agitation in the spray tank, are abrasive to pumps and nozzles (causing premature wear), may produce visible residues on plant and soil surfaces, and can create an inhalation hazard to the applicator while handling (pouring and mixing) the concentrated powder. Wettable powders need to be protected against moisture.

29.1.1.2 Soluble powders (SP)

Soluble powder formulations when mixed with water, these dry formulations dissolve readily and form a true solution. The formulation is dry and consists of the active ingredient and additives. When thoroughly mixed, no further agitation is necessary to keep the active ingredient dissolved in solution. Few formulations of this type are available because few active ingredients are highly soluble in water. Soluble powders are similar to wettable powders, except that the ingredients used completely dissolve in water or another liquid for which the wettable powder is formulated.

29.1.1.3 Emulsifiable Concentrates (EC)

Formulations of this type are liquids that contain the active ingredient, one or more solvents, and an emulsifier that allows mixing with water. Formulations of this type are highly concentrated and relatively inexpensive, easy to handle, transport, and store, require little agitation (will not settle out or separate) and are not abrasive to machinery or spraying equipment. Formulations of this type may, however, have potentially greater phytotoxicity than other formulations and exhibit a potential for over - or under dosing through mixing or calibration errors and are more easily absorbed through skin of humans or animals and contain solvents that may cause deterioration of rubber or plastic hoses and pump parts.

29.1.1.4 Solution concentrates (SC)/Water soluble concentrates (WSC)/Soluble liquids (SL)

Water-soluble concentrates form a true solution when added to water and are applied with water as the carrier. These herbicides usually have an amine (ammonium salt) or mineral salt in the molecule that enables water solubility. These formulations are essentially nonvolatile. Agitation is not necessary to maintain the herbicide in solution. Here, the herbicide is dissolved in a solvent system to provide a concentrate that is soluble in the carrier. Some types of solution concentrates are oil soluble. The herbicide must be soluble in a small enough quantity of concentrate solventto make packaging and shipment of the formulation economical. It must also be completely and rapidly soluble in the carrier at all temperatures and concentrates.

29.1.1.5 Water dispersible granules (WDG, WG, DG) or dry flowables (DF)

Dry flowables are manufactured in the same way as wettable powders except that the powder is aggregated into granular particles. They are mixed with water and applied in a spray exactly like a wettable powder. This dry formulation usually contains 70 to 90 percent active ingredient. The formulation pours easily without the wind blown dust associated with wettable powders and readily disperses in water to form a suspension. Constant agitation is required. Because of their larger particle size, inhalation hazard for the applicator is reduced.

Advantages: 1. easy to store, transport, and handle; 2. Reduced applicator exposure when mixing the dry formulations; 3. concentrate spills are most easy to clean up from porous surfaces.

Disadvantages: 1. Good agitation required; 2.Residues may be visible; 3.Abrasive to sprayers; 4. May be slightly more expensive than other dry formulations; 5. Rapid pouring from large container can cause mixing problem when product mass settles to bottom of the tank.

29.1.1.6 Flowable liquids

Flowable liquids formulations consist of finely ground active ingredient suspended in a liquid. Flowables are mixed with water for application, are easily handled and applied, and seldom clog nozzles. Some of their disadvantages are that they may leave a visible residue on plant and soil surfaces, and typically require constant and thorough agitation to remain in suspension.

29.1.1.7 Granules (GR)

Used exclusively for soil applied herbicides, the active ingredient is formulated onto large particles (granules or pellets). The primary advantages of this type of formulation are that the formulation is ready to use with simple application equipment (seeders or spreaders), and the drift potential is low because the particles are large and settle quickly. The disadvantages of these formulations are that they do not adhere to foliage (not intended for foliar applications), and may require mixing into the soil in order to achieve adequate herbicidal activity.

29.1.1.8 Pellets (P) or Tablets (TB)

Pellets are dry formulation of herbicide and other components in discrete particles usually larger than 100 mm3, tablets are in the form of small flat pellets. Pellets and tablets frequently are used for spot applications. Herbicide concentrations typically are 5 to 20%

29.1.1.9 Fumigants

Fumigants are volatile chemicals applied into confined spaces or into the soil to produce a gas that will destroy weed seeds and act as a soil sterilant.

29.1.1.10 Flowable (F)

Flowable contains an herbicide plus dust diluent that is blended in small quantity of water. *e.g*: Butachlor 50F.

29.1.1.11 Microencapsulated Formulations (ME) or Capsule Suspension (CS)

Microencapsulated formulations are small particles consisting of a herbicide coresurrounded by a barrier layer, usually made up of a polymer shell. They also are refereedto as capsule suspensions because the capsules are suspended in a liquid medium. Microencapsulation greatly reduces the amount of solvent needed.

29.1.1.12 Dust Forms

Herbicides are not used in dust forms for fear of their drift hazards.

29.2 Calibration of Sprayer

The process to find out the exact quantity of carrier (water) required for spraying a given area under a set of conditions is called "calibration". Calibration of a sprayer is a must for spraying herbicide in the field. Care must be taken to see that speed and pressure do not vary significantly from the calibration test to actual field spraying. Filling a known quantity of water into the sprayer and spraying it over a measured area can docalibration.

The quantity of the water required for spraying a given area can be calculated as

$$Q = \frac{(V_1 - V_2)}{a} \times A$$

where,

Q = Quantity of water required for area to be sprayed (liters)

V_1 = Quantity of water taken initially

V_2 = Quantity of water remained after spraying

A = Area to be sprayed m^2

a = measured area sprayed m^2

29.3 Herbicide Application Techniques and Evaluation

Timely application of herbicides at peak periods plays a vital role in ensuring better yields from a crop. The magnitude of this problem is further amplified due to shortage of labour during this time. Hence, mechanization of application is the only viable option in this scenario. Correct Equipment selection for herbicide application is the most important issue we need to address for effective weed control. The choice of equipment depends on its specific use and the need for a particular weed control measure. Studies indicate that seventy percent of the success of herbicidal weed management depends on the effectiveness of its application (Malik R.K. *et.al.*, 2012).In India, improper and ineffective methods of application have resulted in a tremendous wastage of herbicides. This has led to the damage of non – targeted plants and has decreased herbicide effectiveness while posing extensive health hazards for humans and animals alike. Generally, herbicides are applied in the form of solution or granules.

29.3.1 Herbicide Applicators

Non-pressurized applicators include the water can, granular applicator, controlled droplet applicator and direct-contact applicators such as rope wicks and wipers. Hydraulic (pressurized) applicators usually need water to work. Ex: Lever operated knapsack sprayer, motorized or ground driven applicators.

29.3.1.1 Spraying

A spray is defined as liquid discharged in to particles and scattered as dispersed droplets. Water is the most common carriers for herbicide spraying, which will facilitate uniform distribution and even coverage over the area to be treated. In a spray solution the herbicide is in a dissolved, emulsified or suspended form in water. Herbicide application is directed to soil in case of pre-planting and pre-emergence herbicides and to foliage incase of post-emergence herbicides.

29.3.1.2 Spraying Equipment

Herbicides are generally sprayed using hand operated sprayers. Depending upon situations power operated sprayers are also used. Based on the volume of spray liquid used the sprayers are classified.

29.3.2 Classification of Sprayers

Type of sprayers	Volume (l/ha)	Example
High volume sprayers	500 - 750	Knapsack or Backpack sprayer
Medium volume sprayer	250 - 300	Rocker sprayer
Low volume sprayer	150 – 200	Power sprayer
Ultra low volume sprayer	< 5	Controlled droplet applicator

High volume sprayers are mostly used for herbicide applications because it avoids drift hazard. There are two types of sprayers;

(a) Hand operated sprayers.

e.g., Knapsack sprayer, Backpack sprayer and Rocker sprayer etc.

Power driven sprayers

Ex: Power sprayers, power tiller drawn and tractor drawn sprayers

Basic Components of Sprayer

- a tank to hold the spray solution
- a nozzle through which the sprayer solution is discharged
- a pump for generating the necessary pressure
- a filter
- a hose with lance or boom to hold the nozzle and for directing the spray.

29.3.3 Nozzles

It is the most important component of the sprayer because it influences the uniformity in spraying herbicide, spray drift, rate of application, spray pattern, droplet size and site of application. The function of the nozzle is to accelerate and disintegrate the flow of spray fluid passing through it to form a spray. It consists of body, cap, lip or disc and core.

29.3.3.1 Types of Nozzles

Nozzles vary according to sizes and spray patterns as

(i) Cone (hollow or solid) (ii) Fan (iii) Deflector (flood jet)

Cone type nozzle

In cone type nozzle, the liquid is forced through one or more tangential or helical passages in to a swirl chamber through which the liquid passes to a circular orifice at high rotational velocity to form an air chore within the orifice and swirl chamber. The liquid emerges from the orifice as a hollow core or solid core depending upon the disc or core of the nozzles. Cone nozzles are widely used for insecticide and fungicide applications and sometimes for spot application of post emergence herbicides.

Fan type

Herbicide spraying is mostly done with fan and flooding or deflector type nozzles as they produce a more even distribution of spray and uniform coverage than the cone types. Fan nozzles are ideal for spraying herbicides on flat surfaces and for band spraying in row crops. The tip of fan nozzle has a lenticular or rectangular orifice behind which two streams of liquid meet because of the shape of the bore.

Flood or deflector type

It is also known as deflector, anvil or impact nozzle. A jet of liquid passes through a relatively large orifice and strikes a smooth surface at a high angle of incidence to form a fan shaped pattern. The different types of flood jet nozzle are available, their discharge and spray angle under different pressure levels will vary. For best results, the hand operated knapsack sprayer should maintain a pressure of 10 to 15 pounds per square inch (0.7 to 1.05 kg/cm^2). The most suited flood jet nozzle for herbicide spraying is WFN 62. Compressed air (70-275 kPa) forces the liquid from the pressure chamber through the hose to the nozzle. The nozzle turns the liquid into droplets, which aid in uniform coverage of the weed or soil.

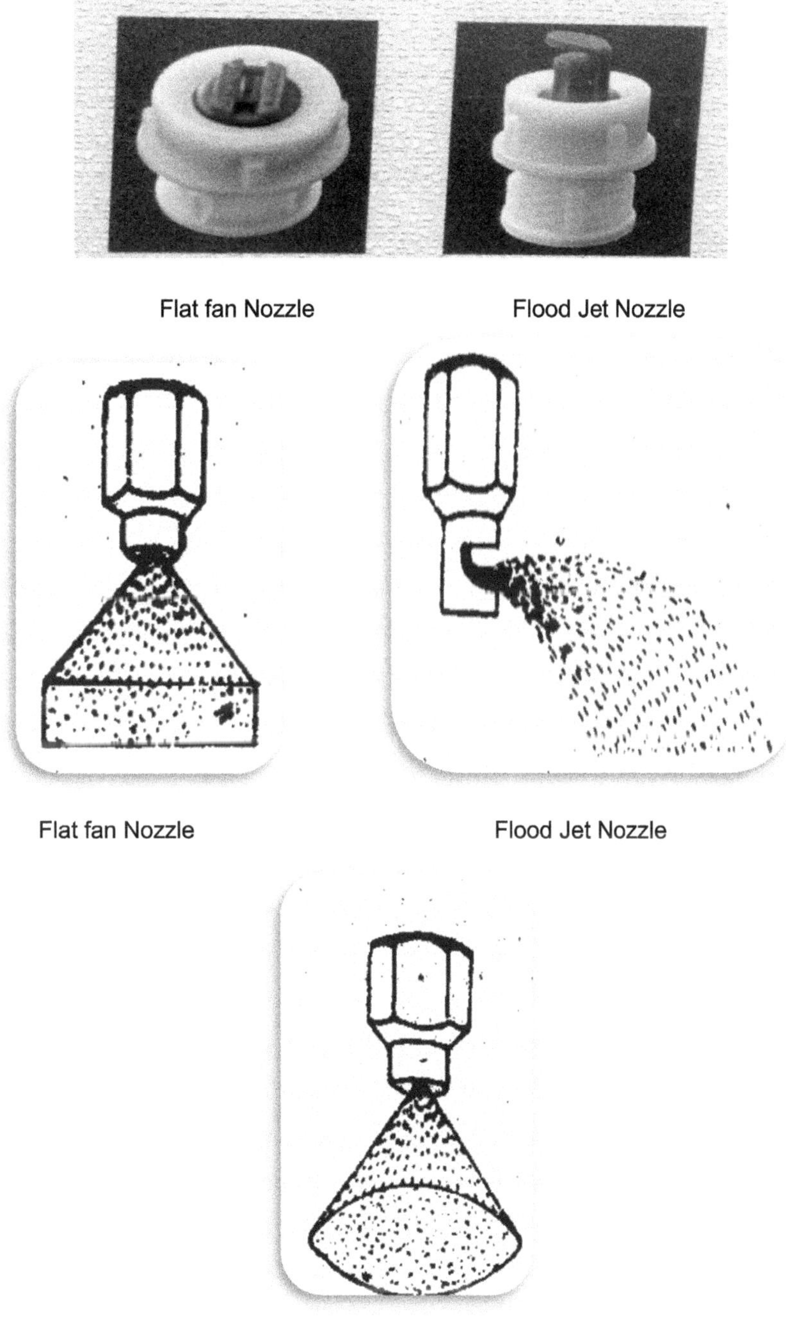

Hollow Cone Nozzle

***Fig. 29.1** Spray pattern of different spray nozzles*

29.3.3.2 Selecting the Nozzle

Impact nozzles cause less drift problems, used for residual herbicides applied ofthe soil. For systemic, translocated herbicides, where through wetting of stems and leaves are not required, an impact nozzle or fan nozzle is recommended. For contact herbicides, use of a hollow cone nozzle, an impact nozzle, or a fan nozzle operated at275 kPa.

29.3.4 Spraying - Precautions

- Do not spray when the wind is too strong. If you must spray when it is windy, hold the nozzle close to the ground (within 50 cm) to prevent droplets from being blown away during calibration of the sprayer and during actual field spraying. Always walk down wind, so that any spray blown off the crop is carried away from you.
- Soil applied herbicides should not be trampled (walk backward while spraying).
- Use soft materials to clean blocked nozzles. Never clean nozzle with a wire or pin.
- A nozzle height of about 50 cm form the target to the ground is ideal. Avoid missing strips or overlapping spray swaths.
- While spraying, maintain a constant walking speed.

29.3.5 Cleaning the Sprayer

- Rinse the sprayer with clean water to remove most of the chemical. A household detergent can be added to water used for cleaning. Pour this out and add clean water again.
- Operate the pump for atleast 10 strokes and pour out the remaining water.
- Repeat these procedures two more times.

29.3.6 Herbicide Drift

It is estimated that two-thirds of pesticide drift problems involve mistakes which could have been avoided. Drift is of concern because it takes the pesticide from the intended target, making it less effective, and deposits it where it is not needed and not wanted. The pesticide then becomes an environmental pollutant in the off-target areas where it can injure susceptible vegetation, contaminate water or damage wildlife. Drift cannot be completely eliminated but the use of proper equipment and application procedures will maintain the drift deposits within acceptable limits.

29.3.6.1 There are Two Kinds of Drift

Particle drift is off-target movement of the spray particles.

Vapor drift is the volatilization of the pesticide molecules and their movement off target.

1. **Wind speed:** When the wind speed was doubled, there was almost a 700 percent increase in drift when the readings were taken 90 feet downwind from the sprayer. Hence the recommendation of spraying in 10 mph winds or less.
2. **Boom height:** When the boom height was increased from 18 to 36 inches, the amount of drift increased 350 percent at 90 feet downwind.
3. **Distance downwind:** If the distance downwind is doubled, the amount of drift decreases five-fold. Therefore, if the distance downwind goes from 100 to 200 feet, you have only 20 percent as much drift at 200 feet as at 100 feet and if the distance goes to 400 feet, you only have 4 percent of the drift you had at 100 feet. Check wind direction and speed when starting to spray a field. You may want to start spraying one side of the field when the wind is lower. Also, it may be necessary to only spray part of a field because of wind speed, wind direction, and distance to susceptible vegetation. The rest of the field can be sprayed when conditions change.

Herbicide drift occurs when the smaller drops in the spray are carried away fromthe target by wind, or when vapor, a volatile herbicide is carried away during or after spraying. Growth regulator herbicides (such as 2, 4-D or MCPA) cause the greatest drift damage. Crops such as tomato, cotton, lettuce, and tree fruits are particularly sensitive to them. Herbicide drift can be avoided by:

- Spraying only when a light wind is blowing away from susceptible crops
- Using large nozzle tips, thus applying larger size droplets and larger spray volumes (more than 100 litres /ha).
- Using minimum pressure required for the nozzles to operate properly.
- Holding sprayer nozzles close to the target.
- Leaving a 10 m wide strip unsprayed between the crop and any field where sensitive crops are growing. Other weed control methods can be applied to the unsprayed strip.*Spray drift can be minimized* Spraying when no wind is blowing
- Using suitable sprayer
- Lowering the spray height from the ground
- Using proper nozzles
- Reducing the delivery pressure
- Use spray shield (hood) to prevent drift completely

29.3.7 Granular Herbicide Application

Most farmers apply granular herbicides by broadcasting. Whatever the method used, safety and uniformity of application are important. Generally herbicides are applied in the form of solution or granules. Of the two, spraying is the most common method of applying solutions of herbicides formulated as wettable powder, soluble powders, emulsifiable concentrate and solution concentrates. When herbicides are formulated as granules they are applied manually by hand or with the help of a granular applicator.

29.4 Why Power Sprayer are not used for Herbicide Spraying?

Power sprayers are discouraged for herbicide application use because of drift hazard as the particle size will be less than 200 micron and the target area cannot be covered and uniform coverage is also not possible.

29.5 What is A Ideal Herbicide ?

An ideal herbicide should have

- High selectivity towards the crop to minimize the crop injury.
- Broad spectrum herbicidal activity
- No carryover to next season.
- Sufficient residual activity to afford season long weed control.
- No hazardous to the operator and corrosive to equipment.
- The price of the chemical should be reasonable so as to provide large margin of monetary gain to the user.

29.6 Qualitative Evaluation

Qualitative evaluation is based on simple tests and assessment is made through visual observations. Scoring is done with reference to control plots.

Assessment of Annual Weeds

For preliminary and observational trials visual assessment is usually quite sufficient. Visual observation is usually based on weed growth; injury to weeds; weed density and weed kill as affected by particular weed control treatment. This can be done by scoring each weed species using the qualitative scale. There are four variations in the scales used for weed control such as 0-1, 1-5, 1-9 and 0-10.

***Table 29.1** Qualitative Scales for Rating Weed Control (European System)*

Rating scale	Effect on weeds
1	Completely kill all the weeds including perennial weeds
2	Very good control
3	Good control of weeds
4	Sufficient in control
5	Medium
6	Fair
7	Poor
8	Very Poor
9	Not effected
	1 to 4 : Positive range; 6 to 9: Negative range

The researcher can use any scale according to his convenience. These scores may be converted into percentage numbers. In the 1-5 and 0-5 scales each number indicates a level commercial acceptability. One to three is unacceptable.

***Table 29.2** Commercial Acceptability of Herbicides with Reference to their Rating Scale*

Rating scale	Commercial acceptability
1	Not acceptable, Poor weed control
2	Not acceptable, poor weed control
3	Not acceptable, poor weed control
4	Minimum acceptable level
5	Excellent control of weeds and readily acceptable

In this system, acceptability is very difficult to define. The acceptability depends upon several factors such as nature of weeds, commercial value of crop, alternative control measures, cost of chemical and environmental conditions. Among all these, the most widely used and simpler rating scale is 0 to 10, where 0 represents no weed control (no effect) and 10 corresponds to complete kill of weeds. The scores can be converted into percentage numbers.

***Table 29.3** Description of Rating Scale in Terms of Weed Control*

Rating	Effect on weeds	Description of weed control
0	None	No weed injury and no control
1	Slight	Very poor control
2	Slight	Poor control
3	Slight	Poor to deficient control
4	Moderate	Deficient control

Table 29.3 *Contd...*

Rating	Effect on weeds	Description of weed control
5	Moderate	Deficient to moderate control
6	Moderate	Moderate control
7	Severe	Satisfactory control
8	Severe	Good control
9	Severe	Good to excellent control
10	Complete	Complete control of weeds

For the conversion of scoring numbers into percentage, 0 represents 0 and 10 represents 100 percent. In a mixed morphological weed population, weeds are to be grouped into 3 categories viz., grasses, sedges and broadleaf weeds. Each morphological group can be assessed separately. This method compares the percent cover and vigour by weed species. Comparisons are made against control plot. In this method, the extent of ground cover by grasses, sedges and broadleaf weeds rates first and percentage infestation of each dominant weed species is estimated later.

Quantitative Evaluation

Quantitative measurements give the exact, accurate and unbiased data. This method is more dependable and the data is more valid particularly for studying the crop growth and yield. It is used in secondary screening trials, advanced trials, special trials for correct assessment. Collection of quantitative data requires more manpower, consumes time and money. Some of the quantitative measurements are weed density, weed biomass, weed height etc.

Assessment of Crop Response

Height of Crop Plants: The effect of weed control treatments is reflecting on crop height. Some herbicides may cause stunted growth and others may help in elongation of plants. Recording crop height helps in knowing the treatmental effects. Usually 10% plants per plot are used for recording this attribute.

Dry Matter Production of Crop: The crop plants are cut at ground level and dry weight is recorded at different intervals to know the growth pattern of forage crops. In grain crops, crop dry weight isto be recorded at the time of harvest. Most often crop dry weight is a better index of weed competition.

Crop Stand: Most of the pre emergence herbicides affect the germination of crop plants. Herbicides, which are having high toxicity, cause mortality of seedlings and loss of stands significantly. Stand count is an important attribute for comparing the herbicidal effects with other practices. Usually, the rows are selected for recording stand establishment.

Standard Yield of Crop: Yield is a final and most important feature of crop performance. The efficiency of herbicide will be judged on the basis of crop yield. Recommendation of herbicides for crop use must be supported by good yield data. Howsoever efficient in the suppression of weeds, the herbicide cannot be accepted in practice, if the yield level is low.

References

1. *Bhan, V.M., V.P.Singh and Anil Dixit, 1996. Weed management – An important tool in improvingthe crop production, NRCWS, Jabalpur, Pp.228.*
2. *Chattopadhyay, S.B. 1980. Principles and procedures of plant protection. Oxford & IBHPublishing Co., New Delhi, Pp.480.*
3. *Gupta, O.P. 1998. Modern weed management. Agro Botanica, Bikanir, RajasthanKwesiAmpong/Nyarko and S.K.DeDatta. 1991. A hand book for weed control in rice. IRRI,Manila. Pp.113.*
4. *Kondap, S.M. and U.C. Upadyay. 1985. A Practical manual on Weed Control. Oxford & IBHPublishing Co., New Delhi.*
5. *Malik, R.K., Pundir, A., Dar, S.R., Singh, S.K., Gopal, R., Shankar, P.R., Singh, N. and Jat, M.L.2012. Sprayers and Spraying Techniques – A manual, CSISA, IRRI and CIMMYT. 20 pp.*

Chapter 30

Integrated Pest Management in Groundnut

T. Murali Krishna, ARK Rao, K. Devaki and K. Raja Reddy

RARS, Tirupati.

30.1 Introduction

The peanut or groundnut (*Arachis hypogaea*) is one of the most important oilseed crops grown in India. It is known by several other names such as African nut, Chinese nut, monkey nut, goober pea, ground pea and ground bean (Johnson, 1964). Groundnut is also regarded as poor man's cashew.

Groundnut is native of Brazil and was introduced into India in the sixteenth century. The most important groundnut growing countries are India, China, Nigeria, Sudan and USA. In the world the crop is grown in an area of 26.62 million hectares by 84 countries with an annual production of 35.66 million tonnes of nuts-in-shell (pods) with a productivity of (1348 kg ha^{-1}). In India, it is grown in 11 states in an area of 4.19 million ha with a production of 5.62 million tonnes of pods per annum. The average productivity of groundnut in India is about (1341 kg ha^{-1}) as against the world's average of (1348 kg ha^{-1}) (Department of Agriculture and Cooperation, Ministry of Agriculture, Govt. of India, 2012). Though India ranks first in area of groundnut cultivation, the productivity is quite low compared to that of USA (3000 kg ha^{-1}), China (2600 kg ha^{-1}), Argentina (2100 kg ha^{-1}) and Indonesia (1550 kg ha^{-1}) (NRCG, 2012). Its cultivation is mostly confined to Gujarat, Andhra Pradesh, Karnataka, Tamil Nadu and Maharashtra accounting more than 80 per cent of the total production and area. The other important states where it is grown are Odisha, West Bengal, Madhya Pradesh, Rajasthan and Uttar Pradesh.

In India, it is grown on 52.64 lakh hectares with a production of 69.64 lakh tonnes (Anonymous, 2012) and contributing to 55 percent of the total oil seed production in the country and ranks first in the world in total area and 2nd in production. In Andhra Pradesh, it is grown in an area of 13.45 lakh hectares with a production of 11.09 lakh tonnes and a productivity of 825 kg ha^{-1} (Directorate of economics and statistics, Govt. of Andhra Pradesh 2013). The reasons for low productivity of groundnut is abiotic stress like drought as the crop is mostly grown under rainfed conditions and insect pests and diseases are the major biotic stresses. Groundnut crop is attacked by about 90 species of insect pests and the major insect pests can be grouped as soil inhabiting insects, foliage feeding insects, sucking insects and insects damaging flowers and growing parts (Wightman,1986).

30.2 Major Insect Pests of Groundnut in India (Faujdar Singh and Oswalt, 1992)

Common Name	Scientific Name
A. Soil Inhabiting insect	
1. Termites	*Odontotermes obesus* Rambur *Odontotermes velonensisWashman;Microtermes thoracalis (Sr)*
2. White grubs	*(Lachnosterna)Holotrichia consanguinea* (Blanch) *(Lachnosterna) Holotrichia serrata* Fabricius
3. False wire worms	*Gonocephalum* sp. *Penthicoides sp.*
4. Earwigs	*Anisolabis annulipes* (Luc), *Euborellia stali*
5. Blind ant	Dorylus orientalis West wood
B. Foliage feeding insects	
1. Leaf miner	*Aproaerema modicella* Deventer
2. Red hairy caterpillar	*Amsacta albistriga* Walker *A. moorei* Butler
3. Bihar hairy caterpillar	*Spilosoma obliqua* Walker
4. Tobacco caterpillar	*Spodoptera litura* Fabricius
5. Gram caterpillar	*Helicoverpa armigera* Hub.
C. Sucking insects	
1. Leaf hoppers (jassids)	*Empoasca motti* *E. kerri* Pruthi
2. Thrips	*Thrips palmi* Karni *Fankliniella schultzei* Trybom *Scirtothrips dorsalis Hood Caliothrips indicus*
3. Aphid	*Aphis craccivora* Koch
D. Flower feeding insects	
1. Blister beetle	*Mylabris pustulata* Thunberg
E. Stem boring insect pests 1. Jewel beetle 2. Spotted pod borer	*Chrysochroa fulgidissima* Tamamushi *Sphenoptera perotetti* *Maruca vitrata*
F. Emerging insect pests	
1. Leaf Bud borer	*Anarsia ephippias*
2. Spotted pod borer	*Maruca vitrata*
3. Green semilooper	*Plusia sp.*
G. Storage pests	
1. Storage bruchid/pod borer	*Caryedon serratus*
2. Rice moth	*Corcyra cephalonica*
H. Non-insect pests (nematodes)	
1. Root knot nematode	*Meloidogyne arenaria* Chitwood
2. Kalahasti malady	*Tylenchorhynchus brevilineatus* Williams

Total annual loss in groundnut due to the pests has been estimated to be about Rs.160 crores (Nandagopal and Basu, 2003). Incidence of various insect pests varies from season to season. Pest complex in *kharif* (rainy season) differs from *rabi/summer* (post rainy season). For instance, serious pests like red hairy caterpillar (*Amsacta albistriga)* occur in *kharif* groundnut and Tobacco caterpillar, *Spodoptera litura* occurs in *rabi/summer crop.*

Insect pests and diseases inflict enormous loss to groundnut. Evidences also indicate a rise in the loss, despite increased use of chemical pesticides. There is a growing concern about the potential adverse effects of chemical pesticides on the human health, environment and biodiversity. These negative externalities, though, cannot be eliminated altogether; but their intensity can be minimized through development, dissemination and promotion of alternative technologies such as safer pesticides and bio-agents as well as good agronomic practices rather than relying solely on chemical pesticides.

Plant protection research has generated many technologies for commercial application, and has proved to provide better pest control and net economic returns, when used in conjunction with other pest control measures. The strategy is often referred to as 'Integrated Pest Management'. *Integrated Pest Management (IPM) means a pest management system that, in the context of the associated environment and the population dynamics of the pest species, utilizes all suitable techniques and methods in a compatible manner as possible and maintains the pest populations at levels below those causing economically unacceptable damage or loss* (FAO, 1967). Thus, IPM is the best combination of cultural, biological and chemical measures that provides the most cost-effective, environmentally sound and socially acceptable method of managing diseases, insects, weeds and other pests.

IPM has grown out of the desire to reduce reliance on chemicals to kill pests. Through the adoption of IPM practices, farmers are able to minimize the pest damage and reduce the costs of production, leading to sustainable increase in crop production and farm income. In recent years, several new molecules with green chemistries have been developed for pest management. With the introduction of these new molecules there is paradigm shift in the use of conventional insecticides. Where the quantities of newer pesticide molecules formulations required per hectare are almost 4-53 times less than the conventional molecules.

Most of the newer insecticides have several advantages over conventional insecticides like high level of selectivity to target pests, excellent efficacy at low rates or dosage, no persistence in the environment, low mammalian toxicity, safer to natural enemies, less likely to cause outbreaks of secondary pests, delayed insecticide resistance and no cross-resistance with the conventional insecticides. All these advantages render their fitness in the integrated pest management programmes. Research has generated several new technologies using naturally occurring enemies of insect pests and plant pathogens (parasitoids, predators and pathogens, antagonistic organisms) for use in IPM. In addition, a number of plant products such as azadirachtin (neem), pyrethrum, nicotine, etc. are also available as bio-pesticides.

Many successes have been reported across the country on the potential use of the natural enemies in combination with the insecticide/fungicide molecules against the pests and diseases of the crops. IPM has been evolving over the decades to address the deleterious impacts of synthetic chemical pesticides on environment ultimately affecting the interests of the farmers. In modern IPM (FAO, 2002) emphasis is given to Agro Eco System Analysis (AESA) where farmers take decisions based on larger range of field observations. The health of a plant is determined by its environment which includes physical factors (i.e. sun, rain, wind and soil nutrients) and biological factors (i.e. insect-pests, diseases and weeds). All these factors can play a role in the balance which exists between herbivore insects and their natural enemies. Understanding the intricate interactions in an ecosystem can play a critical role in pest management (Saroj singh et al., 2014)

It is an approach, which can be gainfully employed by extension functionaries and farmers to analyze field situations with regard to pests, defenders, soil conditions, plant health, the influence of climatic factors and their interrelationship for growing healthy crop. Such a critical analysis of the field situations will help in taking appropriate decision on management practice. The basic components of AESA are:

- Plant health at different stages.
- Built-in-compensation abilities of the plants.
- Pest and defender population dynamics.
- Soil conditions.
- Climatic factors.
- Farmers past experience.

During post rainy season *(rabi/summer) S.litura* has emerged as a major pest of groundnut in India particularly in A.P. This pest alone is causing damage to *rabi/summer* groundnut crop to the extent of 40-60%. The farmers are applying 4-5 rounds of insecticide sprays exclusively to manage this pest during post rainy season. Indiscriminate or misuse of insecticides leads to development of insecticide resistance in *S.litura* (Armes *et.al*, 1997). This pest has acquired resistance to many of the conventional and currently available insecticides (Murali Krishna *et. al*, 2008). Quinalphos and monocrotophos are reported to be not working against *Spodoptera* in groundnut during 2007-08 and 2008-09 *rabi* seasons particularly in Chittoor and Nellore districts of A.P.

30.3 IPM Strategies to Manage Important Groundnut Insect Pests'

30.3.1 Tobacco Caterpillar (Spodoptera Litura)

- Occurrence of S.*litura* a serious pest of groundnut coincides with rabi/summer season viz., October onwards. It lays eggs in clusters of 200-300 and covered with brown hairs and a female can lay about 2000 eggs.

- Just emerged larvae are greenish while 3rd instar onwards the larvae are pale brown in colour with broken dark brown stripes along the body. When it is young it is generally green. It pupates in the soil within 15-21 days; the adult moth comes out after 9-16 days of pupation.
- Cultural: Deep summer ploughing. Planting castor plants as trap crop for egg laying and destroying eggs or first instar stage larvae (on skeletonised leaf) help in reducing the incidence.
- Mechanical: Arrange bird perches 20/acre and pheromone traps @4/acre for monitoring. Collection and destruction of egg mass and just emerged larvae.
- **Biocontrol:** Spray NSKE 5% for stages like egg and 1st instar larvae. Spray *SlNPV* 100 LE/acre or spray *Bt* 1g/lit water.
- Efforts are on to develop *Bt* transgenic groundnut at ICRISAT, Hyderabad. At RARS,Tirupati also research work is initiated on transgenic groundnut.
- **Effective IRM strategies for *S.litura*:** All effective insecticide resistance management (IRM) strategies seek to minimize the selection of resistance to any one type of insecticide. In practice, alternations, sequences or rotations of compounds from different mode of actions (MoA) provide sustainable and effective IRM.
- If defoliation exceeds ETL of 25%, spray Chlorpyriphos 2.5ml/litre of water or Thiodicarb 1g/litre or Novaluron 1ml/litre or Chlorfenapyr 2ml/litre of water (Murali krishna *et al, 2008*). Avoid use of pyrethroids or readily available combination pesticides.
- Indoxacarb 20ml a.i. ha^{-1} or Spinosad 45ml a.i. ha^{-1} is recommended only as a last resort when there is an absolute necessity (Nigam *et al,*2006)
- Poison baiting with rice bran, jaggery and chlorpyriphos (10:1:1) is recommended against 4-5th instar caterpillars which withstand contact insecticides as sprays.

30.3.2 Gram Caterpillar, Helicoverpa Armigera

Larvae feed mostly on tender leaves of groundnut plant. Female moth lays shiny, creamy colored eggs singly on tender parts of the plant. Eggs hatch in 3-4 days. For full development of larva about 3 weeks time is required. Pupation takes in soil for 10-12 days. Insect is very active in post - rainy season (October to March).

Control: IPM as in case of *Spodoptera* except for collection of egg mass. Spray *HaNPV* @ 200LE/acre

30.3.3 Hairy Caterpillars

Amsacta albistriga, A. moorei, Spilosoma (Diacrisia) obliqua.

Adults emerge from soil during June-July with the onset of monsoon rains. A.Albistriga is endemic pest in rainfed groundnut grown in western mandals of Chittoor and Kadapa districts. Spilosoma obliqua has been occurring in North India and in parts of Telangana (Karimnagar district). Full grown larvae devour the entire foliage, flowers and growing points leaving the plant with bare stems. Caterpillars migrate from field to field soon after damaging one field.

- Deep summer ploughing is recommended to expose the pupae in the soil.
- Setting up of bonfires on field bunds between 7-10 p.m in the night from 2nd day of rain during June and July months.
- Cowpea may be planted as trap crop around the groundnut crop
- Collection and destruction of egg masses and hand picking of larvae are effective in reducing the larval population.
- Erection of light traps soon after the monsoon for 20-45 days and collecting and killing of adult moths are found very effective.
- Utilizing Vegetative traps *Calotropis* or *Jatropha* (wild castor) or *Ipomoea* to prevent the migration of the grown up larvae or collect larvae and destruct them on these vegetative traps
- In rainfed area water is scarce and defoliation exceeds ETL of 25%, dusting of Carbaryl or Quinalphos @ 25-30 kg/ha would be effective in the control of young larvae. 200 ml Dichlorvas dissolved in 200 litres of water/acre may be used to control grown up caterpillars or Spray quinalphos 2ml/lit of water.
- The dispersing larvae of hairy caterpillar from one field to another can be checked by digging deep furrow trenches across the direction of their march, and prompt destruction of larvae.
- Poison bait preparations with 10 kg rice bran + 1 kg jaggery + 500 ml quinalphos may be applied in the evening.

30.3.4 Leaf Miner, Aproaerema Modicella

Small blister like mines are seen on the upper leaf surface near mid rib. As the feeding advances, the mines increase in size and the entire leaflet becomes brown. The 3rd and 4th instar larva rolls the same leaf or webs two leaves together as a result it shrivels and dries up. In severe cases the affected crop gives burnt up appearance. Adult catches of 40 per day followed by dry spell for more than10 days favors leaf miner incidence.

- Crop rotation with non-leguminous crops would considerably reduce the leaf miner population. Wherever possible, Soybean may be planted as trap crop Rotation of groundnut with soybean and other leguminous crops should be avoided.
- Set up pheromone traps (Delta sticky or Wota-T traps @ 4/acre) for monitoring

- About 25 species of hymenoptera parasites were found to attack the leaf miner eggs, larvae and pupae on groundnut. *Chelonus sp, Sympiesis sp* and *Stenomesius* sp are prominent among them.
- The insecticidal control should be initiated when the larval mines are 5/plant Spray quinalphos 2ml or monocrotophos 1.6ml /lit of water (Murali Krishna et al, 2009).

30.3.5 Aphid: Aphis Craccivora

Adults and nymphs congregate mostly on growing points and young tender foliage and desap which result in the chlorotic patches and leaf curl. As flowering commences, aphids infest flower stocks and pegs thereby affecting the pod formation

- Coccinellid predators such as *Coccinella septumpunctata, Menochilus sexmaculata* would be effective in reducing the population of aphids.
- Spraying Monocrotophos 1.6 ml or Dimethoate 2 ml/lit of water will provide good control.

30.3.6 Leaf Hopper (Jassid): Empoasca Motti

- Both the adults and nymph suck sap from young leaves mainly from the under surface. The affected leaves of young plant become pale. Older plants on prolonged damage result in a 'V' shaped yellowing on the tips of the leaflets which may spread to the entire leaflet leaving small greenish patches. Later on the yellowing at the tips turns to brown and appears like *Hopper burn.*
- Spraying of monocrotophos 1.6 ml or imidacloprid 0.3 ml/lit of water gives good control.

30.3.7 Thrips: Thrips Palmi; Caliothrips Indicus, Frankliniella Schultzei, Scirtothrips Dorsalis

Incidence first appears from seedling stage and continues at all stages of crop growth. Adults and nymphs rasp the leaf surface and suck the sap. As a result there will be silvery white patches and upward curling of tender leaves. They hide inside the tender folded leaves, rasp and suck the oozing sap. **ETL:** 5thrips/plant when there is no PBND/PSND. Apart from causing direct damage, species like *T.palmi* and others transmit peanut bud necrosis disease (PBND) or peanut stem necrosis disease (PSND) .

- IPM for thrips: Maintain recommended plant population i.e 33/m^2 in *Kharif* and 44/m^2 in rabi/summer
- Seed treatment with imidacloprid 17.8SL @ 2 ml/kg seed or imidacloprid 600 FS @ 2ml + 4ml water/kg seed. Jowar/bajra may be used as barrier crops. Eliminate weeds like *Parthenium* in and around the field.

- Insecticides such as dimethoate 2.0 ml or monocrotophos 1.6 ml or imidacloprid 0.3 ml per litre of water will control thrips effectively.

30.3.8 White Grub: Holotrichia Consanguinea, H. Serrata

The 'C' shaped white, robust grubs feed on rootlets and nodules. The older grubs of 3rd instar devour the entire tap root and such plants wither and die. The maximum damage usually occurs on patches and crops which are nearer to the host plants of adult beetles such as neem. Damaged plants can easily be pulled out.

Control: Adult beetles which emerge in the dusky time and feeding on plants such as neem trees can be sprayed with Carbaryl @ 3g/liter. Spray should be done on a community approach and should be repeated after every rainfall till the middle of July It is more effective and economical, and brings about general reduction in the pest population in the entire area.

- Deep summer ploughing. Light traps will effectively reduce the adult population.
- Apply Phorate 10% granules @ 5 kg/acre at the sowing time. Seed treatment with Chlorpyriphos @ 6 ml/kg seed has been recommended.

30.3.9 Termites: Odontotermes Obesus, Odontotermes sp

- Termites penetrate and hollow out the tap root and cause wilting and premature death of the plant. It also feed on the pod shell and scarify the pods externally rendering them more susceptible in invasion by soil fungi that can attack seeds, usually such seeds rot.
- Drenching spray at the base of the plants with Chlorpyriphos20EC @ 5ml/litre water.

30.4 Management of Nematodes in Groundnut

30.4.1 Kalahasthi Malady

(a) Crop rotation with paddy or other cereal crops like sorghum and maize. Rotation with paddy is very effective in reducing the population of Kalahasthi nematode. Rotation with maize or sorghum reduces root knot nematode population.

(b) Growing green manure crops like sunhemp and incorporation into the soil can help in minimizing the Kalahasthi and root knot nematodes.

(c) Quarantine regulations need to be strictly monitored to check the entry of *Aphelenchoides arachidis* and *Belonolaimus longicaudatus*.

30.4.2 Root Knot Nematode

(a) Deep summer ploughing in states having hot and dry summer help in reducing infestation of plant parasitic nematodes.

(b) Soil amendments with neem cake or castor cakes @ 1 tonne/ha preferably seven days prior to sowing has been found to reduce nematode population.

(c) Farmers of Nellore district of A.P., are recommended to apply Carbofuran @ 1kg a.i/ha to reduce the population of root-knot nematodes infecting groundnut.

30.5 Storage Pests

30.5.1 Storage Bruchid or Pod Borer of Groundnut Caryedon Serratus

The bruchid causes major damage to groundnut in storage. Of late one consignment of groundnut pods which is supposed to be exported from Chennai Air port was rejected due to this pest. It is occurring mostly from March to July. Adult beetles lay eggs on the surface of pod and the grubs on emergence bore into the pods and cause damage to kernels.

The pest can be managed by following suitable preventive measures.

1. The pods has to be stored in polythene layered gunny bag
2. The moisture content of the pods is to be carefully checked. It should not exceed >9-10%
3. The storage place must be dampness free and should be air tight. The store is to be cleaned at regular intervals.
4. Spray malathion @ 5ml/L or deltamethrin @ 2 ml/L at 15-20 days interval
5. If the storage godown is airtight, fumigate the store with aluminium phosphide tablets @ 3/quintal.

30.6 IPM Strategies for Groundnut Insect Pests as a Whole

30.6.1 Cultural Practices

1. Summer deep ploughing to expose pupae of insects,
2. Adopt proper crop rotation to reduce the incidence of nematodes or other pest.
3. Seed treatment to control soil and seed borne diseases should be followed with *Trichoderma* spp. @ 4 g/kg seed or mancozeb @ 3g/kg. seed or tebuconazole @1g/kg seed.

4. Seed treatment with imidacloprid 17.8SL @2 ml/ kg of seed or imidacloprid 600 FS @ 2ml + 4ml water/kg seed. After insecticide seed treatment with *Trichoderma* or fungicide treatment may be taken up.
5. Timely Sowing reduces the incidence of leaf miner.
6. Adopt proper spacing and maintain recommended plant population in both the seasons.
7. The crop should be maintained weed free to reduce the Thrips transmitted viral diseases.
8. Growing trap crop like castor for *Spodoptera* or Soybean for leaf miner or cowpea for red hairy caterpillars on priority of the pest attack in a particular area where groundnut is grown.
9. 5-6 rows of bajra/jowar/maize crops are recommended as barrier crops for thrips. Alternate weed hosts like *Parthenium* be removed and destroyed.

30.6.2 Mechanical Practices

1. Hand picking and destruction of egg mass and just emerged larvae of *Spodoptera*
2. Pheromone traps for monitoring leaf miner, *Spodoptera* and *Helicoverpa*
3. Install 20 bird perches per acre for attracting predatory birds on *Spodoptera* and *Helicoverpa*

30.6.3 Biocontrol Practices

1. Conservation of predators (lacewings, lady bird beetles, spiders) and parasitoids like *Apanteles, Bracon*)
2. Spray *SlNPV* @100 LE/acre at 2nd or 3rd instar stage or Spray Bt @1g/lit water

30.6.4 Chemical Control Measures

1. Avoid mixing of two or more insecticides/tank mixing.
2. Repeated application of same insecticide should be avoided.
3. Avoid using insecticides such as pyrethroids which result in resurgence of sucking
4. Pests and mites.
5. Use neem based formulations NSKE @ 5%.
6. Use selective insecticides
   - ✓ **Up to 20DAS**: for sucking insects like Thrips, Spray monocrotophos 1.6 ml or imidacloprid 0.3 ml/litre water.

- ✓ **20-40 DAS: If** leaf miner (5mines/plant) occurs spray quinalphos @ 2ml/litre water.
- ✓ **40-60 DAS**: If defoliation exceeds 25%, spray Chlorpyriphos 2.5 ml/litre of water or Thiodicarb 1g/lit or new insecticides like Novaluron 1ml/litre or Chlorfenapyr 2ml/litre of water (Murali Krishna *et al, 2008*) for the control of *Spodoptera* and *Helicoverpa.* Emamectin benzoate 0.2g/litre and flubendiamide 0.2 ml/litre water were also found effective.

Groundnut Insect Pests

(A) Soil pests:

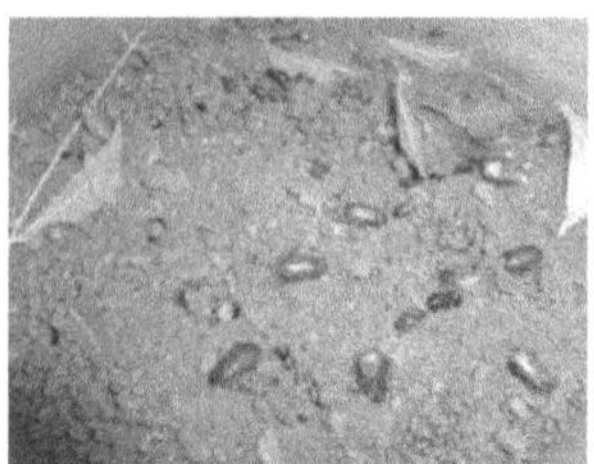

Fig. 30.1 *Root grub and caused damage to groundnut crop*

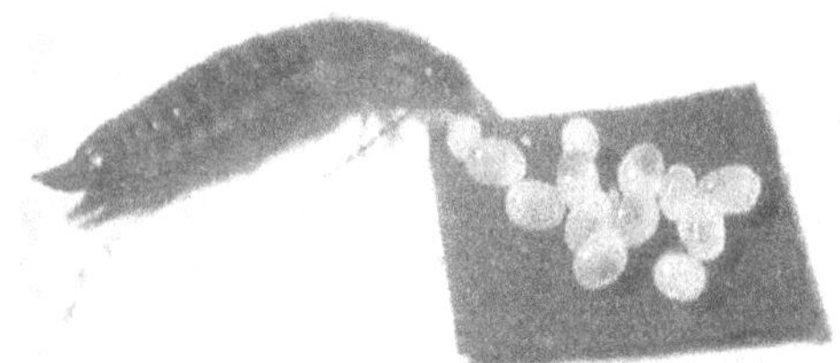
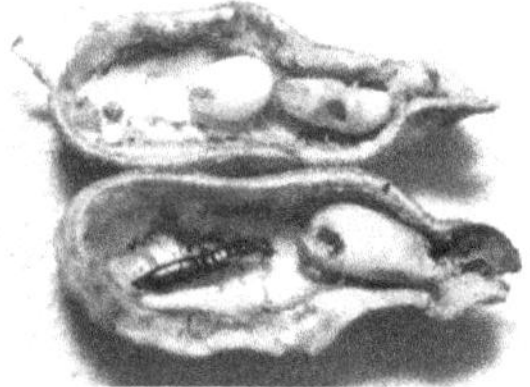

Fig. 30.2 *Earwig and caused damage to groundnut pod*

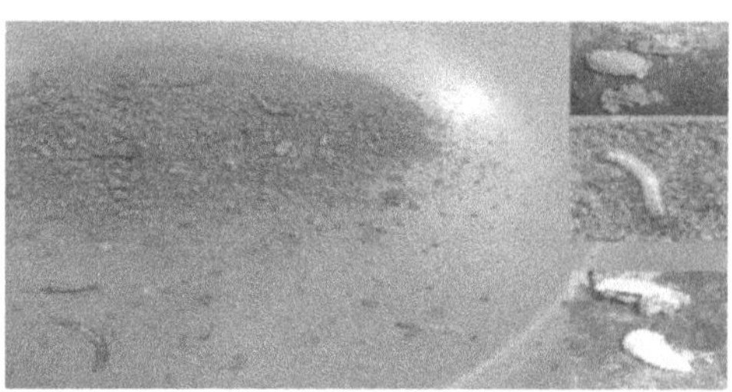

Fig. 30.3 *Click beetle (wire worm adult) feeding on groundnut cotyledon*

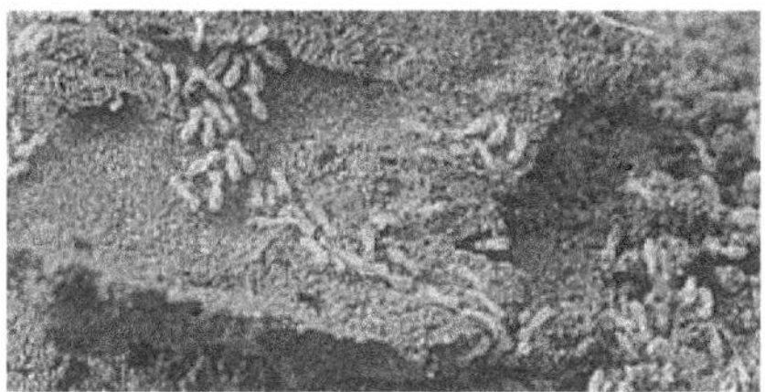
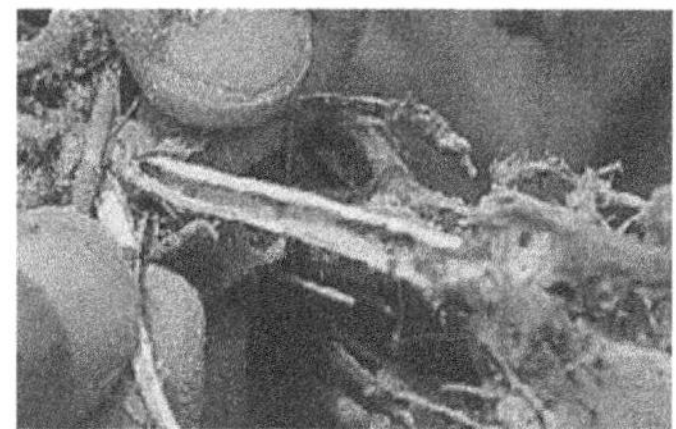

Fig. 30.4 Termites and caused damage to groundnut plant

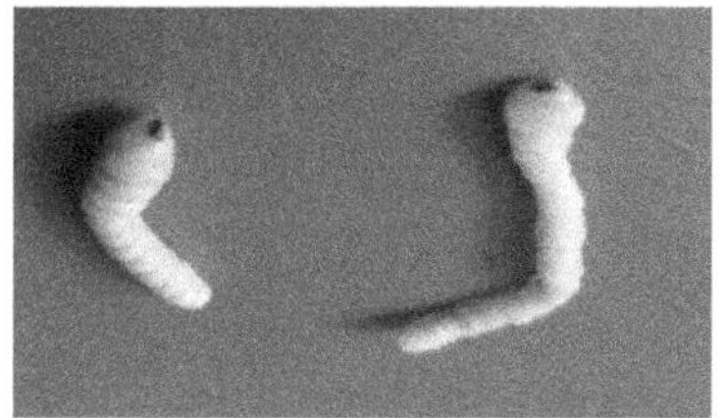

Fig. 30.5 Jewel beetle (stem borer) caused damage to groundnut plant

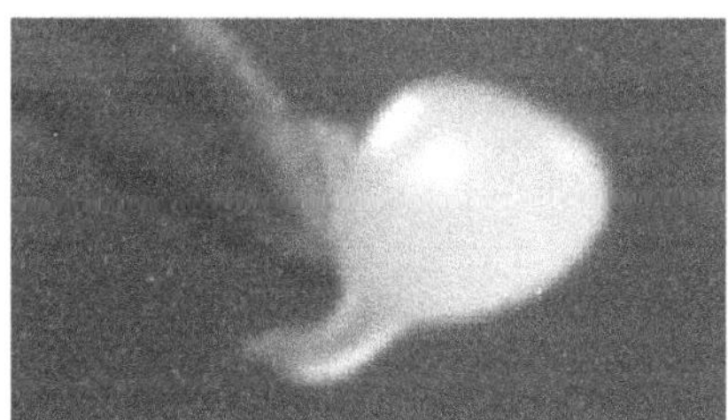

Fig. 30.6 Nematode (M.arenareae) - damage symptoms on groundnut pods

(B) Defoliators in Groundnut

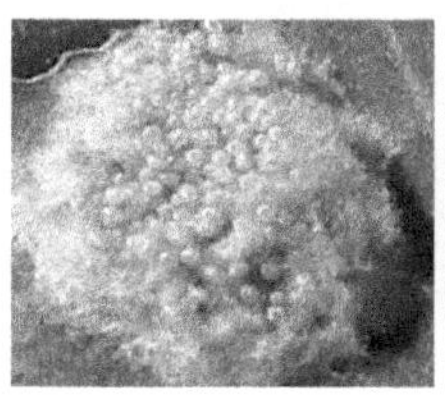
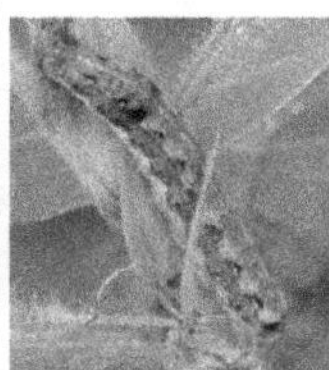
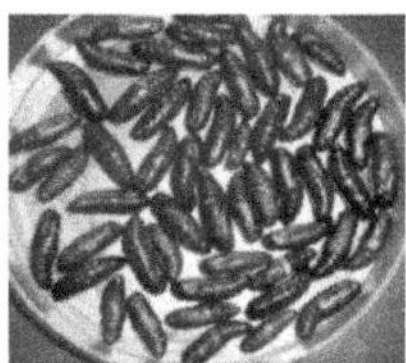

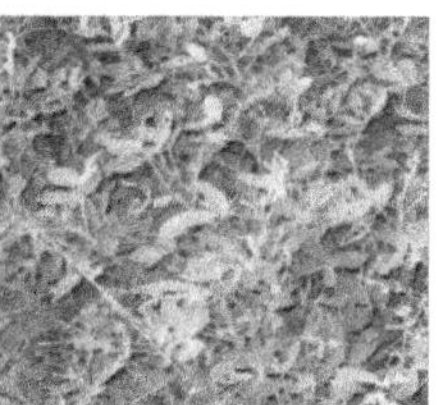

Fig. 30.7 Tobacco cut worm lifecycle and causing damage to groundnut crop

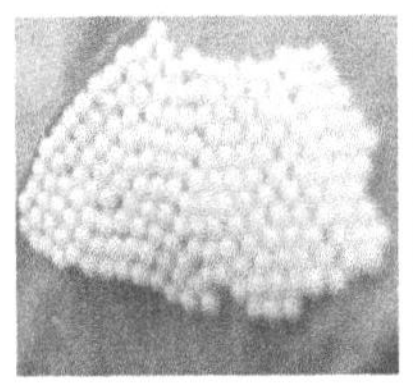

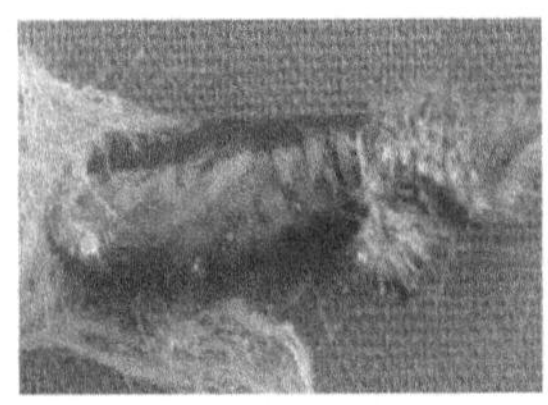

Fig. 30.8 *Lifecycle of Red hairy caterpillar*

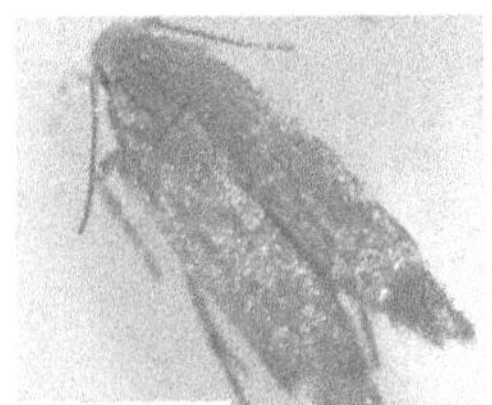
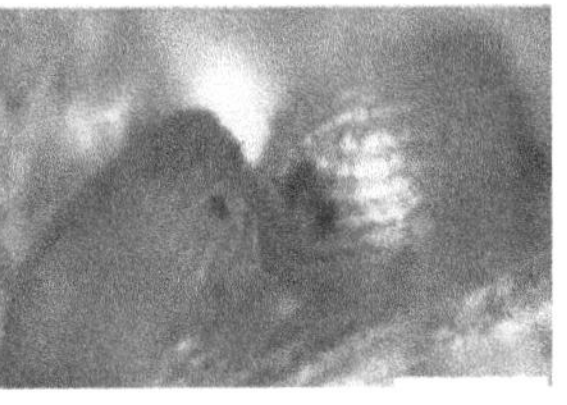
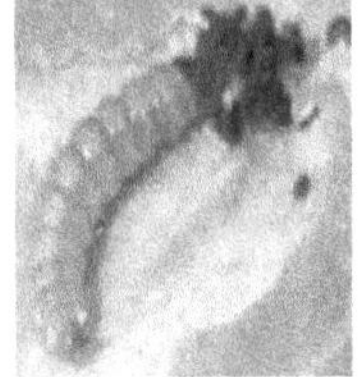

Fig. 30.9 *Groundnut leaf miner*

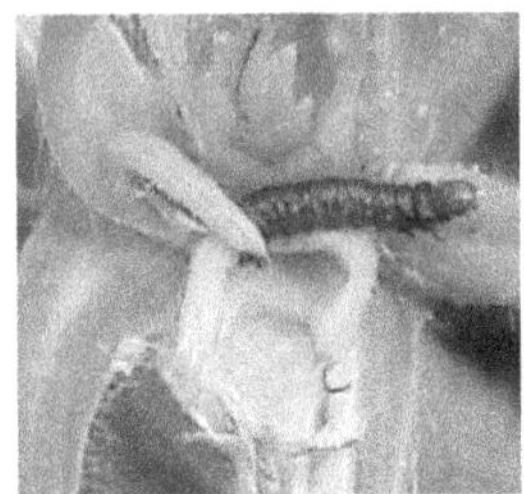

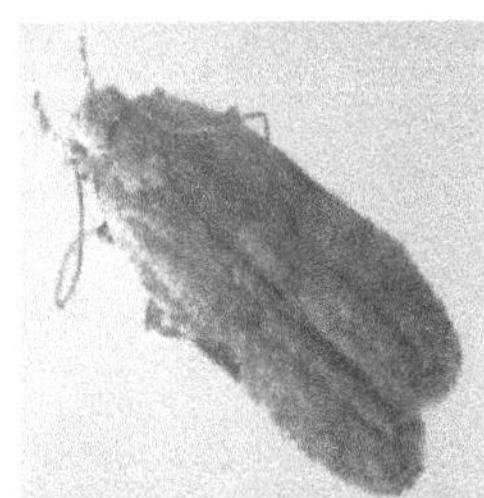

Fig. 30.10 *Bud borer; Anarsia ephippias*

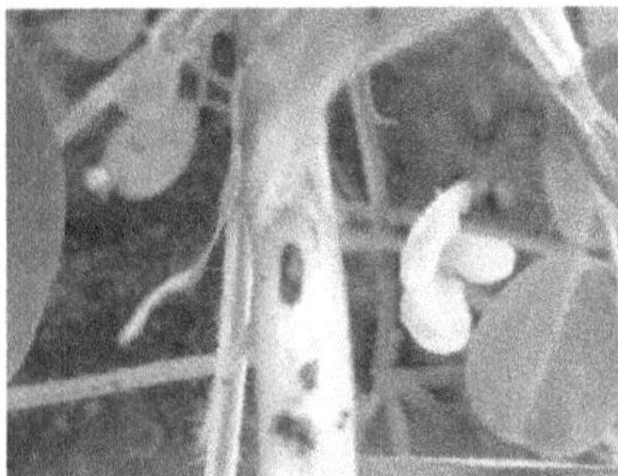

Fig. 30.11 *Spotted pod borer; Maruca vitrata*

(C) Sucking Insect pests

Fig. 30.12 *Thrips and symptoms of damage on groundnut*

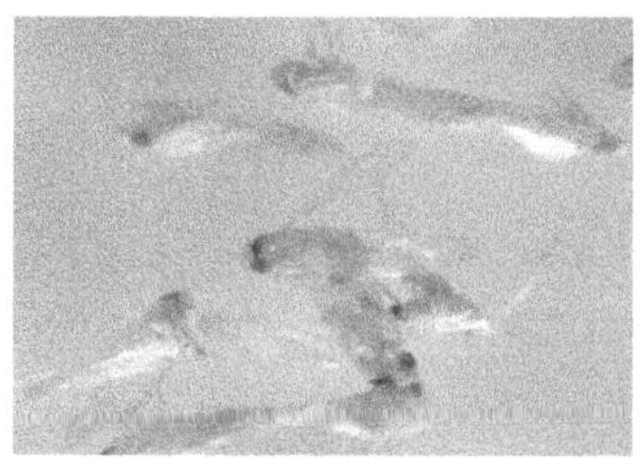

Fig. 30.12 *Leafhopper and symptoms of damage on groundnut*

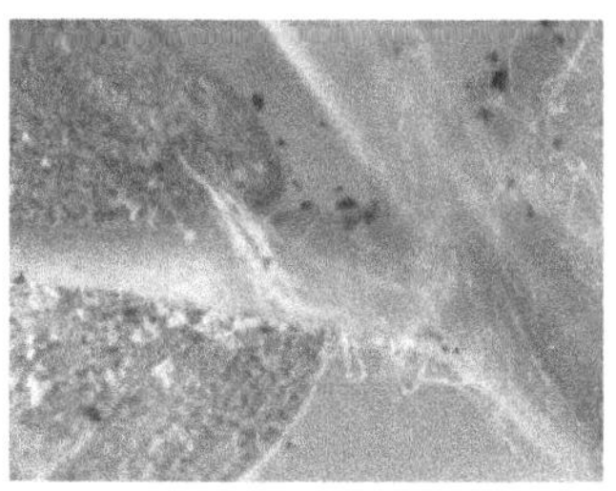

Fig. 30.13 *Mites and symptoms of damage on groundnut*

Fig. 30.14 *Mealybug on groundnut (Maconellicoccus hirsutus) on groundnut pods*

(D) Storage pests

Fig. 30.15 *Symptoms of damage caused by Storage pest bruchid (Caryedon serratus)*

References

1. *Anonymous, 2012 Annual report 2011. AICRP groundnut workshop.P.E-5-6.*
2. *Armes, N.J, Wightman, J.A., Jadhav, D.R and Ranga Rao, G.V. 1997 Status of insecticide resistance in Spodoptera litura in Andhra Pradesh, India Pesticide Science 50(3):240-248.*
3. *Department of Agriculture and Cooperation, Ministry of Agriculture, Govt. of India, 2012. State of Indian Agriculture (2012-2013). P. 171.*
4. *Directorate of economics and statistics, Govt. of Andhra Pradesh, 2013*
5. *FAO, 1967 www.fao.org*
6. *FAO, 2002 www.fao.org*
7. *FAO, 2003 FAO statistical year book 2004-05 www.FAO.org*
8. *Faujdar Singh and Oswalt,D.L 1992 Major insect pests of Groundnut Skill Development series No.7 HRD, India, Patancheru, A.P. India pages:1-30*
9. *Hong Tong, QiSu, Xiaomao Zhou and Lianyang Bai 2013 Field resistance of Spodoptera litura (Lepidoptera: Noctuidae) to organophosphates, pyrethroids, carbamates and four newer chemistry insecticides in Hunan, China J. Pest Sci. 86: 599-609*
10. *Johnson, F.R. 1964 The peanut story Murfreesboro,N.C; Johnson Publishing Co.,*
11. *Muthusamy, Shivakumar, Karthi, Ramkumar 2011 Pesticide detoxification mechanism in field population of Spodoptera litura (Lepidoptera : Noctuidae) from South India Egypt.Acad.J.Biolog.Sci 3(1):51-57*
12. *Murali Krishna,T., Devaki,K., Raja Reddy,K and Venkateswarlu,U 2008 Efficacy of certain new insecticide molecules against groundnut defoliator, Spodoptera litura (Fab,) (Noctuidae:Lepidoptera) Current Biotica 2(2):173-178*

13. *Murali Krishna,T., Devaki,K., Swarna,B Venkateswarlu,U and Raja Reddy,K 2009 Outbreak of groundnut leaf miner, Aproaerema modicella Deventer (Gelechiidae: Lepidoptera) Journal of Oilseeds Research 26(spl issue): 534-538.*
14. *Nandagopal, V and Basu, N.C. 2003 Groundnut insect pests diagnosis and management National Research Center for Groundnut (ICAR), Junagadh Gujarat pp 1-2*
15. *Nigam, SN, Aruna R, Giri DY, Ranga Rao G.V and Reddy AGS 2006 Obtaining Sustainable higher groundnut yields: Principles and Practices of cultivation Information Bulletin No.71 ICRISAT, Patancheru, Andhra Pradesh, India pp.28*
16. *Saroj Singh, P P Thirumalaisamy, G Harish, Datta Ram, S N Sushil, A K Sinha, Ram Asre, K S Kapoor, K Satyagopal, P Jeyakumar, Ajanta Birah, O P Sharma, Someshwar Bhagat, P V Verma, Sunil Kumar, C Chattopadhyay and M S Yadav. 2014. Integrated Pest Management Package for Groundnut. pp. 49. Pub.NCIPM, NewDelhi*
17. *Wightman, J.A. 1986 Insects and related organisms found on Groundnut in the field in Africa and Asia A handbook prepared for trainees at ICRISAT, Patancheru, Andhra Pradesh, India pp.28*

Chapter 31

Seed Related Physiological and Quality Aspects of Groundnut

V. Umamahesh, V. Rajarajeswari, Y. Yohan and A. Reshma

Dept of crop physiology, S.V. Agricultural College, Tirupati.

31.1 Introduction

Groundnut (*Arachis hypogaea* L.) is an important oilseed crop around the world as well as in India. It is ranked as the second most important cultivated grain legume and the fourth largest edible oilseed crop in the world (Shilman *et al.*, 2011). It is grown on nearly 20.88 million ha worldwide with a total production of 34.66 million tons and an average yield of 1660 kg ha^{-1} (FAOSTAT, 2012). Developing countries account for over 97% of world groundnut area and 95% of total production. It is cultivated in more than 100 countries of which China (41 per cent), India (21 per cent), Nigeria (8 percent), and the United States (7 per cent) are the largest producers. India is the second largest groundnut producing nation in the world with an annual production of 5.64 million tons of pods after China, which has an annual production of 15.71 million tons pods (Madhanmohan and Nigam, 2013).

The crop is grown primarily for human consumption and it is a rich source of oil (40–50 percent), proteins (20–50 percent), carbohydrates (10–20 percent), minerals (P, Ca, Mg and K) and Vitamins (E, K and B group) (Savage and Keenan, 1994). Groundnut is an important source of edible oil for millions of people living in the tropics. It provides an inexpensive source of high quality dietary protein and oil. The crop improvement programmes in groundnut aims at high pod yield and oil content. However, for confectionery grade varieties high protein and sugar, low oil and high Oleic/linoleic acid (O/L) ratio are the desirable characters (Dwivedi and Nigam, 1995).

Groundnut is primarily grown for its seeds (kernels). Therefore any agronomic or genetic manipulation in this crop requires a strong fundamental knowledge about physiological and biochemical processes of the plant responsible for better kernel yield. Underground development of pods in groundnut makes it very interesting to learn about the formation and maturation of seeds. Several internal and external factors during the development of seeds will have their effect on seed quality which is manifested in the post harvest life and germination behaviour of this crop. Thus the aim of this review is to critically elucidate various physiological and biochemical processes associated with seed formation, seed maturation, chemical composition, seed quality, seed dormancy and post harvest storage of groundnut seeds.

31.2 Taxonomic Classification

Cultivated groundnut (*Arachis hypogaea* L.) or peanut, is a self-pollinated annual legume crop. Genus *Arachis* belongs to the family Fabaceae (syn: Leguminosae), subfamily Papilionaceae, in the subtribe Stylosanthinae of the tribe Aeschynomeneae. The cultivated groundnut *A. hypogaea* has two subspecies, *viz.*, *hypogaea* and *fastigiata*. Based on their growth habit most commonly known botanical types of groundnut are as follows.

I	Virginia bunch	*Arachis hypogaea* subsps. *hypogaea* var. *hypogaea* (type Virginia)
II	Virginia runner	*Arachis hypogaea* subsps. *hypogaea* var. *hypogaea* (type Virginia)
III	Valencia	*Arachis hypogaea* subsps. *fastigiata* var. *fastigiata* (type Valencia)
IV	Spanish	*Arachis hypogaea* subsps. *fastigiata* var. *vulgaris* (type Spanish)

The comprehensive subspecific classification of *A. hypogaea* is summarized in Table 1 as described by Ramanatha Rao and Murty (1994).

Table 31.1 *Classification of cultivated groundnut (Arachis hypogaea) and traits associated with different botanical varieties*

Characters	Subspecies *hypogaea*		Subspecies *fastigiata*	
	Botanical variety *hypogaea*	**Botanical variety *Hirsute***	**Botanical variety *Fastigiata***	**Botanical variety *vulgaris***
Foliage	Dark green	Dark green	Lighter than *hypogaea*	Lighter than *hypogaea*
Growth habit	Procumbent, decumbent or erect	Procumbent	Erect or decumbent	Erect or decumbent
Branching pattern	Alternate	Alternate	Sequential	Sequential
Main axis	40-50 cm	About 1m	Primary branches shorter than main axis	Primary branches shorter than main axis

Table 31.1 Contd...

Characters	Subspecies *hypogaea*		Subspecies *fastigiata*	
	Botanical variety *hypogaea*	Botanical variety *Hirsute*	Botanical variety *Fastigiata*	Botanical variety *vulgaris*
Floral axes on main stem	Absent	Absent	Present	Present
Inflorescence	Simple	Simple	Usually simple	Compound
Stem hairiness	Less hairy	More hairy	Sub-glabrous to moderately hairy	Hairy
Maturity	Medium to late	Very late	Short duration	Short duration
Seed coat colour	Tan, red, white, purple or variegated	Tan, red, white, purple or variegated	Tan, red, white, yellow, purple or variegated	Tan, red, white or purple
Seed size	Very large to small	Large to small	Medium to small	Medium to small
No. of seeds pod^{-1}	Usually two	Two to four	Two or two to four;rarely five	two
Pod beak	Not very prominent	Strongly beaked and ridged	Absent or prominent or slight	Present or absent
Seed dormancy	Present	Present	Absent	Absent

(After Madhan Mohan and Nigam,2013)

Cultivated groundnut has narrow genetic base due to its allotetraploid origin. However, there is abundant morphological variability present among the varieties in the cultivated species. In groundnut, the sub species *fastigiata,* which includes the *Spanish* and *Valencia* are the most preferred groundnut sub species in the semi-arid tropics, which accounts for about 60% of the world's groundnut production area (Sudharani *et al.*2013). Most of the Spanish bunch varieties are early in duration (105-110 days) where as spreading varieties are medium to long duration (130-150 days).

Some representative examples for different botanical types of groundnut were given below

Spanish bunch	Virginia bunch	Virginia runner
Narayani (TCGS 29),Kadiri 5,Kadiri 6 Kadiri 9, Abhaya (TPT 25), Greeshma	Kadiri 7,Kadiri 8	GG 14,Utkarsh, RG 382

Thus a great variability is observed in morphology of different botanical types of groundnut. However the fundamental processes underlying seed formation, maturation and accumulation of different chemical compounds responsible for seed quality follow a more or less a similar trend.

31.3 Phenological Stages of Groundnut Crop

Expansion, pod formation and seed filling are the three major phenophases in groundnut. However Williums *et al.* (1978) divide the crop's life time in to ten phases *viz.*, : 1- germination and emergence; 2-vegetative growth only; 3-vegetative growth and first flowers;4-active flowering first pods being produced; 5-flowering,peg production, pod production and the start of significant kernel growth; 6-vegetative growth ceased and commencement of rapid kernel growth;7-pod initiation ceased and commencement of rapid kernel growth; 7-pod initiation ceased and rapid kernel growth continued, defoliation usually starts;8-first mature pods present; 9-30-40% of the pods mature;10-50-70% of the pods mature , rapid defoliation and crop lifted at the end of this phase. Further the growth stages of groundnut plants based on visual observations of vegetative and reproductive growth was defined by (Boote 1982) as follows.

Stage	Stage title	Description
Vegetative stages		
VE	Emergence	Cotyledons near the soil surface with the seedling partly visible
V0		Cotyledons are flat and open at or below the soil surface
V1	First tetrafoliate	One of n developed nodes on the main axis, a node is counted when its tetrafoliate is unfolded and its leaflets are flat
Reproductive stages		
R1	Beginning bloom	One open flower at any node
R2	Beginning peg	One elongated peg (gynophore)
R3	Beginning pod	One peg in soil with swollen ovary at least twice the weight of the peg
R4	Full pod	One pod fully expanded to dimensions characteristic for the cultivar
R5	Beginning seed	One fully expanded pod with which seed cotyledon growth is visible when the fruit is cut in cross section
R6	Full seed	One pod with cavity apparently filled by the seed when fresh
R7	Beginning maturity	One pod showing visible natural coloration or blotching of inner pericarp or testa
R8	Harvest maturity	66-75% of all developed pods have testa or pericarp coloration
R9	Over mature pod	One undamaged pod showing orange tan coloration of testa and or natural peg deterioration

31.3.1 Time Line for Different Phenophases

Germination is completed in 7-10 days after the seed is sown (by this time the plant becomes autotrophic)
Root nodules appear at 15-18 days old plant
Flower initiation starts at 20-30 DAS
Pod formation at 45 DAS
Pod achieve maximum number at 55 -70 DAS and then the filling starts.
Seed development is completed by 8-14 weeks after flowering depending upon cultivar.

31.4 Seed Formation

31.4.1 Reproductive Biology

The groundnut plant has an indeterminate growth habit. This means that the flowering and fruiting of the crop occur over a long period of time. Groundnut is a self-pollinated leguminous crop (with less than 1% cross-pollination) but cross-pollination up to a maximum of 10% was also reported. Generally, only one flower opens at each node at any given time at sunrise. There are usually four stages of flowering, the pattern of which depends upon cultivar and environment. At first stage only a few flowers are produced followed by a stage of rapid flowering. A peak in flowering is observed in the third stage followed by a decline in fourth stage (Smith,1954). A flower consists of five petals, ten stamens and a pistil; Stigma becomes receptive to pollen 24 hours before flower opening and remains so after up to 12 hours, while pollen is shed 1 to 8 hours before the flower opening; Anthesis and pollination occur at sunrise with self-pollination taking place within the closed keel of the flower (cliestogamy). About 40% of the flowers fail to begin pod development and another 40% abort before pod development. The ratio of pod produced to flower is generally 1:7.

The number of flowers, pegs, and pods are the most important yield components that affect the yield potential of groundnut (Awal and Ikeda, 2003). After pollination and fertilization the corolla closes, the calyx tube bends, the flower withers, and then the peg is formed. Peg is an expanded intercalary meristem at the base of the basal ovule (ovary) resulting in a lomentiform carpel of 1-5 segments, each containing a single seed with two massive cotyledons and an embryo. When the peg reaches its maximum depth in the soil it stops growing and the pod begins to develop. The pod continues to grow, reaching its maximum size after the penetration of the peg into the soil (Sevgi Calisikan *et al.*, 2008). Pods containing seeds are produced below ground. They vary in shape, size and texture and may contain up to five seeds per pod. The pod expands rapidly in the soil by the development of a large parenchymatous tissue called endocarp lying between ovules and shell. The endocarp recedes with ovules growth and disappears completely when the seeds have matured. The seeds inside start to develop only after the pods attain their full size. Good amount of variation is seen in different cultivars with respect to flowering, maturity period and seed characteristics (Table 31.2)

Table 31.2 *Flowering and maturity period and seed characteristics of five groundnut cultivars*

Characters	Kadiri 3	GAUG 10	ICGS 11	Girnar 1	GG 2
Seed size	**Medium large**	**Large**	**Medium large**	**Medium**	**Medium**
Time to initial flowering (days)	24.0	26.3	25.3	21.0	19.7
Time to 50% flowering (days)	27.3	29.3	27.7	24.0	21.7

Table 31.2 *Contd...*

Characters	Kadiri 3	GAUG 10	ICGS 11	Girnar 1	GG 2
Seed size	Medium large	Large	Medium large	Medium	Medium
Seed length (mm)	11.5-13.0	10.5-16.0	10.0-12.0	11.0-13.0	10.0-14.5
Seed width (mm)	7.0-8.3	5.5-8.2	7.0-8.0	7.0-7.5	6.8-11.8
100-seed weight (g)					
Range	36.0-43.0	44.0-59.0	34.0-46.0	30.0-40.0	30.0-38.0
Average	40	51.5	40	35	34.5
Oil content (%)	49.6	50.0	52.9	53.9	49.0
Protein (%)	28.0	20.1	20.8	21.3	24.1
Sound mature seed (%)					
Range	82.0-93.2	79.0-95.0	80.0-94.3	82.0-88.6	80.0-96.0
Average	87	87	87	85	88
Time to maturity (days)					
Summer season	125	131	117	115	115
Rainy season	115	124	105	104	104
(*Source:* Rajagopal *et al.*, 2002)					

31.4.2 Seed Development

Generally, there are three clearly defined phases of seed development in groundnut. The first is a phase of rapid growth with simultaneous increase in both fresh and dry weight. In the second phase the rate of dry weight increase is often stable, but seed moisture content begins to fall (Fig 1 & 2). The seed then loses moisture in the ripening phase, usually with little change in dry weight. The first and second phases involve division, enlargement and differentiation of seed cells, and the last phase involves changes in seed components, such as storage lipids, fatty acids and other seed storage metabolic substances (Chung *et al.*, 1995). Depending upon cultivar and local environment, stages I and II of seed development may be completed by 8-14 weeks after flowering.

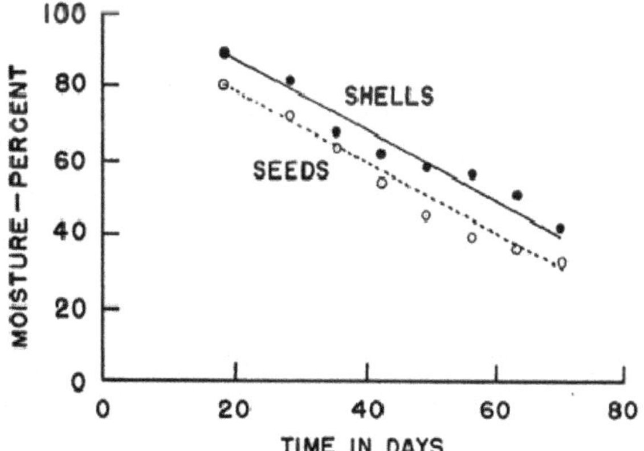

Fig. 31.1 Percentage of moisture in Virginia bunch seed and shell

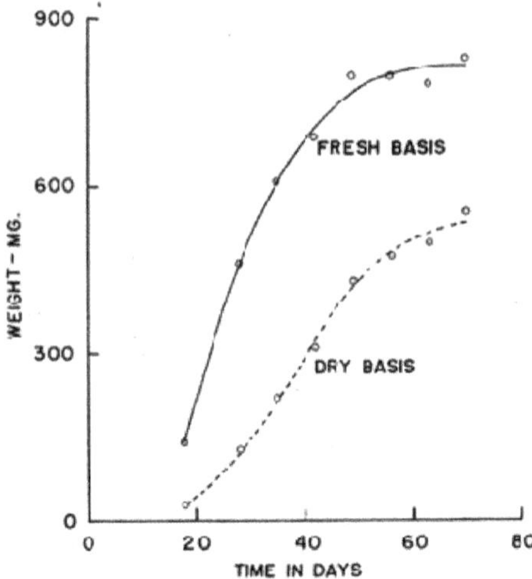

***Fig. 31.2** Average weight of individual Virginia bunch seed*
Source: Pickett, 1950)

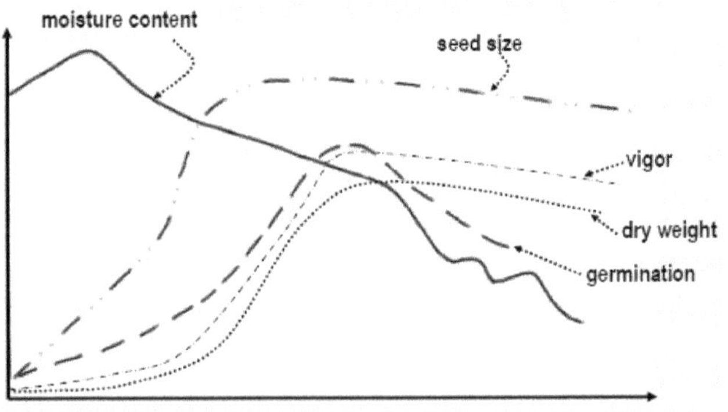

General alterations of seed characteristics during maturation (Carvalho and Nakagawa, 2000)

Fig. 31.3

A seed consists of two large cotyledons (does not contain endosperm), a stem axis and leaf primordial (plumule), hypocotyl primary root (radicle) and seed coat. According to Conagin (1957) the nucellar tissue disappears at an early stage and the endosperm disappears at seed maturity. Cotyledons comprise about 96% of the seed weight and are the major storage tissue for the developing seedlings. (Caliskan *et al.* 2008) reported that dry matter accumulation in each part of the seed continues until maturity, although the accumulation rate differs according to plant age and genotype. Physiological maturity of seed at harvest is usually defined as the time when seed attains its maximum dry weight (ISTA 1987) (Fig 31.3). Fresh weight follows a typical sigmoidal growth pattern (Fig 31.2). From an economic standpoint peanut growers should consider the trend towards weight loss in over-mature seeds. Therefore the stage at which maximum average maturity is achieved before harvest should be identified to maximize yield.

31.5 Changes in Seed Chemical Composition During Seed Maturation

31.5.1 Carbohydrates

With the exception of reducing sugars present in the early stages in small amounts and later only in traces, sucrose is the only soluble sugar in the seed and is considered as the transport sugar. The percentage of sucrose decreases as the seed develops. However, at maturity a slight increase in content of this sugar occurs. Starch also decreases as the seed ripens. The total percentage of sugars and starch approaches a constant value in the fortnight preceding maturity (Fig.4 & 5). In understanding the mutual relation between the physiological process associated with seed maturity and the attributes of groundnut seed, it is clear that sugars are synthesized during the initial development of seed and converted into oil during maturity (Young and Moore, 1972). Starch is not normally considered as a reserve resource in peanut seeds. However, in the early stages of maturity the seed may accumulate starch more rapidly than lipid, the final major reserve. At maturity stage 8 starch and lipid content were approximately equal, 50–55 mg/seed. After this stage starch accumulation became minimal or ceased depending on the harvest period. A possible role of the starch reserves is to provide a sugar source during germination until the glyoxylate system is operating at a sufficient level to supply the germination energy requirements (Pattee *et al.*, 1969).

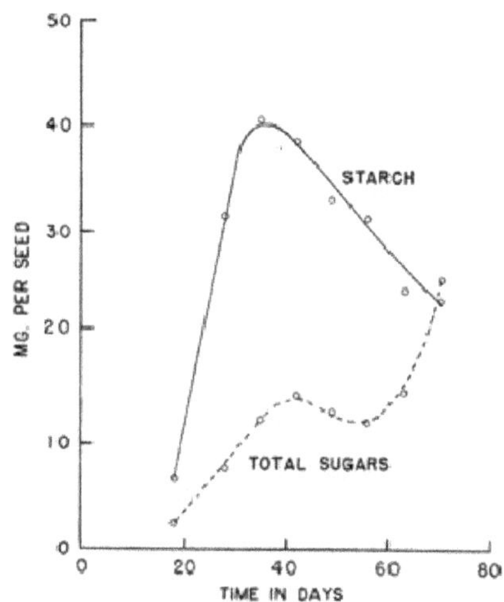

Fig. 31.4 Carbohydrates on Virginia bunch seed on weight per seed basis

Groundnut kernels consist of water-soluble carbohydrates (monosaccharides, disaccharides) and oligosaccharides (including starch, raffinose and stachyose). Together they comprise the total available carbohydrate fraction. Though the Carbohydrate levels are high in newly germinated groundnut kernels but the levels begin to fall from an average of 46% at 5 weeks as the maturing seed begins to store oil. Mature kernels are reported to contain 9.5-19.0% total available carbohydrate as both soluble and insoluble carbohydrate (Abdel Rahman, 1982b; Woodroof, 1983). Water-soluble carbohydrates comprise 3.0 to 8.9% of the available carbohydrate (Sekhon *et al.*, 1970; Wallerstein *et al*, 1989; Rodriguez *et al.*, 1989). Seasonal variation plays a role in the levels of both soluble and insoluble carbohydrates.

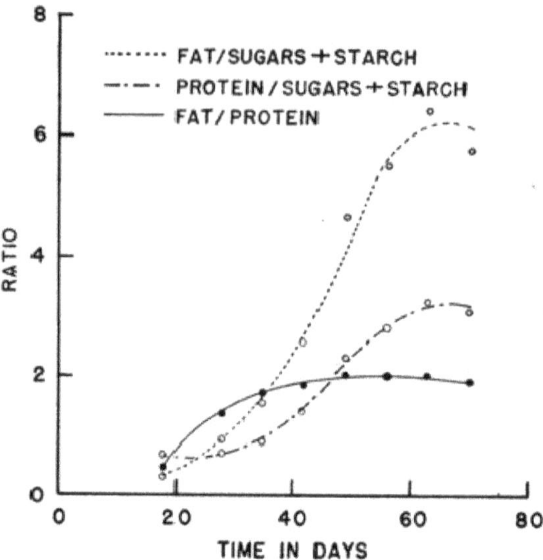

Fig. 31.5 Ratios of fat protein and carbohydrates in Virginia Bunch seed
(***Source:*** Pickett,1950)

31.5.2 Lipids

Fats and oils are important storage forms of reduced carbon in many seeds including those of agriculturally important species such as peanut, soybean, sunflower, canola, and cotton. Oils serve a major storage function.The major constituents of oil seeds are triglycerides while that of cereals is starch, it requires a plant to expend 1.2 g of photosynthate to produce one gram of starch and 3.3 g of photosynthate to produce one gram of triglycerides in oil seeds.

Fats and oils belong to the general class termed lipids, a structurally diverse group of hydrophobic compounds that are soluble in organic solvents and highly insoluble in water. Lipids represent a more reduced form of carbon than carbohydrates, so the complete oxidation of 1g of fat or oil (which contains about 40 KJ of energy) can produce considerably more ATP than the oxidation of 1g of starch (about 15.9 KJ) accordingly the biosynthesis of lipids requires a correspondingly large investment of metabolic energy.

The fatty acids in plants are usually straight chain carboxylic acids having an even number of carbon atoms. The carbon chains can be as short as 12 units and as long as 30 or more, but commonly are 16 or 18 carbons long. Fatty acid biosynthesis involves the cyclic condensation of two carbon units derived from acetyl-Co A. In plants fatty acids are synthesised exclusively in the plastids. The major fatty acids in plant lipids are as follows.

Saturated fatty acids

Lauric acid (12:0) $CH_3(CH_2)_{10}CO_2H$

Myristic acid (14:0) $CH_3(CH_2)_{12}CO_2H$

Palmitic acid (16:0) $CH_3(CH_2)_{14}CO_2H$

Stearic acid (18:0) $CH_3(CH_2)_{16}CO_2H$

Unsaturated fatty acids

Oleic acid (18:1) Linoleic acid (18:2)

Linolenic acid (18:3)

The proportion of fatty acids in plant lipids vary with the plant species. For example Groundnut oil is rich in unsaturated fatty acid, oleic acid (mono unsaturated) and linoleic acid (polyunsaturated) accounting for 38 to 58 percent and 16 to 38 per cent, respectively. Among the saturated fatty acids, palmitic acid is the major one with the proportion of about 10 to 16 percent. Groundnuts characteristically contain high levels of oil and protein and low levels of carbohydrates and ash. The composition of oil is important from the point of view of human nutrition as well as the stability of the oil during storage.

The oil content of the groundnut kernels is modified by the duration of seed development. Oil content increases as seed dry weight increases (from 2nd week onwards), which is most rapid between weeks 7 and 8 (50 % of the oil accumulated by this time), reaching a maximum at 11 weeks (Sanders, 1980). Oil content of the groundnut seeds is increased in direct proportion to the increase in seed dry weight, (Abdel Rahman 1982) found that oil synthesis replaced starch as the dominant reserve accumulation mechanism from 7 weeks onward.

The Virginia type, which is late maturing, and the runner type are generally higher in monounsaturated fatty acids (oleic acid). Further, Virginia cultivars were reported to contain higher levels of linoleic than oleic acid (Sekhon *et al.,* 1973), which is not observed in Spanish cultivars. The early maturing Spanish type has a higher linoleic acid (i.e. polyunsaturated) content but total saturated fatty acids are also higher (Woodroof, 1983) (table.3) One characteristic of groundnut oil is its low linolenic acid (18:3) content (less than 1%). In such small amounts it is able to exert an effect on the oil flavour without contributing to oxidative rancidity.

Oil content of different botanical types of groundnut differs to some extent, as follows (Taira, 1985; Wallerstein *e. tal.,* 1989)

Spanish	42.0-53.8% of dry weight
Virginia	45.0-58.6% of dry weight
Runner	41.23-53.6% of dry weight

Table 31.3 *Fatty acid composition of groundnut oil (%)*

Fatty acid type	Spanish	Virginia	Runner	Unspecified
16:0	8.2-18.4	6.0-13.5	8.6-13.3	6.0-20.12
16:l<o9	0.1-2.3	0-1.1	0.1-0.2	0.1-1.47
18:0	1.1-6.2	1.0-4.9	2.0-3.0	0.8-5.3
18:lco9	36.4-53.1	36.7-61.1	40.9-60.4	35.7-71.5
18:2co6	20.5^10.3	21.6-48.3	21.0-35.2	13.0-41.1
18:3<o6	0.1-0.3	0.1-0.5	0.1-0.2	0-1.5
20:0	1.0-2.1	1.1-1.8	1.0-1.5	0.9-3.5
20:lco9	0.7-1.4	1.1-1.8	0.9-1.9	0.6-2.0
22:0	3.0-4.4	2.7-3.7	2.8-3.9	1.3-5.1
24:0	0.9-2.0	1.5-2.0	1.2-2.6	0.6-5.9
				(After Savage and Keenan, 1994)

Table 31.4 *Fatty acid composition of hexane extracts of groundnut cutivars, post rainy season*

Fatty acid (%)	ICGS 1	ICGS 5	ICGS 11	ICGS 21	ICGS 44	Kadiri 3
Palmitic	12.1	11.9	12.1	11.3	11.9	12.1
Stearic	2.3	2.1	2.3	2.8	2.71	2.9
Oleic	36.7	37.2	36.8	42.9	38.9	38.9
Linoleic	40.5	40.5	40.6	35.1	38.4	38.0
Arachidic	1.4	1.3	1.4	1.4	1.5	1.5
Eicoaenoic	1.4	1.4	1.4	1-3	1.4	1.4
Behenic	32	3.1	3.0	2.9	3.0	2.8
Lignoceric	1.8	1.7	1.8	1.5	1.7	1.8
0/L ratio	0.91	0.92	0.91	123	1.01	1.03
						(after jambunathan,1991)

Strains grown under similar climatic and agronomic conditions still show varietal differences with respect to oil content and composition (Table.4). This was attributed to genetic causes. This makes the manipulation of the oil content and composition through plant breeding techniques a practical objective. Differences in fatty acid composition of groundnut oil is influenced by environmental and seasonal effects variable soil conditions and agricultural practices in addition to varietal differences, genotypic variation and maturity. The environment (especially temperature) during seed formation can modify both the oil content and the fatty acid distribution pattern in a given cultivar (Liu *et al,* 1984). (Taira 1985) showed that the daily mean temperature during ripening correlated negatively with the linoleic acid content.

31.5.3 Protein Content

The protein content of groundnut seeds ranges from 12% to 36.4% (Savage and Keenan, 1994) and may be greatly influenced by environmental factors, such as temperature, irrigation, and fertilization. Protein content increases during ripening and the maximum increase occurs at 9 weeks and then becomes stable (Basha, 1991).

The storage proteins arachin and conarachin account for over 63% and 33% of the total protein content of groundnuts. The ratio of these globulins ranges from 2:1 to 4:1. Conarachin is found in the cytoplasm while arachin, the second major storage protein, is localized in the protein bodies (aleurone grains) of the seeds. Groundnut protein is deficient in some essential amino acids (methionine and lysine) but overall the digestibility of the protein is superior to many other plant protein sources.

31.6 Dormancy

Seed dormancy is the temporary failure of a viable seed to germinate, after a specific length of time and in a particular set of environmental conditions that allow germination after the restrictive state has been terminated by either natural or artificial conditions (Simpson, 1990). The term quiescence is sometimes used incorrectly for dormancy. Quiescent seeds are fully germinable, but do not complete the germination process because of limiting external conditions such as light, temperature, oxygen, or moisture. The term used to describe the transition of dormant seeds to a more readily germinable state is after ripening. After ripening is loss of the dormant state over some period of time through exposure of the seeds to a set of environmental conditions after maturation and separation from the parent plant. Environmental conditions that facilitate after ripening vary with species (Michael E. Foley, 2001).

Different causes for seed dormancy in groundnut were reported by many workers. It is fairly obvious that more than one cause might be responsible for the dormancy of a seed. In a broad view, two types of dormancy can be distinguished *i.e.* (1) "Innate" dormancy where the seeds will not germinate even under favourable conditions and (2) "Imposed" dormancy where seeds will not germinate when conditions are unfavorable.

Dormancy is an important factor in commercial groundnut production. It can be beneficial when dormancy prevents mature seeds from sprouting before harvest. It can be detrimental when dormancy reduces stand or hampers taking a second crop immediately after harvest. The following aspects are considered to be the important factors in influencing seed dormancy in groundnut (Shelar *et al.,* 2014).

1. **Presence of inhibitors and ratio between growth promoter to inhibitor:** Dormancy is determined by the balance between growth inhibitors and growth promoters. Auxins and ethylene are considered to be the promoters and ABA being the inhibitor. (Nagarjun and Gopalkrishnan 1958) reported that non-dormant seeds of TMV-2 groundnut contained a water soluble growth promoting hormone

and the seed extract induced root initiation in the dormant seeds of TMV-3 groundnut. The physiological studies done by Sreeramulu and Rao (1971) revealed that the water soluble hormone was indole acetic acid which has root inducing activity. This auxin was noticed at high levels in the embryonic axis and seedcoat of non-dormant seeds.

2. **Genetic aspects of dormancy:** Dormancy, the physiological phase in seed, is initially determined by the genetic makeup of the seed and varies largely among species and even within a species. Hull (1937) reported that dormancy in groundnut is an inherited character and the rest period extended even up to two years in some varieties. (John *et al.* 1950) pointed out that dormancy is an inherent property of Virginia groundnut. The trait dormancy was found to be partially dominant over the trait non-dormancy. Genetic differences in seed dormancy between strains with different botanical groups were demonstrated by several investigators (Ramachandran *et al.*, 1967 and Lin and Lin, 1971).

3. **Mechanical restriction of the seed coat:** Hardness or impermeability of seed coat is said to be one of the many causes for dormancy. This causes physical restriction to the exchange of gas and water which are essential for the initiation of germination process. The inheritance of hard seed coat varies among and within the species. This dormancy is also mediated by environment prevailing during seed ripening period. Significant morphological differences in the testa among the different cultivars of groundnut were reported.

4. **Presence of ethylene:** Ethylene was involved in the normal regulation of seed dormancy (Toole *et al.*, 1964). (Ketring and Morgan 1969) reported that the embryonic axis of non-dormant seeds of peanut actively produced ethylene during germination, where as ethylene production was low in dormant seeds.

   Taxonomically, the cultivated peanut (*A. hypogaea* L) as mentioned earlier is divided into two subspecies, one with two botanical varieties, and one with four. In the subspecies *hypogaea var. hypogaea* (Virginia and Runner market types) and *var. hirsuta*, the varieties have long duration cycle and seeds are dormant. While in subspecies *fastigiata* with *var. fastigiata* (Valencia market class) and *var. vulgaris* (Spanish market class), the varieties are early-maturing but generally without fresh seed dormancy (Krapovockas and Gregory, 1994).

5. The pods of the subspecies *hypogaea,* are typically two-seeded and these seeds show marked dormancy, ranging from 30 to 360 days (Zade *et al.,* 1986). Their seeds also do not suffer from vivipary when harvesting is delayed. Subspecies *fastigiata* which includes the Spanish and Valencia market types are the most preferred groundnut in the semi-arid tropics, which accounts for about 60% of the world's groundnut production area. Sprouting in this sub-species, which occur before harvest and sometimes beyond, contributes to significant yield losses. Lack of dormancy in bunch types has been described as an inherent property of seed and does not primarily depend on soil conditions (Shelar *et al.*, 2014).Yield losses

due to *in situ* germination in bunch varieties have been estimated to range between 20 and 40% (Reddy *et al.,* 1985). Thus, a short period of seed dormancy is necessary to reduce these losses.

In groundnut, seed dormancy was reported to be controlled by two hormones: abscisic acid which inhibits sprouting and ethylene which is accumulated in storage to break dormancy to allow germination. Abscisic acid may prevent germination and thus cause dormancy by preventing ethylene synthesis by the seeds.

31.6.1 Dormancy in Virginia Type

Seeds of the Virginia botanical type of peanut are dormant for a variable period after harvest. That is, they fail to germinate under conditions normally favorable for germination. Dormancy is an inherent property of Virginia type peanut seeds (John *et al.,* 1950). Dormancies of many seeds have been attributed to inhibitors in the embryo or seed coat or both. (Toole *et al.* 1964) measured the influence of seed coat, seed maturity, and various external factors on dormancy of freshly harvested and stack-cured peanut seeds. According to them, the two seeds in a pod appeared to be about equally mature, but when a difference in maturity was evident the basal seed was almost always more mature than the apical. They further reported a greater degree of dormancy for basal than for apical seed of comparable maturity, as measured by physical appearance. Similarly, (Shibuya 1932) also reported that basal seeds were more dormant than apical, attributed this to the greater degree of maturity of basal seed within a pod. However, (Ketring and Morgan, 1972) have explained this difference as due to the rate of ethylene production ,the more dormant basal seed producing less ethylene than the apical one (Fig 31.6). According to (Shibuya 1932) and (Pettit 1896) the basal seed in a pod matures earlier than the apical.

With seeds of comparable maturity, dormancy was easier to break in apical than in basal seeds. However, physical appearance of peanut seeds does not always give an indication of the relative dormancy. It was reported that after ripening occurs on the plant in the field as the season progresses. Frequently in a seed lot immature seeds were always more dormant than more mature seeds. The more immature the seeds it is more difficult in break the dormancy. Emergence of the radical when ethylene production is near maximum does not establish a causal relationship between ethylene production and the initiation of germination or the breaking of dormancy. However, the facts that endogenous production of ethylene accompanies the natural decline of dormancy, that exogenously applied ethylene can intervene in the gradual breaking of dormancy, and that ethylene can cause the inherently more dormant basal seeds to germinate like the inherently less dormant apical seeds suggest a hormonal role for ethylene. Thus the concentration of C_2H_4 was a more critical factor than the duration of exposure in inducing the inherently more dormant basal seeds to germinate in a manner similar to the inherently less dormant apical seeds.

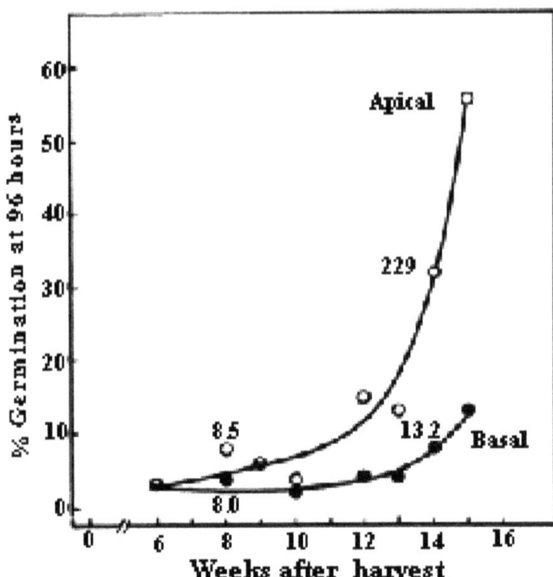

Fig. 31.6 Seed viability test (Ketring and Morgan, 1970)

Inability of freshly harvested peanut seed to germinate is due to dormancy of the embryo and to interference of the seed coat. In a nearly mature seed the outer integument is thick and fleshy and the inner thin. As the seed matures the seed coat loses turgidity and thus decreases in thickness. At full maturity the seed coat is thin and of a brown colour on the inner surface and deep pink on the outer. Thus the seeds if harvested before attaining sufficient maturity were subjected to dormancy.

31.6.2 Factors to be Considered for Breaking of Dormancy in Virginia Types

The effect of removal of the seed coat after it became thin was observed to be influenced by the position of the seed in the pod and the length of time the plants remained in the field before digging. For early diggings, coat removal caused greater numbers of apical than of basal seeds to germinate. The inner layer of the coat appears to exert a greater influence on dormancy than the outer. Germination of more number of seeds was found when the entire coat was removed than when the outer layer alone was removed. This would establish the role of seed coat in inducing dormancy.

Ethylene for 72 hours, on the other hand, caused complete or almost complete germination of peanut seeds of maturity classes 1 to 3 with seed coat intact. A 4-hour ethylene treatment was proved to be highly promotive of germination when given immediately or up to 72 hours after planting. It had no effect on immature seeds with intact seed coats. However, ethylene greatly stimulated the germination of naked embryos of immature seeds. The turgid seed coat of immature seeds appears to be a barrier to the entrance of ethylene. It was observed that ethylene might help in "after-ripening" or "maturing" the dormant embryos.

Although the mechanism of its action is unknown, ethylene triggers prompt and high germination in peanut seeds. The site of reaction or reactions that trigger the initiation of germination may or may not be the same, but all produce an immediate response with high results. Ethylene, 100 ppm at 14 to 28 liters/min., induced complete germination of the more mature classes of seeds with intact seed coat, and a high percentage of immature seeds with seed coat removed.

The CO_2, does not cause complete germination of dormant peanut seeds. However, five per cent CO_2 in air was found capable of stimulating germination of dormant peanut seeds. Thus it was suggested that the only two effective means of breaking peanut seed dormancy are desiccation by heat and ethylene treatment. Ethylene treatment requires less time, and it is effective at low concentrations and at usual seed-germinating temperatures. The first method for breaking the "primary" post harvest dormancy of Virginia-type peanut seeds was heating at 40 to 45^0 C for about 15 days (Bailey *et al.*,1958). Leaching and treatments with natural aromatic compounds were only partially successful in breaking this dormancy (Toole *et al.*, 1964.) (Shibuya 1938) reported that indole-3-acetic in lanolin (0.1 g of IAA/g of lanolin) applied to the exposed radical end of dormant peanut seeds stimulated germination. The three concentrations of gibberellic acid tested were also found to stimulate ethylene production and germination of dormant peanut seeds to some extent (Ketring and Morgan, 1970).

31.6.3 Lack of Dormancy in Spanish and Valencia Sub Species

In groundnut, the sub species *fastigiata,* which includes the *Spanish* and *Valencia* are the most preferred groundnut sub species in the semi-arid tropics, which accounts for about 60% of the world's groundnut production area. Most of the popularly cultivated varieties in southern India are non-dormant and harvesting these cultivars in *rainy season* poses the problem of sprouting in the field and subsequently deterioration of the quality of the kernel to a tune of 20 to 50%. This problem becomes more obvious if harvesting is further delayed.

Development of early groundnut varieties that could cope with drought that may occur at the end of the rainy season period is an important area of groundnut improvement. However, such early varieties belonging to Spanish-type usually lack fresh seed dormancy. Thus rains that may fall at the harvest period could cause *in situ* germination. Moreover, the sprouting is known to increase the susceptibility of peanut to *Aspergillus* infection, which can result in aflatoxin contamination. Thus, a short period of seed dormancy is necessary to reduce these losses.

31.6.3.1 Seed Dormancy Period in Bunch Groundnut Varieties

(Varisai and Dorairaj 1968) screened 206 groundnut varieties under irrigated condition for dormancy. Only 6 bunch varieties had a dormancy period of about 15-20 days. None of the bunch varieties studied was completely dormant. (Reddy *et al*. 1985) studied 17 groundnut varieties belonging to Spanish and Valencia botanical groups and reported that the derivative of the cross, J-11 × Robout-33-was found to possess seed dormancy for a period of about 35 days. (Pandya and Patel 1986) studied seeds of 4 Virginia and 73 Spanish bunch varieties and 5 Spanish × Virginia hybrids for percentage germination after 3-50 days of storage at room temperature. Virginia types and their crosses, particularly G-201, ICGS-6, Robout-33-1and (TMV 10 x Robout-33-1)-2, were found to be more dormant than most of the Spanish types tested. Among the Spanish varieties, RSHY-6, ICGS-21, ICGS-30, ICGS-57, TG-9 and TG-17 were identified as sources of dormancy for breeding.

(Nautiyal *et al.* 1996) screened about 200 Spanish germplasm lines for fresh seed dormancy during *rabi* /summer 1995. They revealed that genotypes showed variation in germination percentage. During rabi /summer 1995, the crop was harvested at 110 DAS and most of the genotypes showed more than 60 per cent fresh seed dormancy, finally they concluded that the dormancy period of most of the genotypes was laid between 30-40 days. However, (Swain and Sahoo 2001) studied the duration of seed dormancy in different bunch type groundnut cultivars and reported wide variation in their duration ranging from 5.4 days to 106.6 days. Further, seeds of *kharif* showed the highest degree of dormancy followed by *rabi* and summer. (Manonmani 2002) studied both dormant and non-dormant cultivars of groundnut, reported that dormant cultivars maintained higher seed germinability in storage for a longer period than the non-dormant cultivar of groundnut varieties. The results indicated that bunch type varieties had less dormancy in comparison to spreading type of cultivars.

31.6.4 Induction of Dormancy in Non Dormant Groundnut Types

31.6.4.1 Genetic Approach

Attempts to select peanut lines from inter subspecies Virginia × Spanish crosses may lead to lines with fresh seed dormancy but matures up to 10 days later than the Spanish parent. Some authors found genetic variability within ssp. *fastigiata* for fresh seed dormancy. Besides, The intensity of dormancy, the duration of dormancy for a given entry was found to be varied more or less over seasons due to non-genetic factors like environmental condition and the mother plant environment. (Varisai Muhammad and Dorairaj, 1968; Pandya and Patel, 1986).

Despite the importance of dormancy in groundnut production, there has been few studies conducted on the inheritance of its fresh seed dormancy, In this regard, (Lin and

Lin 1971) observed monogenic control with complete dominance of dormant over non-dormant seed, whereas (John *et al.* 1950) and (Nautiyal *et al.* 1994) indicated that the character may be quantitatively inherited. (Khalfaoui 1991) concluded that dormancy is a quantitatively inherited trait.

An experiment was conducted at Ghana (Africa) to determine the heritability of fresh seed dormancy in groundnut and to transfer this trait from dormant exotic lines (ICGV-86158 and ICGV-87388) into two non-dormant groundnut varieties (Asibuo *et al.*, 2008). It was reported that seed dormancy is controlled by monogenic inheritance with dormancy dominant over non-dormant, as the results showed that more than 90 per cent of the freshly harvested seeds of the non-dormant parents germinated before 14 days, whereas less than 10 per cent of the seeds of the dormant parents germinated during the same period.

Further, Mathur *et al.* (2000) stated that two cultivars of groundnut *viz.*, PBS-12115 and PBS-12126 were higher yielders and also possess fresh seed dormancy of 21-28 days and 14-21 days respectively. Thus, suggested to use these two genotypes as donor parents for incorporation of fresh seed dormancy in breeding programme.

31.6.4.2 Chemical Methods

Coumarin, phenolic acids (viz., hydoguinone,salicylic acid, trans cinnamic acid etc) and maleic hydrazide (diethanolamine salt of 1,2-dihydroxy-3,6 pyridazine-dione), were some of the chemicals used to induce dormancy in groundnut.

Foliar application of MH-30 at different concentrations *viz.*, 5000 ppm at 70 DAS, 10000 ppm at 90 DAS induced seed dormancy ranging from 30 to 40 per cent (Bhapkar *et al* . 1986) . Seed dormancy period was increased from 5 to 25 days in cv. PI-139915 and PI-169292 when 200 ppm MH was applied as foliar spray one month before harvesting (Abrar and Jadhav, 1991). (Jagatap 2000) studied the induction of seed dormancy in bunchy groundnut genotypes *viz.*, RHRG-12, TAG-24, RHRG-16 and SB-XI. He revealed that seed dormancy could be induced upto 30, 10, 30and 20 days, respectively by foliar application of MH @ 250 ppm than other concentrations (500 and 750 ppm). Induction of seed dormancy in non-dormant groundnut cultivars was also reported by (Nautiyal 2004) using MH at various concentrations and reported that foliar spray of malic hydrazide @ 1000 ppm, 60 days after crop emergence was found to be superior in inducing dormancy in Spanish groundnut cultivars.

Induction of dormancy in TMV-2, a bunch groundnut variety by the foliar application of MH-30 was reported by (Vaithialingam and Rao 1973). Further, (Sudharani *et al.*, 2013) also reported that two foliar sprays of maleic hydrazide at 1250 ppm at 60 and 90 days after sowing would induce dormancy in non-dormant bunch type groundnut cultivar up to 30 days (4 weeks) and protects the seed from *in situ* germination due to untimely rains at the time of harvest of the crop.

(Jaydeva 2008) also tried to induce the dormancy in summer groundnut varieties *viz.*, TG-26, SB-XI & TAG- 24 by spraying Maleic hydrazide at different concentrations. It was observed that foliar spray of MH at 60 and 90 DAS @ 1000 ppm concentration was superior in inducing seed dormancy up to 40 days, irrespective of genotypes as compared to other concentrations of MH. Among the varieties the genotype TG- 26 and SB- XI responded well in getting induce the dormancy up to 30 days, where as in the genotype TAG-24 the dormancy induced was only for 5 days.

Thus maliec hydrazide was extensively used to induce fresh seed dormancy in bunch type groundnut cultivars.

31.7 Seed Deterioration in Storage

The storage potential of seed is a heritable character. It is envisaged that bulk of the carry over seeds would be stored in bags under ambient conditions because in India storage of bulk quantity of seeds under controlled conditions is neither economical nor practicable. Stored seeds of groundnut lose their viability during storage under ambient conditions. Groundnut farmers often store their seeds under ambient conditions where the temperature and humidity can be very high, particularly in countries with warmer climate. Oilseeds are recalcitrant in nature and due to the presence of oil content they develop rancidity and easy spoilage. In India, groundnut is grown in two seasons, and the postrainy season (rabi) harvest often suffers drastic loss of seed viability. Groundnut seeds are known to suffer loss of viability during storage and it depends on several crucial aspects, *viz.*, genotype, characteristics of seeds including their size, nature of storage material (pods or kernels), the chemical constituents of the seed etc. (Delouche *et al.*, 1973; Nautiyal *et al.*, 1990).

Storage potential of seeds of various oilseed crops ranged from 9 to 23 months under ambient condition. In groundnut, erect varieties showed quick increase in germinability up to 91 percent after two months of storage and decreased to 12 percent after eight months of storage. Where as in the semi-spreading group, germinability increased to 97 percent after three months of storage then gradually decreased to 16 per cent after eight months of storage (Swain *et al.,* 2001). (Nautiyal 2004) reported that groundnut seeds could be stored for six months under ambient storage conditions. Further, (Chowdhury *et al.* 2003) elucidated that the validity period fixed for 9 months was found to be appropriate for groundnut.

Seed tissues deteriorate due to ageing and there could be several reasons for such deterioration. The ultimate manifestation of seed deterioration is loss of its ability to germinate, but before that occur different biochemical and physiological changes at sub-cellular level affect the performance of the seed (Ellis and Roberts, 1980). Among the reasons for seed deterioration during the ageing process, lipid peroxidation mediated membrane damage is considered to be most significant (Koostra and Harrington, 1969). During storage, many polyunsaturated fatty acids found in seeds become highly susceptible to peroxidative degradation, in which not only is the lipid itself destroyed, but a complex series of reactions generate a variety of potentially toxic products. The consequences of lipid peroxidation for cellular functioning and survival are therefore severe. The peroxidative changes in the phospholipids also affect membrane integrity. Any loss of structural integrity of the cellular membrane has two major consequences. The cell is unable to respond osmotically, failing to maintain proper turgor, while a substantial efflux of seed metabolites possibly stimulates potentially damaging pathogens. In addition to these various individual effects, the age-induced deficiencies interact to induce cellular debility, which is poorly studied in cultivated groundnut and almost unknown in wild species (Adib sultana, 1994).

31.7.1 Fatty Acid Composition of Seed and Oil Quality in Storage

All fats and oils are composed of fat molecules known as fatty acids. The molecules can be classified into three categories depending on their degree of saturation *viz.,*saturated fatty acids, monounsaturated fatty acids, and polyunsaturated fatty acids. A fatty acid that is missing any hydrogen atoms is classified as being unsaturated. This includes all monounsaturated and polyunsaturated fatty acids. Vegetable oils are promoted as being healthy on the basis of their varying degrees of unsaturation. This can be measured by the number of grams of iodine absorbed across the double bonds by 100 g of oil. Reported iodine values for groundnut oil ranges from 84.5 to 103.0 (Badami *et al.,* 1980). The value is also expressed as percent of free fatty acids calculated as oleic acid. Further, acid values are used to measure the extent to which glyceride in the oil has been decomposed by lipase and other actions such as light and heat.

The oil from mature, freshly ripened groundnuts should contain less than 1% free fatty acids. Therefore the presence of higher levels of free fatty acids in fresh oil may be an indication of tissue damage and the subsequent degradation of lipids by lipolytic enzymes. This usually occurs during storage (Sanders, 1980a). The free fatty acid content of groundnut oil (as % oleic acid) was reported to range from 0.1% to 7.2% (Crawford and Hilditch, 1950).

Groundnut oil is relatively more stable than safflower and sunflower oil, which have higher content of polyunsaturated fatty acids and consequently groundnut oil has a longer shelf life. A major influence on oil storage stability is its fatty acid composition, especially the proportion of unsaturated to saturated fatty acids. For groundnut, an 0:L ratio of 1.6 and above is desirable for longer shelf life (Jambunathan,1991).

31.7.2 Effect of Seed Deteriration

(a) **Loss of Seed Viability and Vigor:** High seed moisture and relative humidity besides temperature were the factors considered detrimental to the viability of stored groundnut seeds, because equilibrium moisture content of seeds at a given temperature increased with relative humidity. (Ketring 1971), while working with Spanish type groundnut observed that RH was more important in deciding seed viability than temperature during storage. He observed that high RH induced various biochemical changes during storage, which lowered the quality of groundnut seeds.

(b) **Membrane Integrity:** High electrical conductivity values of the leachates were reported in aged seeds of groundnut by several researchers (Huang and Fu, 1991; (Chakrahorty *et al.,* 1991) suggesting a damage to the membrane integrity.

(c) **Change in Lipids and Lipid Peroxidation:** The chemical changes in lipids with seed deterioration usually involve breakage of the ester linkage between the acyl chain and glycerol backbone or attack of the unsaturated bonds of fatty acid chain. During ageing certain changes can also happen in the physical properties of lipid (Vertucci, 1992) such as a decrease in the energy associated with the lipid melting. The different explanations of decreased lipid levels and their constituents in aged seeds have been mostly directed towards the effects of lipid peroxidation or degradation by lipolytic enzymes.

(d) **Lipid Peroxidation:** Lipid peroxidation, the oxidative destruction of polyunsaturated fatty acids, is an uncontrolled, autocatalytic process leading to the formation of fatty acid hydroperoxides and to secondary products, including a wide range of aldehyde components. The essential mechanism of lipid peroxidation outlined by (Chessman 1993) indicated that the process starts with the abstraction of H_2 atom from the target fatty acid to form a lipid (fatty acid) radical. This process is known as *initiation.* The hydroxyl (HO) radical, most peroxyl radical (ROO) and most alkoxy radicals (RO) are all capable of oxidizing polyunsaturated fatty acids, while the superoxide radical ($O_2^{\cdot-}$) is not. It is impossible to be certain about the relative significance of the various possible initiating agents. The product of the initiation reaction is a fatty acid radical that rapidly rearranges to form a conjugated diene structure. The extremely rapid addition of oxygen to the fatty acid radical forms a lipid (fatty acid) peroxyl radical (LOO). This is capable of reacting with other polyunsaturated fatty acids, beginning a new chain of oxidation, thus forming a lipid hydroperoxide (LOOH) on the original polyunsaturated fatty acid and generating a new fatty acid radical. In the propagation stage of lipid peroxidation a new chain is initiated by a lipid peroxyl radical and the breakdown of lipid hydroperoxides to more radical intermediates.

Lipid hydroperoxide breakdown is important for two reasons. It generates radicals that propagate lipid peroxidation as it was already stated, and also generates non radical fragmentation such as aldehydes, many of which are biologically active. In biological systems with mixtures of different polyunsaturated fatty acids, lipid peroxidation will generate a mixture of hydroperoxides, the breakdown of each of which can produce a variety of radical species and aldehydes. Unfortunately, it is as yet not clear which of them are actually formed in the biological system, in what quantities they are formed, and what their biological properties are.

The first indication that lipid peroxidation can be a direct cause of seed deterioration came from the report of (Kaloyereas 1958) and since then there was several other findings that substantiate such a claim (Wilson and McDonald, 1986). The peroxidation in stored seeds has been considered to arise either through atmospheric autoxidation or through the agency of lipoxygenase, an enzyme present in many seeds which accelerates the rate of this reaction. Different polyunsaturated fatty acids possess different susceptibilities to peroxidation. The evidence of lipid peroxidation comes mostly from the analysis of the relative changes in the levels of unsaturated fatty acids. In most seeds the lipids that are at risk from autoxidation comprise oleate (18:1), linoleate (18:2) and linolenate (18:3) fatty acyl chains. The degree of un saturation has considerable influence on the rate of degradation. It is stated (Schaich, 1980) that 9, 12-linoleate, with a pair of double bonds that are methylene-interrupted, is degraded about 30 to 40 times faster than 9-oleate, which has only one double bond.

(Koostra and Harrington 1969) were the first to propose the oxidation of membranes as a major mechanism of seed deterioration and since then, considerable research work was carried out to identify the role of lipid peroxidation in seed ageing, most of it was reviewed by (Wilson and McDonald 1986). The various approaches to study seed deterioration included monitoring changes in lipid bond saturation, lipid and phospholipid content, release of free fatty acids and production of lipid peroxides as well as their breakdown products.

The analysis of fatty acid composition by gas chromatography suggested that seeds in dry storage tend to lose polyunsaturation over time (Priestley, 1986). In oilseeds, the common observations have been a decrease in the proportion of polyunsaturated fatty acids like linoleate and linolenate in the stored seeds. In stored groundnut seeds (Uematsu and Ishii 1981) recorded a downward shift in the amount of linoleate which was associated with loss of viability.

In stored seeds, lipid peroxidation has often been studied through analysis of lipid degradation products. For such purpose, a degradation product like malonaldehyde was used as an index of lipid peroxidation in aged seeds. Higher levels of malonaldehyde in aged groundnut seeds was reported by (Chakraborty *et al.* 1991). Seeds subjected to accelerated ageing also showed higher levels of malonaldehyde in soybean (Stewart and Bewley, 1980)

(e) **Enzyme activities:** The major enzymes in lipid degradation in deteriorating oilseeds are lipases and lipoxygenases. It was reported that lipases hydrolyze the ester linkages between fatty acyl chains and glycerol in triglycerides of storage lipids liberating free fatty acids and glycerol (St. Angelo and Ory, 1983). Free fatty acids could be toxic to cells and cause deleterious effects like uncoupling of oxidative phosphorylation in mitochondria (Karnshaw *et al.,* 1970), inhibit hill reaction in chloroplasts (Krogman and flagendorf, 1959) and denature soluble enzymes (Tortora *et al.,* 1978).

Lipoxygenase was found to be responsible for oxidation of polyunsaturated fatty acids and formation of hydroperoxides (St. Angelo and Ory, 1983). Hydroperoxides and their degradation products affect important cellular systems by denaturing proteins and DNA (Benson, 1990).

Phospholipases (an important lipid degrading enzyme) also play an important role during seed ageing and seed deterioration. Phospholipase A cleaves the ester bonds of the glycerol backbone liberating free fatty acids and lysophospholipids. Phospholipase D cleaves the polar head group to leave phosphatidic acid and liberate free fatty acids. These free fatty acids and lysophospholipids are the major components responsible for the increased membrane damage and consequent damage to the seed (Van Bilsen and Hoekstra, 1993).

Increase in different enzyme activities during storage and its association with loss of viability were reported in several oilseed crops. In groundnut (Chakraborty *et al.,* 1991), mustard, corn and soybean (Dey and Mukherjee, 1986) lipase activity showed an increase in the stored seeds along with an increase in free fatty acids.

31.7.3 Estimation of Stored Seed Quality using Acid and Peroxide Values

Hydrolysis of the ester linkages in the presence of lipase liberates free fatty acids which may accumulate leading to lowering of the pH of seed extract. By using pH as an indicator of free fatty acid content, several research workers observed an increase in the acid value with increasing periods of storage. The acid value is defined as the number of milligrams of potassium hydroxide required to neutralize the free fatty acids present in one gram of fat. It is a relative measure of rancidity as free fatty acids are normally formed during decomposition of oil glycerides. Increases in acid value were observed in oilseeds such as groundnut (Subbaraman and Selvaraj, 1989; Chakraborty *et al.,* 1991), sunflower, mustard and soybean (Dey and Mukherjee, 1986). This changed ratio was considered to be responsible for complete loss of viability. They also observed that 20 percent of the total fatty acids in aged seeds were in the free form compared with only 1-2 percent of free acyl units observed in highly viable seeds.

Further Peroxide value was also reported as an indicator of fat oxidation (Gray, 1978). An elevation of peroxide value, which correlates negatively with loss of viability was observed during prolonged storage of groundnut seeds (Mathur *et al.,* 1956). Thus acid and peroxide values could be considered as a means to estimate the seed quality under storage.

References

1. Abdel Rahman, A. H. Y. 1982a. Compositional study on some Egyptian peanut varieties. Rivista Italiana Delle Sostanze Grasse. 59(6): 287-288.
2. Abdel Rahman, A. H. Y. 1982b. Changes in chemical composition of peanut during development and ripening. Rivista Italiana Delle Sostanze Grasse. 59(6): 285-286.
3. Abrar, A. K and Jadhav, B. B. 1991. Effect of growth regulators, chemicals and temperature on dormancy in peanut. Ann. Pl. Physiol. 5(1): 36-42.
4. Awal, M. A and Ikeda, T. 2003. Controlling canopy formation, flowering, and yield in field-grown stands of peanut (Arachis hypogaea L.) with ambient and regulated soil temperature. Field Crops Res. 81: 121-132.
5. Badami, R. C., Shivamurthy, S.C., Joshi, M. S and Patil, K. B. 1980. Yield characters and chemical examination of five varieties of hybrid bunch type peanuts with respect to their parents. Fette Seifen anstrichmittel. 82(10): 400-401.
6. Bailey, W. K., Toole, E. H., Vivian, K., Toole and Drowne, M. J. 1958. Influence of temperature on the after-ripening of freshly-harvested peanuts. Proc. Am. Soc. Hort. Sci. 71: 422-424.
7. Basha, S. M. 1991. Deposition pattern of methionine-rich protein in peanuts. Journal of Agricultural and Food Chemistry. 39(1): 88-91.
8. Benson, E. E. 1990. The detection and characterization of free radical activity in stored seeds and in vitro plant germplasm. Free radical damage in stored plant germplasm. International Board for Plant Genetic Resources. 37-66
9. Bhapkar, D. G., Patil, P. S and Patil, V. A. 1986. Dormancy in groundnut. J. Maharashtra agric. Univ., 11(1): 68-71.
10. Boote, K.J.1982.Growth stages of Peanut (Arachis hypogaea L.), Peanut Science 9: 35-40.
11. Caliskan, S., Caliskan, M. E., Erturk, E., Arslan, M and Arioglu, H. 2008. Growth and development of Virginia type groundnut cultivars under Mediterranean conditions. Acta Agr. Scan. Section B: Plant and Soil Sci. 58: 105-113.
12. Chakraborty, M. K., Chaudhari, A. N., Shukla, Y. M and Patel, K.V. 1991. Biochemical characterisation of groundnut (Arachis hypogea L.) seed In relation to its viability. Indian Journal of Agricultural Sciences. 61(5): 332-334.
13. Chessman, K. H. 1993. Mechanisms and effects of lipid peroxidation. Molecular Aspects of Medicine. Pergamon Press. (14):191-197.
14. Chowdhury, D. S., Mathur and Jayprakash. 2003. Seed research on groundnut under NSP crops. Proceedings of National Workshop on Groundnut Seed Technology. Feb 6-7, 6-19.
15. Chung, C. H., Yee, Y. J. D., Kim, H., Kim, H. K and Chung, D. S. 1995. Changes of lipid, protein, RNA and fatty acid composition in developing sesame (Sesamum indicum L.) seeds. Plant Sci. 109: 237-243.

16. Conagin, C. H. T. M. 1957. Cytology of seed development in the peanut plant. Bragantia 16. 15-33.
17. Crawford, R. V and Hilditch, T. P. 1950. The component fatty acids and glycerides of groundnut oil. Journal of the Science of Food and Agriculture. 1: 372-379.
18. Delouche, J. C., Matthes, R. K., Dougherty, G. M and Boyd, A. H. 1973. Storage of seed in tropical and sub-tropical region. Seed Science and Technology. 1: 671-700.
19. Dey, G and Mukherjee, R. K. 1986. Deteriorative change in seeds during storage and its control by hydration - dehydration pretreatments. Seed Research. 14(l): 49-59.
20. Dwivedi, S. L., Nigam, S. N., Reddy, D. V. R., Reddy, A. S and Ranga Rao, G. V. (1995). "Progress in breeding groundnut varieties resistant to peanut bud necrosis virus and its vector," in Recent Studies on Peanut Bud Necrosis Disease: Proceedings of a Meeting, March 20, 1995, eds Buiel A. A. M., Parlevliet J. E., Lenne J. M., editors. (Patancheru: International Crops Research Institute for the Semi-Arid Tropics/Wageningen: Department of Plant Breeding, Wageningen Agricultural University;) 35-40.
21. Ellis, R. H and Roberts, E. H. 1980a. Improved equations for the prediction of seed longevity. Annals of Botany. 45: 13-30.
22. Gray, J. 1978. Measurement of lipid oxidation: a review. Journal of American Oil Chemists Society. 55: 539-546.
23. http://www.FAO.ORG/FAOSTAT database. accessed on 04/01/2011.
24. Huang, L and Fu, J. 1991. Early events in germinating peanut seeds in relation to the vigour. Acta Scientiarum Naturalium Universitatis Sunyatseni. 30: 98-105.
25. Hull, F. M. 1937. Inheritance of rest period of seeds and certain others characters in the peanut. Florida Agricultural Experiment B, station Bull., 314.
26. International Seed Testing Association (ISTA). 1985a. International rules for seed testing. Seed Science and Technology. 13:299-355.
27. Jagtap, P. B. 2000. Physiological maturity and seed dormancy studies in bunch type groundnut (Arachis hypogaea L.). M.Sc. (Agri.) Thesis, PGI, M.P.K.V., Rahuri.
28. Jambunathan, R. 1991. Groundnut Quality Characteristics. ICRISAT, conference paper CP-631
29. Jaydeva B. 2008. M.Sc. (Agri.) thesis, PGI, MPKV, Rahuri.
30. John, C. M., Seshadri, C. R and Rao, M. B. S. 1950. Dormancy of the seeds in the groundnut. Madras. Agr. J. 35: 159-67.
31. Kaloyereas, S. A. 1958. Rancidity as a factor in the loss of viability of pine and other seeds. Journal of American Oil Chemist's Society. 35: 176-179.
32. Ketring, D. L and Morgan, P. W. 1969. Ethylene as a component of emanations from germinating peanut seeds and its effect on dormant Virginia type sees. Plant Physiol. 44: 326-330.

33. Ketring, D. L and Morgan, P.W. 1970. Physiology of seeds. I. regulation of Dormancy in Virginia –type peanut seeds. Plant Physiol. 45: 268-273.

34. Ketring, D. L and Morgan, P. W. 1971. Physiology of oilseed. II. Dormancy release in Virginia-type peanut seeds by plant growth regulators. Plant Physiol. 47: 488-492.

35. Ketring, D. L and Morgan, P.W. 1972. Physiology of oilseed. IV. Role of endogenous ethylene and inhibitory regulators during natural and induced after ripening of dormant Virginia-type peanut seeds. Plant Physiol. 50: 382-387.

36. Khalfaoui, J. L. B. 1991. Inheritance of seed dormancy in a cross between two Spanish peanut cultivars. Peanut Sci. 18: 65-67.

37. Koostra, P.T and Harrington, J. F. 1969. Biochemical effects of age on membranal lipids of Cucumis sativus (L.) seed. Proceedings of International Seed Testing Association. 34: 329-340.

38. Krapovickas, A and Gregory, W. C. 1994. Taxonomia del genero Arachis (Leguminosae). Bonplandia. 8:1–186.

39. Krogman, D. W and Jagendorf, A.T. 1959. Inhibition of the hill reaction by fatty acids and metal chelating agents. Archives of Biochemistry and Biophysics. 80: 421-430.

40. Lin, H and Lin, C. Y. 1971. Studies on the seed dormancy on peanuts. III. inheritance of seed dormancy of peanuts. J. Agric. Res. 20: 49-53.

41. Liu, S.Y., Chan., K. L and Yang., J. H. 1984. Variation in oil content of the main oil crops of Taiwan. Felle Seifen Anstrichmittel, 86(12). 466-468.

42. Madhan Mohan, M and Nigam, S. N. 2013. Principles and Practices for Groundnut Seed Production in India. Information Bulletin No. 94. Patancheru, Andhra Pradesh, International Crops Research Institute for the Semi-Arid Tropics. 64.

43. Manonmani. 2002. Storability of dormant and non-dormant cultivars of groundnut. Seed Res. 30(1): 158-160.

44. Mathur, P. B., Prasad, M and Singh, K. P. 1956. Studies in the cold storage of peanuts. Journal of Science, Food and Agriculture. 7: 354-360.

45. Mathur, R. K., Manivel, P., Samdur, M. Y. and Bandopadhyay, A. 2000. Screening for fresh seed dormancy in Spanish bunch groundnut. J. Oilseed Res., 17(1): 181-182.

46. Michael E. Foley. 2001. Seed dormancy: an update on terminology, physiological genetics, and quantitative trait loci regulating germinability. Weed Science, 49: 305-317.

47. Nagarajan, S. S and Gopalakrishnan, S. 1958. Root-inducing substance in groundnut seeds. Current Sci. 27: 29-30.

48. Nautiyal, P. C., Ravindra, V and Bandyopadhyay, A. 1994. Peanut seed dormancy, ACIAR-Food Legumes News. 21: 2.

49. Nautiyal, P.C., Ravindra, V and Joshi, Y. C. 1990. Varietal and seasonal variation in seed viability in Spanish groundnut (Arachis hypogaca L.). Indian Journal of Agricultural Sciences. 60(2): 143-145.

50. Nautiyal, P. C. 2004. Issues related to maintenance of seed viability and regulation of dormancy in groundnut. Grountnut research in India.321-335

51. Pandya, R. B and Patel, V. J. 1986. Dormancy in kernels of Spanish and Virginia bunch varieties of groundnut. J. Oilseeds Res. 3(1): 19-27.

52. Pattee, H. E., Purcell. A. E and Johns, E. B. 1969. Changes in carotenoid and oil content during maturation of peanut seeds. Journal of the American Oil Chemists' Society. 46: 629-631.

53. Pickett, T.A. 1950. Composition of developing peanut seed. Plant Physiology. 25: 210-224.

54. Priestley, D. A. 1986. Morphological, structural and biochemical changes associated with seed ageing. Seed Ageing. Cornell University Press. 125-195.

55. Rajgopal, K. L and Chaajlmn, K. 2002. Influence of packaging media on the storability of groundnut (Arachi hypogoea L.) germplasm. Plant Genetic Resources Newsletter. 62-64.

56. Ramachandran, M., Lognathan, N. S., Sridharan, C. S., Chandrasekharan, N. R and Krishnaswami, P. 1967. Evolution of dormant bunch groundnut strains by hybridization. Indian J. Agric.Sci. 37: 428-436.

57. Ramanatha Rao, V and Murty, U. R. 1994. The groundnut crop – A scientific basis for improvement. Botany morphology and anatomy. 43–95.

58. Reddy, P. S., Zade, V. R and Desmukh, S. N. 1985. 1-19, A new Spanish bunch groundnut cultivar with fresh seed dormancy. J. Oilseed Res. 2: 103- 106.

59. Rodriguez, M. M., Basha, S. M and Sanders, T. M. 1989. Maturity and roasting of peanuts as related to precursors of roasted flavour. Journal of Agricultural and Food Chemistry. 37(3): 760-765.

60. Sanders, T. H. 1980b. Falty acid composition of lipid classes in oiis from peanuts differing in variety and maturity. Journal of the American Oil Chemists Society. 57(1): 12-15.

61. Sanders, T.H. 1980a. Effects of variety and maturity on lipid class composition of peanut oil. Journal of the American Oil Chemists' Society. 57(1): 8-11.

62. Savage, G. P and Keenan, J. I. 1994. The Composition and Nutritive Value of Groundnut Kernels. The Groundnut Crop: Scientific basis for improvement. London: Chapman and Hall.173-213.

63. Schaich, K. M. 1980. Free radical initiation in proteins and amino acids by ionizing and ultraviolet radiations and lipid oxidation. Pt. III. Free radical transfer from oxidizing lipids. CRC Crit. Rev. Food Sci. Nut. 13:189-244.

64. Sekhon, K. S., Ahuja, K. L and Sandhu, R. S. 1970. Chemical composition of raw and roasted peanuts. Indian Journal of Nutrition and Dietetics. 7: 243-246.

65. Sekhon. K. S., Ahuja. K. L., Jaswal, S and Bhatia. L. S. 1973. Variability in fatly acid composition in peanut. 2. Spreading group. Journal of the Science of Food and Agriculture. 24(8): 957-960.

66. Sevgi Caliskan, Mehmet Emin Caliskan M E and Mehmet Arslan. 2008. Genotypic Differences for Reproductive Growth, Yield, and Yield Components in Groundnut (Arachis hypogaea L.). Turk J Agric For TUB‹TAK. 32: 415-424.

67. Shelar, V.R., Karjule, A.P., Jayadeva, B. 2014, Induction of dormancy in groundnut- A review. Agicultural reviews.35(3): 216-224

68. Shibuya, T. 1932. Studies on the non germinating seeds of peanut, Arachis hypogaea L. Formosan Agr. Rev. 305: 65-75.

69. Shibuya, T. 1938. Forcing the germination of dormant seeds by means of growth hormone. J. Soc. Trop. Agr. Formosa 10: 1-8.

70. Shilman, F., Brand, Y., Brand, A., Hedvat, I and Hovav, R. 2011. Identification and Molecular Characterization of Homeologous Δ9-Stearoyl Acyl Carrier Protein Desaturase 3 Genes from the Allotetraploid Peanut (Arachis hypogaea). Plant Molecular Biology Reporter, 29, 232–241.

71. Simpson, G. M. 1990. Seed Dormancy in Grasses. New York: Cambridge University Press. 297.

72. Smith, B. W. 1954. Arachis hypogaea. Reproductive efficiency. Amer. J. Bot. 41:607-615.

73. Sreeramulu, N and Rao, I. M. 1971. Physiological studies on dormancy in seed of groundnut. Aust. J. Bot. 19: 273-280.

74. St. Angelo, A. J and Ory, H. L. 1983. Lipid degradation during seed deterioration. Phytopathology. 73: 115-:317.

75. Subbaraman, R and Selvaraj, J. A. 1989. Effect of method of shelling and pod moisture on viability and vigour of groundnut seed in storage. Seeds and Farms. 28: 11-16.

76. Sudha Rani, M., Rao, P. S., Kanaka Durga, K., Bharathi, V., Bharathi, K and Vishnuvardhan Reddy, A. 2013. Induction of dormancy in non-dormant varieties of groundnut. Journal of Cereals and Oilseeds. 4(7): 89-93.

77. Swain, S. K., Sahoo, P and Patnaik, M. C. 2001. Seasonal effect and genotypic variability for storability of seeds in groundnut (Arachis hypogaea L.). Seed Res. 29(1): 109-111.

78. Taira, H. 1985. Oil content and fatty acid composition of peanuts imported in Japan. Journal of the American Oil Chemists' Society. 62(4): 699-702.

79. Toole, V. K., Bailey, W. K and Toole, E. H. 1964. Factors influencing dormancy of peanut seeds. Plant physiol. 39: 822-832.

80. Tortora, P., Hanozet, G. M., Guerritoe, A., Vincenzinni, M.T and Vanni, P. 1978. Selective denaturation of several yeast enzymes by free fatty acids. Biochemica Biophysica Acta. 522: 297-306.

81. Uematsu, T and Ishii, K. 1981. Effects of storage conditions on the quality of peanuts. Bull. Coll. Agric. Vet. Med. 38: 125-134.

82. *Vaithialingam, R and Rao, J. S. 1973. Induction of dormancy in groundnut by pre-harvest foliar spray of M.H. Mad. Agric. J. 6(9-12): 1862-1863.*

83. *Van Bilsen, D. G. J. L and Hoekstra, F. A. 1993. Decreased membrane integrity in ageing Typha latifolia L. pollen: accumulation of lysolipids and free fatty acids. Plant Physiology. 101:675-682.*

84. *Varisai, M. S. M and Stephen Dorairaj, M. 1968. Screening the genetic stock of Arachis hypogaea L. for seed dormancy. Indian J. Agric. Sci. 38: 73-75.*

85. *Vertucci, C. W. 1992. A calorimetric study of the changes in lipids during seed storage under dry conditions. Plant Physiology. 99(1):310-316.*

86. *Waldron, R. A. 1919. The peanut Arachis hypogaea, its history, histology and utility. Penn*

87. *Wallerstein, I. S., Merin, U and Rosenthal. I. 1989. Comparison of kernels of three Virginia-type peanut cultivars. Lebensmitiel-Wissenchaft und Technologie. 22: 179-181.*

88. *Williams, J. H., Hilderbrand, G.L and Tattersfield, J.R. 1978. Effects of weather and genotype x environment interaction on yields of groundnut (Arachis hypogaea L.) Rhod. J. Agric. Res, 16: 193-204.*

89. *Wilson, D. O and McDonald, M. B. 1986. The lipid peroxidation model of seed ageing. Seed Science and Technology. 14: 269-300.*

90. *Woodroof, J. G. 1983. Peanuts Production, processing, products. 3rd edn, Avi Publishing Company, Inc., Westporl, Connecticut.*

91. *Young, J. H and Moore, R. P. 1972. Factors affecting planting quality of seed peanuts. II. Effect of harvest methods. Oleagineux. 27: 89-93.*

92. *Zade, R., Deshmukh, S. N and Reddy, P.S. 1986. Magnitude of dormancy in the released Virginia group cultivars of groundnut. Seed Res.14: 235-238.*

Chapter 32

Management of Viral Diseases of Groundnut

B. V. Bhaskara Reddy

Sr. Scientist, RARS, ANGRAU, Tirupati-517 502.

32.1 Introduction

Groundnut (*Arachis hypogaea* L.) commonly known as peanut, is one of the most important oil seed and food legume crop grown in more than 100 countries. It belongs to sub-family *Faboideae* of the family *Fabaceae.* Groundnut, a native of South America is an important oil seed crop grown in tropical and sub tropical regions of World. The most important groundnut producing countries are China, India, USA, Nigeria, Indonesia, Myanmar, Sudan, Senegal, Argentina and Vietnam. Groundnut crop is grown in an arca of 5.4 m ha with production of 5.5 m t in India (Anonymous, 2009). It is being grown throughout the year in India. The major groundnut producing states are Gujarat, Andhra Pradesh (A.P), Karnataka and Tamil Nadu. The productivity of groundnut was very low (857 kg/ha) in India as compared to world productivity (2149 kg/ha). The main reasons could be not only rainfed nature of crop but also biotic stresses like viral diseases namely stem and bud necrosis diseases caused by Tobacco streak virus (TSV) and Groundnut bud necrosis virus (GBNV) respectively (Reddy *et al.,* 2002). Groundnut is a major legume crop and it is susceptible to unusually broad range of diseases. The range of environments in which it is cultivated makes sure that the crop is constantly challenged by several pathogens like viruses, bacteria, fungal, phytoplasmas, nematodes and parasitic weeds. It is prone to infection by several groups of viruses and there by limiting the crop production and quality. About 31 viruses were reported to naturally infect groundnut around the world and 12 are reported from India (Table 32.1). Nine of them belong to genus Potyvirus, six to Tospovirus, two each to Cucumovirus, Pecluvirus, Sobemovirus and Umbravirus, and one each to Begomovirus, Bromovirus, Carlavirus, Ilarvirus, Luteovirus, Potexvirus, Rhabdovirus and Tymovirus. Of these, 19 were first isolated from groundnut, and remaining first isolated from other hosts but they commonly occur on groundnut. Diseases in groundnut caused by Cucumber mosaic virus (CMV), Groundnut bud necrosis virus (GBNV), Groundnut rosette virus (GRV), Indian peanut clump virus (IPCV), Peanut clump virus (PCV), Peanut mottle virus (PeMoV), Peanut stripe virus

(PStV), Tobacco streak virus (TSV) and Tomato spotted wilt virus (TSWV) are most economically important and are responsible for serious yield losses globally or regionally (Sreenivasulu *et al.,* 2008). They also known to naturally infect several other crops and inflict significant losses in them. In this chapter five viruses (peanut bud necrosis virus, peanut stem necrosis disease, peanut stripe, peanut clump and peanut mottle viruses) were discussed in detail. Eight groundnut infecting viruses, CMV, IPCV, PeMoV, PStV, PCV, Cowpea mild mottle - severe strain (CPPMV), Cowpea aphid-borne mosaic viruses (CABMV) and Peanut stunt virus (PSV) are seed-transmitted and they have quarantine importance (Sivaprasad and Sreenivasulu, 1996). The occurrence of groundnut bud necrosis virus has been reported from India and other Asian countries (Reddy, 1991). Early infection could causes losses up to 100 percent, while late infections might be responsible for a yield loss up to 70% (Narayanasamy, 1993).

***Table 32.1** Viral diseases of groundnut in India*

S. No.	Name of virus/ disease	Genus (Family)	Distribution	Transmission	Reference
1	Groundnut bud necrosis virus (GBNV)	*Topsovirus (Bunyaviridae)*	India, Nepal, China, Sri Lanka,	Sap, thrips (CP)	Ghanekar et al., 1979; Reddy *et al.*, 1992
2	Tobacco streak virus (TSV)	*Ilarvirus (Bromoviridae)*	Brazil, India	Sap, pollen	Prasada Rao *et al.*, 2003
3	Bean golden yellow mosaic virus (BGYMV)/ (Peanut yellow mosaic virus)	*Begamovirus (Geminiviridae)*	India	White flies	Sudhakar Rao *et al.*, 1980
4	Groundnut crinkle virus	*Carlavirus (Flexiviridae)*	China, India, Indonesia,	Sap, white flies, seed	Sivaprasad and Sreenivasulu, 1966
5	Groundnut veinal chlorosis virus (GVCV)	*Rhabdovirus (Unassigned)*	India, Indonesia	Ni	Naidu *et al.*, 1989
6	Groundnut yellow spot virus (GYSV)	*Tospovirus (Bunyaviridae)*	India, Thailand	Sap, thrips (CP)	Reddy *et al.*, 1991; Satynarayana et al., 1996a
7	Indian peanut clump virus (IPCV)	*Pecluvirus*	India, Pakistan	Sap, fungi, seed (up to 6%)	Reddy *et al.*, 1983, Reddy *et al.*, 1999
8	Peanut chlorotic streak virus (PCSV)	*Soymovirus (Caulimoviridae)*	India	Sap	Reddy *et al.*, 1993
9	Peanut chlorotic streak virus-vein banding strain	*Soymovirus (Caulimoviridae)*	India	Sap	Satyanrayana *et al.*, 1994a, 1994b

Table 32.1 ***Contd...***

S. No.	Name of virus/ disease	Genus (Family)	Distribution	Transmission	Reference
10	Peanut green mosaic virus (PGMV)	*Potyvirus (Potyviridae)*	India	Sap, aphids	Sreenivasulu *et al.*, 1981
11	Peanut mottle virus (PeMoV)	*Potyvirus (Potyviridae)*	Worldwide	Sap, aphids (NP), seed (up to 8.5%)	Rajeswari *et al.*, 1983
12	Peanut stripe virus (PSV)	*Potyvirus (Potyviridae)*	Brazil, China, India, US	Sap, aphids (NP), seed (up to 35%)	Demski *et al.*, 1988,

Note: NP-non-persistant transmission, C-circulative transmission, CP-circulative and propagative transmission; Ni- no information available.
(*Source:* Sreenivasulu, 2008; ICTV, 2010).

32.2 Groundnut Bud Necrosis Disease

The disease is widely distributed in India and is known to occur in Andhra Pradesh, Tamil Nadu, Karnataka, Maharashtra and Uttar Pradesh. The early incidence of disease resulted in 100 percent yield losses (Prasada Rao *et al.,* 1978). The TSWV on groundnut in Maharashtra was observed and the incidence ranged from 5- 28 percent in Kharif season (Mali and Patil, 1979). (Gopal and Upadhya 1991) studied the effect of bud necrosis disease on groundnut under field conditions and recorded yield losses upto 50 percent depending upon plant growth at the time of infection. Bud necrosis of groundnut was first reported from India (Reddy et al., 1968). Later this disease was reported from South and Southeast Asia including China, under several different names viz. bud blight, groundnut mosaic, groundnut ring mosaic, bunchy top, chlorosis, ring mottle, ring spot and spotted wilt from India (Reddy, 1988). The causal virus of this disease was initially identified as TSWV in India (Ghanekar *et al.*, 1979). After virus purification and characterization, the virus involved in bud necrosis was identified as a distinct tospovirus and named as GBNV, which is placed in Serogroup IV of the genus Tospovirus (Satyanarayana *et al.*, 1996). It also infects several other crops like mungbean, urdbean, cowpea, soybean, sunflower, carrot, tomato, potato, onion, taro, jute, calatrophis, jasmine (Reddy *et a*l,. 2014). GBNV has become increasingly important in the production of groundnut in South and Southeast Asia. Its incidence in groundnut in India is variable from field to field and from area to area even in a given season (from 70%). Crop losses worth up to US$ 89 million from India were reported (Reddy and Devi, 2003). Basic properties and ecology of GBNV are very similar to TSWV. GBNV is mechanically transmissible and also transmitted by thrips vector, *Thrips palmi* in a persistant manner (Reddy and Devi, 2003). Thrips can acquire the virus only in their larval stage but both nymphs and adults can transmit the virus. It is not seed transmitted in groundnut or other species that it infects. It has extensive host range covering cultivated crops, ornamentals and weeds (Reddy and Wightman, 1988).

32.2.1 Symptoms

The symptoms of bud necrosis caused by GBNV, and spotted wilt caused by TSWV are similar on groundnut. Symptoms first appear on young leaflets as chlorotic spots or mottling that may develop into chlorotic and/or necrotic rings and streaks. When temperatures are above 30°C, the petioles bearing fully expanded leaves with such initial symptoms usually become flaccid and droop. The necrosis from such leaves soon spreads to terminal growing buds. If the plant is young, it may become totally necrotic. Early infection causes stunting and sometimes proliferation of axillary shoots (Fig. 32.1). Leaflets on proliferated axillary shoots are small in size and exhibit puckering, chlorosis, mosaic, oak leaf line patterns and distortion of lamina. Entire plant shows bushy appearance and often confused with infection by other viruses like peanut stunt and groundnut chlorotic rosette. Such highly stunted bushy plants are usually difficult to notice in the fields as they often covered by the canopy of the adjacent healthy plants. The roots become dark, possess few nodules and are stunted. The early-infected plants never produce any worthwhile pods. Any seed produced on early-infected plants are small, shriveled and testae have red, brown or purple mottling. Late infected plants may produce seed of normal size but often testae mottled (Reddy, 1991).

Fig. 32.1 Symptoms of bud necrosis affected groundnut leaves

32.2.2 Management Approaches

The primary sources of virus inoculum include a range of hosts (crops, ornamentals, weeds), which can sustain virus infection and support thrips vector multiplication (Reddy *et al.,* 1983a). The incidence and progress of the disease is dependent on several environmental and cropping practices which influence multiplication and spread of thrips. *Ageratum conyzoides* has been shown to support GBNV and vector multiplication. Primary spread is by thrips coming from other crops or weeds. Secondary spread from infected groundnut plants within a field was not established conclusively. Disease incidence continues to depend on infection by viruliferous thrips that acquire the virus from alternative hosts and thus bud necrosis disease of groundnut is mostly monocyclic type. Management of bud necrosis disease depends upon the control of thrips vectors.

Even though several weeds have been identified as sources of both virus and vectors, their eradication in the tropics is not practical. Insecticidal control of thrips was not effective in reducing the incidence. Cultural practices found to be more promising in reducing the disease incidence (Reddy, 1998). Depending on the arrival of principal thrips vectors the sowing dates in the growing season in endemic areas need to be adjusted so that the good crop canopy can develop before infection occurs and thus disease incidence can be reduced. Intercropping of groundnuts with quick growing cereals like sorghum, pearlmillet or maize can reduce the disease incidence by acting as a barrier for thrips colonization (Bikas *et al.*, 2014). Close spacing was effective for the groundnut crop in peninsular and central India as increasing plant density compensated yield loss due to bud necrosis disease. Optimum plant population will minimize the incidence of the disease. Seed treatments with Imidacloprid @2g/Kg seed may help to reduce incidence at early stage. Spraying once at 20-25DAS with systemic insecticides may reduce the incidence of disease.

Rogueing of early-infected plants in the fields can create gaps, which can lead to increased incidence. Work at ICRISAT has identified some groundnut cultivars and breeding lines that have field resistance to GBNV. They are ICGV 86029, 86031, 86388, 91239, 91245, 91246 and 91249. GBNV resistant groundnut cultivars ICGS 44, ICGS 11 were released in India. (Bera *et al.*, 2014) identified several inter specific material that are immune/resistant to PBNV by phenotyping and genotyping with SSR markers at DGR, Junagadh.

32.3 Peanut Stem Necrosis Disease

PSND is caused tobacco streak virus (TSV) and it was reported on groundnut from India, Pakistan, South Africa and Brazil, but it has become economically important recently in Southern India. It was first recognized in 2000, infecting over 2,25,000 hectares of groundnut in Andhra Pradesh, India, and resulted in yield loss exceeding US $ 65 million. It has a wide host range, infecting more than 200 plant species belongs to 30 dicotyledonous and monocotyledonous plant families and it is also reported from more than 26 countries Worldwide. In India, TSV was first identified on sunflower in (Karnataka in 1997 Singh *et al.,* 1997) and peanut in Ananthapur district of Andhra Pradesh during 1999-2000 (Prasada Rao *et al.,* 2000; Reddy *et al.,* 2002).

32.3.1 Symptoms

TSV infection at early stage of plant growth results in severe necrotic symptoms on leaves, petioles, and stems leading to premature death of the plant. Virus infection at later stages of plant growth (>40 days old) results in partial necrosis on leaves and main stem, proliferation of auxillary buds, and drastic reductions in pod yield. Some plants may not become necrotic, but severally stunted, having small leaves with or without chlorosis. Necrotic spots also occur on the shells of the pods. In severe infection the entire plant becomes necrotic and dies prematurely. In case of late infection, symptoms are confined to only terminal leaflets on a single branch, which may show necrosis. TSV symptoms in groundnut are often confused with those caused by GBNV. Necrosis on stems and petioles and necrotic spots on shells are characteristic of TSV infection (Fig. 32.2). Whereas GBNV induces chlorotic ring spots on leaves and cause proliferation of axillary shoots (Vemana and Jain, 2012). Premature death of plant was the main reason for enormous yield losses during the stem necrosis disease epidemics.

32.3.2 Host range

TSV has a broad host range and known to infect many cultivated plants and weed species belonging to families such as Asteraceae, Brassicaceae, Cucurbitaceae, Rosaceae, Solanaceae, Liliaceae, Malvaceae and other under natural and experimental conditions. The natural host range is listed as *Dahlia* spp, *G.herbaceum, Melilotus albus, Trifolium pretense, P.vulgaris, Glycin max, Rosa setigera, N. tabacum, A. officinalis*, Cotton, Cowpea, Groundnut, Castor, fieldbean, Guar, Onion, Kenaf, Sunflower, Okra, Soybean, Marigold, Cowpea, Mungbean, Safflower, Niger, Sesame, Ixora, Crossandra, *X.strumarium, C. bonplandium, Achyranthus aspera, Acanthosperum hispidium, Commelina benghalensis, T. procumbens, Acalypha indica, Abutilon indicum, Calotropis gigantea, Cleome viscose, Croton sparsiflorus, Digera arvensis, E.hirta, E.geniculata, Lagasca mollis, Leucas aspera, Urdbean, guizotia abyssinica, Commenlina benghalensis, Digera arvensis, Euphorbia hirta, Euphorbia geniculata, Lagasca mallis, Leucas aspera, Parthenium hysterophorus and Tridax procumbens* (Prasad Rao *et al.*, 2003, Reddy *et.al,* 2013, (Sivaprasad *et al.*,2011; 2014).

Fig. 32.2 Peanut stem necrosis disease affected leaves, plants and pods

32.3.3 Transmission

TSV was transmitted by vector, mechanical inoculation, grafting and pollen, but not transmitted by contact between plants and dodder (*Cuscuta campestris*) and also not by contact or soil. The major transmission of TSV is by infected pollen spread by wind or carried by insects. A number of thrips species are known as vectors of TSV. Adults and nymphs of several thrips species were shown to transmit TSV infected pollen from the infected plants by a mechanical mechanism where the virus from pollen carried externally or released from inside the pollen infects plants through feeding wounds caused by the thrips. A similar mechanism was observed in groundnut by three thrips. *F.schultzei* (Tryborn), *S. dorsalis Hood* and *M.usitatus* (Bagnall) (Reddy *et al.,* 2002; Prasada Rao *et al.,* 2003)

Seed transmission of TSV was reported in several crop species elsewhere in the world. Seed transmission is not found in naturally or experimentally infected groundnut, sunflower, parthenium or several other annual crops infected with TSV in India (Prasada Rao *et al.,* 2003, Vemana and Jain, 2012). The seed transmission of TSV to an extent of 2.7-65.7% in cucumber and gherkin in Southern Karnataka was observed (Krishanreddy *et al,.* 2003).

32.3.4 Management Approaches

Diseased plants in the farmers' fields are found to be usually high near the field bunds or waste lands with weeds. Several other crops and weeds have been identified as sources of virus and thrips. Transmission can only occur through infected pollen with the aid of several thrips species (Prasada Rao *et al.*, 2003). Groundnut seed does not carry infectious virus to initiate primary infection cycle. *Parthenium*, a ubiquitous weed, harbors both the virus and the thrips that visit its flowers carry the virus. TSV infected groundnut plants may not serve as a source for secondary spread in the same field or for primary spread to the near by fields because pollen from groundnut flowers is not an efficient source of inoculum. TSV incidence in groundnut fields was correlated with the infected parthenium near the field bunds (Prasada Rao *et al.,* 2003). Removal of the weeds including Parthenium from the vicinity of the groundnut fields is expected to reduce the disease incidence in the fields. Physical barriers that limit the movement of thrips have been shown to reduce TSV incidence. Intercropping of groundnut with maize, pearl millet, pigeonpea, and sorghum (3:1) significantly reduced TSV incidence. Rogueing of early-infected plants may not limit or restrict further spread of the disease. Cultivation of groundnut near sunflower and marigold should be discouraged because they act as a source of virus and/or thrips. Maintenance of sufficient plant density is important to discourage landing of thrips. Seed treatment with systemic insecticides (imidacloprid) may prevent vector infestation at early stages of crop growth. Limited germplasm screening revealed that groundnut varieties ICGV 92267, 99029, 01276 and ICG 94379 show consistently low disease incidence. Efforts are underway to deploy resistance genes from wild species (*Arachis chacoense*) to cultivated species and transgenic resistance to TSV by exploiting its coat protein gene.

32.4 Peanut Clump Virus (Peanut Clump)

Clump disease was reported from Asian countries: India (from Andhra Pradesh, Gujarat, Punjab, Rajastan and Tamil Nadu states) and Pakistan (Sindh and Punjab states); and West African countries (Reddy *et al.,* 1999). It is transmitted through seed and by soil-borne plasmodiophoromycete fungi, *Polymyxa graminis* and by mechanical sap inoculation (Reddy *et al.,* 2005). The viruses survive in resting spores of the fungus and contribute to the spread of the disease. The clump virus that occurs in Indian sub-continent is referred as Indian peanut clump virus (IPCV) where as the virus that occurs in Africa is called as Peanut clump virus (PCV). Both are important because they not only infect groundnut but also several other economically important crops like sugarcane, few cereals and millets and pigeonpea. In groundnut almost 100% crop loss results when the disease occurs in the early growing season, and up to 60% yield loss in late infected plants (Reddy, 1991). The annual loss due to this disease on global scale has been estimated to exceed US$ 38 million (Reddy and Devi, 2003). Further, clump disease also has quarantine importance, as the causal viruses are transmissible through groundnut seed, and also thorough seed of pearl millet, finger millet, foxtail millet, wheat and maize (Delfosse *et al.*, 1999).

32.4.1 Symptoms

The symptoms of clump disease are often confused with green rosette disease. Affected plants usually occur in patches and they are highly stunted and appear green even at crop maturity stage developed. Plants affected by clump disease are conspicuous in the field because of their severe stunting and dark green appearance. Initial symptoms appear on young leaflet as mottling, mosaic and chlorotic rings, but later turn dark green with or without faint mottling as the leaves mature (Fig. 32.3). Early infected plants become severely stunted. Late infected plants may not show conspicuous stunting but appear dark green with faint mottling on younger leaflets. In late infected plants, clumping may be restricted to few branches. Infected plants become bushy and produce several flowers. Early infected plants may not produce any pods and late infected plants may produce poorly developed pods (Reddy *et al.* 1988). These plants often occur in patches and the disease reoccurs in the same area of the groundnut field in successive years.

Fig. 32.3 Groundnut leaves affected by peanut clump virus

32.4.2 Management Approaches

PCV is soil-borne and transmitted by *P. graminis*. The vector in the soil, survives as highly resistant resting spores, responsible for the patchy appearance of the disease and annual recurrence. Eventhough the fungus can transmit the virus to dicot plants, it does not extensively colonize their roots and hence resting spores are seldom detected in such plants. Naturally virus infected groundnut plant roots fail to induce the disease when incorporated into sterile sand, whereas infected sorghum and pearlmillet roots could induce disease. For these reasons, dicot plants are considered as 'fortuitous' hosts that are unlikely to contribute to build-up of virus inoculum. On the other hand monocot crops like maize, pearlmillet and sorghum are regarded as 'preferred' hosts for *P. graminis* because of its high multiplication in them. The virus is also transmitted through the seed of cereal crops. The clump disease often occurs in groundnut crops raised during the rainy season in India. During this season the ambient temperatures range from 25-40°C. The optimum temperature for infection by fungus vector is between 27° and 30°C. Below 23°C infection does not occur and fungus development is delayed, and this could be the reason for the absence of the clump disease in the crops raised during post-rainy season in India, during this season ambient temperatures range between 15-30°C. Rainfall also plays significant role in natural infection by *P. graminis* and *P. graminis* is known to be favored by alternate watering and drainage and thus the disease commonly occurs in sandy soils. Continuous cropping with groundnut, cowpea and pigeonpea is likely to decrease the inoculum in the soil. On the other hand, intercropping with susceptible cereal crops can contribute to increase in incidence and spread of the disease. Hence, it was suggested that pecluviruses are not typical groundnut viruses, but actually graminaceous viruses that possibly coevolved in tropical and sub-tropical areas with wild grasses and field crops such as millets and sorghum (Delfosse *et al.,* 2002). Clump disease is difficult to control as it is a persistant soil-borne disease. However, its satisfactory control can be achieved by adapting cultural practices over a period of time.

The studies on disease epidemiology in India have lead to formulation of the following cultural practices: (a) early sowing of groundnut crop before the onset of monsoon rains; (b) use of pearlmillet as bait crop to reduce the inoculum load in the soil. To achieve this, bait crops are advised to plant soon after the onset of monsoon preferably under irrigation and up rooted in three weeks after germination; (c) avoiding rotation with highly susceptible cereal crops such as maize and wheat; (d) sowing peanut during post rainy season; (e) continuous cropping with dicot hosts to reduce the inoculum in the soil and thus low incidence of disease and; (f) soil solarization during hot summer months. This is achieved by covering well-irrigated soils with a transparent polythene sheet for atleast 3 weeks. Soil biocides are effective in reducing disease incidence but they are not economical to adopt.

32.5 Peanut Stripe Virus (Peanut Stripe)

First reported from China and later (Demski *et al.* 1984) detected potyvirus from peanut seed imported from China and named it 'Peanut stripe virus' (PStV). This virus infection is endemic in Southeast Asia, China and in majority of groundnut growing countries, including India (McDonald *et al.*, 1998). Yield reductions by PStV in groundnut in Georgia were about 7% in experimental plots (Lynch *et al.,* 1988), but yield losses can reach up to 70% in early-infected plants. In northern China annual yield loss estimated at 200,000 tonnes of pods was reported. The seedborne nature of the virus in groundnut is of considerable interest in the international exchange of groundnut germplasm, as this means of virus transmission can lead to the introduction of new strains of the virus into groundnut growing areas. PStV naturally infects not only groundnut but also a few other crops like cowpea, soybean and sesamum and a few legume weeds, *Centrosema pubescens, C. macrocarpum, Calapogonium caeruleum, Crotalaria striata, Desmodium siliquosum and Peuraria phaseoloides* (Sreenivasulu *et al.*, 1992). PStV has moderate experimental host range (Demski *et al.*, 1984). It is sap transmissible and also through groundnut seed upto 37%. However, in naturally infected groundnut plants seed transmission rates are between 0-7%, generally less than 4%. Aphids like *Aphis craccivora, A. gossypii* and *Myzus persicae* transmit the virus in a non-circulative, non-persistant manner.

32.5.1 Symptoms

The virus initially causes stripes and green vein banding symptoms along lateral veins of groundnut leaflets. The virus isolates from different peanut growing regions of the world can induce diverse and distinct symptoms. The first recognized stripe isolate produces discontinuous stripes along the lateral veins on young leaflets of peanut. Older leaflets exhibit striping, mosaic in the form of green islands and an oak leaf pattern. The infected plants are slightly stunted. Most other reported PStV isolates produce initially chlorotic flecks followed by mild mottle, blotch, or chlorotic ring mottle symptoms (Sreenivasulu *et al.*, 2008). The necrotic strain reported from Taiwan produces necrotic lesions on leaves, with the necrosis later extended to the mid-ribs, petioles and sometimes to the stems. This leads to stunting, severe mosaic and systemic foliar distortion or stripe symptoms (Fig 32..4) .

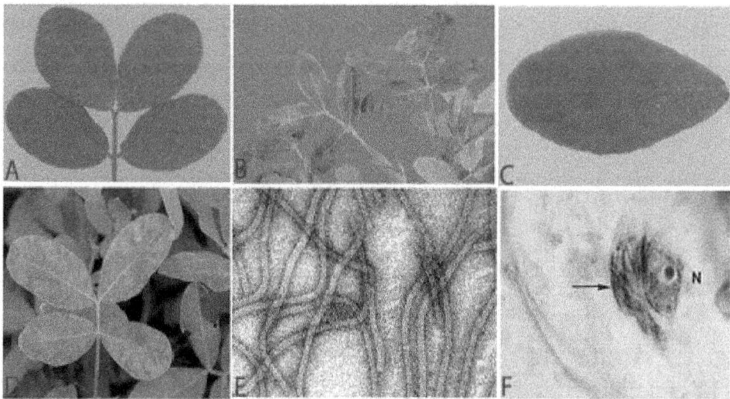

Fig. 32.4 Groundnut leaves affected by peanut stripe virus

32.5.2 Management Approaches

In nature PStV is transmitted in groundnut by seed and aphids in a non-persistant manner. The primary cycle of the disease is initiated through seed-borne infection. In Thailand and USA certain weeds like *Centrosema pubescence, C. macrocarpum, Calopogonium caeruleum, Crotalaria striata, Desmodium siliquiosum Peuraria phaseoloides* and beggar weed have been reported as natural susceptible hosts. In the absence of seed-borne virus, primary cycle can be initiated by aphids that acquire the virus from weed hosts. Secondary spread in the peanut fields is mainly by aphid vectors. Since the primary source of virus inoculum is seed, planting should be done with seed lots obtained from disease-free areas. Attempts made to control PStV by using 10% milk suspension, metasystox or milk alternated with metasystox or pyrimidine carbamate (systemic aphicide) was ineffective (Demski *et al.,* 1993). Enforcing of strict quarantine regulations in countries where the virus is known to be restricted at certain locations is important to avoid introduction of the virus into the virus-free locations. Only certified groundnut seed is to be moved between the locations or countries. Application of plastic film mulch in groundnut fields in China is found to reduce PStV incidence. Resistance to this virus could not be found in cultivated peanut.

32.6 Peanut Mottle Virus

This virus was first reported from USA and is present in all major groundnut-growing countries. It also naturally infects other crops like cowpea, French bean, soybean and white lupine (Brunt et al., 1996). (Demski and Reddy 1997) have considered this virus to be economically important globally. In India in susceptible groundnut cultivars crop losses may reach upto 40 %. Further, it has quarantine importance, as the virus is seed-transmitted. This virus has been identified as a distinct species in the genus Potyvirus, family Potyviridae. Peanut mottle virus (PeMoV) is easily sap transmitted and transmitted through the groundnut seed to variable levels. This virus is also transmitted in a non-persistant manner by *Aphis craccivora, A. gossypii, Myzus persicae, Hyperomyzus lactucae, Rhopalosiphum padii and R. maidis.* When groundnut plants infected with PeMoV and PStV are used as a virus source plants, *A. craccivora* and *M. persicae* are able to transmit either PeMoV or PStV but not both viruses together (Sreenivasulu and Demski, 1988).

32.6.1 Symptoms

PeMoV produces a range of symptoms in groundnut. The widely occurring mild mottle isolate produces mild mottle or mosaic of irregular dark green islands on young leaflets (Fig.32.5). In older leaflets the above symptoms are not obvious but can be seen in transmitted light. In some genotypes, interveinal depressions and inward rolling of margins of leaflets can occur. Infected plants are not usually stunted. The number of root nodules in diseased plants is found to be less, and the

Fig. 32.5 Groundnut leaves affected by peanut mottle virus

number and size of pods are reduced. Different strains of the virus cause varied symptoms. Symptoms in groundnut caused by chlorosis and necrosis strains of PeMoV mimic the symptoms caused by TSWV (Sreenivasulu *et al.,* 1988).

32.6.2 Management Approaches

The low level transmission of the virus through the seed in peanut (0-8.5%) is contributing for the primary spread of the virus. Alternate crops (cowpea, soybean, clover, peas, navy bean, French bean, white lupine) and weeds (*Desmodium, Cassia* sps.), and aphids are facilitating the survival and spread of the virus in nature (Demski, 1975). As the seed is identified as a primary source of virus inoculum, genotypes that do not transmit the virus through the seed, such as ICG 2716 (EC 76446-292), ICG 7013 (NCAC 17133) and ICG 1697 (NCAC 17090), are useful in containing the spread of the virus. These lines were used in conventional breeding programmes to transmit the non-seed transmissible trait to high yielding groundnut cultivars. The seed of advanced breeding lines from these crosses has been tested for frequency of the virus transmission. Two non-seed transmitting high yielding groundnut genotypes (ICGS 65 and ICGS 76) were identified. Usually cultivated groundnut is susceptible to PeMoV. (Kuhn *et al.* 1978) have identified two groundnut genotypes (PI 261945 and PI 261946) that are tolerant to PeMoV infection in USA. Later on in India also, the groundnut genotypes, ICG 5043 and NCAC 2240 were found to be tolerant with no significant effect on yield after virus infection. High yielding groundnut genotype (ICG 89336) with tolerance to PeMoV is available. Identified source of resistance in wild species but its transfer to cultivated species under progress.

References

1. *Bera, S.K, Jignesh H. Kamdar, Anil Kumar Maurya and Pitabash Dash 2014. Molecular diversity and association of simple sequence repeat markers with bud necrosis disease in interspecific breeding lines and cultivars of peanut (Arachis hypogaea L.). Australian Journal of Crop Sciences: 8: 771-880.*
2. *Bhat, A.I., Jain, R.K., Varma, A. and Lal, S.K. 2001 Tospovirus(es) infecting grain legumes in Delhi-their identification by serology and nucleic acid hybridization. Indian Phytopath. 54: 112-116.*
3. *Bhat, A.I., Jain, R.K., Varm, a A. and Lal, S.K. 2002. Nucleocapsid protein gene sequence studies suggest that soyabean bud blight is caused by a strain of groundnut bud necrosis virus.* Curr Sci. *82: 1389-1392.*
4. *Bikas Mandal., Reddy, K.M, Ravi, K.S, Kumar, K, Jain, R.K, Pappu H, 2014. Emerging problems of tospoviruses and their management in Indian sub continent Plant Dis. 86: 4.*
5. *Ghanekar, A.M., Reddy, D.V.R., Izuka, N., Amin, P.W. and Gibson, R.W. 1979. Bud necrosis of groundnut (A. hypogaea) in India caused by tomato spotted wilt virus. Ann. Appl. Biol 93: 175-179.*

6. Krishnareddy M, Lakshmi Raman D, Jalali S, Samuel DK. 2003. Outbreak of Tobacco streak virus causing necrosis of cucumber (Cucumis sativus) and gherkin (Cucumis anguria) in India. Plant Dis. 87:1264-1270.
7. Lava Kumar P, Prasada Rao RDVJ, Reddy AS, Jyothirmai Madhavi K, Anitha K, Waliyar F. 2008. Emergence and spread of tobacco streak virus menace in India and control strategies. Indian J Plant Protect. 36:1-8.
8. Prasada Rao R.D.V.J, Reddy A.S, Chander Rao S, Varaprasad K.S, Thirumala Devi K, Nagaraju, Muniyappa V, Reddy D.V.R. 2000. Tobacco streak ilarvirus as casual agent of sunflower necrosis disease in India. J Oilseed Res. 17:400-401.
9. Prasada Rao R.D.V.J and Reddy A.S. 2005. Viral diseases of groundnut. In: Diseases of Oil Seed Crops (edi. Saharan et al.,), Indus Publishing Co, New Delhi, pp153-175.
10. Reddy AS, Prasad Rao RDVJ, Thirimala Devi K, Reddy SV,Maya MA, Roberts I, Satyanarayana T, Subramaniam K, Reddy DVR. 2002. Occurrence of Tobacco streak virus on peanut (Arachis hypogaea) in India. Plant Dis. 86: 173-8.
11. Reddy, B.V.B, Prasanth, L, Sivaprasad, Y, Sujitha, A and Giridhar Krishna, T. 2013. First report of tobacco streak virus infecting castor bean. Journal of Plant Pathology Doi: 10.4454/JPP.V96I2.004.
12. Reddy, B.V.B, Prasanthi, L, Sivaprasad, Y, Sujitha,A and Giridhar Krishna. T 2013. First report of tobacco streak virus infecting fieldbean (Lablab purpureus L). New Disease Reports 28:21.
13. Reddy, B.V.B, Prasanth, L, Sivaprasad, Y, Sujitha, A and Giridhar Krishna, T. 2013. Genetic variability of tobacco streak virus in southern India based on the analysis of coat protein gene. Archives of Phytopathology and Plant Protection. 2:172-179.
14. Reddy, B.V.B., Sivaprasad, Y. and Sai Gopal, D.V.R. 2011. First report of Peanut bud necrosis virus infecting Calotropis gigantea. Journal of Plant Pathology. S4. 63-S4.89.
15. Reddy, D.V.R. 1991. Bud Necrosis: A Disease of Groundnut Caused by Tomato Spotted Wilt Virus Information Bulletin no. 31, ICRISAT.
16. Reddy DVR, 1991. Groundnut viruses and virus diseases; distribution, identification and control. Rev Plant Pathol, 70: 665-678.
17. Reddy, D.V.R. 1988. Groundnut virus disease occurring in India. In: Monograph of groundnut, pp 508- 525) (eds P.S. Reddy). Indian Council of Agricultural Research, New Delhi, India.
18. Reddy, D.V.R. and Devi, K.T. 2003. Peanuts. In: Viruses and virus-like diseases of major crops in developing countries, pp 397-423. (Eds. Gad Loebenstein and George Thottappilly). Kluwer Acadamic Publishers, Dorderecht.

19. *Reddy D.V.R., Amin, P.W., Mcdownald, D. and Ghanekar, A.M. 1983. Epidemiology and control of groundnut bud necrosis and other diseases of legume crops in India caused by tomato spotted wilt virus. In: Plant virus epidemiology, edi. RT Plumb, JM Thrash, pp: 93-102.*

20. *Reddy, D.V.R., Ratna, A.S., Sudarshana, M.R. and Kiran, K. 1992. Serological relationships and purification of bud necrosis virus, a tospovirus occurring in peanut in India. Ann. Appl. Biol. 120: 279-286.*

21. *Reddy, M, Reddy DVR, Appa Rao A (1968) A new record of virus disease on peanut. Plant Disease Reporter. 52: 449-450.*

22. *Sivaprasad, Y., Bhaskara Reddy, B.V., Sujitha, A, and Sai Gopal, D.V.R. 2014. Sequence diversity of the nucleoprotein gene of* Peanut bud necrosis virus *isolates from South India. Journal of Phytopathology 162:542-547.*

23. *Sivaprasad, Y., Bhaskara Reddy, B.V., Naresh Kumar, C.V.M., Raja Reddy, K. and Sai Gopal, D.V.R. 2011. Jute (Corchorus capsularis): a new host of Peanut bud necrosis virus. New Disease Reports. 23, 33.*

24. *Sivaprasad, Y., Bhaskara Reddy, B.V., Naresh Kumar, C.V.M., Raja Reddy, K. and Sai Gopal, D.V.R. 2011. First report of Groundnut Bud Necrosis Virus infecting Taro (Colocasia esculenta). Australasian Plant Dis. Notes. 6:30-32.*

25. *Sreenivasulu, P, Reddy, C.V.S, Ramesh, B and Kumar, P.L. 2008. Important virus diseases of groundnut. In: Rao, G.P, Paul Khurana, S.M, and Lenardo, S.L (edi), Characterization, diagnosis and management of plant viruses, Studium Press, LLC, Houston, pp 47-98.*

26. *Vemana, K and Jain, R.K. 2012. Biological similarities and differences between Tobacco streak virus and Groundnut bud necrosis virus infecting groundnut (Arachis hypogea). Indian Phytopathology.65:177-182.*

Chapter 33

Preparation of Microbial Biocontrol Agents for Insect Pest Management in Groundnut

K. Devaki, T. Muralikrishna and A. Ramakrishna Rao

Scientist (Entomology), IFT, RARS, Tirupati.

33.1 Introduction

Over reliance on broad spectrum pesticides has been severely condemned in different parts of the world after International Conference on Chemicals Management. Since then, an alternative eco friendly strategy for the management of noxious insect pests has been explored to trim down the harmful effects of chemical insecticides on humanity. Studies of biodiversity in agro ecosystems and the delivery of ecosystem services to agricultural products have usually ignored the contribution of entomopathogens in the regulation of pest populations (Ritu *et al.*, 2012). In recent years, crop protection based on biological control of crop pests with microbial pathogens like virus, bacteria, fungi and nematodes has been recognized as a valuable tool in pest management (Anand *et al.*, 2009; Rao *et al.*, 2006). The appropriate use of eco friendly microbial biopesticide can be engaged in recreation of sustainable organic crop production by providing a stable pest management program.

33.1.1 Status of Biocontrol Agents as Pesticides

Insects, like other organisms, are susceptible to a variety of diseases caused by bacteria, viruses, fungi and protozoans, and these pathogens are exploited for biological control of insect pests through introductory or inundative applications. Microbial pathogens of insects are intensively investigated to develop environmental friendly pest management strategies in agriculture and forestry. In India, mostly imported products of *Bacillus thuringiensis kurstaki* (*Btk*) have been used, which are expensive and there is an urgent need to develop aggressive indigenous *Bt* strains against various pests. Baculoviruses comprising Nuclear Polyhedrosis Virus (NPV) and Granulosis Virus (GV) have been successfully used as insect pathogens because of their high virulence and specificity. NPV and GV formulations are used for lepidopteran pests like *Helicoverpa armigera* (HaNPV) and *Spodoptera litura* (SlNPV) in India, besides *Anticarsia gemmatalis* NPV in

Brazil, *Lymanttria disper* NPV, *Orgyia pseudotsugata* NPV in USA and GV of *Pieris rapae* in China. Lack of easy mass multiplication methods for the commercial production of baculoviruses calls for R & D to develop production in insect tissue cultures. Entomopathogenic fungi like *Beauveria bassiana, B. brongniartii, Metarhizium anisopliae, M. anisopliae* var. *acridium, Lecanicillium* spp., *Hirsutella thompsonii, Nomuraea rileyi* and *Isaria fumosorosea* are gaining importance in the crop pest control in recent years due to the simpler, easier and cheaper mass production techniques (Ramanujam *et al*., 2014) .

The growth rate of the bio-pesticide industry has been forecasted in the next 10 years at 10-15 per cent per annum in contrast to 2-3 per cent for chemical pesticides (Men, 1996). Some of the successful and large-scale pest management programmes using microbial insecticides include management of *Lymantria spp.* outbreak in forestry in Poland and North America with the products of *B. thuringiensis* subspecies *kurstaki (Btk)* (Butt *et al*., 1999), Pine caterpillar management with *B. bassiana* in China (Li *et al*, 2010), locust control with *Metarhizium anisopliae* var. *acridium* in Africa, Australia and China (Lomer et al, 2001), sugarcane spittle bug management by *Metarhizium anisopliae* in Brazil, rubber lace bug control with *Sporothrix insectorum*, corn borer management with *B. bassiana* in Europe and USA (3), and European pine sawfly management with NPV in Europe and North America (Lord, 2005). Several hundreds of commercial products of fungi, bacteria and viruses are available worldwide for the biological control of insect pests in agriculture and forestry. In India, microbial pesticides registered for management of insect pests include 17 products of *B. thuringiensis* var. *kurstaki*, 58 of *B. bassiana,* 49 of *V. lecanii,* 11 of *M. anisopliae,* 18 of HANPV and 3 of SlNPV (Vimaladevi et al., 2009).

Table 33.1 *Status of Production of microbial biocontrol agents used against insect pests in India*

Agent	2005-06	2006-07	2009-2010	
			Solid formulations (T)	Liquid formulations (L)
HaNPV	15639 (L)	7958 (L)	-	12522
S1NPV	3438 (L)	2387 (L)	-	1673
Bacillus thuringenesis	52	0.07	5	
Beauveria bassiana	141	82	244	19478
Metarhizium anisopliae	29	35	20	4730
Verticillium lecanii	45	30	190	10242
Paecilomyces lilacinus	50	3	14	26436
P. fumosoroseus	-		2	4000

(Ramanujan et al., 2014)

33.1.2 Advantages and limitations

The main advantages of these biocontrol agents are their specificity to target pests, safety to the non-target organisms, they do not cause ill effects on environment and human health and can be used against pests which develop resistance to the conventional insecticides, they fit as ideal components in Integrated Pest Management (IPM) and also in organic farming systems. However, these biocontrol agents are reported to be slow in action or kill rate and sensitive to the environment, which lead to the inconsistent and poor success rate in the field. To overcome these problems, attempts are made to identify highly efficient and aggressive strains to improve the field efficacy and to develop suitable formulation technologies for increased field persistence and to withstand harsh environmental factors of radiations and dry weather conditions. In the present topic, various bioagents associated with insect pest control in different crops are discussed herewith.

Entomopathogenic Biocontrol agents

- Bacteria
- Fungi
- Virus
- Protozoa
- Nematode

33.2 Bacterial Biocontrol Agents

33.2.1 Bacillus Thuringiensis

The most successful insect pathogen used for insect control is the bacterium, *Bacillus thuringiensis (Bt),* which presently occupies about 2 percent of the total insecticidal market (Bravo *et al.*, 2011). The year 2011 was marked as a century of research and development of *Bacillus thuringiensis* (*Bt*). *Bt* is a Gram-positive, rod-shaped, spore-forming entomopathogenic bacterium characterized at the species level by the production upon sporulation of a parasporal inclusion body, the crystal which is toxic to insects and other invertebrates and is the most commonly used commercial biopesticide worldwide. *Bt* protein toxins are highly selective to their target insect, are innocuous to humans, vertebrates, plants and are completely biodegradable. Therefore, *Bt* is a viable alternative for the control of insect pests in agriculture and disease spreading vectors in public health. The most widely used bacterial pathogens include subspecies or strains of *B. thuringiensis.* Each one of the strains produces different mix of toxins and specifically kills one or a few related species of insects (*Bt* subspecies *kurstaki* and *aizawai* for lepidopteran larvae) and (*Bt* subspecies *tenebrionis* for coleopteran larvae). Some of these strains are specific to mosquitoes (*Bt* subspecies *israelensis*).

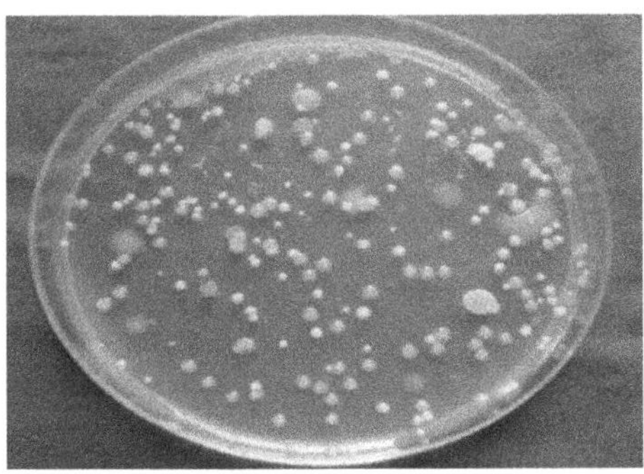

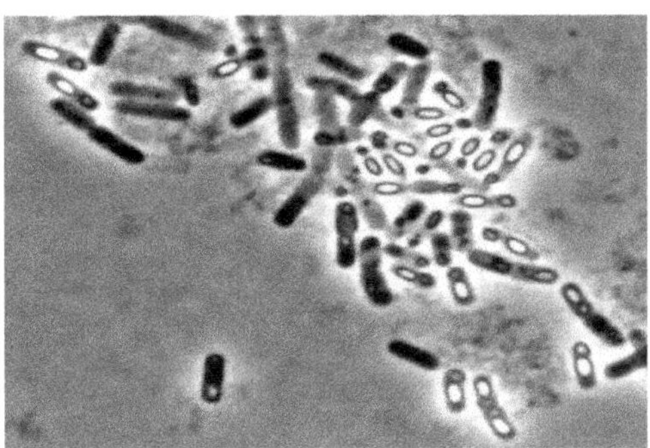

Fig. 33.1 Colonies of Bacillus thuringiensis and Bt cells

33.2.2 Bt scenario in World and India

Table 33.2 *Commercial Bacillus thuringiensis formulations available worldwide*

Manufacturer	Product	Bt strain	Toxic units	Target pests
Valent Biosciences Corporation	Dipel WP (Wettable powder)	*Bacillus thuringiensis*	16000 BIU/mg	Spruoo budworm; gypsy moth; bagworm; spring and fall cankerworms and cabbage looper
	Thuricide 48 LV (Liquid suspension)	*Bacillus thuringiensis berliner* ssp. *kurstaki*	12.7 BIU/L	Bagworm; elm spanworm; fall spanworm; gypsy Moth; spring and fall cankerworm; spruce budworm
	Vectobac-200 G larvicide granules		200 ITU/mg	
	Teknar Granules larvicide		260 AAU/mg	Mosquitoes
	Vectobac 200 G Vectobac 600L aqueous suspension	*Bacillus thuringiensis israelensis*	ITU/mg1986 2007 600ITU/mg	Fungus gnats
	Teknar HP –D larvicide (Aqueous suspension)		3000 AAU/mg	Mosquitoes & black flies
	Novodar flowable concentrate	*Bacillus thuringiensis tenebrionis*	3.6%	Colarado potato and elm leaf beetle

Table 33.2 Contd...

Manufacturer	Product	Bt strain	Toxic units	Target pests
	Foray 48 B		12.7 BIU/L	Spruce budworm (Eastern & Western); gypsy moth; jackpine budworm; eastern hemlock looper; whitemarked tussock moth and forest tent caterpillar White marked tussock moth and forest tent caterpillar satin moth
	Foray 48 B A Low Volume		12.7 BIU/L	
	Foray 76 B (aqueous xoncentrate)		20.0 BIU/L	
	Foray 96 B		25.4 BIU/L	
	Dipel 2 X DF (Dry flowable)	*Bacillus thuringiensis berliner* ssp. *kurstaki*	32000 IU/MG	spruce budworm; gypsy moth; bagworm; spring and fall cankerworm; fall webworms; elm spanworm; tent caterpillar; cabbage looper; leaf roller and diamondback moth
AFA Environment Inc	Aquabac II XT (Liquid suspension) Aquabac 200 G (10/14) (Granules) Aquabac 200 G (10/14) (5/8)	*Bacillus thuringiensis israelensis* Serotype H-14	1.28 BITU/Kg 0.2 BITU/Kg 200 ITU/mg	Mosquitoes
	Aquabac XT	*Bacillus thuringiensis israelensis*	1200 ITU/mg	Mosquitoes & black flies
Abbott Lab. Ltd. 192188	Dipel 176 (Emulsi fiable suspension)		16.9 BIU/L	Forest tent caterpillars; gypsy moth spruce budworms; hemlock looper
Woodstream Canada Corporation Safer's	BTK (Liquid concentrate)	*Bacillus thuringiensis Berliner* ssp *kurstaki*	- 12.7 BIU/L	Gypsy moth; tent caterpillar and cabbage looper
AEF Global Inc.	Bioprotec Aqueous		12.7 BIU/L	Gypsy moth; eastern spruce budworm; western spruce budworm; jack pine budworm; forest tent caterpillar; eastern hemlock looper; bagworm; elm spanworm; fall spanworm; spring & fall cankerworm; satin moth and white marked
	Biological (aqueous suspension)		12.7 BIU/L	
	Bioprotec CAF		17,5000 IU/mg	
	Aqueous Bioprotec HP		12.7 BIU/L	
	Bioprotec ECO		tussock moth	

(ITU), International Toxic Units; (IU), International Units; (BITU), Billion International Toxic Units; AAU, *Aedes Aegypti* Units (1 ITU = 2.5 AAU). IU: it refers to standardized potency (by bioassay) of different marketed *Bt* products against Bt var. *thuringiensis* E-61 standard from Institut Pasteur, Paris, France, assigned a potency of 1000 International Units (IU) per mg. Bioassay is carried out against mediterranean flour moth (*Ephestia kuhniella*) in Europe and in US, a primary reference standard of Bt HD-1-S-1971 strain with an assigned potency of 18000 IU/mg is being used, bioassayed against cabbage looper (*Trichoplusia ni*).

Table 33.3 *Bt strains registered under insecticide act in India*

Bt species	Strain/ Serotype	Formulation type
Bt kurstaki	Strain A97, serotype H-3A	35WP
	Strain DOR –Bt-1 serotype (2a, 3b, 3c)	0.5%WP
	Strain HD-a serotype 3a, 3b	3.5%ES
	Serotype 3a, 3b strain z-52	-
Bt galleriae	Strain R1593m serotype 3a	1.3%FC
Bt israelensis	Strain 164, serotype H-14	WP
	Strain VCR B-17 serotype-H-14	Slow release
	Strain VCR B-17 serotype-H-14	Granule WP
	Serotype-H-14	12AS
	Strain VCR B-17 serotype-H-14	5AS
	Serotype-H-14	5%WP

33.2.3 Preparation of Bt Formulations for Field Application

For mass production of *Bt* solid formulation, powdered Barley is the suitable substrate and most commonly used source. Powdered barley is taken in a 250 ml conical flask. The remaining ingredients (Table 33.4) are dissolved separately in 50 ml distilled water and this is added to already prepared Barley. The p^H of the medium is adjusted to 7.2. Flasks containing media are sterilized at 15 psi for 20 minutes, cooled and inoculated with 2% (v/v) of *B.t.* isolates multiplied on Luria broth and incubated for 48h at 30^0C on a shaker at 200 rpm. The medium with Bt culture from flasks is centrifuged, the pellet subjected to forced air drying under laminar air flow and used for field application. Instead of powdered barley, rice or wheat also can be used.

Table 33.4 *Composition of solid formulations for native B.t. strains*

Components	Quantity
Barley	5 g
Yeast extract	63 mg
$CaCl_2$	24 mg
$MgSO_4$	60 mg
K_2HPO_4	50 mg
KH_2PO_4	50 mg
Water	50 ml

For preparing liquid formulations, forty mL MGM broth was taken in a 250 mL conical flask. Flasks containing media are sterilized at 15psi pressure for 20 minutes, cooled and inoculated with native *B.t.* isolates, along with reference strain (HD1) and incubated for 72 hours in shaker at 200 rpm. This medium then taken at @ 2ml/L was used for field application.

Mass production

Fermenters are used for mass production of Bt in a 5 litre-capacity fermenter as starter inoculums for large scale fermenters. The composition of the culture medium for one litre volume is as follows: yeast extract, 4.0g; glucose, 8.0g; $MgSO_4$-$7H_2O$g, 0.1; KH_2PO_4, 0.1g; $MnSO_4$. $4H_2O$, 0.03g; $CaCl_2$. $2H_2O$, 0.04g; $(NH_4)2SO_4$, 0.1g. The medium has to sterilise in autoclave at 120°C for 20 minutes. Glucose and mineral salts were sterilised separately. The pH has to maintain 7.0 for culture media. Fermentation occurred at 30°C, 400 rpm and aeration rate of 1.5 vvm. After sporulation, the crystal-spore mixture was filtered in a tangential filter, 40-fold concentrated (Microdyn Filter/Module MD 020/ Frings Mycrodyn) and then dried in a forced-air oven. The spray solution is prepared by mixing *Bt* primary powder with detergent, emulsifier, sun block and dispersal agents for easy scattering and slow sedimentation.

33.2.4 Method of Field Application

The concentrated or wettable forms of regular *Bt* can be mixed as per recommended dose of 1g or 1mL/L by adding teepol or robin blue or ujala @ 1ml/L as UV protectant, jaggery @ 2g/L as feeding additive and triton-X @ 1ml/L as emulsifying agent in a normal sprayer then carefully applied to the foliage of affected plants.

Fig. 33.2 Bacillus thuringiensis infected larvae on groundnut and collected

In order for *Bt* to work, the caterpillar has to eat it. Be sure to cover both the tops and bottoms of the leaves, since they feed on both, with an even coating. It's not necessary to drench plants to the point of dripping and thorough coverage is required on foliage. It takes 72 hours for *Bt* to kill the insect after ingestion and the efficacy lasts for few days 7-10 days as Bt is photo degradable. Hence, UV protectants are needed to be added while spraying. There is a chance to develop resistance for *Bt* in insect as insecticides, use it only if absolutely necessary. Use *Bt* as part of Integrated Pest Management program that includes beneficial predators and other natural controls.

33.3 Entomopathogenic Fungi

Entomopathogenic fungi are gaining importance in the crop pest control in recent years, although *Bacillus thuringensis* (*Bt*) is the most widely used microbial agent at present. Fungal pathogens have certain advantages in pest control programmes over other insect pathogens like bacteria and viruses. Mass production techniques of fungi are much simpler, easier and cheaper than those used for *Bt* and NPVs. Fungi unlike bacteria or viruses directly infect through insect cuticle and do not require ingestion for infection and so sucking insects are also infected by entomopathogenic fungi. Entomopathogenic fungi play an important role in the natural pest control in various crops through epizootics. More than 750 sp. of fungi, mostly from hyphomycetes and entomophthorales are pathogenic on insects, many of them offer great potential for pest management. Entomopathogenic fungi like *Beauveria* spp., *Metarhizium*spp., *Lecanicillium* spp. and *Isaria* spp. have been developed as successful mycoinsecticides for various groups of insect pests (De Faria and Wraight, 2007).

Table 33.5 *Occurrence of B.bassiana on different crops in India*

Insect species	Occurrence	Food plant	Reference
Sweetpotato weevil (*cylas formicarius*)	Natural	Potato	Khader khan et al., 1993
Termite (*Odentotermis obesus*)	Natural	Soil	Khader Khan et al., 1993
Rice hispa (*Diclodispa armigera*)	Lab	Rice	Puzari and Harika, 1994.
Nilaparvata lugens (Stal)	Natural	Rice	Ambethagar 1996
Cnaphalocrocis medinalis (Guenee)	Natural	Rice	Ambethagar 1996
Plocaederus ferrugineus (Linn.)	Natural	Cashew	Ambethagar *et al.,* 1998
Sorghum shoot borer (Chilo partellus)	Lab	Sorghum	Umadevi et al., 2001.
Indarbela quadrinotata	Natural	Eucalyptes	Sasidharan and Varma, 2005
Sodoptera litura	Natural	Groundnut	Gurvinder and Padmaja, 2008.
Hairy caterpillar (Spolosoma oblique	Lab	Castor	Ajaykumar pandey 2008
Helicoverpa armigera	Natural	Sunflower	Vimala devi et al, 2008.
Rhizome & Pseudostem weevils of Banana	Natural	Banana	Padmanaban et al., 2009.
Hyadaphis coriander (Aphid)	Lab	Coriander	Selvarj et al., 2010.
Aphid (*Aphis craccivora*)	Lab	Fenugreek	Selvaraj et al., 2010.
Helicoverpa armigera	Lab	Tomato	Ashish et al., 2010.

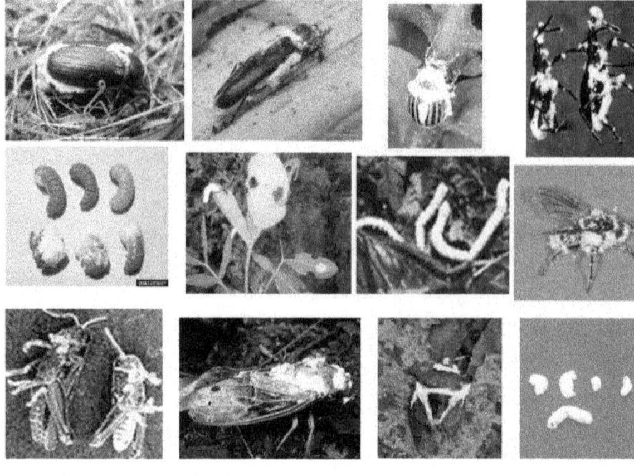

Fig. 33.3 Entomopathogenic fungi infected insects

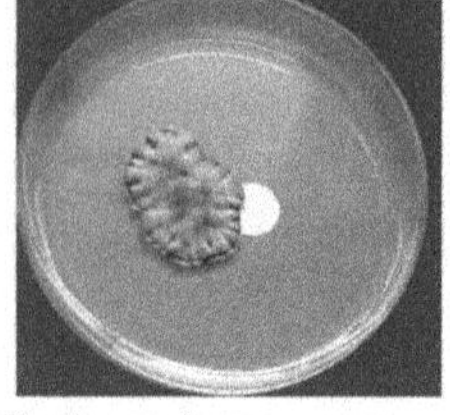

Cladosporium oxysporum

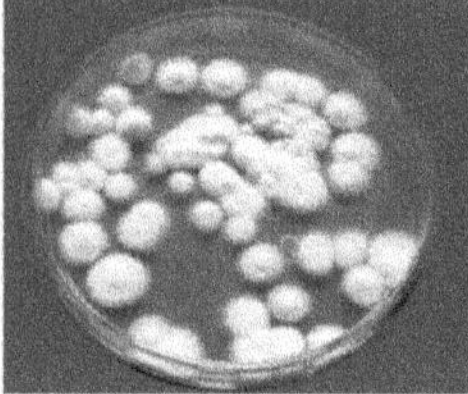

Beauveria bassiana

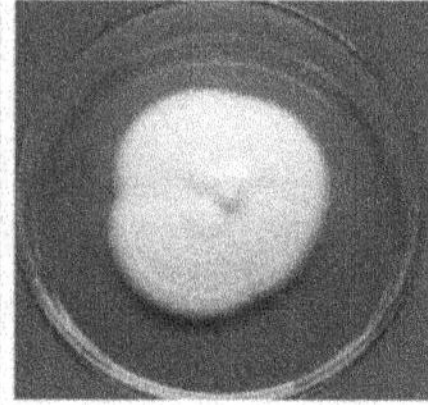

Verticillium lecanii

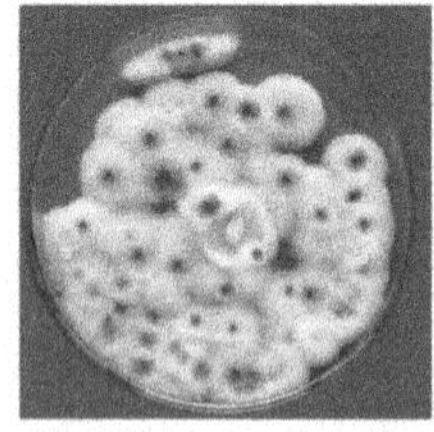

Metarhizium anisopliae

Nomuraea rileyi

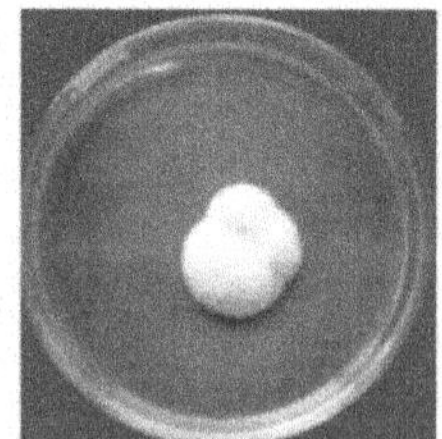

Hirsutella thompsonii

Fig. 33.4 Entomo-pathogenic fungi cultured on growth media

Table 33.6 *Commercial scale production of mycotoxins from different countries*

Fungus	Brand name	Insects	Crops	Manufacturer
Beauveria Bassiana	Mycotrol	Whiteflies/Aphids/Thrips	Field Crops	Mycotech, USA
	Naturalis	Sucking insects	Cotton, Glasshouse crops	Troy BioScience, USA
	Conidia	Coffee berry borer	Coffee	AgrEvo, Germany
	Ostrinol	Corn borer	Maize	NPP (Calioppe), France
	Myco-Jaal	Diamond backback moth	Cabbage	Pest Control India (Pvt) Ltd, India
	Biosoft	*Helocoverpa* & sucking pests	Several crops	Agriland Biotech, India
	Biowonder	Rice pests	Rice	Indore Biotech, India

Table 33.6 Contd...

Fungus	Brand name	Insects	Crops	Manufacturer
B brongniartii	Betel	Scarab beetle Larvae	Sugarcane	NPP(Calioppe), France
	Engerlingspilz	Scarab beetle Larvae	Pasture	Andermatt, Switzerland
	Melocont	Scarab beetle Larvae	Pasture	Pasture Kwizda, Austria
Metarhizium anisopliae	BIO 1020	Black Vine weevil	Glasshouse, Ornamental crops, Nursery stock	Bayer, Germany
	Bio-BlastTM	Termites	Houses	EcoScience, USA
	Bio Magic	Brown plant hopper	Rice	T. Stanes, India
	Multiplex	Root grubs	Several crops	Multiplex, India
	Metarhizium		Sucking pests	
M. flavoviride	Green Muscle	Locusts, Grasshoppers		CABI, UK
	BioGreen	Red-headed cockchafer	Pasture/Turf	Australia
Verticillum lecanii	Vertalec Mycotal	Aphids, Whiteflies and Thrips	Glasshouse crops	Koppert, Holland
	Inovert	Aphids, Scales, Mealybugs -		Inora, India
	Biocatch	Whiteflies	Cotton	T. Stanes, India
	Verticare	Mealybugs & Scales	Citrus	Viswamitra Bio Agro, India
Paecelomycos fumosoroseus	PFR 97TM	Whiteflies/Thrips	Glasshouse crops	Thermo Trilogy, USA
	Prioroty	Mites	Wide range of crops	T. Stanes, India

33.3.2 Mass Production of Entomopathogenic Fungus and Field Application

Dry Sorghum, wheat, maize, pearl millet, barley, pigeon pea and rice grains are some of the suitable solid media for growing entomopathogenic fungi, *N. rileyi* or *B. bassiana*. This grains has to crush in the mixer for about 30 sec to make small pieces. Crushed grains (50 g) with 1 % yeast extract are added to 500 ml capacity glass saline bottles containing 45 ml sterile water and soaked for 24 hrs and autoclaved for 20 minutes at 1.04 Kg/cm2. After cooling, each glass bottle is inoculated with 6 mm circular agar disc from 20 days old fungal culture grown on SMAY plates, and shaken well to disperse the inoculum. The glass bottles are incubated in BOD incubator at 25 + 1°C. After 25 days, fungal mass along with spores from each bottle is harvested along with the substrate and crushed after drying under laminar air flow for 12 hr.

The spore powder along with grain carrier (1g) is suspended in 50 ml sterile 0.02% Tween – 80 solutions (Vyas *et al.*, 1991) and mixed thoroughly using a sterile glass rod.

The suspension can be used for field spray after arriving requiring spore count (Ingle, 2014).

Development of pilot scale production system for *B. bassiana* in fermentor

- Studies on biomass production and CFU counts of *B.bassiana* in Jaggery-Soya Broth (JSB), Jaggery-Soya Yeast Broth (JSYB) and Corn Meal Broth (CMB) in ten-liter fermentor has been studied for seven days at 24 intervals.
- Peak biomass production and CFU counts were observed after 7 days in all the media tested.
- Corn Meal Broth (CMB) supported maximum biomass after 7 days (17.3g/100ml wet wt.) followed by JSYB (12.79 g/100 ml wet wt.) and JSB (12.0g/100ml wet wt.).
- Highest CFU counts were observed with corn meal broth (240×10^{10} CFU/ml).

***Table 33.7** Field application of various entomopathogenic fungi in different crops*

Fungus	Pest & Crop	Field efficacy	Remarks
Beauveria bassiana	Rice Hispa, *Dicladispa armigera*	Spray of *B. Bassiana* spore suspension 10 million spores/ml (1997)	Hazarika and Puzari
	Coffee berry borer *Hypothenemus hampei*	Spray of *B. bassiana* spore suspension (1×107 spores/ml) containing 0.1% sunflower oil and 0.1 percent wetting agent during monsoon reduced 50-60 percent berry borer incidence in Coorg, Karnataka	Anon. (2001)
	Tea looper, *Buzura suppressaria*	Spray of *B. Bassiana* spore suspension (2.5 g/l), gave 88 per cent reduction in West Bengal	Ghatak and Reza (2007)
	Sunflower: *Helicoverpa*	Spray of oil suspension of *B. bassiana* (200 mg/l) *armigera* in Andhra Pradesh	Devi and Hari (2009)
	Green gram: whitegrubs	Soil application @5 × 1013 conidia/ha White grubs effective control achieved in Assam (2008)	Bhattacharyya *et al.*
Beauveria Brongniarti	Sugarcane: white grub, *Holotrichia serrata*	Soil application @ 1kg /acre. Highest yield recorded	Chelvi *et al.* (2010)

Table 33.7 Contd...

Fungus	Pest & Crop	Field efficacy	Remarks
Metarhizium anisopliae	Coconut: Rhinoceros beetle, *Oryctes rhinoceros*	Spraying of Spores in its breeding sites @ 5 X 1011spores/m3 to the compost pits and manure heaps	Anon. (2000)
	Sugarcane: White grub	*M. anisopliae* at $1x10^{13}$/ha gave yield of 91.18 q/ha as compared with chlorpyrifos (93.29 q/ha) in Karnataka	Rachappa *et al.* (2004)
	Pigeon pea: Pod borer, *Helicoverpa armigera*	*M. anisopliae* conidia in an oil formulation was effective in reducing 66.74percent *H. armigera* as compared to 62.58 percent with endosulfan in Maharastra	Nahar *et al.* (2004)
	Potato White grub *Brahmina* sp.	Soil application @ 5 x 1013 conidia/ha along with chlorpyrifos 20 EC at 200 g a.i./ha resulted in the highest tuber yield (155 q/ha) in HP.	Bhagat *et al.* (2003)
	Soyabean: white grub, *Holotrichia longipennis*	Soil application formulation applied @ 5 x 10^{13} conidia/ha, 61.50 percent reduction in grub population	Pandey (2010)
Verticilium lecanii	Coffee green scale, *Coccus viridis*	Spraying spores@ 16 × 106 spores/ml along with Tween-80 twice at 2 weeks interval caused 97.6 percent mortality of the pest	Jayaraj (1989)
	Citrus green scale, *Coccus viridis*	Spraying of spore (2 x 106 spores/ml) along with0.005 per cent quinalphos and 0.05 per cent Teepol was found highly effective killing 95.58 percent and 97.55 percent scales in coffee and citrus respectively	Singh (1995)
	Indian mustard and Rapeseed: Mustard aphid, *Lipaphis erysimi*	Spray @ (106 spores/ml). There was a significant reduction in aphid infestation at 10 DAS	Rana *et al.* (2002)
Nomuraea rileyi	Castor: *Spodoptera litura*	Spraying of spore (10 x 1010 spores/ml) along in AP with 0.02 percent Tween-80	Vimala Devi and Prasad (1997)
	Soybean: *Spodoptera litura Helicoverpa armigera Thysonoplusia orichalcea*	*N. rileyi* spores spraying @ 2 × 108/ml twice at 10 days interval during Kharif in North Karnataka was cheaper than insecticidal treatment and cost effective	Lingappa *et al.* (2002)

33.4 Viruses

Biopesticides based on baculoviruses are ideally suited in integrated pest management programme. Since, baculoviruses are highly specific, known to be completely safe to mankind, animals, and beneficial insects such as bees, predatory insects and parasitoids. Baculoviruses have been described in over 600 insect species. The family Baculoviridae consisting of four Genera: Alpha baculovirus (Lepidoptean specific NPVs), Betabaculoviruses (Lepidopteran specific GVs), Gammabaculoviruses (Hymenopteran specific NPVs) and Deltabaculoviruses (Dipteran specific NPVs) (Jehle *et al.*, 2006). Among the insect baculoviruses, Nuclear Polyhedrosis Virus (NPV) and Granulosis virus (GV) are the most promising for insect control particularly of Lepidoptera and Diptera because of their specificity. NPVs have been successfully used for management of devastating pests like *Heliothis* spp. and *Spodoptera* spp. on cotton, fruit and vegetable crops in several countries (Rosell et al., 2008).

The virus particles (virions) are rod-shaped, 40-70 nm × 250-400 nm, comprising a lipoprotein envelope around a protein capsid containing DNA-protein core. Baculoviruses are large, circular and double-stranded DNA genomes ranging from 81.7 to 178.7 kb . They are pathogenic to Lepidoptera, Hymenoptera, and Diptera.

NPV is known for high epizootic levels and is naturally occurring, self-perpetuating, safe to natural enemies due to host specificity and environmental friendly. Since, NPV is an obligate parasite, it multiply only in living larvae. So, mass production of NPV is tedious job and requires skilled labours. Insect larvae infected with baculoviruses usually die from 3 to 12 days after infection depending on viral dose, temperature and the larval instar at the time of infection. Insects killed by baculoviruses have a characteristic shiny-oily appearance and often crawl to the tops of plants or any other available structure where they die and decompose. The virus particles are attached to the peritrophic membrane lining the midgut. The lipoprotein membrane surrounding the virus fuses with plasma membrane of the gut wall cells and liberates nucleocapsids into the cytoplasm. The nucleotide transports virus DNA into the nucleus of the cell and virus gene expression begins. The virus multiplies rapidly and eventually fills the body of the host with virus particles. The infected larvae are extremely fragile to the touch, rupturing to release fluid filled with infective virus particles.

33.4.2 Production of the Spodoptera Litura Nuclear Polyhedrosis Virus for Groundnut Application

- The virus is effective against the pest on several crops.
- The virus is generally propagated in early fifth instar by surface contamination of semi-synthetic diet.

Production procedure

- The mass production of the NPV is carried out in early fifth instar stage of *S. litura*, which yields maximum amount of the NPV. Therefore, in the host culture laboratory a continuous culture of the insects is maintained with proper handling procedures.
- The larvae are grown in diet held in 5 ml glass vials. When the larvae reach the appropriate stage they are transferred to the virus production facility.
- The NPV is multiplied by feeding the semi synthetic diet coated with a clean inoculum of the NPV that has previously been standardized.
- This is accomplished by placing aliquots of 10 ml of the viral suspension of concentration 1×10^8 Polyhedral Occlusion Bodies (POB) in the centre over the diet surface either in glass vials and spreading the suspension uniformly all over the surface with a polished glass rod.
- Larvae are released singly after 15 min. into each glass vial/cell and incubated at 25°C for 10 days.

The larvae begin to die from 5th day onward. The cadavers are collected individually and transferred to 500 ml plastic containers and frozen immediately until processing

Processing of NPV

The method of processing of NPV requires greater care to avoid losses during processing. The cadavers are brought to normal room temperature by repeatedly thawing the container with cadaver under running tap water. The cadavers are homogenized in sterile ice cold distilled water at the ratio 1: 2.5 (w/v) in a blender or precooled all glass pestle and mortar. The homogenate is filtered through double layered muslin and repeatedly washed with distilled water. The ratio of water to be used for this purpose is 1: 7.5-12.5 (w/v) for the original weight of the cadaver processed. The left over mat on the muslin is discarded and the filtrate can be semi-purified by differential centrifugation. The filtrate is centrifuged for 30-60 sec. at 500 rpm to remove debris. The supernatant is next centrifuged for 20 min at 5,000 rpm. Then the pellet containing the polyhedral occlusion bodies (POB) is suspended in sterile distilled water and washed three times by centrifuging the pellet in distilled water at low rpm followed by centrifugation at high rpm. The pellet finally collected is suspended in distilled water and made up to a known volume, which is necessary to calculate the strength of the POB (5×10^5/ 10 mL suspension) in the purified suspension.

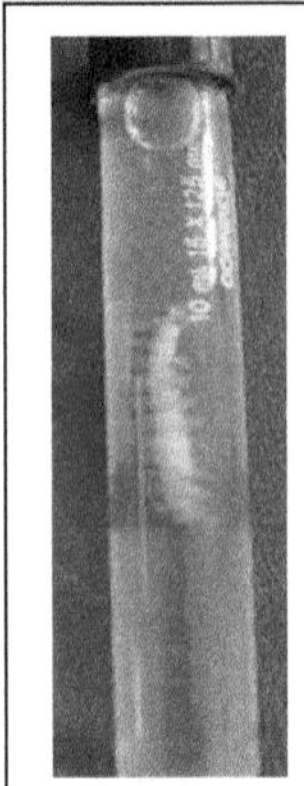

Fig. 33.5 *NPV infected larva*

Fig. 33.6 *NPV larval extract*

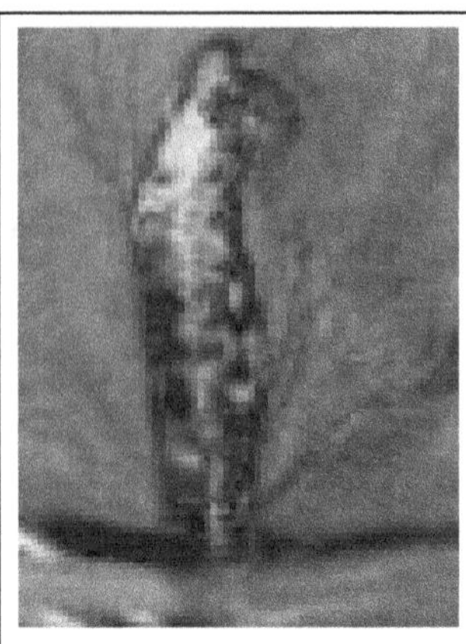

Fig 33.7 *Infected S.litura*

Fig 33.8 *Infected H.armiigera*

Table 33.8 *Baculovirus preparations registered for pest control in various parts of the world*

Insect species I	Trade name of the product I	Country I	Year of registration
Granulosis viruses			
Adoxophyes orana	Capex	Czechoslovakia	1989
Agotis segetum	Agrovir Germany 1990	Germany	1990
Cydia pomonena	Madex	Czechoslovakia	1987
Nuclear polyhedrosis viruses			
Heliothis Spp.	Biotrol VW and Virion H	USA	1973
Heliothis spp.	Elcar	USA	1975
Lymannia dispar	Gypcheck	USA	1978
Mamestra brassicae	Mamestrin	Finland	1988
Neodiprion lecontei	Lecont-virus	USA/Canada	1982/1983
Orgyia seudotsugata	Biocontrol-1 and Virtuss	USA	1976

NPVs recorded in India includes Helicoverpa armigera, S. litura, S. exigua, Amsacta moorei, Agrotis ipsilon, A. segetum, Anadividia peponis, Trichoplusia ni, Thysanoplusia orichalcea, Adisura atkinsoni, Plutella xylostella, Corcyra cephalonica, Mythimna separata and Phthorimaea operculella.

Table 33.9 *Nuclear polyhedrosis viruses recorded in India*

Pest	Crop
H. armiigera	Chickpea and others
S. litura	Tobacco and others
S. exigua	Tom ato and others
Amsaaa moorei	Pulses
Agrotis ipsilon	Potato and others
Anadividia peponis	Gourds
Trichoplusia ni	Potato and others

33.4.3 Method of Application

Spraying of NPV late in the evening and addition of UV blockers are recommended to improve the effectiveness of NPV. For chickpea, pigeon pea and cotton, HaNPV should be used at 250-300 LE, 500 LE and 250 LE/ha, respectively and for groundnut SlNPV should be used at 250 LE/ha.

33.5 Protozoa as Insect Pathogens

Entomopathogenic protozoans are extremely diverse group of organisms comprising around 1000 species attacking invertebrates including insect species and are commonly referred as microsporidians. They are generally host specific and slow acting, producing chronic infections with general debilitation of the host. The spore formed by the protozoan is the infectious stage and has to be ingested by the insect host for pathogenecity. The spore germinates in the midgut and sporoplasm is released invading the target cells causing infection of the host. The infection results in reduced feeding, vigour, fecundity and longevity of the insect host.

Although they are undoubtedly important in natural biological regulation of insect populations, they do not possess the required attributes for a successful microbial insecticide. The most notable entomopathogenic protozoa belong to *Nosema* spp. and *Variomorpha necatrix*. *Nosema locustae* is the only commercially available species of the microsporidian and is marketed for control of grass hoppers and crickets in USA, Canada, Argentina, China and Mali. However, the utility of *N. locustae*as a grasshopper biocontrol agent remains questionable because of the great difficulty in assessing the efficacy in case of a highly mobile insect. *Nosema pyrausta* is another beneficial microsporidian that reduces fecundity and longevity of the adults and also causes mortality of the larvae of European corn borer.

33.6 Entomopathogenic Nematodes

Entomopathogenic nematodes (EPNs) of the families Heterorhabditidae and Steinernematidae are obligate parasites of insects and are used as biological control agents of economically important insect pests. The two major genera are *Heterorhabditis* Poinar, 1976, and *Steinernema* Travassos, 1927, with 85 species described to date. These nematodes possess a symbiotic association with pathogenic bacteria from the *Xenorhabdus* and *Photorhabdus* genera, associated with *Steinernema* and *Heterorhabditis* respectively. Infective juveniles (IJs), considered the only free-living stage of EPNs, enter the host insect through its natural apertures (oral cavity, anus and spiracles) or in some cases through the cuticle. After penetrating the insect's hemocoel, IJs release their symbiotic bacteria, which are the primary agents responsible for host death and also provide the nematodes with nutrition and defence against secondary invaders. The nematodes complete their development and live for two or three generations inside their host. When food is depleted, IJs exit from host cadaver searching for new hosts

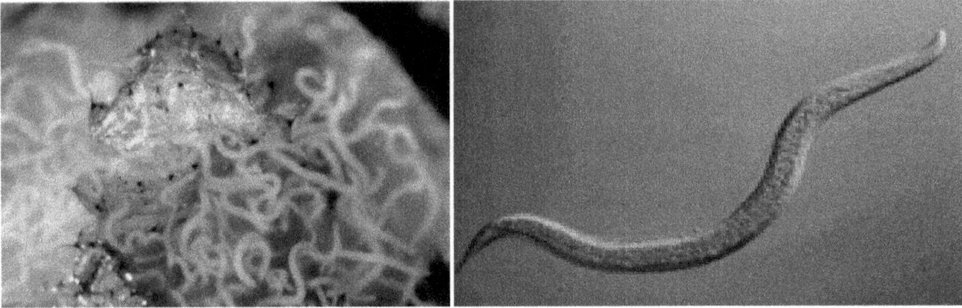

Fig. 33.9 Nematode infestation on larvae

Entomopathogenic nematodes are currently produced by different methods either *in vivo* or *in vitro* (solid and liquid culture) (Friedman, 1990). Each approach has its advantages and disadvantages relative to production cost, technical know-how required, economy of scale, and product quality, and each approach has the potential to be improved. Following production, a variety of formulation and application alternatives are also available.

33.6.1 Production Methods

The general approach to *in vivo* culture is a two dimensional system that relies on production in trays and shelves (Ehlers and Shapiro-Ilan, 2005). *In vivo* production methods for culturing EPNs have been reported by various authors (Shapiro-Ilan et al., 2002a). The systems described in all of these references (with some variation) entail a system that is based on the White trap (White, 1927), which takes advantage of the progeny IJ's natural migration away from the host-cadaver upon emergence. The approach consists of inoculation, harvest, concentration, and (if needed) decontamination. Insect hosts are inoculated on a dish or tray lined with absorbent paper or another substrate conducive to nematode infection such as soil or plaster of Paris. After approximately 2-5 days, infected insects are transferred to the White traps; if infections are allowed to progress too long before transfer the chance of the cadaver rupturing and harm to reproductive nematode stages is increased. White traps consist of a dish or tray on which the cadavers rest surrounded by water, which is contained by a large arena. As IJs emerge they migrate to the surrounding water trap where they are harvested. The scale of the White trap in size and number can be expanded to commercial levels.

In vivo production yields vary greatly among insect hosts and nematode species. Due to high susceptibility to most nematodes, wide availability, ease in rearing, and the ability to produce high yields, the most common insect host used for laboratory and commercial EPN culture is G. *mellonella* (Woodring and Kaya, 1988). Significant research (and commercial application) has also been achieved for production of EPNs in T. *molitor*. Other hosts in which in vivo production has been studied include the navel orangeworm (Ameylois transitella), tobacco budworm (*Heliothis virescens*), cabbage looper (*Trichoplusia ni*), pink bollworm (*Pectinophora gossypiella*), beet armyworm (*Spodoptera exigua*), corn earworm (*Helicoverpa zea*), gypsy moth (*Lymantria dispar*), house cricket (*Acheta domesticus*) and various beetles (Coleoptera)

In vitro culturing of EPNs is based on introducing nematodes to a pure culture of their symbiont in a nutritive medium. In earlier work, to create monoxenic cultures surface sterilized nematodes were added to a lawn of the bacterial symbionts. However, (Lunau *et al.* 1993) suggested that surface sterilization of IJs is insufficient to establish monxenicity because contaminating bacteria survive beneath the nematode's cuticle. Thus, an improved method has been developed where axenic nematode eggs are placed on a pure culture of the symbiont.

Solid culture was first accomplished in two dimensional arenas e.g, petri dishes, using various media (Hara *et al.*, 1981, Wouts, 1981). Subsequently, *in vitro* solid culture advanced considerably with the invention of a three-dimensional rearing system involving nematode culture on crumbled polyether polyurethane foam. A liquid medium is mixed with foam, autoclaved, and then inoculated with bacteria followed by the nematodes. Nematodes are then harvested within 2-5 weeks (Bedding, 1984) by placing the foam onto sieves immersed in water. Media for this approach include various ingredients including peptone, yeast extract, eggs, soy flour, and lard (Han *et al.*, 1993). Large scale production was further advanced through several measures including using bags with gas permeable Tyvac ® strips for ventilation, automated mixing and autoclaving, simultaneous inoculation of nematodes and bacteria, sterile room technology, and automated harvest through centrifugal sifters (Gaugler and Han, 2002).

33.6.2 Method of Application

Entomopathogenic nematodes can be applied with nearly all agronomic or horticultural ground equipment including pressurized sprayers, mist blowers, and electrostatic sprayers or as aerial sprays (Shapiro-Ilan et al., 2006a). The application equipment used depends on the cropping system, and in each case there are a variety of handling considerations including volume, agitation, nozzle type, pressure and recycling time, system environmental conditions, and spray distribution pattern. It is important to ensure adequate agitation during application. For small plot applications, hand-held equipment (e.g., water cans) or back-pack sprayers may be appropriate. When nematodes are applied to larger plots, a suitable spraying apparatus such as a boom sprayer should be considered. Conceivably, applicators could also be using other methods such as through microjet irrigation systems, subsurface injection or baits (Lara et al., 2008). Various formulations for entomopathogenic nematodes may be used for applying EPNs in aqueous suspension including activated charcoal, alginate and polyacrylamide gels, clay, peat, polyurethane sponge, vermiculite, and water dispersible granules (WDG) (Georgis *et al*., 1995).

33.7 Conclusion

Despite their proven efficiency, each and every microbial biocontrol agent has its own limitation in large production as well filed usage. In case of viral biocontrol agents, large-scale commercial application of NPV is low as compared to *Bacillus thuringiensis* based biopesticides. NPVs must replicate in the nuclei of living host cells. They cannot be cultivated in microbial culture media without living cells. This means that the production of NPV isolates as microbial biopesticide, requires a colony of insects or a suitable insect tissue culture. This is more expensive than producing organisms such as the bacterium *Bacillus thuringiensis* in an artificial medium. This hinders its large-scale application in field. The use of *Bt* based biopesticides for longer period invites the problem of developing resistance in insects.

Similarly, for the application of fungal biocontrol agents, environmental humidity and temperature play an important role in the infection and sporulation of these fungi and as such they are highly suitable during cool and humid cropping seasons. Since talc - based formulations of these fungi have limited shelf life of 3-4 months, alternative formulations with longer shelf life (12-18 months) have to be developed besides suitable oil based formulations for dry land agriculture. And few organisms were exploited in biocontrol programme. Still there is a scope to utilize the biodiversity of Entomophthorale group of fungi like *Entomophthora, Zoophthora, Neozygites etc*., which have potential for management of aphids, thrips and lepidopteran pests. Similarly, EPNs also requires specific environmental conditions for their application. The microsporidia are good biocontrol agents but, they are having limited host range. Based on these, it is concluded that, microbial based biopesticides are best suited in integrated pest management programmes as a whole, instead of solo application.

References

1. *Anand R, Prasad B, Tiwary B N (2009) Relative susceptibility of Helicoverpa armigera pupae to selected entomopathogenic fungi. Biological Control 54: 85-92.*
2. *Bravo A, Likitvivatanavong S, Gill SS et al. Insect Biochem Mol* ***41*** *(2011) 423-431*
3. *Butt TM, Harris JG and Powell KA, Microbial biopesticides. In: (Eds: Hall FR and Menn JJ), Methods in Biotechnology, 5 Biopesticides: Use and Delivery, Humana Press, Totowa, NJ, (1999) pp 266-269*
4. *de Faria MR and Wraight SP. 2007. Biol Control* ***43:*** *237-256*
5. *Georgis R, Dunlop DB, Grewal PS. 1995. Formulation of entomopathogenic nematodes. Pp. 197-205 in F. R. Hall, and J. W. Barry, eds. Biorational Pest Control Agents: Formulation and Delivery. Washington, DC: American Chemical Society.*
6. *Hara AH, Lindegren JE, Kaya HK. Monoxenic mass production of the entomogenous nematode Neoplectana carpocapsae Weiser on dog food/agar medium. USDA Advances in Agriculture W-16, 8 pp. 1981*
7. *Ingle. Y.V. 2014. Effect of different growing media on mass production of nomuraea rileyi International journal of Environmental Sciences Volume 4(5). Pp1006-1014.*
8. *Jehle JA, Blissard GW, Bonning BC et al. 2006. Arch Virol* ***15.*** *1257-1266*
9. *Lara JC, Dolinski C, Fernandes de Sousa E, Figueiredo Daher E. Effect of mini-sprinkler irrigation system on Heterorhabditis baujardi LPP7 (Nematoda: Heterorhabditidae) infective juvenile. Scientia Agricola. 2008; 65: 433-437.*
10. *Li Z, Alves SB, Roberts DW et al. Biocontrol Sci Techn* ***20*** *(2010) 117-136*
11. *Lomer CJ, Bateman RP, Johnson DL et al. Annu Rev Entomol* ***46*** *(2001) 667-702*
12. *Lord JC J Invertebr Pathol* ***89*** *(2005) 19-29*
13. *Lunau S, Stoessel S, Schmidt-Peisker AJ, Ehlers RU. Establishment of monoxenic inocula for scaling up in vitro cultures of the entomopathogenic nematodes Steinernema spp. and Heterorhabditis spp. Nematologica. 1993; 39: 385-399.*
14. *Menn JJ J Environ Sci Health* ***B31*** *(1996) 383-389*
15. *Ramanujam, B. Rangeshwaran, R. Sivakmar, G. Mohan. M. and Yandigeri. M S. 2014. Management of Insect Pests by Microorganisms. Proc Indian Natn Sci Acad* ***80*** *No. 2 June 2014 Spl. Sec. pp. 455-471*
16. *Rao C, Devi KU, Khan PAA (2006) Effect of combination treatment with entomopathogenic fungi Beauveria bassiana and Nomuraea rileyi (Hypocreales) on Helicoverpa armigera (Lepidoptera: Noctuidae). Biocontrol Science and Technology 16: 221-232.*

17. *Ritu, A Choudhary Anjali, Tripathi Nidhi, Patil Sheetal and Bharti Deepak. 2012. Biopesticidal Formulation of Beauveria Bassiana Effective against Larvae of Helicoverpa armigera. J Biofertil Biopesticides, 3: 3*
18. *Rosell G, Quero C and Coll J. 2008. J Pestic Sci **33:** 103-121*
19. *Shapiro-Ilan DI, Gaugler R, Tedders WL, Brown I, Lewis EE. Optimization of inoculation for in vivo production of entomopathogenic nematodes. Journal of Nematology. 2002a; 34: 343-350.*
20. *Vimala Devi PS, Rangarao GV, Gopalakrishnan S and Sivakumar G. 2009.*
21. *Environmental Impacts of microbial pesticides. Paper presented at the symposium on 'Biosafety and Environmental Impact of Genetically Modified Organisms and Conventional Technologies for Pest Management' at ICRISAT– Patancheru, (AP). November 20-21, 2009. pp 12-13*
22. *White GF. A method for obtaining infective nematode larvae from cultures. Science. 1927; 66: 302-303.*
23. *Woodring JL, Kaya HK. 1988. Steinernematid and heterorhabditid nematodes: A handbook of biology and techniques. Southern Cooperative Series Bulletin 331, 30 pp, Arkansas Agricultural Experiment Station, Fayetteville, AR.*

Chapter 34

Indigenous Technical Knowledge (ITK) in Groundnut Cultivation

E. Somasundaram

Professor and Head, Department of Sustainable Organic Agriculture, Tamil Nadu Agricultural University, Coimbatore.

34.1 Introduction

Indigenous Technical Knowledge is the local knowledge – knowledge that is unique to a given culture or society. It contrasts with the international knowledge system generated by universities, research institutions and private firms. It is the basis for local–level decision making in agriculture, health care, food preparation, education natural resource management and a host of other activities in rural communities (Warren, 1991).

The term indigenous technical knowledge may be denoted mainly as a tacit type of knowledge that has evolved within the local (grassroots) community and has been passed on from one generation to another, encompasses not only local or indigenous knowledge, but also scientific and other knowledge gained from outsiders. Indigenous knowledge is a product of the adaptation of farming practices to the local environment, creating unique indigenous farming practices and food culture.

For example, *Navathaniyam* - vegetational diversity possessed green manure value, suppressed the weed growth, acted as cultural means of pest management and improved soil properties on incorporation

Organic production system was imperfect and continued to be incomplete for want of an input to replace foliar fertilization, growth-promoting hormones and immunity boosters and ITK based *Panchagavya* is the solution.

Fig. 34.1 Navathaniyam

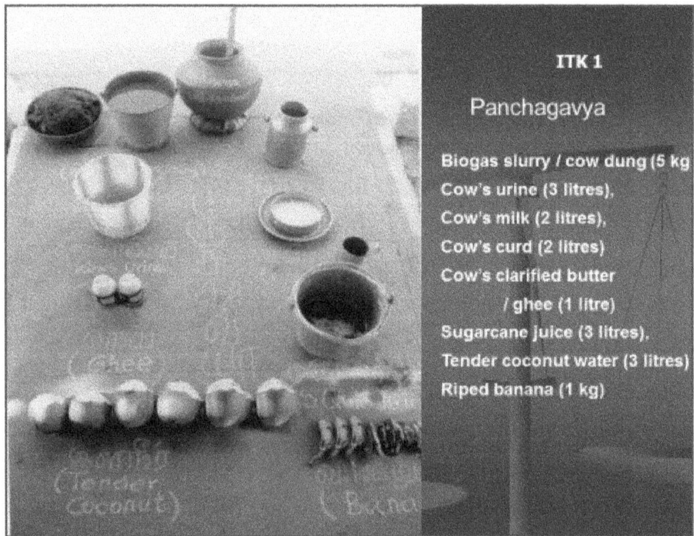

Fig. 34.2 Panchagavya

Many of our ITK's in agriculture and allied fields are obsolete, endangered ones and in extinct particularly during this era of globalization, liberalization and commercialization. There exists a need to systematically document the unwritten body of knowledge.

34.2 Advantages of Indigenous Technical Knowledge

Two powerful advantages:

- low cost and non- monitoring
- Readily available

- ITK system- Socially desirable, economically affordable, sustainable and involves minimum risk to rural farmers and producers- believed to conserve resources.
- ITK- problem solving strategies for local communities' especially small and marginal farmers.
    - ITK is the basis for self-sufficiency and self-determination
    - People are familiar with indigenous practices and techniques
    - ITK draws on local resources
- ITK provides effective alternatives to western know-how
- Indigenous technologies and practices are often cheaper than modern agriculture

Efficiency

- ITK is cost effective
- ITK uses appropriate technology

Effectiveness

- ITK improves chances of adoption
- ITK is integral to local communities

Sustainability

- ITK facilitates mutual adaptation and learning
- ITK empowers local communities

34.3 Principles of ITK

- All ITK's go by the principle of 'Permanence'. It is not so with modern technologies with synthetic inputs.
- All the ills in crop production (soil mining, degradation, pollution, etc.,) are due to decreasing attention to ITK.
- Indigenous practices in agriculture are organic in nature.

ITKs includes traditional varieties, crop management practices (weather forecasting/ rainfall, soil and water management, season, preparatory cultivation, seeds and sowing, manures, legumes and weed management) crop protection practices (pest and disease management, rat control, bird's scaring), post harvest methods, seed storage structures and storage pest control.

34.4 Groundnut Cultivation

Groundnut or peanut is one of the important legume crops of the tropical and semiarid tropical countries, where it provides a major source of edible oil and vegetable protein.

Groundnut kernals contain 47-53% oil and 25-36% protein. The crop is cultivated between 40°N to 40°S of the equator. Groundnut is a self pollinated crop.

The cultivated groundnut (*Arachis hypogaea L.*) belongs to family Fabaceae, subfamily Papilionaceae. In India, it cultivated in 5.31million ha with production of 6.93 million tonne (Anonymous, 2012). In India, it is grown mainly in Gujarat, Maharashtra, Andhra Pradesh, Uttar Pradesh, Tamil Nadu, Rajasthan, Karnataka and Madhya Pradesh. There are two major subspecies of groundnut that also differ in their branching patterns viz., hypogaea and fastigiata. However, most of the commercially cultivated varieties belong to the hypogaea (Virginia or runner), fastigiata (Valencia), and vulgaris (Spanish) botanical variety groups. The groundnut contains high oil (46-52%) and protein (25-36%) content depending on the variety.

34.4.1 Seasons

- Even if the seeds are of good quality, they must be sown in the right time
- Sowing in the right season is better even on a poor land
- The best season for crop- cultivation/sowing are the Tamil months viz., Chittirai (Apr), Aadi (Jul), Aavani (Aug) and Thai (Jan)
- Crops sown on eighteenth day (Aadipeperukku) and new moon day of the Tamil month "Aadi" (July- Aug) yield better
- It is better to avoid sowing on Ashtami (eighth) and Navami (ninth) days from full moon and new moon days
- The crop sown on new moon day escapes from pest and diseases
- Higher yield of tamarind is considered as an indication for good agricultural season and higher yield of mango for poor season.

34.4.2 Preparatory Cultivation

If soil is dried so one ounce become one-quarter ounce, Abundant yields will not require a single handful of fertilizer- Thirukkural

- Achieving fine tilth is better than applying manures
- It is better to have deep ploughing rather than shallow ploughing for good crop growth
- Doing last ploughing across the slope
- Ploughing four times for garden land and seven times for wetland
- Summer ploughing gives good crop in the ensuing season
- Garden and dry lands are ploughed deep to conserve moisture.

34.4.3 Summer Ploughing

Plough the field during summer and two feet height, earthern bunds is made around field which checks run off of the rain water. Advantages of summer ploughing:

- It increases water holding capacity of the soil
- It destroys weeds and pupal case of the pests.
- It conserves soil moisture and reduces the pest incidence
- Water requirement for field preparation of succeeding crop is reduced.
- A piece of raw/fermented areca nut with a betel leaf and a pinch of lime are chewed together and spit on the ploughing spot. If the soil becomes blackish, it indicates that the soil is well tilled. Or if the soil turns reddish, then the test is negative, indicating the need for more tillage (Farmers in Assam).

34.4.4 Seeds and Sowing

- Even if it is quality seed, it should be a dried one because well dried seeds will have higher longevity and keeping quality
- To maintain the seed viability and prevent it from outside damage, the outer shell is not separated
- Shallow sowing is followed in dry lands
- It is better to start planting from 'Sani moolai' (north east) of the field to get higher yields.
- It is better to perform sowing and planting operations during evening hours
- Sandy soil is highly suitable for groundnut crop since it gives less number of ill-filled pods.

34.4.5 Manures

- Organic manures always encourage good crop growth
- Green manure crops are raised and ploughed in- situ before their flowering.
- Green leaf manures like neem, pongam, calotropis, nuna, poovarasu and adothoda used as manures
- Legumes are also used as green manure crop to improve the soil fertility.
- Applying water hyacinth plants either as a compost or as burnt ash to the fields for supplying potash
- Sheep penning results in more yields
- Goat manure gives benefits to crops grown in the same season
- Goat manure is good for the first season and cattle manure and green manure for the second season

- Poultry manure serves as a good source of crop nutrients
- Cow urine is more nutritious
- Soil fertility can be increased by the use of pig dung
- Near the irrigation channel, a pit is dug in which, cow dung, foliage of calotropis gigantean, neem cake powder and cow urine are mixed well and allowed to decompose. Then it is allowed to mix with the irrigation water, to supply the nutrients, and to control pests and diseases.
- Tank silt is applied every year in dry lands for better yields
- Termite mount serve as good manure
- Adding sand from an ant mount to the field gives good yield
- Foliar spray of manures given on full moon day yields better results.
- Bone remains of the dead animals in the farm is a good source of calcium and sulphur

34.4.6 Animal Penning

Penning is an important traditional practice and more than 57% of farmers follow the practice at least once in 2 years. This practice involves camping of cattle, sheep, duck, etc. in the crop fields over night or days of nights after the crop harvest. The farmers prioritize the use of penning with the land patches with their production potential. The paddy, groundnut and vegetables get priority for manuring as they assure food supply and enhancing farm household income.

34.4.7 Sheep Penning

During non cropped season or after harvest, sheep penning is done with 100 sheep's in the field for about 4-5 days. Every day morning the place of penning can be changed. So that the entire field will be covered. At the time of penning, sheep manure enriches the soil fertility.

Fig. 34.3 Sheep penning

Fig. 34.4 Cattle penning

34.5 Identification of Soil Fertility

- If the weed growth is profuse after the rains, it indicates high soil fertility
- Low soil fertility is indicated by the growth of the weed "Aduthinnapalai" (*Aristolochia bracteolata)*

34.6 Plant Growth Regulator

34.6.1 Panchagavya

Panchagavya is a growth promoter produced with the combination of five products obtained from the cow along with few other bio products viz.,

- Cow dung - 5 kg
- Cow's urine - 3 litres
- Ghee - 1 litre
- Cow's Milk - 2 litres
- Curd - 2 litres
- Tender coconut - 3 litres

Cane juice or jaggery - 3 litres (or) 1kg or Honey - ½ kg

Method of Preparation

Add cow dung, cow's urine and ghee in a mud or cement tank of 30 litres capacity. Stir this well in morning and evening for a week. The methane gas gets released from this. To this mixture, add milk, curd, tender coconut, yellow plantain (ripe) and jaggery. After a week's time, this extract can be filtered and used.

Panchagavya showed antimicrobial activity against *Staphylococcus aureus, Salmonella typhi, E.coli* and *Bacillus sp.* and also showed fairly good values for the chemical parameters. Hence it can be said that it is a good bio-pesticide and a good bio-fertilizer.

34.6.2 Cowdung and Green Leaf Mixture

Add 1-2 kg of cowdung- dilutes it with 1 litre of water. To this add 5 kg of green leaves neem, pongam, vitex, lantana and glyricidia (each equal quantity) in a mud pot and pour the diluted cowdung solution into the mudpot. Stir it daily at least once. After 16 days, add 200 gms of Jaggery solution to the mixture and keep on stirring it. After 21 days, filter the solution and spray it on the field. It increases the yield and the vegetative growth too. For 1 litre of this mixture add 5 litres of water.

34.7 Weed Management

It is essential to keep groundnut fields weed free for up to 45 days after crop emergence. Weeds plants may act as alternative hosts of plant parasites allowing the pathogen to survive between seasons and providing a source of inoculums in the new growing season.

- Repeated ploughing will reduce weed population
- Weeds are to be destroyed before their flowering
- Cultivating rice once in three years in garden lands to control 'arugu' (*Cynodon dactylon)* weeds
- Raising and ploughing green manure crops in the field at the time of flowering to reduce the weed population
- To control nut grass (*Cyperus rotundus)* in the field, 50 kg. of neem cake is applied both at the time of ploughing and sowing
- Dissolving 200 g. of salt in 1 litre of water and spraying to eradicate congress and *Parthenium spp.*
- Continuous submergence of the field for some time controls the weeds
- Summer plough reduces weed population (soil solarisation).

34.7.1 Weed Management by Groundnut Farmers

Weeding in groundnut is very important technique and age old also. Farmers are using a 'V' shaped tool called 'hand hoe' to remove the weeds. Weeding is done based on the age and stage of crop. First weeding is done after 15 days of sowing. Second weeding is being practiced at flowering stage (35 days after sowing). There is no adverse effect to crop by this technique. Farmers believe that it improves the yield up to 20%. This ITK is being adopted by groundnut farmers in Dindigul district of Tamil Nadu over 50 years.

Fig. 34.5 Hand hoe used for weeding

34.8 Intercropping in Groundnut

34.8.1 Sorghum Intercropped with Groundnut

Objective of the cropping system is to improve the income and soil fertility. In the village of Ayakudi, Pazhani, Dindigul district, 40% of groundnut growers are used to follow this cropping system for more than 40 years. 6 rows of groundnut with 2 rows of sorghum required 50 kg of groundnut and 5 kg of sorghum seeds to cover up a one acre of land. Farmers are benefitting of Rs. 1500/ac as additional income through this cropping system.

34.8.2 Cotton Intercropped with Groundnut

This indigenous technical knowledge is being practiced by the groundnut farmers in the region of Vilupuram district, Thindivanam taluk, Vada Aalappakam village over 25 years. Line sowing of cotton with 3 feet gap has to be taken up after groundnut sowing. Moisture in the soil is being effectively utilized by cotton after harvest of groundnut. This cost very minimum (Rs. 300/ac) with additional income of Rs. 5000/ac.

34.8.3 Redgram/Maize Intercropped with Groundnut

It's an age old practice to meet out the family requirements.

34.8.4 Chickpea as Border in Groundnut

Objective of the cropping system is to differentiate the groundnut field and to get additional income. This ITK's is being practiced by generations to generations in the groundnut growing areas. In general groundnut is sown during June (Aadipattam) in rainfed areas. Chickpea sowing is done around the field after groundnut. This earns additional income and enhances the soil fertility.

Fig. 34.6 *Groundnut + Maize, Groundnut + Redgram, Groundnut + sorghum*

34.9 Pest and Disease Control in Groundnut

34.9.1 Sucking Pests (Aphids, Thrips etc.)

34.9.1.1 Methodology for Preparing Tobacco Leaf Extract

Ground 1 kg of tobacco leaves well with water and mix it with 10 lit of cow's urine and 1 litre of kerosene and keeps it for 15 days. Filtered the solution and use it for spray. 1 litre of solution has to mix with 9 lit of water and spray during morning hours with hand operated sprayer. Spray can be done twice at 15 days intervals incase of high pest incidence. Farmers of Villupuram district, Tamil Nadu are practicing this ITK against sucking pests over 25 years. This ITK is simple and can be adopted for rice also.

34.9.1.2 Dusting of Ash and Manure

To control aphids, beetles and diseases like leaf spot, leaf burning and potassium deficiency. In the district of Ramanadhapuram, Paramakudi taluk, Ariyanenthal village of Tamil Nadu farmers are adopting this indigenous technical knowledge over 30 years to control pest and diseases in groundnut. 10 kg of well powdered farm yard manure is mixed with 10 kg of ash (collected from brick kiln) and dusted over the groundnut field during morning hours. This technique can be adapted to all pulses and vegetables after 30 days of sowing/planting. This ITK is believed to control 20% of pest and disease and also provide potassium (ash) to crops.

34.9.2 Trap Crops

Crop plants more preferred by the pest for egg laying and feeding are grown as trap crop on the bunds of the main crop or one row after every 10 rows.

- To control red hairy caterpillar in groundnut used cowpea and sunnhemp as trap crop.
- Castor and sunflower used to control *Spodoptera* in groundnut.
- Inter cropping of groundnut with pearl millet reduced the incidence of Thrips, Jassid and leaf miner.
- Grow castor as a border crop (trap crop) to reduce the attack of tobacco cut worms.

***Fig. 34.7** Groundnut + sunflower, Groundnut + castor, Groundnut + pearlmillet*

34.9.2.1 Red Hairy Caterpillar

- Cowpea grown as an intercrop in groundnut fields acts as a trap crop to control red hairy cater pillar.
- Grow cowpea as border crop to attract red hairy caterpillar
- Summer ploughing is practiced to expose and destroy the pupae of red hairy caterpillar pest.
- Grinding well and dissolving 10kg of the leaves of *Aloe vera* in water and spaying for an acre to control red hairy caterpillar in ground nut.

Fig. 34.8 Groundnut + cowpea, summer ploughing to control insects

Others

- During nights, burning heaps of straw in several places near the fields along with a bucket full of water or castor cake dissolved water near the fire to attract and kill the insects.
- Spray lime solution to control leaf roller.
- Spray water to control leaf roller attack.
- To control groundnut ring mosaic, dried sorghum or coconut leaves are powdered and boiled in water to 60° C for one hour, filtered, diluted and sprayed for two times at 10 and 20 days after sowing.
- Mix neem oil with irrigation water during second or third irrigations to prevent root rot in groundnut.
- Spray neem oil @ 6 lit/ac to control root rot.
- Neem oil solution 4% or neem kernel extract 6% is sprayed to control rust disease in groundnut.

34.9.3 Rat Control

Rat damage is the severe menace to crop production. Few indigenous techniques are popular among the farming community over the years. They are

- Keeps the clay ball into the rat holes
- Killing the rat by inserting long iron rod/needle by into rat holes

- Placing the "T" shape sticks (owl stand) in field to attract owl at night time will help in reducing the rat damage
- Strengthen the bund and increase the height of bunds to prevent formation of rat holes.
- Rats do not live in the fields where sheep penning is being practiced
- Pieces of palmyra (*Borassus flabellifer)* leaves are tied on the pole and fixed on the field. The sound produced by the leave scares away the rats.
- To reduce rate population, rat holes are dug and rats are killed after each harvest
- To catch the rats, a trap made up of wire loops on bamboo pegs is being used
- Use of soaked rice as bait attracts more rats
- Planting closely notchi (*Vitex negundo)* and erukku *(Calotropis gigantea)* around the fields as a fence helps to control rat problem.
- Putting the branches of Thanga arali (*Tecoma stands)* around the fields to control rats.
- Big round shaped earthen pots are burried on the field at ground level. Half of the pot is filled with mud slurry on which baiting material is put on a coconut shell. Attracted rats fall inside the pot a*nd* they cannot climb up and get killed.

34.9.3.1 Extract of Ipomoea Fistulosa

Leaves of Ipomoea are boiled in water and filtered. Sorghum grains are boiled in this extract and placed near the rat burrows. It is believed that rats die after eating it.

Others

- Pots are filled with water and buried in the field.
- By leaving space (nearly 1 ft) in all four sides of field
- Cow dung balls are immersed in kerosene and kept in 8-10 places of the field randomly
- Papaya- cut in to pieces and spread around the farm (3 fruits/acre)

34.9.4 Bird's Scaring

- Stones are thrown with the help of 'Kavan' (a 'V' shaped twig with string at top) is made by tying long pieces of gunny thread to both ends of small piece of leather.
- Tying a black cloth to a long pole and placing it in the centre of fields to drive away the crows
- Waste magnetic tapes taken from audio-cassettes are tied in the fields in a criss-cross manner, which the birds will think as a net.

- Pieces of palmyra (*Borassus flabellifer)* leaves are tied on the pole and fixed on the field. The sound produced by the leave scares away the rats
- Beating drums and using crackers

34.9.5 Storage Pest Control

- Groundnut pods are threshed by spreading on a threshing floor and beating with wooden sticks (for oil extraction)
- Drying the seeds and grains of all the crops on the new moon day before storing. It avoids pest attack.
- Crop harvested on full moon day are more prone to storage pest attack
- Mixing the dried leaves of neem with seeds while storing them.
- Generally seeds stored with their outer coat/shell escape from storage pests.
- Coconut oil, peanut oil and sesame oil can be used for protection from bruchids. 5 ml of the oil is mixed with 1 kg of grain legumes. This provides protection up to six months from bruchids.
- Kitchen ash is spread around the groundnut storage bags to prevent insect attack.

34.10 Success Stories

In 1993, Mr Gothi encouraged Shri Bavanjibhai Rathod of Satapar village to take up an experiment on the use of butter milk. Bavanjibhai sprayed butter milk on a twenty-five-day old rainfed crop of groundnut (variety Gujarat-10). To his surprise he found a spectacular effect on the growth and yield of the crop. The yield of pods was higher by about 250 kg/acre while fodder yield was almost thrice compared to the control. He observed that spraying of buttermilk helped both the crops to survive the water stress period of 47 days caused by delay in the last spell of monsoon rains. Jamanbhai could get some yield in the treated plots while the entire crop in the other fields failed completely due to dry spell.

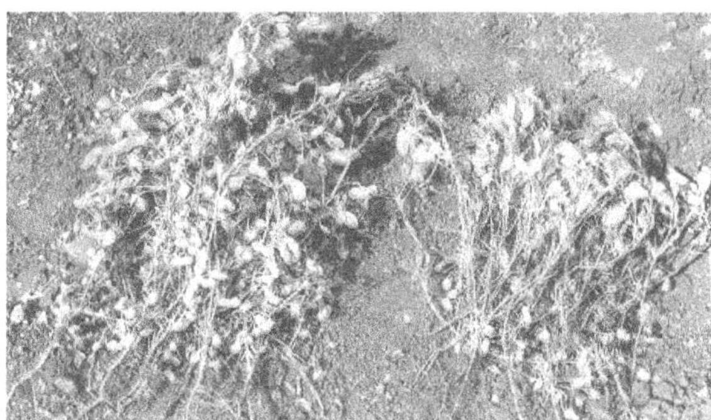

Fig. 34.9 Source: *Honey Bee Vol 7(1) January-March, 1996*

34.10.1 Intercropping for Disease Control in Groundnut

'Tikka' (***Cercospora arachidicala)*** disease of groundnut is caused by a fungus and appears at flowering stage and reduces yield drastically. Intercropping of groundnut and greengram (***Vigna radiata***) in the ratio of 4:1, controls '*tikka*' disease completely. Ashvinbhai encountered this practice accidentally. Once he planted groundnut with greengram and observed that his field was free from '*tikka*' disease while it had affected fields of other farmers in the village. From that year onward he started intercropping of the two. He is practicing it for the last five years and slowly other farmers of his village are also trying it out. (*Source*: Honey Bee Vol. 11(2) April-June, 2000).

34.10.2 Blackgram as Intercrop

Mr. R. Venkatasubbu, Pandasolanallur, Pondicherry. When Black gram is raised in the irrigation channels of the main field with groundnut crop, the incidence of groundnut pests is reduced. Castor is also sown as border crop in the field with groundnut as main crop. Castor acts as a trap crop for control of larval pests of the groundnut crop. Before cultivating the groundnut crop, Bajra is raised. After the harvest of Bajra ear heads, the stumbles and crop residues are left in the field to dry and later incorporated into the soil by ploughing. This is believed to reduce the incidence of nematodes in the succeeding groundnut crop. (Source: Honey Bee Vol. 11(3) July-September, 2000).

34.10.3 Groundnut Blight

The fungal blight (c.o. ***Alternariaalternata***) infects groundnut and characteristic lesions appear on the leaves. Soaking seeds in a mixture of two litre water and 250 g asafetida (***Ferula asafoetida***) per 20 kg seeds for 12 hr, is one of the traditional practices to reduce groundnut blight. Gum resin of asafoetida yields an essential oil which acts both as bactericidal and fungicidal agent (Chopra et.al., 1996). (***Source***: Honey Bee Vol. 11(4) & Vol. 12(1) October 2000 – March, 2001).

References

1. *Anonymous. 2012. Agricultural Statistics at a glance-2012. Directorate of Economics and Statistics, Department of Agriculture and Cooperation, New Delhi. pp 97-98.*
2. *Karthekeyan, C., T. Veeraragavathatham, D. Karpagam and S. Ayisha Pirthous. 2006. Indigenous Technical Knowledge in garden land conditions. Open and Distance Learning Education, Tamil Nadu Agricultural University, Coimbatore- 641 003.*
3. *Warren, D.M. 1991. Indigenous Agricultural systems and Development. Agriculture and Human Values, 8 (1-2).*

Newsletters

1. Honey bee
2. CIKS

Journals

1. Indian Journal of Traditional Knowledge (IJTK)
2. International Journal for Sustainable Development
3. Journal for Sustainable Development

Websites

1. www.ciks.org
2. www.sristi.org

Others

1. www.agritech. tnau.ac.in
2. www.indiaenvironmentportal.org.in

Chapter 35

The Food and Processing Quality Characteristics of Groundnut

J. B. Misra

Technical Advisor, Indian Oilseeds and Produce Export Promotion Council Mumbai-400 021, India.

35.1 Introduction

The ability of any foodstuff to provide any one or more of the major nutrients, vitamins and minerals required for growth and maintenance of human body without being deleterious to health, determines the nutritive value or quality of that foodstuff. There are only a few foodstuffs in nature that are so full of important nutrients as the groundnut kernels are. The protein content of groundnut kernels is more than that in meat and two-and-a-half times that in eggs. Thus groundnutsare one of the cheapest sources of quality protein and have rightly earned the sobriquet of 'a poor man's nut'.

35.2 Nutritive Value of Groundnuts

The nutrient composition of groundnut vis-à-vis major cereals (wheat and rice) is indicated in Table 35.1.

***Table 35.1** Composition of nutrients in groundnuts and cereal grains*

Constituents	Groundnut	Wheat	Rice
Moisture (%)	3.0	12.2	13.7
Protein (%)	25.3	12.1	6.8
Fat (%)	40.1	1.7	0.5
Minerals (%)	2.4	2.7	0.6
Fibre (%)	3.1	1.9	0.2
Carbohydrate (%)	26.1	69.4	78.4
Energy (Kcal/100g)	567	341	345
Ca (mg/100g)	90	48	10

Table 35.1 *Contd...*

Constituents	Groundnut	Wheat	Rice
P (mg/100g)	350	355	160
Fe (mg/100g)	2.5	4.9	0.7
Carotene (μg/100g)	37	29	0
Thiamine (mg/100g)	0.90	0.49	0.06
Riboflavin(mg/100g)	0.13	0.17	0.06
Niacin(mg/100g)	19.9	4.3	1.9
Source: Gopalanet *al.* (1971)			

With about 25% protein, 48% oil, and 3% fiber and high in calcium, thiamine and niacin contents, it has all the potential for use as highly economical food-supplement to fight malnutrition that occurs due to deficiencies of these nutrients in the cereal grains like wheat and rice. Thus groundnut is a nature's gift to man in general and to children, pregnant and nursing women and the poor in particular. The nutritive virtues of groundnuts have attracted the attention of National Aeronautics and Space Administration of the USA. They have identified groundnut as a possible food for advance life-support system for extended space missions.

35.2.1 Energy

Being rich in oil, the calorific value of groundnut kernels is quite high (approx. 570 Kcal/100g). Therefore figure-conscious people sometime avoid consuming groundnuts. The calorific value of the processed groundnut can, however be lowered by partial removal of fat from the kernels. By this technique up to 80% of the oil can be removed.

35.2.2 Carbohydrates

The total carbohydrate content of groundnut kernels may range from 16 to 26%. The carbohydrates are energy giving nutrients (4 kcal/g). Starch (8-14%) and sucrose (3-7%) account for the major fraction of carbohydrates while the reducing sugars (glucose and fructose) comprise a very small fraction (1-2%). Glucose and fructose (monosaccharides), sucrose (disaccharide), and starch (polysaccharide) contribute to the calorific value of groundnut. The oligosaccharides (stachyose andraffinose) and the cellulosic polysaccharides are not digestible hence do not contribute to the calorific value. The cellulose, however, contributes to the fiber-value of kernels. The reducing sugars are associated with browning of kernels that occurs during roasting.

35.2.3 Protein

Amongst the foodstuffs of plant origin, the protein content of groundnuts is second only to soybean. On the criteria of ease of cooking, palatability and digestibility, the groundnut protein, outperforms soybean proteins. As both the content and digestibility of groundnut protein is quite high, groundnut crop can play a crucial role in alleviating protein malnutrition in the developing countries. The groundnut protein comprises two major fractions viz. arachin and conarachin. Although genotypic differences exist in the amino acid composition of protein of different groundnut cultivars, yet these differences are so narrow that they may not be reflected in the overall nutritive quality of groundnut protein.

Compared to FAO pattern of requirements of essential amino acids, the groundnut protein appears to be deficient in lysine, methionine, threonine, and isoleucine. Aspartic acid, glutamic acid and arginine are the most abundant amino acids of groundnut protein and collectively account for about 45% of the total amino acids. The amino acid composition of groundnut is shown in table 35.2.

***Table 35.2** Amino acid composition (g/16g N) of groundnut protein*

Essential amino acid	Cultivar			Non-essential amino acid	Cultivar		
	Kadiri-3	J-11	Mean of five		Kadiri-3	J-11	Mean of five
Isoleucine	3.28	3.33	3.45	Alanine	4.71	4.12	4.11
Leucine	6.30	6.48	6.32	Arginine	11.28	11.47	11.31
Lysine	4.40	4.47	4.30	Aspartic acid	11.18	11.21	11.46
Methionine	1.26	1.21	1.29	Cystine	1.26	1.24	1.23
Phenylalanine	4.96	5.35	5.24	Glutamic acid	21.29	21.88	21.83
Threonine	2.40	2.25	2.52	Glycine	5.52	6.42	5.86
Tryptophan	-	-	-	Histidine	2.21	2.12	2.22
Valine	4.03	4.13	4.32	Proline	6.40	6.30	5.83
				Serine	5.09	5.05	5.32

Source: Jambunathan*et al.*(1992)

35.2.4 Oil

Vegetable oils contain varying degree of unsaturated fatty acids and hence are considered healthier than the fats of animal origin which contain mostly saturated fatty acids. Like most vegetable oils, groundnut oil has a calorific value of 9 Kcal/g. The oil is light yellow in colour, slightly nutty flavour, and low in viscosity. It also contains small quantities of mucilage, albuminous material and free fatty acids, which, however, are absent in the refined oil. The nutritive value of edible oils is by and large determined by their fatty acid composition. Groundnut oil comprises mixed glycerides of about 14 types of fatty acids of which about 80% are unsaturated and 20% are saturated. Nutritionally, however, only eight fatty acids viz. palmitic, stearic, arachidic, behenic, lignoceric, oleic, eicosenoic and linoleic are considered important. Of these, oleic acid– a Monounsaturated Fatty Acid MUFA), and linoleic acid– a Polyunsaturated Fatty Acid (PUFA), are the major fatty acids of groundnut oil, which together constitute about 80% of the total fatty acids.

Among the saturated fatty acids, which comprise the remaining 20% of the total fatty acids, palmitic acid has the largest proportion. Groundnut oil, however, lacks omega-3 fatty acid. There are 0.02-0.6% free fatty acids in groundnut oil which make groundnut and its products susceptible to oxidation (rancidity).Fatty acid composition of groundnut oil is given in table 35.3.

Table 35.3 *Fatty acid composition (%), oleic to linoleic (O/L) ratio and iodine value of oil of groundnut genotypes differing in habit groups*

Fatty acid	Virginia runner		Virginia bunch		Spanish bunch	
	Range	Mean	Range	Mean	Range	Mean
Palmitic (16:0)	8.54-10.77	9.61	9.25-12.12	10.54	9.41-12.61	11.48
Stearic (18:0)	2.03-3.14	2.63	1.55-3.41	2.42	1.74-5.00	3.40
Oleic (18:1)	50.55-54.69	52.18	37.64-49.13	43.42	36.30-49.60	41.89
Linoleic (18:2)	29.70-33.37	31.96	33.82-46.65	39.68	29.96-44.32	39.64
Arachidic (20:0)	0.17-0.53	0.31	0.19-0.79	0.34	0.19-0.80	0.40
Eicosenoic (20:1)	0.51-1.06	0.83	0.57-1.19	0.88	0.33-0.97	0.62
Behenic (22:0)	1.50-2.89	2.37	1.05-4.10	2.77	1.32-4.04	2.67
O/L ratio	1.53-1.83	1.64	0.82-1.45	1.11	0.83-1.53	1.09
Iodine value	99.2-100.9	100.9	101.5-114.2	106.4	93.7-111.2	105.2

Source: Rahejaet *al.* (1987)From the culinary point of view, groundnut oil is considered superior to most vegetable oils as it does not impair the flavour of herbs and spices by contributing its own.

The main dietary essential fatty acid for man is linoleic acid, a PUFA. An intake of 1% of dietary calories in form of PUFA has been found to be adequate. High intake of PUFA is considered not desirable from health point of view. Hence the dietary energy coming from PUFA should not be more than 10% of the total calorific value. Groundnut oil contains much less PUFA (approx. 33%) than the oils from soybean, safflower, sunflower, niger, cottonseed and corn, and hence is in greater conformity with the recommendations on dietary allowances compared to most vegetable oil.

35.2.5 Vitamins

Vitamins are essential for facilitating utilization of major nutrients for promoting growth and development and also repair of human body. So far as vitamin D, K and cobalamin (vitamin B12) are concerned, there is no evidence to suggest their presence in groundnut kernels while the occurrence of vitamin C is disputed. Among the foods of plant origin, the groundnuts are one of the richest sources of thiamine (B1). The kernels are also rich source of niacin (vitamin B3) but very low in fat-soluble vitamins A and D. About 25% of the total thiamine is found in the red skins of the kernels. Hence a large proportion of this vitamin is destroyed during roasting and blanching or by blanching alone. The vitamin content of groundnut is given in table 35.4.

Table 35.4 *Vitamins in groundnut kernels*

Fat soluble (units/100g)		Water soluble (mg/100g)	
Vitamin	**Content**	**Vitamin**	**Content**
Vitamin A (IU)	26	Thiamine	0.99
Vitamin E (mg)	26.3-59.4	Riboflavine	0.13
α-tocopherol (mg)	11.9-25.3	Pyridoxine	0.30
β-tocopherol (mg)	10.4-34.2	Niacin	12.8-16.7
γ-tocopherol (mg)	0.58-2.50	Choline	165-174
		Folic acid	0.28
		Inositol	180
		Biotin	0.33
		Pantothenic acid	2.72
		Vitamin C	5.80

Source: Cobb and Johnson (1973)

35.2.6 Minerals

The total mineral content of groundnut ranges from 2 to 3%. It is good source of some primary dietary minerals such as calcium, magnesium, potassium, sodium, iron, and certain trace elements such as manganese, copper, zinc, and boron in appreciable amounts. Although groundnuts are rich in Ca, Mg and Fe, the contents of these elements do not reflect their nutrient availability correctly as these are present as salts of phytic acid and oxalic acid. As many as 26 inorganic constituents have been found in groundnut kernels as sown in table 35.5.

Table 35.5 *Mineral composition of groundnut kernels*

Mineral	Content (mg/100g)	Mineral	Content (mg/100g)
Potassium	680-890	Boron	2.6-50
Sodium	Trace	Fluorine	0.14
Calcium	20-80	Iodine	0.02
Magnesium	90-340	Strontium	0.8-5.0
Phosphorus	250-660	Barium	8.30
Sulfur	190-240	Vanadium	10-50
Chlorine	Trace	Chromium	1-30
SiO_2	80	Aluminum	100
Zinc	1.7-80	Nickel	3-8
Manganese	0.8-50	Titanium	30-80
Iron	1.8-100	Molybdenum	0.8-3.0
Cobalt	0.03	Tin	0-5
Copper	0.7-30	Lead	0-50

Source: Freeman *et al.* 1954

35.2.7 Fibre

The crude fibre content of groundnuts may range from 2 to 4%. The fibre of groundnut kernels comprises mainly of cellulose, hemicellulose, pectin and small quantities of mucilages, gums and lignin. Considering a recommended intake of 40 g dietary fibre a day, groundnut is not considered a source of dietary fibre, although it may supplement fibre supply.

35.2.8 Noxious components

Like most legumes, groundnut kernels also contain several noxious factors, which can adversely affect the nutritive value or quality of kernels. These factors are groups of unrelated chemical compounds.

(i) **Flatus sugars:** Raffinose and stachyoseare the major oligosaccharides responsible for flatulence (gas formation) in humans. The flatus sugar content of defatted groundnut meal/flour (0.2 to 1.5%) is much lower than that of defatted soybean meal/flour (2-4%). Hence the groundnut mealit is a better supplement for fortifying wheat flour with protein.

(ii) **Saponins:** The bitter flavour that is sometimes encountered while eating groundnuts is due to saponins. In groundnuts, the saponinsare concentrated in the germ of seed. Saponins are soap-like substances characterized by their ability to create foam in water. This ability to behave like a detergent is due to their chemical structure, with one end binding easily to water and the other to cholesterol and fat. Saponins latch on to cholesterol molecules attached to the surface of intestinal cells. Once they attach, they stimulate a reaction that creates pores in the cell surface, producing increased permeability (i.e., leaky gut) and allowing substances to enter directly the bloodstream.

(iii) **Allergens:** Some individual are hypersensitive to groundnut and show swelling, reddening of skin, itching, and choking within an hour of consuming groundnuts and some may even show anaphylactic shock. The allergenicity is spread through both crude arachin and conarachin fractions of groundnut protein. Heat induced conformational changes do not affect the allergenic properties of the major allergen *Ara H 1.* In the USA, one out of every 200 persons is allergic to groundnuts. A clear statistics about the susceptibility of the Indian population is not available.

35.3 Chemical Factors Influencing Shelf-life of Groundnut and its Products

Groundnuts are semi-perishable in nature. Due to ageing, during the post-harvest storage, there is a noticeable deterioration in the organoleptic quality of the stored groundnuts. The deterioration in flavour can also result from chemical reactions, particularly oxidation of fats due to naturally occurring enzymes in groundnuts. The rate and extent of deterioration depends mainly on the fatty acid profiles. When oxidation occurs, the fat in the groundnuts becomes rancid and produces off-flavours.

The rate of enzymatic oxidation of groundnut increases as the groundnutstorage temperature rises. Hence, the storage life of groundnutsincreases dramatically as temperature and relative humidity are lowered. Ideally, groundnuts should be stored within a temperature range of 3-5°C with a relative humidity of 55 to 70%. The lower humidity level produces a lower moisture environment which retards the growth of molds. Temperature reduction not only retards the production of microorganisms but also breaks any insect reproduction cycles.

Under ambient storage conditions, the off-flavours and rancidity develop in a few months only. Compared to in-shell groundnuts the shelf-life of kernels without shell is relatively short. Among the chemical constituents, the profile of fatty acids especially the ratio of oleic to linoleic acid (O/L ratio), free fatty acid and tocopherolcontents, peroxide value, and *p*-anisidine value collectively determine the shelf-life. The sucrose content is the chief determinant of flavour. The attribute of susceptibility to browning upon roasting is due to a reaction between the free amino acids and the reducing sugars at high temperatures. This attribute assumes importance in production of roasted-blanched kernels and also peanut butter.

35.4 The Benefits of Eating Groundnuts

- Groundnuts and its products are very beneficial in the treatment of hemophilia and other such inherited blood disorders. People suffering from nose bleeding also benefit from eating groundnuts and it is also helpful in reducing excessive menstruation bleeding in women.
- Five main components required by the body for its maintenance and repair are food- energy, protein, phosphorous, thiamine and niacin. All these five components are found in good quantities in groundnuts.
- Groundnuts are rich in vitamins and contain at least 13 different types of vitamins that include vitamin A, B, C and E. Along with this, groundnuts are also rich in 26 essential minerals like calcium, iron, zinc, boron, etc. these help in brain function and development and also help to maintain strong bones.
- It is recommended that growing children, expecting women and nursing mothers consume roasted groundnuts with jaggery and goat's milk. It enhances resistance and immunity against dangerous infections like hepatitis and tuberculosis.
- Groundnuts are rich in anti-oxidants and a chemical called resveratrol. These help in reducing the risk of contracting cardiovascular diseases and cancer and also act as anti-ageing factors.

35.5 Quality in Processing and Export Parlance

The Indian groundnuts have a rich nutty flavour, sweet taste, crunchy nature and a relatively longer shelf-life. As of now in India, nearly two hundred groundnut varieties have been notified and of these only about 40-50 are in the seed-chain.

The share of India in the export of groundnut to top ten importing countries during 2013-14 is given in table 35.6.

Table 35.6 *Share of India in the total groundnuts imported by the top ten countries in the year 2013-14*

	Country	Quantity imported (tons)	Exported from India	
			Quantity (tons)	Share (%)
1	The Netherlands*	325,000	692	0.2
2	Indonesia	245,000	227,755	93.0
3	China	180,000	2,687	1.5
4	Germany*	121,000	380	0.3
5	Canada	116,000	88	0.07
6	Mexico	110,000	2,770	2.5
7	Russia	95,000	9,438	9.9
8	UK*	85,000	1,306	1.5
9	Philippines	70,000	60,384	86.3
10	Thailand	66,000	23,116	35.0

*The EU countries; ***Source:*** Oil World 2014 and DGCIS

For determining export-worthiness of the groundnut varieties, besides oil and protein contents, information on a few physical attributes as well as on several other chemical attributes is required. The significance of various physical and chemical attributes in determining the quality of groundnuts kernels is given in table 35.7.

Table 35.7 *Important physical and chemical quality attributes of groundnut kernels*

Factor	Significance
Physical	
Sound mature kernels	Appeal to eyes
Shape of kernels	Appeal to eyes and ease of blanching
Seed size uniformity	Appeal to eyes and ease of processing
Hundred seed mass	Indicator of average seed size
Colour of testa	Appeal to eyes before and to some extent after blanching
Moisture	Determines extent of splitting during decortication and susceptibility to growth of moulds
Chemical	
Oil	Determines flavour and intake of dietary energy

Table 35.7 *Contd...*

Factor	Significance
Free fatty acids	Determine susceptibility to oxidation (rancidity)
Stability index (O/L ratio)	Determines the shelf-life
Tocopherol	Determines the shelf-life
Protein	Important nutrient
Free amino acids	Determine colour and flavour of the roasted groundnuts
Sucrose	Determines taste and flavour of the roasted groundnuts
Reducing sugars	Determine colour and flavour of the roasted groundnuts
Overall flavour	Appeal to taste
P-anisidine value	Determines the shelf-life
Peroxide value	Determines the shelf-life
	Source: Adapted from Misra (2004).

35.5.1 Trade Types: Shape and Colour

Among the physical attributes shape and size of seeds and the colour of testa are considered important. To a very large extent these attributes are determined by the botanical habit group to which a variety belongs. For the purpose of international trade (export/import), the groundnut varieties are classified broadly into only three types on the basis of shape, size and skin-colour of their kernels, viz., Bold (Virginia runner and Virginia bunch), Java (Spanish) and Red Natal (Valencia).

(a) **Bold or Runner:** This includes varieties of both spreading and semi-spreading habit. The seeds of spreading group are attractive bold and of uniform medium size and are preferred in candies. The largest-red-brown-skinned (Virginia bunch) kernels are used for cocktail and salted nuts. The medium-size kernels (Virginia runner) are used for peanut butter, oil and candies. 'GAUG-10' is the most preferred variety among the spreading habit group. The other popular variety is 'M-13'. The seeds of semi-spreading group are the largest and elongated. The varieties of this habit group account for most of the roasted in-shell groundnuts. The larger kernels are sold as salted and roasted groundnuts. 'GG-20' is the most preferred variety in the semi-spreading habit group. Some other semi-spreading varieties are 'TMV-10', 'ICGS-76', 'BAU-13', 'HNG-10', 'Girnar-2' and 'GJG-22'.

(b) **Java or Spanish:** They have small and round kernels covered with a pink or salmon colour skin. The seeds of Spanish types have higher oil content than those of spreading and semi-spreading types. These are used predominantly in coated confections while significant quantities are used for snack-nuts and peanut butter and also for crushing for oil expulsion. A few African countries are the main competitors of India in the export of this type of groundnut. In this group, the most preferred variety by the exporters-importers is 'J-11'. The other important varieties are 'SB-XI', 'SG-99', 'GG-1', 'GG-2', 'ICGS-44', 'ICGV-91114', 'VRI-2', 'TG-

37A', 'TKG-19A', 'AK-159', 'Girnar-3', 'GPBD-4', 'Utkarsh', 'OG-52-1' and 'Kadiri-6'.

(c) **Red Natal or Red-skin Java or Valencia:** Their seeds have a bright-red skin and pods have three or four seeds. The seeds taste sweet and are usually sold as roasted-in-shell. These are much in demand as ingredient of roasted or fried salted-snacks and are also considered excellent for fresh use as boiled groundnuts. 'Gangapuri' is the most well-known Valencia variety. The other varieties of this group are 'Kopergaon-3', 'TMV-11', 'MH-2' and 'MH-4'.

The seeds of Spanish bunch variety 'Narayani', have a bright-red skin. 'Narayani' is quite popular in the southern states of India and its seed often make a good substitute for Valencia types. The seeds of another Spanish variety'JL-24' are quite large (hundred seed mass 53-54 g) and can be considered as a round-bold type.

35.5.2 Count Per Ounce

Within each of the traded type, the seed size is another important determinant of quality, which in international trade is expressed as the count per ounce i.e. is the number of seeds in an ounce (28.35 g). The larger the count the smaller is seed. In the scientific parlance hundred-seed mass (HSM, expressed in grams), however, is used to indicate the average seed size of a genetic variety– the higher the HSM the larger is the seed– which in principle is an inverse of count per ounce. Within each trade type, the prevalent counts along with their HSM equivalent are given in table 35.8.

Table 35.8 *Grades (count per ounce) used in trade and equivalent HSM*

BOLD		JAVA		RED SKIN JAVA	
Counts per ounce	Equivalent HSM (g)	Count per/ ounce	Equivalent HSM (g)	Count per ounce	Equivalent HSM (g)
35/40	70.9-81.0	-	-	-	-
38/42	67.5-74.6	-	-	-	-
40/45	63.0-70.9	-	-	-	-
45/50	56.7-63.0	40/50	56.7-70.9	-	-
45/55	51.5-63.0	45/55	51.5-63.0	-	-
60/70	40.5-47.3	60/70	40.5-47.3	-	-
70/80	35.4-40.5	70/80	35.4-40.5	70/80	35.4-40.5
-	-	80/90	31.5-35.4	80/90	31.5-35.4
-	-	100/120	23.6-25.8	-	-

35.5.3 Moisture Content

The moisture content of the kernels has a direct bearing on the extent of splitting of kernels during decortication. In many processing units there is an undesirable practice of sprinkling the pods with water to enhance moisture content and thus avoid splitting of kernels. The moisture content of a lot to be exported should not be more than 7%. The higher moisture content (>9%) makes the lot susceptible to invasion by *Aspergillus* and enhances the chances of insect activity.

35.5.4 Aflatoxin Contamination and its Impact on International Trade

Aflatoxin is a group of toxic compounds produced by the fungi *Aspergillus flavus* and *A. parasiticus*,when these fungi invade the groundnuts during pre- and post-harvest stages. The four major types of aflatoxin reported to occur in groundnuts are B1, B2, G1 and G2. Aflatoxin B1 is a potent hepatocarcinogen and exerts it toxic effects even if present in microgram quantities in the diet.

The maximum permissible limits for aflatoxins fixed by some countries are: 30 ppb (total) by India; 20 ppb (total) by Brazil, Kenya, Philippines and Saudi Arabia; 15 ppb (total) by the USA, Canada, Indonesia, Australia and the UAE; 10 ppb (total) by Japan, Korea, Taiwan, Vietnam and Malaysia; 5 ppb (total) by Russia, Singapore and Turkey and 2 ppb (B1) and 4 ppb (total) by the European Union.

In 2013-14, India exported 511, 662 metric tons of groundnuts (valued Rs. 3,212 crores) and the bulk of this quantity was imported by Indonesia (227,755 tons), the Philippines (60,384 tons), Malaysia (59,259 tons), Vietnam (40,466 tons), Thailand (23,116 tons) etc. while the export to the EU countries was very small: 1,306 tons to the UK; 692 tons to the Netherlands, 389 tons to Greece, 380 tons to Germany, 377 tons to Croatia and in even smaller quantities to some of the other EU countries.

The European Union is the largest importer of groundnuts in the world and is quite fastidious about the quality of groundnuts it imports. The USA and Argentina continue to be the major suppliers of groundnuts to EU. The Netherlands, a member of EU, is the largest importer of groundnuts in the world which imported 3,25,000 tons in 2013-14. India contributed only a negligible fraction (0.2 percent) to the total groundnuts imported by the Netherlands by supplying only 692 tons.

Because aflatoxins are not an innate constituent of groundnuts, their distribution across the kernels is highly skewed, it is not possible to determine the average value of aflatoxin-load in a consignment of groundnuts but only a probabilistic risk-assessment can be made by adopting a statistically designed sampling technique. Due to such sampling procedures which are followed at the ports of destination as well as origin, the values reported from different laboratories differ quite often and become a matter of dispute. Many a time, a consignment which is declared 'safe' or 'fit' at the port of origin is found 'unsafe' or 'unfit' when sampled at the destination port.

In the last decade, due to aflatoxin contamination several consignments have been rejected at the destination ports and thus the Indian exporters incurred heavy losses. Most of the Indian processing units find it difficult to maintain consistent quality in their export consignments with respect to permissible limits on aflatoxins set by the EU countries.

In groundnut, aflatoxin contamination can occur at various pre-harvest and post-harvest stages. The Good Agricultural Practices (GAP), good storage and transport practices, Good Manufacturing Practices (GMP) and good handling and shipment practices which need to be adopted for managing aflatoxin contamination have been described and published by several agencies including the *Codex Alimentarius*.

In fact, considering the current international export-import scenario, India has a huge potential for exporting high-quality groundnuts to the international buyers among EU countries (the Netherlands, Germany, the UK, Spain, Italy and others), the Russian Federation, Canada and Mexico.

The demand for organic groundnut is also growing in domestic as well as in the international market. The north-eastern states especially Tripura and Nagaland hold a high potential for production of organic groundnut.

The awareness and concern for quality amongst the Indian groundnut farmers, shellers and processors is growing steadily. Multiple sorting and grading are fast becoming a norm. Indian shippers have the capability to prepare and supply edible groundnuts conforming to highest standards.

35.6 Future Perspective

For maintaining quality of the processed groundnuts intended for export, it is necessary that the processors use raw material with consistent quality, lot after lot, and season after season. Although aflatoxin contamination is currently the main non-tariff barrier to export groundnuts from India to EU countries, there are several other factors too that determine the consistency in the quality of the groundnuts to be exported.

Information on export-worthiness of Indian groundnut varieties, however,is not available in a comprehensive manner. The published information is not of much practical value as the varieties included in the studies were mostly grown at a single place and did neither cover all the important attributes nor all the prevalent/recommended varieties. So far there has been a greater emphasis on developing new groundnut varieties from the point of view of enhancing yield through genetic improvement of resistance/tolerance to various biotic and abiotic stresses and the research projects aimed at improving quality from export point of view are yet to deliver results.

It is therefore necessary to generate comprehensive information on the quality attributes of the groundnut varieties popular amongst the Indian farmers and also those which have been recommended by the Central or State Governments for cultivation in the major groundnut growing areas. This will lead to identification of varieties superior in terms of export-worthiness and thus facilitate formulation of a strategy for promoting export oriented cultivation/production of the groundnut crop.

Fresh-seed-dormancy is one of the most desired traits in Spanish types. Most Spanish varieties have only little seed dormancy. Hence, if the mature crop or the harvested lots being cured in field is exposed to unseasonal rains, *in situ* germination of seed occurs which not only adversely affects the organoleptic properties of the seeds but also makes the produce susceptible to invasion by the aflatoxigenic fungi and other microbes.

There is a need for increasing the seed size of Valencia types. Identification of high-yielding groundnut varieties with bold seeds (38/42 count per ounce) which could replace GG 20 is required. Least discolouration upon roasting and ease of blanching are the required traits for exporting blanched groundnuts.

There are evidences to indicate that high-oleic groundnuts (oleic to linoleic ratio value greater than 10) not only have a longer shelf-life but also have improved sensory properties.The Indian groundnut varieties have O/L ratio in the range of 1.2 to 2.5. Thus there is a pressing need for developing high-oleic varieties without compromising on yield. High oleic varieties have already been developed elsewhere and are likely to enter the world markets in the coming years.

References

1. *Ajay BC, Kusuma VP, and Gowda MVC. 2008. Evaluation of groundnut (Arachishypogaea L.) varieties for nutritional traits. Karnataka Journal of Agricultural Science 21: 262-263*
2. *Bangar VD, Gupta DN and Chavan AS. 1997. Nutritive value of promising cultures of groundnut in Kokan region. Journal of Maharashtra Agricultural Universities 22: 154-157.*
3. *Bishi SK, Kumar L, Dagla MC, Mahatma MK, Rathnakumar AL, Lalwani HB and Misra JB. 2013. Characterization of Spanish peanut germplasm (Arachishypogaea L.) for sugar profiling and oil quality. Industrial Crops and Products, 51:46-50.*
4. *Bishi SK, Kumar L, Mahatma MK, Khatediya N, Chauhan SM and Misra JB. 2015. Quality traits of Indian peanut cultivars and their utility as nutritional and functional food. Food Chemistry 167: 107–114.*
5. *Borkar VH and Dharanguttikar VM. 2014. Evaluation of groundnut (Arachishypogaea L.) genotypes for physiological traits. International Journal of Scientific and Research Publications 4: 1-8.*
6. *Casini C, Dardanelli JL, Martínez MJ, Balzarini M, Borgogno CS, and Nassetta M. 2003. Oil quality and sugar content of peanuts (Arachishypogaea) grown in Argentina: their relationship with climatic variables and seed yield.* Journal of Agriculture and Food Chemistry
7. *Cobb WY and Johnson BR. 1973.Physicochemical properties of peanuts.In Peanuts– Culture and Uses.Stilwater USA, American Peanut Research and Education Association. Derbyshire EJ. 2014.*

8. Review of nutritional, organoleptic and biological HO peanut properties. International Journal of Food Science and NutritionEarly Online: 1–10. DOI: 10.3109/09637486.2014.937799p209-261.51:6309-13.
9. Dwivedi SL, Jambunathan R, Nigam SN, Raghunath K, Ravi Shankar K and Nagabhushanam GVS. 1990. Relationship of seed mass to oil and protein contents in peanut (Arachishypogaea L.). Peanut Science 17:48–52.
10. FAO. 1977. Dietary fats and oils in human nutrition. Report of FAO Expert Committee on Food and Nutrtion paper no. 3. FAO, Rome.
11. Freeman AF, Morris NJ and Willich RK. 1954. Peanut butter. US Deptt.Agric. IIC-370.
12. Gopalan C, Ramasastry BV, Balasubrmanian SC, Narasingarao BS, Deosthle YG, and Pant KC. 1971. Nutritive value of Indian foods.National Institute of Nutrition, Hyderabad, India.
13. Jambunathan R, Gurtu S, Raghnath K, Kannan S, Sridhar R, Dwivedi SL, and Nigam SN. 1992. Chemical composition and protein quality of newly released groundnut (Arachishypogaea L.) cultivars. Journal of the Science of Food and Agriculture 59: 161-167.
14. Jambunathan R, Sridhar R, Raghunath K, Dwivedi SL and SN Nigam. 1993. Oil quality characteristics and headspace volatiles of newly released groundnut (Arachishypogaea L.) cultivars. Journal of the Science of Food and Agriculture 61:23-30.
15. Keenan JI and Savage GP. 1994. Mycotoxins in groundnuts with special reference to aflatoxin. In The Groundnut Crop.A Scientific basis of improvement (ed. J Smartt). P 509-551. Chapman Hall, London.ISBN 0 412 408201.
16. Mann SK, Ghuman PK, and Hira CK. 1991.Amino acid composition and protein fractions of groundnut cultivars.Journal of Research, Punjab Agricultural University 28: 421-426.
17. McDaniel KA, White BL, Dean LL, Sanders TH, and Davis JP. 2012. Compositional and mechanical properties of peanuts roasted to equivalent colors using different time/temperature combinations. Jouranl of Food Science 77: 1203-1299.
18. Misra JB, Ghosh PK, Dayal D, and Mathur RS. 2000. Agronomic, nutritional and physical characteristics of some Indian groundnut cultivars. Indian Journal of Agricultural Sciences 70:741-746.
19. Misra JB, Girdhar V, Jain VK and Dhamsaniya NK. 2006.Quality attributes of peanut butter prepared from some Indian peanut cultivars. International Arachis Newsletter 26: 38-40.
20. Misra JB. 1998. Update on transformation of groundnut and potential genes for enhancement of quality. International Arachis Newsletter 17: 18-21.
21. Misra JB. 2001. Variation in Nitrogen-to-Protein Conversion Factor for Peanut. Peanut Science 28: 48-51.

22. *Misra JB. 2004. A mathematical approach to comprehensive evaluation of quality in groundnut. Journal of Food Composition and Analysis 17: 69-79.*
23. *Misra, JB, 2006, Nutritive value of groundnut and composition of Indian groundnut cultivars. In Groundnut Research in India, Ed. Basu, M. S and Singh, N. B., National Research Center for Groundnut, Junagadh, India. pp. 273-291.*
24. *Misra, JB and Nautiyal PC. 2005. Influence of imposition of soil-moisture deficit stress on some quality components of groundnut, ArachishypogaeaL. kernel. Journal of Oilseeds Research 22: 119-124.*
25. *Nagaraj G and Chauhan S. 1987, Seasonal variation in the oil quality of groundnut. Indian Journal of Agricultural Science 57: 522-524.*
26. *Nautiyal PC and Misra, JB. 2005. Effect of drying methods on seed germination and seed protein profile in groundnut, ArachishypogaeaL. Journal of Oilseeds Research 22: 125-128.*
27. *Nile SH. 2013. Fatty acid composition and antioxidant activity of groundnut (Arachishypogaea L.) products.* Food Science and Technology Research *19(06). DOI: 10.3136/fstr.19.957*
28. *Olmedo RH, NepoteV and Grosso NR. 2012. Sensory and chemical stability in coated peanuts with the addition of essential oils and synthetic antioxidants. International Journal of Fats and Oils 63: 5-13.*
29. *Pattee HE, Isleib TG, Giesbrecht FG and McFeeters RF. 2000. Relationships of sweet, bitter, and roasted peanut sensory attributes with carbohydrate components in peanuts. Journal of Agricultural and Food Chemistry 48:757–763*
30. *Pratibha KM and Reddy MU. 1994. Nutrient composition of groundnut cultures* (Arachishypogaea *L.) in relation to their kernel size. QualitasPlantarum-*Plant Foods for Human Nutrition *45:365-369.*
31. *Raheja RK, Batta SK, Ahuja KL, Labana SK and Singh M. 1987.Comparison of oil content and fatty acid composition of peanut genotypes differing in growth habit. QualitasPlantarum-Plants Food for Human Nutrition 37: 103-108.*
32. *Rajgopal RK, Chandran K, Misra JB, Bhalodia PK and Mathur RS. 2000. Evaluation of some bold-seeded groundnut (Arachishypogaea L.) accessions for confectionery attributes. International Arachis Newsletter 20: 18-20.*
33. *Riveros CG, Mestrallet MG, Gayol MF, Quiroga PR, Nepote V, and Grosso NR. 2010. Effect of storage on chemical and sensory profiles of peanut pastes prepared with high-oleic and normal peanuts.* Journal of the Science of Food and Agriculture *2010 Dec; 90(15):2694-9. DOI: 10.1002/jsfa.4142.*
34. *Sharma ND, Mehta, SL, Patil SH, and Eggum BO. 1981. Oil and protein quality of groundnut mutants. QualitasPlantarum-Plant Foods for Human Nutrition 31: 85-90.*

Chapter 36

Kernel Qualitative Traits in Groundnut

Latha .P

Scientist (Crop Physiology), RARS, Tirupati.

36.1 Introduction

After soyabean oil, peanut (*Arachis hypogaea* L.) is one of the most important and economic pellets in tropical and subtropical regions mostly planted to produce 43- 55% oil and protein (Holbrook and Stalker, 2003). Peanut seeds are mainly used in preparation of edible oil, conserve, margarine, soap and the like. Peanut meal is also an invaluable feed for livestock and its nutritional value is the same as soybean meal. After sunflower, it has the highest percentage of sunflower oil among seed crops (Khajehpour, 1999). Peanuts contain high quality proteins with all the amino acids. In this sense, they are richer than soyabeans, which are very famous for their richness in proteins. Peanuts contain nearly 50% fat which almost 70% are unsaturated. Among them are oleic and linoleic (Omega 6). Peanuts contain lipids and carbohydrates which are energy rich compounds, capable of complementing the basic energy demands of the human body. The major carbohydrate present in peanuts is starch which is finally converted to glucose and liberates energy. Peanuts have adequate amounts of carbohydrates amounting to 15% of the RDA value. Peanuts are one of the crops that are susceptible to fungal infection, especially by aflatoxin producing *Aspergillusflavus*. Aflatoxin is a very powerful and dangerous known carcinogen that may cause cirrhosis and cancer.

In peanuts, quality traits might be improved along with increased yields. Hence, it is very essential to know the methodology for estimation of quality traits viz., proteins, oil, carbohydrates, total free amino acids and aflatoxins for screening peanut germplasm with good quality traits which is an essential component of economic yield.

36.2 Proteins

Proteins are made of amino acids arranged in a linear chain joined together by peptide bonds. Many proteins are enzymes that catalyze the chemical reactions in metabolism. Other proteins have structural or mechanical functions, such as those that form the cytoskeleton, a system of scaffolding that maintains the cell shape. Some seeds do not have the optimum quantities of amino acids for human nutrition. For example, corn proteins are generally low in the amino acid lysinebut relatively high in the amino acid methionine. In contrast, soybean proteins are relatively high in lysine but somewhat low in methionine. Peanuts lack the essential amino acid L-methionine, but contain high levels of the other essential amino acids, including L-lysine.

36.2.1 Protein Estimation by Lowry Method

Protein can be estimated by different methods as described by Lowry and also by estimating the total nitrogen content. No method is 100% sensitive. Hydrolysing the protein and estimating the amino acids alone will give the exact quantification. The method developed by (Lowry *et al.* 1951) is sensitive enough to give a moderately constant value and hence largely followed. Protein content of enzyme extracts is usually determined by this method.

Principle

Proteins react with Folin-Ciocalteau reagent to give a blue colored complex. The blue colour developed is due to the reaction of the alkaline cupric tartarate with the protein and the reduction of the phosphomolybdic-phosphotungstic components in the Folin-Ciocalteau reagent by the amino acids tyrosine and tryptophan present in the protein. The blue color intensity is read at 660 nm using spectrophotometer.

Chemicals required

- Sodium carbonate (Na_2CO_3)
- Sodium hydroxide (NaOH)
- Copper sulphate (CuSO4. $5H_2O$)
- Sodium potassium tartarate ($C_4H_4O_6KNa.4H_2O$)
- Folin-Ciocalteau reagent
- Trichloroacetic acid (CCl_3COOH)
- Bovine serum albumin

Preparation of reagents

- Reagent A: 2 % sodium carbonate (Na_2CO_3 anhydrous): Dissolve 2 g of Na_2CO_3 in 0.1 N sodium hydroxide
- Reagent B: 0.5 % copper sulphate ($CuSO_4$, $5H_2O$): 500 mg in 1.0% sodium potassium tartarate (prepare fresh)
- Reagent C: (alkaline copper solution): Mix 50 ml of solution A with 1.0 ml of solution B just before use.
- Folin-Ciocalteau reagent: Take 0.5 ml of Folin-Ciocalteau reagent and 0.5 ml D.D.H_2O for one sample.

- 2 N sodium hydroxide (NaOH): 8 g in 100 ml of D.D.H_2O
- 10% Tri chloro acetic acid (TCA): 10 g in 100 ml (50 ml TCA is cold and 50 ml TCA is normal)

Extraction

1. Take 0.5 g sample and add 5 ml of 10% TCA (normal).
2. Grind the sample in a pestle and mortar.
3. The ground sample is taken in centrifuge tube, add 5 ml of cold TCA mix it and kept it in refrigerator for 15 min.
4. Centrifuge it at 3000 rpm for 15 min. Discard the supernatant.
5. Add 4 ml of 2 N NaOH to the pellet mix it and kept it for overnight.
6. Centrifuge it at 3000 rpm for 15 min.
7. Take 2 ml of supernatant and add 8 ml of distilled water.

Estimation

1. Take 0.1 ml of sample solution and add distilled water to make up the volume to one ml. and 1 ml of distilled water is added for blank.
2. Add 5 ml of reagent C to the sample solution, mix it and keep for 10 min.
3. Add 0.5 ml of Folin-Ciocalteau reagent, mix well and kept it for half an hour in dark.
4. Blue colour will be developed.
5. Read the absorbance at 660 nm.
6. Take 50 mg of BSA (Bovine Serum Albumin) and dissolved in 50 ml of distilled water as stock.
7. Take 10 ml of stock and make up the volume to 50 ml with distilled water (1 ml contains 200 µg proteins).
8. Take 0.2, 0.4, 0.6, 0.8 and 1 ml solution from working standard of BSA, add distilled water and make up to 1 ml.
9. Then proceed as that of the sample and read the color.

Calculation

Draw a standard curve using absorbance versus concentration. Find out the concentration of proteins in the sample using standard regression equation and express as mg per g fresh weight.

The protein content of the kernel can also be estimated by Infratec™ 1241 Grain analyzer and expressed as percentage.

36.3 Kernel Oil

36.3.1 Oil Estimation by Soxhlet Apparatus (SOCS)

The fat or oil is extracted from the moisture free material with ether or petrol of boiling point below 60 ^{0}C. Other substances waxes, resins, organic acids, coloring matter etc., are also extracted with fat (Sadasivam and Manickam, 1992).

Chemicals required:

Petroleum ether

Soxhlet apparatus

Thimbles

Procedure:

1. Oven dry seed for 48 h if seed is fresh.
2. Otherwise dry for 30 min in oven at 100 ^{0}C.
3. Grind 2 g of seed roughly.
4. Weigh 1 g of the ground material and transfer it to thimble.
5. Let the sample weight be W.
6. Mark the condensing flasks with the representative numbers and take its initial weight as W_1.
7. Then place the condensing flasks in the soxhlet apparatus.
8. Pour enough petroleum ether (80 ml) more than sufficient to run the cycle in the extractor.
9. Immerse the thimble into the extractor and place the extractor above the condensing flask.
10. Load all the condensing flasks along with extractors in the apparatus.
11. Connect the top of the extractor to condenser.
12. Check the water level and connect instrument.
13. Open the tap gently. Switch 'ON' instrument.
14. Set the boiling point of solvent as boiling temperature.
15. The boiling temperature may be 10 to 20 ^{0}C more than that of solvents boiling point. Example: Boiling point of ether is 40-60 ^{0}C. Boiling temperature can be 80 ^{0}C.
16. Leave the process about 45-60 min.
17. After the process time, increase the temperature to recovery temperature (maximum boiling point × 2). Example: If the boiling point is 60 ^{0}C, recovery temperature can

be 120 ^{0}C. Now do the rinsing about 2 times in order to collect the remaining oil that may be presented in the sample.

18. Now take out all the condensing flasks from the system and put them in a hot air oven at 80 ^{0}C.
19. After 15-20 min, take out all the beakers and place them in a dessicator about 5 min.
20. Weigh the condensing flask as the final weight of beaker (W_2).
21. The amount of oil present in the sample can be calculated as

$$\% \text{ Oil} = \frac{W_2 - W_1}{W} \times 100$$

The oil content of the kernel can also be estimated by Infratec^{TM} 1241 Grain analyzer and expressed as percentage.

36.4 Total Carbohydrates

Carbohydrates are the important components of storage and structural materials in plants. They exist as free sugars and polysaccharides. The basic units of carbohydrates are monosaccharides which cannot be split by hydrolysis into more simple sugars. The carbohydrate content can be measured by hydrolyzing the polysaccharides into simple sugars by acid hydrolysis and estimating the resultant monosaccharides.

Principle

Carbohydrates are first hydrolysed into simple sugars using dilute hydrochloric acid. In hot acidic medium glucose is dehydrated to hydroxymethyl furfural. This compound condenses with anthrone to form blue-green colored product which is measured at 630 nm (Hedge *et al.*, 1962).

Chemicals required

- Hydrochloric acid (HCl)
- Sulphuric acid (H_2SO_4)
- Anthrone
- Glucose

Preparation of reagents

- 2.5 N Hydrochloric acid (HCl): 20.83 ml of 35% HCl make up to 100 ml of D.D.H_2O.
- Anthrone reagent: Dissolve 200 mg anthrone in 100 ml of ice cold 95 % H_2SO_4. Prepare fresh before use.

Extraction

1. Weigh 100 mg of sample into a test tube.
2. Add 5 ml of 2.5 N HCl. The test tube is then kept in hot water bath for 3 h and cool to room temperature.
3. Add solid sodium carbonate until the effervescence ceases to neutralize it.
4. Make up the volume to 50 ml with distilled water.
5. Centrifuge for 20 min at 2000 rpm.
6. Collect the supernatant.

Estimation

1. Take 1 ml of aliquot for analysis.
2. Cool the aliquot on ice and add 4 ml of ice cold anthrone reagent.
3. Kept in hot water bath for 8 min.
4. Cool rapidly and read the green color intensity in a spectrophotometer using photometric method at 630 nm.
5. Prepare the reagent blank as above by taking 1 ml of distilled water instead of the extract.
6. Dissolve 100 mg glucose in 100 ml of distilled water in a volumetric flask and taken as stock.
7. Take 10 ml of this stock standard and dilute to 100 ml in another volumetric flask for working standard solution.
8. A series of volume from 0.0 to 1.0 ml of this standard solution gives a concentration range 10 to 100 µg.
9. Then proceed as that of the sample and read the color.

Calculation

Draw a standard curve using absorbance versus concentration. Find out the concentration of the total carbohydrates in the sample using standard regression equation and express as g per g fresh weight

36.5 Total Free Amino Acids

The Amino Acids are the organic compounds that form the basic building blocks of proteins. The common feature of the structure of amino acids is having a minimum of two ionizable groups; the acidic carboxyl (-COOH) and the basic amino ($-NH_2$) groups on the same carbon atom called as α-carbon atom. The amino acids also exist in the free form and not bound to proteins are known as free amino acids. They are mostly water soluble in nature. Very often in plant during disease conditions, the free amino acids composition exhibits a change and hence, the measurement of the total free amino acids gives the physiological and health status of the plants.

Principle

Ninhydrin, a powerful oxidizing agent, decarboxylates the alpha-amino acids and carboxylates to give an intensely colored bluish purple product which is colorimetrically measured at 570 nm (TheymoliBalasubramanian and Sadasivam, 1987).

Ninhydrin + alpha- amino acid ---→ Hydrindantin + decarboxylated Amino acid + carbon dioxide + Ammonia.

Hydrindantin + Ninhydrin + Ammonia ---→ Purple coloured product + Water.

Chemicals required

- Ninhydrin
- Stannous chloride
- Citric acid
- Sodium citrate
- Methyl cellosolve
- N- propanol
- Ethanol
- Leucine

Preparation of reagents

- Ninhydrin: Dissolve 0.4 g stannous chloride ($SnCl_2.2H_2O$) in 250 ml of 0.2 M citrate buffer (p^H 5.0). Add this solution to 10 g of ninhydrin in 250 ml of methyl cellosolve (2-methoxyethanol). Prepare freshly and store in brown bottle (carcinogenic).

  0.2 M citrate buffer (p^H 5.0):

  Solution: A 0.2 M citric acid solution

  Solution: B 0.2 M sodium citrate ($C_6H_5O_7Na_3.2H_2O$) solution
- Add solution A 20.5 ml and solution B 29.5 ml diluted to a total of 100 ml with distilled water (p^H 5.0).
- Diluent solvent: Mix equal volumes of water and n- propanol, and use.

Extraction

1. Weigh 500 mg of the plant sample and grind it in a pestle and mortar with 5-10 ml of 80% ethanol.
2. Filter or centrifuge. Save the filtrate or the supernatant.
3. Repeat the extraction twice with residue and pool all the supernatants.
4. Reduce the volume if needed by evaporation and use the extract for the quantitative estimation of total free amino acids.
5. If the tissue is tough, use boiling 80% ethanol for extraction.

Estimation

1. To 0.1 ml of Extract, add 1 ml of Ninhydrin Solution and mix.
2. Make up the volume to 2 ml with distilled water.
3. Heat the tube in a boiling water bath for 20 min.
4. Add 5 ml of the diluents and mix the contents.
5. After 15 min, read intensity of purple color against a reagent blank in a spectrophotometer using photometric method at 570 nm.
6. The color is stable for 1 h. Prepare the reagent blank as above by taking 0.1 ml of 80% ethanol instead of the extract.
7. Dissolve 50 mg leucine in 50 ml of distilled water in a volumetric flask.
8. Take 10 ml of this stock standard and dilute to 100 ml with distilled water in another volumetric flask for working standard solution.
9. A series of volume from 0.1 to 1.0 ml of this standard solution gives a concentration range 10 to 100 μg.
10. Then proceed as that of the sample and read the color.

Calculation

Draw a standard curve using absorbance versus concentration. Find out the concentration of the free amino acids in the sample using standard regression equation and express as μg per g fr.wt.

36.6 Aflatoxins

Aflatoxin is one of the major quality problem in grains of several crops, which hinders domestic consumption as well as the export potential, since the international regulation for minimum standards for aflatoxin contamination is becoming stringent. There were several reports of presence of higher levels of aflatoxin in the cultivated groundnut based cattle and poultry feeds. In India aflatoxins are permissible up to 30 ppb food/grains and 120 ppb in poultry/animal feeds. In this laboratory quantification of aflatoxin levels in peanut kernels by HPLC using Aflatest P columns and PHRED unit (photo chemical reactor for enhanced detection) was standardized (Latha *et al.,* 2011). The reliable and working protocol is given below:

Chemicals required

- HPLC Methanol
- HPLC water
- Aflatest-p-column.
- Sodium chloride.

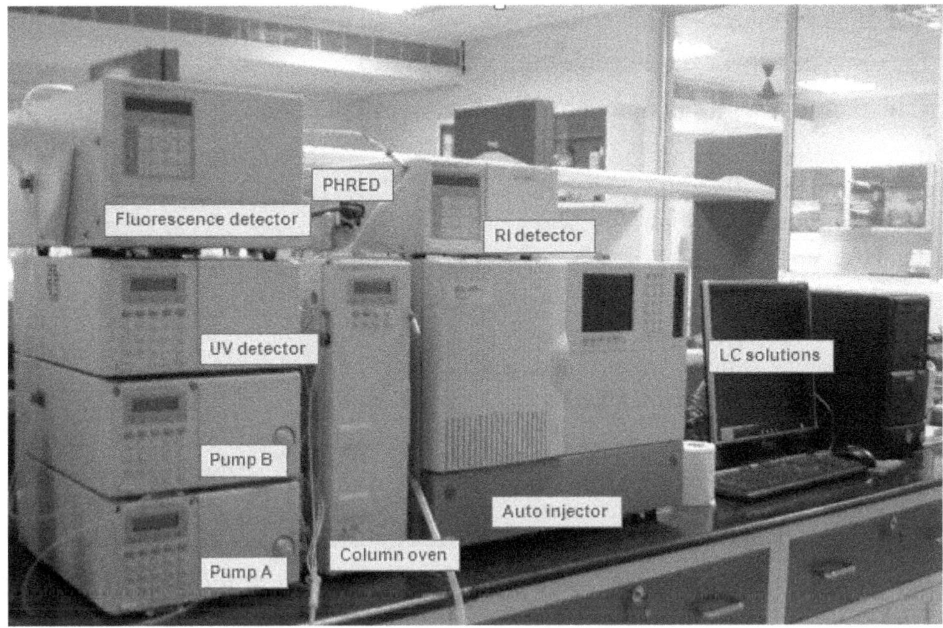

Fig. 36.1 *High performance liquid chromatography*

Extraction of sample

1. Weigh 20 g of sample into a blender jar.
2. Weigh 1 g of sodium chloride salt into blender jar.
3. Add 100 ml of 80% methanol.
4. Cap blender jar and seal with parafilm.
5. Blend at high speed for 1 min.
6. Filter the blender contents through a fluted filter paper into a 250 ml beaker.
7. Pipette 25 ml of filtrate into 50 ml graduated cylinder.
8. Add 25 ml of HPLC water to the cylinder.
9. Filter the content of the graduated cylinder through a glass fiber filter into a 250 ml beaker.
10. This filtrate will be used for passing through aflatest-p-column.

Immune-affinity column clean-up

1. Pipette 10 ml of filtrate, pass through aflatest-p-column and allow absorbing on column.
2. Once 10 ml of the sample has passed through the column, rinse the column with 10 ml of HPLC water.
3. Repeat HPLC water rinse for twice.

4. Add 1ml of methanol to the column and collect all methanol eluent into a test tube.
5. The sample is now ready for injection into the HPLC.

HPLC instrument set up protocol for estimation of aflatoxin

1. Instrument set up protocol for estimation of aflatoxins through HPLC in groundnut kernels was standardized.
2. Standardization was done by changing the mobile phase concentrations, flow rates and injection volume of the standard to be injected.
3. PHRED (Photochemical reactor for enhanced detection) was used as post column derivatization unit.
4. Finally the chromatogram with four sharp aflatoxin peaks (AfB1, AfB2, AfG1 and AfG2) at particular retention time for standard is obtained.
5. The sample is now injected into the HPLC column for detection of aflatoxin in the sample. The retention time of the sample peaks should be similar to that of standard speaks.
6. The instrument set up protocol for aflatoxin estimation is given below.
    - Mobile phase: 63% HPLC water/ 0.02% nitric acid/ 22% HPLC methanol/ 15% HPLC acetonitrile.
    - Flow rate: 1.0 ml/min.
    - Column- symmetry: C-18.
    - Injection volume: 20 µl.
    - Detection: Fluorescence detector having excitation at 365 nm and emission wavelength at 464 nm.
    - Gradient: Isocratic.
    - Run time: 30 min

Fig. 36.2 Groundnut kernels infected by Aspergillusflavus

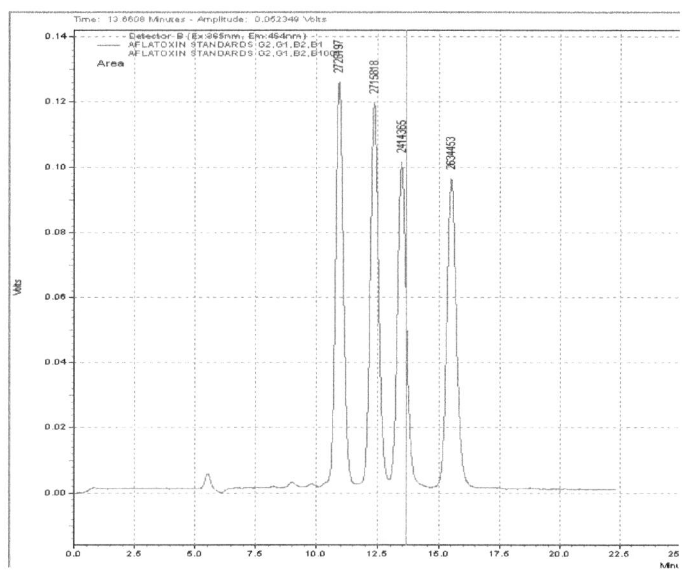

Fig. 36.3 *Chromatogram showing Standard Aflatoxin peaks (AfG2, AfG1, AfB2 and AfB1) eluted at particular retention times.*

References

1. *Holbrook, C.C., and Stalker H.T. (2003). Peanut breeding and genetic resources. Plant Breeding Reviews. 22: 297-356.*
2. *Hedge J E and Hofreiter B T (1962). In: Methods in Carbohydrate Chemistry (eds Whistler, R L and Miller, J N) Academic Press New York.*
3. *Khajehpour, M.R. (2007). Industrial Plants Production, Isfahan SID Technology Publishing, 2nd Edition, p. 350. JahadDaneshgahi Publishing Technology, the second edition*
4. *Latha P, Sudhakar P, Bala Krishna M, Rajiya Begum C and Raja Reddy K (2011). Estimation of groundnut kernel aflatoxins by High Performance Liquid Chromatography using Immunoaffinity column clean up and Post column photochemical derivatization. Legume Research, 34(1):31-35*
5. *Lowry O H, Rosebrough N J, Farr A L and Randall R J (1951). Protein measurement with folin-phenol reagent. J. Biol. Chem. 193: 265 – 275.*

Chapter 37

Problems and Prospects of Export Oriented Groundnut in India

V. Sumathi

Professor, Dept. of Agronomy, S.V. Agrl. College, Tirupati-517502, India.

37.1 Introduction

Groundnut or peanut (*Arachis hypogaea* L.) is considered as one of the important oilseed crops and is grown throughout the world. It has gained a lot of economic and nutritional importance worldwide. The groundnut is particularly valued for its protein content, which is of high biological value and it contains more protein than meat-about two and a half times more than eggs, and far more than any other vegetable food except soybean and yeast. India is one of the largest producer of groundnut in the world and major groundnut exporting country after the China. Indian groundnut is available in different varieties: *i.e.* Bolds or Runners, Javas or Spanish and Red Natals; and have a rich nutty flavor, sweet taste, crunchy texture and a relatively longer shelf life. Groundnuts in India are available throughout the year due to a two-crop cycles harvested in March and October. On an average, India produces 7-8 million tons of groundnuts (unshelled) every year. In fact, considering the current export figures, India has a much larger potential to supply high quality groundnuts to the international buyers. The awareness and concern for quality amongst the Indian groundnut shellers and processors are growing steadily. Indian shippers have the capability to prepare and supply edible peanuts conforming to highest standards.

37.2 Groundnut International Trade

India exports both types of groundnut pods and shelled groundnuts (kernels). It has been a traditional exporter of Hand Picked and Selected (HPS) groundnuts. Indian varieties are available in different varieties: Bold or Runner, Java or Spanish and Red natal. The main groundnut varieties produced in India are Kadiri-2, Kadiri-3, BG-1, BG-2, Kuber, GAUG-1, GAUG-10, PG-1, T-28, T-64, Chandra, Chitra, Kaushal, Amber etc. Apart from raw edible peanuts, India is also in a position to supply blanched peanuts, roasted salted peanuts and dry roasted peanuts and variety of peanut based products. Indian manufacturers have the capability to prepare and supply edible peanuts conforming to highest standards for international market. As per the APEDA, 2015 India has exported 7,08,386 MT of groundnut to the world for the worth of Rs. 4,675 crores, during the year 2014-2015. The major export destinations of Vietnam, Social Republic, Indonesia, Philippines and Thailand.

37.3 Characteristics of Groundnut for Export

Various physical, sensory, chemical and nutritional factors determine the edible quality of groundnut seed. New groundnut cultivars should posses certain physical attributes and chemical composition, maintain some definite processing and end use characteristics to be acceptable to traders, manufacturers and ultimate consumer. High values for sound mature seeds (SMS), 100-seed mass (HSM), kernels with elongated shape, tapering ends and pink to light brown testa colour are considerable traits. (Nigam *et al.*,1989). Large seed size is one of the most important criteria when whole kernels are used for table purpose. Nevertheless, all the large seeded types may not posses the essential confectionery traits. Hence, seed size coupled with the chemical composition will determine the worth of groundnut for direct consumption. Low oil groundnuts are preferred from confectionery point of view due to increasing health awareness among the consumers. The higher the Stability Index, the longer is shelf life of groundnut products (Ahmed and Young, 1982). The higher the protein content, the more is nutritional value. Sucrose is considered desirable from the organoleptic point of view. Sugars, especially reducing sugars and the free amino acids react during roasting to impart characteristic colour of roasted groundnut (Newell *et al.*, 1967) and also the nutty flavour. However, if reducing sugars and free amino acids are present in higher quantities the discolouration of the cooked product occurs due to excessive browning. Some of the important quality attributes of confectionery groundnut for export purpose are presented in Table 37.1.

Table 37.1 *Quality attributes of confectionery groundnut for export purpose*

Trait	Desirable aspects
Seed size	155 to 170 seed/100g
Seed shape	Round or elongated with tapering ends
Colour	Tan, rose tan, red
Flavour (Roasted)	Almond, coffee, nutty, popcorn, smoky and sweet
Texture (Roasted)	Firm and crispy
Biochemical and nutritional	Low oil content, high protein content, high carbohydrates, high O/L ratio, high vitamins B_1, B_2, E, high in minerals like Ca, Mg, Fe; low in anti-nutritional compounds like oxalic and phytic acid
Aflatoxin	Free from aflatoxin

37.3.1 Physical Factors

The price received by the farmer for a lot of confectionery groundnut depends on the size, shape, uniformity in seed mass, integrity of testa, absence of extraneous materials and immature seeds, and integrity of seed during processing and blanching. Extra large seeds also receive a premium in the international market. The price for the produce is reduced based on extent of damaged or split kernels or excessive moisture content.

The groundnut breeder uses the simple equipments and procedures to determine the grade factors for genotypes under evaluation and consideration for release as improved cultivars. These characteristics are genetically controlled in a complex manner. The lines to be released for commercial cultivation must confirm to these varietal parameters so that it could find place in the global market.

37.3.2 Nutritional Factors

The constituents of groundnut kernel, which determine its nutritional value, are described below.

1. **Protein:** It varies from 22-36% that depends on the cultivar, location, season, seed maturity and agronomic practices. Average protein content is higher than that of eggs, dairy products, meat and fish. The digestibility of groundnut protein is very high. Groundnut is deficit in lysine, methionine, theonine and tryptophan. Both the quantity and quality (amino acid profile) of the protein are important breeding considerations to the groundnut breeder, though neither has received much attention because of the primary use of groundnut as an oil seed crop.
2. **Oil:** Oil quality refers to oil content, fatty acid composition, iodine value, and ratio of Oleic to linoleic acid (O/L ratio). Oil content varies from 35 to 55% depending upon variety, season and maturity. Groundnut oil is composed of 80% unsaturated fatty acids. A great deal of variability is present in groundnut germplasm. Several workers have reported genetic manipulation of the proportion of different fatty acids in groundnut oil .Virginia types generally have a higher oleic acid contents and lower linoleic acid contents than Spanish or Valencia types. Varieties for confectionery purpose should have low oil content with high O/L ratio.
3. **Carbohydrates:** Groundnut kernels contain 10- 20% and oil- meal about 38% carbohydrates, which are low in reducing sugars. Sucrose is the most important sugar, which ranges from 3-7%. The oligosachharides, stachyose and raffinose, are responsible for flatulence and bad taste. Genetic differences have been found in groundnut for these intestinal gas- forming sugars. Therefore, sugar content and composition should also be an important component of the selection process in a breeding program.
4. **Vitamins and minerals:** Groundnut is a source of nearly half of the 13 vitamins necessary for normal body growth. It is a good source of B group of vitamins and

tocopherols. However, groundnut is deficient in vitamins A, C, and D Groundnut kernel is a rich source of minerals like P, Ca, Mg and Fe.

5. **Milling quality:** Milling quality is a term used to describe the ease with which groundnut kernels are removed from the shell (shelling) and the seed coat (blanching). Milling quality can be influenced by the environment in which the crop is grown and harvested, and by the shape, structure and composition of pod, seed coat and seed tissue.
6. **Sensory quality:** It is summation of all physical and chemical attributes of groundnut kernels and their products that appeal to human senses and has a bearing on acceptability by consumers. Colour, texture and flavour are the most important components of sensory quality.
7. **Noxious components:** Like most legumes, groundnut kernels also contain several noxious factors, which can adversely affect the nutritive value or quality of kernels. These factors are groups of unrelated chemical compounds like flatus sugars, saponins, allergens, proteinase inhibitors, lectins, goitrogens, phytic acid, aflatoxins etc.

37.4 Factors Affecting Chemical Composition of Kernels

37.4.1 Oil

A negative correlation has been found between oil and protein contents of groundnut genotypes (Dwivedi *et al.*, 1990).

(a) **Habit group:** The oil content was found to be independent of habit group.

(b) **Crop season:** Oil content is generally higher in kernels obtained from the crop grown in summer season (dry, irrigated conditions) than those from rainy season.

(c) **Crop maturity:** The oil content increases with advancing maturity in most cultivars (Nagaraj and Chauhan, 1987).

(d) **Seed mass (seed size):** The maximum percentage of oil was in medium sized seeds.

(e) **Geographical location:** The oil content differs considerably with geographical locations of crop growth in India (Bandyopadhyay and Desai, 2000).

(f) **Effect of fertilizers:** There are conflicting opinions about the individual or combined effect of nitrogen and phosphorus on the oil content. Application of N and /or P did not have any bearing on the oil content of kernels from the crop raised in Vidharba and Kandhesh tracts and Akola of Maharashtra (Puntakumar and Bathkal, 1967, Channawar *et al.*, 1995) and at Tirupati in Andhra Pradesh

(Vali *et al* ., 1978) but (Punoose and George 1974) have reported that application of nitrogen @10 to 20 kg/ha at Vellayani in Kerala, lowered the oil content while application of phosphorus gave a progressive increase in the oil content. However, so far the effect of potash is concerned its application has been found to increase the kernel oil content on red lateritic soil in Vridhachalam, Tamilnadu (Loganathan *et al.*, 1996). In the sandy loam soils of West Bengal, a combined application of N, P and K fertilizers in *kharif* season increased the oil content. (Barik *et al.*, 1994). The positive effect of application of effect of sulphur, whether applied individually or in combination with other nutrients, on the oil content, has been rather unequivocally established in all the major groundnut producing states of India.

(g) **Effect of drought:** A transient mid season drought does not significantly affect the oil content whereas end-of-season drought significantly reduces the oil content (Dwivedi *et al.*, 1996). Imposition of water deficit stress at pod filling or pod maturation phase resulted in decrease in oil content of kernels (Bhalani and Parameshwaran 1992).

(h) **Effect of application of fungicides/ pesticides:** Spray of fungicide, Bavistin, for controlling Tikka disease of groundnut resulted in an increase in the oil content of kernels. While application of sulphur resulted in an increase in free fatty acid pool of oil.

37.4.2 Protein Content

Application of nitrogen or phosphorus or both leads to increase in protein content. Boron and sulphur resulted in a significant increase in protein content and the combined application of B and S had more pronounced effect on protein content. In deficient soils, application of zinc caused an increase in the protein content (Malewar *et al.*, (1992). The end of season drought results in an increased protein content of kernels. The protein content of the kernels obtained from the crop of rainy season is generally higher than that of summer season. Unlike oil content, the protein content of kernels is not much affected by the maturity status of the crop at harvest. (Nagaraj and Chauhan, 1987).

37.4.3 Carbohydrates

Application of zinc in deficient soils, caused an increase in the soluble carbohydrate content.

37.4.4 Minerals

Application of zinc in deficient soils, caused an increase in the total ash content.

37.4.5 Processing

When untreated and salt soaked groundnut kernel samples were roasted at 160^0C for 30 min in a hot air oven or were fried in groundnut oil at 147^0C for 2 min and then stored in sealed polythene bags at 5^0C and 27^0C for up to 150 days, it was observed that the heat treatments significantly reduced the moisture, methionine and tryptophan contents of the kernels, as well as the in vitro protein digestibility (IVPD). Crude oil, crude protein and water soluble protein increased significantly after heat treatments. However, during subsequent storage, the IVPD increased.

37.5 Problems of Confectionery Groundnut for Export

- The attempts made so far to develop non - allergenic groundnut have not been successful. Hence the individuals susceptible to groundnut allergy are left with no choice but to eliminate groundnut and its products from their diet. In the U.S. one person in 200 is estimated to be allergic to peanuts and the figure is thought to be increasing.
- No research appears to have been done in India on the factors controlling the colour of raw or roasted kernels.
- Not much research has been done on the flavour quality of Indian groundnut cultivars.
- Systematic information on the factors determining the texture of groundnut and the effect of cultivar, seed size and cultural practices on the texture is not available.
- In consumer preference driven market, produce of cultivars with superior flavour and texture and low in flatus sugars are likely to fetch a premium price. Lack of cultivars with high blanch ability with enhanced recovery of finished products to the processing industry.
- A perusal of published work on chemical constituents of groundnut shows that various laboratories have not only employed different methods for analysing the nutritional and other constituents of groundnut kernels but also different sampling procedures. Therefore a valid comparison of data across the publications becomes difficult. Thus, there is a need for developing a common protocol of sampling and analytical procedures which may be followed by various laboratories working on quality aspects of groundnut.

37.6 Prospects of Confectionery Groundnut for Export

Breeding of confectionery groundnut varieties coupled with large seed size, uniform appearance and acceptable seed colour appears tenable with the availability of several high yielding, large seeded breeding lines maturing earlier than many Virginia types, and variants with high O/L ratio. Besides, availability of germplasm accessions possessing large seed size, high protein and sucrose content offers immense scope for tailoring large seeded genotypes with desired confectionery traits following conventional breeding approaches. Availability of molecular markers linked to the quality traits would help in marker assisted selection, thus widening the scope of effective selection in the segregating materials in the breeding programme. Along with breeding efforts suitable management practices that are eco-friendly and economically viable should also be developed exclusively for the cultivation of confectionery groundnut.

Suitable regions with low risk of aflatoxin contamination in the groundnut growing belts of the country will have to be identified. These areas with assured irrigation and other inputs can be earmarked for the cultivation of large seeded and confectionery groundnuts. As the demand for the organically grown food is growing in the western markets, opportunities for organic cultivation of groundnut will have to be explored.

Some other important aspects that may be kept under consideration are:

- Developing national and regional groundnut sectoral and seed strategies that lead to increasing exports of edible groundnuts matching end-user needs.
- Implementing integrated aflatoxin management programmes, covering all contamination-related issues, including mould formation and spreading, detoxifying the product and routine monitoring.
- Encouraging regulatory measures to control the flow of contaminated shipments in national and international trade, and technical information and training in aflatoxin control and prevention
- Establishing certified national quality control laboratories, in charge of issuing export quality certificates
- Policy level interventions in terms of providing incentives for the popularisation of large seeded and confectionery groundnut.
- Ensuring availability of good quality groundnut seed to the target group of farmers. For this the seed village concept can be pushed forward, which is already gaining momentum in different parts of the country in case of other crops.
- Proper and timely guidance by the agriculture extension departments or NGOs in crop management especially the water management, which assumes greater significance in containing the aflatoxin menace.
- Creation of proper post harvest processing and storing facilities at the farmers level before the centralised procurement of produce.
- Fine tuning the technologies of grading, packing etc. to cater the needs of the target domestic and export markets.

References

1. Ahmed, E. M. and Young, C.T. 1982. Composition, quality and flavour of peanuts. Peanut Science and Technology. Ed.H.E. Pattee and C. T.Young. pp 655-688. American Peanut Research and Education Society, Yoakum, Texas, USA.
2. APEDA, 2015. Agricultural and Processed Food Products Export Development Authority, Ministry of Commerce and Industry, Govt. of India. Source : http://www.agriexchange.apeda.gov.in
3. Bandyopadhyay, A. and Desai, S., 2000. Groundnut as a source of oil and food-the facts and national perspective. In: National Seminar on Oilseeds and oils – Research Development Needs in the Millennium. Directorate of oilseed Research, Hyderabad 2-4 Feb 72-87.
4. Barik, A. Jana, P.K., Sounda, G and Mukherjee, A.K. 1994. Influence of nitrogen, phosphorus and potassium fertilization on growth, yield and oil content of kharif groundnut. Indian- Agriculturist. 38, 105-111.
5. Bhalani, G. K. and Parameshwaran. M. 1992. Influence of differential irrigation on kernel lipid profile in groundnut. Plant Physiology and Biochemistry. 19, 11-14.
6. Channawar, A.G, Dahatonde. B.N. and Ulemale R. B. 1995. Response of groundnut to nitrogen and phosphorus fertilization. PKV Research Journal 19,71-72.
7. Dwivedi, S.L., Nigam, S.N., Nageswara Rao,R.C., Singh, U., Rao, K.V.S.,1996. Effect of drought on oil fattyacids and protein contents of groundnut Seeds (Arachis hypogaea L). Field Crop Res. 48, 125-123.
8. Dwivedi, S.L.,Thendapani, K. and Nigam, S. N. 1990. Stability of seed mass in confectionery groundnut International Arachis Newsletter 3,14-16.
9. Loganathan, S, Helkiah, J and Thanavelu, S 1996. Effect of FYM, nitrogen and potassium on yield and uptake of nutrients in groundnut. Madras Agricultural Journal (83) : 484-487.
10. Malewar, G.U, Indulkar, B.S and Malewar, V. 1992. Inorganic mineral matter and quality of grains as influenced by zinc levels in peanut cultivars. Annals of Agricultural Research (13) : 109-111.
11. Nagaraj, G and Chauhan, S., 1987, Seasonal variation in the oil quality of groundnut. Indian Journal of Agricultural Sciences., 57: 522-524.
12. Newell, J. A., Mason, M.E. and Matlock, R. S. 1967. Precursors of typical and a typical roasted peanut flavour. Journal of Agricultural and Food Chemistry. 15, 767-772.
13. Nigam, S. N., Dwivedi, S. L., Reddy, L. J. and Vasudev Rao, M. J. 1989. An uptake on groundnut breeding activities at ICRISAT centre with particular reference to breeding and selection for improved quality. Pp. 115-125. Proceedings of the Third Regional Groundnut Workshop, held during 13-18 March. 1988, Lilongwe, Malawi.

14. *Punnoose, K.I and George, C.M. 1974. Studies on effect of nitrogen and phosphorus on yield and quality of groundnut in red loam soils of Kerala. Agricultural Research Journal of Kerala (12) : 151-157*

15. *Puntakumar, S S. and Bathkal, B.G. 1967. Influence of N P fertilisers on composition, growth and yield of groundnut. Indian Journal of Agronomy (12) : 344-350.*

16. *Valli, P.M.. Reddy, M.N. and Reddy, G.H.S. 1978. Effect of different levels and methods of fertiliser application on rainfed groundnut. Mysore Journal of Agricultural Sciences. (12): 408-412.*

Chapter 38

Groundnut Seed Processing Technology

S. Kaleemullah

Professor and Head, Dept. of Agril. Engineering, S.V. Ag C, Tirupati.

38.1 Introduction

Groundnut pods go through a number of post harvest processing operations such as cleaning, grading, separation, drying, storage, packaging, transportation and marketing before it reaches the consumers. Groundnut processing is directed towards conservation of produce and value adding to make the material more readily usable and economically more remunerative. Groundnut is a major oilseed of India. It yields 70% kernel and 28% oil (pod basis). This chapter describes various types of strippers, grader, driers, decorticators, drying and storage technology of groundnut.

38.2 Stripping

The process of removing groundnut pod from the plants or haulms is known as stripping. The most common method of stripping is to pull out the pods from the plants manually, yielding 7-8 kg pods per hour. Stripping manually is a tedious and time consuming operation.

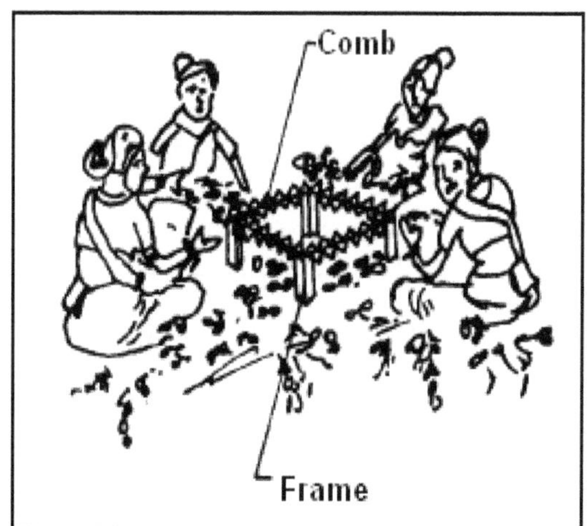

Fig. 38.1 Groundnut stripper (Comb type)

Tamil Nadu Agricultural University (TNAU), Coimbatore has developed two types of manually operated strippers namely; comb type and drum type (Fig 38.1 and 38.2). The comb type stripper consists of a square frame of four vertical standards with a strip of expanded metal fixed on to each of its side in the form of a comb. The stripping of the pod is accomplished by drawing a handful of vines across the comb with slight force. Four women can utilize the comb type stripper simultaneously. Nearly 3 quintals of pod per day can be stripped with 4 women workers. Maximum stripping efficiency of 97% can be achieved by this unit. Drum type strippers consist of a hollow drum formed by two metal discs at the ends, connected on the periphery of 5 1/2 inch M.S. rods inserted inside and covered by thick and soft rubber tubes. This drum is mounted on pedestal bearings and is free to rotate. It is fixed on a framework at a conventional height so that the operator can stand and beat the root portion of handful of plants over the rubber-covered roads of the revolving drum. To avoid scattering of pods, a hood frame is also provided. The roof as well as the three sides, i.e. other than the operator's side is covered with a canvas or gunny bag. One person can strip 20 kg pods per hour and the stripping efficiency is 93%.

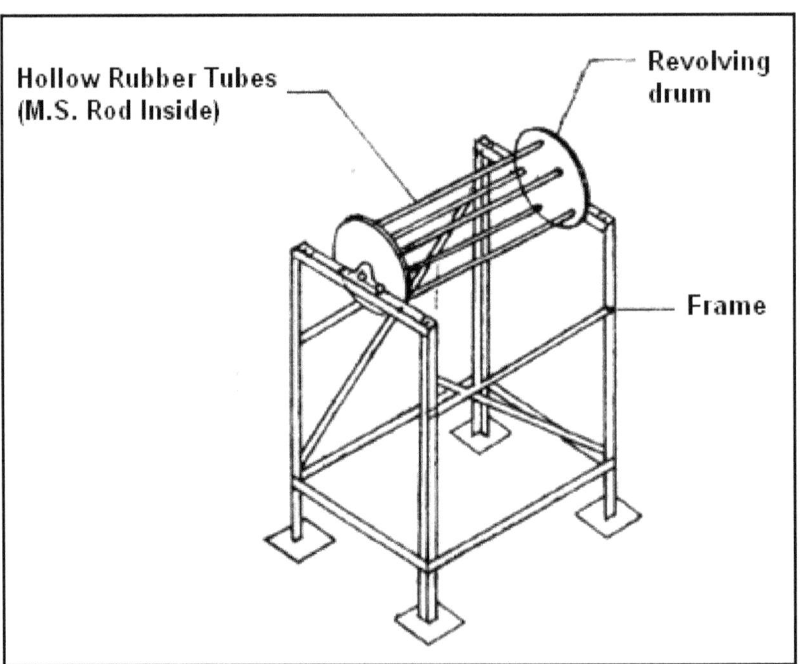

Fig. 38.2 Groundnut stripper (Cylindrical cage type)

In addition, TNAU, Coimbatore and University of Agricultural Science (UAS), Raichur have also developed power operated groundnut strippers (Fig 38.3 and 38.4). TNAU-stripper consists of 400 mm diameter and 900 mm long horizontal rotor (stripping drum) rotating at 300 rpm. It is powered by 1.5 hp electric motor. Nails have been fixed over the periphery of the rotor. A wooden platform is provided at the bottom of the machine so as to stand the labourers on it. Feeding plank made of wood is fitted near the rotor so that the groundnut plants can be kept in position by the labourers for stripping.

A blower is attached to this unit to blow-off the leaves, trash etc. The efficiency of stripping is found to be 90%. Breakage of pods is found to be only 5%.

UAS-Raichur stripper consists of a feeding trough, separating drum, blower, cleaning and collecting chutes assembled on angle iron frame with power transmission through belt and pulley connected to a prime mover. The separating drum is provided with staggered vertical loop type spikes made of galvanized iron rods. The separating drum rotates at 350 rpm. The stripper is also provided with a blower which operates at a speed of 1,750 rpm to remove trash. Two persons are required to feed the crop to the machine. Table 1 presents the specifications and test results of these strippers.

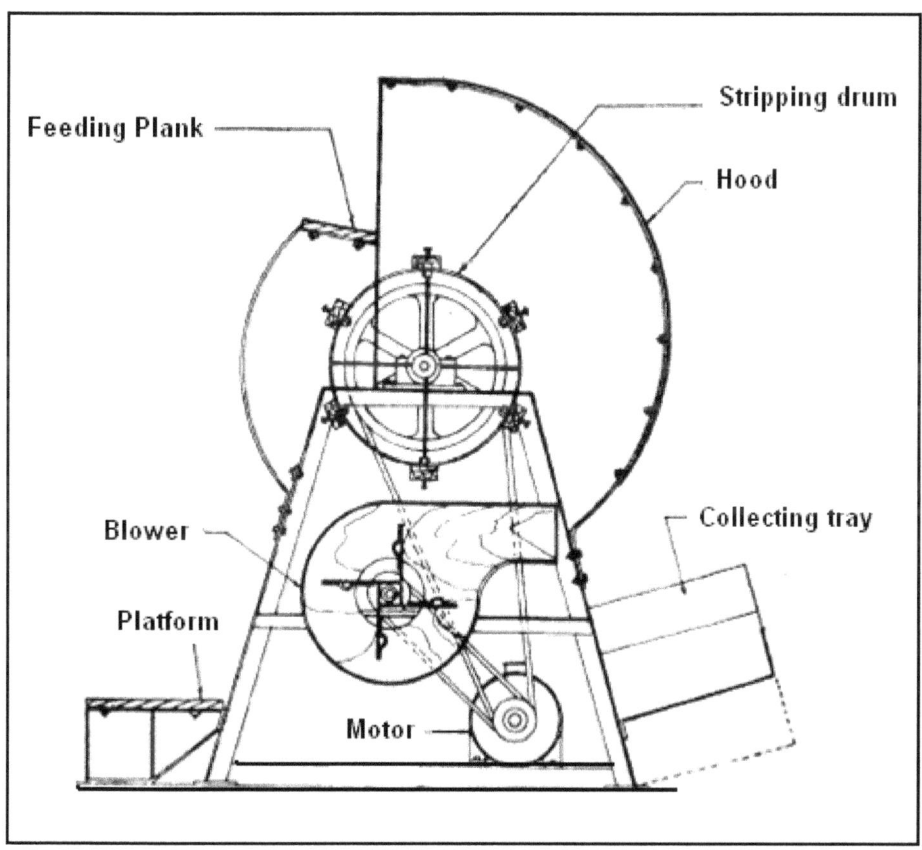

***Fig. 38.3** Power operated groundnut stripper (TNAU model)*

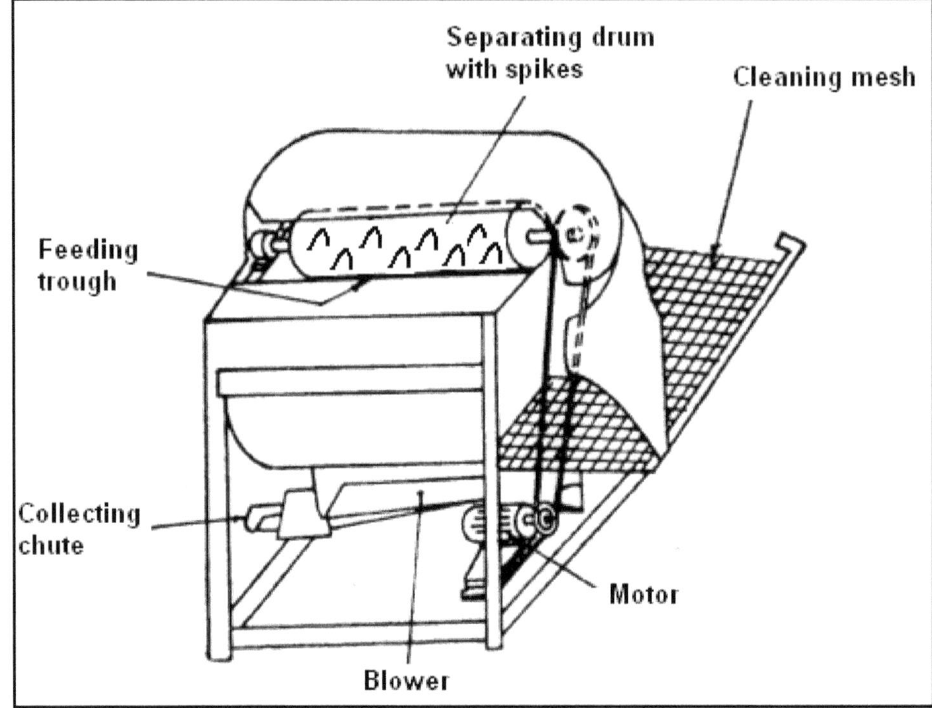

Fig. 38.4 *Power operated groundnut stripper (UAS Model)*

Table 38.1 *Comparative study of groundnut strippers*

Specifications/ Test Result	Manual strippers		Power operated strippers	
	TNAU model	TNAU model	UAS model	TNAU model
Type	Comb	Drum	Loop	Rotor
Capacity, kg/h	50	20	70	100
Power, hp	Manual	Manual	3	1.5
Labour requirement	4	1	2	4
Stripping efficiency,%	97	93	90	93

38.2.1 Dry Pod Groundnut Thresher

The harvested crop needs to be dried before threshing. It is operated either with 10 H.P diesel engine or electric motor (Fig 38.5). It consists of frame, feed hopper, hammer type threshing cylinder, concave, oscillating sieves and a blower. A concave is provided under the threshing cylinder for separation of pods from plants. In order to separate all the unwanted material after threshing from the pod, two sieves have been provided below the concave. It has the feed rate of 500 kg/h and output capacity of 200 kg/h with 95% threshing efficiency. Its cost is around Rs. 1,40,000 including 10 H.P diesel engine.

Fig. 38.5 Dry pod thresher for groundnut

38.2.2 Wet Pod Groundnut Thresher

It is used for separating pods from plants immediately after harvesting the groundnut crop (Fig 38.6). It is operated with PTO shaft of 35 H.P. tractor. It consists of frame, feed hopper, drum type threshing cylinder, concave, oscillating sieves and a blower. A concave is provided under the threshing cylinder for separation of pods and stripped plants. An outlet is provided at the rear portion of cylinder for stripped plants. Two sieves have been provided below the concave in order to separate all the unwanted material after threshing. A centrifugal blower with spiral casing has been provided in between the two cleaning sieves for blowing of light weight plant material coming along with threshed pods from the concave. It has the feed rate of 750 kg/h and output capacity of 300 kg/h with 96% threshing efficiency. It can be easily transported from one place to another place as it is provided with pneumatic tyres. Its cost is around Rs. 2,00,000.

Fig. 38.6 Wet pod thresher groundnut

38.3 Drying

Groundnut is dried either by the natural or the artificial methods. The equilibrium between the pod moisture content and atmospheric relative humidity during the drying process has been investigated by several workers. The results showed that the shell, the testa (skin of kernel) and the kernel have different equilibrium moisture contents at the same humidity (Table 38.2). The rate at which pods lose moisture to the air during drying and the rate of change of moisture content during storage, depend on the physical structure of the pods as well as the temperature, velocity and relative humidity of the air.

Table 38.2 *Equilibrium moisture contents of groundnut at various levels of RH*

Components of pods	Percentage moisture content (wet basis) at different percentage RH						
	44	53	64	70	75	86	92
Unshelled	6.2	6.9	8.2	8.2	9.0	12.8	19.3
Kernel	5.2	5.8	-	6.7	7.1	11.3	17.2
Shell	9.6	11.5	12.4	-	14.5	16.5	20.1
Skin (testa)	13.9	14.3	15.1	-	17.9	19.9	52.8

Source: *Blatchford and Hall, 1963*

38.3.1 Natural Methods

38.3.1.1 Scattered on Ground

In this method, plants are placed directly on the ground, foliage downwards, so that the pods are exposed to the Sun. Conversely, in some areas, foliage is placed upwards with the pods in contact with the moist-soil and protected from the direct sunrays. Plants are left in this position for varied periods of time, which often depends on the individual farmer. The moisture content of groundnut will be reduced from 40% to 6% within 15-20 days depending upon the intensity of solar radiation. If groundnuts are left too long on the ground to dry, dew and sunlight tend to discolour the nuts. Moulds may develop and cause breaking of pods while overheating may cause splitting of the kernels during shelling.

38.3.1.2 Windrows

This method is used for curing groundnuts prior to further drying in stacks in South Africa and Israel. After harvest plants are dried in inverted windrows for 2 to 3 days. In these windrows, the pods on top may be exposed to the weather or they may be underneath, next to the ground covered by the foliage. Loose fluffy windrows seem to permit good air circulation, which ensure uniform and fast curing of the pods. Groundnuts 'dried' in fairly large windrows with the pods protected from full sunlight by the haulms have been shown to lose moisture more slowly and suffer no apparent damage as compared to the pods exposed to the sun in small, thin windrows. If there are rains before the pods are dry enough to be picked-up, they may become damaged by the moulding and consequent blackening of the shell, together with some blackening of kernels.

Heat and too rapid drying usually damage groundnuts exposed to direct sunlight at the top of windrows for more than two days. Pods dried following thin windrows produce hard kernels, excessive skinning and breakage during shelling. The seed also loses their viability rapidly during storage. Consequently, in colder areas freshly dug and windrowed groundnuts, which become frosted, will have impaired viability and reduced vigour of seedlings. When windrow drying is used as a preliminary to all drying, curing may be done for three to four days to attain moisture content between 15 and 20 percent. It is recommended that when windrow drying is used before drying in sacks, the plants should only be left in windrows to wilt for 18 to 24 hours (Blatchford and Hall, 1963).

38.3.1.3 DOR Method

(Directorate of Oilseeds Research 1983), Hyderabad, India, developed a method for drying the pods under shaded conditions for maintaining seed viability. In this method two big heaps one-meter in diameter are tied near the base with a rope. One heap is inverted over the other in such a way that haulms of the upper heap cover the exposed peripheral pods from the direct sunlight. Seed of the pods dried following this method retains about 80 percent germiability, even after 8 months of storage. In the regions, where groundnut is grown only in the summer season, pods dried following the DOR method could be stored for the sowing of the next summer season. There are two major drawbacks to this method. First, in case of cultivars with less height and foliage, if the upper bundle does not completely cover the peripheral pods of the lower bundle by its haulms completely, the peripheral pods are exposed to sunlight and become prone to lose seed viability. Second, if there are rains within 1 or 2 days after the harvest, the seed quality is damaged seriously.

38.3.1.4 NRCG Method

A new method for the drying of groundnut pods was developed at National Research Centre for Groundnut (NRCG) to overcome the drawbacks of the DOR method. In this method a tripod type structure (pyramid shape) can be raised in the field with the help of three bamboo poles of about 1.5 m long. A coir rope can be wound around the structure starting from the bottom to the top, while maintaining a space of 6 to 8 inches between two loops. Immediately after harvest groundnut plants are hanged on the rope of the structure in inverted position, pods up and haulms down and the structure is filled with groundnut plants in a way that the pods of an upper ring covered the haulms of the lower ring thus forming a sloping structure like the roofing of a thatched house. The plants can be arranged bottom ring upwards (Fig. 38.16). Groundnut pods along with the plants can be allowed to dry in the structure in the field for five days.

Seed dried following this method maintains more than 80 percent germinability, even after 10 months of storage. In India this method was found very useful in maintaining the viability and quality of the seed in the areas where rain-showers are frequent during drying period i.e. the Northeastern part states of Orissa and Bengal. Seed obtained from the pods dried following NRCG method retained more natural testa colour than the conventional, windrows and DOR methods. If there are heavy rains during the drying period the structures may be covered with polyethylene sheeting, to serve as protection from rainwater.

38.3.1.5 Stacks

(Blatchford and Hall, 1963) defined the term 'stack' as structure formed by grouping a number of plants together. They described four types of stacks: ordinary, ventilated, poled and ventilated poled stacks.

The ordinary stack is the simplest type of stack and is formed by gathering the plants into heaps, the dimensions of which often range from 3 feet in diameter and 2 to 3 feet high to about 12 feet in diameter and 5 feet high. The pods may either be scattered throughout the stacks or lie at the stack centre or around the outside of the stack, depending on the area and the type of groundnut grown. In several groundnut-growing areas, pods are picked from the plant after drying for about two to four weeks in these stacks and are made ready for selling or storage. The drying period in these stacks also varies, frequently lasting from 10 to 15 days but in a few cases only two to three days. The small stacks are gathered together into large stacks with the haulms towards the outside and the pods towards the inside. These larger stacks are usually built at the edges of fields and some farmers choose shaded sites for stacking to avoid over drying by the sun.

In ventilated stacks the plants are grouped together so that the centre of the stack remains open to assist ventilation. The pods, as with ordinary stacks, may be scattered throughout the stack or all lie at the stacks centre or around the periphery of the stack. Farmers in Rhodesia and Nigeria build stacks of this type locally known as 'cocks'. The stacks should be built on small mounds of earth and the pods placed at the inside of the stacks. The central funnel should be covered with grass to prevent the pods at the top from being scorched by the sun. It is also suggested that pieces of plastic sheeting cover the stack would help to shed rain to outer edges.

When stacks are built near the farm buildings on an area of cleared ground, 'ventilation tunnels' constructed with the help of poles available locally is recommended. In most humid parts of Kenya plants are often built into a stack called 'poled-stacks'. In this method plants are grouped around a centre pole, which supports the stack and prevents it being blown down by the wind. The pods may be scattered throughout the stack or may lie the stack centre or round the periphery. It is reported that drying in such stacks takes about three to five weeks. The losses are less in the stacking method than in the scattered or windrows methods.

After lifting, plants can be stocked by hand to dry, on some kind of simple frame or pole as used in some parts of Central Africa. Stacking groundnut normally produce excellent quality hay and pods, but are confined to less-developed areas, where machinery is not easily available or where labour is still plentiful. When it is necessary to leave the plants to dry in the field after lifting, some method of keeping the pods off the ground is necessary. This will reduce losses from rotting and termites etc. and allow air to circulate more freely. The simplest method is to invert the plants in lines, so that pods are upward resting on the foliage.

38.3.1.6 Platforms

Platforms of various heights may be built to raise the plants off the ground during curing and so reduce moisture damage in the bottom layer of pods and avoid damage by the cattle also. In Gambia stacks of groundnut plants are often put straight on to raised platforms immediately after harvesting. In Guinea after curing the plants in small stacks, local farmers sometimes put plants to dry in larger stacks on raised platforms.

38.3.1.7 Racks

The curing of pods on racks has been referred to in a number of countries. In North Rhodesia to prevent termite damage to groundnuts during the curing period, a horizontal rack is used. The rack consisted of crossed pieces of local wood 36 inches long, 18 inches apart and rose 18 inches off the ground. Plants hung on the rack are protected from termites and could be arranged so that the nuts are shaded from the direct rays of the sun. The moisture content of pods on the rack comes down from 21 percent to 6 percent during the first 7 days of curing.

38.3.1.8 On Ground Surface

Drying of pods by spreading them in a thin layer on the soil or woven matting or tarpaulin material is a common practice in many parts of India. In India, the threshed pods being spread out in the farmer's threshing yard on a hardened mud, cement, or stone slab floor. A layer 1 and 1/2 inches deep needs no stirring, while one 3 inches deep needs stirring on alternate day.

Two major disadvantages were noticed in drying the pods by leaving them spread in a layer on the ground or on areas of concrete or on matting, etc. Initially, there is the problem of moisture in the ground in contact with the pods together with restricted air movement within the produce. The second difficulty is the time and effort required to gather the pods together, cover them during rain showers and re-spreading the pods as soon as possible to continue drying. A new type of plastic sundryer, which help the farmers to overcome these difficulties, has been designed in the Tropical Stored Produced Liaison Department (Pest Infestation Laboratory) and known as Allegate sundryer.

38.3.1.9 Trays

In some countries farmers are encouraged to spread their produce on trays, which they leave exposed to sun-drying during the day and shifts into the house at night. In Uganda for example trays, which hold one hundred kilograms of produce consist of a metal mesh base and wooden sides with handles at both ends. These trays can be raised off the ground by supporting the four corners on sticks. The Government subsidizes these drying trays to the farmers.

38.3.1.10 Platforms

Well-cured pods after removal from the plants are practiced to heap on platforms to complete drying. Very often the pods are left on such platforms for an indefinite period of time and may, in some cases, even be stored there.

38.3.1.11 In Sacks

In cases where the moisture content of threshed unshelled and shelled pods is too high, the pods are sometimes bagged and every day the bags are brought out to the storeroom and left in the open. This is common practice at agricultural stations in Africa.

38.3.1.12 Racks

Several trials have been conducted on the natural drying of threshed groundnuts in bags in Australia. In the suspended bag trial, groundnuts of 30 percent initial moisture content in open weave bags were suspended vertically from a horizontal wooden rack supported at both ends by strong vertical posts. A galvanized iron roof provided protection from the rain. Staggered hanging of bags at centre distances of 22 inches, two bags rows deep is reported to have permitted safe drying of groundnuts from approximately 30 percent moisture content to safe storage moisture content in 10 days. As a result of this trial it was considered that in humid conditions, it might be necessary to use supplemental heat to achieve safe storage moisture content.

38.3.1.13 Artificial Methods

The artificial drying of groundnuts on a commercial scale has not been widely adopted, but research on the wide applicability of the artificial drying methods is still being conducted in many developing countries.

Intermittent drying of threshed groundnuts has been investigated and shown to reduce the total time of drying when compared with drying without 'rest' intervals. The theory is that by drying the shells and then allowing the nuts to 'rest' for some time, moisture will move from the kernel to the shells; the moisture differential between the shell and the drying air will thereby be greater and drying will consequently be more efficient.

In general, attempts to dry green (uncured) pods on the plant have given poor results in relation to the quality of both the kernels and the haulms. The total moisture percent was such that if drying was at a moderate rate, mould quickly occurred and if drying was at a fast, quite considerable breakage of the kernels resulted. Unsatisfactory results were also obtained when the haulms were clipped at various periods before curing (Teter, 1954). It has been noticed that an initial period of curing which reduces moisture from about 50 to 60 percent to about 25 percent is necessary, if good quality groundnuts are to be produced by subsequent artificial drying. After partial windrow curing, groundnuts on the haulms have been successfully dried artificially in the United States of America. The best results have been obtained using a temperature of 27 to 32°C and an airflow rate of 10 to 12 c.f.m.cu-1 ft and discontinuing drying when moisture content reaches below 7 percent.

38.3.1.13.1 Batch dryers

These consist briefly of a fan and heater unit (oil furnace) connected by suitable ducting to an air chamber, which forms the bottom part of a platform, tray, bin or silo. The threshed groundnuts are contained above the air chamber and the drier is so designed as to ensure that the heated air has to pass through the groundnuts in escaping into the atmosphere. To avoid the possibility of tainting (off flavors) from products of combustion an indirectly fired heater is normally used. In Tanzania test were carried out using a 10-aperture platform or sack drier with uncured groundnuts at about 48 percent moisture content. It was found that in practice using a single layer of sacks and air at 54°C moving at 110 cu.ft per minute (c.f.m) per sack the moisture content would be reduced from 48 to 8 percent in about 16 hours. During these tests it was also found that the maximum temperature should be limited to 54°C since at 60°C a distinctive smell appeared which may be associated with some form of deterioration. Further drying of nuts at the higher temperature i.e. above 38°C renders the seed unsuitable for the sowings due to loss of viability.

38.3.1.13.2 In bulk

Tests were conducted in Israel on the artificial drying of groundnuts in three rectangular wooden bins, 2 m high with a slatted and wire netting floor. Air for drying was supplied by a fan to a plenum chamber below the bins and additional heat could be supplied to the air stream by indirect heating. These tests showed that there was no decrease in the quality of the product if it was dried in bulk in layers of 1.8 m high at an air velocity of about 0.5 m per second (m sec-1) which is equivalent to 98.4 ft per minute (f.p.m.) and an air temperature less than 40°C. Artificial drying had no influence on the colour and appearance of the shells or on the germination of seed. A bin-type drier used in Alabama to dry rain-soaked nuts consisted of an underground air duct and drying bins placed over vents in the air duct. The drying temperature was 46°C and the results were satisfactory (Smyth 1959).

38.3.2 Drying of Pods in India

In India, groundnut is traditionally dried in sun which requires 4-6 days for reducing the moisture content of groundnut pods from 26% to 13% on mud floor. On cement floor, the drying takes place faster and reduces about 1/3 of total drying time in comparison to earthen floor (IIT, Kharagpur). As per studies conducted at TNAU, Coimbatare, about 15 h are required to reduce the moisture content from 29% to 9% in mechanical drier at an airflow rate of 27 cfm and bed thickness of 15 cm for POL-2 variety pods at 35°C temperature. In case of TMV—7 variety, 21 h are required to reduce the moisture from 29.8 to 7.3%. For such reduction in moisture content, sun drying requires 48 to 72 h. Though drying methods do not effect the viability of seed, however, it is reported that about 1% oil content is reduced when groundnut is dried by a mechanical drier compared to shade dried seeds.

For mechanical drying of groundnut, 2 driers have been designed, developed and tested. The specifications and test results of these driers are given in Table 3 while Fig 38.7 and 38.8 show these driers.

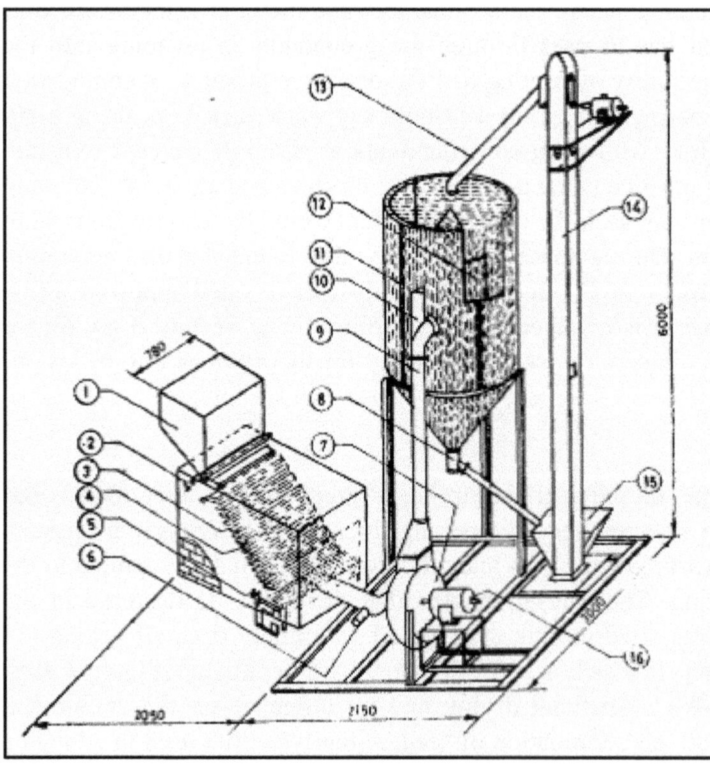

1. Hopper 2. Pulley 3. Inclined Grate 4. Furnace Space 5. Rotating Lever 6. Secondary inlet 7. Blower 8. Discharge Gate 9. Hot Air Duct 10. Flue Gas Exit By Pass11. Outer Cylinder 12. Inner Cylinder 13. Delivery Pipe 14. Bucket Elevator 15. Seed Loading/ Discharge Hopper 16. Motor

Fig. 38.7 *Recirculating batch drier (Dimensions are in mm)*

1. Drying Chamber 2. Fuel Tank 3. Blower 4. Burner 5.Motor 6.Thermometer

Fig. 38.8 Bin drier

Table 38.3 Comparative study of groundnut driers

Specifications/Test results	Driers	
	Recirculating batch drier	Bin drier
Developed at	IIT, Kharagpur	TNAU, Coimbatore
Capacity	1250 kg/batch Shelled groundnut	(80 kg/h)
Power requirement, hp	5	2
Labour requirement	2	1
Suitability to other crops	Paddy, wheat, maize sorghum, millets, pulses etc.	Paddy, millets

PKV, Akola has developed a waste fired drier which is fabricated with locally available materials. The drier (Fig 38.9) can dry about 3.5 t pods/d. The best drying temperature for groundnut is 36-58°C.

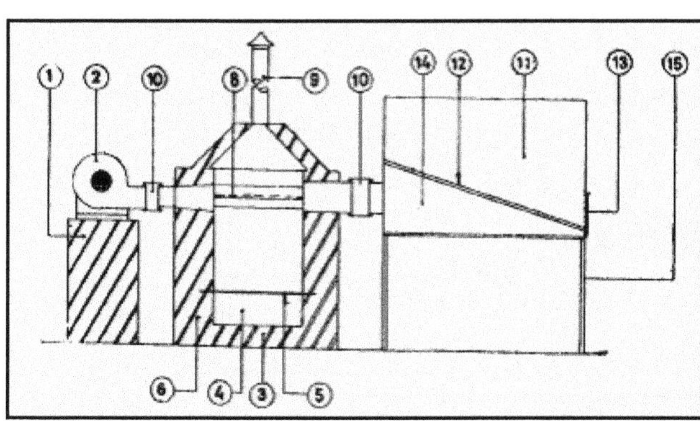

1. Blower 2. Foundation for Blower 3. Foundation for Furnace 4. Ash Collection 5. Grate 6. Brick Wall 23 cm. Thick 7. Furnace 60 cm. x 60 cm. 8. Heat Exchanger 9. Cnimney 10. Canvas Duct 11. Dry Bin 12. False Bottom 13. Grain Outlet 14. Planum Chamber 15. Stand

Fig. 38.9 Waste fired dryer

Gujarat Agriculture University (GAU), Junagadh has developed a plastic enclosure (Fig 38.10) with a floor area of 6 m x 4 m and 2 m height for sun drying of groundnut. This structure needs 8-9 days to reduce the moisture content of pods from 43.5 to 18% (wb). Under this enclosure, the pods are protected from the unfavourable weather conditions, animals, birds and insects. However, this structure, when not in use for drying purpose, could also be conveniently used for storage or as a green house.

GAU, Junagadh has also developed a flat plate solar collector type batch dryer (capacity 200 kg/batch) with floor size of 4 m × 2.5 m for freshly harvested groundnut vines and pods. It is reported that the duration for drying of freshly harvested whole groundnut plants form 65% to 15% m.c. is reduced by half (2 d) using solar dryer by supplying ambient air as compared to the traditional method.

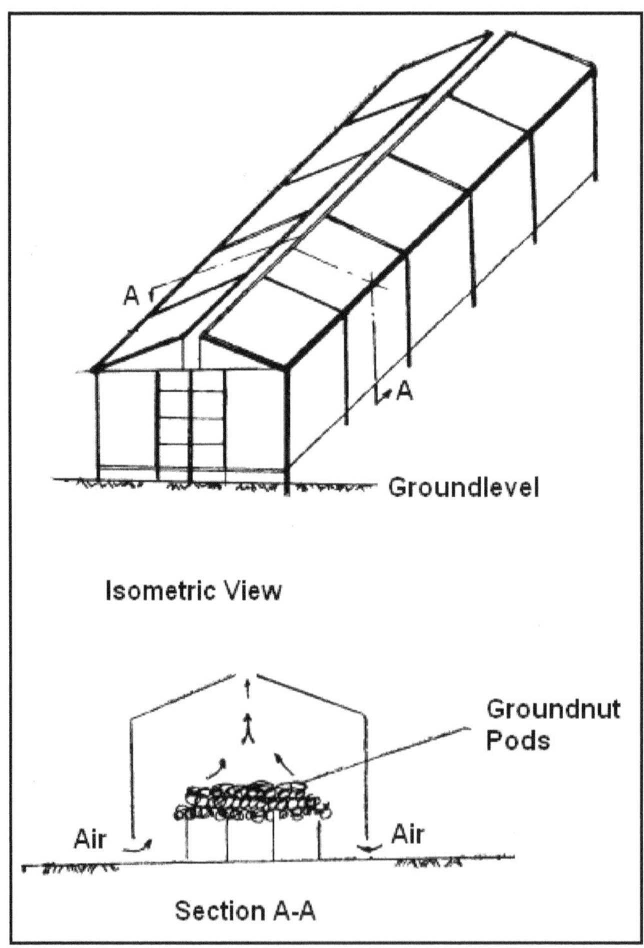

***Fig. 38.10** Details of the plastic enclosure construction for groundnut solar drying*

38.3 Grading

A power operated groundnut grader was developed at TNAU, Coimbatore (Fig. 38.11). It mainly consists of a feed hopper, two slotted oscillating sieves, an eccentric mechanism and a frame. The two oscillating sieves have slot sizes of 10.75 mm × 50 mm and 9.50 mm × 50.50 mm, respectively. Sieves are oscillated by the eccentric mechanism. The machine grades groundnut pods/kernels into three distinct grades according to size. Oscillating sieves could be replaced by different grades depending upon the groundnut varieties to be graded.

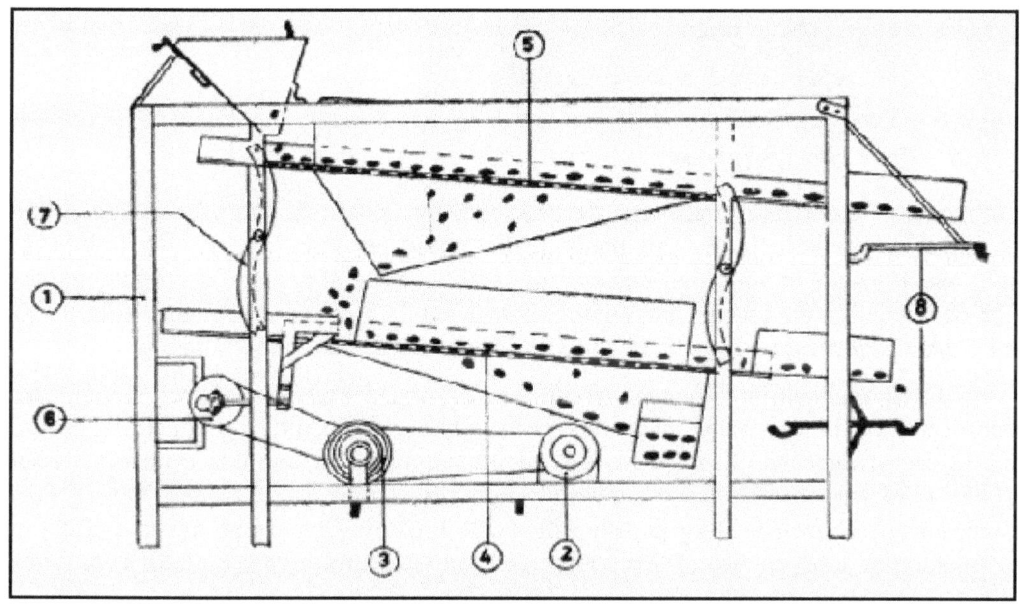

1. Frame 2. Motor 3. Cone Pulley 4. Bottom Sieve 5. Top Sieve 6. Ecentric 7. Bell Crank Lever 8. Bag Holder

Fig. 38.11 Groundnut grader (TNAU model)

38.3.1 Groundnut Kernel Grader

A groundnut kernel grader developed at ARS, Anantapur (Fig 38.12) has three sizes of sieves to grade groundnut kernel based on three sizes i.e >1.24 cm, 1.06-1.24 cm and <1.06 cm. Its capacity is 300 kg/h without any kernel damage and operated with 2 H.P electric motor. It's cost is around Rs.50,000/-

Fig. 38.12 Groundnut kernel grader (ARS, Anantapur)

38.4 Decortication

Traditionally groundnut pods are decorticated by hand shelling for removal and separation of the kernels. By this method, the output per man hour is very low (1-2.5 kg/h) which results in very high unit cost of shelling. Several types of decorticators have been developed in the country but among them, the CIAE manually operated (Fig 38.13) and TNAU decorticators (Fig 38.14 and 38.15) have been found most suitable.

CIAE, Bhopal has modified its decorticator giving two new prototypes viz. (i) mini decorticator useful for rural women who are usually accustomed to work in sitting posture and (ii) decorticator with feeder-cum-separator attachment. The mini decorticator is small in size (250 mm × 500 mm × 630mm), light in weight (5.7 kg) and low in cost. Its capacity is about 35-40 kg pods/h with 1-2% broken. There is no adverse effect on germination of seeds by use of this equipment. The other prototype (Fig 38.16) provided with a feeder and separator attachment has a sieve of 250 mm × 500 mm size (dia of holes 11 mm). The feeder is reported to facilitate the easy operation of the equipment with an increased capacity of 50-55 kg pods/h, (25-37% more than the previous design). Damage to the kernels reduced from 2 to 1% giving 70-75% separation efficiency and 100% shelling efficiency. Rubber tyre has been reported useful in design of groundnut shellers in Thailand. Based on the experience of Thailand, a motorized rubber tyre groundnut sheller has been developed at CIAE Bhopal (Fig 38.17). This sheller is operated by a 2 HP electric motor. A three layer sieve has been provided for grading of kernels. The speed of grading sieve is 250 rpm with 50 mm stroke.

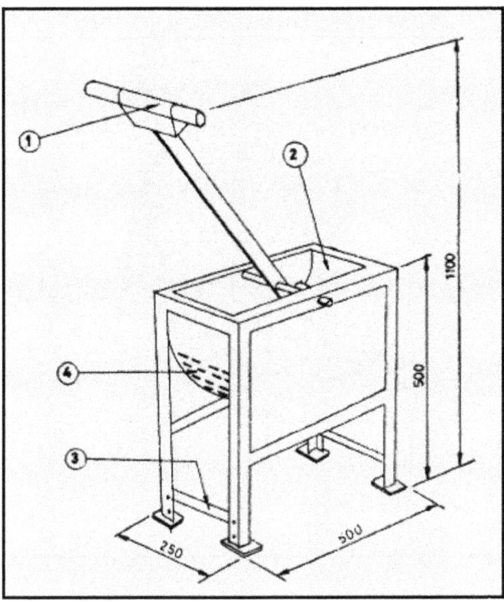

1. Handle 2. Hopper 3. Foot Rest 4. Sieve

***Fig. 38.13** Manual Groundnut Decorticator (CIAE Model)*

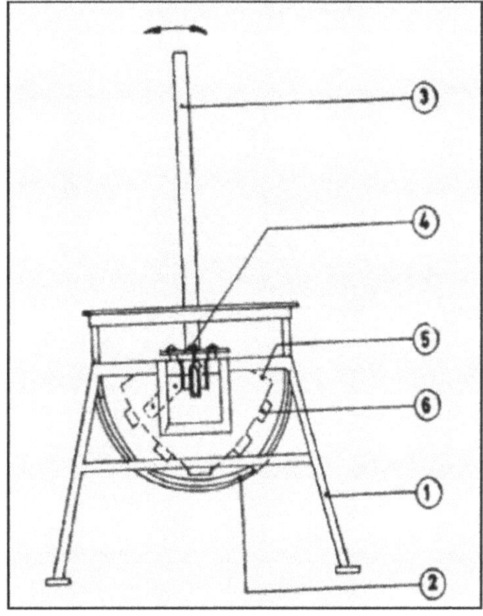

1. Stand 2. Sieve 3. Handle 4. Clearance Adjusting Bolt 5. Oscillating Sector 6. Pegs

***Fig. 38.14** Improved Hand Operated Groundnut Decorticator*

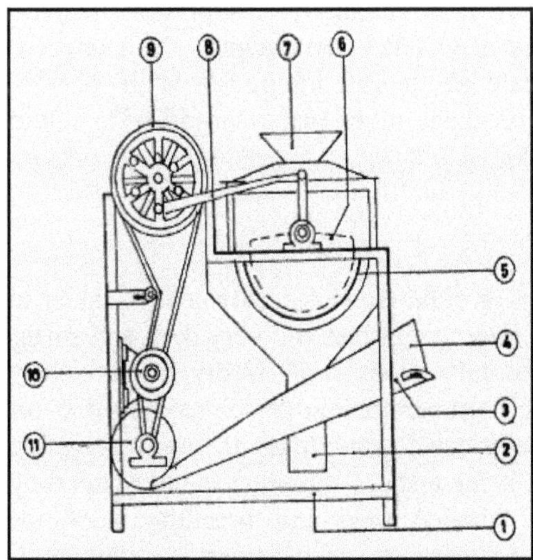

1. Frame 2. Kernel 3. Chute 4. Hull 5. Concave Sieve 6. Occillating Rod 7. Feed Hopper 8. Connecting Rod 9. Fly Wheel 10. Electric Motor

***Fig. 38.15** Power Operated TNAU*

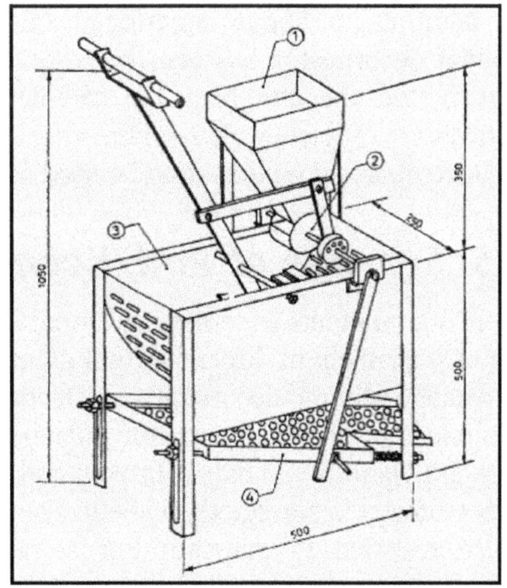

1. Hopper 2. Feeder 3. Decorticating Unit 4. Separator

***Fig. 38.16** CIAE Groundnut Decorticator Attached with Feeder &*

Model Groundnut *Separator (Dimensions in mm)*

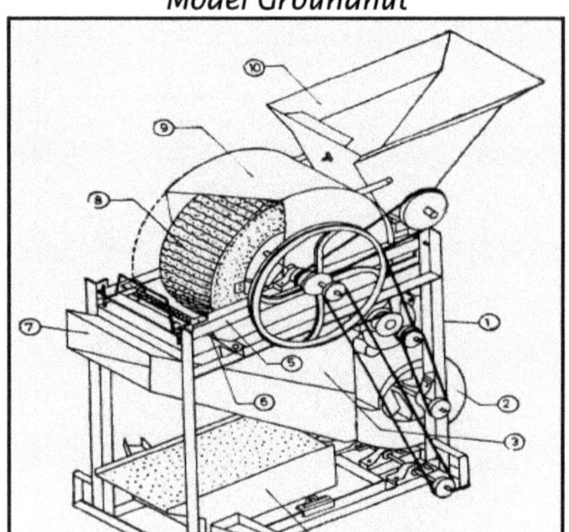

1. Main Frame 2. Blower 3. Trough 4. Grading
5. Rubber Tire Frame 6. Concave 7. Blower Chute
8. Rubber Tyre Assembly 9. Cover 10. Feed Hopper

Fig. 38.17 *Motorized rubber tyre-sheller*

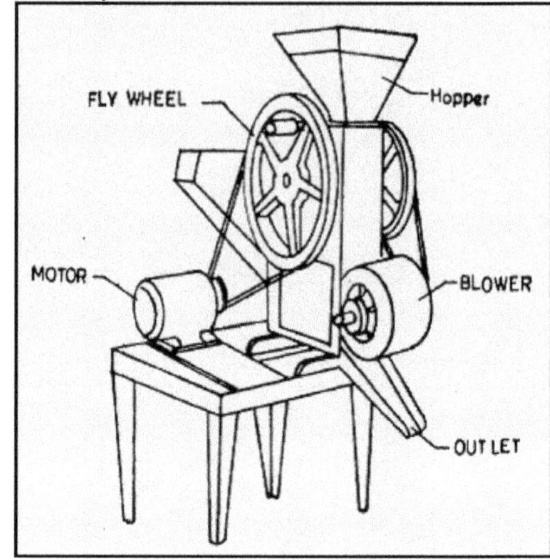

Fig. 38.18 *Power operated mini groundnut decorticator*

In addition to above mentioned decorticators, a manually-cum-power operated groundnut decorticator has also been developed at O.T.R.I., Anantapur. With one HP electric motor, the unit (Fig 18) can decorticate about 150 kg pods/h with 94-96% decortications efficiency. It is claimed that no breakage takes place and 57-63% sound kernels are obtained which could be used as seed.

38.5 Storage of Pod/Kernels

Unshelled groundnuts are normally stored in bags or bulk. However, care is to be taken so as to not store them in direct sunlight otherwise they would become very dark and suffer high damage during decortication. Decorticated nuts store well, if dry but are very susceptible to insect and vitamin damage. High temperature adversely affects the oil quality and viability. Sound, unbroken and undecorticated groundnuts are less susceptible to insect attacks in storage than shelled nuts but if the testa of the latter is unbroken they are also resistant to damage. For storage of shelled seed and breeding stock, of decorticator is an air conditioned store is almost essential if nuts are to retain their viability for more than a season. A temperature of 2-4°C with a R.H. of 65% allows storage for at least 2 years with little loss of viability. For less critical uses, low oxygen or nitrogen atmospheres can improve shelled groundnut storage.

In Indian conditions, small amounts of shelled dried seed, treated with some insecticide could be successfully stored for one year without significant loss of viability in laminated polyvinyl bags (Ramamoorthy, 1979). Studies conducted at P.K.V., Akola (1974) showed that out of various storage structures used for groundnut, polyethylene bags were found most suitable recording minimum storage loss of only 5.8% as compared to 8.7% in PKV bin, 9% in gunny bags/rectangular metal bins, 9.3% in mud plastered Nirgudi (local) bins and 10.7% in Pusa and Hapur bins. The insect species recorded were *Coeyra Cephalonica Staint* and *Tribolium Castaneum*. Oil content of pods decreased slightly during storage.

The storage of shelled groundnut was studied at UAS, (Raichur 1977) to reduce the space requirements and labour involved in storage of groundnut pods. Six types of different storage structures, namely; plywood, metal, plastic, earthen pots, gunny bags and polyethylene bags were evaluated for this purpose. The physical condition of groundnut kernels in plywood, metal and plastic bins after 9 months was quite unsatisfactory. Considering various parameters *viz.* seed damage, germination percentage and protein content, higher ratings were given to earthen pots and polyethylene bags as shown in Table 4.

Table 38.4 *Comparative performance of various storage structures for groundnut kernels*

Storage structure	Fat acidity mg/KOH%	Oil content %	Protein content% (Oil free basis)	Seed damage%	Germination %
Fresh sample	15.00	48.11	52.76	-	-
Plywood	57.75	46.57	52.69	17.31	31.00
Metal	64.00	41.44	49.88	33.09	7.16
Plastic	61 .68	45.88	50.07	15.82	11.50
Earthen pots	45.25	46.72	51.35	6.79	59.00
Gunny bags	54.41	46.16	50.74	9.17	35.83
Polyethylene bags	46.71	47.77	50.49	4.40	44.83

The safe moisture content limit for storage of groundnut kernels and pods are 4-6% and 7-9% respectively. However, studies conducted at TNAU, (Coimbatore 1978) showed that groundnut variety SBX-1 could be safely stored for 9 months at a moisture level of 9-10%. Due to efficiency of the improved storage structures, the free fatty acids formation, which is an indication of seed quality deterioration, was minimum in the plywood storage bin as compared to other structures as shown in Table 5.

A similar study has been conducted at IIT, Kharagpur where groundnut pods of TMV-2 variety (m. c. 5.8%) were stored in 4 types of storage structures for 120 days. These structures were earthen pot, tar painted polyethylene lined bamboo bin (TPB), bamboo cement bin and metal bin. Moisture content of groundnut increased with storage time for each type of storage structure, however, the degree of rise in moisture was highest for seeds kept in tar painted polyethylene lined bamboo bin and least in metal bin (Fig 19). As the seeds became aged, there was a continuous loss in dry weight of groundnut in all structures. This loss in weight was maximum in tar-painted polyethylene lined bamboo bin and minimum in metal bin (Fig 20).

***Table 38.5** Evaluation of different storage structures for groundnut pods*

Particulars	Hapur bin	Plastic silo	Ply wood bin	Double walled PE lined bamboo bin	Coal tar drum bin	Gunny bags
Capacity, kg	300	300	300	300	80	150 (5 bags)
Moisture content, % wb						
Initial	10.0	10.0	10.0	10.0	10.0	10.0
Final	10.2	10.7	10.4	10.8	10.2	10.8
Grain temp., °C						
Initial	31.0	31.0	31.0	31.0	31.0	31.0
Final	31.2	31.2	30.4	30.5	31.2	31.0
Loss in weight due to insect infestation, %						
Initial	Nil	Nil	Nil	Nil	Nil	Nil
Final	14.73	14.27	8.8	12.51	12.51	22.32
Germination, %						
Initial	76.0	76.0	76.0	76.0	76.0	76.0
Final	70.0	72.0	72.0	68.0	70.0	64.0
Biochemical changes						
(a) Proteins, %						
Initial	24.66	24.66	24.66	24.66	24.66	24.66
Final	24.70	24.37	24.25	24.60	24.20	24.75
(b) Carbohydrate, %						
Initial	15.18	15.18	15.18	15.18	15.18	15.18
Final	15.25	15.29	15.59	15.55	15.40	15.75
(c) FFA, % mg of KOH/g						
Initial	1.03	1.03	1.03	1.03	1.03	1.03
Final	3.36	2.25	2.22	1.99	2.59	1.95
(d) Oil content, %						
Initial	49.15	49.15	49.15	49.15	49.15	49.15
Final	48.80	48.25	49.95	49.00	48.90	48.95

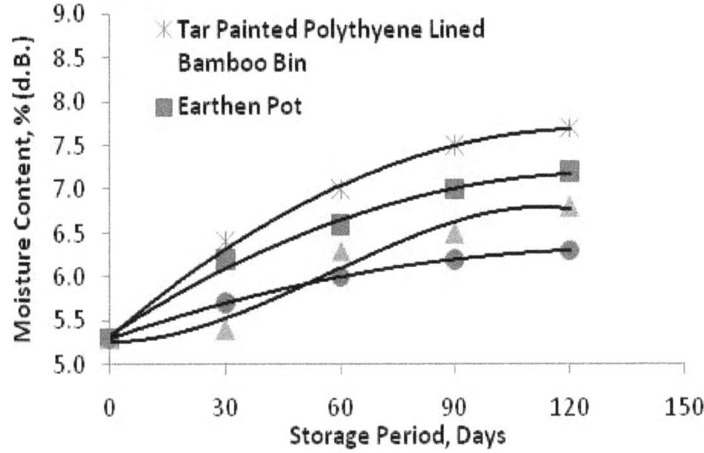

Fig. 38.19 *Moisture content of groundnut (Kernel) stored in different storage structures*

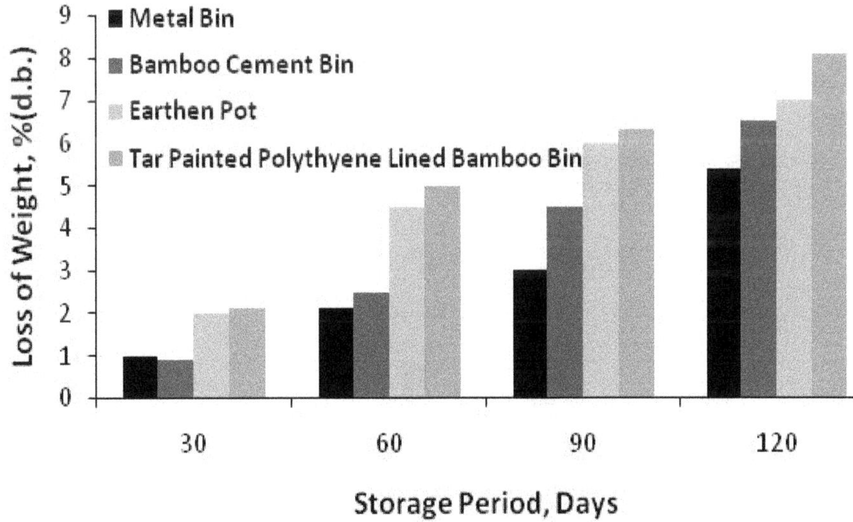

Fig. 38.20 *Loss of weight of groundnut samples stored in different storage structures*

Fig 38.21 shows the influence of storage period on germination and root/shoot lengths which decreased with increase in storage time. However, metal bin was found to be the best structure and TP bin as worst structure in terms of germination. Appreciable changes were observed in shoot length of groundnut as shown in Fig 38.22.

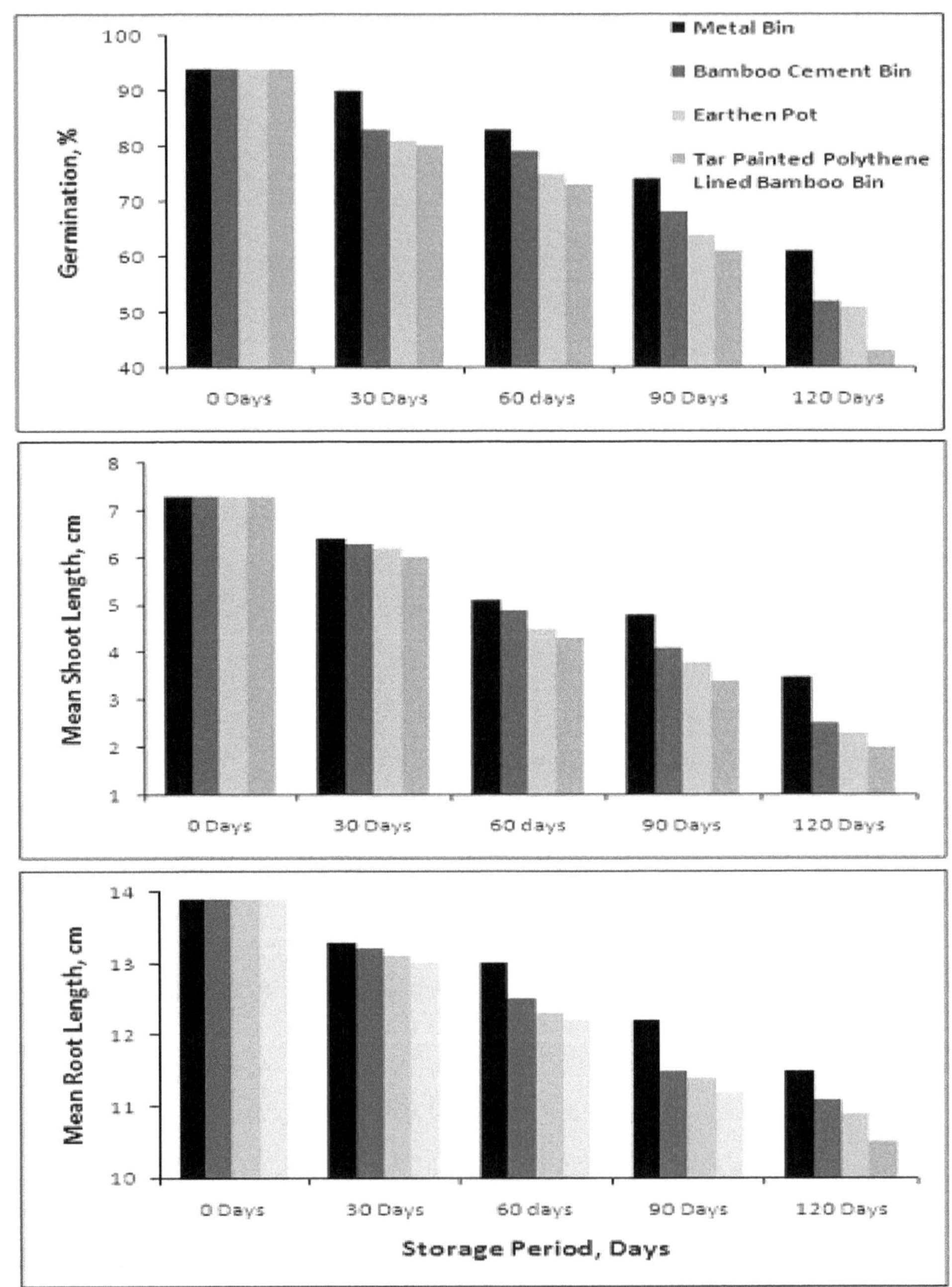

***Fig. 38.21** Influence of storage period on percent germination, shoot length and root length of groundnut samples stored in different storage structures*

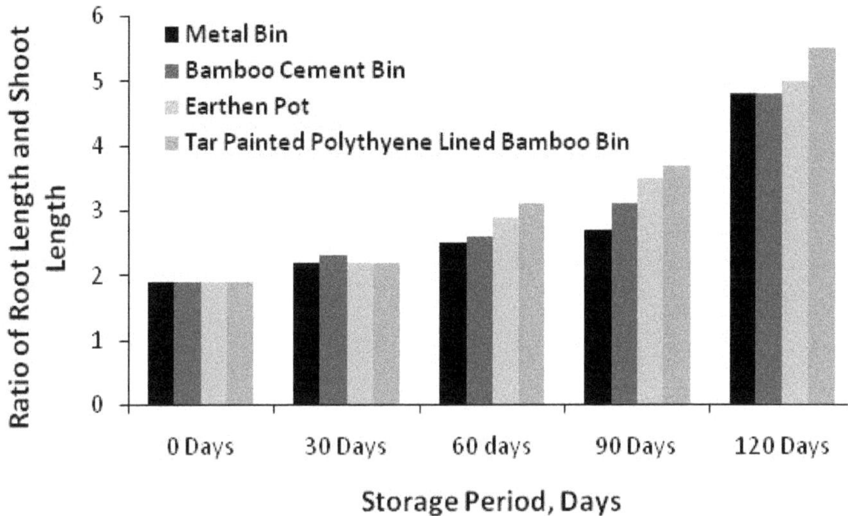

Fig. 38.22 Ratio of root length and shoot length of groundnut samples stored in different storage structures during the storage period of 120 days

The variations in total oil content and free fatty acid content after 120 days are shown in Table 6. Development of free fatty acidity, presented in Fig 4.23 showed that both metal bin and bamboo-Cement bin could resist the formation of appreciable amount of FFA up to 90 days. On the other hand FFA rose very sharply between 30 to 60 days of storage and then decreased again sharply to a very lower value during 60 to 120 days in earthen pot and tar-painted PE lined bamboo bin. This indicates that groundnut deteriorates very fast in these two structures after 60 days of storage.

Table 38.6 Variations in Oil and FFA contents of groundnut stored in different storage structures

Storage period	Total oil content, (% db)				Free Fatty Acid, (% db)			
	MB	BCB	EP	TPB	MB	BCB	EP	TPB
0	53.90	53.90	53.90	53.90	3.55	3.55	3.55	3.55
30	53.78	53.12	52.85	52.93	4.64	4.58	7.33	8.30
60	53.92	53.00	52.52	52.08	5.85	5.93	10.82	11.50
90	53.75	52.95	52.10	51.98	6.70	6.82	10.22	10.89
120	53.69	52.51	52.03	51.46	6.92	7.25	5.42	3.20

MB – Metal bin　　BCB – Bamboo cement bin

EP – Earthen pot　　TPB – Tar painted polyethylene lined bamboo bin

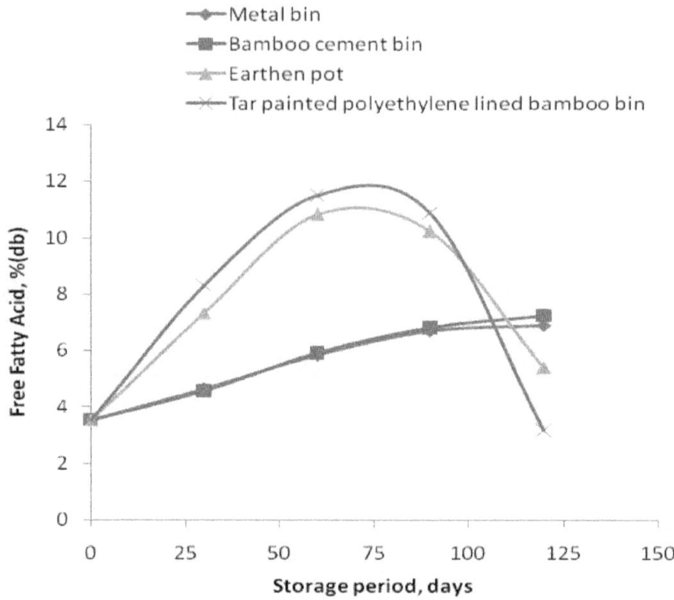

***Fig. 38.23** Effect of storage period on changes in free Fatty Acid content of groundnut stored in different storage structures*

Studies conducted so far have revealed that storage of groundnut at temperatures below 20°C is not safe if the relative humidity is not brought below 75% as it leads to decrease in nutritional quality and increase in aflatoxin content even if the initial groundnut moisture content is within the safe storage limits. Temperature of 30°C and above is safe for storage when the initial moisture content of the pod is within safe limit and relative humidity is below 65%. Thus the period for which the groundnuts can be stored without effecting the germiability and quality as well as aflatoxin development depends on the temperature and relative humidity during storage.

38.5.1 Storage for Seed Purpose

Only sound, mature, clean and well filled pods from 'true to type' plants should be selected for seed purpose. Well-dried pods with about 5% moisture content should be stored to avoid fungal and insect pests attack in storage. The pods should be stored in polythene-lined gunny bags or in some other safe storage structure in a well-ventilated and rodent free room, which is not in general use and out of bounds to children. The bags should be placed on wooden planks (not more than five in a stack) (Fig 38.24) and away from walls to avoid damage from dampness and should be protected from storage pests by dusting the bags with 5% lindane or 5% malathion dust. In case of pest outbreak in storage, the bags should be fumigated with celphos (tablets) @ 3 g bag-1 (40 kg bag) and under cover with polythene sheet for 4-5 days.

Fig. 38.24 Groundnut bags stored on wooden planks

38.6 Production of Oil

Oil contains high amounts of energy and fat-soluble vitamins (A, D, E, and K) and essential fatty acids. The oil content of the kernels is between 45% and 55%. The peanuts are prepared for the oil extraction process by being shelled and cleaned. Oil production requires some type of press with which to extract the oil form the groundnuts and filtering equipment. The following flow chart explains the production of oil from groundnut (Fig. 38.25).

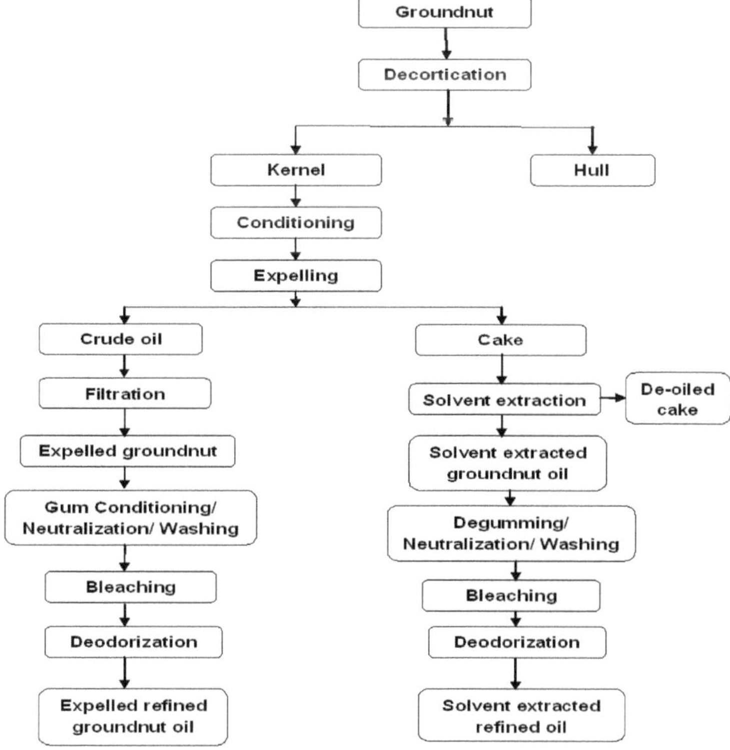

Fig. 38.25 Process flow chart for production of oil from groundnut

38.7 Uses of Groundnut Shell

Of the several million tonnes of groundnut produced each year, shell form about 25 percent of the total mass produced and their utilization thus becomes very important. At present in the developing countries the majority of groundnut hulls are either burned, dumped in forest areas or left to deteriorate naturally. The shell can be used as cattle feed, soil amendment and manure, mulch, particleboard, briquette and pelletization.

38.7.1 Briquette and Pelletization

Energy shortage in rural areas has several far-reaching ill consequences, the security of fuel wood forces people to use animal dung and crop residue as fuel, reducing the soil fertility and productivity.

Following is the method for briquette and pelletization of groundnut shell for fuel purpose. The process of briquette and pelletization is based on the principle that at set pressure fraction is related logarithmically to the height of the layer of compressed material. During the pelleting of fine-ground material, fraction increased as the diameter of the cylinder decreased and as pressure increased. It was calculated that the length of the holes in roller type disc dyes and ring dyes should be increased as the diameter of the holes and particles size of the material increased. Wafers of groundnut husk and bark could be formed by pressure <11MPa. The moisture content of the material should be between 10 to 16 percent and particle density 1.0 to 1.4 g cm^{-3}. (Esaki, *et al.*, 1986). In India small-scale industries are forming briquette from groundnut husk, which is being used in other industries as a fuel for boilers (Fig 38.26). The briquette industries require a strong support and encouragement from the governments.

Fig. 38.26 *Small-scale industries preparing groundnut Briquettes for fuel purposes*

References

1. Blatchford, S.M. and D.W. Hall, 1963. Methods of drying groundnuts, Part I: Natural Methods, part II: Artificial Methods, Printed by Secretary Indian Central Oilseeds Committee, Hyderabad.
2. Esaki. H, T. Satake. and, K. Guo. 1986. Studies on pelleting biomass II. Compression characteristics during formation of pellets and wafers. Journal of the Japanese Society of Agricultural Machinery 48: 1, 83-90.
3. Nautiyal, P.C. 2002. Groundnut: Post-harvest operations. Food and Agriculture Organization, Rome.
4. Shukla B.D., Srivastava, P.K. and Gupta R.K. 1992. Oil Seeds Processing Technology, Central Institute of Agricultural Engineering, Berasia Road, Bhopal.
5. Smyth, M.J. 1959 Drying and storage of peanuts in Queensland. For Department of Primary Industry, The peanut marketing Board, Melbourne,: Williams F. Yann, duplic., 50pp.
6. Teter, N.C. 1954. Peanut curing as related to mechanization. Agric. Engg., St. Joseph, Mich., 35, No. 8, 568-569.

Chapter 39

Value Addition to Groundnut

Swarnalatha and Veeraiah
KVK, Utukur, YSR Dist.

39.1 Introduction

Ground nut is a native of South America. It originated in the region of southern Bolivia and north-western Argentina but is now cultivated in all tropical and sub tropical countries in the world. It was introduced into the Malabar Coast of South Western India in the late 15th century by the Portuguese. Groundnut is an annual herbaceous legume with erect, semi erect or spreading growth habit. It is an unusual plant because it flowers above ground and pods, containing one to five edible seeds, are produced below the ground due to the geotropic nature of its gynophores.

Due to its high nutrient content and good taste ground nut has been an important food in various cultures throughout history. In a country like India, where people are basically vegetarian & production of pulses are low, ground nut seems to be one of suitable alternative to the pulses. The UNICEF, according to its recent report, has made a mention about the wide spread protein energy malnutrition (PEM) in Andhra Pradesh & 50% of the children within the age group of 4 are under weight. Besides 30% of the children are born with low birth weight. The situation is alarming & it requires corrective measures to combat the menace of malnutrition within the available resources. A project on "promoting groundnut as food crop for sustainable nutritional security is taken up by a novel joint venture by ICAR/TMOP/UNDP at Andhra Pradesh had a great significance.

Ground nuts are a whole some food and a treasure house of minerals and vitamins essential for healthy living. The pronounced flavour, crunchy texture and high protein render groundnuts especially suitable for use in confections. It is called the poor man's cashew nut. It forms an important ingredient in a variety of traditional foods. At present, because of readymade food culture, the age old home made groundnut recipes are disappearing.

39.2 Nutrition Facts

- Groundnut kernels, on average, contain 48% fat, 26% protein, 17% carbohydrates, 2% fibre, 2% ash and 1% or less vitamins & minerals and the remaining is moisture. It is the world fourth most important source of edible oil and the third most importance source of vegetable protein. The oil content of groundnut is approximately 50% monosaturates and 30% polyunsaturates. Just one ounce of ground nuts contains nearly half of the 13 vitamins necessary for growth of the human body and its maintenance and 35% of the 20 minerals needed.
- WHO recommends that about one ounce of pulses, nuts and seeds should be eaten daily.
- ***Caution:*** Those who have 'nut allergy' keep away from groundnut & other nuts. Always use healthy & aflotoxin free kernels in preparation of food products.

Table 39.1 *Nutrient contents in one ounce (28.3 g) of raw groundnut kernels % daily*

Nutrients	Amount	Value	Uses in the body / Remarks
Calories	161.0	n/a	Energy rich food due to its fat content. A very high proportion of unsaturated fats and high satiety value make groundnut part of a healthy diet.
Protein	7.3g	14.2%	A powerhouse of less expensive vegetable protein
Total Carbohydrates	4.6g	1.5%	Good for diabetic diets to its low glycaemic index (a measure of the rate at which carbohydrates from a particular food breakdown and release glucose in particular food breakdown and release glucose in blood stream).
Dietary Fiber	2.4g	9.4%	Reduce risk of some types of cancer, helps control blood sugar levels and may help reduce the levels of cholesterol in blood.
Total Fat	14.0g	21.8%	Concentrated source of energy, provides essential fatty acids, carries fat soluble vitamins such as A,D and E and helps maintain healthy skin. Suitable for Indian style of cooking due to its high smoking point (240°C)
Saturated Fat	1.9g	9.5%	A low proportion of saturated fat (bad fat). Saturated fat intake should be less than 10% of the total daily intake of calories.
Monounsaturated Fat	6.9g	n/a	The monounsaturated fats help to remove cholesterol including LDL cholesterol from the blood, thus giving protection from heart attack.
Polyunsaturated Fat	4.4g	n/a	Along with monounsaturated fats, polyunsaturated fats are healthy and necessary for the healthy body.

Table 39.1 ***Contd...***

Nutrients	Amount	Value	Uses in the body / Remarks
Vitamin – E	2.4 mgAT	17.5%	Vital antioxidant which protects vitamin A and the body's cells and tissues from damage. It is important for the immune system and might aid in the prevention of tumor growth. It plays a role in preventing coronary heart disease.
Folate	68 mcg	16.5%	Important for the development of new cells in the body, particularly during growth and pregnancy. It helps to prevent birth defects.
Niacin	3.26 mg	16.3%	Functioning in more than 50 of the body processes, niacin is primarily important in the release of energy from the food that we eat as well as maintenance of healthy skin, the nervous system and the digestive tract.
Thiamin (B1)	0.18 mg	12%	Needed to endure normal functioning of the nervous system, appetite and digestion.
Riboflavin (B2)	0.04 mg	21.3%	Releases energy from the food we eat, helps skin stay healthy and assists in the normal functioning of the eye.
Vitamin B6	0.10 mg	5.7%	Makes and breaks down proteins in the body makes red blood cells used to transport oxygen in the body.
Zinc	0.93 mg	5.9%	Aids in the formation of protein, wound healing, blood formation, taste perception, appetite, night vision and general growth and maintenance of all tissues.
Copper	0.32 mg	15.2%	Important in the formation of haemoglobin, health of bones, blood vessels and nerves.
Selenium	2.0 mcg	2.8%	A trace element required in small quantities for normal functioning of the immune system.
Magnesium	48 mg	12.5%	Important in the building of bones and teeth, creation of protein, transmission of nerve impulses and maintenance of body temperature.
Phosphorus	107 mg	10.6%	Component of all soft tissues that are fundamental to growth, maintenance and repairs of bones and teeth.
Potassium	200 mg	5.3%	Needed to endure water balance in the body and in the creation of protein, It also helps release energy from nutrients and aids in never impulse transmission.
Calcium	26 mg	3.5%	Needed for development and maintenance of healthy bones and teeth.
Sodium	5 mg	0.22%	Naturally low sodium food.
Iron	1.3 mg	8.1%	Aids in transport and distribution of oxygen in the body's cells.

Table 39.1 *Contd...*

Nutrients	Amount	Value	Uses in the body / Remarks
Boron	1.0 mg	100%	Major factor in the metabolization of calcium in the body and plays significant role in development and maintenance of strong and healthy bones.
Cholesterol	0.0 mg	--	Free from cholesterol.
Arginine	0.88 mg	n/a	Improves wound healing and immunity.
Total Phytosterols	62.4 mg	n/a	These phytochemicals help to prevent diseases and enhance health.
Resveratrol	Present	--	Ounce for ounce groundnut have about half of the amount of resveratrol in wine (160 mcg/g), 73 mcg/g without skin. Resveratrol, a phytochemical, has a possible role in reducing cancer and can inhibit build up of platelets in blood vessels. It is a potent antioxidant which can reduce the oxidation of LDL cholesterol.
Beta - sitosterol	18.4 mg	n/a	Has anticancer properties and prevents cholesterol uptake.

39.3 Value Addition to Groundnut

39.3.1 Groundnut Pakodi

Ingredients

Rice flour - 1/4kg
Bengal gram (chickpea) flour - 1kg
Groundnut seeds - 1kg
Chili powder - 2 ts
Baking soda - Small quantity
Salt to taste - Oil to fry

Method: Take groundnut in vessel. Add rice flour, Bengal gram flour, small quantity of oil, baking soda, chilli powder and salt. Add water and prepare batter. Heat oil in a kadai and deep fry the groundnuts (dipped in the batter).

Makes 6 serving 783 calories each.

39.3.2 Groundnut Holige

For the covering

Sorghum flour - 1cup
Maida (refined wheat flour) - 1cup
Groundnut oil - 4 ts

For the filling:

Roasted and skinned ground	-	1cup
Roasted gingelly (sesame) seed	-	1cup
Poppy seeds, roasted	-	2 ts
Grated coconut	-	3 ts
Cardamom	-	2 cups

Method: Mix the ingredients for the covering. Add water and knead into a dough as for chapaties. Powder groundnut seed and gingelly seeds together. Add jiggery, roasted poppy seed powder and great coconut to the filling mixture. Divide the dough and filling into equal number of portions. Roll out each portion of the dough into a small round. Place a portion of filling in it. Bring up the edges and seal. Roll out into thin rounds and roast on a hot flat pan. Serve the holige hot or cold with fresh melted ghee.

39.3.3 Ground Nut Burfi

Ingredients

Roasted and skinned groundnuts	-	¼ kg
Powdered jaggery	-	¼ kg
Ghee	-	½ cup
Cashew nuts	-	50 gm
Grated copra	-	2 tp
Clove powder	-	Pinch

Method: Mix jaggery with one cup of water and place on fire, stirring till it begins to thicken. Add roasted groundnuts and continue to cook. Add cashew nuts and clove powder and spread in a greased plate or tray. Cool and cut into pieces.

39.3.4 Groundnut Burfi -2

Ingredients

Sugar	-	1 kg
Powdered jaggery	-	1 kg
Roasted and skinned groundnuts	-	2 kgs

Method: Mix sugar and jaggery with 3 cups of water. Place fire stirring till it begins to thicken. Add roasted groundnuts and continue to cook till the mixture leaves the sides of the pan and forms a thick lump. Spread on a greased plate or metal tray cool and cut into pieces.

39.3.5 Groundnut Halwa

Ingredients

Milk	-	1 Lit
Coconut	-	1
Roasted and skinned groundnut	-	100 gm
Sugar	-	300 gm

Method: Grate the coconut and mix with milk and sugar. Place on fire and stir till it begins to thicken. Add the groundnut powder and continue to cook till the mixture leaves the sides of the pan and forms a thick lump. Remove from fire, cool it and keep it in a refrigerator for 2hours and serve chilled. Makes 8 servings.

39.3.6 Peanut Candy

Ingredients

Skinned groundnuts	-	4 cups
Ghcc	-	1/4 cup
Powdered jaggery	-	1 ½ cup
Coconut	-	½ cup
Cardamom powder	-	1 tp

Method: Slice the coconut into thin small square pieces. In a kadai, add the jaggery with ½ cup water and coconut pieces. When the mixture boils and starts thickening, add on teaspoon of jaggery Mix to the cup of water to test for doneness. (The jaggery should not dissolve in water. When you roll it into a ball, the jaggery should be soft.) Add cardamom powder to the jaggery mix. In a broad plate, place the cleaned peanuts and pour the jaggery mix over it one ladle at a time. Mix them evenly with a thin edged spatula. Make lemon sized balls with your hand while it is still hot. Dip hands in water or rice flour periodically while making the candy. The peanut balls start hardening as they cool down. The same peanut – jaggery mixture can be poured into different moulds. They are cut into small pieces according to requirement. For garnishing, dried grated coconut dipped in various food colours is used.

39.3.7 Peanutty Bar Delights

Ingredients

Flour	-	1 ½ cup
Baking powder	-	1tp

Salt	-	1tp
Granulated sugar	-	1 cup
Butter	-	¼ cup
Egg	-	1
Whole roasted peanuts	-	½ cup
Brown sugar	-	1 cup
Egg whites, stiffly beaten	-	2
Vanilla extract	-	1tp

Method: Preheat oven to 375 F. Cream butter and sugar in a large mixing bowl, add eggs and vanilla. Beat until fluffy. Mix flour with soda in a large bowl, add oats and toss to coat with flour mixture. Add to first mixture a cup at a time until completely mixed. Stir in peanuts and chocolate bits. (Butter will be stiff). Drop from a teaspoon onto a cokkie sheet. Bake for 10 – 12 minutes. Yield 6 dozen cookies of 232 calories each.

39.3.8 Peanut Punch - 1

Ingredients

Shelled roasted peanuts	-	8 ounces
Water	-	1 ½ pints
Condensed milk	-	1 can
Vanilla essence	-	½ tp
Nutmeg, freshly grated	-	½ tp
Honey	-	1tp

Method: Place the shelled peanuts in a blender. Pour water in gradually while blending the peanuts. Blend thoroughly. Strain mixture through a sieve lined with muslin, pouring in a little at a time. Discard the contents. Add milk and other ingredients, to sweeten the taste. Makes 4 servings of 520 calories each.

39.3.9 Peanut Punch- 2

Ingredients

Smooth peanut butter	-	½ to 2/3 cup
Cold whole milk	-	3 cups
Sugar	-	4 tp
Cinnamon to taste		
Nutmeg to taste		

Method: Place all ingredients in a blender on high speed for about 30 seconds or until thick and frothy. Pour into glass(es) and serve immediately.

Makes 1 serving of 1,392 calories (can be divided in to 3 servings.)

39.3.10 Groundnut Milk

Ingredients

Roasted and skinned groundnut seed	-	100 gm
Sugar	-	75 gm
Water	-	1 lit
Cardamom	-	5
Essence	-	Few drops

Method: Remove the germs from the roasted groundnut cotyledons. Add a little water and grind the groundnut seeds to smooth paste. Add 1 litter of water to the paste and mix it. Filter the milk. Add sugar, cardamom powder and essence and boil the milk.

39.3.11 Ground Nut Pakodi

Ingredients

Rice flour	-	¼ kg
Bengal gram(chickpea) flour	-	1 kg
Ground nut seed	-	1 kg
Chili powder	-	2 tp
Baking soda	-	Small quantity

Salt to taste

Oil to fry

Method: Take ground nut in vessel. Add rice flour, Bengal gram flour, small quantity of oil, baking soda, chilli powder and salt. Add water and prepare batter. Heat oil in a kadai and deep fry the groundnuts (dipped in the batter).

39.3.12 A Once up on a Time Spicy Peanut

Ingredients

Peanuts	-	125 gm
Grams (chickpea) flour	-	½ cup
Rice flour	-	¼ cup

Red chilli powder	-	½ tp
Fresh coriander leaves, chopped	-	½ cup
Onions, chopped	-	¼ cup
Water	-	¾ cup
Oil to deep fry		
Salt to taste		

Method: Roast for 10- 15 minutes in a pan on the stove top. Make a batter by mixing together all the remaining ingredients . Remove the skin of the roasted peanuts carefully. Add roasted peanuts to the batter. Heat oil in a wok or kadai. Drop spoonful of the peanut butter mixture into the hot oil. Fry till crisp. Drain on clean kitchen napkins/ towel/ absorbent paper. Serve hot with chutney. Makes 2- 3 servings of 524 calories each.

39.3.13 Whole Bringal Curry

Ingredients

Tender brinjal	-	250 gm
Ripe tamarind	-	lemon size ball
Roasted and skinned groundnut	-	50 gm
Chili powder	-	1 tp
Cumin seeds	-	1 tp
Salt to taste	-	
Grated copra	-	20 gm
Garlic	-	6 flakes
Onion	-	1 medium sized
Coriander leaves, chopped	-	2 tp
Chopped ginger	-	2 cm piece
Turmeric powder	-	½ tp
Groundnut oil	-	¼ cup

Method: Lightly fry all the ingredients, except brinjal, tamarind and chilli powder, using very little oil and grind them to a smooth paste. Slit the whole brinjal horizontally into four parts and fill the brinjal with roasted masala paste. Heat the oil in a kadai, Saute onion till golden brown, then add the chilli powder and cook it for a while. Now add the whole brinjals filled with masala paste and tamarind juice. Cook on a low flame, till the brinjal are soft. garnish with coriander. Makes 4 servings of 140 calories each.

39.3.14 Groundnut Rasam

Ingredients

Ripe tamarind	-	Lemon sized ball
Coriander leaves chopped	-	2 tp
Roasted and skinned groundnut	-	50 gm
Rasam powder	-	½ tp
Cumin seeds	-	1tp
Red chilies	-	2

Method: Lightly roast groundnut seeds, cumin seeds and red chillies and powder them. Heat oil in a kadai, season with onion and red chillies. Add tamarind juice, salt, rasam powder, coriander leaves and groundnut powder. Boil the rasam for 5 minutes before serving. Makes 4 serving of 122 calories each.

39.3.15 Groundnut Chutney

Ingredients

Green chilli	-	5
Roasted and skinned groundnuts	-	250 gm
Coriander seed	-	1tp
Ripe tamarind	-	lemon size balls
Onion	-	1
Salt to taste	-	3 pods

Method: Lightly garlic fry groundnut seed, green chillies, sliced onion and tamarind with a little oil. Add a little amount of water and grind the mixture to a smooth paste. Season with groundnut oil and garlic. Makes 4 servings of 418 calories each.

39.3.16 Tamarind Leaf Powder

Ingredients

Roasted and skinned groundnut	-	100 gm
Red chillies	-	10
Horse gram	-	25 gm
Dried tamarind leaves	-	50 gm
Salt to taste		
Sugar to taste		

Method: Heat oil in a pan and roast the grated copra, red chilies, horse gram and groundnut seeds. Remove them from the pan. Fry the dried tamarind leaves in the same pan. Mix all ingredients, and salt and sugar and powder them.

39.3.17 Groundnut Chutney Powder

Ingredients

Roasted and skinned groundnut	-	20 gm
Popped chana dal	-	20 gm
Grated copra	-	10 gm
Chili powder	-	2 tp
Salt to taste		

Method: Heat ghee or oil and roast groundnut seed, popped chana dhal and grated COPRA. Add chilli powder and salt and powder them.

References

1. *Swaminathan.M, "The use of oil seeds & nuts in the feeding of infants & pre-school children in the developing countries. PL. Fds.IIVM.Nutri.I.205.1969.*
2. *Gopalan, C.Rama Sastri, B>V. And Balasubramanian, S.C. Nutritive value of Indian*
3. *Foods, I.C.M.R., New Delhi, 1971.*
4. *Mrs. B.N.Sulochana, Mrs. K.Bhargavi, Dr. T.Yellamanda Reddy, Dr.M.S.Basu – Culinary preparations with Groundnut. UNDP sub programme on promoting Groundnut as food crop for sustained nutrition security" by NRC for Groundnut, Junagadh, Gujarat.*
5. *K. Ravishankar, K.P.Sreejith "Groundnut Gourmet"by ICRISAT.*
6. *Dr. M.Swaminathan, Ädvanced text book on Food & Nutrition" Volume-I, BAPPCO 1999.*

Chapter 40

Statistical Tools for Sustainable Agriculture

P. Lavanya Kumari

Scientist, Division of Statistics, RARS, ANGRAU, Tirupati.

40.1 Introduction

Statistical analysis is a vital component in every aspect of contemporary research. Statistics, as a science of data collection, analysis, interpretation, explanation and presentation will help the researchers for proper characterization, summarization, presentation and interpretation of research findings based on objectives framed in the study. This chapter is intended to explore the need of Statistics in Agricultural Research by explaining the appropriate tests according to type of measurement and number of groups (samples) under study. Statistical tools can be viewed as two groups; 'tests of differences' and 'tests of relations or associations' for better understanding. Statistical analyses may be invalid if the assumptions behind those tests are violated.

This chapter contains four sections in which the first section discusses a step wise procedure to conduct *Scientific Research in Agriculture* while the second section deals with various *Designs of Experiment* and their purposes, whereas the third section explains important aspects of *Tests of hypothesis* including Posthoc tests and the analysis of *Repeated Measures*. Finally the fourth section guides on the Software based Statistical Analysis using MS-Excel and SPSS.

40.2 Scientific Research in Agriculture

40.2.1 The Scientific Method of conducting Research

In a broad sense, the traditional Scientific Research consists of four interrelated stages viz., Problem definition, Data gathering, Data analysis, and Data interpretation in which Statistical tools are used in data gathering, data analysis and interpretation stages. But in detail, there are eight phases in conducting research in a scientific manner.

1. Identification of the problem
2. Review of literature

3. Formulation of the hypothesis
4. Design and conduct experiment
5. Data Collection
6. Data Analysis
7. Draw conclusions
8. Make recommendations

1. Identification of the problem

The first step in conducting agricultural research is identifying the problem to be investigated. Typically, this can be stated in the form of a question. The research problem should grow out of the researcher's areas of interest and personal experience in agriculture. By choosing a problem in an area of interest, the researcher will find the experimentation process to be more enjoyable and more relevant.

2. Review of literature

The next step in the scientific method is to research the problem. The researcher should utilize various sources to gain knowledge about the problem to be tested. This information will allow the researcher to move to the next step in the scientific method that is forming a hypothesis.

3. Formulation of the hypothesis

The third step in the process is to predict the results of the experiment, called a hypothesis.

A *hypothesis* is a tentatively accepted theory that explains the relationship between two variables. Generally, hypotheses should indicate the nature of the relationship between variables. The researcher should utilize background knowledge and experience in this area to formulate the hypothesis.

Once the hypothesis is made, it is subjected to a test that attempts to isolate the important factors. This test is called an *experiment.*

4. Design and conduct experiment

(a) The fourth step in the scientific method is to design and conduct the experiment. Factors to consider in the design of the experiment include observation and measurement procedures, control measures, and selection of variables.

- A *variable* is a characteristic by which an object of phenomenon may be described. Variables change from time to time and may have more than one value.
- An *independent variable* is the characteristic that the researcher believes will affect another variable. Researchers manipulate an independent variable by managing its presence in the experiment.

- The manipulation of the independent variable is known as the *treatment*.
- *The dependent variable* is the characteristic that will be observed. It is expected to change as a result of the independent variable.

(b) Measurement of the dependent variable is critical to the success of the experiment. During this phase, the researcher must determine the most precise and appropriate way to measure the effect of the independent variable.

(c) It is critical that the design of the experiment match the stated hypothesis. This means that the researcher must make sure that the experiment answers the question stated earlier. The research design also should provide for replication.

(d) *Replication* is the exact duplication of the experiment which allows the results to be verified across numerous trials, resulting in greater confidence in the results. After all design procedures have been followed, then it is time to conduct the experiment.

5. Data collection

(a) The next stage in the experiment is data collection. The researcher should consistently and precisely follow the previously determined methods outlined in the experiment design in the measurement and collection of data.

(b) A pattern or format of collecting and recording data should have been developed in the design stage of the experiment.

(c) Data summary charts are usually used to record experimental data. These charts should be as simple as possible, yet provide for all data to be completely and consistently recorded.

6. Data analysis

(a) Once the data is collected, the next step is to analyze the data. In this stage, data that was collected during the experiment is analyzed so it can be used later to draw conclusions. The research will follow the procedures outlined in the experiment design stage in organizing the data. This can be a very exciting stage of the research process because results of the experiment become evident.

(b) Once the raw data has been organized into a usable form, the researcher can draw conclusions from the information. The researcher must answer the question, "is my hypothesis correct or not?" This is done by looking for patterns in the data or by interpreting calculations that were performed on the data.

(c) During this process, the researcher must also determine if there is any evidence to suggest that the data might be inaccurate or in any way misrepresent what actually happened in the experiment. At this point any limitations of the experiment also must be noted.

(d) Depending upon the nature of the experiment, the researcher may then be able to formulate recommendations based on the findings of the experiment.

7. Draw conclusions

Findings should be explained in detail through conclusions, written discussion, and recommendations. Results of this experiment should be compared to results obtained from similar experiments.

8. Make recommendations

Finally, based upon the results of the completed experiment, the researcher should identify new research questions to explore through further experimentation.

Through this continuing cycle of questioning, experimentation, further questioning, and further experimentation, agricultural research is done.

40.2.2 Research need to have Controlled Conditions

(i) Research must be done in a controlled and precise way to ensure that the results of the experiment are valid and reliable.

(ii) The results in an experiment must be accountable. Results are accountable if the experiment has an independent variable, dependent variable, controlled extraneous variables, and constants.

- *Extraneous variables* are the factors which may impart the effect on the dependent variable. The extraneous variables must be limited to the greatest degree. If extraneous variables are present in an uncontrolled manner, the results may not be accurate or repeatable.
- A controlled experiment must have *constants* or conditions in an experiment that remain the same. *Control* is the process of holding factors constant in an experiment.

40.3 Overview on Design of Experiments

The Purpose of designing an experiment in scientific way is to increase the precision of the experimental results. In order to increase the precision, the experimental error (i.e., plot to plot variation) needs to be minimized.

40.3.1 Terminology of Experimental Designs

1. An *experiment* is a planned inquiry to obtain new knowledge or to confirm or deny previous results.
2. The *population of inference* is the set of ALL entities to which the researcher intends to have the results of the experiment be applicable.
3. The *experimental unit* is the smallest entity to which the treatment is applied and which is capable of being assigned a different treatment independently of other experimental units if the randomization was repeated. This is the most seriously misunderstood and incorrectly applied aspect of experimental design.

4. A *factor* is a procedure or condition whose effect is to be measured.
5. A *level* of a factor is a specific manifestation of the factor to be included in the experiment.
6. *Experimental error* is a measure of the variation which exists among experimental units treated alike. *It can be minimized but cannot be eradicated completely.*
7. A *factor level* is said to be replicated if it occurs in more than one experimental unit in the experiment.
8. Prof. R.A. Fisher proposed three principles of Good Experimentation viz., *Replication, Randomization and Local Control* in order to minimize the experimental error as much as possible.
9. *Randomization* in the application of factor levels to experimental units occurs only if each experimental unit has equal and independent chance of receiving any factor level and if each experimental unit is subsequently handled independently. In other words Randomization means random allocation of treatments to different experimental units. For this purpose random number tables can be used.
10. *Replication* means repetition of the trial in the same conditions. Replication is the source of error.
11. *Local control* is nothing but grouping of homogenous experimental units into blocks. Experimental units under each block need not be in a single row or single column, sometimes they may be scattered around the experimental area.

    Note:

    (a) Replication and Randomization help us to estimate the Experimental Error. But Local Control will help us to minimize the Experimental error.

    (b) *There is negative correlation between application of Local control and experimental error.*

    (c) A satisfactory precision can be obtained by maintaining error degrees of freedom not less than 12.

40.3.2 Various Experimental Designs at a Glance

1. **Completely Randomized Design (CRD):** In this design all treatments will be allocated completely at random to all experimental units so that each experimental unit has the same chance of receiving any one treatment. It utilizes only two principles viz., Replication and Randomization.
2. **Randomized Complete Block Design (RCBD):** It is the arrangement of 't' treatments in 'n' blocks such that each treatment should appear once in any block. It utilizes three principles of experimentation. It is the most suitable design for Agriculture.

Fig 40.1 is showing different ways of laying out six treatments (T1–T6) in one replicate of a Randomized Complete Block Design. If the soil gradient goes from the top to the bottom then the complete block should be as homogeneous as possible in terms of soil and other environmental factors. The researcher may not know the direction of the soil gradient on the experimental area and may allocate the treatments in the wrong direction. However, it is clear that with the smaller the block size less chance exists of having heterogeneity of soil within the complete block. When the block size is large, the chances of having heterogeneity soil conditions within the block increase. As the number of treatments increases, the chances of finding a homogeneous complete block dramatically decreases and another design is required. It is recommended that the block or (replicates) be as compact as possible.

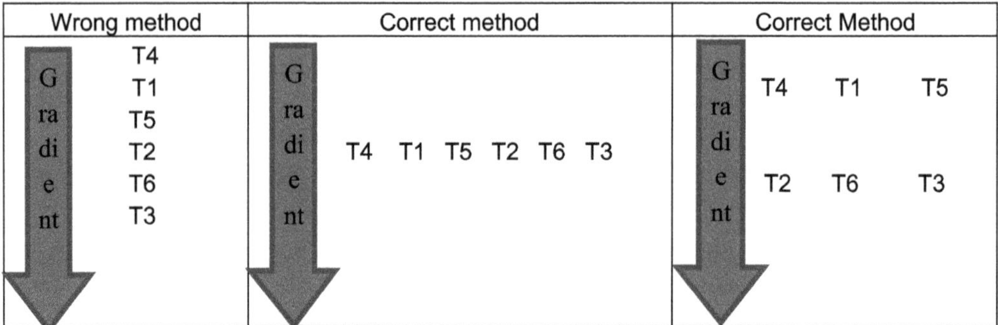

***Fig 40.1** Layout of treatments in a complete one block*

3. **Incomplete Block Designs – Lattice *(from Cochran and Cox, 1957)***

These designs are arranged in blocks that are smaller than a complete replication, in order to eliminate heterogeneity to a greater extent than is possible with randomized blocks; this reduction in the size of block was achieved by sacrificing all, or part of, the information on certain treatment comparisons. They may be 'balanced or partially balanced'.

Fig 40.2 shows the possible layout in the field of a Randomized Complete Block Design (RCBD) and the layout of an incomplete block design of four incomplete blocks of size 4. The layout of the RCBD covers a band of the field that accounts for soil gradient on the upper and lower parts of the field. The two possible layouts of the 4 × 4 incomplete blocks will control local variability in a much more efficient manner.

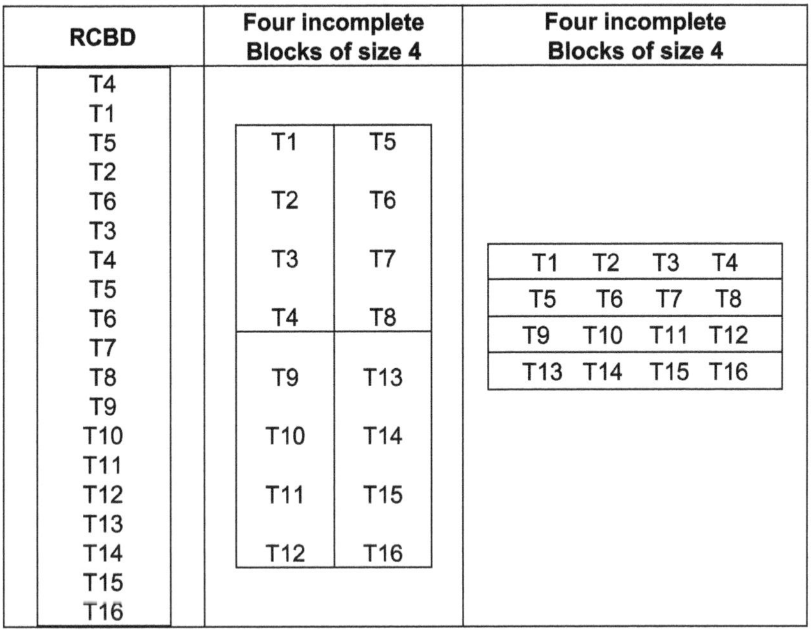

<table>
<tr><th>RCBD</th><th>Four incomplete Blocks of size 4</th><th>Four incomplete Blocks of size 4</th></tr>
<tr>
<td>T4
T1
T5
T2
T6
T3
T4
T5
T6
T7
T8
T9
T10
T11
T12
T13
T14
T15
T16</td>
<td><table>
<tr><td>T1
T2
T3
T4</td><td>T5
T6
T7
T8</td></tr>
<tr><td>T9
T10
T11
T12</td><td>T13
T14
T15
T16</td></tr>
</table></td>
<td><table>
<tr><td>T1 T2 T3 T4</td></tr>
<tr><td>T5 T6 T7 T8</td></tr>
<tr><td>T9 T10 T11 T12</td></tr>
<tr><td>T13 T14 T15 T16</td></tr>
</table></td>
</tr>
</table>

Fig 40.2 *Lay out in the field of a Randomised Complete Block Design (RCBD) and the layout of an Incomplete Block Design of four incomplete block size of 4*

4. Balanced Incomplete Block Designs (BIBD)

The balanced designs will be illustrated first by simple examples of the experimental plans. Consider table 40.1 below which compares 9 treatments in incomplete blocks of 3 Experimental Units (EU) with 4 replications. Every pair of treatments will be found to occur once, and only once, in the same block.

For instance, treatment 1 occupies the same block with treatments 2 and 3 in the first replication, with treatments 4 and 7 in the second replication, treatments 5 and 9 in the third replication, and treatments 6 and 8 in the fourth replication. When the results are analyzed by the method of least squares, all pairs of treatments are compared with approximately the same precision, even though the differences among blocks may be large.

Table 40.1 *Balanced design for 9 treatments in blocks of 3 units*

Block	Rep. I			Block	Rep. II			Block	Rep. III			Block	Rep. IV		
(1)	1	2	3	(4)	1	4	7	(7)	1	5	9	(10)	1	8	6
(2)	4	5	6	(5)	2	5	8	(8)	7	2	6	(11)	4	2	9
(3)	7	8	9	(6)	3	6	9	(9)	4	8	3	(12)	7	5	3

The above design belongs to the group known as 'balanced lattices', so-called because the plans are conveniently written down by drawing a square lattice, with the treatment numbers at the intersections of the lines. In the balanced lattices, the number of treatments must be an exact square while the number of units per block is the corresponding square root. Balanced designs can be constructed for other numbers of treatments and units per block. Table 40.2 shows 7 treatments arranged in blocks of 3 units.

***Table 40.2** Balanced design for 7 treatments in blocks of 3 units*

Block	Rep. I			Block	Rep. II			Block	Rep. III			Block	Rep. IV		
(1)	1	2	4	(3)	3	4	6	(5)	1	5	6	(7)	1	3	7
(2)	2	3	5	(4)	4	5	7	(6)	2	6	7				

Again every pair of treatments occurs once within same block, In this case, however, the blocks cannot be grouped in separate replications, since 7 is not divisible by 3. Designs of this type are known as 'balanced incomplete blocks'. For certain numbers of treatments and units per block, both of the types above can be laid out in a kind of Latin Square formation to allow the elimination of variation arising from the grouping. The appropriate rearrangement for the first example (Table 40.1) is shown in the Table 40.3.

***Table 40.3** Balanced design for 9 treatments in 4 lattice squares*

Rep. I				Rep. II				Rep. III				Rep. IV			
	Columns				Columns				Columns				Columns		
Rows	(1)	(2)	(3)	Rows	(4)	(5)	(6)	Rows	(7)	(8)	(9)	Rows	(10)	(11)	(12)
(1)	1	2	3	(4)	1	4	7	(7)	1	6	8	(10)	1	9	5
(2)	4	5	6	(5)	2	5	8	(8)	9	2	4	(11)	6	2	7
(3)	7	8	9	(6)	3	6	9	(9)	5	7	3	(12)	8	4	3

Every pair of treatments now occurs once in the same row and also once in the same column. All comparisons between pairs of treatments are of nearly equal precision. This design is known as a 'lattice square'.

5. **Partially balanced designs**

Although a balanced design can be constructed with any number of treatments and any number of units per block, the minimum number of replications is fixed by these 2 variables. In most cases this number is too large for the usual conditions of experimentation and less replicates are used. Simple examples of such designs are the lattices. These are constructed in the same manner as balanced lattices, except that there are fewer replications.

For example, the design with the first 2 replications for a balanced design with 9 treatments is called a 'simple lattice', and that with 3 replications a 'triple lattice' (Table 40.3). Similarly, with a lattice square, as in Table 40.2 above, may be used with less than the full number of replicates necessary for balance. For all these designs the number of treatments must be a perfect square.

Partially balanced designs are less suitable than balanced designs as the statistical analysis is more complicated. When the variation among blocks (or rows and columns) is large, some pairs of treatments are more precisely compared than others, and several different standard errors may have to be computed for tests of significance. These difficulties increase as the design departs more and more from the symmetry of the balanced design.

The simplest of the partially balanced designs are those with two associate classes. In these, some pairs of treatments occur in the same block λ_1 times, while other pairs occur in the same block λ_2 times, where λ_1 and λ_2 are whole numbers. In the lattice designs, which are of this type, some pairs of treatments never occur together in a block, i.e., $\lambda_1 = 0$; other pairs occur once in the same block ($\lambda_2 = 1$).

Note: Incomplete block designs are no more difficult than randomized blocks. Some extra planning is involved in drawing up and randomizing the experimental plan, especially if care is taken to make the best possible grouping of the experimental units.

6. **Other important designs**

   Latin Square Design (LSD): It is the arrangement of 't' treatments in 't' rows and 't' columns such that each treatment should appear once in a row or once in a column. It also utilizes three principles of good experimentation but local control will be applied twice. This design will be used when no.of treatments are ranged between 5 to 12.

   Factorial Experiments (FEs): When several factors are investigated simultaneously in a single experiment, such experiments are known as factorial experiments. The term factor refers to a set of related treatments. If the factors under investigation will have same no.of levels, then it is called symmetric factorial experiment, otherwise asymmetric factorial experiment.

   Split plot design: It is also a 2-factor experiment. One factor is called Main plot and other factor is called Sub plot. Main plot is assigned with factor which is of less importance to the resultant (less precision) and subplot is assigned to the factor which is of more importance.

   Strip plot design: This is also a 2 factorial experiment. There are certain factors like Spacing, Methods of ploughing, Levels of irrigation and dates of sowing, which requires plots in larger sizes. Land constraint can be overwhelmed with the help of this design.

Augmented design: (Pinney 1991) has made use of Augmented design for On-farm trials or prototype evaluation trials. On-farm trials are an essential means of technology transfer and can be viewed as a means s of proving feed-back from the farm to the research station. It is a very powerful and appropriate strategy for agroforestry research, especially for the applied type. The technology developed at research stations is teste under farm conditions to obtain information about its performance and acceptability by the farmers.

40.3.3 Sample Layouts of the Designs

Readymade layouts depending on the requirements of the researchers are available in the following URL of IASRI.

http://www.iasri.res.in/design/Basic % 20Designs/basicdesign.aspx

Table 40.4 *Model Layout for CRD with 6 treatments and 3 replicates.*

Randomized Layout																	
Experimental Unit																	
1	2	3	4	5	6	7	8	9	10	11	12	13	14	15	16	17	18
T3	T5	T3	T3	T5	T4	T4	T2	T6	T2	T1	T1	T6	T4	T1	T6	T5	T2

Outline of ANOVA	
Source	Degrees of Freedom
Treatments	5
Error	12
Total	17

Table 40.5 *Model Layout for RCBD with 6 treatments and 4 blocks*

Randomized Layout					
Block - 1					
1	2	3	4	5	6
T4	T2	T3	T1	T5	T6
Block - 2					
1	2	3	4	5	6
T2	T6	T1	T5	T3	T4
Block - 3					
1	2	3	4	5	6
T6	T3	T2	T4	T5	T1
Block - 4					
1	2	3	4	5	6
T2	T3	T6	T5	T1	T4

Outline of ANOVA	
Source	Degrees of Freedom
Blocks	3
Treatments	5
Error	15
Total	23

Table 40.6 *Model Layout for LSD with 5 treatments*

Randomized Layout					
Rows	Columns				
	1	2	3	4	5
1	T1	T3	T2	T4	T5
2	T3	T2	T4	T5	T1
3	T2	T4	T5	T1	T3
4	T4	T5	T1	T3	T2
5	T5	T1	T3	T2	T4

Outline of ANOVA	
Source	Degrees of Freedom
Rows	4
Columns	4
Treatments	4
Error	12
Total	24

Table 40.7 *Model Layout for Factorial CRD with two factors with 2 and 3 levels each in 3 replications*

Randomized Layout																	
Experimental Unit																	
1	2	3	4	5	6	7	8	9	10	11	12	13	14	15	16	17	18
T3	T5	T5	T4	T6	T6	T1	T1	T3	T5	T4	T2	T1	T2	T4	T6	T2	T3

Treatment Mapping		
Treatment	Factors	
	1	2
T1	0	0
T2	0	1
T3	0	2
T4	1	0
T5	1	1
T6	1	2

Outline of ANOVA	
Source	Degrees of Freedom
Treatments	5
Main Effects	
Factor 1	1
Factor 2	2
Two factor Interactions	
Factor 1 * Factor 2	2
Error	12
Total	17

Table 40.8 *Model Layout for Factorial RRD with two factors having 2 and 3 levels each in 4 blocks*

Treatment Mapping		
Treatment	Factors	
	1	2
T1	0	0
T2	0	1
T3	0	2
T4	1	0
T5	1	1
T6	1	2

Outline of ANOVA	
Source	Degrees of Freedom
Blocks	3
Treatments	5
Main Effects	
Factor 1	1
Factor 2	2
Two factor Interactions	
Factor 1 * Factor 2	2
Error	15
Total	23

40.4 Testing of Hypothesis

40.4.1 Important aspects in Testing of Hypothesis

1. **Statistical hypothesis**

A Statistical Hypothesis is an assertion or conjecture (tentative conclusion) either about the form or about the parameter of the distribution.

Example:

(i) The average yield of a variety-X is 1200 kg/ha.

(ii) Almost 50% of the area is covered under variety-X in southern zone.

In a statistical hypothesis, all the parameters of a distribution may be specified completely or partly. *A statistical hypothesis in which all the parameters of a distribution are completely specified is called simple hypothesis, otherwise, it is known as composite hypothesis.*

For example, in case of normal population, the hypothesis

(i) Mean (μ) =20, variance (σ^2) = 5 (Simple hypothesis)

(ii) μ= 20, σ^2>1 (Composite hypothesis)

(iii) μ =20 (Composite hypothesis)

Null hypothesis: The statistical hypothesis under sample study is called null hypothesis. It is usually that the observations are the result purely of chance. It is usually denoted by H_0.

Alternative hypothesis: In respect of every null hypothesis, it is desirable to state, what is called an alternative hypothesis. It is complementary to the null hypothesis. In other words "The desirable attitude of the researcher about the hypothesis is termed as alternative hypothesis". It is taken usually that the observations are the result of real effect plus chance variation. It is usually denoted by H_1 . For example, if one wishes to compare the yield per hectare of a new line with that of standard line, then, the null hypothesis:

H_0 : Yield per hectare of new line (μ_1) = Yield per hectare of standard line (μ_2)

The alternative hypothesis corresponding to H_0 , can be the following:

(i) $H_1 : \mu_1 < \mu_2$ (Left tail alternative)

(ii) $H_1 : \mu_1 > \mu_2$ (Right tail alternative)

(iii) $H_1 : \mu_1 \neq \mu_2$ (Two tailed alternative)

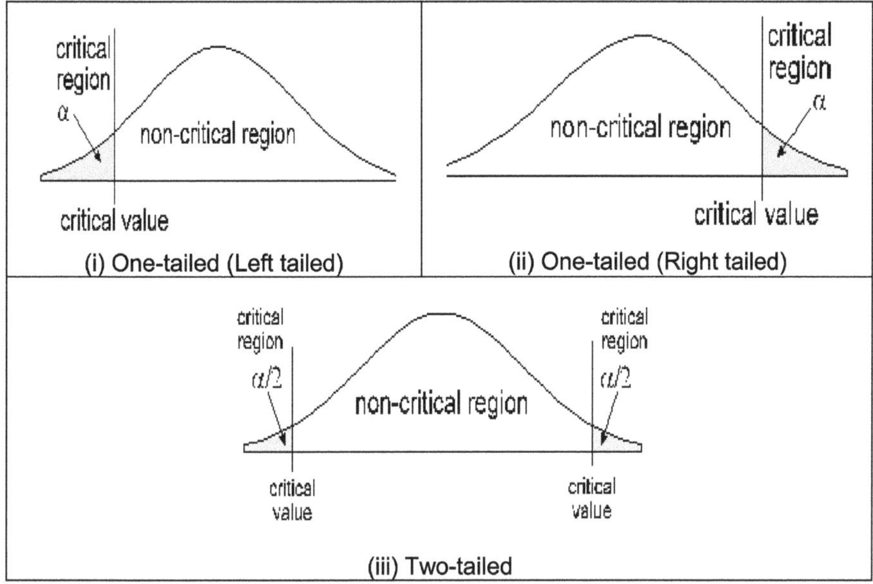

Images (i) and (ii) are called one tailed test and (iii) is a two tailed test. Whether one sets up a one tailed test or a two-tailed test depends upon the conclusion to be drawn if H_0 is rejected. The location of the critical region will be decided only after H_1 has been stated.

For example in testing a new pesticide, one sets up the hypothesis that it is no better than similar pesticides now on the market and tests this hypothesis against the alternative hypothesis that the new pesticide is superior. Such an alternative hypothesis will result in a one tailed test (right tail alternative). If we wish to compare a new cultivating technique with the Farmer's Practice, the alternative hypothesis should allow for the new approach to be either inferior or superior to the Farmer's Practice. Hence the test is two-tailed. Consider the following examples to fix up the idea.

Example-1

An Agricultural Officer claims that 70% farmers in Southern Zone are cultivating Rice. To test this claim, a large sample of farmers has to be interviewed; the proportion of the farmers who cultivates Rice is recorded and used as our test statistic.

If the test statistic is substantially higher or lower than Proportion (p) = 0.7, we should reject the officer's claim. Hence, we should make the test H_0 : p = 0.7 and H_1: p ≠ 0.7. After framing the null hypothesis and alternative hypothesis, researcher has to select an appropriate procedure that will enable to decide whether to reject null hypothesis (H_0) or not. Such a rule or procedure is called **test of hypothesis.**

2. **Critical region**

   It is region of rejecting null hypothesis when it is true if sample point belongs to it and is denoted by 'C'. Suppose that if the test is based on a sample of size 2, then the outcome set or sample space is the first quadrant in a two dimensional space and a test criterion will enable us to separate our outcome set into two complementary subsets, C and C^c*(C-compliment).* If the sample point falls in the subset C, H_0 is rejected, otherwise, H_0 is accepted.

3. **Types of errors**

   The terms acceptance and rejection are, however, not to be taken in their literal senses. Thus acceptance of H_0 does not mean that H_0 has been proved true. It means only that so far as the given observations are concerned, there is no evidence to believe otherwise. Similarly, rejection of H_0 does not disprove the hypothesis; if merely means that H_0 does not look plausible in the light of given observations. It is now known that, in order to establish the null hypothesis, we have to study the sample instead of entire population. Hence, whatever, decision rule we may employ, there is every chance of committing errors in the decision for rejecting or accepting the hypothesis. Four possible situations, which can arise in any test procedure, are given in the following table.

		Situation	
		H_0 is true	H_0 is false
		(H_1 is false)	(H_1 is true)
Decision	Reject H_0	Type-I error	Correct decision
	Accept H_0	Correct decision	Type-II error

From the table, it is clear that the errors committed in making decisions are of two types.

Type I error: Reject H_0 when H_0 is true.

Type II error: Accept (do not reject) H_0 when H_0 is false.

For example, a Scientist would like to decide whether a treatment is effective or not. The statistical hypothesis in this case is:

H_0 : Treatment (new practice) is not effective;

H_1: Treatment (a new practice) is very effective

In this situation, two types of errors which the Scientist may commit are:

Type I error: New practice which is really ineffective is treated as effective.

Type II error: New practice which is really effective is treated as ineffective.

Since, it is more dangerous to misguide the farmer by promoting a useless practice. Therefore, Type I error is more serious than the Type II error.

Probabilities of the errors:

Probability of Type I error = P (Reject H_0 / H_0 is true) =α

Probability of Type II error = P (Accept H_0 / H_1 is true) =β

In quality control terminology,

Type I error amounts to rejecting a lot of grains when it is good and

Type II error may be regarded as accepting a lot when it is bad.

P (Reject a lot when it is good) = α (producer's risk)

P (Accept a lot when it is bad) = β (consumer's risk)

4. Level of significance

The probability of Type I error (α) is called the level of significance. It is also known as size of the critical region. Although 5% level of significance has been taken as a rough line demarcation in which deviations due to sampling fluctuations alone will be interpreted as real ones in 5% of the cases. Hence, yet the inferences about the population based on samples are subject to some degree of uncertainty. It is not possible to remove this uncertainty completely, but it can be reduced by choosing the level of significance still lesser like 1%, in which chances of interpreting the deviations due to sampling fluctuations, as real one, is only one in 100 cases.

Power of the Test: The probability of rejecting H_0 when H_1 is true is called power function of the test.

$$\begin{aligned}\text{Power of the test} &= \text{P (Reject } H_0/\ H_0 \text{ is true)}\\ &= 1\text{- P (Accept } H_0\ /\ H_1 \text{ is true)}\\ &= 1 - \beta.\end{aligned}$$

The value of this function plays the same role in hypothesis testing as the mean square error plays in estimation. It is usually used as our standard in assessing the goodness of a test or in comparing two tests of same size. The value of this function for a particular point is called the power of the test.

5. How to control two types of errors

In testing of hypothesis, the ideal procedure is to minimize the probabilities of both types of errors. Unfortunately, for a fixed sample size n, both the probabilities cannot be controlled simultaneously. A test which minimizes the one type of error maximizes the other type of error.

For example, if there is a critical region which makes the probability of Type I error zero, will be of the form always accept H_0 and that the probability of Type II error will be one. It is, therefore, desirable to fix the probability of one of the error and choose a critical region which minimizes the probability of the other error. As Type I error is consider to be more serious than Type II error. Therefore, fix the probability of Type I error (α) and minimize the probability of Type II error (β) and hence, maximize the power of the test.

p-value: In technical terms, a p- value is the probability (chance) of obtaining an effect at least as extreme as the one in the sample data, assuming the truth of the null hypothesis.

For example, suppose that a pesticide study produced a p-value of 0.04. This p-value indicates that if the pesticide had no effect, one would obtain the observed difference or more in 4% of studies due to random sampling error.

All hypothesis tests ultimately use a p-value to weigh the strength of the evidence (what the data are telling about the population). The p-value is a number between 0 and 1 and interpreted in the following way:

- A small p-value (typically ≤ 0.05) indicates strong evidence against the null hypothesis, so you reject the null hypothesis.
- A large p-value (> 0.05) indicates weak evidence against the null hypothesis, so researcher fails to reject the null hypothesis.
- p-values very close to the cutoff (0.05) are considered to be marginal (could go either way).
- *Always report the p-value so that readers can draw their own conclusions.*

6. Reporting the decision on p-value

Almost all statistical tests can be inferred based on p-value obtained through the test subject to the following conditions

(i) If p-value < 0.01 then it can be concluded that the difference/association is significant at 1% level and '**' (double asterisk marks) are to be marked beside the corresponding test value i.e., t-value or chi-square value or F-value.

For example: t-value =17.465**, p-value = 0.0021 (since < 0.01)

(ii) If p-value < 0.05 then it can be concluded that the difference/association is significant at 5% level and '*' (single asterisk mark) is to be marked beside the corresponding test value i.e., t-value or chi-square value or F-value.

For example: t-value = 9.654*, p-value = 0.042 (since < 0.05)

(iii) If p-value >0.05 then it can be concluded that the difference/association is not significant. In this case NS (Not Significant) is to be marked beside the corresponding test value i.e., t-value or chi-square value or F-value.

For example: t-value = 1.245^{NS}, p-value = 0.086

7. Difference in decision making between One-tailed and two tailed tests

Case	Null Hypothesis	Alternative hypothesis	Type of test	LOS at p-value is to be compared	LOS where Decision is taken
1	H_0: $T_1 = T_2$	H_1: $T_1 \neq T_2$ or H_1: T1 > T2 & H_1: T1 < T2	Two tailed (both sides)	0.05 or 0.01	5% or 1%
2	H_0: $T_1 = T_2$	Vs. H_1: T1 > T2	One-tailed (Right tailed)	0.10 or 0.02	5% or 1%
3	H_0: $T_1 = T_2$	Vs. H_1: T1 < T2	One-tailed (Left tailed)	0.10 or 0.02	5% or 1%

Advantage of One-tailed over Two-tailed test (if applicable)

If p-value = 0.065	NS	(in case of two tailed)
If p-value = 0.065	Sig @ 5%	(in case of one-tailed)
If p-value = 0.02	Sig @ 5% level	(in case of two-tailed)
If p-value = 0.02	Sig @ 1% level	(in case of one-tailed)

If the LOS is low Precision is high in Results

40.4.2 Steps in Testing of Hypothesis

1. Set up the null hypothesis H_0 and alternative hypothesis H_1 by keeping objective of the study in mind.
2. Choose an appropriate test based on framed hypothesis and type of the variable under study and check for assumptions underlying it.
3. Proper conclusions will be drawn based on p-value of the test at the respective levels of significance.

40.4.3 Parametric and Non Parametric Statistical Tests

1. **Assumptions for parametric tests**
   - Each sample is an independent random sample
   - The distribution of the response variable follows a normal distribution
   - The population variances are equal across responses for the group levels. This can be evaluated by using the following *rule of thumb*: if the largest sample standard deviation divided by the smallest sample standard deviation is not greater than two, then assume that the population variances are equal.
2. **Test for Normality**
   - One sample Kolmogrov Smirnov Test (a Non Parametric test)
   - P-P Plot

- Q-Q Plot
- Mean, Median and Mode will coincide
- Quartiles will be in equidistant

  i.e., $Q_3 - Q_2 \approx Q_2 - Q_1$

3. Transformations to achieve Normality and for variance stabilization using MS-Excel

- =ASIN(SQRT(D2/100))*360/(2*3.14159)
- =SQRT(D2)
- =Ln (D2)

Note: where D2 is cell address where value to be transformed is available

Value	Arc sin transformation
71.67	57.84181
81.36	64.42198
63.9	53.07048
65.53	54.04774
68.16	55.64847
65.66	54.12614
71.05	57.44892

40.4.4 Choosing Appropriate Test as per the Objective

Table 40.9 *Different Statistical techniques as per the hypothesis under study*

Goal	Type of Data		
	Measurement (from *Normal* Population)	**Rank, Score, or Measurement (from *Non- Normal* Population)**	**Binomial (Two Possible Outcomes)**
Describe one group	Mean, SD	Median, interquartile range	Proportion
Compare one group to a hypothetical value	One-sample t-test	Wilcoxon test	Chi-square or Binomial test
Compare two independent groups	Independent sample t-test	Mann-Whitney test	Fisher's test (Chi-square for large samples)
Compare two paired groups	Paired t-test	Wilcoxon test	McNemar's test
Compare three or more independent groups	One-way ANOVA	Kruskal-Wallis test	Chi-square test
Compare three or more related groups;	Repeated-measures ANOVA	Friedman test	Cochrane Q

Table 40.9 *Contd...*

Goal	Type of Data		
	Measurement (from *Normal* Population)	Rank, Score, or Measurement (from *Non- Normal* Population)	Binomial (Two Possible Outcomes)
Quantify association between two variables	Pearson correlation	Spearman Correlation	Contingency coefficients
Predicting values from one variable	Simple linear Regression or Nonlinear Regression	Nonparametric Regression	Simple logistic Regression
Predict values from several variables	Multiple linear Regression or Multiple nonlinear Regression	-	Multiple Logistic Regression

40.4.5 Appropriate Statistical Techniques based on Level of Measurement

Statistical analysis may be invalid if the assumptions behind those tests are violated. If the data are normally distributed, and other assumptions are met, parametric tests are the most powerful. If the data are non-normal but other criteria are met, non-parametric statistics provide valid analysis.

Levels of Measurements at a glance

There are four levels of measurements viz., Nominal, Ordinal, Interval and Ratio.

(i) ***Nominal:*** A variable whose values are not numerical. Examples include type of variety (Narayani, Dharani K6..etc), colours (green, yellow, brown). A variable with just two categories is said to be dichotomous, whereas one with more than two categories is described as Polytomous.

(ii) ***Ordinal:*** A categorical variable in which the categories have an obvious order (for example in ground nut sowing, emergence, vegetative stage, flowering stage, pegging stage, pod stage…..). The simplest ordinal scale is a ranking. If you are asked to rank 5 types of crops from most profitable to least profitable. But your profitable crop may be far superior to the second preferred crop but, to another respondent with the same top and second crop, the distance may be subjectively small. An ordinal scale only lets us interpret gross order and not the relative positional distances.

(iii) ***Interval:*** A scale of measurement which can be sued to measure the difference, or distance, between two general states or points. For example use a ruler to measure length, use a stop-watch to measure a time interval, use a 7 point rating scale to measure satisfaction etc. It is an interval scale because it is assumed to have equidistant points between each of the scale elements. This means that we can interpret differences in the distance along the scale. We contrast this to an ordinal scale where we can only talk about differences in order, not differences in the degree of order

(iv) ***Ratio:*** A ratio scale is the top level of measurement by which the difference or distance is measured between a state or point of interest and a standard state or point. The factor which clearly defines a ratio scale is that it has a true zero point. The simplest example of a ratio scale is the measurement of length.

Appropriate statistical techniques as per the type of variable under study are summarized in the table 40.10 to table 40.13. But table 40.9 provides a summarized form of statistical techniques according to hypothesis to be tested under the study.

Table 40.10 *Statistical techniques to test differences if Measurement level is Nominal*

One Sample	Two samples		More than two samples	
	Independent	Paired (matched)	Independent	Paired(matched)
One-sample Chi-square test	Chi-square test	McNemar's test	Chi-square test	Cochran's Q test

Table 40.11 *Statistical techniques to test differences if Measurement level is Ordinal*

One sample	Two samples			More than two samples		
	Independent	Paired (matched)	Mixed	Independent	Paired (matched)	Mixed
Kolmogorov One sample	Rank-sum or Median test	Signed Rank test	Rank-sum for Gains	Kruskall-Wallis One-way ANOVA	Friedman's Two-way ANOVA	Kruskall-Wallis for Gains

Table 40.12 *Statistical techniques to test differences if Measurement level is Interval*

One sample	Two samples			More than two samples		
	Independent	Paired (matched)	Mixed	Independent	Paired (matched)	Mixed
Single sample t-test	Pooled t-test	Paired t-test	Pooled t-test for Gains or ANCOVA	One-way CRD or Factorial ANOVA	Repeated Measures ANOVA	ANOVA for Gains or ANCOVA

Table 40.13 Statistical techniques to test Associations or Relations

Nominal	Ordinal	Interval
Chi-square Analysis; Odds Ratio/Relative Risk; Log-linear Models ; Multiple Categorical Procedures; Cohen's Kappa (Agreement Measure)	Spearman's Rho – Rank-order correlation coefficient Or Kendall's Tau	Pearson's r- Product moment Correlation coefficient; Simple Linear Regression (one independent variable); Multiple Linear Regression (more than one independent variable)

40.4.6 Flowchart in Deciding an Appropriate Statistical Technique

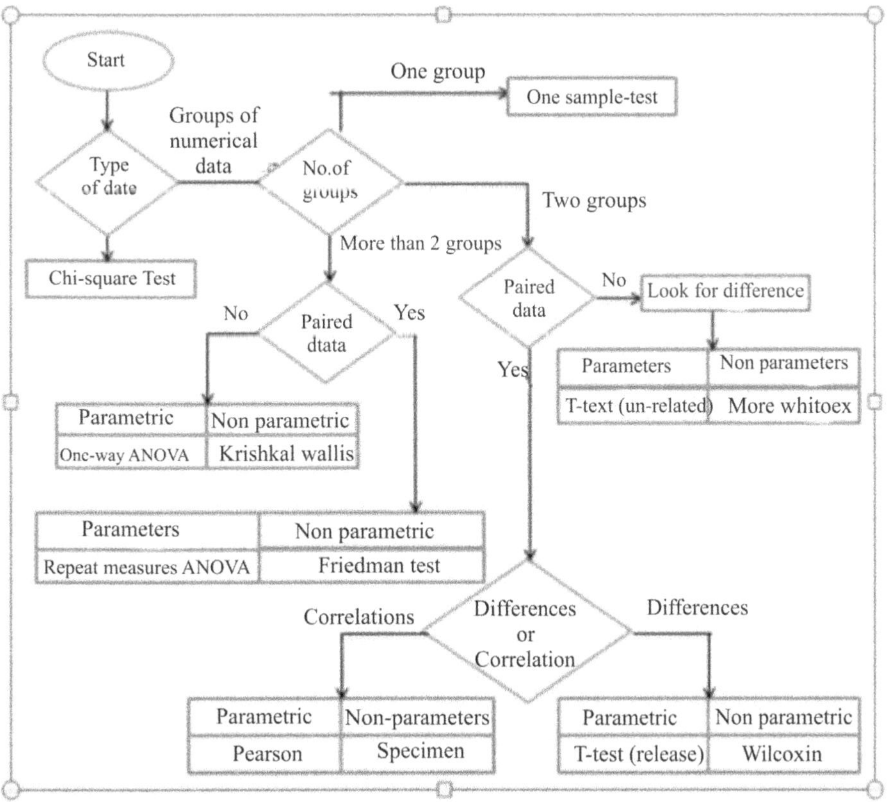

40.4.7 LSD (CD) Versus other Multiple Comparison Tests

1. **Ned of CD/LSD**

   After ANOVA (Analysis of Variance) test and get a significant result, that means at least one of the groups tested differs from the other groups. Further, if we wish to know *which* group differs more significantly, Fisher developed the least significant difference test in 1935, which is only used when you reject the null hypothesis as a result of your hypothesis test results.

The LSD calculates the smallest significant between two means as if a test had been run on those two means (as opposed to all of the groups together). This enables you to make direct comparisons between two means from two individual groups. Any difference larger than the LSD is considered a significant result. Here the important remarks regarding CD/LSD. But LSD post hoc comparison does not control *experiment wise error* rate (e.g. *comparison wise error rate* (= 0.05 per comparison). Hence, this is considered a liberal post hoc analysis.

Points to remember

LSD/CD test will be used only when ANOVA is significant with 0.05for each comparison which inflates type-I error if treatments are more than 5 and will become invalid. Hence it can be used only when the no.of treatments are below 5 or for specified and limited pair wise comparisons.

2. **Comparison wise error rate vs Experiment wise error rate**
   - *Comparison wise error* rate means (α, alpha) 5% for each comparison. For example if no. of treatments are 5, then possible pair wise comparisons are 5_{C2} (=10) viz., (A,B), (A,C), (A,D), (A,E), (B,C), (B,D), (B,E), (C,D), (C,E) and (D,E). Then Comparision wise error rate (LOS) will be 10* 0.05 = 0.50 = *50%.*
   - *Experiment wise error rate* - the probability of making a Type I error for the set of all possible comparisons is denoted by $\alpha_e = 1 - (1- \alpha)^c$ 'where α is LOS of each comparision and c is no.of comparisons. Then the experiment wise error rate without any adjustments would be: $\alpha_e = 1 - (1-\alpha)^c = 1-(1-0.05)^{10} = 1 - 0.95^{10} = 1 - 0.5987 = 0.4013 \simeq$ *40%*

3. **Other Multiple Comparison Tests or Post-Hoc tests**

   There are several multiple comparision tests of having their own advantage and disadvantages. Depending on test criteria and interest on type-I and type-II errors, researcher has to select an appropriate post hoc test.

   (a) ***Bonferroni test:*** This approach can be thought of as "alpha-splitting." If c comparisons are required then LOS α* for each comparison will be α/c, where α is your desired overall significance level. This method is completely general: it applies to any set of c comparisions, not only to multiple comparisons following ANOVA. The main advantage of this approach is that it is very easy, as well as very widely applicable. The main disadvantage is that it often is unnecessarily conservative (weak): α* is smaller than it needs to be.

   (b) ***Tukey's test:*** It is very popular, but too conservative in that it results in a low degree of Type I Error but too high Type II Error (incorrectly rejects H1). Makes comparisons without inflating the experiment wise error rate.

(c) ***Scheffe's test:*** It is too conservative and results in a low degree of Type I Error but too high Type II Error (incorrectly rejects H_1) also too hard to find significant results.

(d) ***Neuman-Keuls:*** Like Fisher's LSD, it allows the family wise $\alpha > 0.05$. Pretty crappy test for that reason.

(e) ***Duncan's multiple range test:*** (Duncan 1955) developed a procedure for obtaining all pair wise comparisons among sample means. Although this procedure makes use of the Studentised range, his error rate is neither on an experiment wise basis (as with Tukey's) nor on a per comparisons basis. It can be applicable even if ANOVA is insignificant.

(f) ***Dunnet's test:*** It is specifically designed for comparing a control group with several treatment groups.

(g) ***Ryan/Einot/Gabriel/Welsch (REGWQ) Procedure (Good compromise of Type I/ II Error):*** Like Tukey's HSD, but adjusts α (like the Dunn-Šidák Test) to make the test less conservative.

40.4.8 Pooled Analysis using Repeated Measures ANOVA

Repeated measures ANOVA is the equivalent of the one-way ANOVA, but for related, not independent groups, and is the extension of the paired t-test. A repeated measures ANOVA is also referred to as a within-subjects ANOVA or ANOVA for correlated samples. All these names imply the nature of the repeated measures ANOVA which is meant to detect any overall differences between related means. This particular test requires one independent variable and one dependent variable. The dependent variable needs to be continuous (plant height) and the independent variable categorical (either nominal or ordinal just like treatments).

When to use a Repeated Measures ANOVA

The Repeated Measures ANOVA can be used in two types of studies that investigate either ***changes in mean scores over three or more time points***, in which researcher might be investigating the effect of 'dates of sowing' on the 'yields of groundnut' and want to measure yields of groundnut in 3 separate years.

Another situation where the ***differences in mean scores under three or more different conditions*** in which, investigator might have same experiment with various treatments and 'plant height' is noted three different time periods in the same experiments. The important point with these two study designs is that the same sample are being measured more than once on the same dependent variable, that is the reason why it is called repeated measures. In this procedure, time period influence can be identified and separated from the error term, increasing power.

40.5 Software Based Statistical Analysis

The statistical analysis of data has been dramatically affected by computers. In fact, the impact has been so great that it is difficult to imagine performing an adequate statistical analysis without the use of computers. Statistical analysis continues to be one of the most active areas of continuing change and development. Statistical packages are expanding to address more of the tasks of data manipulation, storage, and presentation.

Statistical packages like Statistical Package for the Social Sciences (SPSS), the Statistical Analysis System (SAS), Microsoft Excel, Minitab, Systat, and many other packages were conceived, created, refined and distributed wide among the research community.

Statistical Analysis using MS-Excel

MS Excel is a Windows-based spreadsheet package designed to perform calculations, create graphs and to make reports. This is a member of the MS-Office family.

40.5.1 Features of Excel Worksheet

(a) Every sheet contains a 256 or 16384 columns labelled as A, B,....Z, AA, AB,...XFD.

(b) The number of rows in a sheet is 65536 or 1048576 numbered as 1,2,3..................

(c) ***Auto fill:*** We can create a series of values in a column or row by using Fill option of the Edit menu.

Procedure: Type first two values of the series → hold them and drag down to get complete series

(d) ***Paste special:*** We can copy a portion of data and paste it as a Transpose (Rows and Columns interchanged) at a different location.

Procedure: Select → Copy → Choose another location → Right Click → Paste Special → Select Transpose option → OK.

(e) ***Freeze panes:*** To keep the first row (headings row) and the first column always visible, (like serial number), we use Freeze Panes option as from the main menu follows.

Procedure: Put mouse on B2 → Click on View → Click on Freeze Panes.

40.5.2 Statistical Functions available in Excel

(i) Excel has many library functions called Paste functions. Such functions can be pasted into any cell by simply clicking the icon f_x found in the toolbars.

Insert → Function → Statistical.

(ii) Working procedure with library (Statistical) functions

A function can be pasted in a cell by putting the mouse in the cell and selecting the required function from the function window.

Procedure:

- Insert → Function → Statistical → Click on required function
- In resultant dialogue box input the parameters for evaluating the function.

40.5.3 Data Analysis Pak

Excel has a built-in Statistical package for carrying out data analysis called ***Data Analysis Pak***. To start with Data Analysis we have to click on **Tools → Data Analysis** in Ms-Office versions upto 2003. In case of Office 2007 and above versions, it is available at **Data → Data Analysis.** We get a menu with the following analysis tools. If not available it can be retrieved from the system with few steps as mentioned below.

How to retrieve data analysis pak

This feature is usually hiden and can be brought out to the menu by clicking the button sequencing.

Office button → excel options → add- ins → analysis tool pack (as shown in MS-Excel Screen-1) → ok.

MS-Excel Screen-1

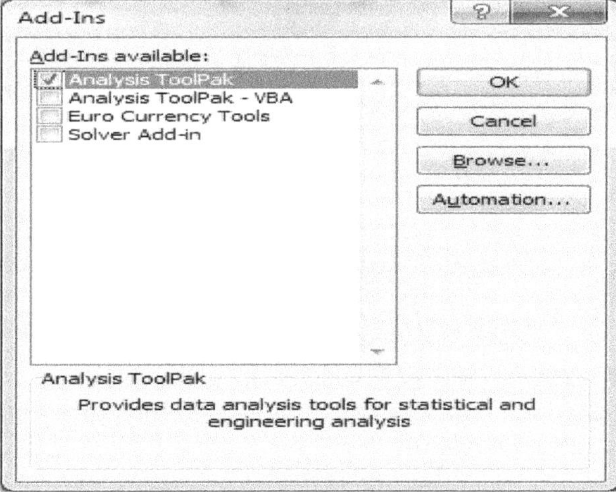

Automatically **Data Analysis Pak** will be included in the menu of data.

Goto Data >Data analysis, then the available list of tests will appear as shown in MS-Excel Screen-2.

MS-Excel Screen-2

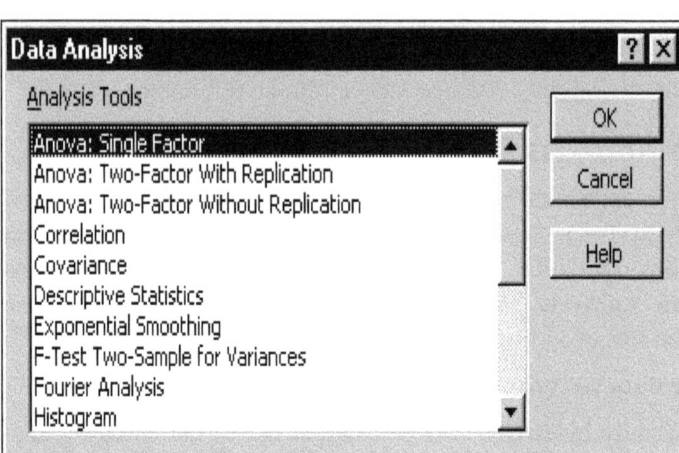

Table 40.14 *Statistical Tools Available in Data Analysis Pak (DAP)*

Anova: Single factor	Moving Average
Anova: Two-Factor with Replication	Random Number Generation
Anova: Two-Factor without replication	Rank and Percentile
Correlation	**Regression**
Covariance	Sampling
Descriptive Statistics	**t-test: Paired Two Sample for Means**
Exponential Smoothing	**t-test: Two-Sample Assuming Equal Variances**
F-Test Two-Sample for Variances	**t-test: Two-Sample Assuming Unequal Variances**
Fourier Analysis	
Histogram	Z-test: Two-Sample for Means

Note: Tests indicated by bold letter are most commonly used tools in Agriculture Research.

40.5.4 Student's T-test using MS-Excel (Working Example)

The following data refers to 100 pod weight (in grams)of variety-1 and variety-2 of Groundnut experimentation. Test whether these two groups differ significantly in their means.

Variety-1	20.43	22.51	18.99	20.49	23.12	25.63	18.08	20.63	22.55	22.43	22.77	23.23
Variety-2	17.7	21.4	20.7	19.3	21	17.9	18.6	18.5	18.2	20.3		

Null hypothesis: The mean weights of 100 pods of the two varieties are same

Procedure:

1. Enter the data in two columns (or rows) with proper headings.
2. Select Tools and Choose Data Analysis
3. Choose the option't-test: Two Sample Assuming Equal variance' and press OK.
4. Choose the appropriate data range with mouse.
5. Select Labels option
6. Set the Hypothesized mean difference as zero.
7. Click OK. The results appear in a separate sheet.
8. If the p-value of the two sided t-test is less than 0.05 conclude that the difference is significant.

The Excel output appears as given below.

t-Test: Two-Sample Assuming Equal Variances		
	Variety-1	Variety-2
Mean	21.73	19.36
Variance	4.319	1.89
Observations	12	10
Pooled Variance	3.229	
Hypothesized Mean Difference	0	
df	20	
t Stat	3.090	
P(T<=t) one-tail	0.0028	
t Critical one-tail	1.7244	
P(T<=t) two-tail	**0.0057**	
t Critical two-tail	2.0858	

***Conclusion*:** Since the p-value (=0.0057) is less than 0.01, hence it can be concluded that there is significant difference between two varieties with regard to 100 pod weight in which variety-1 have more weight (21.73g) than variety-2 (19.36 g).

Note: *Other tests can also be done in similar way by proving required inputs.*

40.5.5 Method of Data Entry in Excel to be Analyzed using SPSS

Table 40.15 *Data entry procedure when we have one factor; Fertilizer dosage (100% RDF, 125% RDF, 150% RDF) with three replications*

Factors : 1 (Fertilizer doses) independent

Replications : 3

Variables : 4 (Pod yield, pod weight, Kernel weight and Shelling %)

	A	B	C	D	E	F
1	Fertilizer dosage	Replication	Pod yield kg/ha	100 pod weight (g)	100 kernel weight (g)	Shelling percentage
2	1	1	941	108	49.9	75.74
3	1	2	1447	106.3	47.6	81.84
4	1	3	1011	112.9	54.1	79.09
5	2	1	689	78.1	35.5	72.21
6	2	2	1038	82.2	36.2	82.11
7	2	3	789	76.8	32.8	78.64
8	3	1	1163	91.2	42.3	83.44
9	3	2	1318	90.7	41.5	88.75
10	3	3	1117	90.4	42.3	87.16
11	4	1	1088	116.5	49.3	79.57
12	4	2	1551	113.5	46.3	78.85
13	4	3	1240	113.7	50.6	81.09

***Table 40.16** Data entry procedure when we have two factors (Date of Sowing (DOS1, DOS2, DOS3), Varieties (V1, V2, V3)) with three replications*

Independent factors : 2 (Date of sowing and varieties)

Replications : 3

Variables : 2 (100 seed weight and Shelling %)

	A	B	C	D	E	F
1	Sample No	Date of Sowing	Varieties	Replication	100 seed weight	Shelling percentage
2	1	1	1	1	24.8	61.75
3	2	1	1	2	29.6	58.10
4	3	1	1	3	22.7	46.92
5	4	1	2	1	19.6	60.27
6	5	1	2	2	20.9	60.22
7	6	1	2	3	26.6	60.15
8	7	1	3	1	21.9	59.51
9	8	1	3	2	25.6	62.05
10	9	1	3	3	25.5	54.72
11	10	2	1	1	22.1	67.52
12	11	2	1	2	22.8	61.50
13	12	2	1	3	22.7	63.38
14	13	2	2	1	24.0	51.97
15	14	2	2	2	28.5	60.89
16	15	2	2	3	23.6	60.32
17	16	2	3	1	31.7	57.42
18	17	2	3	2	21.2	62.49
19	18	2	3	3	24.5	69.16

Table 40.17 *Data entry procedure when we have two factors (Treatments (T1, T2, T3, T4), Spacing (S1, S2, S3)) with three replications for two years (2011-12 and 2012-13)*

Total Factors	:	3 (Year, Treatments, Spacing)
Independent factors	:	2 (Treatments and Spacing)
Related or dependent factor	:	1 (Year)
Replications	:	3
Variables	:	4 (Plant height, No. of branches, 100 seed weight, Shelling %)

	A	B	C	D	E	F	G	H	I	J	K	L
1	Sample No.	Treatment	Spacing	Replication	Plant height (2011-12)	Plant height (2012-13)	No. of branches (2012-13)	No. of branches (2012-13)	100 seed weight (2011-12)	100 seed weight (2012-13)	Shelling percentage 2011-12	Shelling percentage (2012-13)
2	1	1	1	1	110	85.0	7.14	6.6	24.8	27.0	61.75	62.05
3	2	1	1	2	129	124.0	7.45	11.4	29.36	31.45	58.10	68.15
4	3	1	1	3	153	156.0	7.58	8.2	22.7	30.50	46.92	67.75
5	4	1	2	1	145	96.6	6.80	5.6	19.6	33.15	60.27	60.25
6	5	1	2	2	183	113.0	7.21	8.8	20.9	33.25	60.22	65.90
7	6	1	2	3	201	158.0	7.45	7.2	26.6	30.85	60.15	68.95
8	7	1	3	1	142	138.0	6.80	7.3	21.9	33.35	59.51	71.10
9	8	1	3	2	176	112.0	6.70	6.3	25.6	30.40	62.05	70.60
10	9	1	3	3	174	132.0	6.98	7.3	25.5	30.30	54.72	66.85
11	10	2	1	1	148	126.0	6.00	6.5	22.1	32.10	67.52	69.85
12	11	2	1	2	185	95.0	6.70	5.8	22.8	30.55	61.50	67.35
13	12	2	1	3	200	136.6	6.75	8.0	22.7	32.65	63.38	70.30
14	13	2	2	1	120	123.0	7.06	6.0	24.0	33.15	61.97	67.65
15	14	2	2	2	138	115.2	7.28	7.2	28.5	26.05	60.89	59.60
16	15	2	2	3	150	131.0	8.27	6.1	23.6	26.50	60.32	62.55
17	16	2	3	1	145	95.4	6.90	6.9	31.7	23.50	57.42	60.15
18	17	2	3	2	186	116.0	7.34	5.4	21.2	27.00	62.49	61.75
19	18	2	3	3	201	154.	7.65	8.4	24.5	28.25	69.16	67.20
20	19	2	1	1	138	98.0	6.75	6.8	30.9	30.9	64.74	63.20

Statistical Analysis using SPSS

40.5.6 About SPSS and its Advantages Over MS-Excel

SPSS is a windows based program that can be used to perform data entry and analysis and to create tables and graphs. SPSS is capable in handling large amounts of data and familiarity with this program should serve users well in the future. SPSS is updated often.

Have a glance at the advantages of SPSS over MS-Excel.

	Limitations of Excel	Advantages of SPSS
1	Data should be numeric and labels cannot be attached to the codes used in the data.	Data labels can be defined for all the variables so that the output becomes more readable.
2	The number of available statistical tools is limited and helps in simple analysis only.	Advanced Statistical tools are available
3	Many analytical tools require the data to be given in separate rows or columns but not as a mixed collection of records.	SPSS can automatically pick up cases from data file of mixed records.
4	Analysis variance can be computed in Excel only if the data is grouped and arranged in separate columns.	ANOVA can be carried for several dependent variables at a time with one or more factors.
5	Multivariate procedures like Stepwise regression, Factor Analysis, MANOVA are not available in the Data Analysis Pak.	Multivariate tools like Stepwise Regression, Generalized Linear Models, Factor Analysis, Discriminant Analysis and MANOVA can be handled with large sets of data.
6	Graphs of general nature like Bar charts, Pie Charts and line charts etc are available in Excel.	Graphs like P-P plot and Q-Q plot which have applications in survival analysis are available.

40.5.7 Importing Data from MS-Excel to SPSS

Enter the collected data in the Excel spreadsheet as discussed in 40.5.5 section and open it in SPSS sheet using following steps.

- Open SPSS sheet
- Go to file > Open > data (a window will appear like below)
- In the resultant window, select the Excel file (where it was stored) and file type as Excel then click on open.
- We can view the Excel data in SPSS.

40.5.8 Labeling or Decoding in SPSS

- Open SPSS sheet
- Go to variable view (which is in left bottom corner)

- Click on the cell under values against corresponding variable (variety, fertiliser dose, date of sowing …etc)

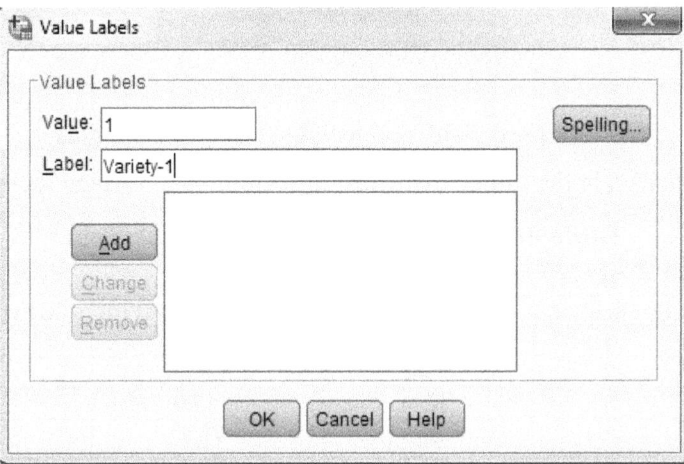

- Type '1' against *value* and type 'Variety-1' against *Label* and click add. Likewise enter all available codes and its labels defined under that variable. Finally click on Ok.

40.5.9 About Analyze Menu in SPSS

This is one of the Main menus which will help us to conduct almost all statistical analysis required in Agricultural Research.

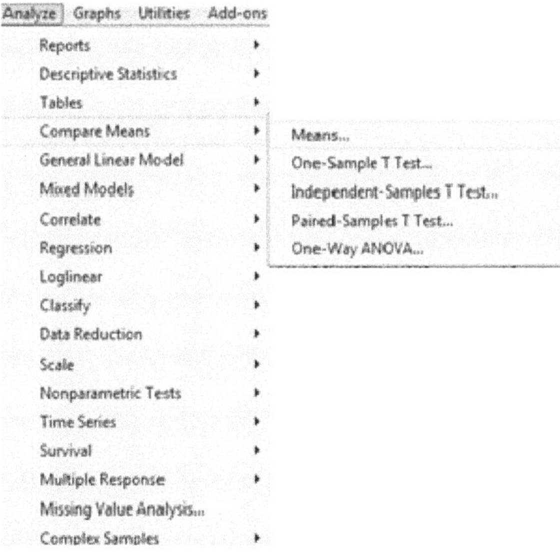

40.5.10 Procedure to Conduct Analysis of a Simple RBD Data in SPSS

Consider the model data

Factors	:	1 (varieties) independent
Replications	:	3
Variables	:	4 (Pod yield, pod weight and Kernel weight)

Step-1: Enter the data in the format explained in section 40.5.5 which looks like below.

Table 40.18 *Model data enter in Excel sheet*

	A	B	C	D	E
1	Varieties	Replication	Pod yield (jg/ha)	100 pod weight (g)	100 kernel weight 9g)
2	1	1	941	108	49.9
3	1	2	1447	106.3	47.6
4	1	3	1011	112.9	54.1
5	2	1	689	78.1	35.5
6	2	2	1038	82.2	36.2
7	2	3	789	76.8	32.8
8	3	1	1163	91.2	42.3
9	3	2	1318	90.7	41.5
10	3	3	1117	90.4	42.3
11	4	1	1088	116.5	49.3
12	4	2	1551	113.5	46.3
13	4	3	1240	113.7	50.6

Step-2: Import the excel data into SPSS sheet as explained in section 40.5.7 and it will be as shown in Fig 40.3.

Step-3: Similarly provide labels for each code given under variable as explained in section 40.5.8 and it will be as shown in Fig 40.4.

1:

	varieties	Replication	Podyieldkgha	weight100pods	weight100kernel
1	1	1	941	108	49.9
2	1	2	1447	106	47.6
3	1	3	1011	113	54.1
4	2	1	689	78	35.5
5	2	2	1038	82	36.2
6	2	3	789	77	32.8
7	3	1	1163	91	42.3
8	3	2	1318	91	41.5
9	3	3	1117	90	42.3
10	4	1	1088	117	49.3
11	4	2	1551	114	46.3
12	4	3	1240	114	50.6

Fig. 40.3 *SPSS sheet after opening Excel data*

1 :

	varieties	Replication	Podyieldkgha	weight100pods	weight100kernel
1	Variety-1	1	941	108	49.9
2	Variety-1	2	1447	106	47.6
3	Variety-1	3	1011	113	54.1
4	Variety-2	1	689	78	35.5
5	Variety-2	2	1038	82	36.2
6	Variety-2	3	789	77	32.8
7	Variety-3	1	1163	91	42.3
8	Variety-3	2	1318	91	41.5
9	Variety-3	3	1117	90	42.3
10	Variety-4	1	1088	117	49.3
11	Variety-4	2	1551	114	46.3
12	Variety-4	3	1240	114	50.6

***Fig. 40.4** SPSS sheet after giving Labels to varieties*

Step-4: Then Analyse > General Linear Model > Univariate> then the resultant window will be as shown in Fig 40.5

- Send 'pod yield' to *dependent variables* category.
- Send 'replication' and 'varieties' to *Fixed factor* as shown in Fig 40.6.
- Click on *model* and click on *customs*.
- Send *replication* and *varieties* from left side box to right side *model* and click on *continue.*
- Click on options and select *'Descriptive Statistics' and select Duncan's and Dunnet's test from Posthoc tab.*
- *Click on OK.*

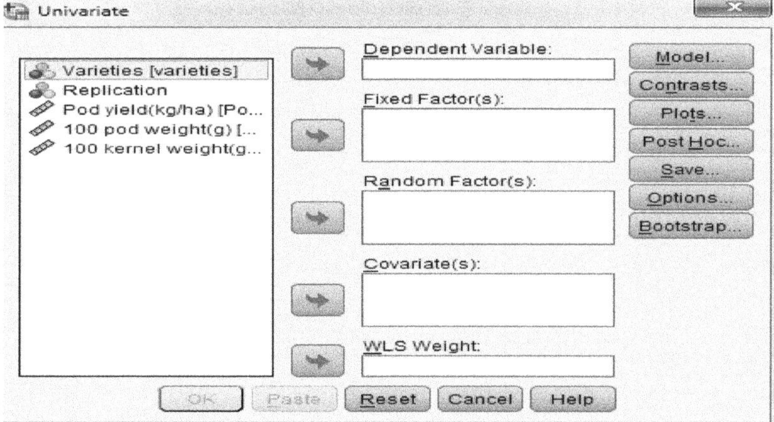

***Fig. 40.5** SPSS sheet after opening Excel data*

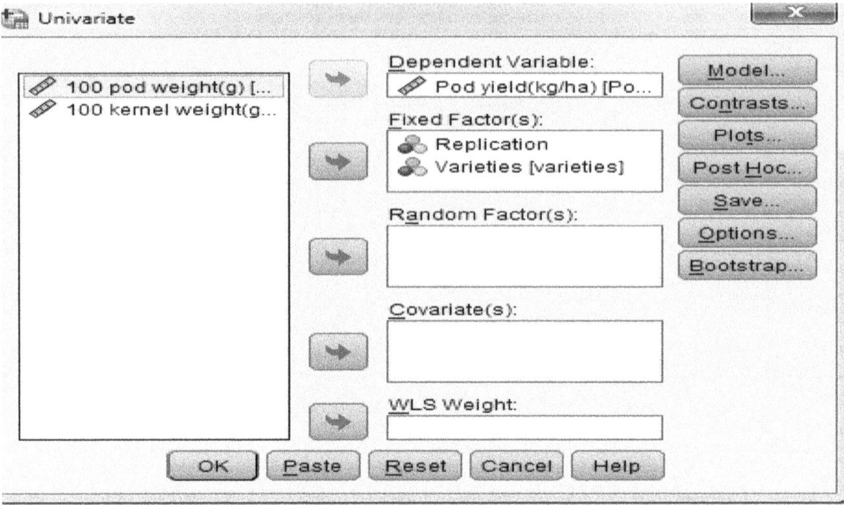

***Fig. 40.6** SPSS sheet after giving Labels to varieties*

***Table 40.19** SPSS Output of Simple RBD*

Replication		Mean	Std. Deviation	N
1	Variety-1	941.0		1
	Variety-2	689.0		1
	Variety-3	1163.0		1
	Variety-4	1088.0		1
	Total	970.2	208.9	4
2	Variety-1	1447.0		1
	Variety-2	1038.0		1
	Variety-3	1318.0		1
	Variety-4	1551.0		1
	Total	1338.5	221.8	4
3	Variety-1	1011.0		1
	Variety-2	789.0		1
	Variety-3	1117.0		1
	Variety-4	1240.0		1
	Total	1039.2	191.2	4
Total	**Variety-1**	**1133.0**	**274.1**	**3**
	Variety-2	**838.6**	**179.7**	**3**
	Variety-3	**1199.3**	**105.3**	**3**
	Variety-4	**1293.0**	**236.0**	**3**
	Total	**1116.0**	**251.3**	**12**

Source	Type III Sum of Squares	df	Mean Square	F	p-value
Corrected Model	652988.2	5	130597.6	18.673	.001
Intercept	14945472.0	1	14945472.0	2136.908	.000
Replication	306559.5	2	153279.7	21.916	.002
varieties	346428.7	3	115476.2	16.511	.003
Error	41963.8	6	6993.9		
Total	15640424.0	12			
Corrected Total	694952.0	11			

Duncan Pod yield(kg/ha)			
Varieties	N	Subset	
		1	2
Variety-2	3	838.67	
Variety-1	3		1133.00
Variety-3	3		1199.33
Variety-4	3		1293.00

CV = SQRT(6993.9)/1116*100 = 7.49

Presentation of Results:

Table 40.20 Summary of Statistical Analysis Pod yield(kg/ha)

Variety	Mean	Sd
Variety-1 a	1133.0	274.1
Variety-2 b	838.6	179.7
Variety-3 a	1199.3	105.3
Variety-4 a	1293.0	236.0
Total	1116.0	251.3
F-value: 16.511**(0.003<0.01) and CV = 7.49		

**significant at 1% level
same alphabets indicates insignificant difference (DMRT)

Findings

From the results it can be concluded that variation due to varieties is significant at 1% level since p-value (0.003) <0.01. CV is also in considerable limit which shows the consistency of the experiment. Further Duncan's Multiple Range Test (DMRT) reveals that variety-2 differs significantly from other varieties with low yield of 838.6 kg/ha while Variety-1, Variety-3 and variety-4 produce yields at the same level.

40.6 Conclusion

There is nothing magical about Statistics; however, like any tool, the appropriate use of statistics depends upon basic knowledge of their capacities and limitations. Fortunately, one need not be a Statistician to take advantage of this important tool of scientific inquiry. If the data are normally distributed, and other assumptions are met, Parametric tests are the most powerful. If the data are non-normal but other criteria are met, Non-Parametric statistics provide valid analyses. When neither set of assumptions has been met, both tests should be implemented to see if they agree. The most commonly used Non-Parametric tests, and their Parametric equivalents explained in this chapter will guide the research community in selecting appropriate Statistical tools for data analysis. Several issues related to the interpretation and value of statistical significance will enhance the quality of conclusions drawn by the researchers.

References

1. *Agarwal B.L (2003), Programmed Statistics, New Age International Publishers, Second edition.*
2. *Cochran, WG. and Cox, DR. (1957) Experimental designs (Second Edition). John Wiley and Sons, Inc., New York.*
3. *IASRI website and study material.*

4. *Hollander, M. and Wolfe, D.A. (1973), Nonparametric Statistical Methods, New York: John Wiley & Sons, Inc. SAS Institute (2003). SAS OnlineDoc, Version 8, Cary, NC: SAS Institute, Inc. 1999.*
5. *Lavanya Kumari. P (April 2012). Significance of Statistics in Social Science Research International Journal of Research SHODH PRERAK, ISSN.2231-413X, Vol.No.11, Issue.2, pp.278-283.*
6. *Lentner, M. and Bishop, T. (1993) Experimental design and analysis (Second Edition). Valley Book Company, Blacksburg, Virginia*
7. *Peter Sedlmeier, (2006), the role of scales in teaching statistics for social science*
8. *Rajagopalan. V (2006), Selected Statistical Tests, New Age International Publisher.*
9. *Richard A.Johnson (1995), Miller & Freund's Probability and Statistics for Engineers, 5th Edition, Prentice-Hall of India, New Delhi. sciences (2nd Ed.). New York, NY: McGraw-Hill.*
10. *Siegel, S., & Castellan, N.J. (1988). Nonparametric statistics for the behavioral students, ICOTS-7, Sedlmeier.*
11. *Zimmerman, D.W. (1998). Invalidation of parametric and nonparametric statistical tests by concurrent violation of two assumptions. Journal of Experimental Education, Vol: 67, pp.55-68.*

Chapter 41

Correlation and Regression

B. Ravindra Reddy

Asst. Professor (Statistics & Maths), S.V. A C, ANGRAU, Tirupati.

41.1 Introduction

Scientists in the Agricultural and biological fields who are involved in research constantly face problems associated with planning, designing and conducting and finally analysis of experiments. Basic familiarity and understanding of statistical methods that deal with issues of concern would be helpful in many ways. Any research endeavour may follow these phases – **Conception, Design, Data collection, Analysis and Dissemination**. After collection of data (univariate/bi-variate/multivariate) various tools and techniques may be used for valid results and conclusions.

In Agriculture experimentation we may deal with a number of variables at a time, instead of single variable. And it is our common experience that all these variables may not be independent of each other; rather they tend to vary side by side. Most of the growth and economic variables are found to follow the above characteristics. For example while dealing with yield component analysis for any crop it is found that yield is an ultimate variable contributed or influenced or effected by number of other factor. If we consider the yield of paddy; then one can find that the factors like number of hills per square meter, number of tillers per hill, number of effective/panicle bearing tillers per hill, length of the panicle, number of grains per panicle etc., are influence the yield. Variation in one or more of the above mentioned factors resulted in variations of the yield. Thus, yield may vary because of variation in number of hills per square meter or variations in number of tillers per hill or so on. Again, yield may vary because of variation in number of hill per square meter and, or number of tiller per hill and other factors. When we consider variations in one variable due to variation in any other variable then it becomes a bi-variate case. On the other hand when the variations of more than two variables are considered at a time it becomes a muti-variate case.

The change or variation in one variable due to change in other variable(s) may follow a linear or non-linear (curvilinear) relationship or any other form. The problem lies in measuring the change in one variable due to the change in other variables and vice-versa *(Sahu P.K. 2007)*.

In this section we will consider the linear relationship among the variables.

41.2 Correlation Analysis

In case of bivariate or multivariate normal distributions, we may be interested in discovering and measuring the magnitude and direction of the relationship between two or more variables. For this purpose we use the statistical tool known as correlation.

Definition

The covariation that exists between two interrelated variables of a bivariate distribution is called correlation. If the change in one variable produces corresponding changes in the other variable then the two variables are said to be correlated and the degree of association ship (or extent of the relationship) is known as correlation.

Example: The relationship between variables like Rainfall and Yield, Income & Expenditure, Prices & Demand etc., is called correlation.

Multiple Correlation

Sometimes there is interaction between many variables and the value of one variable may influenced by many others, then we are interested to study the joint effect of a group of variables upon a variable not included in that group is called multiple correlation.

Example: The study of variation in yield due to various factors viz., rainfall, temperature, fertilizer, IPM practices etc., is called multiple correlation.

Types of correlation

(a) ***Positive correlation:*** If the two variables deviate in the same direction, i.e., if the increase (or) decrease in one variable results in a corresponding increase (or) decrease in the other variable, correlation is said to be direct or positive.

Examples: Heights and Weights, Household income and expenditure, amount of rainfall and yield of crops, Prices and supply of commodities, Feed and milk yield of an animal, Soluble nitrogen and total chlorophyll in the leaves of paddy.

(b) ***Negative correlation:*** If the two variables constantly deviate in the opposite direction i.e., if increase (or) decrease in one variable results in corresponding decrease (or) increase in the other variable, correlation is said to be inverse or negative.

Examples: Price and demand of a goods, Volume and pressure of perfect gas, sales of woolen garments and the day temperature, Yield of crop and plant infestation

(c) ***No or zero correlation:*** If there is no relationship between the two variables such that the value of one variable change and the other variable remain constant is called no or zero correlation.

Methods of studying correlation

1. Scatter Diagram
2. Karl Pearson's Coefficient of Correlation
3. Spearman's Rank Correlation
4. Regression Lines
5. Probit and
6. Logit Analysis

Scatter diagram: It is the simplest way of the diagrammatic representation of bivariate data. Thus for the bivariate distribution (x_i, y_i); $i = j = 1, 2, \ldots n$, If the values of the variables X and Y be plotted along the X-axis and Y-axis respectively in the xy-plane, the diagram of dots so obtained is known as scatter diagram. From the scatter diagram, if the points are very close to each other, we should expect a fairly good amount of correlation between the variables and if the points are widely scattered, a poor correlation is expected. This method however, is not suitable if the number of observations is fairly large.

If the plotted point shows an upward trend of a straight line then we say that both the variables are positively correlated

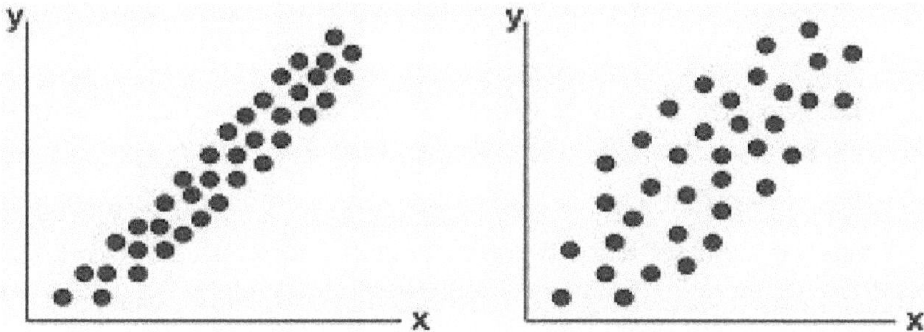

***Fig. 41.1** Positive Correlation*

When the plotted points shows a downward trend of a straight line then we say that both the variables are negatively correlated

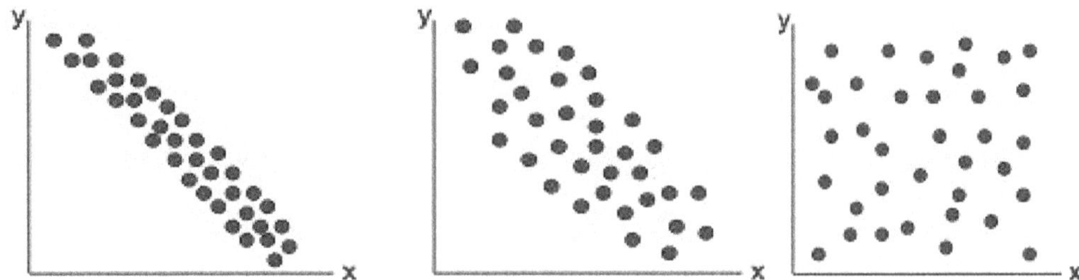

***Fig. 41.2** Negative Correlation* ***Fig. 41.3** No Correlation*

If the plotted points spread on whole of the graph sheet, then we say that both the variables are not correlated.

Karl pearson's coefficient of correlation: Prof. Karl Pearson, a British Biometrician suggested a measure of correlation between two variables. It is known as Karl Pearson's coefficient of correlation. It is useful for measuring the degree of linear relationship between the two variables x and y. It is usually denoted by r_{xy} or 'r'.

$$r_{xy} = \frac{\text{cov}(x, y)}{\sqrt{V(x)V(y)}} = \frac{\frac{1}{n}\sum_{i=1}^{n}(x_i - \bar{x})\,(y_i - \bar{y})}{\sqrt{\frac{1}{n}\sum_{i=1}^{n}(x_i - \bar{x})^2}\sqrt{\frac{1}{n}\sum_{i=1}^{n}(y_i - \bar{y})}^{\,2}}$$

$$\text{After simplification; } r = \frac{\Sigma xy - \frac{(\Sigma x)\,(\Sigma x)}{n}}{\sqrt{\left\{\Sigma x^2 \frac{(\Sigma x)^2}{n}\right\}}\sqrt{\left\{\Sigma y^2 \frac{(\Sigma y)^2}{n}\right\}}}$$

Remarks

- The correlation coefficient never exceeds unity. It always lies between –1 and +1 (i.e. $-1 \leq r \leq 1$)
- If $r = +1$ then we say that there is a perfect positive correlation between x and y
- If $r = -1$ then we say that there is a perfect negative correlation between x and y
- If $r = 0$ then the two variables x and y are called uncorrelated variables
- No unit of measurement.

Test for significance of correlation coefficient

If 'r' is the observed correlation coefficient in a sample of 'n' pairs of observations from a bivariate normal population, then Prof. Fisher proved that under the null hypothesis

$$H_0 : \rho = 0$$

the variables x, y follow a bivariate normal distribution. If the population correlation coefficient of x and y is denoted by ρ, then it is often of interest to test whether ρ is zero or different from zero, on the basis of observed correlation coefficient 'r'. Thus if 'r' is the sample correlation coefficient based on a sample of 'n' observations, then the appropriate test statistic for testing the null hypothesis $H_0 : \rho = 0$ against the alternative hypothesis H_1: ρ # 0 is

$$t = \frac{|r|}{\sqrt{1-r^2}}\sqrt{(n-2)}$$

t follows Student's t – distribution with (n-2) d.f.

If calculated value of $|t|$ > table value of t with (n-2) d.f. at specified level of significance, then the null hypothesis is rejected. That is, there may be significant correlation between the two variables. Otherwise, the null hypothesis is accepted.

Rank Correlation

The correlation between ranks assigned with respect to two characteristics is called rank correlation. The numerical measure of Rank Correlation coefficient is denoted by 'ρ' and is given by:

$$\rho = 1 - 6\ \Sigma d^2/n(n^2 - 1)$$

Here, $d = x_i - y_i$ and n = no. of observations

***Example* 1:** From a paddy field, 12 plants were selected at random. The length of panicles in cm (x) and the number of grains per panicle (y) of the selected plants were recorded. The results are given in the following table. Calculate correlation coefficient and its testing (Example from study material).

y	112	131	147	90	110	106	127	145	85	94	142	111
x	22.9	23.9	24.8	21.2	22.2	22.7	23.0	24.0	20.6	21.0	24.0	23.1

Solution: Correlation coefficient $r_{xy} = r = \dfrac{\Sigma xy - \dfrac{(\Sigma x)(\Sigma y)}{n}}{\sqrt{\left\{\Sigma x^2 - \dfrac{(\Sigma x)^2}{n}\right\}\left\{\Sigma x^2 - \dfrac{(\Sigma x)^2}{n}\right\}}}$

Where n = number of observations

Testing the correlation coefficient:

Null hypothesis H_0: Population correlation coefficient 'ρ' = 0

Under H_0, the test statistic becomes

$$t = \frac{r\sqrt{n-2}}{\sqrt{1-r^2}} \sim t_{(n-2)} \text{ d.f.}$$

Calculations:

X	Y	x2	y2	xy
112	22.9	12544	524.41	2564.8
131	23.9	17161	571.21	3130.9
147	24.8	21609	615.04	3645.6
90	21.2	8100	449.44	1908.0
110	22.2	12100	492.84	2442.0
106	22.7	11236	515.29	2406.2
127	23.0	16129	529.00	2921.0
145	24.0	21025	576.00	3480.0
85	20.6	7225	424.36	1751.0
94	21.0	8836	441.00	1974.0
142	24.0	20164	576.00	3408.0
111	23.1	12321	533.61	2564.1
Σx =1400	Σy = 273.4	Σx^2 =168450	Σy^2 = 6248.20	Σxy = 32195.6

$$r = \frac{32195.6 - \frac{(1400)x273.4)}{12}}{\sqrt{\left(168450 - \frac{(1400)^2}{12}\right)}\sqrt{\left(6248.20 - \frac{(273.4)^2}{12}\right)}}$$

$$= \frac{298.9333}{71.5308x19.2367} = 0.95$$

The value of coefficient of determination (r^2) = 0.90

Now to test the null hypothesis we have to calculate t-statistic as given below:

$$t = \frac{0.95\sqrt{12-2}}{\sqrt{1-(0.95)^2}} = \frac{3.0042}{0.3122} = 9.6$$

't' critical (table) value for 10 d.f. at 5% LOS is 2.23

Conclusion:

Since calculated value i.e. 9.6 is > t table value i.e., 2.23, it can be inferred that there exists significant positive correlation between (x, y).

Procedure for obtaining correlation coefficient by using MS-Excel

1. Enter the data in Excel worksheet for x and y in two columns
2. Select 'Tools', 'Data Analysis' then scroll down the list, select 'Correlation' and click OK.
3. A new window will appear where the following information needs to be entered

   ***Input range*:** Highlight the two columns of data that are the paired values for the two variables. The cell range will automatically appear in the box. If column headings are included in this range, tick the Labels box.

   ***Output range*:** Click in this box then select a cell on the worksheet where the user wants to displayed the output.
4. Click OK and a table will be displayed showing the correlation coefficient (r) for the data.

(OR)

= CORREL (array1, array2) also returns the correlation coefficient between two data sets.

Note: If we wish to find out the value of the correlation coefficient for each pair of the variables in the case of more than two variables, we proceed on similar lines by entering the data pertains to available variables in columns and by selecting the same as input range. However, Excel does not give the p-value/t-statistic regarding the significance of the observed r. We have to calculate the t-statistic value manually as explained in the above example or use advanced statistical packages like SPSS, SAS etc.,

Procedure for obtaining correlation coefficient by using *SPSS*

1. Enter the data either directly in SPSS data editor by declaring the variables in variable view or we can import the Excel file by entering the data on variables in columuns.
2. Once the data are entered or import the Excel file, go to *Analyze, Correlation, Bivariate*. Then, you will get the dialogue box and move the variables (quantitative only) that you wish to correlate into the variables box and hit OK.
3. You will get correlation matrix of required no. of variables. It gives results of pair wise correlations(r) along with p-value and N.
4. The r-value indicates strength and direction (±) of the correlation. The '*' means we can reject the null hypothesis (H_0) at 5% level of significance (LOS) and '**' means reject the null hypothesis at 1% LOS (i.e., The correlation of corresponding pair of variables significant)
5. The p-value is the probability value. Reject H_0 if $p \leq 0.05$ at 5% LOS and $p \leq 0.01$ at 1% LOS.

41.3 Regression Analysis

The term `regression' literally means "stepping back towards the average". It was first used by a British Biometrician Sir Francis Galton. The average relationship between the independent and dependent variables may be expressed as a function. Such functional relationship between two variables is termed as regression.

In regression analysis independent variable is also known as regressor or predictor or explanatory variable while dependent variable is also known as regressed or explained variable. Similarly, in regression analysis one of the two variables is identified as cause and the other is effect. Hence, regression models are also known as *cause and effect models (Sarma KLAP et. all 2013)*. Further in regression analysis, the value of the dependent variable for a given value of the independent variable can be estimated.

When only two variables are involved the functional relationship is known as simple regression. If the relationship between two variables is a straight line, it is known as simple linear regression, otherwise it is called as simple non-linear regression. The regression analysis can be carried out as follows:

The regression equation of Y on X is given as

$$Y = a + bX$$

where, Y = dependent variable; X = independent variable and
a = intercept b = the regression coefficient (or slope) of the line.

a and b are also called as constants the constants a and b can be estimated with by applying the 'least squares method'. This involves minimizing $\sum_{i}^{n} e^2 = \sum_{i}^{i} (y - a - bx)^2$.

This gives

$$b_{yx} = b = \frac{Cov(XY)}{v(X)} = \frac{\Sigma XY - \frac{(\Sigma x)(\Sigma y)}{n}}{\Sigma X^2 - \frac{(\Sigma X)^2}{n}}$$

and $$a = \overline{Y} - b\overline{X}$$

where b is called the estimate of regression coefficient of Y on X and it measures the change in Y for a unit change in X.

Similarly, the regression equation of X on Y is given as

$$X = a^1 + b^1 Y$$

where, X = dependent variable and Y = independent variable

$$b_{xy} = b^1 = \frac{Cov(XY)}{v(Y)} = \frac{\Sigma XY - \frac{(\Sigma x)(\Sigma y)}{n}}{\Sigma Y^2 - \frac{(\Sigma Y)^2}{n}}$$

and $a^1 = \overline{X} - b^1\overline{Y}$

where, b^1 is known as the estimate of regression coefficient of X on Y and 'a' is intercept

Properties of regression coefficient

1. Correlation coefficient is the geometric mean of the two regression coefficients.
2. If one of the regression coefficient is greater than unity, the other must be less than unity.
3. Units of 'b' are same as that of the dependent variable.
4. Regression is only a one-way relationship between y (dependent variable) and x (independent variable).
5. The range of b is from $-\infty$ to ∞. $-\infty$ for negative b and $+\infty$ for positive b.
6. If r = 0, the two variables are uncorrelated, the lines of regression become perpendicular to each other
7. If r = ± 1, in this case the two lines of regression either coincide or they are parallel to each other
8. If the regression coefficients are positive, r is positive and if the regression coefficients are negative, r is negative

Multiple Linear Regression

In simple regression analysis studies involved two variables only, but in many cases we required to work with multivariable data. In such case, the value of one variable may be influenced by many others. For example the yield of crop depends on the plant character, climatic factors, soil parameter, quality of seed, type of irrigation etc. Thus to predicting the value of one variable from the knowledge of several other variables we use multiple regression analysis.

Suppose we have 'k' independent variables (explanatory variables) X_1, X_2, X_3 ...X_k then the multiple liner regression equation is given by:

$$Y = \alpha + \beta_1 X_1 + \beta_2 X_2 + \beta_3 X_3 + \ldots\ldots\ldots\ldots\ldots \beta_k X_k + \varepsilon$$

Where, α = intercept and $\beta_1, \beta_2, \beta_3 \ldots \beta_k$ are partial regression co-efficients.

ε = error term

***Example* 2:** The following data give the yield per plant (gm) (Y) and days to flowering (X) of 11 pigeonpea plants. Fit regression equations of Y on X and X on Y. Also estimate Y when the flowering period is 149 (Example from study material).

***Solution*:**

The regression equation of Y on X is given by;

$$Y = a + bX$$

Where, Y = dependent variable, X = independent variable

and a = intercept, b = the regression coefficient of Y on X

The constants a and b can be estimated with the help of 'least squares method'

Where, $$b = \frac{\Sigma XY - \frac{(\Sigma x)\ (\Sigma y)}{n}}{\Sigma x^2 - \frac{(\Sigma x)^2}{n}}; \qquad a = \overline{Y} - b\overline{X}$$

Similarly, the regression equation of X on Y is given by; $X = a^1 + b^1Y$

Where X = dependent variable and Y = independent variable

Where $$b^1 = \frac{\Sigma XY - \frac{(\Sigma x)\ (\Sigma y)}{n}}{\Sigma Y^2 - \frac{(\Sigma Y)^2}{n}}; \qquad \text{and} \qquad a^1 = \overline{X} - b^1\overline{Y}$$

***Calculations*:**

Y	X	Y^2	X^2	XY
24.72	91	611.08	8281	2249.52
20.25	76	410.06	5776	1539.00
38.56	98	1486.87	9604	3778.88
74.72	136	5583.08	18496	10161.92
72.75	128	5292.56	16384	9312.00
78.45	119	6154.40	14161	9335.55
69.8	142	4872.04	20164	9911.60
80.4	135	6464.16	18225	10854.00
160.2	147	25664.04	21609	23549.40
165.75	145	27473.06	21025	24033.75
77.56	122	6015.55	14884	9462.32
ΣY = 863.16	ΣX = 1339	ΣY^2 = 90026.91	ΣX^2 = 168609	ΣXY = 14187.94

$$b = \frac{114187.94 - \frac{(1339)(863.16)}{11}}{168609 - \frac{(1339)^2}{11}} = \frac{9117.83}{5616.18} = 1.62$$

$$\overline{X} = \frac{\Sigma X}{n} = \frac{1339}{11} = 121.73 \quad \text{and} \quad \overline{Y} = \frac{\Sigma Y}{n} = \frac{863.16}{11} = 78.47$$

Therefore, a = 78.47 – 1.62 × 121.73 = – 118.74

The regression equation of Y on X is Y = – 118.74 + 1.62X

To estimate Y when X = 149

$$\hat{Y} = -118.74 + 1.62 \times 149 = 122.64$$

Now,
$$b^1 = \frac{114187.94 - \frac{(1339)(863.16)}{11}}{90026.91 - \frac{(863.16)^2}{11}} = \frac{9117.83}{22295.53} = 0.41$$

$$a^1 = \overline{X} - b^1\overline{Y} = 121.73 - 0.41 \times 78.47 = 89.56$$

The regression equation of X on Y is X = 89.56 + 0.41Y

Result

1. The regression equation of Y on X is Y = a + bX = Y = – 118.74 + 1.62X
2. The regression equation of X on Y is X = 89.56 + 0.41Y
3. Estimated Yield is 122.64

Procedure for Regression analysis by using MS-Excel *(Microsoft Office 2003)*

1. Enter the data in Excel worksheet for x and y in two columns and if multivariate case enter the data many no. of columns as depends no. of independent variables.
2. Select 'Tools', 'Data Analysis' then scroll down the list, select 'Regression' and click OK.
3. A new window will appear where the following information needs to be entered
   Input Y Range and Input X Range: Input Y range Refers to the spreadsheet cells containing the dependent variable Y and the Input X range to those containing independent variables X (in multivariate case X = X1, X2, …). Select the data from respective columns to this boxes.
4. If we do not want to force our model through the origin we leaves the 'constant is zero' box unchecked.
5. Select the 'Confience level' either 0.05 or 0.01 as required level of significance.

6. *Output range*: Click in this box then select a region on the worksheet where the user wants the data table displayed. The minimal output has two parts 'Regression Statistics' and 'ANOVA'. Checking the appropriate boxes in sub-blocks 'Residuals' and Normal Probability' will expand the default output information.
7. Click OK and a table will be displayed showing the required output.

Procedure for regression analysis by using *SPSS*

1. Enter the data either directly in SPSS data editor by declaring the variables in variable view or we can import the Excel file by entering the data on variables in columns.
2. Once the data are entered or import the Excel file, go to *Analyze, Regression, Linear.* Then, you will get the dialogue box and move the dependent variable Y *Dependent* box (one variable only) that you wish to what you predict and move explanatory variables called dependent variables (more than one in multiple regression) into *Independent(s)* and hit OK.
3. You will get output as four tables, but you need only two viz.. Model Summary and Coefficients. The Model summary table gives you the R-value and the R^2 value. The Coefficients table gives you the **a** (intercept/constant) & partial regression co-efficient **b's** (b's according to no. of independent variables). The multiple R gives the amount of relation between independent variable(s) with dependent variable. The R^2 gives the amount of variation in dependent variable explained by the independent variable(s). The Adjusted R^2 used as an indicator of an adequacy of the model, since it takes into account not only deviations, but also numbers of degrees of freedom. Standard error is an estimate of the standard deviation of experimental values of the dependent variable with respect to those predicted by the regression model.
4. The p-value to check for significance. We reject H_0 if $p \leq 0.05$ at 5% LOS. This means the relationship is reliable and can be used to make predictions. The Multiple R, R^2, Adjusted R^2 closer to 1 and the smaller p-value (≤ 0.05) is better for model.

41.4 Diagnostic Analysis of Regression

The R^2 value is the commonly used measure of the adequacy of the linear model. We can test for its statistical significance with the help ANOVA and understand the p-value. Now the question arises for the researcher. If the regression model is used to predict the values of Y at the know values of X, will these estimates match with the actual Y values or will there be any difference? If so what would be the average error ? The Software's (Excel/ SPSS) resolves these issues as a part of the regression analysis by selecting the *Residuals and residual plots, Line fit plot, and the Normal probability plots (Q-Q plots).*

Residuals and residual plots: The difference between the actual and the predicted value of Y form a given model is called the residual. If the model is a good fit to the data, we can expect these residuals to be close to zero. However they do differ and software shows these differences in the form of residual output. Software also provides the plot of these standardized residuals.

***Line fit plot*:** The plot of the actual and predicted values obtained from the regression model is shown in line fit graph. This is useful to judge the adequacy of the model. If many of the predicted values differ from the actual values we may consider the model is not good representative of the relationship.

Normal probability plots (Q-Q plots): The Normal probability plots are used to test the normality of a data which is one of the assumptions in regression analysis. It is a plot of the observed data against the normal quartiles. If the data has truly come from a normal population, we can expect a linear fit between normal quartiles and observed values. This method works well when the number of values is moderately large (*Kachigan Sam Kash 1986)*.

41.5 Probit Analysis

The idea of probit analysis was originally published in *Science* by Chester Ittner Bliss in 1934. He worked as an entomologist for the Connecticut agricultural experiment station and was primarily concerned with finding an effective pesticide to control insects that fed on grape leaves (Greenberg 1980). By plotting the response of the insects to various concentrations of pesticides, he could visually see that each pesticide affected the insects at different concentrations, i.e. one was more effective than the other. However, he didn't have a statistically sound method to compare this difference. The most logical approach would be to fit a regression of the response versus the concentration, or dose and compare between the different pesticides. Yet, the relationship of response to dose was sigmoid in nature and at the time regression was only used on linear data. Therefore, Bliss developed the idea of transforming the sigmoid dose-response curve to a straight line. In 1952, a professor of statistics at the University of Edinburgh by the name of David Finney took Bliss' idea and wrote a book called *Probit Analysis* (Finney 1952).

Today, probit analysis is still the preferred statistical method in understanding dose-response relationships. Probit analysis is a type of regression used to analyze binomial response variables. It transforms the sigmoid dose-response curve to a straight line that can then be analyzed by regression either through least squares or maximum likelihood.

Probit analysis can be carried out by either using tables to estimate the probits and fitting the relationship by eye or use hand calculating the probits, regression coefficient and confidence intervals or having statistical package such as SPSS or SAS (*Kim Vincent* date unknown).

Definition: Probit Analysis is a specialized regression model of binomial response variables.

Remember that regression is a method of fitting a line to your data to compare the relationship of the response variable or dependent variable (Y) to the independent variable (X).

$$Y = a + b X + e$$

Where

a = y-intercept

b = the slope of the line

e = error term

Also remember that a binomial response variable refers to a response variable with only two outcomes. For example; The effectiveness or toxicity of pesticides: Death/no death, Flipping a coin: Heads or tails and testing beauty products: Rash/no rash.

Applications

Probit analysis is used to analyze many kinds of dose-response or binomial response experiments in a variety of fields. However, because background knowledge of probit analysis stems only from toxicology, hence the examples considered from toxicology.

Probit Analysis is commonly used in toxicology to determine the relative toxicity of chemicals to living organisms. This is done by testing the response of an organism under various concentrations of each of the chemicals in question and then comparing the concentrations at which one encounters a response. As discussed above, the response is always binomial (e.g. death/no death) and the relationship between the response and the various concentrations is always sigmoid. Probit analysis acts as a transformation from sigmoid to linear and then runs a regression on the relationship (*Ranga Swamy R. 2007*).

Once a regression is run, the researcher can use the output of the probit analysis to compare the amount of chemical required to create the same response in each of the various chemicals. There are many endpoints used to compare the differing toxicities of chemicals, but the LC50 (liquids) or LD50 (solids) are the most widely used outcomes of the modern dose-response experiments. The LC50/LD50 represent the concentration (LC50) or dose (LD50) at which 50% of the population responds.

For example, consider comparing the toxicity of two different pesticides to aphids, pesticide A and pesticide B. If the LC50 of pesticide A is 50 ug/L and the LC50 of pesticide B is 10ug/L, pesticide B is more toxic than A because it only takes 10 ug/L to kill 50% of the aphids, versus 50ug/L of pesticide B.

How does probit analysis work? How to get from dose-response curve to an LC50?

The probit analysis can be carried out with various methods. The easiest by far is to use a statistical package such as SPSS, SAS or R.

SPSS uses maximum likelihood to estimate the linear regression. To run the probit anaylsis in SPSS, follow the following simple steps:

Input a minimum of three columns into the Data Editor

- Number of individuals per container that responded
- Total of individuals per container
- Concentrations

For example in the following screen, a mort is the number of individuals that responded per container, a total is the total number of individuals per container, and a_conc are the concentrations. Row 7 in the following example is data from the control where 0 out of 10 responded at a concentration of 0.

***Screen* 1:**

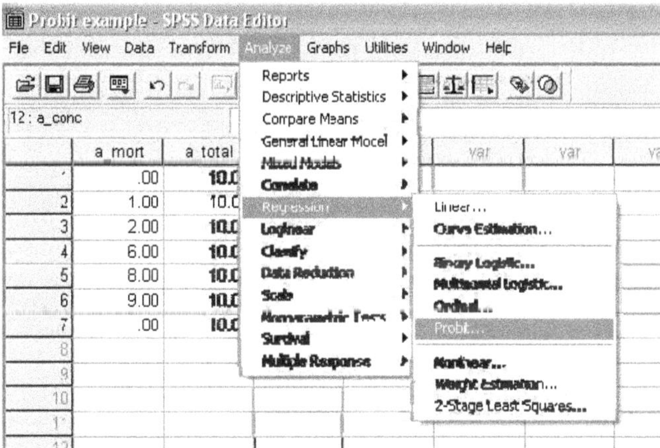

After you columns are set, simply go to analyze, regression, probit:

***Screen* 2:**

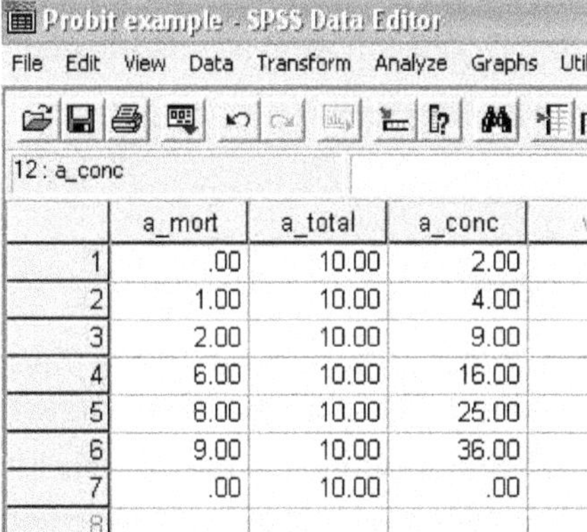

	a_mort	a_total	a_conc
1	.00	10.00	2.00
2	1.00	10.00	4.00
3	2.00	10.00	9.00
4	6.00	10.00	16.00
5	8.00	10.00	25.00
6	9.00	10.00	36.00
7	.00	10.00	.00
8			

Then set your number responded column as the "Response Frequency", the total number per container as the "Total Observed", and the concentrations as the "Covariates". Don't forget to select the log base 10 to transform your concentrations.

***Screen* 3:**

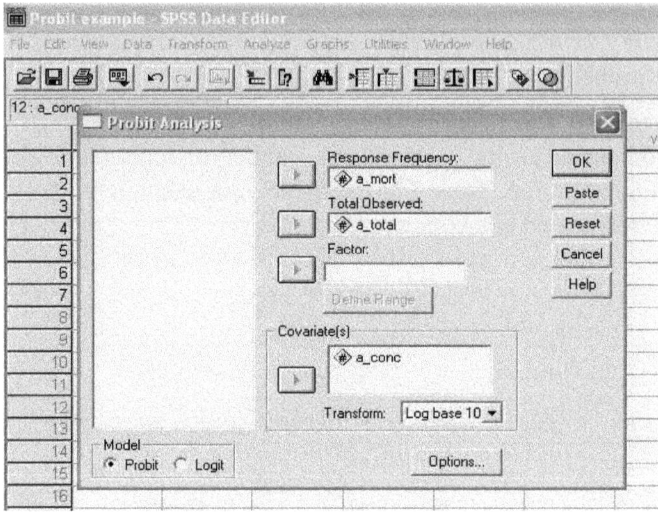

If you run the above example, you will see that SPSS determines an optimal solution after 14 iterations.

Find the LC50

Method A**:** Using your hand drawn graph, either created by eye or by calculating the regression by hand, find the probit of 5 in the y-axis, then move down to the x-axis and find the log of the concentration associated with it. Then take the inverse of the log and voila! You have the LC50.

Method B**:** The LC50 is determined by searching the probit list for a probit of 5.00 and then taking the inverse log of the concentration it is associated with.

Determine the 95% confidence intervals

SPSS and other computer programs calculate this automatically

41.6 Logit vs. Probit

Logit is another form of transforming binomial data into linearity and is very similar to probit. Logit functions by taking the log of the odds: logit(P) = log P/ (1-P). Yet, the relationship between logit and probit is almost in distinguishable: Logit $\approx (\pi / \sqrt{3})$ x probit. In general, if response vs. dose data are not normally distributed, Finney suggests using the logit over the probit transformation (Finney, 1952). The Logit provides a better fit in the presence of extreme independent variable levels and conversely that probit better fit random effects models with moderate data sets (*Kim Vincent* date unknown).

References

1. *Finney, D. J., Ed. (1952). Probit Analysis. Cambridge, England, Cambridge University Press. Greenberg, B. G. (1980). "Chester I. Bliss, 1899-1979." International Statistical Review / Revue International de Statistique **8**(1): 135-136.*
2. *Kachigan Sam Kash (1986). Statistical Analysis-An Interdisciplinary Introduction to Univariate and Multivariate Methods, Radius Press, New York.*
3. *Kim Vincent (date unknown). Probit Analysis Retrieved June 23, 2015 from http://userwww.sfsu.edu/efc/classes/bio1710/probit/.pd*
4. *Microsoft MS-Excel & SPSS13.0. Online help.*
5. *Rangaswmy, R.(2007). A Text book of Agricultural Statistics. New Age International(P) Limted, Publishers, New Delhi, India.*
6. *Sahu, P.K. (2007). Agriculture and Applied Statistics Vol-I, Kalyani Publishers, New Delhi, India.*
7. *Sarma, K.L.A.P, Ravindra Reddy, B and Pullaiah, T (2013). Biostatistics. Daya Publishing House, Astral International Pvt.Ltd., New Delhi.*

Chapter 42

Strategies for Mitigation of Phytonematode Problems in Groundnut

V. Krishna Rao

Professor and University Head of Plant Pathology and Coordinator-Farmers Call Centre, Professor Jayashankar Telangana State Agricultural University, Rajendranagar, Hyderabad, Telangana.

42.1 Current Status of Phytonematode Problems

In the present scenario of agriculture production system, vagaries due to various biotic stresses have increased considerably due to intensification in crop production. Diseases caused by various pathogenic fungi, bacteria, viruses nematodes etc., have been found to contribute 25% of the total losses inflicted by pests and pathogens. Among the diseases, greatest losses are caused by fungi (42%) followed by bacteria (27%), viruses (18%), and nematodes (13%). The existence of nematodes known for 4 billion years has contributed for their rich biodiversity, prevalence, occurrence and distribution under varied agro-eco-systems. Phytonematodes have assumed alarming stature by their adverse effect on crop health, directly as pathogens as well as indirectly by pre-disposing the plants to fungal/ bacterial diseases and vectoring viruses resulting in disease complexes, wherein nematodes play pivotal role as aggravators, incitants, predisposers. Besides causing quantitative losses, nematodes are known to reduce vitamins, minerals in edible plant parts and market value. There is hardly any flora including oilseeds that are unaffected by these tiny hidden enemies which often remain ignored/unnoticed/misdiagnosed because of the lack of awareness and above ground non-specific symptoms. In India on an average, a national loss of Rs. 21,068.73 million has been estimated due to plant parasitic nematodes including oil yielding crops i.e., groundnut causing about 10-15% losses in crop productivity.

42.2 Nematodes as Phytoparasites

Unlike insects, viruses, fungi or bacteria, nematode populations increase rather slowly and the diseases caused by them are generally not of sudden epidemic type resulting in obvious destruction of crop and yield but cause a rather slow decline in yields. Nematodes have six stages in their life cycle- the egg, four larval stages and the adult. Length of life cycle of plant parasitic nematodes can be anywhere from 20-40 days (on average 25 days at 22^0 C) or may require several years depending upon a great variety of conditions and circumstances. The emerging second-stage juvenile (J_2) is the most common infective stage. Nematodes are omnipresent and can be transmitted through a variety of methods. Most nematodes migrate only a few feet a year through soil but dispersal may occur through several ways viz., soil adhering to farm implements, irrigation water, heavy rain causing soil erosion, infested planting material etc.

42.3 Phytonematode Threat to Groundnut Production

Groundnut is attacked by various types of nematodes. On a global basis, 91 nematodes are reported to be associated with groundnut. The yield losses due to phytonematodes in groundnut on world basis have been estimated to be around 12 percent, while monetary losses are estimated at over one billion dollars.

Meloidogyne spp. are internationally important nematode pests of groundnut followed by Pratylenchus spp. Root knot nematodes are distributed throughout groundnut growing areas of India. However, the detailed information on nematode problems associated with groundnut in India is very scanty. At least 26 species belonging to 20 genera of nematodes are known to infect groundnut and major nematodes which attack groundnut in India are root knot nematodes (M. Arenaria Northern root knot nematode), M Javanica (Javanese root knot nematode) & M Incognita (Southern root knot nematode), stunt nematode (Tylenchorynchus brevilineatus) and lesion nematode (Pratylenchus brachyurus).

42.3.1 Root Knot Nematode

Root knot nematodes are distributed throughout major groundnut growing states of India and the most important nematode species limiting yields in groundnut. Because of conspicuous galls and warts they produce on the root system of the plants, the seriousness of the damage has been recognized by farmers. The infection at early stage impairs growth of tap root and effect the growth more severely than at later stage. Inoculum level of 500-1000 J_2/kg soil cause significant reduction in yield. Yield losses to the extent of 59-100% by M. Arenaria and 38-54% by M. Javanica were reported. The principal species of root knot nematodes attacking groundnut in India are M. Arenaria, M. Javanica and M. Incognita

42.3.2 Stunt Nematode

Kalahasty malady caused by stunt nematode, Tylenchorynchus brevilineatus was first reported from Kalahasty area of Chittoor dist in Andhra Pradesh during 1976 and also observed during post rainy season cultivation of groundnut in Chittoor, Nellore and Prakasam districts of Andhra Pradesh resulting in considerable yield losses.

The occurrence of this nematode has also been reported from Tamil Nadu during 1996. This is one of the most economically important nematode diseases of groundnut which occurs severely on pods reducing not only the yields by 20-60% but also the market value of pods.

42.3.3 Lesion Nematode

First reported in India (Varaprasad and Sharma, 1990) from Andhra Pradesh and later reported from other parts of the country. This nematode is minor problem in India and yield losses were not estimated properly.

42.4 Symptoms of Nematode Damage

General Symptoms caused by phytonematodes in groundnut

Phytonematodes are obligate parasites and typically attack underground parts of the plant with most of the population occupying the top 15 cm of soil and the crop damage depends on plant species or cultivar, plant age, nematode species, level of soil infestation, and prevailing ecological factors and associated root pathogens.

Misdiagnosis of the problem

The damage caused by phytonematodes often goes unnoticed as they do not produce any easily distinguishable above ground symptoms on their host plants and it is often difficult to diagnose on the basis of aboveground symptoms alone. Field symptoms are of generalized nature giving an impression that the crop is suffering from shortage of water and fertilizers even though both of these are in adequate amounts. Symptoms of a mineral deficiency or toxicity, low soil pH, several soil borne diseases, or drought stress may also easily be confused with those of nematode injury leading to misdiagnosis of the problem. Moreover, the debilitation symptoms (plants looking pale, unthrifty, stunting, mild yellowing of foliage) caused by nematodes is often confuse with nutritional deficiency, as a result fertilizers are applied in place of nematicides which prove ineffective and non-economic.

General field symptoms caused by phytonematodes

General symptoms of nematode infestation in the field appear as scattered irregularly distributed patches of dull green or pale yellow foliage with stunted growth, reflect the area "hot spots" in the field ranging in size from a few feet to several acres, symptoms of water and nutrient stress; excessive wilting during dry (mid day wilting) hot conditions, but recover at night. The patches continue to enlarge if the cultivation of the susceptible host plant continues. Wilting, defoliation and consequent death may be associated with root knots/root rot/root lesions in hot weather in spite of abundant water. Accelerated root decay may occur due to invasion by secondary soil borne pathogens (fungi, bacteria) and an incurable stage is reached finally resulting in severe reduction in plant growth and yield

Root knot nematodes

In general, symptoms of root knot nematode infection are similar for all species. Galls are formed on pegs, pods and roots (2-200 mm in diameter) Heavily galled pegs often rot and then break when the peanuts are dug, leaving numerous pods scattered across the soil surface M. arenaria is the most economically important nematode ,drastically reducing yields and causing nuts to be of very poor quality. Symptoms on foliage may be expressed at any time during the growing season generally about 100 days after planting. Plants may die especially if the nematode is interacting with other root pathogens. Leaves become chlorotic and scorched along their margins. Reduction in number of feeder roots and root stunting. Nematode infection weakens the peg and break during harvest, contributing to additional yield losses. Root-knot is generally more severe in sandy soils and under adverse environmental conditions such as drought and high temperatures. Nematode interact with root pathogens like Sclerotium rolfsii and cause more severe symptoms or break resistance to the disease agent, or allow weakly parasitic and non-pathogenic organisms to cause disease resulting more plant death.

Stunt nematode

T. brevelineatus induces the disease "Kalahasti malady" which is characterized by reduction in pod size, brownish to black discolouration on the pod surface and appearance of small brownish yellow lesions on pegs and developing pods. Affected pods become small, shriveled, and show ugly external appearance called "shutti kaya" in telugu. Brownish to black necrotic lesions on root surface, pod stalks, pod surface, and reduced pod size. The margins of the lesions become slightly elevated because of cellular proliferation around the site of nematode infection, reduction of pod stalks. The surface of heavily infested pods become completely discoloured without effecting on seeds with in pods. Plants become stunted, with dark green foliage. Disease is generally distributed n patches within field and more severe in sandy soils.

Lesion nematode

Fibrous root development of peanut is greatly reduced by the lesion nematodes. Damaged roots are usually discolored, decayed, and stunted. Numerous small, brown to black spots, often give damaged pods a speckled appearance. The most obvious symptoms are: small purplish brown to black lesions on groundnut shell with distinct boundaries. First appear as small brownish tunnels in the shell and later coalesce to form large lesions. Soil fungi often invade and rot the feeding sites of the lesion nematode on peanut roots and pods. Nematode usually enters the elongation zone of the root, feeding on the cells of the cortex and inflicting extensive injury. Lesions start as small brown flecks and increase in size until large parts of the root and pods are destroyed. Once the lesions have appeared, secondary soil borne pathogen like Aspergillus spp can easily enter the damaged roots/

pods leading to greater necrosis and rotting of pegs and pods. The nematode does not induce distinct field symptoms.

42.5 Detection of Phytonematodes

Since many of the symptoms are similar to those caused by bacteria, fungi etc. an assay of soil and root sample is often needed to accurately detect the occurrence of most phyto nematodes and is usually the only way to confirm a nematode diagnosis. Detection of nematodes is recommended, the year before any field goes into peanuts, regardless of previous cropping history. Particular attention must be paid to those fields with continuous peanut cropping. Fields in summer fallow, particularly those with a good stand of volunteer peanuts, are also likely targets of nematode attack. Even trace to low numbers of some nematodes, especially root knot, could cause serious damage on next year's peanut crop. Collect soil samples from nearby healthy plants as well as from soil around plants showing typical decline symptoms but not from around dead or dying plants or from soils that are dry or extremely wet.

42.6 Phytonematode-Fungal Disease Complexes

In nature, nematodes interact with other soil organisms like fungi and bacteria wherein nematodes play a pivotal role as aggravators, incitants, predisposers, and vectors of some phytopathogens. The damage due to nematodes is many a time a result of interaction between nematodes especially Fusarium and Aspergillus resulting in disease complexes causing losses exceeding those caused by either of the pathogens alone Different pathogens on same plant interact, which may results in disease complexes leading to susceptibility by predisposition against a particular pathogen. The fungus- nematode and bacterium nematode interaction are numerous and weak fungal or bacterial pathogens can cause considerable damage once they gain entry into plant roots in the presence of feeding nematodes. The combination of a fungus and nematode often results in synergistic interaction wherein the crop loss is greater than expected from either of the pathogen alone or an additive effect of the two together. The result for a cultivar sensitive to the interaction can be total crop failure.

Groundnut is one of the crops in which nematodes reduces their vigour and also reduce crop yield directly by damaging pegs, pods and kernals and indirectly by feeding on roots and weakening the plants. Most important are the secondary infections caused by avenues caused due to penetration. The infections of other plant pathogens like bacteria and fungi are on the rise and further weaken the plant. Nematodes make the plant vulnerable to damage from pathogens which are otherwise weak. The concomitant occurrence of phytonematodes in general with Fusarium spp. or Aspergillus spp. Sclerotium rolfsii, Pythium spp, Aspergillus spp. Rhizoctonia spp. results in disease complex due to synergetic effect thereby causing symptoms worse than those produced by either pathogen alone.

42.7 Typical Life Cycles of Phytonematodes

All the phytonematodes have a relatively simple basic life cycle consisting of an egg stage, four juvenile stages and an adult stage and these diverse life cycles are important for designing nematode management strategies. The first stage juvenile develops in the egg and the first molt usually occurs within egg shell, giving rise to second stage juvenile which emerges free into soil or plant tissue. Once the nematode begins feeding on tissues of favourable host, the second, third and fourth molts occur giving rise to third, fourth and fifth or adult stages respectively. Between molts, there is further growth and development of nematode, with concurrent development of reproductive systems in two sexes. Upon maturity, the female deposits eggs and the life cycle is repeated.

All species of Meloidogyne are sedentary endoparasites and have the same type of life cycle and J2 is the infective stage. M. Arenaria M Javanica and M Incognita generally completes life cycle in 21, 42 and 29 days respectively. Four races of M incognita, two of M. arenaria two of M. arenaria, and one each of M. javanica and M. hapla have been identified. Groundnut is generally highly resistant to M. incognita, M. javanica, M. arenaria - race 2, but susceptible to M. hapla. and M. arenaria-.race-1.

The species of Tylenchorynchus are migratory ectoparasites and life cycle completes in 25-28 days. All the stages from J_2 to adult feeds on plant roots. The species of Pratylemchsu are migratory endoparasites. All stages are infective to roots. During period of drought, the nematode may remain quiescent and the length of life cycle is usually from 4 to 5 weeks.

42.8 Host Range of Phytonematodes in Groundnut

Meloidogyne species feed on more than 3,000 plant species including vegetables, legumes, cereals and grasses. Some less common species are very host specific viz., M arenaria on groundnut, wheat, watermelon, tobacco, grapevine, weeds etc. Pratylenchus species feed on 500 hosts including beans, blackgram, corn, grasses, mungbean, papaya, pea, pepper, redgram, rice, sorghum, sugarcane, sunflower, tomato, tobacco, turmeric, watermelon, weed hosts etc. Tylenchorynchus species feed on bean, blackgram, chickpea, cluster bean, corn cotton, pea, bhendi, ginger, horsegram, muskmelon, brinjal, onion, papaya, pea, bajra, pepper, redgram, rice, soybean, sugarcane, sunflower, tobacco, tomato, turmeric, weeds etc.

42.9 Major Determinants of Field Wide Distribution

There is a general lack of awareness on survival of nematodes among the farmers. This can result in unintentional movement of nematodes through soil, irrigation water, equipment etc. Severe nematode infestations may develop with intensive peanut production and thrive in sandy soils where damage is usually greatest. The damage caused by phytonematode depends on various factors viz., previous cropping pattern, initial inoculums, and movement of infested soil through water etc. A number of other factors such as nematode species and population, susceptibility of hot plant, soil factors and presence of other organisms govern the damage caused by nematodes. The nematodes may be additive stress factors and can induce predisposition of their hosts to attack by other pathogens such as fungi and bacteria. The higher temperatures and longer growing season induce more nematode generation and consequently an increased population. Though farmers practice crop rotation, and cropping systems that suppress nematode population naturally, nematodes like root knot and lesion nematode multiply rapidly and cause decline in yields due to their polyphagous feeding and wide host range.

42.10 Phytonematode Management Strategies in Groundnut

Management of nematode diseases is difficult particularly in developing countries like India where nematicides are very expensive. Several management practices have been evaluated and recommended for the control of phytonematodes. Though nematicides are found to be effective, indiscriminate use of chemical pesticides has been cautioned on account of their deleterious effect on the ecosystem. Keeping in view the dangerous effects of nematicides on environment and ultimately on human health, emphasis is being given to alternative methods of management i.e., non- chemical management or INM in the context of sustainable agriculture. Biointensive integrated approaches utilizing cultural methods, resistant cultivars, bioagents, judicious application of nematicides at a recommended dose and time should form the main nematode management strategy to reduce nematode population in a field leading to successful crop production and protection. Hence, integrated nematode management appears to be an alternative since no single method will give an effective and economical management of nematodes for a long time. The success and sustainability of INM strategy, especially with resource poor farmers, greatly depends on their involvement to generate solutions suitable for their particular farming systems and integrating control components that are ecologically sound and readily available to them.

42.10.1 Cultural Practices

In crops where nematicide usage is not profitable due to low returns, cultural practices are the only available alternatives to manage nematode problems.

42.10.1.1 Crop Rotation

Crop rotation with certain non-hosts, graminaceous poor hosts and certain antagonistic crop like Marigold (Tagetes sp.) for one or two years have been found effective in reducing the population of nematodes. Agronomic and economic factors are also important in the selection of a rotation crop. An adequate weed control programme is absolutely necessary for a crop rotation scheme to be effective because many weed species serve as source of survival.

Management Meloidogyne spp is difficult because of adoptability of nematodes to a wide host range (> 3000 plant species). In general monocotyledons like sorghum, maize, bajra, are good crops to rotate with groundnut. Breaks of minimum three years of rotation between susceptible crops should give better control of root knot nematodes. Sesame, castor, are good crops to reduce M. arenaria populations. Root knot nematodes are mainly managed by rotation (two years) with cotton and maize with the exception of M. Arenaraia. M. Arenaria can also feed on soybean, tomato and maize and should be avoided from rotation. Rabi groundnut preceded by summer rice and Kharif rice or summer rice and kharif sun hemp or kharif rice or groundnut-rice - rice -millets were found effective. Peanut-soybean rotations should be avoided, particularly in fields infested with the peanut root-knot nematode. Similarly, cotton, tobacco, sorghum, maize should be avoided from rotation particularly in case of lesion Nematode

42.10.1.2 Fallowing and Ploughing during Summer Months

Hot and dry conditions in India are favorable for controlling the nematodes since these are sensitive to heat and drying action of the sun and the wind. Keeping the land fallow during the summer months and deep ploughing of soil infested with nematodes during the month of May-June two to three times at an interval of 10 to 15 days and exposing the nematodes in soil and infected tissue to solar heat was found to be an excellent method to reduce nematodes.

42.10.1.3 Organic Amendments

The addition of organic matter to soils is known to improve soil structure, aids in water retention, provide nutrients and better plant vigour which in turn can make the plant stress caused by plant parasitic nematodes less severe. Animal manures, poultry litter, crop residues, Farm Yard Manures (FYM) etc. are the typical examples of soil amendments used for management of nematodes. Application of organic amendments encourage the growth of natural enemies of nematodes and the decomposition products of these amendments are also toxic to nematodes. Further, improved agronomic practices in terms of optimum irrigation, fertilization and other necessary inputs at appropriate time during the crop growth period help the plants to tolerate the attack of nematodes. Application of organic amendments viz., neem or castor cake @ 1 tonne/ha to soil or, green manuring alongwith seed treatment carbosulfan (25 DS) @ 3% a.i. (w/w) was found to reduce root-knot nematode damage with improved yields in groundnut.

42.10.2 Trap and Antagonistic Crops

The use of trap crop is a method whereby a highly susceptible crop is grown in the nematode infested fields and allowed to grow over a time period so as to trap the penetrating juveniles of root knot nematodes in the roots. Later the crop is ploughed up or destroyed prior to the development of the penetrating juveniles to attain adult stage. Further, plants that release certain root exudates in the rhizosphere having nematicidal values, when inter-cropped with the susceptible crop or grown as cover crop such as marigold, mustard, sesame and asparagus (Asparagus officinalis) can also help in reducing the population of root-knot nematodes.

42.10.3 Resistant and Tolerant Varieties

This is one of the most practical and economical methods of controlling nematodes. Though, there are practical difficulties, the development of multi-resistant strains of plants to nematodes and fungi is the best choice since the main drawback sometimes is the development of resistant breaking nematodes which make the process futile. This resistance must be combined with desirable agronomic traits of recommended cultivars and should give higher yields over the recommended varieties under nematode infested conditions. Kalahasti is a short duration high yielding variety and resistant to kalahasty malady and recommended for post rainy season (November –December/ March –April) especially in kalahasty malady endemic areas. It is also suitable for rainy season cultivation in high rainfall areas of North Coastal and Northern Telangana districts .The other varieties found resistant to kalahasty malady are Tirupat-2, Tirupati-3, Prasuna.

42.10.4 Botanical Nematicides

Certain substances of plant origin are known to be toxic to nematodes. A few commercial products based on neem seed kernel, oil, leaves, etc., have become available. These can be applied to seed for reducing nematode damage to plants at the early growth stage.

42.10.5 Present Status and Practical Applications of BCAs in India

Biological control assumes major importance in sustainable agriculture where soil borne pathogens including nematodes are serious and can serve as an important tool in organic farming systems to improve organic matter status of the soil through the use of FYM, crop residues, etc., which provides the necessary substrate for rapid growth of antagonistic organisms. The use of Bio-Control Agents (BCAs) for effective nematode management offers economical, long lasting and eco-friendly non chemical options. Currently, a few commercial formulations of Paecilomyces lilacinus, Trichoderma viride and T. Harzianum Pochonia (Verticillium) chlamydosporium and Pasteuria penetrans are available in the market for controlling nematodes. The antagonistic plant growth-promoting rhizobacteria like, Pseudomonas fluorescens, Bacillus, etc., has also received prominent attention because of the dual role of these bacteria in plant growth promotion and disease control.

However, research on biological control of phytoparasitic nematodes is less explored particularly in India. However, the major setback with the application of BCAs is inconsistency in experimental results from field conditions. Many of the known fungal enemies of nematodes are lacking in some of the key desirable characteristics of bio control agents and this may account for lack of success where they have been tested. Similarly, the weakness with opportunistic fungi is related to their biology as the control is opportunity dependent and they are known to cause allergy in humans. Even though P. penetrans appears to be a potential biocontrol agent, commercial exploitation is prevented by lack of large scale production methods. The control achieved through biocontrol agents is often slow acting, less robust, species specific, than achieved with chemicals implying the use of biocontrol agents integration with other management strategies.

42.10.6 Chemical Methods

This is the most effective method but the main drawbacks are the prohibitive costs and the destruction of normal micro-flora and fauna of the soil, environmental pollution and development of pesticide resistance. Carbofuran (**Furadan** 3% granular formulation) is one of the largest selling granular insecticides in India. Till date, there is no recorded instance of resistance development to carbofuran in any insect species worldwide, thus making it a versatile choice to growers to manage the crop pests. It is a broad spectrum insecticide and nematicide controlling a wide range of foliar pests, soil pests including nematodes. Granules of carbofuran are low-strength formulations meant to be applied without dilution and especially well suited for soil applications.

42.11 Integrated Nematode Management

- If climate is suitable, summer ploughing to expose the soil to solar radiation
- Soil application of neem cake @ 1 t/ha + neem oil @ 5litre/ha seven days prior to sowing followed by seed treatment with carbosulfan 25 DS @ 3% a.i. (w/w) (AICRP on nematodes) or soil application of carbofuran 3G @ 6-12 kg/Ac (once in 4 years) at 25-30 days after sowing after giving irrigation (ANGRAU/PJTSAU).
- Seed treatment with neem seed kernel powder @ 10 g/kg seed in combination with bioagents (Trichoderma viride/Pseudomonas fluorescens @10g/kg).
- Sowing of groundnut with castor as an intercrop (row ratio 2:1) In situations where, for economic reasons if there are difficulties in adopting crop rotation, rotation with resistant cultivar can be followed.
- Rabi groundnut preceded by summer rice and kharif rice or summer rice and kharif sun hemp or kharif rice or groundnut-rice - rice -millets
- In areas where cultivation is on a large scale, it is economically possible to adopt chemical control.

- In areas where cultivation is on small scale, the only means of management is by good husbandry like removal of crop residues and volunteer plants and efficient weed clearance.

42.12 Future Research Strategies

- The following thrust areas needs focus in the management of phytoparasitci nematodes in general and groundnut in particular.
- The management strategy should be taken in total considering all pests and diseases including nematodes infesting a crop
- There is a need for systemic survey and crop loss assessment for nematodes in groundnut to be done in farmers fields.
- Documentation of changing disease scenario in different zones and to identify high risk hot spots based on GIS (Geographical Information System) to develop disease warning systems
- Special attention should be given to reduce population of most economically important nematodes and cultivation of their susceptible crops viz., all crops for Meloidogyne spp., maize, tobacco, sugarcane for Pratylenchus brachyurus, sugarcane, tobacco, for Tylenchorynchus brevilineatus should be rotated with tolerant crops.
- Field sanitation with regard to removal of crop debris and weeds which may support nematode survival and multiplication in the absence of main crop must be adopted, Over flooding of water from adjoining fields should be avoided. Use of biopesticides should be invariably adopted as seed dressing or application to the soil. If nematode infestation is detected during crop growth, judicious doses of nematicides may be applied, as at this stage no other management options shall give economic returns.
- Spread of important phytoparasitic nematodes remains a global concern for low productivity.
- Nematodes must be detected at an early stage and eradicate them to prevent further spread.
- Carbofuran, the only insecticide found effective against nematodes doesn't have label claim against plant parasitic nematodes on a number of economically important crops. Under such situation there is dire need to identify new molecules for nematode control or expand the label claims of existing insecticides that are having nematicidal properties.
- Application of novel management approaches viz., agro-homeopathy- to exploit the enormous potential of homoeopathic drugs as crop Protector and Booster in the management of crop diseases and nano based delivery system in crop protection

- Development of simple, efficient and economically viable mass production, and formulation/ delivery system of bio control agents against economically important phytoparasitic nematodes under field conditions and advice for the users.
- Role of phytoparasitic nematodes to be investigated particularly in complexe diseases i.e., groundnut pod discolouration and an alternative, eco friendly and economically viable INM strategies to be identified to reduce further build up of inocula of complex pathogens.

42.13 Conclusion

A large number of nematodes have been found associated with groundnut but the research activities has been confined to root knot and stunt nematode where in both the nematodes cause sizeable amount of reduction in yield and market value. However, In India, the information on problems of other nematodes and their damage in groundnut is scanty. Although nematicides have been found quite effective but their use is avoided because of high cost and toxicity. The nature and magnitude of important nematode diseases of groundnut indicate that nematode problems can no longer be ignored and needs immediate and greater research emphasis. It would be wise to realize and exploit economically and environmentally safe non chemical/INM strategies with emphasis on cultural methods, tolerant hosts, potential biological agents needs to be practiced and popularized in a wider perspective so as to phase out or reduce the use of synthetic molecules due to pesticide residue problems. Awareness among growers and users about harmful effects of agricultural chemicals and benefits of non chemical methods are essential to develop a culture of bio- organic farming in the country. A relationship between nematode numbers in the soil and crop performance must first be established, since density-dependent damage relationship for phytoparasitic nematodes is the foundation to accurately advise the farmers of nematode hazards.

References

1. *Anonymous. 2010. Economically important plant parasitic nematodes distribution – ATLAS. Directorate of Information and Publication of Agriculture, ICAR, New Delhi.*
2. *Faujdar Singh and D.L. Oswalt. 1992 Major Diseases of Groundnut Compiled by ICRISAT. Human Resource Development Program ICRISAT, Patancheru.*
3. *Jain, R. K., 2010. In: Proceedings of the National Symposium on Innovations in Nematological Research for Agricultural Sustainability – Challenges and a Road map Ahead, Coimbatore.*
4. *Mani A., Kumar G.Ratna.1990. Plant parasitic nematodes associated with groundnut in Andhra Pradesh. Indian Journal of Nematology. 20: 44-48*
5. *Mujebar rahman khan, Shamim Jairajpuri, M 2010. Nematode infestations. Part II: industrial crops. The National Academy of Sciences, India.*

6. *Prasad D and Rangappa T. 1994. Distribution and estimation of avoidable loss due to nematodes associated with kalahasty melady of groundnut. Annals of plant protection Sciences. 2: 37-4*
7. *Rao, M. S., Parvatha Reddy, P. and Walia, R. K., 2001. In: Proceedings of the National Nematology Congress – Centenary Celebrations, IARI, New Delhi.*
8. *Ravichandra N G. 2008. Plant Nematolgy. I K International Publishing House Ltd New Delhi.*
9. *Anonymous. 2004. Training modules- training programme on 'emerging trends in biological control' 2- 8, PDBCC, Bangalore.*
10. *Reddy, D. D .R., Subramanyam, P., Sankar reddy, G H , Raja Reddy C., Siva rao. D. V.1984.A*
11. *Nematode disease of peanut caused by Tylenchorynchus brevilineatus .Plant Disease.68: 526- 529*
12. *Reddy, P. P., 2008. Diseases of Horticultural Crops: Nematode Problems and their Management, Scientific Publishers (India), Jodhpur.*
13. *Sharma , S. B. 1985 A world list of nematode pathogens associated with chickpea, groundnut, pearlmillet, pigeonpean and sorghum, ICRISAT Patancheru, India.*
14. *Sharma, S.B. 1988. Preliminary survey for nematode caused disease of groundnut in some districts of Andhra Pradesh, India. International Arachis newsletter. 3: 11-13, ICRISAT, Patancheru.*
15. *Sharma, S B, Varaprasad, K S. Balasubramanian, P. Sivakumar, C.V. 1996 Kalahasty malady on groundnut outside Andhra Pradesh, India. ICRIAST, Patancheru.*
16. *Sudheer John, M., Kalaiarasan, P, Senthamarai, M, Prabhu, S, Prasad Rao, G M V, Priya, P, Harinathanaidu, P, Haritha, V, Reddykumar, M, Sailajarani. 2008. Diversity and community structure of major plant parasitic nematodes in selected districts of Andhra Pradesh, India. Indian Journal of Nematology. 38: 68-74*
17. *Varaprasad, K S and Giribabu, P. 2014 Emerging nematode pest challenges in vegetable oil production in India. Directorate of Oil Seeds Research, Rajendranagar, Hyderabad.*

Chapter 43

Fungal Disease Management in Groundnut

M. Reddy Kumar

Scientist (Plant Pathology), IFT, RARS, Tirupati.

43.1 Introduction

Groundnut (peanut), *Arachis hypogaea* L., an annual legume native to South America, is primarily cultivated in areas of the world bounded by latitudes 40°N and 40°S. Approximately 80% of the world groundnut crop is produced in developing countries where the yields are usually very low. Diseases are major constraints to groundnut production throughout the world. A large number of diseases caused by fungi, bacteria, viruses, mycoplasma, nematodes, and parasitic flowering plants have been reported, and with some exceptions, they are widely distributed. All parts of the groundnut plant are susceptible to diseases. This chapter is designed to assist agricultural scientists and extension workers in the preliminary field diagnosis of common diseases of groundnut caused by fungi, bacteria, viruses, mycoplasma, nematodes, and parasitic flowering plant.

The most characteristic field symptoms of each disease are illustrated and described. It is important to understand the diagnosis of diseases based on their pattern of occurrence and macroscopic symptoms can in most cases be only tentative, and that for a critical differentiation of diseases with nearly similar field symptoms, microscopic examination of the diseased tissues and fruiting structures of the pathogen is required.

43.2 Seed and Seedling Diseases

43.2.1 Aspergillus Crown Rot/Collar Rot: Aspergillus niger van Tieghem

Symptoms: In a moist soil environment seeds may be attacked and killed leading to a pre-emergence rotting. If seeds that did not emerge are carefully removed from the soil they may be covered with masses of black conidia which give them a sooty appearance. Post-emergence infection often culminates in death and rapid decay of seedlings. The first symptom in emerged seedlings is usually rapid desiccation of the entire plant. Affected tissue is covered by the sporulating fungus at the soil surface. As the infection progresses, the entire collar region becomes dark brown and shredded. Seedling infection commonly occurs in the cotyledonary-hypocotyl region shortly after emergence. The infected areas become water-soaked and light brown, and are soon covered with black fungal spores (Fig 43.1). Mature plants are also attacked. Lesions develop on the stem just below the soil surface and then spread upward along the branches. Because of the woodiness of mature plants, symptoms are generally not observed until either the branches wilt permanently, or wilting of the entire plant is apparent. The dead and dried branches are easily detached from the disintegrated collar region. The fungus sporulates on the surface of mature pods resulting in patches of black sooty spores.

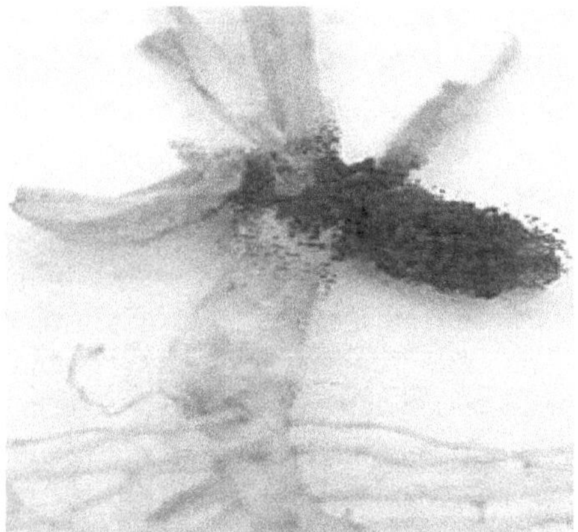

Fig. 43.1 *Groundnut plant and pods affected by Crown Rot*

Control

- Avoid deep sowing, avoid injuries to seedlings
- Soil application of neem cake or castor cake @ 500 kg/ha
- Seed treatment with Mancozeb @ 3 g/Kg or Tebuconazole @ 1.5 g/kg or *Trichoderma formulation* @ 4 g/kg seed and their soil application as per local recommendation.
- Foliar spray with Mancozeb @ 2.5 g/lt of water

43.2.2 Yellow Mold: Aspergillus Flavus Link ex Fries

Symptoms: Seeds and nonemerged seedlings attacked by the fungus are rapidly reduced to a shriveled, dried brown or black mass covered by yellow or greenish yellow spores. Decay is most rapid when infected seeds are sown and the fungus becomes active as the seeds hydrate. In some cases the emerging radical and hypocotyls become infected and rapidly decay. After seedling emergence there might not be any new infection, but previously infected cotyledons have necrotic lesions with reddish-brown margins covered by yellow or greenish-yellow spores (Fig 43.2). This phase of the disease is very similar to the crown rot caused by *Aspergillus niger,* and in some diseased plants both fungi may be present. When the strain of *A. flavus* causing the seedling disease produces aflatoxin, the plants may be severely stunted with chlorotic or pale green leaves with vein-clearing of leaflets, that are smaller than usual and have pointed tips (Fig 43.2). Root development is reduced by infection. This symptom is known as 'aflaroot'.

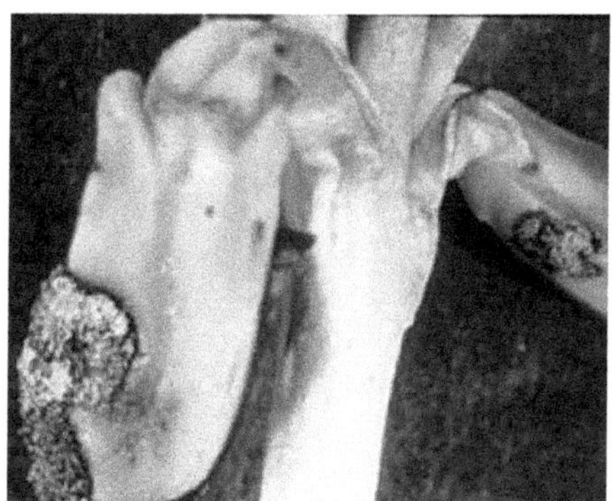

Fig. 43.2 *Yellow mold affected groundnut seedling and plant*

43.2.3 Diplodia Collar Rot: Lasiodiplodia Theobromae (Pat.) Griffon & Mubl.

Symptoms: Seedlings or maturing plants are infected at or near the soil surface and the fungus quickly invades the stem. Wilting of lateral branches or the entire plant is the first field symptom. Infected plants usually die within a few days. The base of the infected plants and the tap root become slate grey to black and easily shred (Fig 43.3). Black pycnidia develop as pimple-like dots on the necrotic tissues.

Fig. 43.3 *Diplodia collar rot affected seedling*

43.2.4 Rhizoctonia Damping-off: Rhizoctonia Solani Kühn

***Symptoms*:** Pre-or post emergence invasion by seed or soil-borne inoculums results in seedling death. In emerged seedlings, lesions are most frequently observed on the hypocotyl as sunken, elongate, dark brown areas just below the soil surface (Fig 43.4). The lesions enlarge, become black, and girdle the hypocotyl leading to typical 'damping-off' symptoms. Similar lesions develop on tap roots, and extend to the entire root system leading to dry root rot, which subsequently kills the plant. The decayed areas are often covered by a light brown mycelia mat. Dark brown, pinhead-sized sclerotia may develop on the dead tissues.

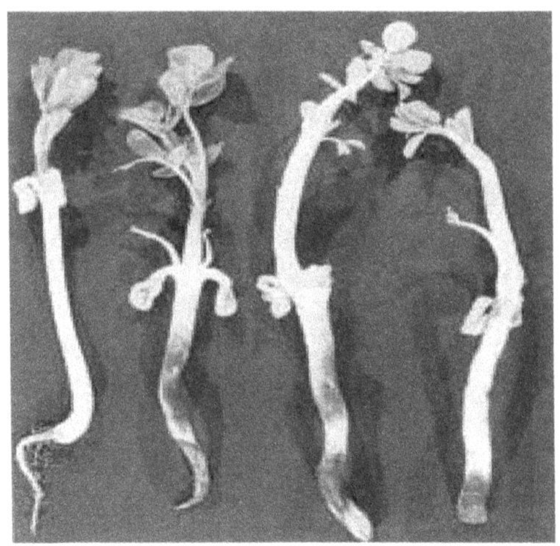

Fig. 43.4 *Groundnut seedling affected by Rhizoctonia*

43.3 Foliar Diseases

43.3.1 Early Leaf Spot: Cercospora Arachidicola Hori

***Symptoms*:** Lesions are sub-circular, dark brown on the upper leaflet surface where most sporulation occurs, and a lighter shade of brown on the lower leaflet surface [Fig 43.5, three leaflets showing upper surface, one leaflet (top right) turned to show lower surface]. A chlorotic halo commonly surrounds the lesion on the upper surface. Spores of *C. arachidicola* are produced mainly on the necrotic tissue of lesions on the upper leaflet surface. Lesions are also produced on petioles, stems, stipules, and pegs; these are oval to elongate and have more distinct margins than leaflet lesions (Fig 43.5). When the disease attack is severe, affected leaflets become chlorotic, then necrotic, and lesions often coalesce, resulting in premature senescence and shedding of leaflets (Fig 43.5).

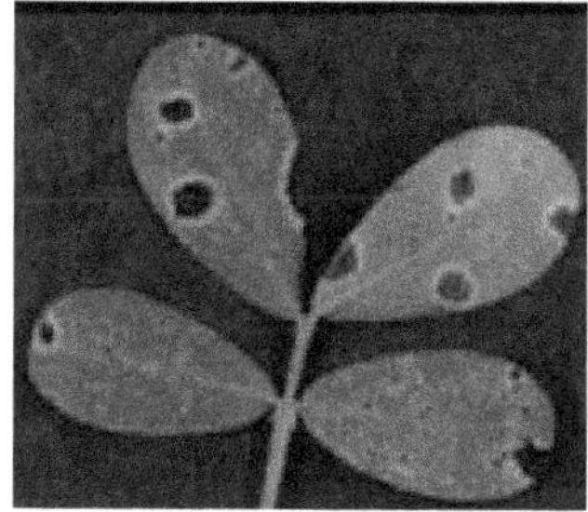
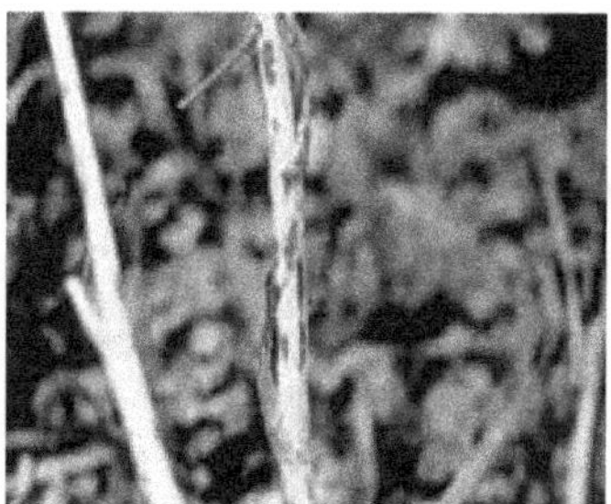
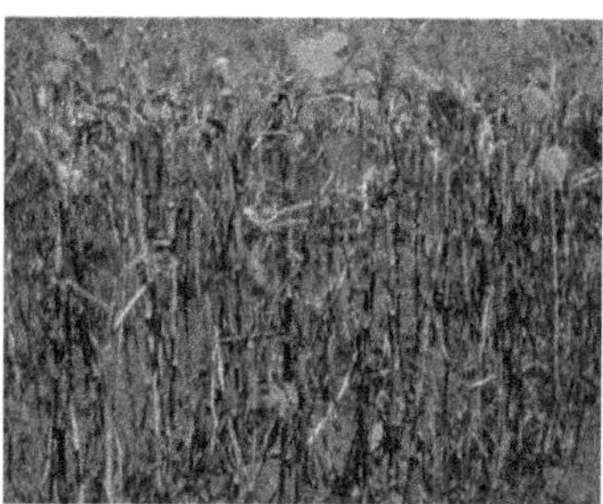

Fig. 43.5 Early leaf spot

43.3.2 Late Leaf Spot: Phaeoisariopsis Personata (Berk. & M.A. Curtis) Van Arx; Cercosporidium Personatum (Berk. & M.A. Curtis) Deighton

***Symptoms*:** When the lower leaflet surface is viewed, late leaf spots (Fig 43.6, right) are nearly circular and darker than early leaf spots (Fig 43.6, left). On the lower leaflet surface where most of the sporulation occurs the lesions are black and slightly rough in appearance (Fig 43.6, right, 6x). Circular rings of fruiting structures on the lesions of the lower leaflet surface can be seen with the aid of a hand lens. Late leaf spot lesions can be distinguished from those of early leaf spot that are lighter in color and may be surrounded by a chlorotic halo (Fig 43.6, left, 6x). When disease attack is severe, affected leaflets become chlorotic, then necrotic, and lesions often coalesce, resulting in premature senescence and shedding of the leaflets (Fig 43.6). Oval to elongate lesions similar to those of early leaf spot (Fig 43.5) are also formed on petioles, stems, stipules, and pegs. Late leaf spot attack is usually coincident with that of rust.

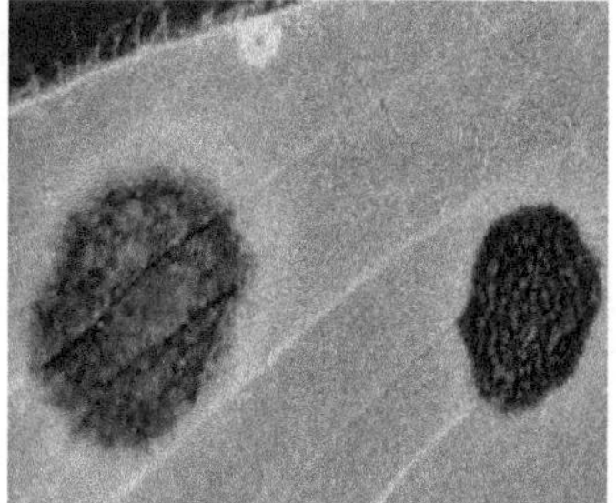

Fig. 43.6 Late leaf spot

Control measures:

- Deep burying of crop residues in the soil
- removal of volunteer groundnut plants

- resistant / tolerant varieties like TCGS-1043 (Dharani), ICGV-87160, ICGV-86590, TAG-24, Kadiri-4
- Intercropping Pearl millet or sorghum with groundnut (1:3 or 1:5)
- Foliar application of neem leaf extract (2-5%) or 5% neem seed kernel extract at 2 weeks interval
- Foliar spray of Carbendazim @ 1 g + Mancozeb @ 2.5 g per litre of water 2-3 sprays from the initiation of the disease.

43.3.3 Rust: Puccinia Arachidis Spegazzini

Symptoms: Orange-colored pustules (Fig 43.7, 7x) appear on the lower surfaces of leaflets and rupture to release masses of reddish-brown spores. In susceptible (Fig 43.7) cultivars the original pustules may be surrounded by colonies of secondary pustules. Pustules may appear late on the upper surface of the leaflets opposite those on the lower surface. Lesions can be formed on all aerial plant parts except flowers. Those on the stems are elongate (Fig 43.7). In contrast to the rapid defoliation associated with leaf spots, leaves infected with rust become necrotic and dry up, but tend to remain attached to the plant (Fig 43.7). Rust disease usually occurs at the same time as late leaf spot.

Fig. 43.7 Groundnut affected by Rust

Control:

- Early sowing in the first fortnight of June to avoid disease incidence.
- Foliar application of aqueous neem leaf extract @ 2.5%
- Spray Mancozeb or Chlorothalonil @ 2.5 g/litre of water or Calixin @ 1 ml.

43.3.4 Alternaria Leaf Spot and Veinal Necrosis: Alternaria Alternata (Fries) Keissler

Symptoms: Small, chlorotic, water-soaked lesions appear randomly over both surfaces of the leaflets. They enlarge, become irregular in shape, and are brown with darker brown margins (Fig 43.8); the central portions later become pale, dry rapidly, and disintegrate. Veins and veinlets adjacent to the lesions become necrotic. Affected leaflets are chlorotic and in severe attacks they prematurely senesce. When many lesions are present they coalesce, giving the leaf a ragged and blighted appearance. Profuse sporulation occurs on older lesions on the upper surfaces of the leaflets. This pathogen also causes leaf blight.

Fig. 43.8 *Alternaria leaf spot and veinal necrosis*

43.3.5 Powdery Mildew: Oidium Arachidis Chorin

Symptoms: Large spots cover the upper surfaces of leaflets. These spots are covered with superficial sporulating fungal growth which gives them a powdery white appearance (Fig 43.9). The centers of the spots later become brown and necrotic.

Fig. 43.9 *Powdery mildew*

43.3.6 Myrothecium Leaf Blight: Myrothecium Roridum Tode ex Fries

Symptoms: Lesions are round to irregular with grey brown colored centers and brown margins surrounded by chlorotic halos. They occur on both surfaces of leaflets. When lesions coalesce leaves appear blighted. Abundant olive green to black fruiting bodies, often arranged in concentric rings, are formed on the necrotic areas of both leaflet surfaces (Fig 43.10).

Fig. 43.10 Myrothecium leaf blight

43.3.7 Zonate Leaf Spot: Cristulariella Moricola (Hino) Redhead

Symptoms: Necrotic spots ranging from 2 to 12 mm in diameter appear on leaflets (Fig 43.11). Small lesions have a light brown center surrounded by a brown ring of necrotic tissue, but larger lesions show a zonate pattern on both leaflet surfaces. Spores may be produced on necrotic tissues of both leaflet surfaces.

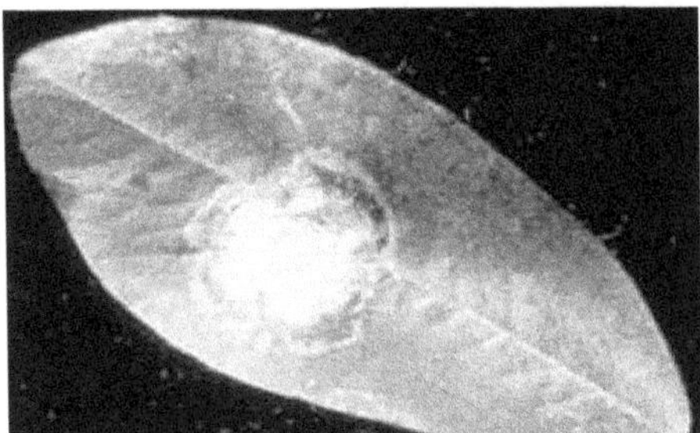

Fig. 43.11 Zonate leaf spot

43.3.8 Pepper Spot and Leaf Scorch: Leptosphaerulina Crassiasca (sechet) Jackson & Bell

Symptoms: The pepper spot phase of the disease is recognized by the appearance of minute necrotic spots on lower leaves (Fig 15). The spots are numerous, dark brown to black, irregular to circular and occasionally depressed. Discrete lesions appear on both sides of the leaflet; but are more often found on upper surfaces. When abundant, lesions coalesce giving the leaflet surface a netted appearance. Leaf scorch, the most common symptom, frequently develops on the tips, and occasionally on the margins, of leaflets. The wedge-shaped lesions have a bright yellow zone along the periphery of their advancing margins (Fig 43.12, below). The necrotic tissue becomes dark brown and tends to fragment along the leaflet margins, so that the leaves look tattered.

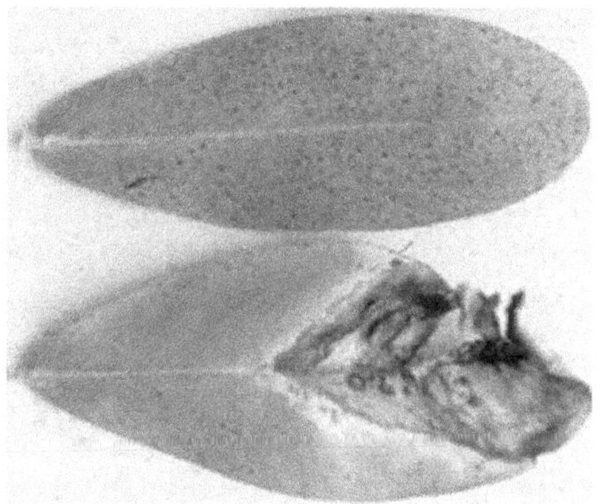

Fig. 43.12 Pepper spot and leaf scorch

43.3.9 Anthracnose: Colletotrichum Arachidis Sawada; Colletotrichum Dematium (Pers.) Grove Colletotrichum Mangenoti Chevaugeon

***Symptoms*:** Wedge-shaped lesions which resemble those caused by *Leptosphaerulina crassiasca* appear on the leaflet tips (Fig 43.13). Lesions may also develop on the leaflet margins leading to marginal blight. The periphery of the advancing margins of the lesion is surrounded by a bright yellow zone. The necrotic tissue becomes dark brown and tends to fragment along the leaflet margins. The disease may also extend to stipules and stems. Fruiting bodies, visible through a hand lens, are abundant on diseased tissues.

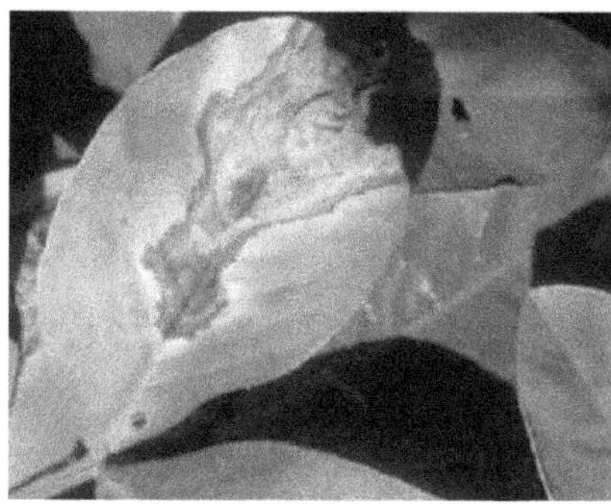

Fig. 43.13 Anthracnose

43.3.10 Alternaria Leaf Blight: Alternaria Alternate (Fr.) Keissler; Alternaria Tenuissima (Kunze ex Pers.) Wiltshire; Alternaria Arachidis Kulkarni

***Symptoms*:** Light to dark brown irregular spots appear towards the tips or margins of the leaves. These rapidly enlarge and coalesce to form wedge-shaped lesions. A distinct chlorotic zone commonly develops along the edge of necrotic tissue (Fig 43.14). In the later stages of disease development, blighted areas become dark brown and brittle and tend to fragment. Dark spore masses appear on necrotic tissues. Defoliation occurs when the disease is severe. The scorch symptoms resemble those caused by *Leptosphaerulina crassiasca* and *colletotrichum* spp, and it is often difficult to distinguish between these diseases in the field. *Alternaria alternata* also causes leaf spots and veinal necrosis as described earlier.

Fig. 43.14 Alternaria leaf blight

43.3.11 Pestalotiopsis Leaf Blight: Pestalotiopsis Arachidis Satya

***Symptoms*:** Dark brown circular lesions surrounded by faint yellow haloes appear on infected leaves. Lesions enlarge and coalesce leading to severe necrosis of leaflets, especially on the margins (Fig 43.15). Very small black spherical fruiting bodies visible through a hand lens can be found in the centers of the lesions or in blighted areas of leaflets. Symptoms of this disease resemble those of alternaria leaf blight, except that the yellow border surrounding the blighted areas is less conspicuous.

Fig. 43.15 Pestalotiopsis

43.3.12 Stem, Root and Pod Diseases Stem Rot: *Sclerotium rolfsii Saccardo*

***Symptoms*:** The pathogen attacks all parts of the plant, but stem infection is the most common and destructive. Yellowing and wilting of branches near the base of the plant is the first symptom. Sheaths of white mycelium develop at or near the soil line around the affected areas of the stem which become shredded (Fig 43.16). Abundant sclerotia, white initially and later turning brown, develop on the infected areas. The entire plant can be killed, but in some cases only two or three branches are affected. Infection of pegs, pods, and roots occurs either independently or together with stem infection. Lesions on the developing pegs may retard pod development. Orange or brown colored lesions may be found on the pods. Severely infected pods are completely covered with a white mycelia mat, and eventually decay. In some cases the seeds from the diseased pods show a characteristic bluish-grey discoloration of the testa known as 'blue damage' (Fig 43.16).

Fig. 43.16 *Stem and pod diseases stem rot*

Control

- Deep Ploughing
- Non dirting cultivation in combination with minimizing defoliation by leaf spot and leaf eating caterpillar control.
- Deep burial of surface organic matter and crop debris by ploughing it to a depth of 8-10 inches.
- Early sowing and close planting
- Crop rotation with Cotton, Wheat, Maize, Jowar, Onion and Garlic
- Use of tolerant/resistant varieties like Dharani, DH-8, ICGV-86590 and OG-52-1.

- Seed treatment with *Trichoderma formulation* @ 4g/kg seed.
- Soil application of castor cake or neem cake or mustard cake @ 500-1000 kg/ha and *Trichoderma formulation* as per local recommendations or Hexaconazole @ 2 ml/lt of water.

43.4 Pod Rot: Pythium Myriotylum Dreschler; Rhizoctonia Solani Kuhn; Fusarium Solani (Mart.) Saccardo f. sp Phaseoli (Burkholder) Snyder & Hans.; Fusarium Oxysporum Schlechtend. Emend Snyder & Hans.; Macrophomina Phaseolina (Tassi) Goidanich

Symptoms: Pod rots are characterized by the presence of brownish-black lesions on the shell (Fig 43.17). The lesions enlarge and coalesce, and the entire pod surface becomes discolored, and in advanced stages the shell tissues disintegrated and the kernels decay. The texture and color of the rotted tissues vary depending on the organisms involved and other edaphic factors. Pods rotted by species of Fusarium have pink-or purple stained shells. *Sclerotium rolfsii, Sclerotinia* spp, *Verticillium* spp, and *Botrytis cinerea* cause pod rotting in addition to the other damage described earlier. *Rhizoctonia solani* in addition to causing pod, root, and stem rots can also invade lower leaves in contact with soil causing a leaf blight (Fig 43.17).

Fig. 43.17 Pod Rot

References

1. Feakin, S.D. (ed.) 1973. Pest control in groundnuts. PANS Manual 2, London, UK: Centre for Overseas Pest Research.197 pp.
2. Garren, KH., and Jackson, C.R. 1973. Peanut diseases. Pages 429-494 in Peanuts _ culture and uses. Yoakum, Texas, USA: American Peanut Research and Education Association, Inc.
3. Jackson, C.R., and Bell, D.K 1969. Diseases of peanut (groundnut) caused by fungi. Research Bulletin no:56. Athens,
4. Georgia, USA: University of Georgia college of Agriculture Experiment Station. 127 pp.
5. Klesser PJ. 1966. Tomato spotted wilt virus on Arachis hypogaea. South Afr. J. Agric. Sci. 9:731–36.
6. Kolte, S.J. 1985. Diseases of annual oilseed crops. Peanut diseases. Vol.1. Boca Ration, Florida, USA: CRC Press, Inc. 143 pp.
7. McDonald, D., and Raheja, A.K. 1980. Pests, diseases resistance, and crop protection in groundnuts. Pages 501-514
8. In advances in legume science (summerfleld, R.J., and Bunting, A.H. eds.). Kew, Surrey, UK: Royal Botanic Gardens.
9. McDonald, D., and Subrahmanyam, P. 1992. Rust of groundnut. Pages 272-284 in Plant diseases of international importance. Diseases of vegetables and oil seed crops vol. II. (Chaube H.S., Kumar, J., Mukhopadhyay, A.N., and Singh, U.S., eds.). Englewood Cliffs, New jersey 07632, USA: Prentice- Hall, Inc.
10. McDonald, D., Subrahmanyam, P., Gibbons, R.W., and Smith, D.H. 1985. Early and late leaf spots of groundnut. Information Bulletin no: 21. Patancheru, A.P. 502324, India. International Crops Research Institute for the Semi-Arid Tropics. 24 pp.
11. Porter, D.M., Smith, D.H., and Rodriguez-Kabana, R. 1982. Peanut plant diseases. Pages 326-410 in Peanut science and technology (Pattee, H.E., and Young, C.T., eds.). Yoakum Texas USA: American Peanut Research and Education Society, Inc.
12. Porter, D.M., Smith, H.H., and Rodriguez-Kabana, R. (eds.). 1984. Compendium of peanut diseases. St. Paul Minnesota, USA: American Phytopathological Society. 73 pp.
13. Prasada Rao, R.D.V.J., Reddy, D.V.R., Nigam, S.N., Reddy, A.S., Waliyar, F., Yellamanda Reddy, T., Subramanyam, K., John Sudheer, M., Naik, K.S.S.., Bandyopadhyay, A., Desai, S., Ghewande, M.P., Basu,
14. M.S., and Somasekhar. 2003. Peanut Stem Necrosis: A New Disease of Groundnut in India. Information Bulletin no. 67. Patancheru 502324, Andhra Pradesh, India: International Crops Research Institute for the Semi-Arid Tropics. 16 pp.
15. Reddy, D.V.R. 1988. Virus diseases. Pages 508-526 in Groundnut (Reddy, P.S., ed.). New Delhi, India: Indian Council of Agricultural Research. 565 pp.

16. Reddy, D.V.R. 1991. Groundnut viruses and virus diseases: distribution, identification and control. CAB International, Review of Plant Pathology 70:665-668.

17. Reddy, D.V.R., Buiel, A.A.M., Satyanarayana, T., Dwivedi, S.L., Reddy, A.S., Ratna, A.S., Vijaya Lakshmi, K., Ranga Rao, G.V., Naidu, R.A., and Wightman, J.A. 1995. Peanut bud necrosis disease: an overview. Pages 3-7

18. In Recent studies on peanut bud necrosis disease: proceedings of a Meeting, 20 Mar 1995, ICRISAT Asia Center, India (Buiel, A.A.M., Parlevliet, J.E., and Lenne, J.M., eds.). Patancheru 502324, Andhra Pradesh, India: International Crops Research Institute for the Semi-Arid Tropics; and P O Box 386, 6700 AJ Wageningen, The Netherlands: Department of Plant Breeding, Agricultural University of Wageningen.

19. Reddy, A.S., Prasada Rao, R.D.V.J., Thirumala- Devi, K., Reddy, S.V., Mayo, M.A., Roberts, I., Satyanarayana, T., Subramaniam, K., and Reddy, D.V.R. 2002. Occurrence of Tobacco streak virus on peanut (Arachis hypogaea) in India. Plant Dis. 86:173-178.

20. Smith, D.H., Pauer, G.D.C., and Shokes, F.M. 1992. Cercosporidium and cercospora leaf spots of peanut (groundnut). Pages 285-304 in plant diseases of international importance. Diseases of vegetables and oil seed crops vol. II. (Chaube, H.S., Kumar, J., Mukhopadhyay, A.N., and Singh, U.S., eds.). Englewood Cliffs, New Jersey 07632, USA: Prentice-Hall, Inc.

21. Subrahmanyam P., and McDonald, D 1983. Rust diseases of groundnut. Information Bulletin no: 13. Patancheru, A.P. 502 324, India. International Crops Research Institute for the Semi-Arid tropics. 20 pp.

22. Subrahmanyam P., and Ravindranath, V. 1988. Fungal and nematode diseases. Pages 453-507 in groundnut (Reddy, P.S., ed.). New Delhi, India. Indian Council of Agricultural Research. 565 pp.

23. Subrahmanyam, P., Reddy, D.V.R., Sharma, S.B., Mehan, V.K., and McDonald, D. 1990. A world list of groundnut diseases. Legumes Pathology Progress Report no.12. Patancheru, A.P. 502 324, India: Legumes Program, International Crops Research Institute for the Semi-Arid Tropics. 14 pp. (Limited distribution).

24. Wongkaew, S. 1985. Compendium of groundnut diseases in Thailand. Bangkok, Thailand: Publishing Co.76 pp.

Chapter 44

Dissemination of Groundnut Research Technology through Farmers Call Centre: A Critical Analysis

S. Hemalatha and A. Pratap Kumar Reddy
Farmers Call Centre, ANGRAU, Andhra Pradesh.

44.1 Introduction

Technology transfer in agriculture is an uphill task and sometimes persuasion may be necessary to convince the farmers to switchover from conventional approaches to new technology. Direct interaction with individual farmer is an extension strategy that is gaining wide popularity in present-day agriculture. Though this method is expensive and requires a sizeable number of technical experts, the results are often tangible. An innovation in this direction is to address farmers through a cost-effective approach through live communication over wire. The establishment of Farmers Call Centres (FCCs), and Kisan Call Centres (KCCs) based on ICT was one such innovation in India during 2003 (FACET, 2012).

In Andhra Pradesh FCC was established in 2003 and is first of its kind in India, serving as a role model for several call centres run by private and public sectors (ICAR). The FCC of A.P. is a collaborative innovation initiated by the Acharya N. G. Ranga Agricultural University and State Department of Agriculture, Govt. of A.P. This FCC caters to the needs of farmers on agri-related issues. Multi-faceted benefits can be derived through this live, two-way communication between technical experts located at the FCC and farmers (Rajasri *et al.* 2014). Another advantage of the FCC is that solutions to farmers' queries are being given in the local dialect (Telugu) so as to avoid any gaps in communication.

In this chapter, we have compiled the information on farmers' queries pertaining to groundnut production from Andhra Pradesh and Telangana states during 2014-15. The periods of analysis were June 2014 to May 2015, covering *Kharif* 2014, *Rabi* 2014-15 and Summer 2015 groundnut crops.

44.2 Analysis of Groundnut Queries

During the year 2014-15, in the United Andhra Pradesh, region-wise, queries (%) on groundnut cultivation were more from Rayalaseema region (58%), followed by Telangana (31%) and Coastal Andhra (11%) (Fig 44.1).

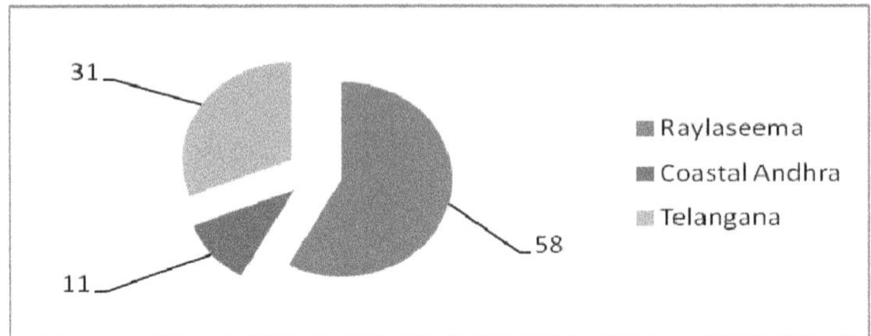

Fig. 44.1 Queries relating to groundnut cultivation, received at Farmers Call Centre, Hyderabad during 2014-15.

44.2.1 Groundnut Queries from Andhra Pradesh

In Andhra Pradesh State, during June 2014 to May 2015, covering the calendar groundnuts of *Kharif*, *Rabi* and Summer crops, the following is the analysis on district-wise queries raised by farmers. The per cent telephone queries received on groundnut were highest from Anantapur district (27.6%), followed by Kurnool (21.8%). More than 10% of the queries were also from districts such as Nellore (14.4%) and Kadapa (13.2%). For the remaining districts, the per cent groundnut calls ranged up to 6.6. The details of the per cent groundnut queries district-wise in Andhra Pradesh are given in Fig 44. 2. From East Godavari district, there were no telephone calls from farmers during 2014-'15 (Fig 44.2).

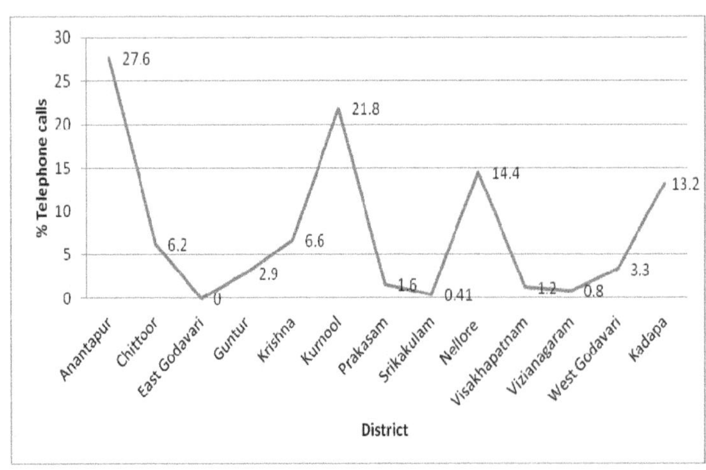

Fig. 44.2 District-wise telephone queries from Andhra Pradesh, on groundnut cultivation, during 2014-'15.

Month-wise, the farmers' queries were on increasing note from June 2014 to November 2014 (for *Kharif* crop) and then decreased thereafter up to May 2015 with February 2015 as an exception. During *Kharif* 2014, highest calls were received (16.7%) during November 2014 followed by December 2014 (12.1%). The telephone calls from Andhra Pradesh ranged from 2.5 to 16.7% in *Kharif* (up to November). During *Rabi* and Summer crops, next highest farmers' calls were during February 2015 (10.1%), followed by January 2015 (9%), March 2015 (8.14%). The farmers' telephone calls during April and May 2015 were 3.44 and 3.07% respectively (Fig 44.3).

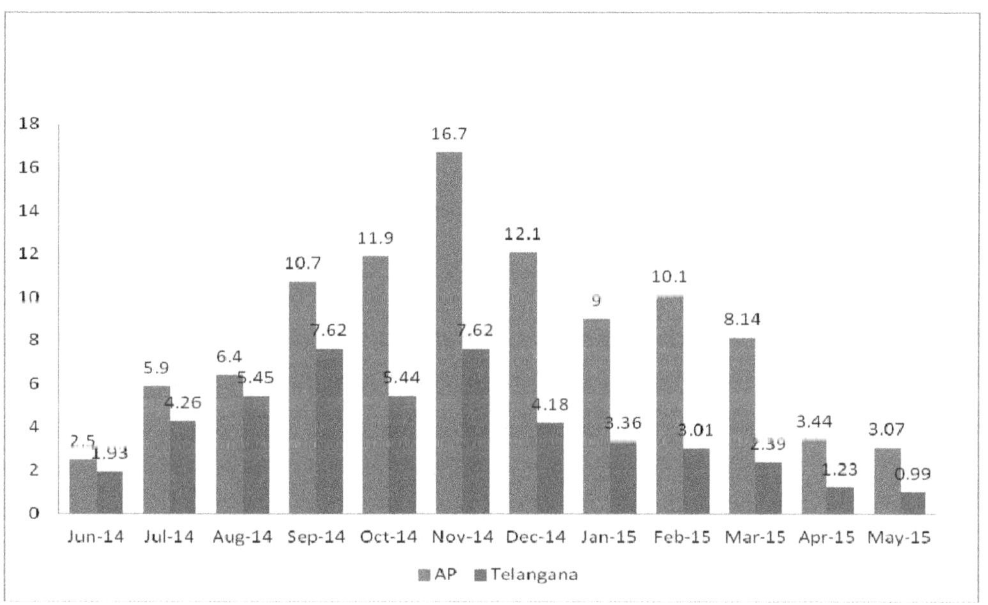

***Fig. 44.3** Month-wise farmers' queries to Farmers' Call Centre on groundnut cultivation during 2014-'15.*

44.2.2 Groundnut Queries from Telangana

In Telangana, on groundnut cultivation, farmers' calls were categorized district-wise and the results are presented in Fig 44.4. Highest per cent of queries were received from Mahaboobnagar district (48%), followed by Warangal (28%). These two districts account to 76% of the total queries on groundnut cultivation in Telanagana State during 2014-15 (Fig 44.3). Among the remaining districts, queries were up to 9% in Ranga Reddy, 6% in Adilabad, 5% in Nalgonda and 4% in Khammam. Per cent queries from Medak were 3%, whereas Hyderabad and Karimnagar recorded 1% queries each on groundnut cultivation.

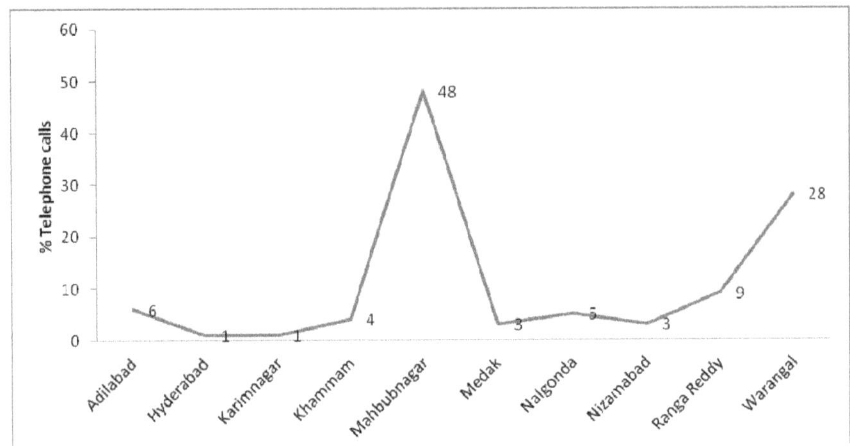

***Fig. 44.4** District-wise telephone queries from Telangana, on groundnut cultivation, received during 2014-'15.*

Month-wise, the farmers' queries were on increasing note from June 2014 to September 2014 (for *Kharif* crop) and then decreased in October 2014. However, in November, there was an increase up to 7.62% from the previous months' queries of 5.4% (Fig 44.3). From November 2014 to May 2015, a decreasing trend on queries was observed. During *Kharif* 2014, highest calls were received during September and November (7.62% each). The telephone calls from Telangana ranged from 1.93 to 7.62% in *Kharif* (up to November). During *Rabi* and Summer crops, thereafter November, next highest farmers' calls were during December 2015 (4.18%), followed by January 2015 (3.36%), February 2015 (3.01%). The farmers' telephone calls during April and May 2015 were 1.23 and 0.99% respectively (Fig 44.3).

***Table 44.1** Telephone queries from farmers of AP and Telangana on plant protection aspects of groundnut cultivation received during (2014-'15).*

S.No.	Pest	AP	Telangana
	Pests		
1	Leaf eating Caterpillars	36.8	40
2	Red hairy caterpillar	3.9	8
3	Plant hoppers	3.3	4
4	Rat damage	1.3	-
5	Leaf folder	25	44
6	Sucking pests	9.21	4
7	Storage pests	2.6	-
8	Root grub	3.9	-
9	Nematodes	5.3	-

Table 44.1 *Contd...*

S.No.	Pest	AP	Telangana
Diseases			
10	Stem rot	18.8	-
11	Root rot	13	33.3
12	Rust	7.3	6.7
13	Leaf spot	53.6	53.3
14	Powdery mildew	2.9	-
15	Wilt	1.4	-
16	*Trichoderma* usage	1.4	-
17	Bud necrosis	1.4	6.7

Compiled results of farmers' queries on groundnut during 2014–15 at the FCC indicated that leaf eating caterpillars (*Spodoptera* and *Helicoverpa*), leaf folder are rampant on groundnut. Among diseases, Leaf spot and Stem rot were more prevalent than others in both Telangana and Andhra Pradesh states (Table 44.1).

44.3 Discussion

Based on queries received on groundnut at FCC, predictions on the incidence of pests and diseases were made district-wise and month-wise. Our observations and data base generated over the period under report are important tools for forecasting the pest and disease outbreaks in various districts of Andhra Pradesh and Telangana for both *Kharif* and *Rabi*. For instance, farmers in the respective districts during September/October (*Kharif*) and January/February (*Rabi*) can be advised well in advance on possible outbreak of these pests and diseases. This enables them to save the crop from biotic stress and reap greater benefits of groundnut cultivation. The success of sustainable agricultural practices in plant protection can also be dependent on the intensity with which the pest or disease outbreak is reported (Suresh et al, 2013).

44.4 Conclusion

Data generated at the FCC based on live communication with farmers will be an important tool in prioritizing research and for diagnostic surveys by scientists. Pooled data, especially on the incidence of pests and diseases in various crops over many years, enables the prediction of risk and sensitive areas based on queries. Over the years, the FCC has witnessed an overwhelming response from farmers. Therefore popularising the toll-free numbers (1100/1800-425-1110 for Telangana and 1800-425-4440 for Andhra Pradesh) is of pivotal importance so as to exploit the services of FCC to its fullest potential.

References

1. FACET. (2012). Briefing Paper: Call Centres for Agricultural Information. Supported by USAID's Fostering Agriculture Competitiveness Employing Information Communication Technologies 1-3.
2. Rajasri, M., Hemalatha, S., Suresh, K., Vijay Krishna Kumar, K., Suneetha, Y. and Lava Kumar Reddy, M. (2014) The Role of Farmers Call Centre (FCC) in Technology transfer to farmers of Andhra Pradesh, India: A case study on Rice pests, diseases and PGPR published in the proceedings of 3rd Asian PGPR on Recent Advances in Biofertilizers and Biofungicides (PGPR) for Sustainable Agriculture pp.468-480.
3. Suresh, K., Hemalatha, S. Rajasri, M, Vijay Krishna Kumar, K. and Suneetha, Y. 2013 Services of Farmers' Call Centre (FCC)-ANGRAU in dissemination of agricultural technology in Andhra Pradesh, India Published in International Journal of Applied Biology and Pharmaceutical technology 4(2): 92 April-June.

Chapter 45

Recent Advances in Agriculture Transfer of Technology

V. Anitha and R. Sudhakar Reddy

Professor Jayashankar Telangana State Agricultural University, Rajendranagar, Hyderabad.

45.1 Introduction

Agriculture, by far is the mainstay of Indian economy even today as more than 60 percent of the population is dependent on farming and allied fields. The Indian agriculture sector has made considerable progress in the last few decades with its large resources of land, water and sunshine. However between 1970 and 2011, the GDP share of agriculture has fallen from 43 to 16%. This is not because of reduced importance of agriculture, or a consequence of agriculture policy. This is largely because of the rapid economic growth in services, industrial output, and non-agricultural sectors in India during 2000 to 2010. This alarming downfall of growth in agriculture sector imposes greater responsibilities on the Universities and research organizations to explore the possibilities to take the fruits of new innovations and developments in agriculture research to the uncovered areas and people with integration of teaching, research and extension.

Several emerging challenges confront Indian farmers today. These include limited land and water availability, which is further exacerbated by degradation of natural resources; climate changes; changes in demand and consumption patterns, shift toward high-value agriculture; increasing population pressure; and liberalization of trade. Recent global food price increases and high levels of inflation have provided an opportunity to increase farmers' profitability. However, to realize the benefit of higher process, farmers need to access a wider range of information, related not only to production technologies but also to post harvest processes, access to remunerative markets, price information etc. The role of agricultural extension and technology transfer assumes importance in this context.

The diverse nature of the Indian subcontinent, with its wide variety of agroclimatic regions and broad range of socioeconomic conditions in the rural population, calls for agricultural extension approaches that are context-and situation-specific. With more than 81 percent of Indian farmers cultivating an area of 2 hectares or less (India, Directorate of Economics and Statistics 2009; NSSO 2006), there is an increasing need for stronger intermediaries that can facilitate information access for diverse smallholder farmers.

In addition to needing different types of information and using different information sources, different farmers will have different search behaviors. Factors such as literacy or access to resources will have a large impact on information needs, searching behavior, access, and use. Swanson (2008) described various target groups; rural and farm women, small and marginal subsistence farmers, medium-scale farmers, commercial farmers, and rural youth. In India, for example, the information needs of the 360 million farmers (expected to rise to 600 million by 2020) who operate under rainfed conditions and contribute 45 percent of production will be different from the needs of farmers in well-endowed irrigated areas. In most rainfed areas, due to higher temporal and spatial variability in rainfall, nutrient-poor soils, ad poor socioeconomic conditions, many farmers are not able to produce beyond the quantity needed for self-consumption. Additionally, livestock plays a greater role in these areas that crop production does, and rainfed regions also employ a greater number of women than irrigated areas do.

A holistic approach to agricultural extension today goes beyond technology transfer for major crop and livestock production systems. It also includes goals for human capital development, in terms of enhancing the management and technical skills of farm households relating to production and postharvest handling of high-value crops, livestock and fisheries, sustainable natural resource management, family health and nutrition, and leadership and organizational skills, in addition to social capital development, that is, organizing producer groups (Swanson 2008).

45.2 Transformation of Extension

The Training and Visit System (T&V) introduced in the 1970s to foster the Green Revolution in the country was successful with its top down approach largely due to the supply led approach at that point of time. Widespread changes in agriculture have rapidly transformed old staple based subsistence farming into a high value information intensive commercial enterprise which has necessitated agricultural extension in our country to transform into a demand driven pluralistic model with several public/private and Civil society organizations taking part in transfer of technology to the farmers (Table 45.1). Farmers participation in the development and dissemination of technology has emerged as an important consideration in this dynamic process.

Table 45.1 *Pluralistic dimension of the Indian agricultural extension system (adapted from Anderson and Feder (2004))*

Source of Finance	Service providers					
	Public Sector	Private sector			Civil Society	
		Input supplier	Processors/ Traders	Private RAS providers	NGO	Farmer Cooperatives
Public	ATMA, KVK, MANAGE, DAATTC, State DOA, Kisan Call Centre,	-	-	-	-	-
NGO/Donors	ATMA	-	-	Mobile applications	BAIF PRADAN BASIX	-
Private companies	-	Agribusinesses Agriclinics,	Contract farming schemes	Mobile applications	PRADAN BASIX	-
Farmers/ Farmer Cooperatives	Indian Diary Cooperative	-	-	-	-	230 million members of Indian agricultural cooperatives

The aim of all these extension service providers is to reach farmers in the most remote areas in as little time as possible. Emerging digital technologies (ICTs) can play an important role in supporting extension in this regard. ICTs is an umbrella term that includes computer hardware and software; digital broadcast and communication technologies such as television, radio, mobile phones, electronic information repositories such as World Wide Web and those found on CD ROMs.

45.3 ICT Initiatives in India

In recent times, efforts are being made to develop personalized information dissemination systems in agriculture to improve the utilization and performance of agricultural technology among Indian farmers. Hitherto dissemination of information was through print media such as newspaper, magazines, books and communication media like radio which disseminated information in an ad-hoc and generalized manner. Recognizing the need to develop improved information dissemination methods by exploiting the recent developments in Information and Communication the GOI has stepped up its efforts in investing on ICT infrastructure.

In a developing country like India where the introduction of ICT tools for agricultural development is still new, challenges have been many both in the implementation mechanism and on the farmers or the end users side. Proper understanding of its use, its credibility and the access are the major bottlenecks in ICT application and usage, but wherever applied, ICT had shown promising results, like IKSL, kisan Kerala, Kisan Call Centres, mobile services, kisan portal, agropedia, eAqua, and web services. Many expert system viz. Decision Support Systems, GIS, financial services in mobile baking etc., are helping many, in various aspects of farming. The ICTs have also increased the Public Private Partnership in agricultural development and the collaboration among them in input delivery, information delivery, etc., but the sustainability of the ICT projects after specified project period and continuing them with same enthusiasm is still a challenge.

45.4 ICT use in Agriculture

- E-commerce for direct linkages between local producers, traders, retailers and suppliers;
- The facilitation of interaction among researchers, extension (knowledge) workers, and farmers;
- Question-and-answer services where experts respond to queries on specialized subjects ICT services to block- and district-level developmental officials for greater efficiency in delivering services for overall agricultural development;
- Tele-education for farmers-up-to-date information, supplied to farmers as early as possible, about subjects such as packages of practices, market information, weather forecasting, input supplies, credit availability, etc;
- Creating of databases with details of the resources of local villages and villagers, site-specific information systems, expert systems, etc.;
- Provision of early warning systems about disease/pest problems, information regarding rural development programmes and crop insurances, postharvest technology, etc.;
- Facilitation of land records and online registration services;
- Improved marketing of milk and milk products;
- Services providing information to farmers regarding farm business and management;
- Increased efficiency and productivity of cooperative societies through the computer communication network and the latest database technology;
- Websites established by agricultural research institutes making the latest information available to extension (knowledge) workers and obtaining their feedback.

Some of the successful ICT initiatives are :

45.4.1 ikisan

ikisan is the ICT initiative of the Nagarjuna group of companies, in the state of Andhra Pradesh, a largest private entity, meeting farmers' agricultural needs. iKisan was set up with two components, the iKisan.com website, to provide agricultural information online, and technical centres at village level to assist the member farmers for a nominal fee of Rs. 20/month.

45.4.2 E sagu

In E *sagu*, rather than visiting the crop in person, the agricultural scientist delivers the expert advice by getting the crop status in the form of digital photographs and other information provided by local coordinators at the village generally experienced local farmer. The E. *sagu* centre in the village is equipped with computers and dial up internet connection to transmit information to and from the agricultural expert.

45.5 Digital Green

The Digital Green approach which started in 2008 involves the use of videos for dissemination of information about agricultural techniques that farmers can adopt in order to increase their productivity which is unique in that it involves

(a) a participatory process for local video production
(b) a human mediated instruction model for dissemination and training
(c) a hardware and software technology platform for exchanging data in areas with limited internet and electrical grid connectivity and
(d) an interactive model to progressively better address the needs and interests of the community with analytical tools and interactive phone based feedback channels.

45.5.1 Farmers Portal www.farmer.gov.in

Farmers Portal is an initiative of the Department of Agriculture & Cooperation, Government of India to facilitate a single point access to information on the Web. The portal provides access to information about Seeds, Fertilizers, Pesticides, Market Price, Agromet advisory, farm machinery, package of practices, extension activities, beneficiary list under NHM and NFSM, programmes and schemes, Insurance, Storage, Credit, Farmer friendly literature, through a single window to farmers and stakeholders.

45.5.2 M-kisan www.mkisan.gov.in/skoch.aspx

This project developed by the Department of Agriculture & Cooperation has widened the outreach of scientist, experts and Government officers posted down to the block level to disseminate information and to provide advisories to farmers through their mobile telephones. SMS portal was launched in July 2013 and since its inception over 72 crore messages have been sent to farmers across the country. These messages are specific to farmers' specific needs & relevant at a particular point of time. mKisan SMS portal has empowered central and state government organizations in Agriculture and allied sectors including state agriculture universities, KrishiVigyanKendras, Agromet forecasts units of India Meterological Department, ICAR institutes, organization in Animal Husbandry,

Dairying and Fisheries etc. to give information/services/advisories to farmers by SMS in their language, preference of agricultural practices and locations.

45.5.3 Kisan Call Centre

Kisan Call Centre (KCC) was launched by the Department of Agriculture and Cooperation, Ministry of Agriculture, Government of India in 2004 with an objective to respond to the issues raised by the farmers across the country, in their local language, Queries related to agriculture and allied sectors are being addressed by subject matter specialists from the State Agricultural Universities and the State Departments concerned. As on today, there are 25 such Call Centres operational across the country in different states.

The KCC operates at three levels viz., Level-I, Level-II and Level-III. In order to access this service, the farmer is required to dial a toll-free number 1551 (from landline) or 1800-180-1551 (from mobile). Once the farmer's call lands at the designated Call Centre for the state, it is attended to by the Level-I functionaries, who record the basic information of the farmer and respond to the farmer's query. If the queries cannot be answered at Level-I, those relating to Crop Technology are escalated to designated Subject Matter Specialists of SAUs/ICAR, while those related to programs/schemes are escalated to technical officials of State Departments of Agriculture, Animal husbandry, Horticulture, Fisheries, Marketing etc. functioning at Level-II. If the queries cannot be answered even at Level-II, these are escalated to a Central Institute (functioning in that state) designated as Level-III by DAC, MoA, GoI for responding to such queries within 72 hours of their receipt.

45.5.4 Agmarknet www.agmarknet.nic.in

AGMARKNET is the Agricultural Marketing Information Network in the country, under the Central Sector Scheme "Marketing Research and Information Network" (MRIN) of the Directorate of Marketing and Inspection, Ministry of Agriculture, Government of India. The project networks over 3000 Agricultural Produce Markets across the country. This is a 'farmer centric' project for speedy collection and diffusion of market information.

45.5.5 Seednet www.seednet.nic.in

The Seed net portal gives access to seed related information. Research institutes, certifying agencies and seed testing organizations can get seed related information such as testing procedures, new crop information etc through the portal. The varieties database provides information on various available varieties of each seed, information about newly released seeds etc. The seed management database provides information on availability, supply and demand for different varieties of seeds.

45.5.6 Fertilizer Monitoring System (FMS) http://www.urvarak.co.in/

The Fertilizer Monitoring System (FMS) launched in 2007 by the Department of Fertilizers is an attempt to provide comprehensive information of fertilizers. The site currently focuses on various fertilizers, phosphatic and urea, and provides information on the fertilizer companies dealing with these fertilizers, the rate of concession on each fertilizer, its MRPs and product wise/state wise details of dispatch and receipts of fertilizers at different destinations across the country.

45.5.7 Rice Knowledge Management Portal (RKMP) www.rkmp.co.in:

Rice Knowledge Management Portal developed by the Indian Council of Agricultural Research (ICAR) serves as an information highway to share rice knowledge among the stakeholders. The Extension and Farmers domains provide production know how, package of practices, FAQs etc., in English and local languages. In the research domain, various services are provided such as AICRIP Intranet, archives of AICRIP data (27000 datasets!), communities of practice (CoP), bio-informatics suite, approach papers, India Rice Research Repository (i3R), status papers on rice etc.

45.5.8 Mass Media Support to Agricultural Extension

The "Mass Media Support to Agricultural Extension "scheme of Department of Agriculture Cooperation (DAC), GOI, was initially launched on 21st January, 2004 and the revised scheme on 24. 02. 2005 wherein the existing infrastructure of Doordarshan and All India Radio is being utilized to make the farmers aware about modern farm technologies. Farm related programmes of 30 minute duration are being telecast 6 days a week through National, Regional Kendras (18) and High Power /Low Power Transmitters (180) of Doordarshan. Similarly, 96 Rural FM Radio Stations of All India Radio are being utilized to broadcast 30 minutes of programme for farmers 6 days a week. The Free Commercial Time (FCT) available under KrishiDarshan and Kisanvani programme is being utilized for dissemination of Advisories on Rabi /Kharif and popularizing the central scheme like contingency plan developed by State Governments and emergent issues like Drought, Flood credit and insurance, Kisan Call Card National Food Security Mission (NFSM), MSP.

45.5.9 Community Radio Initiative

The advent of Community Radio (FM radio owned & operated by community) has brought lot of relevance in current Indian agriculture. Prime Minister Modi has called for establishing one community radio in each agricultural college to transform the way local agricultural knowledge and experiences are shared.

Examples: Bangladesh Community Radios, Farm Radio International, Farmer Voice Radio, Mali Shambani (Kenya).

In most cases, community radio can provide agricultural information that is topical, relevant and of interest to farmers' problems. Further follow-up, interviews with farmers who have successfully improved their farms etc., will speed up modern farming.

45.5.10 Village Knowledge Centres (VKCs)

The Village Knowledge Centre programme was started in 1998 in Pondicherry by the M.S. Swaminathan Research Foundation (MSSRF) with the support of IDRC and CIDA. In order to ensure access to all, the VKCs are located in public buildings (Panchayats, SHG building, Community Hall, School, Farmers and Fishermen Associations, Women Self-Help Groups, Community Based Organization, etc.). The content is developed in close interaction with the local people in a botton-up manner. VKCs are connected with hub centres through hybrid wired and wireless network. These centres provide information on agriculture, health, employment, weather, education, government entitlements, micro enterprises training etc.

45.5.11 E-choupal Experience of ITC

Use of ICT platform to enhance transactional efficiency in the agro-sector is very high. ITC Ltd implemented a project on electronic market place for the soyabean farmers in the state of Madhya Pradesh. The project owes its success to the factors such as utilization of local leadership in the villages, a sustainable business model and collaboration between the local authorities and the corporate implementer. The technology embarked was easy to replicate and easily scalable, and it was customized according to the needs of the local farmers. The project has helped the farmers developing sustainable income levels. Elimination of the middlemen, developing easy access to the market place and shared ownership of the project.

45.5.12 E-governance in Fisheries

A novel example in e-governance was that set-up by an NGO in the fishermen colony in Pondicherry. The project started by MSSRF was successful since the entire project was implemented in the phased manner, it was customized according the needs of the rural users, and it helped in developing a systematic knowledge base for sharing within the community. The Fisher Friend Mobile Application (FFMA) is a unique, single window solution for the holistic shore-to-shore needs of the fishing community, providing vulnerable fishermen immediate access to critical, near real-time knowledge and information services on weather, potential fishing zones, ocean state forecasts, and market related information.

45.6 Vikaspedia www.vikaspedia.in

The Agriculture section of Vikaspedia portal developed by CDAC aims to provide relevant information and knowledge based content related to agriculture and allied sectors, contributed by volunteers in these sectors. As a stakeholder in this platform, you can express your views, share your experiences, participate in relevant discussions, update your knowledge from the available resources and disseminate the knowledge to others.

The major focus of Agriculture sector of Vikaspedia portal is pertaining to the areas of Crop Production, Post Harvest Technologies, Livestock, Poultry, Fisheries, Agro enterprises, Best practices, Policies and schemes, Agri Credit, Agri Insurance, Agri directory, Discussion Forum

45.7 Agrisnet

Government of India launched a Central Sector Scheme titled, 'Strengthening/Prompting Agricultural Informatics & Communications' of which one component in AGRISNET operational in all the States. Telangana State has also established AGRISNET project with the following goals

- Providing information to farmers on inputs so that they get alerted on quantity on the input and its supply chain.
- Providing information of Govt .schemes implemented in the state.
- Providing recommendations based on soil test results and reduce indiscriminate use of fertilizers soil health
- Updates farmers in latest technology in Agriculture sector for Productivity and quality farmers.

45.7.1 Mobile Telephones

- Mobile phones can be defined as Basic (with talk and text), Featured (with talk, text and some other applications) or Smart – As above plus they can access the Internet, Mobile phones have emerged as a widely available option; providing multiple opportunities for improving communication.
- Mobile phones can be used for
- Voice to voice – call and ask specific questions (e.g., market prices, input availability, talk back radio, call centres)
- Voice to machine – call and get an automated response
- Text to pre-packaged response
- Text to call center for individualized text or recorded voice response
- Video screens

- Accessing the internet
- Mobile apps (e.g., Diagnostic Tools, Nutrient Managers)
- There are several examples from India and elsewhere as how mobile phones could be used. Some of them include - Banking: m-pesa; IFFCO Kisan Sanchar Limited (IKSL), Market information: Esoko, Technical information; e-Krishak; iCow; mKisan

Ideally speaking mobile phones can eliminate certain bottlenecks of language and illiteracy that exist in farmers communication. However mobile phone coverage can be limited, costs can be high, the amount of information that can be downloaded can be limited, screen size can limit usefulness of video, some applications require availability of experts or expert systems which may not be the case. Nevertheless novel and exciting business prospects are evolving, boding well for agricultural extension.

45.7.2 Social Media

Face book, Twitter and Whatsapp are all examples of Social media which can engage a large number of people particularly the young farmers and stakeholders in problem solving in agriculture. Krishi India, Google baraza are some examples.

45.8 ICT Initiatives at Professor Jayashankar Telangana State Agriculture University (PJTSAU)

45.8.1 Electronic Wing

Professor Jayashankar Telangana State Agricultural University, fore front in the "Service of the Farmers" is always striving to deliver the latest technologies developed in the University to the farming community through different communication channels. Realizing the importance of the electronic media as the most powerful tool to reach a large audience in the most effective manner in the quickest possible time with maximum accuracy, PJTSAU through its ElectronicWing, the first of its kind in the country caters to the Mass Communication needs relevant to Electronic Media that includes Doordarshan and private TV channels telecasting agricultural programmes for the state. It is also involved in the production of Video capsule programmes, Digital Video discs, Quickies, knowledge material for Information Kiosks and Agricultural Portals. The electronic wing has developed interactive and descriptive DVDs (64) on various crops and technologies depicting seed to seed production processes. It also provides visual and technical support to as many as 14 TV channels for the agricultural programmes broadcast throughout the week. The 60 – 30 minute Phone in live programmes encouraging farmer queries to scientists on different TV channels, broadcast on different days of the week are also facilitated by the electronic wing.

45.8.2 Interactive Information Dissemination System (IIDS)

The IIDS model is in operation in select KVKs of PJTSAU which addresses the problems of farmers right from the choice of inputs in the farming system to marketing of the farm product in a location specific manner. Farmers are required to be registered with their farm and other details. The agriexpert provides personalized solutions based on the input such as smart phone supported images/videos provided by the farmers and his available farm profile.

45.8.3 KisanSms Portal

The farmers portal of the Department of Agriculture and Cooperation, Government of India is a platform for farmers to seek any information related to Agriculture. Detailed information on farmers insurance, agriculture storage, crops, extension activities, seeds, pesticides, farm machinery, market prices, package of practices, programmes, welfare schemes etc., is provided to farmers through this portal. It is implemented by Zonal Project Directorate (ZPD) of ICAR as the nodal agency. All the KVKs under ZPD are to register 1000 farmers with the portal and they will be provided timely alerts and Agricultural Messages by the nodal agency. PJTSAU is a part of this programme, with its six KVKs registering 6000 farmers form Telangana and providing need based information.

45.8.4 Smart Information Kiosks

The Smart Information Kiosks is a net based information house maintained by all the KVKs and DAATTCs of PJTSAU, which offers information in text form, pictures and videos related to Agriculture and allied fields. Farmers visiting these centres can access information on their own on a touch screen. It also provides a platform to share and upload information among all the centres.

45.8.5 Telangana Yuva Raithu Sagubadi

Over the past few years, rural youth have been shying away from agriculture and globally there is an increasing interest in finding ways of engaging youth in agriculture Young farmers can play an important role in ensuring food security if they are encouraged to involve in farming by providing incentives and addressing the many challenges they face. A step in this direction is the *Telangana Yuva raithu sagubadi* programme launched by PJTSAU in 2014 for improving the entrepreneurial skills of practicing young farmers. A *Telangana Yuva raithu sagubadi* **Whatsapp** group consisting of the participants of this programme has also been formed to facilitate sharing of information and innovative ideas.

45.8.6 Aakasavani

University is actively participating in all agricultural programmes of Aakasavani. Radio scripts for*vyavasayavaartalu*aired for 10 minutes every day from 6.50-7.00PM. a thirty minute *Vyavasayapatasala*programmeaired every week is also being co-ordinated by electronic wing.

Morning	6:30 – 6:45 :	PolamPanulu
Afternoon	1:20 – 1:30 :	Paadipantalu
Evening	6:40-6:50 :	Gramaseemalu (Every Sunday)
Evening	6:50 – 7:00 :	Vyavasayasuchanalu
Night	7:15 – 7:45 :	Illu, Vakili
Monday:	7.15 -7.45 PM-	VyavasayaPatashala
Thursday:	7.15-7.45 PM-	Phone in live programme

45.8.7 ChenuKaburlu

Christened recently by PJTSAU has a noble intention of linking the innovations of the to the farming community of Telangana, utilizing the students' creativity. This Radio-Club formed by the participation of the students of faculty of Agriculture, Agricultural Engineering and Home Science of Telangana State proposes to create, promote and broadcast the factual information on quality life and the best farming practices for farming families through radio communication. The uniqueness of this programme is the amalgamation of culture and science. ChenuKaburluit upgrades students' knowledge on the subject, enhances their communication-skill, builds confidence, inculcates commitment towards interaction with clientele and also updates the farming community on latest farming advances in tune witheir own cultural identity.

45.9 Conclusion

With ICTs many unreached parts of the country have been reached and many milestones have been achieved but there is still a long way to go to harness the full potential of ICTs in agricultural development. Agricultural extension workers should be empowered to become knowledge workers through ICTs to enable the bottom-up, demand driven technology generation, assessment and transfer. ICT should be used to harness knowledge for strong organizations of small producers where strategic planning can be used to provide members with least-cost inputs, better storage facilities, improved transportation links and collective negotiations with buyers for enhancing overall farm productivity and living standards.

References

1. Anderson. J. and G. Feder 2004. Agricultural extension: good intentions and hard realities. The World Bank Research Observer 19 (1): 41-60
2. Bhupal Raj G and Praveen Rao V 2015. Role of ICT in Agriculture and its use in Professor Jayashankar Telangana State Agricultural University for Transfer of Technology. The Best of India.1(2) 54-56
3. Ferroni M and Zhou Y 2012. Achievements and Challenges in Agricultural Extension in India. Global Journal of Emerging Market Economies. 4(3) 319-346
4. Glendenning C. J., Babu .S and Okyere K. A. 2010: Review of Agricultural Extension in India. Are Farmers' Information Needs Being Met ? IFPRI Discussion Paper 01048. Eastern and Southern Africa Regional Office. pp 35
5. Meera S N 2015. ICT as Great Levellers: An Overview of Agriculture, Gender and ICTs. Training Manual on Gender, Short course on Transformative Approaches in Gender Mainstreaming, Gender Budgeting and Women Empowerment in Agriculture, IIRR(June 8-7, 2015) 135-143
6. Sharma V.P and Murthy L 2015. Information and Communication Technologies (ICTs) in Agricultural Extension: Indian Initiatives. The Best of India.1(2) 59-60
7. Stefanie Kaegi 2015. The experiences of India's agricultural extension system in reaching a large number of farmers with rural advisory services. Background paper to the SDC face-to-face workshop "Reaching the Millions!" in Hanoi, Vietnam, March 2015.

Chapter 46

Technology Assessment and Refinement (TAR) in Groundnut - The Concept and Practice

K. Mohan
Scientist (Agricultural Extension), RARS, Tirupati.

46.1 Introduction

Technology development is a non-going response of scientific knowledge to changing requirements of society. It focuses on a target group keeping in mind the resource base, socio-cultural factors, and government policies to exploit the available opportunities and match scientific knowledge with requirements.

A good development process should be flexible and offer options to the target group for successful adaptation and adoption of any new technology. Generation of new scientific knowledge is essential to upgrade the existing technology, so a strong and well-focused research program is a prerequisite for any technology development. This research program may involve both on-station and off-station research which complement each other and help in developing the most appropriate technology Developments in scientific knowledge do not necessarily address social, cultural, and economic issues so it is left to the technology development process (putting scientific knowledge to practical use by the society) to address these issues and improve the quality of life and the environment. Verification of the technology in the targeted system is an important part of the technology development process. (Nigam and Gowda, 1996).

TAR is an prime activity of the extension system in India which links the research extension and farmer bridged by the technology. During 2014-15, in India, 2,174 crop technological interventions were assessed across 4159 locations by laying out 23,568 trials on the farmers' fields on various crops under different thematic areas and a total of 143 farm-women specific appropriate technological interventions were assessed at 225 locations through 1,848 trials under various thematic areas. As far as refinement is concerned a total of 1897 trials were conducted at 309 locations to refine 253 technologies under different thematic areas of crop production and 17 women specific income generation technologies were also refined by conducting 158 trials in 20 locations. (ICAR, 2015)

Before understanding the TAR and its applications it is important to comprehend some of the concepts underlying in TAR process. These concepts will enable for better implementation of TAR process by the extension system.

46.2 Concept of Technology Development, Assessment and Refinement

Technology

Technology is any systematic knowledge and action applicable to any recurrent activity. Technology involves application of science and knowledge to practical use, which enable man to live more comfortably. The Merriam-Webster dictionary offers a definition of the term: "the practical application of knowledge especially in a particular area" and "a capability given by the practical application of knowledge".

Technology can be most broadly defined as the entities, both material and immaterial, created by the application of mental and physical effort in order to achieve some value. In agriculture, the term technology often confuses practitioners. This is because agricultural technology is a complex blend of materials, processes and knowledge. Swanson (1997) has classified agricultural technologies into two major categories:

1. Material technology, where knowledge is embodied into a technological product; and
2. Knowledge based technology, such as the technical knowledge, management skills and other processes that farmers need for better farm management and livelihood support.

Scientists need to have clarity over the technologies which they are assessing and refining in response to a specific problem in a specific microlocation. For example, a SMS may be assessing the efficacy of a particular management practice on a crop's yield or growth in the KVK district. Such management practices can be broadly classified as Knowledge based technology. Alternatively, all technological products tested and demonstrated under OFT and FLD fall under material technology. Ex: Seeds, pesticides, fertilizer, farm machinery, irrigation systems etc.

Technology Development

Technology Development (also called technology innovation) is a process consisting of all the decisions and activities which a scientist does from recognition of a need/ problem with planning, testing, conducting research, verification, testing and dissemination for adoption. During the same time, some problems on the technology might get back to the scientist for solution thus resulting in refinement of the same. Thus, technology development is a continuous process. KVKs act as a crucial player in testing, adaptation and integration of proven technologies. The KVK scientists have to equip themselves for 'technology application' - a process which includes the above mentioned processes; thus contributing their part in the overall process of technology development.

Understanding Technology Development Process

Understanding technology development process in agriculture and its components is vital for success of scientists. Technology development basically constitutes seven processes. They are:

1. Technology generation
2. Technology testing
3. Technology adaptation
4. Technology integration
5. Technology demonstration
6. Technology dissemination
7. Technology adoption

(Items 2–4 are bracketed: Technology Assessment and Refinement (TAR); items 5–7 are bracketed: Front Line Demonstration (FLD))

Technology generation, the starting point of technology development process is mainly a function of research system. Testing, adaptation and integration processes constitute technology assessment and refinement which KVK system executes through OFTs. The feedback is passed over to research system. KVK system also involves in technology demonstration through FLDs. Feed-forward from successful OFTs and FLDs is communicated to the extension system for mass popularization in the district. Technology adoption; the final act, occurs among the members of client system i.e. farmers.

A conceptual model of technology development process in agriculture depicting the components and actors involved is explained below:

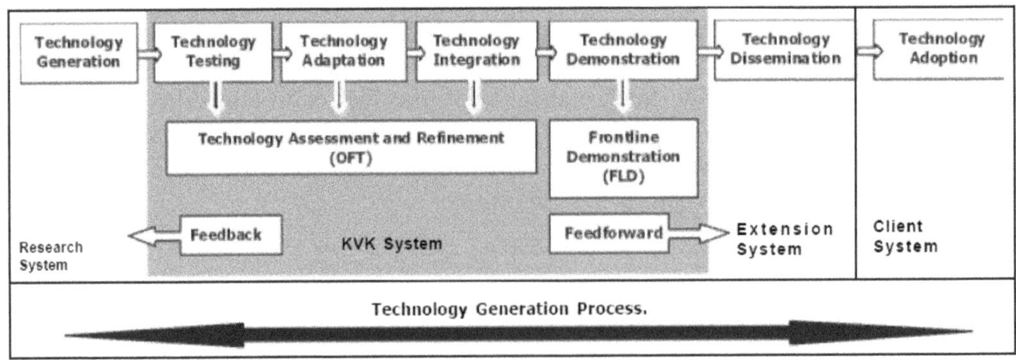

Fig 46.1 Conceptual Model of Technology Generation Process (Developed by Sajeev, M.V and V. Venkatasubramanian

Research system generates new technologies. In Indian agriculture, research system comprises of ICAR institutes, SAUs, departments like DBT, DST, other Science and Technology Institutions and Commodity boards. NGOs, Corporate and farmer innovators also contribute to technology generation.

Extension system comprises of State departments of agriculture, animal husbandry and veterinary, fisheries, sericulture etc. SAUs, ICAR institutes, commodity boards, NGOs and Corporate sector also contribute to extension system.

Earlier, due to the primary focus on vocational training, KVKs were categorized under extension system itself. But today, with mandates being focused on assessment, refinement and demonstration of frontier technologies, the KVK system positions itself clearly between the research and extension systems thus acting both as a feedback and feed forward mechanism. Client system comprises of the ultimate end-user i.e. the farmer. Although KVK system does assessment, refinement and demonstration of new technologies as part of technology development process, some technologies get refined or rejected even in the last stage at farmer fields. Hence, client system even though being the final actor in technology development process, plays the ultimate decisive role.

Technology Management

Technology management can be defined as the integrated planning, design, optimization, operation and control of technological products, processes and services. A better definition for KVK scientists would be "the management of the use of any technology for farmer advantage." The KVK role under technology management is very huge where-in it selects latest technologies, tests them for suitability in different micro-locations of the district and demonstrates the proven ones to farmers and extension system.

Technology Fatigue

Linkages between the laboratory and farmer fields have weakened and extension services often have little to extend by way of specific information and advice on the basis of location, time and farming system. Good quality seeds at affordable prices are in short supply and spurious pesticides and biofertilizers are being sold in the absence of effective quality control systems. Farmers have no way of getting proactive advice on land use, based on meteorological and marketing factors. No wonder the prevailing gap between potential and actual yields, even with technologies currently available, is very wide (NCF, 2007). In case of KVKs, it was found utilizing old and obsolete technologies for OFTs, FLDs and training programmes thus resulting in poor feed-forward to the extension system. A knowledge deficit as mentioned above coupled with the usage of obsolete technologies and package of practices together leads to a situation called 'technology fatigue'. Indian agriculture, particularly agriculture by resource poor farmers in rural areas is now bearing the brunt of technology fatigue. The KVK role lies in providing timely supply of proven technologies specific to various micro-locations of the district thus alleviating the technology fatigue existing in the district.

Technology Gap

Technology Gap is the gap between the level of recommendation and the extent of adoption (against recommendations). Technology gaps are a major source of concern for extension system. The success of traditional transfer of technology (TOT) models were mainly evaluated on the basis of the extent of narrowing down in technology gaps achieved by them. KVK system being primarily focused on assessment, refinement and demonstration of new technologies, its role lies in feeding proven technologies to the main extension system. Thus, the primary focus of KVK should not be mistaken as reduction of existing technology gaps. Rather, they are meant at alleviating "technology fatigue" by providing timely supply of proven technologies specific to various micro-locations of the district. Alleviation of technology fatigue is accomplished through processes of technology and methodology backstopping.

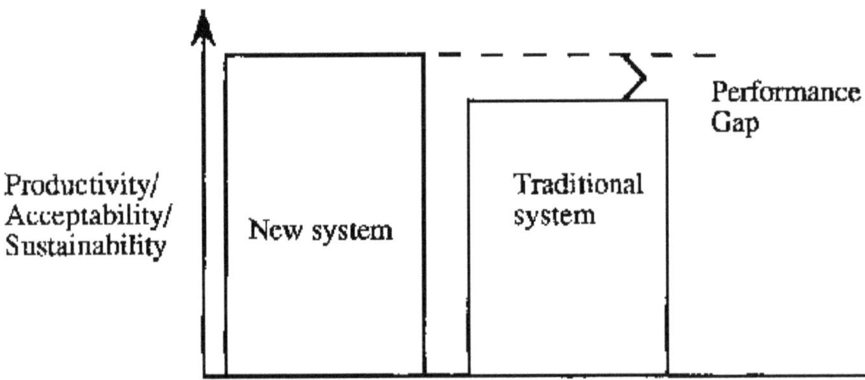

46.3 Role of Tar in Research - Extension - Farmer Relationship

The role of OFR as a link between research and extension is much simpler when dealing with single component technologies. For such technologies, the effects show up quickly, and farmers' interest and participation can be expected to rise if early results are promising. An active involvement of the farmer in experimental OFR on such technologies can lead directly to extension work without necessarily passing through the developmental OFR process. This is in line with traditional thinking on the research/extension linkage. However, the process will depend on the existence of an effective extension service which can link up with the research group for the dissemination of the technology.

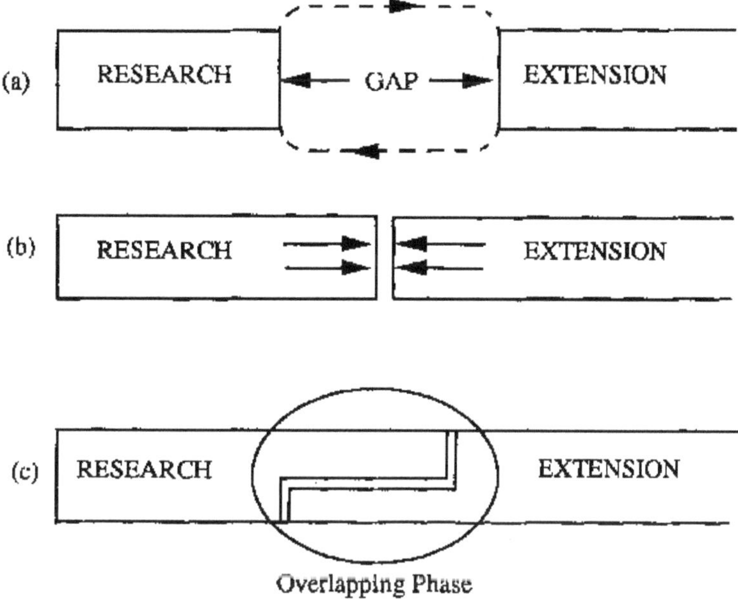

(a) Illustrates the gap that usually exists between research and extension in developing countries.

(b) Illustrates a back-to-back research/extension linkage. Suggested for simple technologies such as fertilizer-use.

(c) Illustrates the overlapping phase between research and extension in which the two units operate together in a developmental OFR activity. Suggested for composite technologies such as alley farming.

Illustration: *Overview of linkage between research and extension*

The research-extension linkage is more complex in the case of composite technologies. For OFR to lead to extensive adoption and extension, farmer involvement will have to be developed to the point where farmers have management control over the system under trial. In view of the complexity, flexibility, and long waiting time for benefits to show up, composite technologies would need to go through the three stages of developmental OFR.

A feed-back mechanism exists among all three processes such that problems identified during developmental OFR can receive further research attention in either on-station or experimental OFR trials.

For composite technologies, the interface between research and extension is more likely to be provided through a developmental OFR approach rather than through experimental OFR. Results from the experimental OFR trials will be worked into the later stages of the developmental OFR process (i.e., the pilot project stage). The direct involvement of extension agents during the developmental phase will facilitate this process. Meanwhile the involvement of the researcher will ensure that problems and opportunities arising as a result of farmer modifications and adaptation are taken into account in further development and fine-tuning of the system.

Experimental OFR

This is the more commonly known and practiced of the two types of OFR. It is performed for big-physical, technical, and economic assessment of alternative systems or treatments within the framework of standard experimental designs. Bio-physical assessment aims at determining the system's biological and physical yield and productivity, while economic assessment inquires into the availability of labor, cash, and other resources for meeting the projected needs of the alternative system, and looks into the level and dependability of profit.

Experimental OFR trials emanate directly from on-station research. Their structure and design are very similar to those used on-station. Generally, however, on-farm experimentation is kept as simple as possible to ensure effective farmer understanding of issues and meaningful involvement and contribution. Depending on the nature of farmer/ researcher involvement in the trials, experimental OFR may be further classified into three different types, namely:

- researcher-managed trials,
- researcher/farmer-managed trials,
- farmer-managed trials.

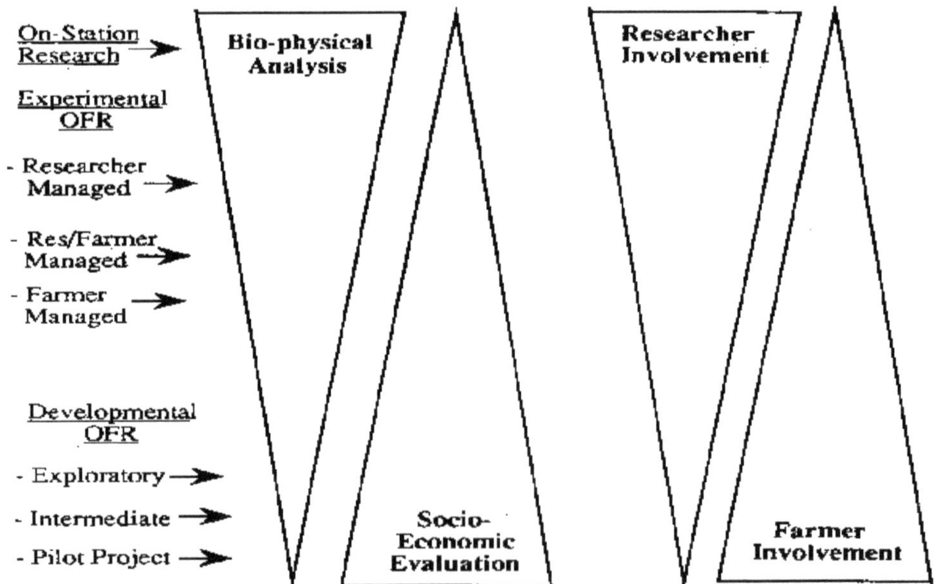

***Illustration:** Research objectives and levels of farmer involvement vary in the different types of on-farm research*

Technology Backstopping

Backstopping refers to any precaution taken against an emergency condition. Accordingly, technology backstopping can be defined as any technology precaution taken to combat technology fatigue. In simple terms, technology backstopping is the process of making available ready to use technologies for farm families through assessment, refinement and demonstration processes in order to combat the existing/forecasted technology fatigue.

Methodology Backstopping

This is a process almost similar to technology backstopping but differs with respect to the kind of technology solution offered. Instead of material technology, methodology backstopping aims at assessment, refinement and demonstration of knowledge based technologies often referred to as methodologies/package of practices. It provides detailed procedures to carry out the technology application functions by the extension personnel in the field. It includes methodologies for conducting OFT, which includes TAR, demonstrations, training, conducting surveys, impact assessment and evaluation etc.

Typology of Technology Passage through Extension System

Extension system has successfully established itself between the research and extension systems. Technology development process as explained earlier, invariably has assessment, refinement and demonstration components. Hence, there is a passage of technologies through various stages in a extension system. This passage doesn't follow a uniform pattern. For example, a technology may go through assessment stage and demonstration stage but not through refinement stage. Based on the stages passed, five different typologies of technology passage can be identified for an extension system. A proper understanding of these typologies will help extension personnel in deciding whether a particular technology will go for OFT and FLDs or both. The typologies identified by us are:

1. **Source - Demonstration**

   In this type the technology from any source/provider directly goes to demonstration by an extension system. This happens when the extension unit is completely sure that the technology is fully suited for the district and can go directly for FLD. Here, the technology doesn't pass through assessment and refinement stages.

2. **Source - Assessment**

   In this type the technology from any source/provider goes for assessment by extension system. This happens when the extension unit is not sure that the technology is fully suited for different micro-locations of the district. Here the technology fails at assessment stage itself and hence doesn't move to refinement or demonstration stages.

3. **Source - Assessment - Refinement**

   This type is a variation of type 2. Here, the extension system is not sure that the technology is fully suited for different micro-locations of the district. The technology goes for and succeeds in assessment but needs refinement and hence moves to refinement stage. Here, the technology fails in refinement stage and hence doesn't move to demonstration stage.

4. **Source - Assessment - Demonstration**

   This type follows type 3. This happens when the extension unit becomes sure that the technology is fully suited for different micro-locations of the district. The technology fully succeeds in assessment and hence moves to demonstration stage. Here, the technology doesn't require refinement and hence move to demonstration stage.

Source - Assessment - Refinement - Demonstration

This type also follows type 3. This happens when the extension system becomes sure that the technology is fully suited for different micro-locations of the district. The technology succeeds in assessment and refinement and moves to demonstration stage. Here, the technology is successfully refined by extension unit and taken to demonstration stage *i.e.*, FLD.

FLDs are supposed to be taken up on proven technologies only. Hence, it makes obvious that once demonstrated it will go to the extension system and client system. Rarely FLDs may fail thus preventing the technology passage. But extension units are not supposed to demonstrate such technologies which are not fully proven. The failure of FLD can be due to some extraneous factors rather than technological factors.

46.4 Conducting Technology Assessment and Refinement

Technology Assessment and Refinement (TAR) refers to a set of procedures whose purpose is to develop recommendations for a particular agro-climatic situation/ location through assessment and refinement of recently released technology through participatory approach. It refers to the process or a set of activities before taking up new scientific information for its dissemination in a new production system. *OFTs conducted by KVKs are based on this concept and thus distinguish it from agronomic and research trials.* As detailed earlier, the process of TAR has three components. They are technology testing, technology adaptation and technology integration. TAR should be:

- Site specific
- Holistic
- Farmer participatory

- Technical solution to existing problems
- Inter-disciplinary
- Interactive

This process involves Scientist-Farmer linkage in terms of:

- Sufficient understanding of the farming situations
- Adequate perception of farmers' circumstances and their needs
- The variability of conditions on the research status as compared to farmers' fields
- Problem orientation instead of disciplinary approach

Technology assessment in agriculture by KVKs is the study and evaluation of new technologies under different micro locations. It is based on the conviction that new discoveries by the researchers are relevant for the farming systems at large, and that technological progress can never be free of implications. Also, technology assessment recognizes the fact that scientists at research stations normally are not trained field level workers themselves and accordingly ought to be very careful while passing positive judgments on the field level implications of their own, or their organization's new findings or technologies. Considering the above factors, the ICAR has envisaged On Farm Trials (OFTs) through its vast network of 562 KVKs covering almost the entire geographical area of the country.

On Farm Trials (OFT): Concept

An On-Farm Trial aims at testing a new technology or an idea in farmer's fields, under farmers' conditions and management, by using farmer's own practice as control. It should help to develop innovations consistent with farmer's circumstances, compatible with the actual farming system and corresponding to farmer's goals and preferences. On-farm-trial is not identical to a demonstration plot, which aims at showing farmers a technology of which researchers and extension agents are sure that it works in the area. *It should be noted that OFTs are strictly to be conducted in collaborating farmer fields and not in extension unit land.*

Stakeholders of On-farm Trials

There are various stakeholders in an on-farm trial. Understanding them and their roles can help extension system to develop better OFTs. The stakeholders are:

1. The farmers who are the clients for the out-coming results,
2. The SMS who should help the farmers to overcome their problems and improve their economical situation. On farm trials can give them valuable information in this respect,
3. The Scientist who needs to apply promising on-station results under farmers' conditions before releasing the technology to the extension service,

4. The extension system and government itself, who is interested in seeing an efficient and participatory technology development model evolving, since most top-down approaches have failed miserably.

Typology of on Farm Trials

We can distinguish three types of OFTs in India according to the stakeholder who is going to take the lead role:

- ***Type 1, Research driven:*** Research system designed and managed (with the assistance of extension)
- ***Type 2, Extension driven:*** Extension System or KVK system designed and managed by farmers
- ***Type 3, Farmer driven:*** Farmers designed and managed, with the assistance of Extension system/KVK system.

Broad Steps in Conducting OFTs

The process of On Farm Trial broadly has five steps. They are:

1. Diagnosis
2. Planning
3. Conducting
4. Assessment and
5. Extrapolation/Diffusion.

Extension Centres identify the farming systems/system interactions through the following sources/methods:

- Secondary data/ sources.
- PRA methods to be exercised by SMS in collaboration with local people.
- Interview with local officials/ opinion leaders.
- Informal farm surveys.
- Farmers' interview using structured/ semi-structured schedules and
- Field observations.

46.5 Participatory Technology Development (PTD)

Extension centres aiming at farmer driven OFTs (Type III) which are a step above the present extension driven ones have to resort to Participatory Technology Development (PTD) process. It is an approach, which involves encouraging farmers and other stakeholders to engage in experiments in their own fields so that they can learn, adopt new technologies and spread them to other farmers. In its purest form PTD is a process in which:

- problems for which solutions have been sought are identified and prioritized by farmers,
- alternative solutions to be tested are defined by farmers
- design of experiment is decided by farmers,
- the implementation is done by farmers,
- monitoring is decided and executed by farmers and
- evaluation is done by farmers

A growing number of documented examples in recent years reveal that PTD has got accepted as a research approach in agriculture, animal husbandry and Natural Resource Management (NRM). It has been recognized that research is effective in improving farmers' livelihoods if farmers play a vital role in the process.

The objectives of PTD approach are to:

- Empower clients to develop and use livestock technologies
- Develop appropriate livestock technologies suited to the farm holdings
- Empower stakeholders, especially the marginalized ones, on their own decision making so that their capacity to make effective demands on research and extension organizations is strengthened and
- To improve the functional efficiency of formal research

The steps in PTD process involves:

1. Getting started in field
2. Understanding problems and potentials
3. Looking for readiness to try
4. Designing experiments
5. Trying out
6. Evaluation of the experiments
7. Sharing results with others
8. Sustaining PTD process and
9. Scaling up/ Phasing out

Obstacles to PTD as viewed by farmers are as follows:

- ➢ Research would take too much time
- ➢ Too great a risk in their production systems
- ➢ Fear that they may be punished, if the technology fails

- ➢ Lack of intention to experiment an option, which they consider risky, insignificant, or having delayed benefits
- ➢ Lack of communication link with scientists and
- ➢ A general feeling that their problems are unimportant to scientists

Obstacles to PTD process from Scientists' view point is as follows:

- ➢ Results could be spoiled by mismanagement of factors outside the researchers' control
- ➢ Failure of technologies in farmers' fields may be construed as inadequacies of scientists
- ➢ Risk of exposing farmers to too many uncertainties
- ➢ Lack of skills and experience in participatory research and
- ➢ Lack of time, motivation and communication skills to approach farmers

Benefits of PTD process in research/extension include:

- Encourages farmer experimentation
- ITK of farmers can be gainfully tapped
- Farmers' knowledge and creativity are harnessed to develop appropriate technology
- Farmers are motivated when their views are respected
- Technologies are more rigorously tested under users' conditions.
- Technologies to suite diverse agro-ecological & socio-economic situations
- Technologies are more likely to be adopted
- Technologies are in users hands more rapidly
- Complements station-based research

The success of an OFT should not be confused with success of the technology tested. A negative result of a technology tested shows that the technology is not suited for the specific micro-location of the district. This finding also refers to the success of the trial. Some technologies may not need refinement thus qualifying directly for frontline demonstrations. Some may successfully undergo refinement and reach the demonstration stage while some technologies fail to get refined in the farmer field. The technologies which successfully come out of On Farm Trials are then recommended for Frontline Demonstrations (FLDs).

46.6 Evidences of Tar Programmes of Groundnut in Sourthern Andhra Pradesh

TAR in Groundnut is the play pivoted role in better application research generated technologies. Groundnut technologies perform to their potential well where in a definite farming systems which are suitable and in order to tap the potential the research generated technologies need to be assessed and suitable refinement of the technology need to be carried out in the given location.

Entire Southern Andhra Pradesh comprises two Agro-climatic zone of Andhra Pradesh viz., Southern Zone (Chittoor, YSR and SPSR Nellore districts) and Scares Rainfall zone (Ananthapuramu and Kurnool districts) are quite suitable of cultivation of Groundnut. The Regional Agricultural Research Station, Tirupati and Regional Agricultural Research Station, Nandyal along with the Agricultural Research Station, Kadiri, Agricultural Research Station, Ananthapuramu, Agricultural Research Station, Utukur Kadapa are predominantly focusing on development of groundnut production technologies including new varieties, crop protection technologies, best suited agronomic practices and post harvest management activities. Many technologies generated in the research stations are being tested in the KVKs and District Agricultural Advisory and Transfer of Technology Centre. TAR trials by the extension centres in the zone not only helping to improve the technologies to upscale their adoption but also is serving as an effective feedback system to the research stations for take up the research programmes based on the refined technologies addressing the problems comprehensively.

Thematic wise TAR activities in Groundnut

1. Thematic area: Crop Improvement

TAR 1: Assessment of drought tolerant groundnut variety, Dharani (FLD):

TAR was conducted in KVK, Kalikiri during the year 2012-14 to assess the drought tolerant groundnut new variety Dharani compared with Narayani variety in 2 ha area in 5 locations.

Table 46.1 *Results of assessment of drought tolerant groundnut variety, Dharani*

Variety	Yield (kg/ha)				Cost of cultivation (Rs/ha)	Gross returns (Rs/ha)	Net returns (Rs/ha)	B:C ratio	% increase
	2012	2013	2014	Average					
Dharani	810	1550	925	**1095**	27500	38550	11050	1.4	6.3
Narayani	775	1465	850	**1030**	27500	33700	6200	1.2	

(Source: Annual Report 2014-15, KVK, Kalikiri)

TAR 2: Varietal evaluation of Groundnut with Dharani and K-6 against local variety (TMV-2)

Groundnut variety Dharani was evaluated in the OFT programmes of KVK, Kalyandurgam during the year 2012-14 in rainfed alfisols farming situation in 2 ha are in 5 locations.

Table 46.2 Results of Varietal evaluation of Groundnut with Dharani

Assessment of yield	Yield (kg/ha) (Rs. 40/kg)			Haulm yield (Kg/ha) (Rs. 2/kg)		
	Dharani	K-6	Local	Dharani	K-6	Local
Average yields	615	569	533	812	765	608
Percent increase over control %	15%	7%		33%	26%	
Additional returns	3688	1754				
B:C ratio	1:1.17	1:1.08	1:1.07			

(Source: Annual Report 2014-15, KVK, Kalyandurgam)

Farmers reaction:

- Farmers are satisfied with the yields of Groundnut Dharani variety over K-6 and local variety.
- More no. of good filled pods was observed in Dharani variety (8-10) than K-6 (7) and local variety (TMV -2) (6-8).
- Higher shelling percentage was observed in K-6 (74%) and Dharani (74%) than local variety (TMV-2 (65%)).
- Highest 100-kernel weight was observed in dharani variety (35.7g) than K-6 (30.5 g) and local variety (22.80 g)

TAR 3: Performance evaluation of Groundnut variety Dharani (TCGS-1043)

Assessment groundnut variety Dharani was evaluated in the OFT programmes of KVK, Kadapa during the year 2012-14 in rainfed alfisols farming situation in 2 ha are in 5 locations.

Table 46.3 Performance evaluation of Groundnut variety Dharani (TCGS-1043)

Technology Intervention	Average Yield (kg/ha)		Feedback
	Trial	Control	
Performance evaluation of Groundnut variety Dharani (TCGS-1043)	1705	1620	Dharani *var.* has recorded higher yield than K-6 *var.* even under severe dry spells up to 45 DAS. Manual picking of Dharani comparatively easier than K-6 because all pods come out from soil

(Source: Annual Report 2014-15, KVK, Kadapa)

2. Thematic Area: Crop Production

TAR 1: Weed Management with Herbicides in Groundnut

Assessment of weed control technology was conducted in KVK, Kadapa on rainfed black and red soils during Kharif 2012-14 compared to farmers practice and at two different stages of crop growth in 2 ha areas in 5 locations.

Table 46.4 *Results of Weed Management with Herbicides in Groundnut*

Treatments	Average Yield (kg/ha)	Net returns (Rs/ha)	B:C ratio
T1:Farmers practice	468.8	401	1:1
T2: Pre emergence application of pendimethalin + manual weeding at 30 DAS	491.7	3236	1.2:1
T3: Pre emergence application of Pendimethalin + Post emergence application of Quizalophas ethyl at 22 DAS	439.5	3864	1.3:1

(Source: Annual Report 2014-15, KVK, Kadapa)

3. Thematic area: Crop Protection:

TAR 1: Management of PBND in groundnut

Assessment of PBND IDM practices was conducted KVK, Kalikiri during 2012-14 in rainfed alfisols during kharif season in 2 ha area in 5 locations in the Western Chittoor district.

Table 46.5 *Results of Management of PBND in Groundnut*

Treatments	Percent damage by PBND	Yield (kg/ha)				Cost of cultivation (Rs/ha)	Gross returns (Rs/ha)	Net returns (Rs/ha)	B:C ratio
		2012	2013	2014	Average				
Farmers practice	8-9%	950	1150	750	950	25000	34625	9625	1.3
IDM practices includes*	5-6%	1125	1215	900	1075	25000	40375	15375	1.6

(Source: Annual Report 2014-15, KVK, Kalikiri)

***IDM Practices include:** 1. Seed treatment with imidacloprid @ 2 ml/kg and mancozeb @ 3g/kg of Seed. 2. Border cropping with sorghum or maize or bajra, 3. Inter cropping of groundnut + redgram at 7:1 ratio, 4. Spraying of monocrotophos @1.6 ml/lit at 25-30 DAS.

TAR 2: Assessment of seed treatment chemicals in groundnut

Efficacy of seed treatment chemicals in groundnut was assessed in KVK, Kalyandurg during 2012-14 in rainfed alfisols during kharif season in South western mandals of Ananthapuramu district.

Table 46.6 *Assessment of seed treatment chemicals in groundnut*

Details	Duration (Days)		Yield (kg/ha)			Haulm yield (kg/ha)			Collar rot incidence percentage		
	Demo (T1 & T2)	Control	T1	T2	FP	T1	T2	FP	T1	T2	FP
Average Treatments	105	105	643	616	573	982	819	790	12.5%	18%	20%
Percent Increase over control			12.2%	7.5%		24.3%	3.6%				

Table 46.6 ***Contd...***

Treatments	Additional cost on seed treatment chemicals	Additional returns	Initial plant population/m^2	Final plant population/m^2
T1 = Imidacloprid @ 2 ml + Tebuconazole @1g/kg seed	Rs. 828/ha	Rs. 1272/ha	32	28
T2= Imidacloprid @ 2 ml/kg + Mancozeb@ 3 g/kg seed	Rs. 742/ha	Rs. 548/ha	28	23
FP (Mancozeb @ 3 g/kg seed)	-	-	25	20

(Source: Annual Report 2014-15, KVK, Kalyandurgam)

Farmers reaction: Farmers realized that the importance of seed treatment in controlling the collar rot disease

TAR 3: Assessment of seed treatment chemicals in rainfed groundnut

Assessment of seed treatment chemicals in rainfed groundnut was tested in 05 locations and recorded 15-30% less collar rot disease incidence when compared to farmers practice. It is recorded an average yield of 800kg/ha which is 14.2% higher than farmers practice (700kg/ha) (Source: Annual Report 2014-15, KVK, Kalikiri).

TAR 4: Management of groundnut leaf miner by Delta 'T' traps

Delta 'T' traps (Mass trapping) were assessed in controlling leaf miner in Nellore district during 2012-14 in 5 locations which resulted in substantial reduction in incidence and damage of leaf miner in groundnut.

Table 46.7 *Results of management of groundnut leaf miner by Delta 'T' traps*

Technology Intervention	Yield (kg/ha)		Feedback
	Trial	Control	
Management of groundnut leaf miner by Delta 'T' traps	1702	1606	Installation of Delta 'T' traps 4 or 8 traps (mass trapping) helps in reducing the incidence of leaf miner which is a major problem and causing substantial yield loss to the farmers

(Source: Annual Report 2014-15, DAATTC, Nellore)

4. Thematic area: Farm Mechanization

TAR 1: Sowing of Groundnut with tractor drawn Anantha planter against farmers gorru practice

Drought mitigating farm mechanization technology assessment was conducted in KVK, Kalyandurgam where Anantha planter was used for saving time of sowing and utilize the available soil moisture immediately after receipt of rains. In this assessment the efficacy of Anantha planter was assessed compared to gorru practice of sowing South Western mandals of Ananthapuramu district in 2 ha area in 5 locations.

Table 46.8 Results of Sowing of Groundnut with tractor drawn Anantha planter against farmers gorru practice

Details	Duration (Days)		Final plant population/m^2		Cost of operation (Rs/ha)		Yield (kg/ha)	
	Anantha planter	Gorru sowing	Anantha planter	Gorru sowing	Anantha planter	Gorru sowing	Anantha planter	Gorru sowing
Average	105	105	30	24	600	1500	563	507
% increase							11%	
Additional returns							2,128	

Farmers feedback: (i) Farmers realized that there is 11 % increase in yield (ii) Rs. 900 per ha is saved on sowing operation (iii) 25 kg of seed per hectare was saved with Anantha seed drill

Conclusion

Generation of technologies is based on the identified problems and to address the same in a holistic manner. It is imperative to assess the applicability, performance, potential and suitability of the generated technologies in different agro-climatic situations and carrying out necessary refinement that suits best to the given location. Also it is TAR activities acts a connecting link between the research -extension -farmers. Further, the conduced TAR trials in Groundnut suggest that the technologies assessed before mass demonstrations for wide spread adoption resulted promising benefits to the farmers.

Suggested Reading

Atta-Krah, A.N., and P.A. Francis 1987. The role of on-farm trials in the evaluation of composite technologies: The case of alley farming in southern Nigeria. Agricultural Systems 23 (2): 133-152.

Merrill-Sands, D., P. Ewell, S. Biggs and J. McAllister 1989. Issues in institutionalizing on-farm client-oriented research: A review of experiences from nine National Agricultural Research Systems. Quarterly Journal of International Agriculture 28 (3/4): 279-300.

Venkatasuramanian, V, Sanjeev M V and Singha A KConcepts, 2003 Approaches and Methodologies for Technology Application and Transfer. Chapter III. Indian Council of Agricultural Research, New Delhi. Available at http://icarzcu3.gov.in/book_cover/Concepts 2ndEdn/4Chapter%203.pdf

References

1. *Annual Report 2014-15 District Agricultural Advisory and Transfer of Technology Centres. ANGRAU. Kadapa*

2. *Annual Report 2014-15 District Agricultural Advisory and Transfer of Technology Centres. ANGRAU. Nellore*
3. *Annual Report 2014-15 District Agricultural Advisory and Transfer of Technology Centres. ANGRAU. Chittoor*
4. *Annual Report 2014-15 Krishi Vigyan Kendra, Kadapa*
5. *Annual Report 2014-15 Krishi Vigyan Kendra, Kalikiri*
6. *Annual Report 2014-15 Krishi Vigyan Kendra, Kalyandurgam*
7. *Annual Report 2014-15 Krishi Vigyan Kendra, Nellore*
8. *Atta-Krah, A.N., and P.A. Francis 1987. The role of on-farm trials in the evaluation of composite technologies: The case of alley farming in southern Nigeria. Agricultural Systems 23 (2): 133-152*
9. *Francis, P.A., and A.N. Atta-Krah. 1989. Sociological and ecological factors in technology adoption: Fodder trees in southeast Nigeria. Experimental Agric. 25: 1-10.*
10. *ICAR 2007-08 Annual Report 2007-08. Indian Council of Agricultural Research, New Delhi. http://www.icar.org.in/files/reports/icar-dare-annual-reports/2007-08/Technology%20 Assessment.pdf*
11. *ICAR 2015 Annual Report 2014-15. Indian Council of Agricultural Research, New Delhi*
12. *Kang, B.T., L. Reynolds, and A.N. Atta-Krah. 1990. Alley Farming. Advances in Agronomy 43: 315-359*
13. *Nigam S. N. and Gowda C. L. L. 1996. Technology development and transfer in agriculture. (In En. Summary in Ch.) Pages 183-193 in Achieving high groundnut yields: proceedings of an international workshop, 25-29 Aug 1995, Laixi City, Shandong, China (Renard, C., Gowda, C.L.L., Nigam, S.N., and Johansen, C. eds.). Patancheru 502 324, Andhra Pradesh, India: International Crops Research Institute for the Semi-Arid Tropics*
14. *Venkatasuramanian, V, Sanjeev M V and Singha A K 2003. Concepts, Approaches and Methodologies for Technology Application and Transfer. Chapter III. Indian Council of Agricultural Research, New Delhi. Available at http://icarzcu3.gov.in/book_cover/Concepts 2ndEdn/4Chapter%203.pdf*

Chapter 47

Planting Density -Impact on Groundnut Production

Dr. Mallu Srinivasa Reddy, Dr. P.V.M. Reddy, G. K. Reddy, P. M. Reddy and P. Bala Hussain Reddy

ANGRAU, A.P.

47.1 INTRODUCTION

Groundnut (*Arachis hypogaea* L.) is an important food legume and an oil seed in the world and presently grown in about 90 countries over an area of 26.4 M ha worldwide with a total production of 38.2 M t (FAO, 2010-11). Globally India ranks first in area and second in production of groundnut with an area of 6.16 M ha with a total production of 7.17 M t and constitutes about 50 per cent of total oilseed production (CMIE, 2009-10 and Agrostats, 2009-10). However, in India the productivity of groundnut is low (1164 kg ha^{-1}) and its productivity in Andhra Pradesh (876 kg ha^{-1}) is much lower compared to national average. In spite of cultivation of many alternative oil seed crops, the area under groundnut production is static due to the demand for the kernel both for table and oil purpose and familiarity of its cultivation with the farming community.

47.2 DIFFERENT METHODS OF SOWING OF GROUNDNUT

47.2.1 Broadcasting

Literally means scattering the seeds. Broadcasting is done for many crops but it is **very unusual for groundnut due to high cost of seed.** Broadcasting is mostly followed for small sized to medium sized crops. This is the largest method of sowing followed in India since it is the easiest and cheapest and requires minimum labours.

To have optimum plant population in unit area certain rules should be followed.

(i) Only a skilled person should broadcast the seeds for uniform scattering

(ii) The ploughed field should be in a perfect condition to trigger germination The seeds are broadcasted in a narrow strip and the sowing is completed strip by strip. To ensure a good and uniform population, it is better to broadcast on either direction. This is called criss-cross sowing. If the seed is too small, it is mixed with sand to make a bulky one and for easy handling. In certain cases the person sowing will be beating the seeds against the basket for uniform scattering. After broadcasting the seeds are covered gently either using a country plough with a very shallow ploughing or some wooden planks (boards/levelers) are used to cover the surface. In some cases tree twigs or shrub branches are used. **If the seeds are large, levelers collect the seeds and leave in the other side**. Comb harrow is the best used one.

Disadvantages

1. All the seeds broadcasted do not have contact with the soil. 100% germination is not possible.
2. Seed rate is not sufficient. Enhanced seed rate required
3. Seeds cannot be placed in desired depth. Desired depth ensures perfect anchorage. Lodging (falling down) is common in broadcasting for tall growing crops but never for groundnut.

*Please remember broadcasting is almost never followed in groundnut. The above information is only subjective for other crops.

47.2.2 Dibbling

Line sowing: Inserting a seed through a hole at a desired depth and covering the hole. Dibbling on plain surface and ridges and furrows or beds and channels. This types of sowing is practiced only under suitable soil condition.

Dibbling is followed in groundnut under scarcity of seed especially in seed production of nuclear and breeder seed. Advantages of line sowing and dibbling

(i) Uniform population
(ii) Better germination
(iii) Reduced seed rate.

47.2.3 Sowing behind the Plough

Sowing behind the plough (line sowing) is done manually or mechanical means. Seeds are dropped in the furrows opened by the plough and the same is closed or covered when the next furrow is opened. The seeds are sown at uniform distance. Manual method is a laborious and time consuming process. Seeds like redgram, cowpea and **groundnut** are sown behind the country plough. Major sown crop is groundnut. Seeds are sown by mechanical means by ‘Gorus’ – seed drill. A seed drill has a plough share and hopper.

Seeds are placed on hopper. Different types of seed drill are available, e.g., simple Goru – Guntakas. Advantages – i) The seeds are placed at desired depth covered by iron planks, except very small, very large seeds most of the seeds can be sown, e.g. maize, sorghum, millets, sunflower, etc.

47.2.4 Drill Sowing (or) Drilling

Drilling is the practice of dropping seeds in a definite depth covered with soil and compacted. In this method, sowing implements are used for placing the seeds into the soil. Both animal drawn gorrus and power operated (seed drills) implements are available. Seeds are drilled continuously or at regular intervals in rows. In this method, depth of sowing can be maintained and fertilizer can also be applied simultaneously. It is possible to take up sowing of inter crops also. It requires more time, energy and cost, but maintains uniform population per unit area Seeds are placed at uniform depth, covered and compacted

47.3 Plant Population or Plant Density

Number of plants per unit area in the cropped field is the plant population.

47.3.1 Optimum Plant Population

Optimum plant population – It is the number of plants required to produce maximum output or biomass per unit area.

Any increase beyond this stage results in either no increase or reduction in biomass.

47.3.2 Crop Geometry

The arrangement of the plants in different rows and columns in an area to efficiently utilize the natural resources is called crop geometry. It is otherwise area occupied by a single plant e.g. Groundnut –30 cm × 10 cm for Spanish bunch for Kharif. This is very essential to utilize the resources like light, water, nutrient and space.

Different geometries are available for crop production. Importance of plant population/ crop geometry

1. Yield of any crop depends on final plant population
2. The plant population depends on germination percentage and survival rate in the field
3. Under rain fed conditions, high plant population will deplete the soil moisture before maturity, where as low plant population will leave the soil moisture unutilized.
4. When soil moisture and nutrients are not limited high plant population is necessary to utilize the other growth factors like solar radiation efficiently

5. Under low plant population individual plant yield will be more due to wide spacing.
6. Under high plant population individual plant yield will be low due to narrow spacing leading to competition between plants.
7. Yield per plant decreases gradually as plant population per unit area is increased, but yield per unit area increases upto certain level of population
8. That level of plant population is called as optimum population
9. So to get maximum yield per unit area, optimum plant population is necessary. So the optimum plant population for each crop should be identified.

47.3.3 Factors Affecting Plant Population

47.3.3.1 Size of the Plant

- The volume occupied by the plant at the time of flowering decides the spacing of the crop
- Plants of red gram, cotton, sugarcane etc., occupy larger volume of space in the field compared to rice, wheat, ragi
- **Even the varieties of the same crop differ in size of the plant. Groundnut is a classical example of variation in size of the plant with varieties. The variety TAG-24 is significantly of small size compared to Narayani and K6.**

47.3.3.2 Elasticity of the Plant

Variation in size of the plant between minimum size of the plant that can produce some economic yield to the maximum size of the plant that can reach under unlimited space and resources is the elasticity of the plant. The elasticity is due to tillering and branching habit of the plants. **In case of groundnut the elasticity is mainly due to production of branches and the length of the branches.**

47.3.3.3 Foraging Area or Soil Cover

- Should cover the soil as early as possible so as to intercept maximum sunlight
- Higher the intercepted radiation more will be the dry matter produced
- Close spaced crops intercept more Solar radiation than wide spaced crops Groundnut has great light interception and covers the soil with in 40 DAS.

47.3.3.4 Dry Matter Partitioning

Dry matter partitioning is related to amount of photosynthates translocated to the economical portions of the plant. It varies from crop to crop and also with in varieties.

47.3.4 Environmental Factors

47.3.4.1 Time of Sowing

- The crop is subjected to various weather conditions when sown at different periods.
- Among weather factors, day length and temperature influence the plant population. As low temperature retards growth, high plant population is required to cover the soil

47.3.4.2 Rainfall / Irrigation

- Plant population has to be less under rainfed than irrigated condition
- Under more plant densities, more water is lost through transpiration
- Under adequate rainfall / irrigation, high plant population is recommended.

47.3.4.3 Fertilizer Application

- Higher plant population is necessary to fully utilize higher level of nutrients in the soil to realize higher yield.
- Nutrient uptake increases with in plant population
- High population under low fertility soil leads to nutrient deficiency symptoms leading low yield

47.3.4.4 Seed Rate

- Quantity of seed sown/unit area, viability and establishment rate decides the plant population.
- The seed rate of groundnut for small seeded varieties varies from 125 kg ha^{-1} to 375 kg ha^{-1} adopted by the farmers for bold seeded varieties.

One of the main constraints for lower productivity of groundnut is non maintenance of optimum plant population. Determination of the optimum plant population density is a major agronomic goal for optimizing the yield. Sowing at precise seed rate results in optimizing the plant population density and also reduces the seed costs. Two general concepts are frequently used to explain the relationship between row spacing plant densities and yield.

First, maximum yield can be obtained only if the plant community produces enough leaf area to provide maximum insolation interception during reproductive growth.

Secondly equidistant spacing between plants will maximize yield because it minimizes inter plant competition. The optimum plant density and planting pattern at one site may not apply at other locations because of regional variations in weather and soil factors which mean further trails are needed at each site to validate general recommendations.

A field experiment was conducted during *rabi* season, 2012 to study the influence of different levels of plant densities on growth parameters and yield parameters of groundnut varieties. The experiment was conducted with three varieties and four levels of plant densities replicated thrice. Among the varieties tested TCGS-29 (Narayani) significantly produced tallest plants over K6 and TAG-24. With respect to LAI, except at 60 DAS the LAI of the varieties was not significant. However at 60 DAS significantly highest LAI was recorded with K6 and the lowest was recorded with TAG-24. The dry matter produced by TAG-24 was statistically lowest at 30 and 60 DAS. However the same variety produced significantly highest dry matter during harvest over the other varieties tested. For all the varieties tested, the plant height was most influenced by different plant densities. However the LAI and dry matter production significantly increased with increased population levels. With regard to pod production, haulm yield and harvest index all the varieties produced statistically equal yield. However variation in density of planting resulted in production of significantly lower pod yield with 33.3 plants m^{-2} and the rest of the treatments produced statistically equal yield. However superior haulm yield was recorded with 66.6 plants m^{-2}.

References

1. *Agrostats, 2009-10. http://www.agrostats.com/world-statistic/world-peanut.html*
2. *CMIE. 2009. Centre for monitoring Indian Economy.URL: http://www.CMIE.com*
3. *FAO. 2010. Food and Agricultural Organization.URL: http://www.fao.org/*

Chapter 48

New Paradigms in Improvement of Drought Tolerance in Crop Plants

CVCM Reddy, PVRM Reddy and P. Bala Hussain Reddy*

Department of Biotechnology, West Virginia State University, WV, USA

Drought stress has a tremendous effect on agriculture, thus negatively influencing human activities (Pareek *et al.*, 2010). During crop domestication, it is assumed that plants were selected on the basis of different traits, water limitation tolerance being unlikely one of them. More recently drought episodes have resulted in extensive famines in India as well as in other regions of the world; also producing famine, contagious diseases and deaths. According to the World Health Organization, drought season can be predicted; however, erratic precipitation is modulated by climate changes as well as by the increase in the global temperature. All these alterations are related directly to human activities (Hoerling and Kumar, 2003). Since it is predicted that the world precipitation pattern will change, either heavy rain or drought are similarly expected in the long term (Allen and Ingram, 2002). Therefore, drought will continue being a problem to be solved for crop production since water limitation causes stress in plants and limits the production of crops worldwide (Boyer,1982 and Zhang *et al.*, 2004).

The need for new alternatives for a sustainable agriculture such as drought tolerant plants will provide a practical solution to alleviate the problem of water limitation. Most of the alternatives are based either on accelerating the selection of natural varieties and inserting genes from other plants varieties or species with the capacity to provide drought tolerance (Pareek *et al.*, 2010). To achieve this goal, the biological base for drought tolerance needs to be understood.

48.2 Physiological Adaptations to Water Deficit

In nature, terrestrial plants have already developed a number of strategies, which are genetically encoded (Monneveux and Belhassen, 1996). One of the most striking adaptations consists in accumulating water to delay or escape from the stress. In contrast, drought tolerant plants, such as resurrection plants, are able to cope with the stress by diminishing their metabolic functions, which are resumed once water potential increases (Bartels and Sunkar, 2005). Other plants are able to maintain their biological functions at low water potentials, although with limited development (Ramanjulu and Bartels, 2002). Other strategies to limit the loss of water to air include abscisic acid mediated regulation of stomatal closure. Stomata, highly specialized cells involved in gas exchange, can account for a high water loss through leaf transpiration; the adaptation to drought consists in stomata closure during stress (Blum, 1996). This adaptation implies the accumulation of gases such as CO_2 which diminish photosynthesis (Bohnert and Sheveleva, 1998). The result is a general energy imbalance, in which a net flow of electrons occurs toward oxygen giving rise to oxygen reactive species and oxidative processes (Levine, 1999). Regarding root development, an interesting general adaptation is hygrotropism, in which roots detect a water gradient and redirect its growth towards it (Lambers*et al.,* 2000).

48.3 Biochemical Responses

Adaptive responses are based on complex changes to cope with stress mainly focused to maintain water potential in key tissues. At the cellular level, the cell membranes as well as the endomembrane system change dramatically their disposition and limit organelle function as well as integrity in response to the stress (Gigon *et al.,* 2004). The cell wall, considered as a barrier, also provides protection; although limited due to its deforming properties (Murphy and Ortega, 1995). When a cell suffers dehydration, the rigidity of the cell wall can provide mechanical protection; however, this organelle is permeable, thus allowing desiccation if stronger stress is applied (Verslues *et al.,* 2006). A common biochemical adaptation is osmotic adjustment which is the result of certain newly-synthesized metabolites (Bartels and Sunkar, 2005). These are hydrophilic highly soluble molecules, able to produce a salvation surface, that capture water molecules to be later available during water limitation. Examples of these osmolytes are aminoacids, glycine, betaine, sugars and sugar alcohols which are non toxic molecules at high concentrations, thus not interfering with cellular metabolism. Interestingly, other hyphothesis consider osmolytes may have additional functions besides turbidity maintenance, such as coping with oxidative stress by quelling reactive oxygen species (Bartels and Sunkar, 2005). Sugars accumulated under drought stress or likely to be stabilize membranes and prevent membrane fusion together with other macro molecules such as LEA proteins. Glycine betaine has been also described as osmo protectant, maintaining water equibrillium in plant organs (Chen and Murata, 2002).

48.4 Responses at the Molecular Level

Drought tolerance is considered as quantitative trait, involving the participation of a complex set of genes. The studies that have established this have been performed on model plants as well as in drought tolerant species (Montalvo – Hernandez *et al.*, 2008). When drought stress perceived by the plant, changes in the expression pattern have been monitored, ranging from genes whose products are involved in early response such as signal transduction, transcription and translation factors; to late response genes, such as water transport, osmotic balance, oxidative stress and damage repair. (Ramanjulu and Bartle, 2002). Adaptive responses are observed as a consequence of such changes, including early flowering, growth inhibition, among others (Bray, 2002).

48.5 Drought Sensing and Signal Transduction

Although, the sensor for drought stress is still unknown, it is generally accepted that the organ with such ability is the route. The plant regulator abscisic acid (ABA) is a key endogenous messenger mediating this stress response (Raghavendra *et al.*, 2010). Diverse hypotheses suggest that a redox imbalance could trigger the response to drought; Another possibility is change in cell wall membrane integrity (Kacperska, 2004). In Arabidopsis, two histidine kinase receptors (ATHK), belonging to the two component family have been described as induced in early drought stress (Urao *et al.*, 1999). Since drought and salinity induce high levels of ABA, together with major changes in gene expression and adaptive physiological responses (Christmann *et al.*, 2007), it is considered that ABA has a central role in early plant response to drought. The recent discovery of the ABA receptor RCARs/PYR1/PYLs, which inactivates type 2C protein phosphatases (ABI1 and ABI) (Kang *et al.*, 2010, provides insights on ABA dependent gene expression and ion channels. Other kinases identified as induced by stress (Conley *et al.*, 1997) are MAPKs (Jonak *et al.*, 1996), which are also induced by other abiotic factors, suggesting that reversible phosphorylation plays a key role in drought signaling. Calcium dependent protein kinases (CDPKs) are also induced when intra or extra cellular Ca^{2+} concentration increases, as a result of ABA mediated stomatal closure (Ramanjulu and Bartle, 2002).

48.6 Induced Genes at Transcriptional Level

Plant gene expression is controlled at different levels and a significant number of drought-induced genes appear to be controlled at the transcriptional level. Massive trancriptome analysis using RNA microarrays have established that ABA dependent and independent signal transduction pathways operate in drought stress plants.

48.7 Drought Induced Proteins

Translational control is another mechanism regulating plant responses to abiotic stress. Synthesized proteins have direct functions in membrane and protein protection. They involve the acquisition of water and ions, as well as their transportation and homeostasis maintenance in basal cell functions. Late embryogenesis abundant protein (LEA) family belongs to this group and was described as highly accumulated in plant embryos (Galau *et al.*, 1986). LEA proteins are expressed at basal levels and can be induced to high levels during osmotic and drought stress. Heat shock proteins (HSP), highly accumulated during stress and also know as molecular chaperones, are widely distributed in nature; they are involved in protein folding and assembly, as well as in the removal and disposal of non functional proteins (Wang *et al.*, 2004). HSP are induced by drought and *in vivo* evidence suggests that HSPs prevent protein thermal aggregation, thus facilitating the recovery of cell functions after abiotic stress. Turnover of damage macromolecules in stress is a basic process to maintain cell homeostasis.

48.8 Oxidative Stress

One of the main effects of the dehydration on plants is the production of reactive oxygen radicals (ROS) such as single oxygen, superoxide anion radicals, hydroxyl radicals and hydrogen peroxide (Bartels, 2001) Since ROS are mainly produced in chloroplast, the photosynthetic activity id compromised during stress; drought tolerance is unequivocally related to efficient antioxidant cellular processes. Factors participating in antioxidative activity are nonenzymatic and enzymatic (Bartels and Sunkar, 2005); among the former are vitamins C and E, glutathione, flavoineds, alkaloids, carotenoids, polyamines, etc. The study of molecular, biochemical and physiological mechanisms the plants employ to respond to drought stress has provided scientific knowledge for plant breeding. Despite drought tolerance is being a quantitative trait, key genes can significantly contribute to ameliorate the damage produced by water limitation. A number of genetically improved drought tolerant crops have been developed by different approaches, such as conventional breeding, marker-assisted breeding and transgenic approach. For modern agriculture, a combination of techniques will likely be needed to produce new varieties showing drought tolerance in the field.

48.9 Conventional Breeding

Conventional improvement to obtained new individual is based on their genetic variation and uses the selection to incorporate better characteristics into the progeny. For this purpose plants possessing desirable traits are selected and then crossed to exchange their genes, so that the offspring has new genetic arrangements. Individual plants are tested for the expression of the desirable characteristics and its maintenance in future plants generations (McCouch, 2004). In practice drought tolerance is selected together with plants plant productivity; non commercial varieties display drought tolerance are crossed with susceptible, higher-yield plants.

48.9.1 Marker assisted Breeding

Genetic improvement can be assisted by using recognizable tags linked to targets gens: these are known as molecular markers, which are based on polymorphisims that occur naturally in the DNA sequence. Different methods are employed to detect linked markers such as RFLPs (Restriction Fragments Length Polymorphism), SSRs (Simple Sequence Repeats), the SNPs (Single Nucleotide Polymorphism), among others (Van Berloo *et al.*, 2008). The genetic factors responsible for part of the phenotypic variation observed for quantitative characteristics are named QTLs (Quantitative Trait Loci). The use of molecular markers assisting the selection of new varieties has an enormous potential to accelerate this process.

48.9.2 Association Mapping

The phenotypic variation of drought tolerant important traits is influenced by multiple QTLs, their interaction, the environment and the QTL x environment interactions. Association mapping, also known as Linkage Disequilibrium (LD) mapping, is a relatively new and promising tool for dissecting complex traits. Association mapping in comparison to the traditional linkage mapping has major advantages due to increased mapping resolution through exploitation of historical and evolutionary recombination events at the population level (Nordborg and Tavare 2002). The prerequisites to perform association mapping include a dense genetic linkage map, passport information and phenotypic data, an understanding of population structure, and contrasting genotypes for beneficial traits (Kresovich *et al.*, 2002). The marker-trait association approach relies on the assumption that an allele responsible for a phenotype and the associated flanking markers are inherited as a block, and therefore neutral marker-based selection will be predictive of allelic content at critical genes determining favorable phenotype. Such marker-trait associations in a collection of plant genetic resources would allow the assessment of the genetic potential of specific genotypes prior to phenotypic evaluation and identification of superior drought tolerant trait alleles in germplasm collection (Gebhardt*et al.,* 2004).

48.10 Plant Breeding through Genetic Engineering

The capacity to stably insert a whole collection of drought related genes to plant genomes has opened amazing opportunities for crop improvement. Recombinant DNA technology as well as plant regeneration is now applicable to a significant number of crop species (Gosal *et al.*, 2009). Key genes observed as induced under abiotic stress in other plants or microorganisms have been employed to generate transgenic plants.

The generation of novel plant varieties displaying tolerance to abiotic stress is highly expected to cope with the unfavorable environmental challenges. Despite the current knowledge on the scientific bases of drought tolerance, more information is needed to understand and be able to manipulate such complex quantitative trait. As described above, traditional breeding have provided an extent number of tolerant varieties; however new approaches are also available to obtain improved plant. The genetic engineering has already produced plants with important tolerance and is likely to be the best approach to obtain superior lines bearing abiotic tolerance, among other agronomic ally important traits.

References

1. Allen, M.R. and W.J. Ingram, 2002. Constrains on future changes in climate and the hydrologic cycle. Nature, 419: 224-232.
2. Bartels, D. and R. Sunkar, 2005. Drought and salt tolerance in plants. Crit. Rev. Plant. Sci., 24: 23-58.
3. Bartels, D., 2001. Targeting detoxification pathways: An efficient approach to obtain plants with multiple stress tolerance. Trends Plant Sci., 6: 284-286.
4. Blum, A., 1996. Crop responses to drought and the interpretation of adaptation. Plant Growth Regul., 20: 135-148.
5. Bohnert, H.J. and E. Sheveleva, 1998. Plant stress adaptations-making metabolism move. Curr. Opinion Plant Biol., 1: 267-274.
6. Bray, E., 2002. Abscisic acid regulation of gene expression during water-deficit stress in the era of the Arabidopsis genome. Plant Cell Environ., 25: 153-161.
7. Chandler, J. and D. Bartels, 1999. Plant Desiccations. In: Plant Responses to Environmental Stresses: from Phytohormones to Genome Reorganization, Lerner, H.R. (Ed.). Marcel Dekker, New York, USA, pp: 575-590.
8. Chen, T.H.H. and N. Murata, 2002. Enhancement of tolerance of abiotic stress by metabolic engineering of betaines and other compatable solutes. Curr. Opin. Plant. Biol., 5: 250-257.
9. Christmann, A., E.W. Weiler, E. Steudle and E.Grill, 2007. A hydraulic signal root to shoot signaling of water shortage. Plant J., 52: 167-174.
10. Conley,T.R., R.E.Sharp and J.C. Walker,1997. Water deficit rapidly stimulates the activity of a protein kinase in the elongation zone of the maize primary root. Plant Physiol., 113:219-226.
11. Galau, G.W., D.W. Hughes and L. Dure, 1986. Abscisic acid induction of cloned cotton Late Embryogenesis Abundant (LEA) messenger RNAs. Plant Mol. Biol., 7: 155-170.
12. Gebhardt C, A. Ballvora, B. Walkemeie, P. Oberhagemann, K. Schuler, 2004. Assessing genetic potential in germplasm collections of crop plants by marker-trait association: a case study for potatoes with quantitative variation of resistance to late blight and maturity type. Mol Breed 13: 93–102.
13. Gigon, A., A.R. Matos, D. Laffray, Y. Zuily-Fodil and A.T. Pham-Thi, 2004. Effect of drought stress on lipid metabolism in the leaves of Arabidopsis thaliana (Ecotype Columbia).Ann. Bot., 94: 345-351.
14. Gosal, S.S., S.H. Wani and M.S. kang, 2009.Biotechnology and drought tolerance. J. Crop Improvement, 23: 19-54.

15. Hoerling, M. and Kumar, 2003.The perfect ocean for drought. Science, 299: 691-694.
16. Jonak,C., S. Kiegerl, W. Ligterink, P.J. Barker, N.S. Huskisson and H.hirt, 1996. Stress signaling in plants: A mitogen-activated protein kinase pathway is activated by cold and drought. Proc.Natl.Acad.Sci. USA., 93: 11274-11279.
17. Kacperska, A., 2004. Sensor types in signal transduction pathways in plant cells responding to abiotic stressors: do they depend on stress intensity? Phyisilogya Plantarum, 122: 159-168.
18. Kresovich, S., A.J. Luongo and S.J. Schloss, 2002. Mining the gold: finding allelic variants for improved crop conservation and use. In: Engels JMM, Rao VR, Brown AHD, Jackson MT (eds) Managing Plant Genetic Diversity. CAB International, Wallingford,UK, pp 379–386.
19. Lambers, H., F. Stuart-Chapin III and T.L. Pons, 2000. Plant Physiological Ecology. Springer- Verlag, New York, USA, pp: 540.
20. Levine, A., 1999. Oxidative Stress as a Regulator of Environmental Responses in Plants. In: Plant Responses to Environmental Stresses: from Phytohormones to Genome Reorganization, Lerner, H.R. (Ed.). Marcel Dekker, New York, USA, pp: 247-264.
21. McCouch, S., 2004. Diversifying selection in plant breeding. PLoS Biol., 2: e347-e347.
22. MCue, K.F. and A.D. Hanson, 1990. Drought and salt tolerance: Towards understanding and application. Trends Biotechnol., 8: 358-362.
23. Monneveux, P. and E. Belhassen, 1996.The diversity of drought adaptation in the wide. Plant Growth Regul., 20: 85-92.
24. Montalvo – Hernandez,L., E. Elias Piedra-Ibarra, L. Gomez-Silva,R. Lira-Carmona and J.A. Acosta-Gallegos, 2008. Differential accumulation of mRNAs in drought – tolerant and susceptible common bean cultivars in response to water deficit. New Phytol., 177: 102-113.
25. Murphy, R. and J.K.E. Ortega, 1995. A new pressure probe method to determine the average volumetric elastic-modulus of cells in plant tissue. Plant Physiol., 107: 995-1005.
26. Nordborg, M. and S. Tavare, 2002. Linkage disequilibrium: What history has to tell us. Trends Genet 18: 83–90.
27. Pareek, A., S.K. Sopory, H.J. Bohnert, and Govindjee 2010. Abiotic Stress Adaptation in Plants. Published by Springer, P.O. Box 17, 3300 AA Dordrecht, The Netherlands.
28. Raghavendra, A.S., V.K. Gonugunta, A. Christmann and E. Grill, 2010. ABA perception and signaling. Trends Plant Sci., 15: 395-401.
29. Ramanjulu, S. and D. Bartels, 2002. Drought and desiccation-induced modulation of gene expression in plants. Plant Cell Environ., 25: 141-151.

30. Urao, T., B. Yakubov, R. Satoh, K. Yamaguchi-Shinozaki, M. Seki. T. Hirayama and K. Shinozaki, 1999. A transmembrane hybrid-type histidine kinase in Arabidopsis functions as an osmosensor. Plant Cell, 11: 1743-1754.

31. Van Berloo, R., A.Zhu, R.Ursem, H.Varbakel, G.Gort and F.A. Van Eeuuwijk, 2008. Diversity and linkage disequilibrium analysis within a selected set of cultivated tomatoes. Theor Applied Genet, 117: 89-101.

32. Verslues, P.E., M. Agrawal, S. Katiyar-Agrawal, J.Zhu and J.K. Zhu, 2006. Methods and concepts in quantifying resistance to drought, salt and freezing, abiotic stresses that affect plant water status. Plant J., 45: 523-539.

33. Wang, W.X., B. Vinocur, O. Shoseyov and A. Altman, 2004. Role of plant heat-shock proteins and molecular chaperones in the abiotic stress response. Trends Plant Sci., 9: 244-252.

34. Zhang, J.Z., R.A. Creelman and J.K. Zhu, 2004. From Laboratory to Field. Using Information from Arabidopsis to Engineer Salt, Cold, and Drought Tolerance in Crops. Plant Physiol 135(2): 615-621.

www.ingramcontent.com/pod-product-compliance
Ingram Content Group UK Ltd.
Pitfield, Milton Keynes, MK11 3LW, UK
UKHW061533190726
13845UKWH00005B/87